Process Safety Management Mentoring 2013

(PSM2)

Topical Conference at the 2013 AIChE Spring Meeting and 9th Global Congress on Process Safety

San Antonio, Texas, USA
28 April - 2 May 2013

ISBN: 978-1-62748-450-3

Printed from e-media with permission by:

Curran Associates, Inc.
57 Morehouse Lane
Red Hook, NY 12571

Printed by Curran Associates, Inc. (2013)

For permission requests, please contact AIChE
at the address below.

AIChE
3 Park Avenue
New York, NY 10016-5991

Phone: (203) 702-7660
Fax: (203) 775-5177

www.aiche.org

Additional copies of this publication are available from:

Curran Associates, Inc.
57 Morehouse Lane
Red Hook, NY 12571 USA
Phone: 845-758-0400
Fax: 845-758-2634
Email: curran@proceedings.com
Web: www.proceedings.com

TABLE OF CONTENTS

Advancing Process Safety in the Petroleum Refining & Petrochemical Industry

Lara Swett
American Fuel & Petrochemical Manufacturers (AFPM)
1667 K St., NW, Suite 700, Washington, DC 20006
lswett@afpm.org

Ron Chittim
American Petroleum Institute (API)
1220 L Street, Northwest, Washington, DC 20005
Chittim@api.org

Jerry Forest
Celanese
1601 W. LBJ Freeway, Dallas, TX 75234
Jerry.Forest@Celanese.com

Prepared for Presentation at
American Institute of Chemical Engineers
2013 Spring Meeting
9th Global Congress on Process Safety
San Antonio, Texas
April 28 – May 1, 2013

UNPUBLISHED

Advancing Process Safety in the Petroleum Refining & Petrochemical Industry

Lara Swett
American Fuel & Petrochemical Manufacturers (AFPM)
1667 K St., NW, Suite 700, Washington, DC 20006
lswett@afpm.org

Ron Chittim
American Petroleum Institute (API)

Jerry Forest
Celanese

Keywords: AFPM, API, Advancing Process Safety, APS, Event Sharing, Hazard Identification, Regional Networks, 754 Metrics, Training and Certification, Site Assessments

Abstract

In April of 2012, AFPM (American Fuel and Petrochemical Manufacturers) and API (American Petroleum Institute) launched six new programs to further advance process safety improvements in the refining and petrochemical industry. The programs are industry event sharing, hazard identification documents, process safety regional networks, process safety training & certification, process safety performance metrics and process safety site assessments. Active industry participation in the programs currently exceeds 95% of the U.S. refining capacity and includes several petrochemical companies. This paper will provide background on why these programs were developed, review each program in detail, explain how industry is utilizing each program and how knowledge is being shared across the industry at the site and company levels.

1. Background and Program Development

The petroleum refining and petrochemical manufacturing industry is committed to protecting the health and safety of our workers, contractors and neighbors as well as the environment in which they operate. For decades, industry has worked together to develop safe work practices and standards aimed at reducing risks of major industry hazards and process safety incidents. Despite these efforts, process safety incidents occur. As a result, there has been increased focus and attention on process safety by industry, government, Non-governmental Organizations (NGO's) and the media.

In 2010, a group of refining and petrochemical industry executives belonging to the then NPRA (now AFPM) tasked process safety professionals within their companies to develop a working group to explore ways to collectively improve process safety performance across the entire refining and petrochemical industry. The goal was to develop programs that would create a significant improvement in process safety performance. Shortly after work began, API was asked to join the effort to leverage the strengths and resources of both organizations to successfully develop and implement the programs. Over a two year period, the group developed a suite of six programs currently being utilized by the petroleum refining and petrochemical industry. Together these programs are called the Advancing Process Safety Programs.

1.1 Program Development and Governance

Several factors at the outset of this project were essential to the development of the Advancing Process Safety Programs and led to its successful launch in April 2012. These contributing factors fostered the development and completion of high quality programs in a condensed time frame.

1.1.1 Executive Leadership

The industry leaders that initiated this project played an integral role in development of the programs. They provided clear guidance on program development, provided industry resources, gained support from their peers, and took steps to ensure the quality and effectiveness of the programs to achieve process safety performance improvements across the entire U.S. refining and petrochemical industry. Their oversight and support illustrated how leadership commitment can enhance the process safety culture.

1.1.2 Clear Guidance

Guiding principles were developed by the executive leadership at the outset of the project and are the corner stone of each program.

Guiding Principles:

- *Focus is "Working on ourselves"*
- *Start small with clear easy steps; grow as needed to drive significant process safety improvements*
- *Don't prescribe how, focus on the desired outcomes*
- *Focus on mitigating real risk, not low risk items - move the needle*
- *Should be applicable to all refiners and petrochemical sites/companies, (i.e. small, medium, and large sites)*

The guiding principles allowed each program development team to focus on what was important and design high-quality programs to improve process safety performance across the entire industry. As a result, the Advancing Process Safety programs are applicable to all refining and petrochemical manufacturing sites regardless of their size or company type.

1.1.3 Desire of Industry to Improve

In the early development of these programs it was apparent that industry had a clear desire to improve process safety performance. Industry motivation advanced each of these programs through the development process and enhanced their quality. They developed programs they knew would be constructive and beneficial to all of industry. This "push" by industry enabled these programs to be developed with the assistance of over 150 industry volunteers with a breadth of knowledge and resources and allowed for quick adoption of these programs by industry. Within 8 month of release approximately 95% of the U.S. refining capacity was actively participating in one or more of the programs. Figure 1 shows the breakdown of AFPM membership participation by refining capacity, companies, and sites.

Figure 1: AFPM Member participation in Advancing Process Safety Programs (as of 3/8/2013)

	AFPM Participating Refining Members	% of member participation
Refining Capacity bbl/day	16,650,089	99%
Member Companies	32	84%
Member Sites	118	94%
NOTE: Total US refining capacity equals 17,264,178 barrels per day.		

1.2 Program Development and Governance

Development of the Advancing Process Safety Programs began in July 2010 with a group of process safety professionals named the Process Safety Workgroup (The Workgroup). The Workgroup began with 15 volunteers and has grown to almost 30. Shortly after development work began, Subgroups were formed for each of the proposed programs. The Workgroup was tasked to oversee the program development by the Subgroups and report back to the AFPM and API executive leadership on a frequent basis. In addition to the program Subgroups, a Subgroup of company legal representatives was established to assist with legal issues that arose during the program development process. Each program was assigned a legal representative. With the support of company leadership, the legal Subgroup quickly worked through issues while maintaining the quality and effectiveness of the programs.

The Workgroup developed the concept for each program in seven months, and then spent another year developing the tools for preparation of the program launch in April 2012.

The governance that led to the development of these programs remains in place and serves an important role managing each program and providing feedback for continuous improvement (see figure 2.)

Figure 2: Oversight and governance of the Advancing Process Safety Programs

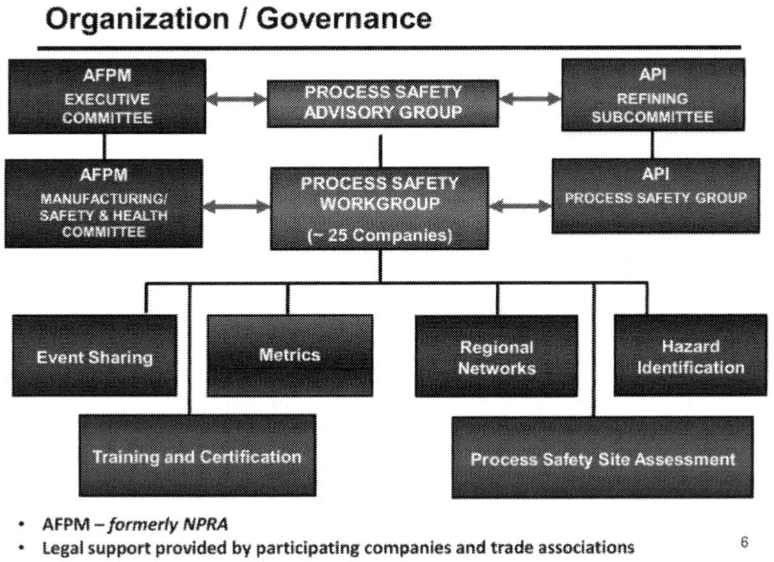

2. Program Descriptions and Product Offerings

Following are detailed descriptions of each program and the products or tools that are available to industry.

2.1 Process Safety Metrics & Analysis Program

In April 2010, an industry standard was published that created a single definition of a process safety event to allow industry to collect, analyze and benchmark process safety performance. The Process Safety Metrics & Analysis program was developed to support industry implementation of the *ANSI/API RP 754 Process Safety Indicators for the Refining and Petrochemical Industries* [1] standard. The Subgroup is tasked with helping sites report consistent and accurate Tier 1 and Tier 2 events (as defined in RP 754) that help measure the process safety performance of the refining and petrochemical industry, provide benchmarking opportunities, and analyze data to help identify opportunities for improvement.

2.1.1 Expert Input of Event Categorization

The Metrics & Analysis Subgroup is comprised of several industry representatives that contributed to the development of API 754 and are therefore in a unique position to assist industry in answering questions related to the standard. Since the release of the standard, this group helped sites with API 754 interpretations, answered questions to clarify categorization of specific events and collected unique API 754 event examples to further enhance understanding of API 754 definitions. API 754 Tier 1 and Tier 2 definitions are located in Appendix A. The Subgroup has posted a Frequently Asked Questions (FAQ) page on the API web site [2] which can be accessed from the references section of the paper.

2.1.2 Quarterly Conference Calls

The Metrics & Analysis Subgroup host conference calls on a quarterly basis for process safety professionals to discuss new FAQs, share industry experiences and lessons learned from effective Tier 3 and Tier 4 metrics implementation, clarify mis-interpretations of the 754 standard, and hold an open question & answer forum. The calls have proven to be effective in helping industry report consistent PSE metrics to the trade associations. For example, based on calendar year 2010 PSE reporting, it was noticed the many sites were not reporting contractor hours and PSE's from pressure relief devices (PRD) consistently. Both of these issues were discussed in detail on a quarterly conference call. As a result, the 2011 PSE reporting showed a significant improvement in the accuracy of this specific information. In addition to quarterly conference calls, this group presents at industry forums, helps solicit support for API 754 with international trade associations and puts on industry workshops.

2.1.3 Analysis of Process Safety Event Data

AFPM and API collect industry PSE data on an annual basis to develop a benchmarking report for industry. The Metrics & Analysis Subgroup is responsible for analyzing the report to identify industry trends and opportunities for process safety performance improvements. In addition, this group provides feedback to other programs within Advancing Process Safety in order to help them focus their efforts on real risks and events that are occurring in industry.

Figure 3: Example of Benchmarking Tool with Sample Data

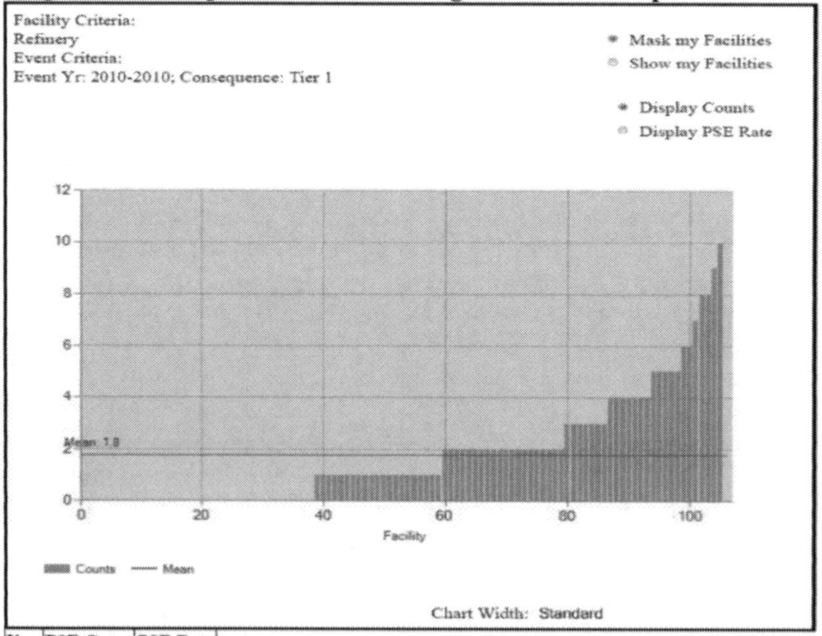

Year	PSE Count	PSE Rate
2010	189	0.1301894

In addition to the Annual Report, the Subgroup developed a Process Safety Metrics data analysis and benchmarking tool on the AFPM Safety Portal as shown in figure 3. This tool allows industry representatives to drill down into the API 754 categories and benchmark their company's performance against the rest of refining and petrochemical industry.

Figure 3 is an example of a company performing a search with the Process Safety Metrics data analysis and benchmarking tool. The search results reflect all Tier 1 Process safety event submissions from 2010. The Y-axis is the count of process safety events and the X-axis represents the number each site submitted.

2.1.4 Industry Participation in Process Safety Metrics

In calendar year 2011, 105 Refining Sites and 98 Petrochemical sites reported their Tier 1 and Tier 2 process safety event data to participate in the annual AFPM Process Safety Event Report. In addition, many companies have developed implemented and Tier 3 & Tier 4 metrics.

One participating company that collects and reports Tier 1 & 2 PSE metrics also developed a full suite of Tier 3 & 4 metrics to track their performance across all sites. After collecting several years of data, they were able to perform a "deep dive analysis" on process safety events and make significant recommendations to improve programs that ultimately reduced risk and improved their performance. Over the years, this company has observed significant improvements in their Tier 1 and Tier 2 process safety performance.

Another participating company has a corporate wide focus on Tier 3 Safe Operating Limit Exceedences (SOLEs) that helped to highlight the critical importance of implementing a robust operating limits program at their refineries. Their program includes a clear definition of a safe operating limit versus a reliability operating limit, a mandatory policy, procedure or internal company standard that defines operating limit implementation schedules and milestones, site procedures, complete operating parameter and limit assessments, complete alarm strategies and configurations, and operating limit guideline documents for all major process units, fired heaters, rotating equipment, offsite, and utilities. To track the Tier 3 metrics, this company established a Tier 4 metric to monitor implementation progress and investigate Tier 3 SOLEs for trend analysis and learning.

These companies and others share lessons learned and process safety achievements from their metrics programs in detail on quarterly API 754 conference calls and presentations at Regional Network meetings.

2.2 Process Safety Event Sharing Program

Learning from incidents is fundamental for safety management and incident prevention. The Event Sharing Program was created to provide a mechanism for industry to share causal factors and lessons learned information from API 754 Tier 1 process safety events as well as other Tier 2 events, key findings and near misses considered to have high learning value for industry (HLVE).

The Event Sharing Subgroup developed an Event Sharing Database that allows industry representatives to input process safety events and near misses. The categories utilized by the database are taken from API 754 so that companies can easily enhance their API 754 data to include causes and learnings.

The Event Sharing Database was launched in April 2012 with approximately 50 events. One year later, the database contains over 100 events from more than 20 process units in petroleum refining and petrochemical manufacturing.

2.2.1 Event Sharing Database

The Event Sharing program database is located in the AFPM Safety Portal. The database allows users to search by:

- Keywords
- Refining and Petrochemical Process
- Mode
- Equipment Point of Release
- Equipment Initiating Release
- Material
- Cause
- Classification (Tier 1, Tier 2 HLVE, Near Miss HLVE, Key Finding HLVE)
- HLVE Event

The events can be viewed on the screen or opened in a PDF or Excel file. This program is continuously improved based on feedback from users. In 2013 the database will be upgraded to include the ability to upload a file or picture to the event submission. The Subgroup is also investigating the ability to add sub-categories to the existing API 754 categories based on a recommendation from both the Metrics & Analysis Subgroup and the Event Sharing Subgroup. After reviewing the API 754 data and the submissions in the Event Sharing Database, the Subgroup would like to investigate further into each category, specifically piping systems, where the Subgroups were seeing a majority of the events submitted. By understanding more about events stemming from piping systems, industry can learn and prevent similar events in the future.

2.2.2 Safety Bulletins

Figure 4: Process Safety Bulletin

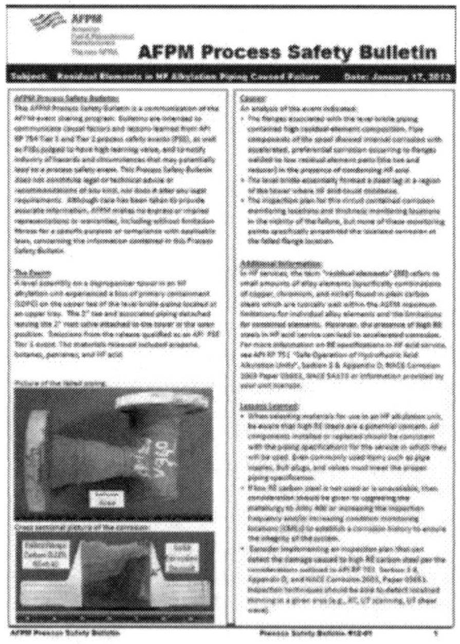
AFPM Process Safety Bulletin

The Event Sharing Subgroup is responsible for maintaining the integrity of the Event Sharing Database as well as distributing event information from the database through the AFPM Process Safety Bulletins, as shown in figure 4. The January 2013 bulletin focused on "Residual Elements in HF Alkylation Piping Caused Failure" and was released to over 1000 industry representatives. Based on the information in the bulletin, one company upgraded its inspection program enterprise-wide to investigate for the failure mechanism described in the event. Another company distributed the bulletin to their sites with directions to inspect the units for the hazard described. Based on the inspection results, preventative actions were taken to reduce the physical risk around residual element hazards. This event was also discussed in detail during several regional network meetings.

2.3 Hazard Identification Program

The Hazard Identification program was developed to create resources for industry on process safety hazards that are not currently addressed in existing industry standards (API, ASME, and NFPA). The Subgroup is developing a library of documents designed to provide information and ready guides for potentially overlooked and not widely known process safety hazards. The library will also share process safety hazard lessons learned from industry related incidents in order to improve process safety awareness and retain institutional knowledge on process safety hazards for future generations. Hazard Identification documents can be used for operator certification, PHA preparation, MOC hazard assessment, tailgate topics, emergency drills and more.

2.3.1 Hazard Identification Documents

Hazard Identification Documents are located in the AFPM Safety Portal and are accessible to all AFPM members. Completed documents include Hot Taps, LPG Storage, Flare Operations, Process Sampling for QA/QC, Shift Turnover, Vacuum Trucks, Critical Crane Lifts, and Atmospheric Tank Preparation for Out of Service Maintenance.

The Hazard Identification Document sections include:

1. Purpose and Use
2. Scope
3. Examples of Inherent Concerns and/or Hazards
    a. Specific hazards organized by task
4. References and Resources
    a. Standards, regulations, public reports, and other public reference material
5. Industry Incidents
    a. A listing of industry incidents with links to public reports that are associated with the hazards described in the document.

Document topics are picked from industry input, hazards identified from recent industry events, and feedback from other Advancing Process Safety Programs. An online form is available on the Safety Portal for industry representatives to suggest topics for future development.

The documents are often discussed in detail during the Regional Network meetings. The network participants are asked to confirm that their site's practices and procedures address all of the hazards included on the Hazard Identification document. Some sites have implemented updates to a practice or procedure based on a specific Hazard Identification document.

For example, the Shift Turnover document was discussed during a Regional Network meeting. It was noted that many of the sites represented did not have a formal process for shift turnover. As a result, participants were asked to talk to their sites about developing a shift handover procedure and provide feedback during the next network meeting.

Another site distributed several of the documents to their contractors to ensure they had complete knowledge of potential hazards when working on tanks or vacuum trucks at their site.

2.4 Process Safety Regional Networks

The Process Safety Regional Networks Program was developed to allow informal collaboration between process safety practitioners at the plant level in an effort to impact overall safety performance. There were several trade association groups sharing process safety information across the industry, however, many of those groups did not include plant level practitioners. This program was modeled after the success of the Western States Petroleum Association (WSPA) Process Safety Group in California and the Rocky Mountain Process Safety Group in Montana. Currently there are eight Process Safety Regional Networks in the U.S. as shown in Figure 5. The two on the West Coast are facilitated by WSPA and the other six are facilitated by AFPM. The agenda items focus on process safety practices, lessons learned from industry events, industry standards, regulatory challenges, hazard identification, and other topics viewed as a priority by the network members.

The Regional Network Program has been successful in utilizing the Advancing Process Safety Programs to help facilitate sharing and conversation. Network participants have provided important feedback and suggestions for improvement of the Advancing Process Safety Programs.

Figure 5: Regional Network Map

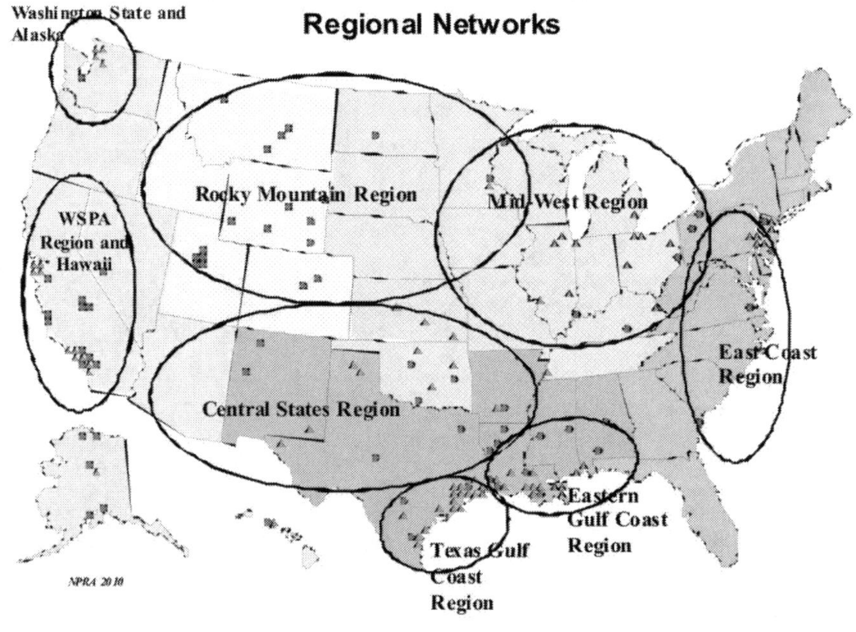

Examples of topics discussed at the Network meetings include: Culture Surveys, Hot Taps and Live Flare Work procedures, process safety training for new employees, electronic MOC's, Process Safety Metrics, Incident Sharing, Process Control Systems, Process Safety Information, Emergency MOC procedures, Mechanical Integrity, Operational Discipline, Employee Participation, and more. The Network's informal sharing allows sites to compare and contrast various methods to achieve the same result.

To facilitate sharing among regions, AFPM developed a module on the Safety Portal that allows all regional network participants to view and search agendas, meeting minutes, documents, and presentations shared in other regions. For example, an individual can search on "MOC" and view the presentation given during a Central States Regional meeting on electronic MOC's and a presentation given during a Rocky Mountain Regional meeting on emergency MOCs. Contact information for regional participants is also provided to allow greater information sharing.

Currently there are 150 process safety practitioners from companies representing 95% of the U.S. refining capacity participating on the eight Regional Networks in the United States. This number includes professionals from several petrochemical sites. The networks meet several times a year in the U.S. with the common goal of collectively enhancing process safety performance.

2.5 Process Safety Training & Certification

Figure 6: Process Safety Training

The goal for the Process Safety Training and Certification program is to help training vendors, industry and other 3rd parties create their own programs based on criteria developed by industry, to ensure training topics and delivery are consistent and meets industry's expectations. To date, training criteria has been developed for: What is Process Safety?, Hazard Recognition, Process Safety Leadership, Conduct of Operations and Risk Assessment. Training vendors, companies or other 3rd parties can submit their training programs (based on the industry developed training criteria) to API for certification under API's Training Provider Certification Program thereby ensuring that the training meets industry's needs.

One company recently incorporated the approved training criteria into their own program titled, "I am a Process Safety Leader" (See Figure 6). The feedback from employees has been positive and effective in bringing elements of process safety into their individual jobs and responsibilities. In the picture shown above, the company utilized the Swiss Cheese Model to demonstrate the critical layers of protection and each individual's role in making sure there are never any "holes" in the cheese.

2.6 Process Safety Site Assessments

The Process Safety Site Assessment Program developed by API focuses on higher risk process safety-related activities in the refining and petrochemical industry. The assessments were at first focused on the petroleum refining industry, but are now available to petrochemical manufacturing and chemical facilities. The assessments are conducted by independent, industry qualified third party teams of expert assessors and focus on both the quality of a site's written programs and the effectiveness of field implementation.

Protocols for the assessments were developed by industry experts in accordance to the API standards development process. The protocols were widely distributed for review and comment and are available through API by request. The protocols contain sections on Process Safety Leadership, Management of Change, Mechanical Integrity (fixed equipment), Safe Work Practices, Operating Practices, Facility Siting, Process Hazard Analysis, and HF Alkylation (API RP 751 [3]). The assessment does not qualify for a three year PSM audit, however, the full API RP 751 assessment can be used for the three year RP 751 audit.

The assessors are first screened to ensure adequate process safety experience, undergo industry led assessor training, and are further interviewed by industry representatives before potentially being selected to the assessor "pool." Assessor teams have on average 35 years of experience per person - this experience level is one of the elements that make this assessment program unique from other industry assessments. In addition, the assessors are trained to help the facility get better rather than focusing on what is wrong with the site. Teams are staffed with enough assessors such that the assessment can be completed in one week.

Seven assessments have been completed since April 2012 and more are being scheduled for 2013 and 2014. This program is designed to promote learning from sharing of experiences and industry proven practices, provide benchmarking opportunities for a site through the use of industry proven protocols, and serve as a feedback mechanism for analysis of performance data to identify trends and patterns as shown in the sample data analysis of figure 7.

Figure 7: Sample Data Analysis

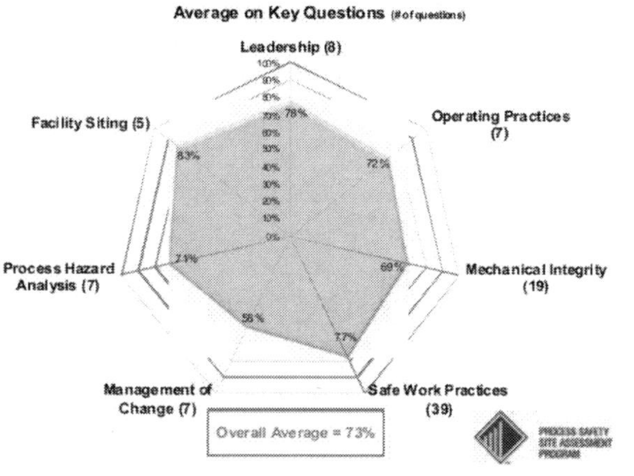

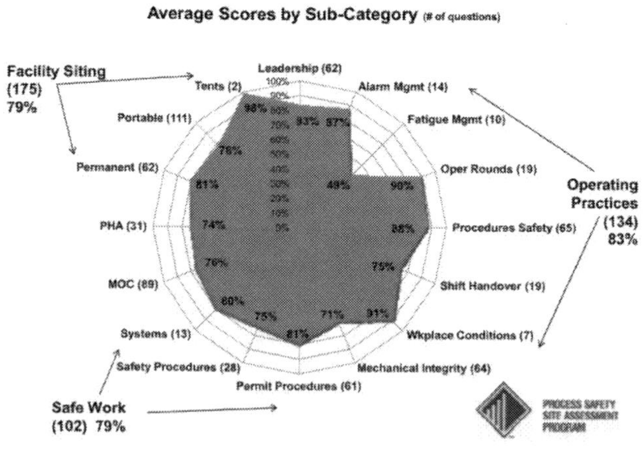

<u>NOTE: The above data is very preliminary and is being shown only as examples of the types of analysis that can be accomplished. No conclusions or actions should be taken as a result of this data.</u>

3. Communication and Continuous Improvement

It was clear from the beginning that development of the Advancing Process Safety programs was only the first step. Additional processes and tools were necessary to enhance industry communication and allow for continuous feedback. The Workgroup developed several tools to help industry stay informed and up to date on the programs. These tools provide processes and feedback mechanisms of the programs.

3.1 AFPM Safety Portal [4]

It was acknowledged early in the development of these programs it would be necessary to provide a single place to house information on all of the products and tools developed for each program. The Safety Portal was developed to be an interactive database where member companies enter safety statistics and events, search on that data, receive notifications when new information is entered and analyze existing industry data. A snapshot of the portal is shown in figure 8.

Figure 8: AFPM Safety Portal

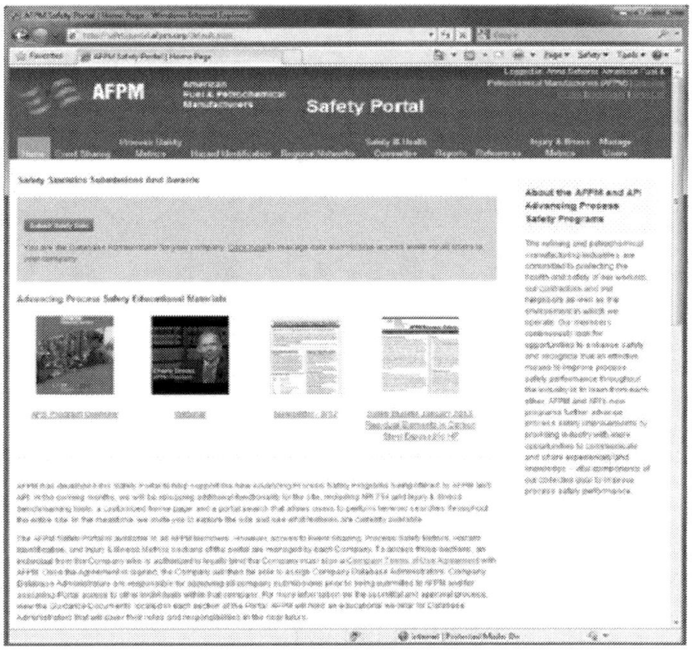

The portal is organized by the following tabs Event Sharing, Process Safety Metrics, Hazard Identification documents, Regional Networks, Safety & Health Committee, Reports, References, and Injury & Illness metrics. The portal has the ability to keyword search across all of the tabs. For example, if someone is searching for Hot Taps, the search results will provide the following results; all events in the event sharing database that had a keyword of Hot Taps, all hazard identification documents that deal with hot taps, the Mid-West Regional Network presentation on Hot Taps, and all AFPM conference papers and presentations on Hot Taps. Within a few seconds, the user can have a tremendous amount of information that previously would have taken hours to perform through internet searches and several segregated database searches.

In addition to searching, the Safety Portal has a customized homepage and sends email notifications to users when new data is uploaded. Based on the feedback received, the portal has allowed individuals within industry to participate in the Advancing Process Safety Programs without spending a lot of time or resources searching and analyzing the information.

3.2 *API Website – Program Information*

The API website contains information on three of the Advancing Process Safety Programs – Metrics & Analysis/API RP 754; Training and Certification; and Process Safety Site Assessments. All website links are located in the references section of this paper.

3.2.1 Metrics & Analysis – API RP 754

Individuals can learn more about API RP 754 [5] as well as access the document (read-only) free of charge on the API website [6]. In addition, a series of four API 754 webinars were developed to help industry become familiar with the document. The webinars cover: 1) Business Case and Leadership Overview; 2) Tier 1 and Tier 2 Process Safety Events; 3) Tier 3 and Tier 4 Process Safety Indicators; and 4) Implementation of API 754. For each of the topics, there is the final webinar, slide pack and notes [7].

3.2.2 Training and Certification

Information for the Training and Certification program can also be found on the API website. The actual training criteria that has been developed for What is Process Safety?, Hazard Recognition, Process Safety Leadership, Conduct of Operations and Risk Assessment are posted online [8].

3.2.3 Site Assessments

As mentioned above, API has developed a new 3rd party Process Safety Site Assessment program. The following information about this program is available on the API website [9]:

- Program description
- Program Guidance Document
- Assessment Protocols (by request)
- Assessor Qualification Process
- Assessor Qualification Requirements

3.3 Advancing Process Safety Newsletter

The Advancing Process Safety Newsletter is the latest communication tool developed by The Workgroup. The newsletter updates industry representatives on a trimester basis on new reports, program updates, and activities within the Advancing Process Safety Programs. The newsletter is distributed to approximately 1000 industry representatives. This tool has been successful in increasing awareness and participation at sites that were not previously participating as well as notifying Safety Portal users of updated information and reports.

3.4 Presentations at Conference and Workshops

Process Safety Workgroup members are always looking for opportunities to speak at conferences, workshops and other industry forums to share information on the Advancing Process Safety programs and new industry learnings that have resulted from the programs. Other international trade associations in Canada and Europe have contacted Workgroup participants to request information and guidance documents on the Advancing Process Safety Programs in order to use them as a model for their own programs.

3.5 Integration of Programs and Continuous Improvement

Developing the programs and communicating them broadly is not enough to ensure continuous improvement in process safety performance. Each program needs to be evaluated on a continuous basis so that it can be updated and learnings captured to share throughout the refining and petrochemical industry. This ensures that industry resources are focused on items that pose the greatest risk. The Workgroup is responsible for identifying integration opportunities.

Here are a few examples of the integration of the programs:

- Site Assessments revealed that some sites may have inconsistent Hot Taps and Live Flare Work procedures compared to others in the industry. The Site Assessment Subgroup submitted these as topics for consideration through the online form for the Hazard ID Subgroup. This resulted in Hot Taps being added as a Hazard ID Document and Live Flare Work was placed in the queue for 2013 development. The new Hazard ID documents are shared at the Regional Network Meeting and with site contractors.
- A site submitted a Tier 1 event to the event sharing database, the Event Sharing Subgroup recommended that the hazard be widely communicated throughout industry. A Process Safety Bulletin was developed and distributed. The Regional Network participants discussed the bulletin in detail during their meetings. Based on this knowledge, sites can inspect their process units for similar hazards and therefore prevent a possible similar event.
- An analysis of Process Safety Tier 1 & Tier 2 PSEs and the Event Sharing Database, showed a significant amount of events occurring in Piping Systems. The Event Sharing Subgroup decided to update the event sharing form to include sub categories for piping systems to better understand the hazards. Once data is collected, it will be shared with the Hazard ID Subgroup to potentially develop a Hazard ID document and the Site Assessment Subgroup to update their protocols.

The integration of these programs allows each program to develop tools and products that are based on actual hazards seen in industry and allows individual sites to focus their resources on items that will improve their process safety performance.

4. Conclusion

AFPM and API successfully completed the development and implementation of six process safety programs to assist the U.S. refining and petrochemical manufacturing industries improve their process safety performance. The programs are process safety metrics & Analysis, event sharing, hazard identification documents, regional networks, training & certification, and site assessments. There is significant support from almost the entire refining industry which includes active participation from 95% of the U.S. refining capacity and several petrochemical companies and we are actively recruiting new participants. Executive leadership, clear guidance, and the desire from the industry to improve were key factors that led to the preliminary success of these programs.

The Workgroup developed processes to evaluate and improve the Advancing Process Safety programs to ensure their effectiveness in improving process safety performance in the future. It is too early to clearly understand the correlation between these programs and the actual process safety improvements; however, initial industry feedback indicates a connection may be apparent in the future.

The Advancing Process Safety initiative driven by an energetic, supportive leadership and designed by the vastly, knowledgeable process safety experts from the refining and petrochemical manufacturing industries provides an amazing step forward to enhance a process safety culture that will lead to process safety performance improvements.

5. Appendix A: API RP 754 Tier 1 and Tier 2 Definitions

A Tier 1 or Tier 2 Process Safety Event begins with an unplanned or uncontrolled release of any material, including non-toxic and non-flammable materials from a process that results in one or more consequences described in the RP.

Tier 1 Consequences	Tier 2 Consequences
an employee, contractor or subcontractor "days away from work" injury and/or fatality	An employee, contractor or subcontractor recordable injury
a hospital admission and/or fatality of a third-party	
an officially declared community evacuation or community shelter-in-place	
A fire or explosion resulting in greater than or equal to \$25,000 of direct cost to the Company	A fire or explosion resulting in greater than or equal to \$2,500 of direct cost to the Company
A release of material greater than the threshold quantities described in Table 1 of the standard in any one-hour period	A release of material greater than the threshold quantities described in Table 12of the standard in any one-hour period
A pressure relief device (PRD) discharge to atmosphere whether directly or via a downstream destructive device that results in one or more of the following four consequences: - liquid carryover; - discharge to a potentially unsafe location; - an on-site shelter-in-place; public protective measures (e.g. road closure); - and a PRD discharge quantity greater than the threshold quantities in Table 1 of the standard in any one-hour period	A pressure relief device (PRD) discharge to atmosphere whether directly or via a downstream destructive device that results in one or more of the following four consequences: - liquid carryover; - discharge to a potentially unsafe location; - an on-site shelter-in-place; public protective measures (e.g. road closure); - and a PRD discharge quantity greater than the threshold quantities in Table 2 of the standard in any one-hour period
A release of material greater than the threshold quantities described in Table 1 of the standard in any one-hour period	A release of material greater than the threshold quantities described in Table 2 of the standard in any one-hour period

6. References

[1] *Process Safety Performance Indicators for the Refining* and Petrochemical Industries, API Standard 754, first edition, 2010

[2] API 754 Frequently Asked Questions [Online] Available: http://www.api.org/environment-health-and-safety/process-safety/process-safety-standards/measuring-process-safety-rp-754.aspx

[3] *Safe Operation of Hydrofluoric Acid Alkylation Unit*, API Standard 751, third edition, 2007

[4] AFPM Safety Portal [Online] Available: http://safetyportal.afpm.org/

[5] API 754, additional information [Online] Available: http://www.api.org/environment-health-and-safety/process-safety/process-safety-standards/standard-rp-754.aspx

[6] API 754, read only copy [Online] Available: http://publications.api.org/

[7] API 754, educational webinars [Online] Available: http://www.api.org/environment-health-and-safety/process-safety/rp-754-webinars/rp754-webinars.aspx

[8] Training and Certification program information and criteria [Online] Available: http://www.api.org/certification-programs/training-provider-tpcp/advancing-process-safety.aspx

[9] Process Safety Site Assessment program information [Online] Available: http://www.api.org/environment-health-and-safety/process-safety/advancing-process-safety-programs/api-process-safety-site-assessment-program.aspx

Process Safety: From Concept to Commissioning

Jennifer F. Mize
Principal Chemical Engineer
Eastman Chemical Company
P.O. Box 511, Building18
Kingsport, TN 37662
jfmize@eastman.com

Prepared for Presentation at
American Institute of Chemical Engineers
2013 Spring Meeting
9th Global Congress on Process Safety
San Antonio, Texas
April 28 – May 1, 2013

UNPUBLISHED

Process Safety: From Concept to Commissioning

Jennifer F. Mize
Principal Chemical Engineer
Eastman Chemical Company
P.O. Box 511, Building18
Kingsport, TN 37662
jfmize@eastman.com

Keywords: Process safety, inherently safer design, relief design, Safety Instrumented Systems, hazard analysis

Abstract

Process safety is an essential component in all phases of project design, from initial research and development to commercial process design and commissioning. Data obtained in the laboratory and piloting efforts feed into the engineering design of the commercial facility. Integration of process safety into the design process will ensure the optimal utilization of inherently safer design, relief and effluent handling systems, and Safety Instrumented Systems in the final design.

I. Introduction

Process safety is an essential component in all phases of project design, from initial research and development to commercial process design. During research and development efforts, a process safety review can help identify the hazards associated with the planned chemistry and initiate the investigation of inherently safer chemistry. As the project progresses into the piloting stage, experiments can be designed to identify potential process upsets and generate data to facilitate the design of relief systems and Safety Instrumented Systems (SIS). Experiments can also determine the vessel design conditions that can eliminate or mitigate as many potential relief scenarios as possible in the commercial facility. The information generated in the pilot plant will feed into the process safety effort during the engineering design of the commercial facility allowing critical design decisions to be made early in the project to ensure the optimal utilization of inherently safer design, and effluent handling relief systems, and Safety Instrumented Systems in the final design.

There are multiple roles within the process safety discipline that will need to be filled at each design phase. A relief designer, Layers of Protection Analysis (LOPA) practitioner, SIS designer, and Process Hazard Analysis (PHA) study leader are needed in the development of the process safety design. Consultation with chemists, process experts, operators, equipment designers, control systems engineers, and industrial hygienists is essential to a comprehensive evaluation of process hazards.

II. Conceptual Design

2.1 Laboratory and pilot plant work

Process safety efforts start with laboratory and pilot plant work. Hazard identification should be included as a routine part of the planning of laboratory experiments and development of piloting efforts. Identifying the hazards will allow for the design, construction, and operation of a safe laboratory unit as well as a safe scale-up to a pilot unit and then to commercialization. Prior to any formal hazard identification process, a thorough review of existing literature and standard chemical references should be conducted to obtain basic information about the physical and chemical hazards of the chemicals that will be used. Standard references such as Kirk-Othmer [1], Bretherick [2], Sax [3], and others will provide additional understanding of the reactive potential of a given system.

Prior to starting the hazard identification process, all available information on the chemicals should be gathered. In some cases, due to the experimental nature of the chemicals produced, Safety Datasheets (SDS) may not be available. Hazards may have to be assessed based on chemical structure. Testing may be needed to identify chemical properties so that they can be safely handled.

The hazard assessment should be documented at each stage so the information can be used as the process is developed and scaled up. The method and level of detail for the hazard assessment will vary at each stage of the process development. A checklist may be sufficient to guide the discussion for the laboratory and pilot plant work, while the more rigorous methods set forth by OSHA 1910.119 (e)[4] including Checklist, "What if", Hazard and Operability Analysis (HAZOP), and Failure Modes and Effects Analysis (FMEA) should be used during the Front-End Engineering and Detailed Design efforts.

The hazard assessment conducted during the conceptual design should include an evaluation of the following:

- Reactive Hazards
- Fire and Explosive Hazards
- Respiratory, Skin, and Eye Hazards
- Materials Handling

Pilot plants are typically operated to achieve steady-state conditions and generate data that will support scale-up to a commercial plant. Another important task in the piloting efforts is the identification of potential safety concerns during upset conditions. A working knowledge of the intended chemistry is not sufficient to ensure reactive hazards are adequately addressed. The unintended chemistry, such as side reactions, decompositions, and unidentified intermediates, can lead to unsafe conditions in the unit. Increasing or decreasing the concentration of reactants can lead to reactive hazards that may not have been identified in the development of the original intended chemistry. For example, the accumulation of byproducts and impurities in a particular system due to closing of the recycle streams could lead to hazards in the pilot unit that were not experienced at the laboratory scale. Furthermore, reaction kinetics may be sensitive to changes in operating conditions such as temperature and pressure. Reactants and products may be unstable or subject to thermal decomposition. Incompatibilities with seal fluids, heating media, lubricants, or other materials could lead to additional safety concerns. [5]

The process safety engineer can assist with designing experiments to identify hazards and generate data for the process safety evaluations during the design of a manufacturing facility. Material generated in the laboratory and/or pilot plant can be used for Differential Scanning Calorimetry (DSC) and other screening tests to identify reactivity concerns. Potential reactions, instabilities, or incompatibilities can be addressed by substituting other solvents or reactants that do not exhibit unstable behavior or reactivity.

2.2 Conceptual Safety Review / Process Safety Design Philosophy

Once the conceptual design of a commercial facility starts, a safety review should be conducted with process engineers, chemists, and pilot plant operators with knowledge of the process chemistry and operation. The goal of this review is to document the main hazards of the process and develop the general process safety design philosophy for the facility. Data from the laboratory and pilot plant work are crucial to this effort.

The conceptual safety review and documentation of the process safety design philosophy will allow the process safety engineer to communicate the hazards and mitigations to the other disciplines (process design, equipment design, vessel design, etc.), to be included in their design deliverables. The best opportunity to achieve inherently safer design occurs at this point in the design process. When inherently safer alternatives are not available, relief systems and SIS can then be implemented to address the hazards. The process safety engineer can utilize each of these options to optimize the safety design.

2.2.1 Inherently safer design

An inherently safer design is one that eliminates or reduces hazards rather than mitigating or controlling them. The process flowsheet should be reviewed to identify opportunities to implement inherently safer design, which can be achieved, for example, by selecting a vessel or system with a Maximum Allowable Working Pressure (MAWP) that is greater than the maximum possible pressure that can be generated during upset conditions. Alternatively, the supply pressure and/or temperature of the utilities can be selected to eliminate them as potential sources of overpressure. As previously mentioned, potential reactions, instabilities or incompatibilities can be addressed by substituting other solvents or reactants that do not exhibit unstable behavior or reactivity.

The process safety engineer should consider the following issues when looking for opportunities to eliminate or reduce the hazards present in the preliminary design. When identified early in the conceptual design, these strategies can be integrated into the design process and validated in the pilot plant to confirm their viability. If a given hazard cannot be eliminated, the implementation of inherently safer options can be used to minimize the size of relief systems and/or reduce the required reliability of SIS to effectively mitigate it.

- Vapor pressure of components at maximum utility temperatures

In order to eliminate loss of cooling and abnormal heating as relief scenarios, the vapor pressure of all components, including cleaning solvents, must be less than the system rated pressure at the maximum temperature of the heating media. This can be achieved by selecting a heating media with the lowest practical operating temperature and designing the system to operate at this temperature. The system MAWP is then selected based upon the maximum vapor pressure that can be experienced with the selected heating media. The relief designer must consider all operating modes, including maintenance and cleaning operations, when selecting the system MAWP. The designer must consider the use of all cleaning agents, such as water or solvents, which may generate sufficient vapor pressure to overpressure the system.

- Maximum utility pressure

Utilities directly connected to the vessel, such as inert gas, water, or steam, can be a source of overpressure if the supply pressure is greater than the system MAWP. In order to eliminate tube rupture as a potential relief scenario, the supply pressure of the heating and cooling media must be less than the system MAWP.

- Reactivity concerns

Potential reactions, instabilities, or incompatibilities can be addressed by substituting other solvents or reactants that do not exhibit unstable behavior or reactivity. Evaluation of alternate solvents and reactants should be considered during laboratory and pilot plant work. Utility supply temperature can be limited to ensure the system operates significantly below the initiation temperature of identified exothermic reactions or thermal decompositions. Heating and cooling media that are compatible with the process materials can be selected to eliminate the potential for reactions due to mixing of process and utility streams due to leaks or tube failures.

- Liquid overpressure

In order to eliminate liquid overpressure as a potential relief scenario the maximum supply pressure of all liquid feeds to the vessels should be less than the system MAWP. This can be readily achieved with pressure vessels designed per ASME Code, [6] which have pressure ratings above 15 psig. Alternatively, relief systems with discharge to a properly designed effluent handling system can be used to address liquid overpressure. Low pressure storage tanks designed to API 650 [7] cannot be filled above the top weld line, so relief cannot be provided for liquid overpressure by a device located on the top head. Relief for liquid overpressure can be provided by a relief device located on the top head of API 620 [8] storage tanks, which can be rated for 2.5 psig to 15 psig. Designing an API 620 storage tank for 15 psig allows the use of a relief valve to address liquid overpressure. Alternatively, a Safety Instrumented System can be used to shut off liquid feeds into the tank on high level. If there are multiple feed lines, it can be difficult to achieve the required reliability in the SIS.

- Vacuum

All vessels that are not rated for full vacuum require vacuum protection. The installation of a vacuum vent introduces a source of air inflow to the process that may lead to the development of a flammable atmosphere inside the process vessel. There may also be operational concerns associated with air entering the process vessel. For this reason, a blanketing system is typically used in conjunction with a vacuum vent to allow inert gas to enter the vessel to relieve vacuum prior to opening the vacuum vent as a secondary source of vacuum protection. Required air and/or inert gas flowrates to relieve vacuum due to pumping out a vessel are easily supplied with a small vacuum vent and blanketing system. Low pressure storage tanks should be requested with a Maximum Allowable Working Vacuum (MAWV) of at least 1 oz/in^2 to allow use of standard conservation vents to relieve vacuum.

Vacuum generated due to collapsing vapor in a distillation column or other system with a high mass of vapor can require very large air and/or inert gas flowrates. This may require the installation of a large vacuum vent and an accumulation system to store inert gas so it can be introduced quickly to relieve vacuum. All pressure vessels should be rated for full vacuum unless this requires significant additional expense to achieve. If a full vacuum rating cannot be obtained, the vessel manufacturer should be requested to calculate and rate the vessel for the highest possible vacuum rating achievable with the requested MAWP.

- External fire

External fire can be addressed by building layout and/or fire protection including fire resistant insulation, vessel specific deluge, and/or foam fire systems. Locating equipment greater than 30 feet above a surface that can sustain a pool fire, segregating equipment and piping containing ignitable materials from equipment and piping with no ignitable materials, and designing underground tanks can eliminate the need to design relief systems for external fire impingement. Utilizing aboveground buried tanks severely limits the potential fire exposure for storage systems. Providing fire resistant insulation and/or vessel specific deluge will reduce the size of relief systems when designing for external fire impingement. Refer to NFPA 30 Flammable and Combustible Liquids Code [9] and OSHA 1910.106 Flammable and combustible liquids [10] for additional information.

2.2.2 Relief and Effluent Handling Systems

The relief designer should determine which vessels need individual relief devices and which vessels will be relieved as a system, such as distillation units. The need for normal venting devices should also be determined. The appropriate type of relief device (relief valve, rupture disk, and/or conservation vent) should be selected for each installation. Evaluation of the relief design philosophy at this point in the design will allow representation of the correct relief system/device and discharge location on the Piping and Instrumentation Diagrams (P&IDs) as they are developed.

The safe discharge location must be identified for each individual relief device. In order to determine what constitutes a safe discharge location the toxicity, flammability and volatility of the material(s) involved must be considered. Both OSHA 1910.106 and NFPA 30 refer to discharging to a "safe location". OSHA 1910.106 is essentially the 1971 version of NFPA 30. Since the regulations do not provide prescriptive information on discharging to a safe location, documents such as API 521 [11] and books such as Guidelines for Pressure Relief and Effluent Handling [12], published by the Center for Chemical Process Safety (CCPS) of the American Institute of Chemical Engineers, provide good engineering practice for the design of relief systems. OSHA 1910.119 Appendix C, Section 3, (part of the OSHA PSM Standard) provides additional sources for good engineering practices.

Secondary containment, such as blowdown tanks, vent stacks, scrubbers, and flares, should be designed to handle credible simultaneous relief flow from connected equipment. Higher relief flowrates will result in larger secondary containment systems. The overall process safety design philosophy will optimize the use of relief, Safety Instrumented Systems, inherently safer design, and other mitigations to "right-size" the relief and effluent handling systems.

2.2.3 Safety Instrumented Systems

The process safety engineer should determine where Safety Instrumented Systems will be used to address hazards identified in the conceptual safety review. General design considerations can include the use of SIS for liquid overpressure/overfill protection, abnormal heating, tube failure in shell and tube heat exchangers, and pump and compressor deadhead and isolation protection. These can be represented on the P&IDs using a standard template for the each interlock. Non-SIL rated interlocks and interlocks implemented in the basic process control system (BPCS) may also provide mitigation. Layer of Protection Analysis should be utilized to determine the required reliability for the SIL interlocks and evaluate the effectiveness of other interlocks for each scenario.

2.2.4 Difficult to resolve issues

There are some hazards that can prove to be difficult to address with either relief systems and/or Safety Instrumented Systems. Addressing these hazards may require a combination of relief systems or Safety Instrumented Systems with inherently safer design (selection of vessel MAWPs) and/or specific equipment design.

- High pressure to low pressure

High pressure to low pressure is characterized by the ratio of the operating pressures. Typically, a ratio of high operating pressure to low pressure MAWP of greater than 2.5 is considered to be of significant concern. This ratio is used as an estimate of the transition point from elastic to plastic deformation in typical pressure vessel design. Maintaining the ratio below 2.5 will prevent catastrophic failure of the vessel but may result in vessel damage due to overpressure above the test pressure. A risk assessment should be conducted to ensure sufficient layers of protection are in place to address overpressure.

Generally, the transition from a section of the process that operates at high pressure to a section that operates at low pressure is accomplished by the use of a pressure reducing device like a control valve. If this valve fails to achieve the designed pressure reduction due to human error (manual bypass or commanding the valve to open), mechanical failure, or failure of the basic process control system, the high pressure equipment can de-inventory quickly into the low pressure vessels with potential for catastrophic failure of the low pressure vessels. Due to the high pressure ratio, the flowrate across a failed open valve may require large relief devices to protect the low pressure vessels. For very high flowrates, multiple relief devices may be required.

If emergency pressure relief cannot be provided for this case, a SIS may be used to shut off the flow from the high pressure equipment to the low pressure equipment. However, without mitigation credit from an adequately designed emergency relief system, the required reliability of the SIS may be difficult to achieve and maintain.

This hazard can be addressed by selecting the vessel MAWPs to achieve a high operating pressure to low pressure MAWP ratio of less than 2.5 and/or designing the pressure reducing device to limit the maximum flowrate in a failed state. Limiting the maximum flowrate can allow design of a reasonably sized relief and effluent handling system to provide adequate protection.

- Tube rupture, especially high pressure gas/vapor into low pressure liquid (high pressure ratio)

A tube leak or rupture in a shell and tube exchanger can allow high pressure material to enter the low pressure side. For pressure ratios above 2.5, a rupture disk (non-reclosing relief device) must be used to provide relief. A relief valve (reclosing device) will not operate quickly enough to prevent failure of the low pressure side of the exchanger.

All tube failures (except high pressure vapor into liquid) resulting from a tube leak or a tube leak that has propagated into a full tube break can be addressed with a Safety Instrumented System. For high pressure vapor into liquid an engineering analysis should be conducted to verify if a SIS has sufficient time to operate. All tube failures resulting from a full, instantaneous tube break need to be addressed with a relief device; a SIS cannot operate quickly enough to prevent vessel failure.

Rupture disks may be undesirable to protect liquid-filled exchangers because of the potential for disk failure due to water hammer. Proper effluent handling must be provided based on the hazards of the relieving fluid, which is not the normal contents of the side of the exchanger that requires protection. Reaction due to mixing of incompatible materials in the exchanger must also be evaluated in the relief design. Designing the exchanger to limit the operating ratio below 2.5 should be strongly considered.

- Introduction of Volatiles

Potential hazards exist when materials of significantly different volatilities are present in the process. Leakage of a low boiling liquid into a hot vessel can result in a high vapor generation rate as the low boiler flashes. Low boiling liquids can enter a hot vessel due to a tube rupture or process error allowing reverse flow or flow through the wrong path. Relief device sizing for this scenario can result in very large relief devices. This concern can be addressed by selecting a system MAWP above the vapor pressure of the low boilers at maximum process and utility temperatures or by the use of SIS to prevent low boilers from entering hot vessels.

- External fire

For many vessels, external fire is the governing relief scenario because the unmitigated heat input due to a pool fire can require large relief devices to provide emergency pressure relief. External pool fire can be mitigated by the use of fire resistant insulation and/or vessel specific deluge to reduce the required size of the relief devices. Due to the high temperatures that can be experienced during a pool fire, vessels containing high boilers (boiling point above the vessel rated temperature) and gas-filled vessels can fail due to overheating of the vessel walls before reaching the set pressure of the relief devices. A depressurization system can be designed to de-inventory these vessels prior to failure. Safety Instrumented Systems cannot be used to mitigate external fire because the components can be damaged or destroyed by the fire.

Development of the process safety design philosophy and documentation of process hazards during the conceptual design will ensure that this information is available to the equipment designers and control system engineers as they start their efforts during Front-End Engineering. This will ensure that equipment specifications will include the requested MAWP as well as nozzles for the installation of relief systems and the instrumentation needed for the SIS. Safety Instrumented Systems can be addressed in the development of the interlock narratives and any non-SIL rated or BPCS interlocks used for hazard mitigation can be included in control systems design.

III. Front-End Engineering

Information from the conceptual safety review and general safety design philosophy will feed into the development of the Piping and Instrumentation Diagrams (P&IDs). Since the relief and Safety Instrumented System design efforts may be conducted in parallel with the Preliminary Hazard Analysis (PrHA), it is critical that the preliminary safety interlocks and relief systems are represented on the P&IDs that will be used during the PrHA. This will allow the team to assess the adequacy of the proposed mitigations and controls and will minimize recommendations that simply identify the need for mitigation that will have to be designed after the conclusion of the study. It is more effective and efficient for the team to assess the adequacy of the proposed mitigations and controls during the study. Representation of the preliminary safety interlocks and relief systems on the P&IDs will also allow the interlock and relief system designers to start their design efforts prior to the conclusion of the PrHA.

3.1 Preliminary Hazard Analysis

The PrHA should be conducted early enough in the design phase of the project to allow the hazards to be effectively addressed as part of the process design efforts. Early identification of hazards and the required controls also lowers cost by preventing scope changes late in the project. Waiting for the PrHA team to identify the need for all interlocks and relief systems forces the process safety design to take place later in the project schedule increasing the likelihood of delays and scope changes and may eliminate the capability to efficiently implement inherently safer options.

In order for the PrHA to be effective, preliminary Process Safety Information (PSI), including P&IDs, Heat and Material Balances (HMBs), and relief system and Safety Instrumented Systems design must be available for review. At least two review cycles should have been completed on the P&IDs by the project team and operations representatives. This will allow sufficient time for a thorough review of the P&IDs (first cycle) followed by incorporation of the markups from the review into the P&IDs used for the PrHA (second cycle). Plant layouts and site plans should be available for review, along with the preliminary control system design and the process safety design philosophy.

Since the PrHA is being conducted in the design phase of the project, maintenance, operability, and other design issues will be identified during the study. The study leader should ensure that any design discussions do not interfere with the hazard analysis. If it is determined that the design is still in development such that the PrHA is not effective, it should be stopped and resumed once the design is better developed.

A team report should be generated at the conclusion of the PrHA to provide documentation of hazards and proposed controls for use by the project team. Recommendations made during the study should be incorporated and addressed during the remainder of the design phase.

3.2 Relief Design

The relief scope should be developed and documented during the design effort. The scope should include preliminary sizing, proposed set pressures, and discharge location for each device. Materials of construction should be identified for the relief devices due to the potential for long-lead times on specialty alloys. The relief designer should review the P&IDs to evaluate the need for relief devices on utility supplies, piping systems, pumps, and downstream of control valves or pressure reducing systems. Proposed relief systems with discharge location should be represented on the P&IDs used for the PrHA.

Once the PrHA has been completed, the relief designer should review the team report to ensure that all scenarios requiring relief are included in the sizing basis. The relief scope should include specific guidance for completing relief calculations for the project, including codes, standards, regulations, and good engineering practices that must be incorporated. A plan should be developed for review and approval of all relief calculations generated for the project.

3.3 Safety Instrumented Systems

The SIS scope should be developed and documented during the design effort. The scope should include identification of the initiating event, proposed initiator and actions, and required reliability. LOPA or other risk assessments should be used to determine the required reliability for proposed interlocks. This will allow the required instrumentation, field elements, logic solver, and other components for the SIS to be included in the overall control system design. The interlock designer should review the P&IDs to evaluate the need for SIS on pumps or other rotating equipment. Proposed SIS should be represented on the P&IDs used in the PrHA. Once the PrHA has been completed, the interlock designer should review the team report to ensure that all scenarios requiring mitigation are included in the SIS design.

The SIS scope should include specific guidance for completing the interlock design for the project, including codes, standards, regulations, and good engineering practices that must be incorporated. A plan should be developed for review and approval of all LOPAs and Safety Integrity Level (SIL) verification calculations generated for the project.

IV. Detailed Design

Information developed in the Front-End Engineering effort will feed into the detailed design efforts for equipment design, instrumentation and control systems, piping systems, relief systems, and Safety Instrumented Systems.

4.1 Initial Process Hazard Analysis

The intent of the initial PHA is to provide a comprehensive safety review of the near-complete design package. In order for the study to be accurate, complete, and representative of the process as-built, it should take place near the end of detailed design prior to construction of the facility. Ideally, the design package should be at least 80% complete when the PHA is performed. The study may focus primarily on the process changes that take place between the completion of the PrHA and the initiation of the PHA. A change log must be maintained for significant changes that occur during detailed design to allow revalidation of the PrHA as the basis for the formal PHA. If this change log does not adequately document the changes, the study leader may choose to review the design in total at the time of the PHA. Special emphasis should be given to reviewing process design details, operating procedures, emergency response plans, and startup/shutdown procedures that were not available for review during the PrHA. Additionally, recommendations made in the PrHA which could not be resolved until detailed design should be reviewed during the PHA for satisfactory resolution. Any recommendations made in the initial PHA must be resolved before the process can be commissioned.

Once the initial PHA has been completed, a formal Management of Change (MOC) process must be implemented. MOCs that are generated after the completion of the initial PHA will be reviewed during the normal revalidation of the process PHA.

4.2 Relief Design

The relief designer will be responsible for generating and/ or reviewing relief calculations and evaluating piping layouts. Relief device datasheets will be generated for purchase.

4.3 Safety Instrumented Systems

The safety interlock designer will be responsible for generating and/ or reviewing SIL verification calculations. Instrument datasheets will be generated for purchase.

V. Commissioning

Integration of process safety from concept to detailed design will facilitate safe commissioning and start-up. Thorough safety reviews provide the opportunity for comprehensive reviews of the P&IDs, equipment design, and control systems design that can minimize the difficulties associated with starting up a new facility.

A Pre-Startup Safety Review (PSSR) should be conducted prior to the introduction of process chemicals. This review should verify:

- All interim and final field inspections are complete with all findings documented and tracked to completion
- Safety, operating, maintenance, and emergency procedures are in place
- Initial Process Hazard Analysis is complete, with all recommendations resolved or implemented
- Appropriate training has been completed

All Process Safety Information generated during the project design efforts should be archived and managed according to site procedures and practices. This documentation will be needed for the PHA revalidations that must be conducted every five years for OSHA PSM covered processes.

VI. Conclusion

Process safety is an essential component of process design. Integration of process safety at all stages of project design, from laboratory and piloting efforts to detailed design, allows for early identification of process hazards and more efficient mitigation of those hazards as the design progresses. The use of inherently safer design, relief and effluent handling systems, and Safety Instrumented Systems will result in an optimized process safety plan to achieve a safe and effective final design.

VII. References

[1] Kirk-Othmer, Kirk-Othmer Encyclopedia of Chemical Technology, 5th Edition (Wiley, 2007)

[2] Urben, Peter, Bretherick's Handbook of Reactive Chemical Hazards, 7th Edition (Academic Press, 2006)

[3] Lewis, Richard J., Sr., Sax's Dangerous Properties of Industrial Materials, 11th Edition (Wiley-Interscience, 2004)

[4] Occupational Safety & Health Administration, U.S. Department of Labor, Occupational Safety and Health Standards: 1910.119, Process safety management of highly hazardous chemicals

[5] Browning, Christine E. and Jennifer F. Mize, " Identification of Process Hazards in Laboratories and Pilot Plants", Proceedings of the 8th Global Congress on Process Safety, April 1-4, 2012

[6] American Society of Mechanical Engineers, Boiler and Pressure Vessel Code (BPVC), Section VIII, Division

[7] American Petroleum Institute, API Standard 650: Welded Tanks for Oil Storage

[8] American Petroleum Institute, API Standard 620: Design and Construction of Large Welded Low Pressure Storage Tanks

[9] National Fire Protection Association, NFPA 30: Flammable and Combustible Liquids Code

[10] Occupational Safety & Health Administration, U.S. Department of Labor, Occupational Safety and Health Standards: 1910.106, Flammable and combustible liquids

[11] ANSI/API Standard 521, Pressure Relieving and Depressuring Systems / ISO 23251, Petroleum, petrochemical, and natural gas industries – Pressure-relieving and Depressuring Systems

[12] Center for Chemical Process Safety (CCPS), Guidelines for Pressure Relief and Effluent Handling Systems (CCPS, 1998)

Additional references not cited:

[13] Mize, Jennifer F., "Relief Design for Laboratories and Pilot Plants", Proceedings of the 8th Global Congress on Process Safety, April 1-4, 2012

9TH GLOBAL CONGRESS ON PROCESS SAFETY

PSM OSHA Citations and the Litigation Process: What Are the Risks?

Michael Taylor, Esq.
Jackson Lewis LLP
10701 Parkridge Boulevard, Suite 300
Reston, Virginia 20191
michael.taylor@jacksonlewis.com

Prepared for Presentation at
American Institute of Chemical Engineers
2013 Spring Meeting
9th Global Congress on Process Safety
San Antonio, Texas
April 28 – May 1, 2013

UNPUBLISHED

PSM OSHA Citations and the Litigation Process: What Are the Risks?

Michael Taylor, Esq.
Jackson Lewis LLP
10701 Parkridge Boulevard, Suite 300
Reston, Virginia 20191
michael.taylor@jacksonlewis.com

Keywords: OSHA, Citations, Litigation, OSHRC

Abstract

The federal Occupational Safety and Health Administration launched the National Emphasis Program for the oil refining industry and most recently the chemical manufacturing industry. This has resulted in record-breaking citations and proposed penalties. OSHA has also publicly announced on numerous occasions that it is going to continue to maintain a "strong and credible enforcement presence" in the process safety management community. Accordingly, Process Safety Engineers should know how to identify and evaluate the risks associated with OSHA citations alleging a violation of the Process Safety Management standard, 29 C.F.R. § 1910.119, and have a basic understanding of the litigation process. With this knowledge, Process Safety Engineers can help evaluate whether to contest the citations, which must take place within fifteen days of receiving the citations, and develop a plan of action to minimize the risks.

1. Introduction

This paper identifies and analyzes the many risks associated with OSHA citations alleging a violation of the Process Safety Management standard, 29 C.F.R. § 1910.119. Those risks involve: (1) administrative liability; (2) civil liability; (3) criminal liability; (4) abatement costs; (5) negative publicity; (6) follow-up inspections; (7) Severe Violator Enforcement Program; (8) workers' compensation costs; (9) contract work; (10) collective bargaining; (11) enterprise-wide relief; and (12) safety culture. This paper also discusses the litigation process before the federal Occupational Safety and Health Review Commission ("OSHRC"), the federal agency in charge of adjudicating workplace safety and health disputes between OSHA and private industry.

2. The Risks of Process Safety Management OSHA Citations

Process Safety Engineers should know how to identify and evaluate the risks associated with OSHA citations alleging a violation of the PSM standard. With this knowledge, Process Safety Engineers can help evaluate the whether to contest the citations and develop a plan to minimize the risks.

There are several risks in which Process Safety Engineers should identify and evaluate. First, a Process Safety Engineer should identify and evaluate administrative liability. Under the Occupational Safety and Health Act of 1970 ("OSH Act"), 29 C.F.R. § 651 *et seq.*, a citation item may be characterized as serious, willful, repeat, or failure to abate. A serious violation has a maximum penalty of $7000, a willful or repeat violation has a maximum penalty of $70,000, and a failure-to-abate violation has a maximum penalty of $7,000 per day.

Employers may file a notice of contest because a citation item is characterized as willful, repeat, or as a failure-to-abate. The goal would be to lower the financial risk associated with these types of characterizations along with the stigma attached to them. Employers may also file a notice of contest because the current citations put them at risk of receiving willful, repeat, or failure-to-abate citations in the future. OSHA proposed penalties can be quite costly as evident in the following chart:

Table 1. Top Enforcement Cases Based on Total Issued Penalty

	Company Name	Inspection #	Issuance Date	Total Issued Penalty
1	BP Products North America	311962674 308314640	10/29/2009	$81,340,000
2	BP Products North America	308314640 308314988	09/21/2005	$21,361,500
3	IMC Fertilizer/Angus Chemical	107607863 107607871	10/31/1991	$11,550,000
4	Imperial Suger	310988712 311522858	07/25/2008	$8,777,500
5	O&G Industries, Inc.*	109179937 314295460	08/03/2010	$8,347,000
6	Samsung Guam, Inc.	107329740 106196801	09/21/1995	$8,260,000
7	CITGO Petroleum	110416880	08/29/1991	$8,155,000
8	Dayton Tire	109061648	04/18/1994	$7,490,000
9	USX (aka U.S. Steel Corp.)	100504950 018252858 102873288	10/26/1989 11/02/1989	$7,275,300
10	Keystone Construction Maintenance*	109179952 314295445	08/03/2010	$6,623,000

See http://www.osha.gov/dep/enforcement/top_cases.html

Second, a Process Safety Engineer should identify and evaluate civil liability. In general, state workers' compensation exclusivity provisions prohibit employees from filing a civil cause of action against their employers. *See Larsen's Workers' Compensation Law § 100.01.* These provisions do not prohibit employees from filing a cause of action against another employer that may have caused their injuries. In this regard, approximately thirty two states authorize the use of OSHA standards as evidence of negligence at trial, and approximately fourteen states authorize the use of OSHA standards as negligence per se. In a state that authorizes the use of OSHA standards as evidence of negligence, a plaintiff may argue, for example, that the PSM standard is evidence that the employer had a duty of care. The plaintiff may also argue that noncompliance with the terms of the PSM standard is evidence that the employer breached its duty of care. The employer can still present evidence, however, that it acted reasonably and therefore did not breach its duty of care. In a state that authorizes the use of OSHA standards as negligence per se, a plaintiff may argue that the PSM standard *proves* that the employer had a duty of care. The plaintiff may also argue that noncompliance with the terms of the PSM standard *proves* that the employer breached its duty of care. The employer is prohibited from presenting evidence, however, that it acted reasonably and therefore did not breach its duty of care. Experienced plaintiff counsel will use the OSHA inspection (i.e., employee statements, documents, and photographs) and citations in the civil case. Employers may file a notice of contest because the citations put them at risk in the civil case. Employers may also file a notice of contest in order to minimize the risk the citations may have in a civil case in the future.

Third, a Process Safety Engineer should identify and evaluate criminal liability. Section 17(e) of the OSH Act states that an employer who willfully violates the law and that violation caused the death to an employee shall be guilty of a misdemeanor with a maximum of six months in prison. Moreover, under the Sentencing Reform Act of 1984, 18 U.S.C. § 3551 *et seq.*, an individual may receive a maximum criminal penalty of $250,000, and the organization may receive a maximum criminal penalty of $500,000. The United States Department of Justice handles the criminal prosecution. There may be State criminal prosecution as well.

In the last few years, there has been an increase of federal criminal prosecutions. Employers may want to file a notice of contest because the citations put them at risk in the criminal case. Employers may also want to file a notice of contest in order to minimize the risk the citations may have in criminal prosecution in the future.

Fourth, Process Safety Engineers should evaluate abatement costs. Abatement may be extremely expensive and have minimal or no safety or health benefit, if at all. When evaluating the potential cost, employers should analyze the initial and on-going capital costs needed in order to implement the abatement. When evaluating the potential safety or health benefit, employers should consult with internal or external experts. Abatement costs are often a key factor in deciding whether to file a notice of contest and therefore proceed with litigation.

Fifth, Process Safety Engineers should evaluate the impact of a negative press release. Currently, OSHA has been issuing negative press releases with very harsh language shortly after issuing citations and proposed penalties to an employer. OSHA calls such press releases "regulation by shaming" and asserts that press releases can be "very effective" in doing the right thing. In essence, the employer is presumed guilty prior to being able to defend itself during litigation.

A negative press release can harm an employer in many different ways. A negative press release may, for example, damage an employer's relationships with current customers and employees. A negative press release may also damage an employer's ability to obtain relationships with potential customers. A negative press release may also damage an employer's reputation within the local community. A negative press release can also impact civil and criminal litigation. Simply put, a negative press release can create an impression that the worksite is a very dangerous place to work. By filing a notice of contest, an employer can defend itself by achieving a favorable settlement agreement or winning at trial. In either case, the employer may want to consider issuing its own press release to minimize or neutralize the impact of OSHA's press release.

Sixth, Process Safety Engineers should evaluate a potential follow-up OSHA inspection. In general, employers who tend to "rollover" after receiving significant citations and notifications of proposed penalties often receive more follow-up inspections (and therefore more citations and notifications of proposed penalties) than employers who "politely push back" after receiving the same. It is a myth that if an employer accepts the citations and pay the penalties as proposed, OSHA will not come back to the worksite for a follow-up inspection. Employers who tend to fight everything at no cost often find themselves having repeated visits from OSHA. Accordingly, employers may want to file a notice of contest and "politely push back" to minimize the risk of OSHA conducting a follow-up inspection.

Seventh, Process Safety Engineers should evaluate the Severe Violator Enforcement Program ("SVEP"). On June 18, 2010, OSHA launched the SVEP. Certain criteria must be met before an employer can be put in the SVEP. For example, three or more willful, repeat, or failure-to-abate violations of the PSM standard would put an employer in the SVEP. Once an employer is in the SVEP, OSHA may conduct nationwide inspections of the employer at the same time or over a certain period of time. This can pose a significant risk to an employer and cause major business interruptions. An employer can get out of the SVEP by having the qualifying criteria withdrawn or re-characterized through a settlement agreement or after a trial. Thus, an employer may want to file a notice of contest to try to get out of the SVEP or minimize the likelihood of getting on the SVEP in the future.

Eighth, Process Safety Engineers should evaluate workers' compensation costs. Most states have statutes or regulations that authorize the increase in workers' compensation payments based on OSHA citations. This can be of tremendous significance when there is a fatality of a young employee. An employer may want to file a notice of contest in order to minimize the risk the citations may have in the workers' compensation proceeding.

Ninth, Process Safety Engineers should evaluate whether the citations may impact the ability of the employer to gain private or government contract work. In the private sector, for example, employers may evaluate the OSHA citation history for contractors prior to accepting their bid for work. This causes some contractors to file a notice of contest in order to minimize the risk the citations may have on their ability to gain contract work. In the government sector, for example, the U.S. Army recently eliminated a contractor from a competition for energy-efficient construction work up to $800 million for safety infractions.

Tenth, Process Safety Engineers should evaluate the likelihood that a union may use citations as leverage in the organizing and collective bargaining process. Many unions use citations and penalties as leverage in the organizing process in an attempt to persuade the employees for the need of a union. Many unions also use citations and penalties as leverage in the collective bargaining process in attempt to force an employer to agree to develop and implement other things related to safety or health in the workplace. For example, in 2012 the United Steelworkers used the Refinery NEP citations as leverage to obtain safety improvements in the collective bargaining contracts.

Eleventh, Process Safety Engineers should evaluate enterprise-wide relief. For example, in July 2010, OSHA filed a complaint against the U.S. Postal Service seeking enterprise-wide relief regarding alleged electrical hazards at its Providence, Rhode Island facility. OSHA requested that the OSHRC order the U.S. Postal Service to correct electrical hazards at its other 350 facilities nationwide, even though OSHA had not formally inspected the other 350 facilities nationwide. The OSHRC has not determined whether it has this type of remedial authority. Employers would be subjected to enormous risk (i.e., abatement costs; administrative liability, negative publicity) with enterprise-wide relief being enforced. Thus, employers may want to file a notice of contest if they believe that OSHA intends to use this type of enforcement tactic.

Finally, Process Safety Engineers should evaluate the impact the citations will have on the safety culture in the workplace. This applies to the corporate, regional, and local level. Sometimes, employer written program contains safety and health rules go beyond what is required in a standard or regulation; the employer effectively communicates those rules to employees; the employer takes affirmative steps to discover violations of the rule through internal and external auditing; and the employer disciplines employees accordingly when they discover violations of the rules. As a result, employers often file a notice of contest because they have pride in their safety and health program and will not admit that they violated the law.

After identifying and evaluating the risks, Process Safety Engineers should consider whether to have an informal conference with OSHA. Employers can have an informal conference with OSHA on or before the deadline for filing a notice of contest with OSHA. A notice of contest is a legal challenge to the validity of the citation items. Employers have fifteen working days from the date they received citations to file a notice of contest with OSHA. If employers do not file within the fifteen-day period, they lose their right to challenge the validity of the citation items.

In general, OSHA does not make significant concessions at the informal conference stage, particularly when the citations allege violations of the PSM standard. Depending on the Area Office, OSHA may reduce the total proposed penalty by a few dollars and re-characterize a citation item from serious to another-than-serious violation. Typically, OSHA does not make significant concessions until after an employer files a notice of contest and engages in discovery. The informal conference can be a useful discovery tool, however. That is, employers can use the informal conference as a fact-finding mission to determine the facts in which OSHA relies to substantiate the validity of the citation items. This, in turn, will enable employers to better prepare for litigation and settlement discussions as the case proceeds.

3. The Litigation Process

Process Safety Engineers should also have a basic understanding of the litigation process before the OSHRC. With this knowledge, Process Safety Engineers can help evaluate the risks of litigation and determine whether to contest the citations.

Process Safety Engineers should be aware that after the notice of contest has been filed, OSHA is required to forward the notice of contest to the federal Occupational Safety and Health Review Commission in Washington, DC. The OSHRC is a federal agency in charge of adjudicating workplace safety and health disputes between OSHA and private industry The OSHRC assigns a docket number to the notice of contest. This starts the litigation process.

After the OSHRC assigns a docket number to the case, the Chief Administrative Law Judge assigns the case to an administrative law judge. There are three administrative law judge offices: (1) Washington, DC; (2) Denver, Colorado; and (3) Atlanta, Georgia. The administrative law judge assigned to a case typically depends on the location of where the citations were issued. For example, if citations are issued in New Jersey, there is a strong possibility that an administrative law judge from the Washington, DC office will be assigned to the case.

The Chief Administrative Law Judge also assigns the case to Simplified or Conventional Proceedings. A case qualifies for Simplified Proceedings if any of the following conditions exists: (1) there are relatively few citation items; (2) an aggregate penalty of not more than $20,000; (3) no allegation of willful or repeat violation; (4) no fatality involved; (5) a hearing is expected to take less than 2 days; or (6) a small employer. An employer is entitled to receive the following from OSHA: (1) OSHA 1A and 1B Forms; (2) Photographs or Videotape; and (3) Exculpatory Evidence. Employers must provide any documents related to any affirmative defense. After the exchange of information, there is an informal trial. Certain OSHRC procedural rules apply. The Federal Rules of Evidence do not apply.

The administrative law judge assigned to the case may remove the case from Simplified Proceedings without a motion being filed. A party may file a motion to remove the case from simplified proceedings and must show good cause for removal. The parties may file a joint motion to remove the case from Simplified Proceedings and are not required to show good cause for removal. Few cases in Simplified Proceedings actually proceed to a trial. Most cases settle before trial.

In a Conventional Proceeding, OSHA files a Complaint, and the employer files an Answer to the Complaint. The parties may serve interrogatories, requests for admissions, and document requests. The parties may agree to take depositions. At the conclusion of discovery, there is a formal trial. The OSHRC procedural rules, Federal Rules of Civil Procedure, and the Federal Rules of Evidence apply.

If the proposed penalty is $100,000 or greater, the Chief Administrative Law Judge must assign a settlement judge (not the trial judge) to the case, and the parties are required to attend a settlement conference. Typically, the settlement conference is held in person after discovery is complete. Some settlement judges are more formal than other settlement judges. Some settlement judges require the parties to submit a position paper that explains the strengths and weaknesses of their case and what it will take to actually settle the case. The settlement judge does not share this information with the other party. Instead, the settlement judge will use the information and try to persuade each party to reach a middle ground during the conference. Other settlement judges do not require a position paper. These types of settlement judges take a more "hands-off' approach and do not express views regarding the strengths and weaknesses of the case. Experience indicates that the latter type of settlement judge tends to be less successful in resolving the case at the settlement conference.

Like cases in Simplified Proceedings, very few cases in Conventional Proceedings actually proceed to trial. When a case actually settles depends, for the most part, on the attorney representing OSHA. Some attorneys for OSHA attempt to initiate settlement discussions early in the case - after receiving the inspection file. Some attorneys for OSHA will only talk settlement after discovery is complete. There is value in this approach because it is not until after discovery is complete that counsel on both sides can fully evaluate the strengths and weaknesses of their case and advise their clients where a middle ground might be reached. Some attorneys for OSHA will only talk settlement at the eve of trial. This is designed, for the most part, to gain the most leverage as possible, with these types of cases sometimes settling on the courthouse steps.

Even though very few Simplified or Conventional cases actually proceed to a trial, employers must be prepared in the event that the case does not settle. This includes, among other things, drafting a pre-hearing statement, preparing fact and expert witnesses for examination, and assembling exhibits to be used at trial.

After the trial is completed, the administrative law judge will request post-hearing briefs. After reviewing the post-hearing briefs, trial transcript, and exhibits, the administrative law judge will issue a written decision with findings of fact and conclusions of law. The administrative law judge will affirm, modify, or vacate the citations, characterizations, or proposed penalties.

The written decision becomes a Final Order in thirty days from the date of docketing unless within that period of time one of the parties files a Petition for Discretionary Review with the Chairman and the Two Commissioners in Washington, DC, and the Chairman or one of the Two Commissioners grants the Petition for Discretionary Review. Thus, in order to give the Chairman and the two Commissioners time to review and analyze the merits of the Petition for Discretionary Review, a party should file the Petition for Discretionary Review soon after the administrative law judge issues the written decision.

If the Petition for Discretionary Review is granted, the parties will be asked to file briefs. In general, the petitioning party is required to file the first brief, and the brief must be filed within forty days of the briefing notice. The responding party is required to file a brief within thirty days of receiving the brief of the petitioning party. The petitioning party must file a reply brief within fifteen days of receiving the brief of the responding party. An oral argument may, but is not often, be heard.

History shows that oral argument is a rare occasion and takes place only when there is a significant issue that affects the regulated community as a whole.

After reviewing all of the briefs, trial transcript, and exhibits, the OSHRC will issue a written decision. A majority will affirm, modify, or vacate the citations, characterizations, or proposed penalties. A party has sixty days from the date of the Final Order to file an appeal with the appropriate United States Court of Appeals.

An employer is authorized to file an appeal in the circuit in which the violation is alleged to have occurred, where the employer has its principle office, or in the Court of Appeals for the District of Columbia Circuit. OSHA is authorized to file an appeal in the circuit in which the violation is alleged to have occurred, where the employer has its principle office, but not in the Court of Appeals for the District of Columbia Circuit.

After oral argument and reviewing all of the briefs, trial transcript, and exhibits, the United States Court of Appeals will issue a written decision. The citation item, characterization, or proposed penalty will be affirmed, modified, or vacated.

OSHA and Employers can appeal to the United States Supreme Court. Many Years Since the Supreme Court has granted reviewed an OSHRC decision.

NOTE: The litigation process for State plans is substantially similar to the litigation process before the OSHRC.

4. Conclusion

Because of recent explosions at oil refineries and chemical manufacturing plants, OSHA has been keenly focusing on the process safety management community. This has led to National Emphasis Programs for the oil refining and chemical manufacturing industries and has resulted in record-breaking OSHA citations and proposed penalties. Thus, Process Safety Engineers should know how to identify and evaluate the risks associated with OSHA citations alleging a violation of the PSM standard and have a basic understanding of the litigation process. With this knowledge, Process Safety Engineers can help evaluate the whether to contest the citations, which must take place within fifteen days of receiving the citations, and develop a plan of action to minimize the risks.

5. References

[1] "Process Safety Management standard," 29 C.F.R. 1910.119

[2] "Occupational Safety and Health Act (OSHA)," 29 C.F.R. 651, 1970

[3] "Top Enforcement Cases Based on Total Issued Penalty," http://www.osha.gov/dep/enforcement/top_cases.html (accessed Feb. 16, 2013)

[4] "Larson's Workers' Compensation Law," 100.01 (1952)

[5] "Occupational Safety and Health Act (OSHA)," 29 C.F.R. 651. 17(e), 1984, http://www.osha.gov/pls/oshaweb/owadisp.show_document?p_table=OSHACT&p_id=3371 (accessed Feb. 19, 2013)

Safety Culture - The Foundation of Sustainability

Robert Smith/Principal Author
ABB, Inc
3700 W. Sam Houston Pkwy S.
Houston, TX 77042
Robert.Smith@us.abb.com

Prepared for Presentation at
American Institute of Chemical Engineers
2013 Spring Meeting
9th Global Congress on Process Safety
San Antonio, Texas
April 28 – May 1, 2013

UNPUBLISHED

Safety Culture - The Foundation of Sustainability

Robert Smith/Principal Author
ABB, Inc.
3700 W. Sam Houston Pkwy S
Houston, TX 77042
Robert.smith@us.abb.com

Keywords: process safety, asset integrity, sustainability, risk profile, strategies, operational improvement, occupational safety

Abstract

A safe operating environment is cornerstone to any company's operations and working environment. By understanding a company's goals and objectives with a detailed understanding of the current culture, a process can be determined to close the gaps and align each and every employee with the overall goals and objectives of the organization; making a safer and more sustainable operation.

This presentation provides a background on safety and performance management processes while addressing practical approaches to forming strategies for a safety culture, leadership in a safety culture, gaining employee involvement through safety contact and observation, and measuring the success of safety initiatives. All of this is done in the context of continuous improvement.

1. Background

Before we can improve safety, imbed it in the operational culture and sustain it we need to first look at and understand the critical elements that make up the different types of safety. As it relates to this topic they include Occupational Safety, Process Safety – as it relates to safety in high risk manufacturing, such as refining, chemical plants and offshore production assets we also need to look at Asset Integrity Management, which ties to Maintenance & Reliability. All of this is driven ultimately by the organizations risk profile. Lastly to understand where we are going we need to understand the elements of Sustainability.

2. Occupational Health and Safety

Occupational health and safety is a cross-disciplinary area concerned with protecting the safety, health and welfare of people engaged in work or employment. The goals of occupational safety and health programs include fostering a safe and healthy work environment.[1] OSH may also protect co-workers, family members, employers, customers, and many others who might be affected by the workplace environment.

Occupational Safety and Health (OSH) can also is important for moral, legal, and financial reasons. All organizations have a responsibility to ensure that employees and any other person who may be affected by the company's operations remain safe at all times.[2] Moral obligations would involve the protection of employee's lives and health. Legal reasons for OSH practices relate to the preventative, punitive and compensatory effects of laws that protect worker's safety and health. OSH can also reduce employee injury and illness related costs, including medical care, sick leave and disability benefit costs. OSH may involve interactions among many subject areas, including occupational medicine, occupational hygiene, health, safety, industrial engineering, chemistry, health physics, ergonomics and occupational health psychology.

Occupational safety and health primarily covers the management of personal safety, however, well-developed management systems also address process safety issues. The tools, techniques, programs etc. required to manage both process and occupational safety can sometimes be the same, for example a permitting system, and in other cases may have very different approaches. Layers of Protection Analysis (LOPA) or Quantified Risk Assessment (QRA) for example, focus on process safety whereas Personal Protective Equipment (PPE) is very much an occupational safety issue.

3. Process Safety

Process safety focuses on preventing fires, explosions and accidental chemical releases in chemical process facilities or other facilities dealing with hazardous materials such as refineries, and oil and gas (onshore and offshore) production installations.

A common tool used to explain the various different but connected systems related to achieving process safety is described by the Swiss cheese model. In this model, barriers that prevent, detect, control and mitigate a major accident are depicted as slices, each having a number of holes. The holes represent imperfections in the barrier, which can be defined as specific performance standards. The better managed the barrier, the smaller these holes will be. When a major accident happens, this is invariably because all the imperfections in the barriers (the holes) have lined up. It is the multiplicity of barriers that provide the protection.

Figure 1: Swiss Cheese Model

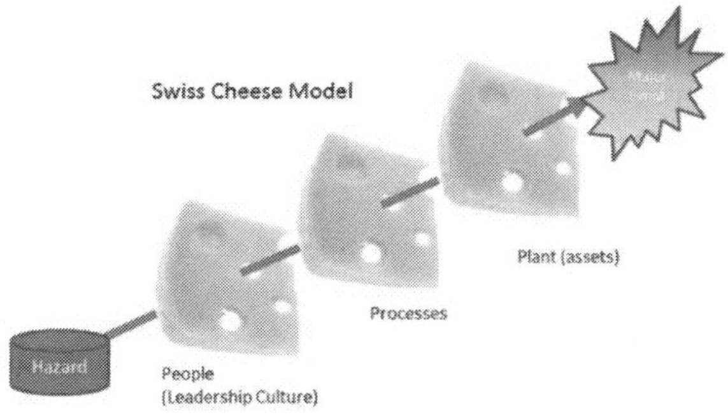

4. Process Safety Management

Process safety generally refers to the prevention of unintentional releases of chemicals, energy, or other potentially dangerous materials, including steam, during the course of chemical processes that can have a serious effect to the plant and environment. Process safety involves the prevention of leaks, spills, equipment malfunction, over-pressures, over-temperatures, corrosion, metal fatigue and other similar conditions. Process safety programs focus on design and engineering of facilities, maintenance of equipment, effective alarms, effective control points, procedures and training. It is sometimes useful to consider process safety as the *outcome* or *result* of a wide range of technical, management and operational disciplines coming together in an organized way.

In short the elements of Process Safety are:

OSHA 29 CFR 1910.119 Requirements	
• Process Safety Information	• Hot Work Permits
• Process Hazard Review	• Management of Change (MOC)
• Operation Procedures	• Incident Investigations
• Training	• Emergency Response Planning
• Contractors	• Compliance Safety Audits
• Pre-Startup Safety Review	• Mechanical Integrity
• Employee Participation	• Trade Secrets

The Center for Chemical Process Safety (CCPS) of the American Institute of Chemical Engineers (AIChE) has published a widely used book that explains various methods for identifying hazards in industrial facilities and quantifying their potential severity.[4] Appendix D of the OSHA's PSM regulations endorses the use of the methods explained in that book.

5. Asset Integrity

Asset Integrity Management improves plant reliability and safety while reducing un-planned maintenance and repair costs.

Asset Integrity Management Systems (AIMS) outline the ability of an asset to perform its required function effectively and efficiently whilst protecting health, safety and the environment and the means of ensuring that the people, systems, processes and resources that deliver integrity are in place, in use and will perform when required over the whole lifecycle of the asset.

An Asset Integrity Management System should address the quality at every stage of the asset life cycle, from the design of new facilities to maintenance management to decommissioning. Inspections, auditing and overall quality processes are just some of the tools designed to make an integrity management system effective.

The AIMS should also strive to maintain the asset in a Fit-For-Service (FFS) condition while extending its remaining life in the most reliable, safe, and cost-effective manner. AIM programs in the U.S. typically target meeting meet API-580, API-581. There is also a PAS 55 standard that is the British Standards Institution's (BSI) Publicly Available Specification for the optimized management of physical assets - it provides clear definitions and a 28-point requirements specification for establishing and verifying a joined-up, optimized and whole-life management system for all types of physical assets. More can be learned about PAS 55 at www.PAS55.net

An asset integrity management system does not exist in isolation. In order to successfully implement an asset integrity management system in a dynamic operating environment, it is essential that all stakeholders have a consistent and a unified understanding of what the essentials of asset integrity are and how they can be applied in their day to day operations, yet this is often cited as among the most significant challenges in achieving an integrity culture within an organization.

A well-managed asset integrity program help operators identify and reduce safety risks before they escalate, but focusing on asset integrity can also play a major role in both achieving operational excellence and extending the life of ageing assets.

6. Sustainability

The concept of sustainability centers on a balance of our society, economy and environment for current and future health. Sustainability is the intersection of people, planet and profit, commonly referred to as the "three-legged stool."

Adopting a culture of sustainability is the ultimate goal for a campus concerned with the impact it has on the community. When a campus embraces sustainability in all aspects of academics, research and operations it will truly be sustainable. A sustainable campus promotes socially just and vibrant communities, provides a means for community members to meet economic needs and conserves resources.

Figure 2 – Three Spheres of Sustainability

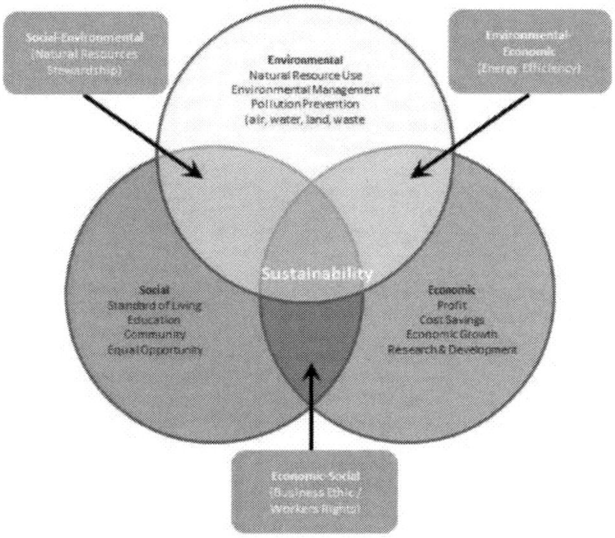

7. Risk Profile

A corporate risk profile is a document that outlines the potential risks to a business operation. Typically the risk profile will outline the number of risks, type of risk and potential effects of risks. This outline allows a business to anticipate additional costs or disruption to operations. It also describes the willingness of a company to take risks and how those risks will affect the operational strategy of the company.

Having and using a Risk Profile is a key step in building an overall safe operating environment. This document should be the foundational element that all subsequent programs and measure are mapped against. From observation in talking to literally hundreds of operations, typically it is found that many organizations either do not have or have not reviewed their own risk profile and it becomes another "binder on the shelf". This is followed by the third group that does not have anything in place.

8. Strategies for Creating a Safety Culture

Fundamentally there are four key pillars that create a safe environment that in turns leads to happy and healthy employees which in turn impacts the attainment of financial goals.

8.1. Leadership and Safety

Management leadership provides the motivating force and the resources for organizing and controlling activities within an organization. In an effective program, management regards worker safety and health as a fundamental value of the organization. Ideally, this means that concern for every aspect of the safety and health of all workers throughout the facility is demonstrated.

The safety and health systems should incorporate:

- The Policy - Reasons for establishing a safety and health program
- The end goal - Where you want to end up and
- The path - how is your business or operation going to get there and what are the objectives.

These are some of the actions recommended by OSHA[8] to ensure that management leadership is in place.

Other recommended actions for management leadership include visible management involvement, assigning and communicating responsibility, authority and resources to responsible parties and holding those parties accountable. In addition, management should ensure that workers are encouraged to report hazards, symptoms, injuries and illnesses, and that there are no programs or policies which discourage this reporting.

8.2. Lead by Example

Managers should look for methods that fit their style and workplace. Successful top managers use a variety of techniques that visibly involve them in the safety and health protection of their workers. Some methods include:

- Getting out where you can be seen, informally or through formal inspections.
- Being accessible.
- Being an example, by knowing and following the rules employees are expected to follow.
- Being involved by participating on the workplace Safety and Health Committee.

OSHA provides some recommendations for leadership teams on how organizations can create a leadership based safety culture.

- Define safety responsibilities for all levels of the organization, e.g., safety is a line management function.
- Develop upstream measures, e.g., number of reports of hazards/suggestions, number of committee projects/successes
- Align management and supervisors by establishing a shared vision of safety and health goals and objectives vs. production.
- Implement a process that holds managers and supervisors accountable for visibly being involved, setting the proper example, and leading a positive change for safety and health.
- Evaluate and rebuild any incentives and disciplinary systems for safety and health, as necessary.
- Ensure the safety committee is functioning appropriately, e.g., membership, responsibilities, functions, authority, meeting management skills, proper meeting protocol.
- Provide multiple paths for employees to bring forward suggestions, concerns, or problems. One mechanism should use the chain of command and ensure no repercussions. Hold supervisors and middle managers accountable for being responsive.
- Develop a system that tracks and ensures timeliness in hazard correction. Many sites have been successful in building this in with an already existing work order / Work Management systems.
- Ensure reporting of injuries, first aid cases, and the near misses. Educate employees about the accident pyramid and importance of reporting minor incidents. Prepare management for an initial increase in incidents and a rise in rates. This will occur if underreporting exists in the organization. It will level off, then decline as the system changes take hold.
- Evaluate and rebuild the incident investigation system as necessary to ensure that investigations are timely, complete, and effective. They should get to the root causes and avoid blaming workers.

8.3 Employee Involvement

Health and safety at work is about securing the health, safety and welfare of people at work by reducing risks, and protecting them and others from harm or illness arising out of work activities by taking the right precautions. It is the responsibility of employers and its management to manage health and safety sensibly in their business, and involving employees is at the very heart of this.

Current legislation provides a framework of requirements for employee involvement in health and safety. Employers are required, under these regulations and other health and safety regulations, to give employees information, instruction and training, and engage in consultation with them about matters affecting their health and safety. Communication and consultation are the most basic forms of employee involvement in health and safety management.

Consultation and workforce involvement is not something you should be daunted by or discouraged about. It does not have to be lengthy and bureaucratic. If you have a small business, consultation can be an informal process of talking to your employees regularly and considering their views when you take decisions about health and safety. It can sometimes be more effective to have a simple process rather than something overly complicated. However, it is equally important that you do make time to talk and think together about issues more formally where this is necessary.

Managers still must manage and make the final decision, but talking to your employees is an important part of successfully managing health and safety.

Don't over think getting employees involved in safety. It does not have to be complex or difficult. At the heart of the issue, executives and managers need to be proficient at creating and maintaining working relationships with their teams. Some simple ways to achieve employee involvement include:

- Post the company's written safety and health policy for all to see.
- Involve employees in policy making on safety and health issues.
- Take an active part in safety activities.
- Hold meetings that focus on employee safety and health.
- Lead by example - Abide by all safety and health rules.
- Invest time, effort, and money in your safety and health program.

8.4 Measuring Program Success

As the saying goes you can only mange what you can measure. Implementing any safety program will require that the program is measured in the form of metrics that will be balanced between leading and lagging metrics, as well as take a balanced look across the organizations operations. For example, one popular measurement structure is the Balanced Scorecard. The Balanced Scorecard was originated by Drs. Robert Kaplan (Harvard Business School) and David Norton as a performance measurement framework that added strategic non-financial performance measures to traditional financial metrics to give managers and executives a more 'balanced' view of organizational performance.

Typically a Balanced Scorecard will look at four key areas in your business, these areas include; Financial, Customer, Employee, and Business Processes – these four areas in turn directly tie to the vision and strategy of the business. It is important to note the difference between leading and lagging indicators.

Lagging indicators are typically "output" orientated, easy to measure but much more difficult to improve or influence. Leading indicators are typically "input" orientated and more difficult to measure, but provide some hint of future performance.

EHS Key Performance Indicators can be found almost everywhere you look, including company press releases and annual reports. Although valuable, these KPIs alone do not drive improved performance since most of the data indicate what has already happened, and you've missed the chance to alter those outcomes.

It is imperative to collect good data, and evaluate the metrics that give a good idea of where the company is headed. Incident rates, Worker Compensation costs and other lagging indicators provide for historical trends, but do little to show what might be in the future. Instead, you need some Leading indicators such as:

- Corrective Actions / Preventative Actions (CAPA) action items overdue
- Audits and inspections completed on time
- Number of audits with compliance issues
- Behavioral Based Safety (BBS) Participation Rate
- Job Safety Analysis (JSA) occurrence count

There are many more potential measures so it is critical to evaluate which metrics will help your business. Literally there are thousands of data points and the key is to define the most important Key Performance Indicators, and displaying those KPIs on intuitive dashboards.

The key to a successful program is to first pick the measurement framework that works best for your organization, then look carefully at the measures that are put in place. The key to success is to start out simple and focused. From observation, many companies start out trying to measure either too many metrics or the wrong metrics or some combination of both which leads to frustration and ultimately the program initiative dissipating over a relatively short amount of time.

8.5 Continuous Improvement and Safety Culture

Continuous improvement is an ongoing effort to improve products, services or processes. These efforts can seek "incremental" improvement over time or "breakthrough" improvement all at once.

Among the most widely used tools for continuous improvement is a four-step quality model—the plan-do-check-act (PDCA) cycle, also known as Deming Cycle or Shewhart Cycle:

- **Plan:** Identify an opportunity and plan for change.
- **Do:** Implement the change on a small scale.
- **Check:** Use data to analyze the results of the change and determine whether it made a difference.
- **Act:** If the change was successful, implement it on a wider scale and continuously assess your results. If the change did not work, begin the cycle again.

Figure 3 Plan, Do, Check, Act (PDCA) Cycle

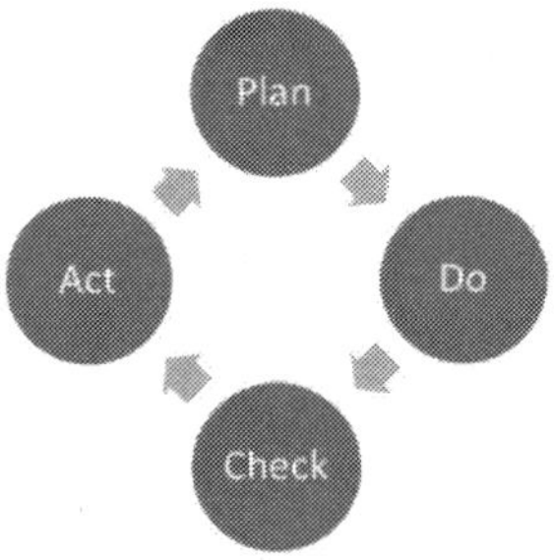

8.6 The Keys to Success

There are several keys to success that should be considered when looking at continuous improvement initiatives. Some of these include:

- **Customize how and where continuous improvement is applied.** One size of continuous improvement doesn't fit all parts of the organization. The kind of rigor required in a manufacturing environment may be unnecessary, or even destructive, in a research or design shop. Sure it's important to inject discipline into product and service development, but not so much that it discourages creativity.

- **Question whether processes should be improved, eliminated, or disrupted -** Too many continuous improvement projects focus so much on gaining efficiencies that they don't challenge the basic assumptions of what's being done. For example, a six sigma team in one global consumer products firm spent a great deal of time streamlining information flows between headquarters and the field sales force, but didn't question how the information was ultimately used. Once they did, they were able to eliminate much of the data and free up thousands of hours that were redeployed to customer-facing activities.

- **Assess the impact on company culture** - Take a hard look at the cultural implications of continuous improvement. How do they affect day-to-day behaviors? A data-driven mindset may encourage managers to ignore intuition or anomalous data that doesn't fit preconceived notions. In other cases it causes managers to ask execution-oriented, cost-focused questions way too early, instead of percolating and exploring ideas through messy experimentation that can't be justified through traditional metrics.

Figure 4. Continuous Improvement Chart

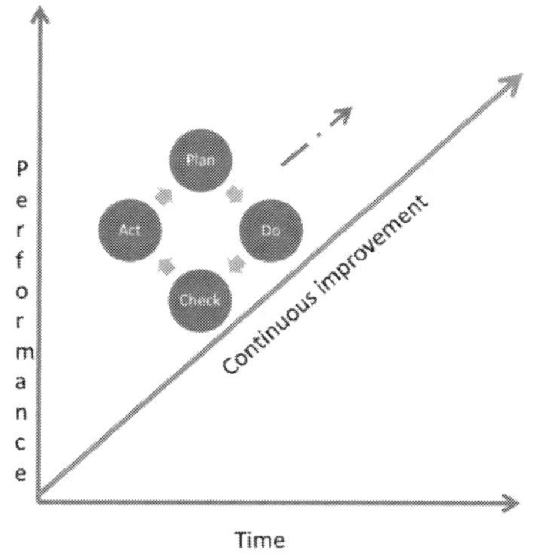

9. Conclusion

Creating a safety culture takes time, requires executive management to buy in both financially and more importantly by action and instill the culture throughout the organization. People outside the organization can always assist in the process, however, the culture of the organizations are born from the actions of its leaders.

To improve the safety culture, an overall safety program must be established, shared and measured. Once the proper metrics and KPIs have been selected a process of improvement and continuous improvement can be implemented. However, the sustainability program will be driven by employee involvement and the culture they embrace on a day to day basis. Having a program is not enough, finding and nurturing the culture of safety is the only way to make safety sustainable, which in turn is one of the key factors in driving operational success.

References

1. "Employers Safe Working Practices, Health & Safety Policy". Citation.co.uk. Retrieved 2013-02-15.
2. "Occupational Health Services And Practice". Ilo.org. Retrieved 2013-02-15.
3. OSHA's PSM regulations at 29CFR1910.119
4. ^ Center for Chemical Process Safety (1992). *Guidelines for Hazard Evaluation Procedures, with Worked Examples* (2nd Edition ed.). Wiley-AIChE.ISBN 0-8169-0491-X.
5. Asset Integrity Engineering http://www.assetintegrityengineering.com/Asset-Integrity/index.html
6. Employee involvement - Involving your workforce in health and safety Good practice for all workplaces First published 2008 ISBN 978 0 7176 6227 2
7. http://www.hss.doe.gov/sesa/corporatesafety/feosh/feosh_employees.html
8. http://www.osha.gov/SLTC/etools/safetyhealth/comp1_mgt_lead.html
9. About the Balanced Scorecard; http://www.balancedscorecard.org/BSCResources/AbouttheBalancedScorecard/tabid/55/Default.aspx. Retrieved 2013-01-07
10. Malcolm Baldrige Quality Criteria for Manufacturers; http://www.nist.gov/baldrige/enter/manufacturing.cfm
11. OSA Safety Health Management; http://www.osha.gov/SLTC/etools/safetyhealth/comp1.html. Retrieved 2012.10.15
12. American Society of Quality. http://asq.org/learn-about-quality/continuous-improvement/overview/overview.html. Retrieved 2012-10-10

Journey to Incident-Free Workplace

Kief Hess
Chemtura Corporation
Philadelphia Pennsylvania
Kief.Hess@Chemtura.com

Prepared for Presentation at
American Institute of Chemical Engineers
2013 Spring Meeting
9th Global Congress on Process Safety
San Antonio, Texas
April 28 – May 1, 2013

UNPUBLISHED

Journey to Incident-Free Workplace

Kief Hess
Director, Process Safety
Chemtura Corporation

ABSTRACT

Safety Excellence is an essential element to achieving manufacturing excellence. This paper describes proven methods of crafting a journey towards achieving safety excellence in the chemical industry. The elements described in this paper helped Chemtura to steer the direction and guide the Corporation on a journey towards safety excellence. The leadership vision, associated techniques for investigating incidents, communication of investigation results, along with the development of tracking tools are described herein. Critical success factors and pitfalls are also illustrated. This paper presents a systematic approach to communicate the vision, collect and assess data, and share risk awareness information across an organization that can lead to an incident-free workplace.

1.0 Introduction

Employee injuries and process accidents cost the chemical industry millions of dollars each year in medical, environmental cleanup, business interruption, legal liability, and equipment replacement costs. Ensuring excellence in safety performance and process safety is essential to minimizing these costs and maximizing manufacturing excellence. Accidents, injuries, and incidents divert vital resources and distract organizations from excelling in manufacturing.
In order to accomplish excellence in safety, companies must provide a clear, simple vision that is understood by all employees so that they know the expectations and the ultimate goal. The vision must be considered achievable, otherwise, employees will not buy in.

Continuously reinforcing this vision with subsequent learning programs throughout the organization help to keep employees engaged and focused on "incident-free."

There are many elements that contribute to the path of success, but for Chemtura, the journey was simple:

Thoroughly investigating all incidents and the generation of corrective actions to address the cause(s),sharing learnings and corrective actions with <u>all</u> employees, expanding investigations and sharing to include critical near-misses, and identifying a series of "leading metrics" to establish that fundamental management systems are in place and fully functional.

Let's take a closer look at our Journey.

Organizations must provide a vision that will ultimately challenge the employees to the end-goal. The path on the journey will change with time based on signals and signs of actual performance and incident rates. For Chemtura, the following provided the direction on the Journey.

2.0 Creating the Vision

The vision is the starting point on the journey to safety excellence. It is a statement set by the corporation to clearly and concisely convey the goal of the company. The vision must be ever-lasting and a frequently communicated statement across the entire organization.

The statement must do three things: engage, stimulate, and energize the workforce.

The vision must be simple, understandable, and, above all, attainable. The least successful visions are those that are too complicated. These are visions that the workforce does not believe can be achieved.

Additionally, the corporation must always uphold the vision to strive towards that end. Focus must be maintained every single day.

A number of years ago, Chemtura Corporation realized that its safety performance record was not acceptable. Chemtura wanted a vision and assembled a broad cross-section of leadership. After using the guidelines stated above, the vision was crafted. It reads, "All employees actively leading the journey of living incident-free."

This vision statement contains several key elements:

First, it is a short and simple statement that can be remembered by all employees.

Secondly, it contains "all employees actively leading…,"which infers that each and every employee must be engaged and active to help lead the efforts toward achieving the goal.

Thirdly, it establishes a journey that has no end date. Chemtura is not striving to be incident-free in one or two years, but, instead, is calling upon employees to work together to keep safety top-of-mind.. It is and can be achievable. Everyone must learn proper safety and must implement new, safer procedures and practices that build an incident-free workplace where injuries, accidents, and incidents do not occur.

So, with this vision, employees are *engaged* in investigations and work task improvement, stimulated to use their "learnings," and energized toward continuous improvement to sustain progress towards the goal.

3.0 All Significant Incidents Must Be Thoroughly Investigated

This is an important step. All incidents must be reported, documented, and posted to a common portal that all employees can access. When an injury, accident, or incident occurs, it must be thoroughly investigated to the minutest detail. Having this drill-down helps to understand not only the factors that lead up to the event, but also shows how well the organization is adhering to rules and procedures and recognizing hazards. The investigation must identify all the failures or "causes," not just a single root cause. A review of several major investigation reports from the chemical industry has shown many incidents may have as many failures that ultimately contributed or led directly to the event. Identifying all the contributing causes and ensuring there are corrective actions for each will help "mis-take proof" the event from occurring again.

At Chemtura, an incident investigation technique called "causal-mapping" was identified as the preferred technique to be used throughout the organization. This simple but very powerful technique allows investigation teams to recreate the details of the incident to identify all the causes and effects that lead to the incident, which, in turn, enables the team to develop specific corrective actions for each cause. Having all significant incidents investigated in this manner helps all employees to better understand how incidents occur and why they occur. This also helps to identify "signals" that may have gone un-noticed had the technique not been used.

Causal mapping starts with identifying the incident and working backwards to identify all the causes necessary to have the effect. For example, in order to have a fire, three elements—oxygen, fuel, and an ignition source—are necessary. Then, each one of these becomes a root for further evaluation. For example, what circumstances made fuel present? What had to occur to have oxygen? By breaking down each component or cause and determining what is needed for the effect, findings all of the influences and "failures" that lead to or contributed to the incident are identified. By knowing all the contributing factors and implementing effective corrective actions for each and many of those factors, future repeat incident should be eliminated.

The causal map allows the team to drill-down into very small details, such as employee actions, positions, and practices, which may otherwise be lost if not used. This drill-down data proves valuable in gaining the "string-of-failures," yielding greater understanding of the activities that led to the event. These are considered "learnings."

In the example below, one can see how the causal maps were developed, with the white boxes to the left indicating the potential effect and the boxes on the right identifying the cause. Each cause, then, is actually an effect for which additional causes can be determined. The pink box is the "evidence" box that contains information that supports or denies the cause-effect relationship. The causal map should have a number of green boxes to represent solutions or possible solutions, which would prevent this cause from occurring again and having a follow-on effect. In this example, there are a number of evidence boxes and solution boxes that help lead to the "causes" of the event. The investigation team will follow the causes all the way to the end until either the evidence will not support further drill-down or it leads to adequate detail about a given cause.

Figure # 1 Example of Incident Causal Map

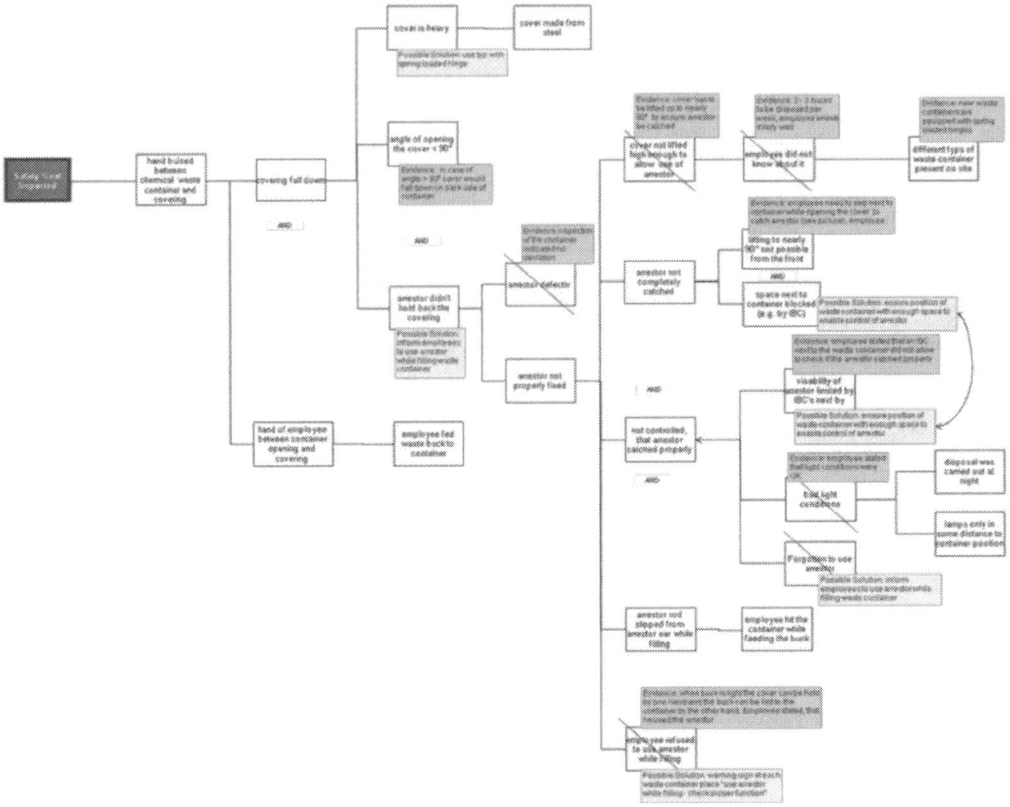

Figure 1 – Causal Map

4.0 All Corrective Actions are Closed

The results of the investigation are an important step; however, it is also critical that the organization take specific corrective action. As part of the investigation, corrective actions for each and every cause are identified (possible solutions or green boxes are discussed above). Of all those possible corrective actions, the feasible/desired corrective actions are selected for implementation. Each of the corrective actions selected must then be closely tracked until closure. Corrective actions are frequently identified during investigations, but following the commissioning of equipment and processes, employees often get distracted, and corrective actions are left incomplete. This can derail the organization towards its goal of instilling appropriate layers of protection to avoid future incidents. If corrective actions are not completed promptly, the employees will quickly decide that corrective actions are not important, and thus the time and effort becomes less valued (and hence, faith in the vision wanes). Firm, specific corrective actions must be identified, resourced, completed, and fully functional following the incident. The organization must have strong will and robust systems to ensure tracking of all corrective actions to prompt closure. If the targeted closure date is not met, then an appropriate escalation to higher management levels in the organization must be communicated to obtain the necessary resources for closure. Persons responsible to close the corrective action must understand that they will indeed be held accountable for prompt closure. As corrective actions are implemented, the organization must continue to implement "layers of protection" to prevent future incidents for long-term sustainability. In this way, not only are the causes of the incident addressed, but the aforementioned vision is given support among the employees.

5.0 Learnings Must be Shared

Sharing the insight of investigations is key to staying on the path to the journey of incident-free. All employees must be acutely aware of incidents and be able to internalize those learnings in their own workplace- no matter where they work in the organization. This is particularly important in a global corporation with multiple locations. Without the sharing of incidents, causes and corrective actions, employees don't become aware of the potential hazards in their workplace and cannot take appropriate action to prevent it from occurring. The potential learning from one incident in one facility, in essence, becomes unused in any of the other plants.

To accomplish this sharing, Chemtura communicates in various ways. First, a corporate and plant leadership call is held to discuss the early facts of the incident. Secondly, a preliminary written report is distributed among the organization leaders. Third, a bi-weekly global conference call is held with any interested member from all sites to share the incident learnings and recommendations. A representative from each site where a significant incident or near-miss occurred will present a "safety alert" during the global conference call, emphasizing the learnings of the incident. The call is recorded for those that miss it. Fourthly, a standard template is used for all safety alerts and is posted on an organizational shared drive for all global employees to access. The safety alerts shared from the global calls can then be used for local safety meetings, tool-box talks, and during site safety committee meetings to ensure a broad communication among all employees. The global discussions have shown that many incidents have occurred previously at other locations, further emphasizing that institutional learnings and corrective actions have not been fully integrated and utilized as an asset throughout the organization.

6.0 Drill Down to Near-Miss Incidents

As thorough investigations are conducted and corrective actions are identified, tracked, implemented, and communicated, the number of incidents will begin to decrease. With a lessening of actual incidents, learning opportunities also fade. For continuous improvement, the organization must recognize this reduction in learning opportunities and support improvement with new, continuing data. To do so, an organization can shift focus to near-misses (NMs) and less significant incidents.

Generally, every organization records near-miss data, but they often deal with the more significant near-misses. At Chemtura, there were several sites that were doing a good job in collecting near-miss data; however, overall near-miss reporting was generally lacking. Reporting failed to identify lesser misses or overlooked the smaller details, namely lack of sign-offs, a diversion from procedure and protocol, failure to complete an MOC form—all of which can represent some form of risk. Near-miss reports can provide a continuous stream of learning opportunities, even if a NM does not lead to an incident. These additional learnings will foster increased understanding of the conditions and behaviors that may lead to accidents and incidents.

Emphasizing near-miss reporting may be difficult at first. Employees may view it as additional work to an already hefty workload. Some may hesitate to report actions of their co-workers. Others may fail to document their own unsafe practices. Also, early on, employees may not be able to recognize all of the potential NMs around them. To overcome this issue, the organization must have strong top-level leadership to reassure the organization of the importance of NM reporting and the valued learnings they can have on further enhancing safety. Hopefully, with employee buy-in and engagement and recognition of the benefits of steps 1-4 of this journey, there will be a bottoms-up support for work on NMs as well.

One approach to control the resource impact on an organization is to select those NM reports which have the greatest risk potential. The NMs that have a high risk potential should be investigated as if had it resulted in an actual significant incident, including using the causal map and sharing those learnings. A good tool to risk-rank NMs is to use a "risk-ranking" chart. This chart has severity (A through F) on one axis and likelihood (1 through 6) on the other. The NMs with a higher potential severity and potential likelihood result in a red risk ranking. Those with red risk rankings are investigated, corrected, and communicated globally. As the red NMs potentials begin to drop off, the organization will investigate and share the yellow risk NMs, or those with a medium potential severity and potential likelihood. See the chart below.

Figure #2 Risk Matrix

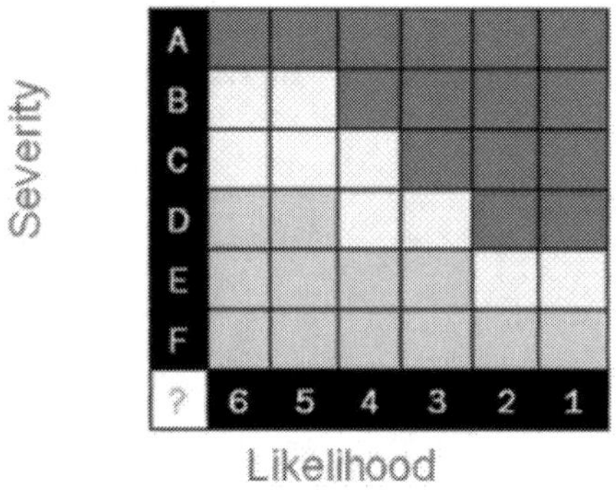

Although an incident did not occur, the learning from these NMs helps to continue the sharing of failures (learning assets) so employees can continue to learn and recognize risks. Understanding the learnings also enhances employees' abilities to see or spot potential hazards before they occur. As more NMs are reported, more learning and risk awareness is acquired. To date,

Chemtura reports nearly thousand NMs per month, with several hundred reviewed and shared globally annually.

There is generally a good correlation of increased NM reporting with lower accidents, incidents, and injuries. However, the organization must be cautious. NMs not only have to be reported, but there must be follow-up in some form of investigation into the cause and subsequent learnings in order for the reports to be beneficial.

At Chemtura, it was initially noted that an increase in NM reporting did not result in improved safety statistics. A standard metric (the number of NM/employee) was established for the sites to report. In this case, the NMs being reported were only done to meet for the metric and not for the value in learning opportunities. NMs were not being reviewed or discussed. We caution that local site leadership must overcome the concern that a large number of NM reporting may reflect poorly on the plant. Leadership must understand and recognize the real value in reporting and sharing across the organization and report as NMs are observed and evaluated. It's not necessarily the number of NMs reported, but the discussions and the understanding of what went wrong that will drive the sites performance to zero incidents.

7.0 Understand trends with Near-Miss Data

The near-miss reporting will yield a tremendous value of information, particularly when analyzed from a causal perspective. At Chemtura, a subcommittee of global employees developed a list of all causes from causal mapping of prior incidents, using a simple "Fish-Bone" diagram technique. This familiar diagram was used to identify specific causal codes and enable sites to select one or more of the 18 highlighted elements as the main cause(s) of the near-miss. Those causes that appear in the greatest number can then be targeted by the organization or local site to drive improvement. For example, if a large percentage of NMs occur due to poor maintenance, then the site will review and make the necessary changes to equipment preventative maintenance. The focus on a specific cause factor from the "fish-bone" will help lead to further reduction in accidents and incidents.

Figure # 4 Causal Factor – Fish-Bone

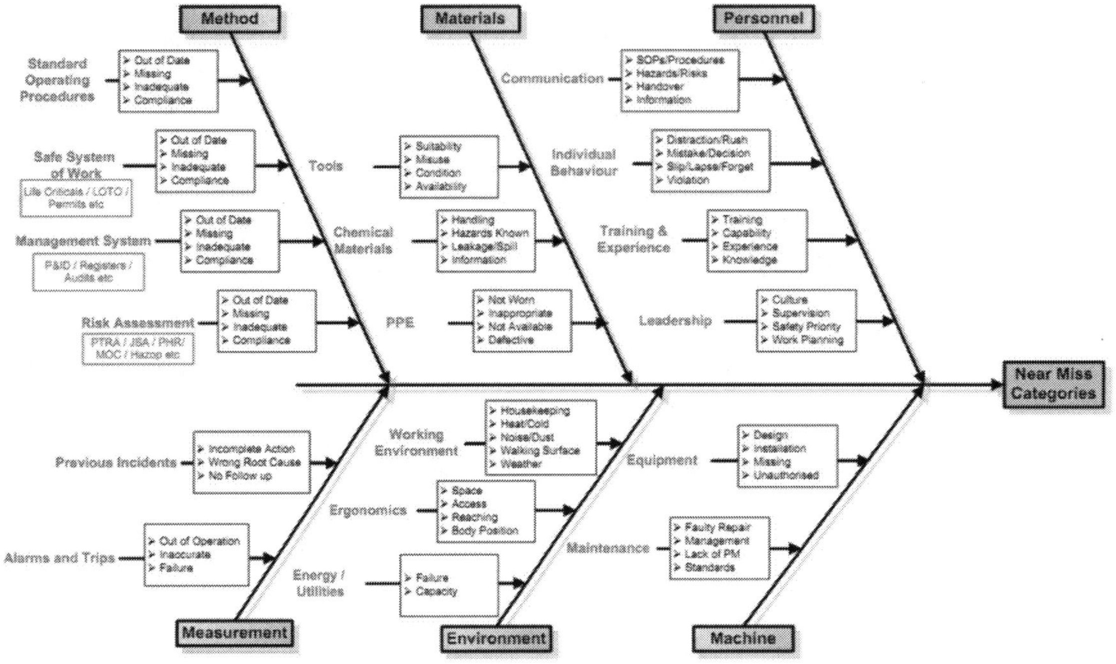

The data can be analyzed on a total organization perspective as well as geographical/regional data to determine if specific regions or even specific facilities have differing causal factors. The organization must use the data and let the data lead and guide future efforts in achieving safety excellence.

8.0 Leading Metrics Help Create the Road on the Journey

The last important compass used on the journey is to use predictive information to determine where failures might occur. Incidents can be measured because they are factual events; however, having the ability to see in advance where potential failure of practices, equipment, and people may occur can be beneficial.

Chemtura wanted to identify which system(s) is the foundation to sustain safety excellence. Once known, having a set of metrics to monitor the implementation and effectiveness of this critical system would help ensure the organization has the proper foundation to build success.

Leading metrics might represent elements of management systems that are in place that help drive improvement in performance. Examples of leading metrics might be: critical inspections, on-time process hazards analysis, past due findings/inspections/trainings, Standard Operating Procedure reviews, number of behavioral observations, audits conducted, actions closure, etc. Using the Chemtura "Layer of Protection," as seen below, the corporation identified PHA and

MI (mechanical integrity) as critical management systems that, if executed efficiently, would provide an optimum first layer of protection to ensure process safety and personnel safety. Why? If the plant does not understand the potential hazards and consequences associated with chemical, equipment, or operational practices, then it is reasonable to believe that incidents will occur. Ensuring the hazard studies are thoroughly conducted and all recommendations are completed on time is considered a critical leading metric. Additionally, the mechanical integrity program is another critical management system. If the organization cannot ensure that all the equipment is properly "fit-for-service" and operated as designed and installed, then it is reasonable to believe that poor mechanical integrity will lead to incidents, accidents, and injuries.

Figure # 5 Chemtura's Layer of Protection Chart

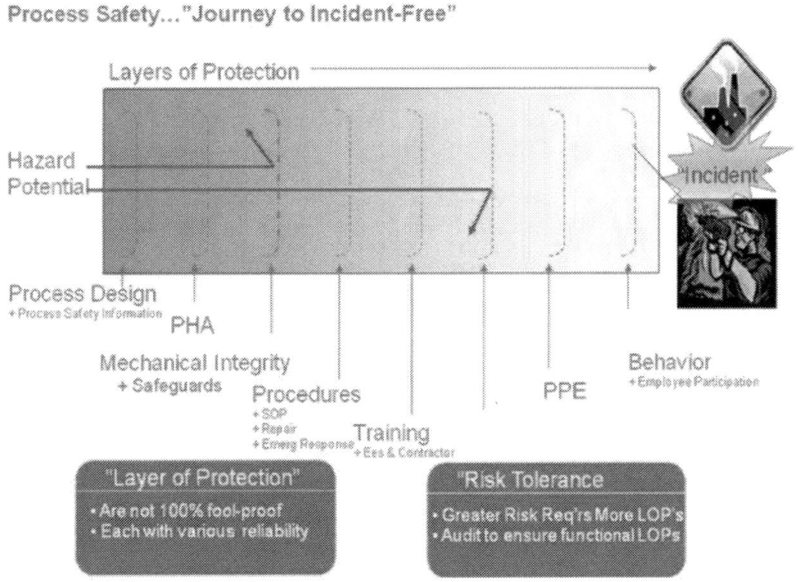

Every day, there are many layers of protection that are accurately and correctly implemented that prevent an incident. An organization must understand the ones they believe are important and monitor their effectiveness.

In the Chemtura organization, leading metrics make up an important element of monthly and quarterly reports. At the end of each month and quarter, each site will enter leading metric data for PHA and mechanical integrity and post on a portal. Following each quarter, the leading metrics are presented to senior management and the Board of Directors. Presenting the data to these groups stresses the importance to all employees the importance of the management systems being monitored (PHA and MI) have on safety excellence.

These simple metrics allow the site management teams to understand how well the PHA and mechanical integrity program is being executed and sustained.

For PHA's, the leading metric are:

% PHA conducted on time – Government Required
% PHA conducted on time – Non-government required
% Recommendations Completed
% Recommendations Past Due >30 days, > 90 days, or > 180 days.

For Mechanical Integrity Inspections, eight various systems are identified which leading metrics are required for each:

Pressure Vessels
Atmospheric Tanks
Relief Devices
Safety Instrumented System
Critical Transfer Systems
Critical Sensors, LEL, Toxic, Etc.
Piping
Electrical

These major equipment categories are believed to be critical to the reliable and safe operation of the facility. Included in each of the equipment categories are:

% inspection complete
% inspections past due in the > 30 days , > 90 days, or > 180 days,
% of "deficiencies" detected or discovered during the inspection.

The recently added metric for "deficiency" represents how well the equipment or system is maintained during normal operation and in between preventative maintenance. Having a leading metric of a very low number of deficiencies may indicate the mechanical maintenance and operability of the equipment is good. However, a leading metrics showing a high number of deficiencies may indicate an incorrect application, a lack of repair, or poor preventative maintenance procedures.

Table #1 Simplified PHA Leading Metric Data Entry

	PHA- Regulated Processes						PHA - Non-Covered Processes						
Site	Total Number	% Completed	% Reco's Closed	Percentage of "Overdue"			Total Number	% Completed	% Reco Closed	Percentage of "Overdue"		
				High	Med	Low				High	Med	Low
Site 1												
Site 2												
Site ...												

Table # 2 Simplified Mechanical Integrity Leading Metric Data Entry

	Mechanical Integrity																	
Site	Pressure Vessels			"Overdue Inspections"			Safety Relief Devices			% "Overdue Inspections"			Critical Interlocks			%"Overdue Inspections"		
	Total	% Up-to-date	Deficiencies	> 30 days	>90 days	>180 days	Total	% Inspected per site plan	Deficiencies	> 30 days	>90 days	>180 days	Total	% tested per site plan	Deficiencies	> 30 days	>90 days	>180 days
Site 1																		
Site 2																		
Site ...																		

9.0 Conclusion

In closing, to achieve manufacturing excellence, an organization must have a robust safety program that includes proper design, well-trained employees, and the proper layers of protection to achieve an incident-free workplace. The journey that will lead to safety excellence must first come from a clear vision that promotes a learning organization. Learning from incidents and near-misses, along with monitoring vital leading metrics, will lead to safety excellence. The examples provided in this paper can have an impact on safety practices towards the goal of having an incident-free workplace.

Engaging employees to actively lead toward the goal, *stimulating* them with data and its analysis, and *energizing* them toward continuous improvement will lead to a sustaining world-class safety performance..

Precise Process Safety Information- A Key for Successful Process Safety Management Program

Sandipan Laskar
Sr. Process Safety Manager
Clean Harbors Inc.
2027 Independence PKWY S
La Porte, TX 77571
laskar.sandipan@cleanharbors.com

Prepared for Presentation at
American Institute of Chemical Engineers
2013 Spring Meeting
9th Global Congress on Process Safety
San Antonio, Texas
April 28 – May 1, 2013

Precise Process Safety Information- A Key for Successful Process Safety Management Program

Sandipan Laskar
Sr. Process Safety Manager
Clean Harbors Inc.
2027 Independence PKWY S
La Porte, TX 77571
laskar.sandipan@cleanharbors.com

Keywords: Process Safety Management (PSM), Process Safety Information (PSI), SMP.

Abstract

Process safety has become one of the inevitable elements in keeping businesses in various industries sustainable. Various industries have suffered painful lessons from incidents and have taken steps to prevent similar tragedies from re-occurring. The Bhopal incident in 1984 and the more recent Deepwater Horizon incident are few examples which show the catastrophic impact of process safety related incidents. Although eliminating risk entirely is not feasible, especially in handling highly hazardous chemicals, recent progress in process safety has improved safety performance through recognizing the risk and implementing safety measures to prevent or mitigate catastrophic consequences.

A safety management system provides a systematic way to identify hazards and control risks while maintaining assurance that these risk controls are effective. The main objective of any safety management program is to reduce the risk to a level that is as low as is reasonably practicable. Although all of the elements of a safety management program are important, one of the core requirements is maintaining accurate process safety information. If these data are not maintained, discrepancies cannot be identified during the Process Hazard Analysis or Pre Startup Safety Reviews, Procedures and Training elements will be based on incorrect information, and loss of containment or other catastrophic events could result from problems with Hot Work, Mechanical Integrity, and other elements. This paper proposes a new methodology to ensure proper documentation of the critical systems and introduces a three phase system to ensure correct, up-to-date process safety information. The paper illustrates how up-to-date process safety information leverages overall process safety management program.

The paper will assist facility personnel to utilize good engineering practice in design and documentation of critical systems, to meet both onshore and offshore process safety requirements, and to develop consistent company best practices for process safety management. In addition to the documentation methodology, the three phase approach furnishes a company with an effective tool for designing, developing and implementing a PSI program that will not only be in compliance with regulation, but will provide a basis for maintaining process, production and facility equipments in a reliable and safe operating condition.

1. Introduction

Process safety information (PSI) refers to the complete and accurate written information about the process chemistry, equipment and technology. Process safety information might be considered as the foundation of PSM Program in that it tells you what you are dealing with from both the equipment and the process standpoint. Much of the required information on the process technology and equipment can be included on the P&IDs. API RP 14J provides an industry consensus guideline for the minimum safety (and environmental) information that should be maintained. Developing and maintaining this information should be adequate for meeting the intent of 1910.119 as well as SEMP. The compilation of the process safety information is to enable the employer and employees involved in operating the process to identify and understand the hazards posed by those processes involving highly hazardous chemicals. This information includes: a simplified process and instrument diagram (P&ID), process design information, relief valve sizing information, process safety information, layout drawings, fire protection and safety equipment information, hazards analysis information, and material safety data. A detailed list delineating the regulatory requirements of process safety information to be in compliance with OSHA PSM is discussed in later part of the paper. The many facets of PSI are shown below. The figure below represents how accurate process safety information leverages other PSM elements.

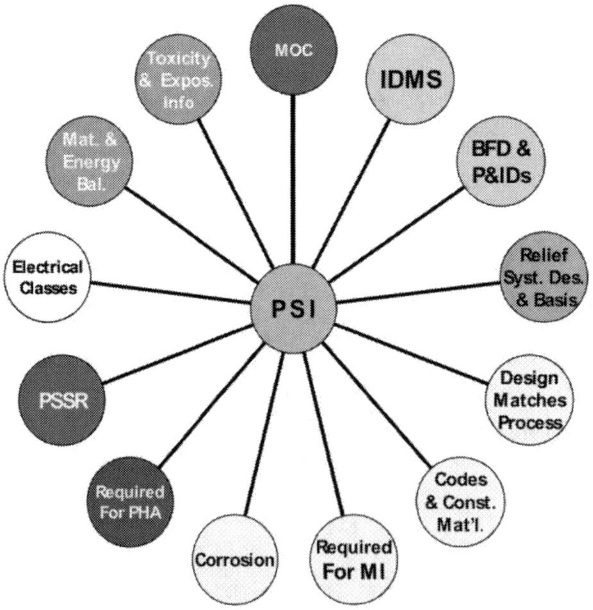

Figure 1- Different Elements of Process Safety Information (OSHA's 29 CFR 1910.119, Para d)

Although all of the elements of a PSM program are important, one of the core requirements is maintaining accurate process safety information. If these data are not maintained, discrepancies cannot be identified during the Process Hazard Analysis or Pre Startup Safety Reviews,

Procedures and Training elements will be based on incorrect information, and loss of containment or other catastrophic events could result from problems with Hot Work, Mechanical Integrity, and other elements. Underscoring the interlinked nature of the PSM program is the fact that design according to RAGAGEP becomes inherent – if a relief device datasheet is to be correct and accurate, it means that the systems the device protects (vessels, pumps, etc.) must have correct and accurate information, and that the design methodology to size the relief device has been properly executed. In other words, correct design with inadequate documentation is as much a problem within the PSM program as correct documentation of an inadequate design. Either situation can and has led to significant incidents in the past.

2. Regulations

2.1. Safety and Environmental Management Program (SEMP)

For offshore facilities, a more holistic approach applying to both production and drilling and without a threshold volume was published originally in 1998 by the API, with the most recent update in 2004 (API 2004). The recommended practice 75 (API RP 75) establishes methods to develop a Safety and Environmental Management Program (SEMP) covering 12 target elements, which addresses all the major aspects of an OSHA PSM program. Although the MMS and industry groups ran a number of audit surveys during the initial implementation period of API RP 75 (BOEMRE 2010), recent compliance status data are not publicly available.

The MMS published an Advance Notice of Proposed Rulemaking (ANPR) in the May 2006 US Federal Register related to the creation of an enforceable PSM/SEMP, and provided a report in 2009 on the comments and data gathered during the voluntary PSM/SEMP period (MMS 2009). Data analysis was performed on a number of MMS incident reports, noncompliance notices, and spill reports, showing a preponderance of incidents resulting from failures in one or more of four main program areas.

Historically the MMS has not been regarded as a strong enforcement agency, though with the reorganization announced by the Interior Secretary in May 2010 (Straub 2010) splitting the field development, enforcement, and revenue functions of the former MMS body, in combination with the promulgation of an enforceable PSM/SEMP and the recent major catastrophe in the Gulf of Mexico, one might have expected a more stringent regulatory environment with greater emphasis on design review and prevention as well as increased penalties and fines should incidents occur. Following a flurry of activity in the second half of 2010, the SEMP as described in API RP 75 became mandatory 15 November 2010 (BOEMRE 2010).

API RP 75 is basically a performance based and non prescriptive standard does not provide much detail to do with the implementation of the elements. API RP 75 does reference a large number of other standards. Most of these are from API itself but some are from other professional organization such as the ASME (American Society of Mechanical Engineers).

2.2. Process Safety Management (PSM)

PSM is abbreviated to Process Safety Management, which is defined as an analytical tool used to prevent release of highly hazardous chemicals. The U.S. Occupational Safety and Health Administration (OSHA) 1910.119 define all elements of process safety management plan. The 14 PSM elements are shown below. Process Safety Management (PSM) refers to a set of inter-related approaches to manage hazards associated with the process industries and is intended to reduce the frequency and severity of incidents resulting from releases of chemicals and other energy sources (US OSHA 1993). These standards are composed of organizational and operational procedures, design guidance, audit programs, and a host of other methods. The different elements of the PSM program, is as shown below.

Figure 2- Elements of Process Safety Management Program

These guidelines are performance based and as such do not provide explicit techniques to achieve the goals of the program. For instance, with respect to compilation of process safety information described in subpart d, the standard requires the compilation of a number of relevant data types in a current and accurate state, but does not specify format or style of presentation, define these data types, or describe methods to validate content. The OSHA PSM guidelines are enforceable throughout the United States and specific US holdings for facilities having a threshold volume of hazardous chemicals, but are not applicable where another federal agency has occupational safety and health jurisdiction, including in federal waters controlled by Bureau of Ocean Management Regulation and Enforcement (BOEMRE, formerly Minerals Management Service (MMS)). Oil and gas drilling and servicing operations and normally unmanned remote facilities are specifically excluded from the standard, but typical facilities in the upstream industry such as production facilities, gas oil separation plants, and gas plants are covered if the threshold chemical volumes, typically hydrocarbons, are present.

Though OSHA had defined all the 14 elements of PSM plan but these elements cannot work as stand alone. All of those elements mentioned above are interlinked and interdependent. There is a tremendous interdependency of the various elements of PSM. All elements are related and are necessary to make up the entire PSM picture. Every element either contributes information to other elements for the completion or utilizes information from other elements in order to be completed. Hence the methodology mentioned in this paper will not only satisfy Process Safety Information element of PSM but also it will facilitate to comply other elements (Process Hazard Analysis, Operating Procedures, Management of Change) of PSM.

In order to be in compliance with the OSHA PSM regulations the process safety information should include information pertaining to the hazards of the highly hazardous chemicals used or produced by the process, information pertaining to the technology of the process and information pertaining to the equipment in the process.

1) Information pertaining to the hazards of the highly hazardous chemicals in the process should consist of at least the following:
   a) Toxicity information
   b) Permissible exposure limits
   c) Physical data
   d) Reactivity data
   e) Corrosivity data
   f) Thermal and chemical stability data
   g) Hazardous effects of inadvertent mixing of different materials that could forseeably occur.
2) Information pertaining to the technology of the process should include at least the following:
   a) A block flow diagram or simplified process flow diagram
   b) Process chemistry
   c) Maximum intended inventory
   d) Safety upper and lower limits for such items as temperatures, pressures, flows or compositions
   e) An evaluation of the consequences of deviations, including those effecting the safety and health of the employees.
3) Information pertaining to the equipment in the process should include following:
   a) Materials of construction
   b) Piping and instrument diagram (P&ID's)
   c) Electrical classification
   d) Relief system design and design basis
   e) Ventilation system design
   f) Design codes and standards employed
   g) Material and energy balances for processes built after May 26 1992
   h) Safety system (for example interlocks, detection or suppression systems)

4) The employer should document that equipment complies with recognized and generally accepted good engineering practices (RAGAGEP)
5) For existing equipment designed and constructed in accordance with codes, standards or practices that are no longer in general use, the employer should determine and document that the equipment is designed, maintained, inspected, tested and operating in a safe manner.

The table below illustrates different elements of regulations for safety management program as applied to downstream and upstream facilities.

Table 1- PSM vs. SEMP Regulation Comparison

OSHA's PSM	SEMP
1. Employee Participation 2. Process Safety Information 3. Process Hazard Analysis 4. Operating Procedures 5. Training 6. Contractors 7. Pre-startup Safety Review 8. Mechanical Integrity 9. Hot Work 10. Management of Change 11. Incident Investigation 12. Emergency Planning and Response 13. Compliance Audits 14. Trade Secrets	1. Safety and Environmental Information 2. Hazard Analysis 3. Operating Procedures 4. Training 5. Pre-Startup Review 6. Assurance of Quality and Mechanical Integrity of Equipment 7. Safe Work Practices 8. Management of Change 9. Investigation of Incidents 10. Emergency Response and Control 11. Audit of Safety and Environmental Management Program Elements 12. Records and Documentation

Generally the OSHA guidelines rely on recognized and generally accepted good engineering practice (RAGAGEP), but allows companies to utilize alternate methods as long as they can prove that performance resulting from that method meets the standard. Good engineering practices are defined in the applicable codes, standards, industry publications and in-house engineering standards. In general, in-house engineering standards supplement the public domain references and should be applied on a case-by-case basis.

The specifications and techniques that form RAGAGEP are generally understood to mean recommended practices from the American Petroleum Institute (API), codes from the American Society of Mechanical Engineers (ASME), standards from the International Standards Organization (ISO), certification agency standards (DNV, Lloyd's, ABS), and other sources in regular use. A list of most frequently used equipments along with the available RAGAGEP for design, inspection and testing is shown below.

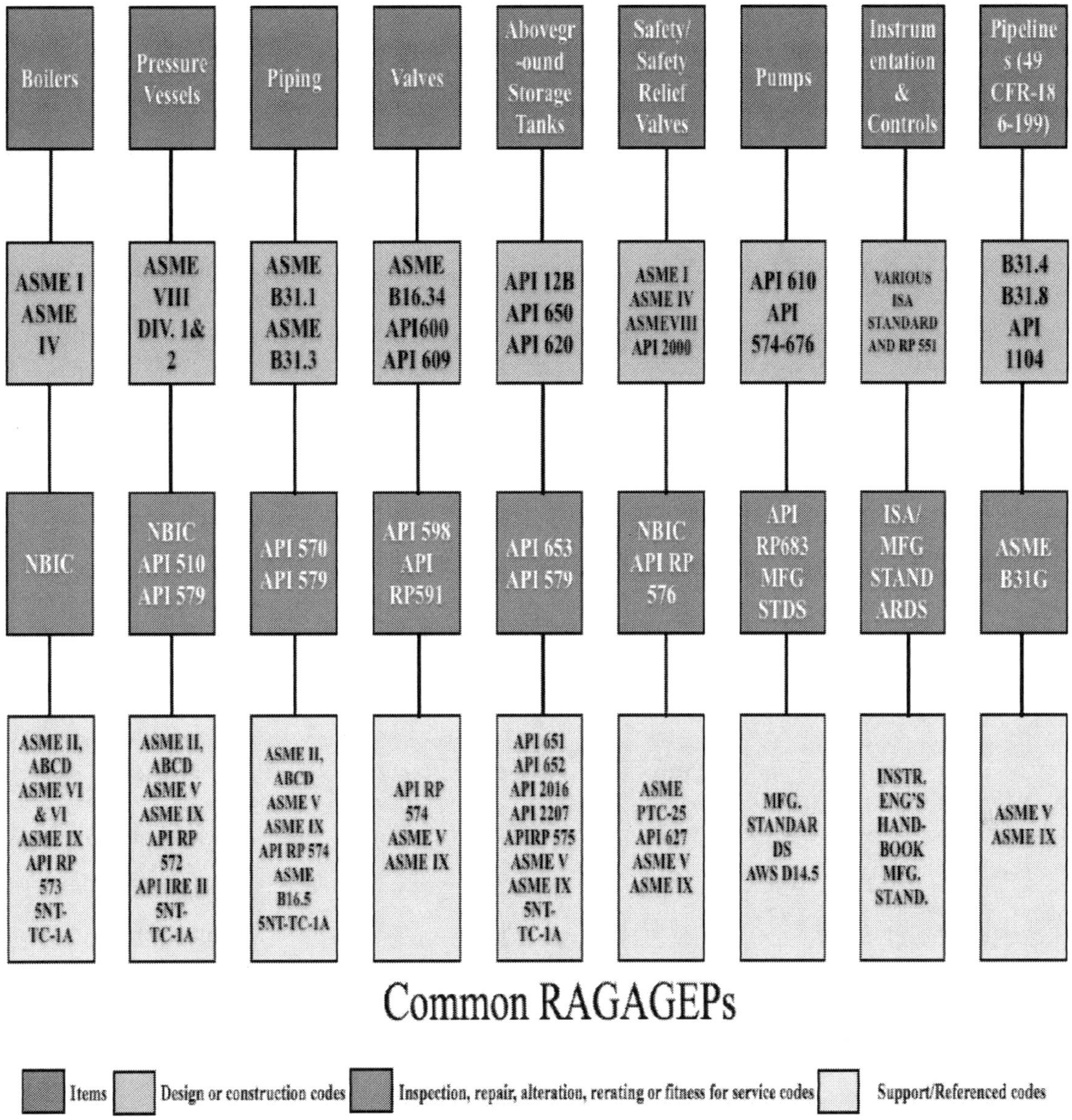

Figure 3- Common List of RAGAGEPs

3. Refinery National Emphasis Program (NEP)

OSHA had already identified numerous violations and issued a significant number of citations. The result of refinery inspections demonstrated that OSHA has adopted a very aggressive enforcement stance and is seeking to impose significant civil penalties and related sanctions for noncompliance. The most recent announcement of expanding the PSM NEP to include chemical plant increased the chances of including the production facilities and gas plants as well as a part of its overall enforcement activities.

Initially, OSHA stated that its PSM NEP inspections would focus on implementation (not documentation) to ensure that substantive workplace hazards are being appropriately addressed. This would have represented a departure from OSHA's oft-perceived focus on "paperwork" violations. However, based on the first-year inspections, it appears that the compliance officers are, in fact, focusing on both implementation and documentation (Table 2). Furthermore, the OSHA inspectors are typically reviewing every element of the complete PSM program. OSHA's apparent continued focus on documentation (and not solely implementation) of relevant standards is demonstrated by the large number of citations related to operating procedure. Furthermore, OSHA compliance officers are seeking to validate whether the operating procedures have been implemented at each individual step specified in the relevant OSHA compliance guidance.

With regard to implementation, compliance officers have been targeting the follow-up portion of several PSM elements, including obligations related to the closure of process hazard analysis (PHA) findings and incident investigation corrective actions. Other implementation areas of focus include execution of inspections on process equipment (for which a willful-violation citation was issued), and updated proves and instrumentation diagrams (P&IDs). Violations were also frequently issued for failure to provide: the relief system design basis (d)(3)(i)(d), safe work practices for employees and contractors (f)(4), written procedures related to employee participation (c)(1) and written procedures associated with equipment inspections (j)(2).

Table 2-The most commonly cited PSM elements during refinery NEP inspections

PSM Element	PSM standard Section 1910.119	Number of Citations
Process Safety Information (PSI)	(d)(3)	49
Operating Procedures	(f)(1)	47
Process Hazard Analysis (PHA) criteria	(e)(3)	40
Mechanical Integrity (MI) Inspection and Testing	(j)(4)	32
PHA Recommendation Follow-Up	(e)(5)	16
MI-Operating Deficient Equipment	(j)(5)	16
Management of Change (MOC)	(l)(1)	15

4. Importance of Process Safety Information (PSI)

Process safety information is the information about the process chemistry, equipment and technology of production facilities or chemical plant. It is collected from many places inside and outside the respective company: research and development, engineering, operations, and also suppliers of raw materials, process technology, and equipment. For most personnel the first exposure of the PSI may have been at a Process Hazard Analysis (PHA). PSI is essential to the implementation of an effective process safety management program as well as a process hazard analysis. The compiled information will help an organization to assess and determine the following:

- Fire and explosion risks
- Reactivity hazards
- Safety and health hazards to workers
- The corrosion and erosion effects on the process equipment and tools
- Proper inventory levels of flammables and chemicals to maintain safe use and storage conditions

The process safety information allows a team to assess if engineering practices around process design as well as any process equipment used to date is compliant with established codes and standards. Correct and up-to-date process safety information provides the fundamental building block for a strong safety management program. PSI will help document process equipment, facility instrument data and will provide ongoing evidence of successful testing and inspection, and stable operation. For example, a new valve must meet the specifications on the PSI from the plant engineering design. Some other examples of PSI include piping and instrumentation drawings, hazardous area classification drawings and the reports from process hazard analysis, management of change reviews, incident investigations, personal protective equipment requirements, operating and maintenance procedures. A complete list of elements of the PSI program is shown in figure 1.

5. Methodology

Process safety information (PSI) is essential to safe plant operation and maintenance, but it is valuable only if it is correct, up to date, and used. Some of the challenges of PSI program includes but are not limited to ready availability, data accuracy between different repositories, document control and information validation. The three phase methodology delineated in this paper will help to eradicate the challenges and help to set up a well structured process safety information program and thereby leveraging the existing safety management program.

5.1. Process Safety Information Program Design – Phase I

The first step in designing of process safety information program is to understand and outline the company objectives. The compliance philosophy of the company must be defined here. The objective and compliance philosophy should be recorded and kept with the program manual for future reference. This information is very useful when new processes and equipments are being considered for addition into the PSI program. Once the program objective is defined then the PSI team should be developed and a team leader should be designated. At this time resources must be identified and allocated for the program. As shown in figure 2, a commitment from management is the foundation for a successful process safety information program. The team leader in most instances is more effective when his or her responsibilities include not only the initial design, development and implementation of the program but the continuous operation or management of the program as well. The PSI team should consist of subject matter experts (Engineering and Information Technology), an advisory group both internally at site and externally at corporate level, plant manager, production manager, safety and health and the upper management of the company.

If an existing process safety information program is in place, program and schedules should be reviewed to verify that the available process information is up to date and accurate in different repositories. As shown in figure 1 and 2, process safety information is directly integrated with all other PSM elements. During review of the existing system or design of new program, verification of this interface should be made.

5.2. Process Safety Information Program Design – Phase II

The next step in developing a process safety information program is the team to identify the covered process and generate a rationale for applicability of PSM to a specific process. The team should define clearly PSM applicability as well as non-applicability of all areas within a facility and should also state the threshold limit. The OSHA Process Safety Management Regulation (29 CFR 1910.119) and EPA Risk Management Programs Regulation (40 CFR Part 68) may help drive this decision for covered processes. It is recommended that a set of piping and instrumentation drawings for the process be color-coded to designate covered equipments. This will facilitate efficient identification of the covered equipments in the process.

The PSI team should evaluate different repositories where the process safety information should be stored and different tools available. All process safety information databases should have a standardized structure. If a company has different facilities then each facility should have similar process safety structure. A typical folders structure for multiple facility process safety information is shown in figure 4.

The main PSM folder is sub-divided into multiple facilities. Each facility should have individual folder all 14 elements of the process safety management program. The PSI folder is further sub-divided into three folders, drawings, equipments and relief devices. The drawing folder should have an index file listing all the drawings. A typical drawing index list is shown in table 1. This folder should list all the drawings including piping and instrumentation drawings, electrical drawing, process flow diagrams and plot plans. The equipment folder should contain all equipment information as listed below.

- U-1A forms
- Equipment manuals
- Equipment drawings
- Design datasheets
- Pump curves

- Testing information
- Actual construction drawings
- Equipment nameplate
- Any MOC's related with the specific piece of equipment

An example of a standardized process safety information folder structure for a company with multiple facilities is shown in the figure below.

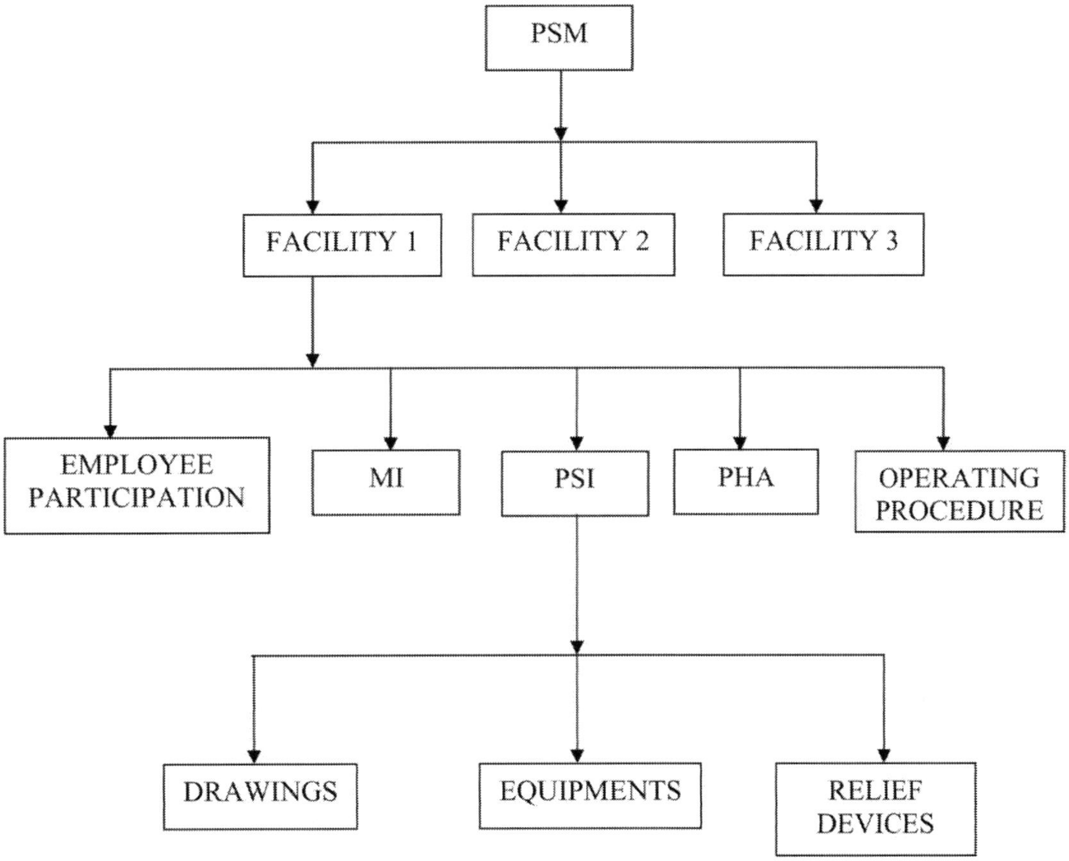

Figure 4- Typical Folder Structure for a Process Safety Information Database

The drawing index list should list drawing number, revision, notes and comments. It should have a provision to check in and checkout drawings from the repository which serves an effective way to control the access of document in a facility. The drawing index list provides a comprehensive list of all the drawings namely Process and Instrumentation Diagram (PID), Process Flow Diagram (PFD) and Electrical one line drawing.

The other element of the process safety information folder is the equipments sub folder. A typical equipment index list is shown below.

Table 2- Typical Index List for Process Equipment

SR No.	PID	Equipment Tag#	Description	Type	PSM	Design Code	Comments
1	H-201-PI-103-D	T-72	Waste Storage Tank	TNK	Yes	API 650	
2	H-201-PI-103-D	T-77	Waste Storage Tank	TNK	Yes	API 650	MOC 23x pending
3	H-201-PI-104B-D	T-70	Waste Storage Tank	TNK	Yes	API 650	
4	H-201-PI-104B-D	T-75	Waste Storage Tank	TNK	Yes	API 650	
5	H-201-PI-105A-D	T-27	Waste Storage Tank	TNK	Yes	API 650	
6	H-201-PI-105B-D	P-7	Pump inlet line 2"WL-P05-79-2" ST, discharge line2"-WL-P05-80	P	Yes	API 610	
7	H-201-PI-105B-D	P-8	Pump inlet line 2"WL-P05-79-2" ST, discharge line2"-WL-P05-80	P	Yes	API 610	
8	H-201-PI-106-D	P-204	Pump discharge to 2"TOX-POS	P	Yes	API 610	
9	H-201-PI-103-D	T-34	Waste Storage Tank	TNK	Yes	API 650	
10	H-201-PI-103-D	T-35	Waste Storage Tank	TNK	Yes	API 650	
11	H-201-PI-104B-D	T-37	Waste Storage Tank	TNK	Yes	API 650	
12	H-201-PI-104B-D	T-44	Waste Storage Tank	TNK	Yes	API 650	
13	H-201-PI-105A-D	T-45	Waste Storage Tank	TNK	Yes	API 650	

The relief valve folder should contain relief valve datasheet, calculations, inspection information, manufacturer manual and design basis. The critical system design calculations should also contain the scenario identification, sizing basis and the protected equipment design documents. The relief valve index list should provide information regarding valve tag number, equipment protected, inlet and outlet size, orifice designation and codes used for designing of the relief valve. A typical relief valve index list is shown below in table 3. The folder structure shown above is for the implementation of a simple PSI program, in case of more complex program with huge amount of data an electronic management system should be in place to manage and control data access. The electronic management system helps in managing the data more efficiently and also control the data access.

Table 3- Typical Index List for Relief Valve

Sr No	Tag #	Location	PID	Model	Inlet	Outlet	Set Pressure (psig)	Design Codes	Comments
1	PSV 2072	on Waste storage tank (T-72)	H-201-PI-XX	Crosby	2	3	50	API 520/521	
2	PSE 2074	on Waste storage tank (T-72)	H-201-PI-XX						
3	PSV 2077	on Waste storage tank (T-77)	H-201-PI-XX	Crosby	2	3	50	API 520/521	
4	PSE 2078	on Waste storage tank (T-77)	H-201-PI-XX						
5	PSE 2079	on Waste storage tank (T-78)	H-201-PI-XX						
6	PSV 2080	Waste tank (T-78)	H-201-PI-XX	Crosby	2	3	50	API 520/521	

The process safety information program development phase also includes developing a robust document control system which tracks the modifications. The document control of the process information is very critical as this defines the accuracy of the information over a period of time. The primary function of document control is to track the changes as well as maintain a history log of all the changes made to the information of a specific piece of equipments or drawings. The document control provides an easy way to verify all the changes and also it can provide information about this document is most recent.

A faulty document control will generate multiple copies of the same process information which will leads to an in accuracy in the data between different repositories and this leads to the third and final phase of the process safety information methodology.

5.3. Process Safety Information Program Design – Phase III

The last phase of the process safety information program design involves update and validation of the process safety information on a regular basis. A typical example of steps involved in validation of piping and instrumentation drawing is shown below.

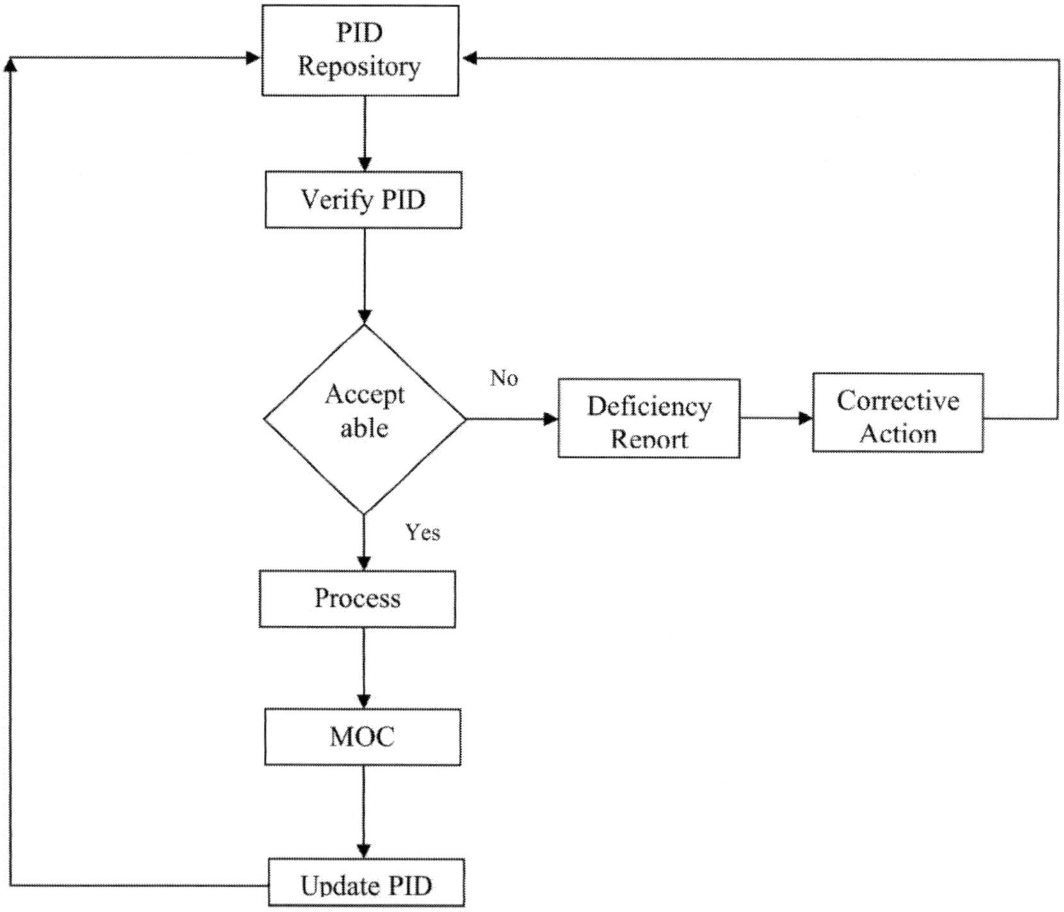

Figure 5- PID Validation Work Process

The validation of the existing process information is an important step for the overall success of the process safety information program. Mostly all chemical plants and refineries re-validate their critical system design over a 3 to 5 years basis but during this there could numerous changes in the process and associated process equipments. So to capture those changes in the updated critical system analysis the up to date process information is of utmost importance.

6. Conclusion

The three step process mentioned in this paper will ensure proper documentation of the critical system. The methodology mentioned above will help to standardize the process safety information, accurate up to date process data and thereby lay the foundation for a strong process safety management program. The proposed methodology ensures process safety without creating a large paperwork burden. The database provides operating and maintenance staff, supervisors and foreman with the information, training and motivation they need to conduct their business in a safer manner. It is also intended that the information presented will help in proper documentation and thereby enable a faster response in case of an emergency. The above mentioned methodology and the database ensures most cost effective process for implementing process safety management program in a manner that meets OSHA's expectations while contributing to a safer production facilities.

7. References

[1] "API RP 75, Recommended Practice for Development of a Safety and Environmental Management Program for Offshore Operations and Facilities", third edition. 2004. Washington, DC: API

[2] "Oil and Gas and Sulphur Operations in the Outer Continental Shelf--Safety and Environmental Management Systems", Bureau of Ocean Energy Management, Regulation and Enforcement Implementation (BOEMRE), *Final Rule*, 30 CFR Part 250, Volume 74, Number 199, (October 15, 2010)

[3] "Petroleum Refinery Process Safety Management National Emphasis Program", U.S. Occupational Safety and Health Administration (US OSHA), Directive CPL 03-00-004, (June 7, 2007)

[4] "PSM Covered Chemical Facilities National Emphasis Program", U.S. Occupational Safety and Health Administration (US OSHA), Directive CPL 09-06-002, (July 27, 2009)

[5] "Survey for API RP 75", Bureau of Ocean Energy Management, Regulation and Enforcement Implementation (BOEMRE), (September. 17, 2010)

(43a) Learning From High Potential Near Misses in the Refining Sector

Ramesh VM, Research Scholar, Bharathiar University, 641046, India and Dr. Sakthivel R, Management Studies, Govt Arts College, 641018

Abstract: Safety is the first priority for any industry. Though steps are taken to ensure an accident-free environment, "Near Miss" is one phenomenon whose possibility of occurrence cannot be ruled out completely. What matters here is the way in which such instances are tackled preventing accidents. Moreover, these incidents can be used as an effective tool in education and training. Each near miss thus offers a free learning tool and a whole new experience with a scope to preach and practice efficient safety mechanisms. Near misses can be classified into two types- Low potential near miss and High potential near miss. All near misses whether it is of Low or high potential require in-depth analysis leading to identification of the root cause so as to avoid the recurrence of such incidents in future. The outcome of every near miss is to reach down the line so as to spread the useful information to all working in the industry. Many workers do not report "near misses" due to fear of negative actions from the management that can deprive them of their future progressions in the Industry. Steps need to be taken to overcome this fear as capturing of all near misses will definitely help the industry in bringing out innovative ideas to ensure safety. The exercise of accident prevention will bear fruits only if the effort to spread the message of usefulness of near miss reporting is spread down the line. Some of the near misses that occurred in the refinery sector are identified to be of High potential and major mishaps were averted only because of the timely action in the second of need. The ability of the concerned to do the right job at the right hour can be attributed to the training given to every employee on fire fighting which includes the analysis of near miss as well as the analysis of various other accidents happened in the other industries.I wish to share few high potential near misses which can be a learning to all working the refinery sector.

Felt Leadership: An Essential Ingredient in Successful PSM Implementation

Brian D. Rains
DuPont Sustainable Solutions
974 Centre Road; CRP-730
Brian.D.Rains@DuPont.com

Prepared for Presentation at
American Institute of Chemical Engineers
2013 Spring Meeting
9th Global Congress on Process Safety
San Antonio, Texas
April 28 – May 1, 2013

UNPUBLISHED

Felt Leadership:
An Essential Ingredient in Successful PSM Implementation

Brian D. Rains
DuPont Sustainable Solutions
974 Centre Road; CRP-730
Brian.D.Rains@DuPont.com

Keywords: Felt Leadership, PSM, High Performing Work Organizations

Abstract

"Felt Leadership" is a term that is frequently mentioned in today's PSM conferences and informal conversations as a quality that must be demonstrated by management if PSM implementation is going to be successful. Interestingly, these two words were first introduced by the DuPont Company over 20 years ago and the context was very different then and not even directly related to safety. Felt Leadership was one of the six attributes of a "High Performing Work System." As one of these 6 attributes, it was deemed essential to achieve outstanding operations and business results through people. This paper will not only describe "felt leadership" in meaningful, practical terms but will also deliver insights as to why it is both appropriate and necessary for effective PSM implementation as well as high performing work organizations.

1. Introduction

One of the clear lessons learned from the Baker Panel Report of the BP-Texas City incident was the role of leadership in successfully implementing and sustaining excellent PSM performance. Of the Panel's ten recommendations, five spoke directly to changes that leadership needed to make. One of the panel members, Dr. Paul V. Tebo, former DuPont Vice President of Safety, Health, and Environment, has actually stated that in his opinion, all ten recommendations speak to leadership. He went on to say that the fundamental, underlying issue at BP-Texas City was "leadership, leadership, leadership."

- **RECOMMENDATION # 1 – PROCESS SAFETY LEADERSHIP ****
- **RECOMMENDATION #2 – INTEGRATED AND COMPREHENSIVE PROCESS SAFETY MANAGEMENT SYSTEM**
- **RECOMMENDATION #3 – PROCESS SAFETY KNOWLEDGE AND EXPERTISE**
- **RECOMMENDATION #4 – PROCESS SAFETY CULTURE ****

- **RECOMMENDATION #5 – CLEARLY DEFINED EXPECTATIONS AND ACCOUNTABILITY FOR PROCESS SAFETY ****

- **RECOMMENDATION #6 – SUPPORT FOR LINE MANAGEMENT ****

- **RECOMMENDATION #7 – LEADING AND LAGGING PERFORMANCE**
- **NDICATORS FOR PROCESS SAFETY**

- **RECOMMENDATION #8 – PROCESS SAFETY AUDITING**

- **RECOMMENDATION #9 – BOARD MONITORING ****

- **RECOMMENDATION #10 – INDUSTRY LEADER**

**** Direct Leadership Responsibility**

As helpful and appropriate as these corporate recommendations were at the time and continue to be, the subject of PSM leadership can, at times, need to be brought down to a very individual and even personal level. In recent presentations at PSM Conferences, the term "felt leadership" has been used to attempt to describe a brand or style of leadership that is also necessary in complex chemical processes and operations. It has been used to describe a quality of leadership that can be sensed or "felt" rather than just heard or seen. It suggests leadership with passion and authenticity and that is genuine and even humble.

2. Origin of Felt Leadership in DuPont

The term "felt leadership" was first used in the DuPont Company in a systemic way in 1991. And the context was very different. In fact, it was not used in connection with safety in any direct way.

A group of Plant Managers, of which I was one, had been formed into a body called the "Plant Managers Board." We had a very specific task and that was to find and identify best practices related to "High Performing Work Systems" or HPWS for short. High Performing Work Systems were described then as "high involvement systems of accomplishing work in which all employees have developed the capability to connect with and drive the quality of business results to be world class."

This board was made up of approximately 10 Plant Managers from all regions of the world. Some were quite young whereas others were more experienced. Some managed very large plants whereas others managed small plants. (I was personally leading a site organization of 20 DuPont employees.) Some spoke English as a second (or third) language whereas others only spoke English. Some were engineers whereas others had educational backgrounds spanning the disciplines. We came together face-to-face at least quarterly for roughly two years. And we almost always visited different operating sites to observe and analyze. Sometimes the facilities were DuPont's whereas in other cases, we visited such diverse facilities as the Honda car assembly plant in Marysville, Ohio, USA.

After a visit to a DuPont facility in rural Illinois, this group met together in a hotel conference room and attempted to synthesize our collective observations into something usable and concise; something that we could attempt to package and leverage across DuPont Operations. We filled tear sheet after tear sheet of paper until we literally covered the four walls in the room. We had seen and observed much, even too much as there was so much data.

Then, the key moment arrived. One of our members, who later rose to become DuPont's Sr. Vice President of Operations (and interestingly spoke English as a second language) took the floor and articulated in simple terms what became to be known as the "Six Attributes of HPWS." The first attribute was "Vision, Mission and Strategic Intent Known by All." The second was "Committed and Dedicated Leadership 'Felt' within the Organization", or simply "Felt Leadership."

Six Attributes of High Performing Work Systems in DuPont

1. Vision, mission, strategic intent known by all.
2. Committed and dedicated leadership "felt" within the organization.
3. Every person knows the paying customer and how their work impacts the customer.
4. Teamwork and people working in teams is part of the structure.
5. Everyone learns, and teaches. "Learn-Teach-Learn".
6. Principle driven.

Phrases that were used at the time to further clarify the meaning of Felt Leadership included:

- Creates vision and energy
- Stretches goals
- Invites positive push back
- Results and output oriented
- "Felt" sense of purpose

Further insight into what was the intended meaning of Felt Leadership was captured in a "Six Attributes of HPWS Self-Assessment Tool." The questions asked were:

- Do the leaders have a clear vision for change based on needs of all stakeholders?
- Are the leaders visible, knowledgeable and open with people?
- Do the leaders value upgrades from anywhere?
- Do the leaders continually set objectives that demand that people develop and grow skills needed to move toward the vision?
- Do the leaders engage people in a process that builds accountability, willingness, and confidence to act in an empowered way?

3. Felt Leadership and PSM

With all this as a backdrop, what really does "Felt Leadership" have to do with Process Safety Management?

A more descriptive term for what process safety is all about might be "Operational Risk Management." Operational Risk Management implies that:

- It is the operation or the materials processing activities that must be controlled.
- It is the reduction of risk to an acceptable level with the ultimate elimination of significant process incidents that is the goal.
- It is the management of these risks that is one of the organization's key responsibilities and challenges.

The degree of success that is achieved in the management of operational risks can be equated with the quality of decision making by and within the operations organization. And leadership is all about influencing and improving the quality of decision making. Thus, Felt Leadership can best be described in the context of PSM as the approach that leaders follow to insure the highest quality of decision making by the organization in the management of the existing operational risks. Decision making links Felt Leadership to PSM Performance. Felt Leadership becomes a necessary ingredient in successful PSM implementation and execution.

4. Demonstrating Felt Leadership

Within the organization of one of our global clients, inspired leaders captured the vision of their roles with respect to PSM and embraced the concept of providing Felt Leadership. But they did not know how to act differently or more effectively in the management of their significant operational risks. They understood the concept of Felt Leadership as it related to PSM but with no role models, they struggled to visualize or project themselves effectively into their/environment or culture in a uniquely additive way. They needed help and they asked for help. They wondered what supportive but probing questions they should even ask to strengthen a culture of PSM Excellence through Felt Leadership. We provided the following as initial guidance:

1. While visiting a control room and speaking to panel board operator, the focus of the discussion could be on safe operating limits for the process:
    - Does he/she know the safe operating limits, and the consequences of operation outside those limits?
    - Is he/she currently operating with any interlocks by-passed? If so, was the by-pass procedure appropriately followed?
    - How long has this situation (interlock bypassed) existed and how long is the by-pass condition expected to last?
    - How many panel alarms are there per shift? Is there any program to reduce the number of panel alarms?

- Is he/she aware of any work going on in the plant, so that he can alert people to any process emergency?

2. While visiting a plant, question the operations supervisor to see what work permits have been prepared for the day. Take time and review them, looking for the quality of the risk assessment and the precautions that have been put in place to mitigate the identified risks.

3. While visiting a plant, a review of a broad range of PSM elements can occur, such as asking operations personnel about:
   - Any incidents that have occurred in other similar operations.
   - The frequency and nature of emergency and disaster mock drills.
   - The number and nature of near misses that have occurred recently and the results of investigations into these and any other incidents that have occurred in the plant.
   - The results of recent PSM audits and what weaknesses were identified.
   - Whether there are any tests or process changes active in the plant.

4. In conversation with operators in the field, discuss the procedure governing the activity underway and how well it is written and followed.

The coaching of these leaders was actually very similar to what had occurred previously inside DuPont as we strived to strengthen our PSM Performance and the quality of Felt Leadership. Simple "Felt Leadership Cards" were printed to help Operations Leaders consider how best to demonstrate their commitment and passion for PSM.

These cards are still in use today and one is duplicated on the next two pages.

Personal "Felt Leadership" – PSM

What is my spoken and unspoken message on PSM expectations to the organization – integral to core values and business success or a "necessary evil"?

What have I done this month to review PSM metrics and performance indicators for the site?

What am I doing to reinforce Operational Discipline?

Is my organization providing adequate "rewards and recognition" for PSM accomplishments?

How is my organization performing versus annual PSM COT's and objectives?

Where do we need to enhance short and long term PSM resources and organizational capabilities?

What is the status of PSM integration improvements at recently acquired sites or businesses?

What are my direct reports doing to reinforce the value, expectations and accountabilities for PSM on an ongoing and consistent basis?

Are we providing enough support and resources to maintain site equipment and infrastructure?

What have I done personally to engage new leaders in my organization that plays key PSM roles to ensure effective MOC- Personnel?

Have I established clear measures on overall PSM program performance (not just a focus on incidents)?

The Goal is Zero

Ten Key PSM Questions for the Line Organization

1) What trends are your PSM tracking systems showing – do you have any overdue PSM action items within the work group?

2) What recent improvements or accomplishments in PSM does your organization feel good about?

3) What is your biggest concern about the area's PSM program right now and what are you doing about it?

4) What additional focus or support do you need, if any, to improve and maintain PSM?

5) What are your first party PSM auditing and incident trend analysis showing across operations?

6) What are you doing to improve your people's knowledge and capability in PSM to eliminate hazards and incidents?

7) How are you driving a consistent "no leak" and "no alarm" operating philosophy in the operating area?

8) What are you personally doing to demonstrate "felt leadership" in PSM on a routine basis and is it effective?

9) How do you ensure all your people training and written operating procedures up to date and complete?

10) Can you show the status of mechanical integrity tests and PSM work orders in the tracking system and are they up to date?

Other words have been used to describe how Felt Leadership applies to PSM Performance. The following ten basic characteristics have been drafted to summarize this thinking.

Ten Characteristics of PSM Felt Leadership

1. Be visible to the organization.
2. Be relentless about time with your people.
3. Recognize your role as a teacher/trainer.
4. Develop your own safety functioning skills and pass them along to the organization.
5. Behave and lead as you desire others to do.
6. Maintain a self-safety focus.
7. Confirm and reconfirm safety as the Number 1 VALUE.
8. Continuous emphasis and clarity around SHE expectations.
9. Show a passion for ZERO injuries, illnesses, incidents.
10. Celebrate and recognize "ZERO" successes.

Some additional guidance for each of these ten characteristics is included in the appendix.

Additional specific actions that management can take to demonstrate Felt Leadership and their commitment to safety include:

- Leading by example in following all safety rules and procedures;
- Showing a passion for safety & a genuine care for individuals;
- Leading safety meetings with direct reports;
- Including safety in all routine meetings and discussions;
- Participating in safety observations regularly;
- Reviewing/critiquing all incident investigations and personally getting involved in high potential and serious investigations;
- Meeting with your organization to express concern and reinforce the learnings from incident investigations; and
- Using company events to recognize/celebrate good safety performance by individuals and groups.

Other attributes of Felt Leadership can be demonstrated by always following this guidance:

- I always practice what I preach with reference to safety.
- I know all potential hazards in my area.
- I audit for hazards.
- I do not permit unsafe shortcuts in order to get a job completed.
- I check on the use of PPE.
- I correct unsafe practices without delay.
- I make sure workers have proper tools.
- I make sure tools are in good condition.

- I encourage reports on unsafe conditions.
- I promptly respond to workers' safety concerns.
- I investigate incidents promptly and fully.
- I encourage safe suggestions.
- I encourage housekeeping for safety.
- I consider my own health and safety.
- I regularly attend safety meetings.
- I set high standards and try to be the example for everyone else.
- I ensure that all workers are properly trained in safety.
- I stop work if necessary to correct unsafe practices.
- I remove unsafe workers from the job.
- I encourage safety communications and follow-up on corrective actions.

5. Conclusion

The successful implementation of robust Process Safety Managing Systems in complex chemical operations requires a high degree of order and high quality decision making. Leadership is fundamentally charged with creating a culture where all members of the organization respond to these needs and perform at a very high level. Although the term Felt Leadership was initially used in DuPont in the context of an essential ingredient of a High Performing Work System, its use as a descriptor of the necessary leadership qualities in Process Safety Management is both appropriate and accurate.

Felt Leadership is essential to the successful implementation and execution of Process Safety Management.

6. Appendix

Ten Characteristics of PSM Felt Leadership

1. **Be visible to the organization:**
    - "Walk the Talk".
    - Provide a positive example to your employees.

2. **Be relentless about time with your people:**
    - Find ways to spend time with employees/contractors.
    - This will encourage you people to do the same.

3. **Recognize your role as a teacher/trainer:**
    - Teach safety through your:
        - Own behavior
        - Interactions with your employees/contractors
    - You are always 'on stage'.

4. **Develop your own safety functioning skills and pass them along to the organization:**
    - Continuously develop and hone skills such as safety observations, incident investigations, improving leadership skills (Felt Leadership), etc.
    - Make sure your people have the time and resources to develop their safety skills.
    - Reinforce and support what they learn when they return from training.

5. **Behave and lead as you desire others to do:**
    - What people see you do is what they believe.
    - When you notice an unsafe situation and do something about it, you are behaving as you want all of your employees to behave.

6. **Maintain a self-safety focus:**
    - Assess yourself in terms of how 'good' or 'bad' a job you are doing.
    - Stop and think about – 'How did that go?'; 'What might I have done differently to get an even better result?'

7. **Confirm and reconfirm safety as the Number 1 VALUE:**
    - Make comments about safety as you walk throughout the work area.
    - Identify and reinforce what is being done right.
    - Identify unsafe conditions and behaviors and correct them before they cause injuries.
    - Learn what safety issues exist on the shop floor.
    - Learn how the issues are being handled.

- Determine how well employees understand and apply the core principles in their work.

8. **Continuous emphasis and clarity around SHE expectations:**
    - Establish and publicize SHE expectations.
    - Include expected safety behavior.
    - Continuously emphasize their importance by:
        - Including them in communications.
        - Weaving them into most, if not all, conversations you have with your organization.

9. **Show a passion for ZERO injuries, illnesses, incidents:**
    - Make clear your expectation that 'zero' is the goal.
    - Put safety upfront on all interactions.
    - Ensure that your safety performance expectations of your contractors is the same as your own employees.
    - Require that everyone work on a steadily improving path in our journey to ZERO.
    - Periodically focus special attention on an aspect of needed safety improvement; be specific.

10. **Celebrate and recognize "ZERO" successes:**
    - Live safety, not just preach it.
    - Recognize and reward improvement and success.
    - Recognize the achievement at all levels and make sure leadership recognizes and shows appreciation.
    - Recognize specific aspects of safety that have improved as a result of targeted efforts.

Localization of control rooms using a MINLP approach, based on accidental explosion scenarios

Jaime E. Cadena
Chemical Engineering Department, Universidad de los Andes
Cra 1 Este No 19A – 40, Bogotá, Colombia
je.cadena21@uniandes.edu.co

Lina M. Sierra, Sergio E. Rodríguez, Manuel A. Ramos, Jorge M. Gómez, Felipe Muñoz
Chemical Engineering Department, Universidad de los Andes
Cra 1 Este No 19A – 40, Bogotá, Colombia
lm.sierra443@uniandes.edu.co, se.rodriguez68@uniandes.edu.co, ma.ramos43@uniandes.edu.co, jorgomez@uniandes.edu.co, fmunoz@uniandes.edu.co

Prepared for Presentation at
American Institute of Chemical Engineers
2013 Spring Meeting
9th Global Congress on Process Safety
San Antonio, Texas
April 28 – May 1, 2013

UNPUBLISHED

Localization of control rooms using a MINLP approach, based on accidental explosion scenarios

Jaime E. Cadena
Chemical Engineering Department, Universidad de los Andes
Cra 1 Este No 19A – 40, Bogotá, Colombia
je.cadena21@uniandes.edu.co

Lina M. Sierra, Sergio E. Rodríguez, Manuel A. Ramos, Jorge M. Gómez, Felipe Muñoz
Chemical Engineering Department, Universidad de los Andes

Keywords: facility layout, optimization problem, uncertainty reduction, Monte Carlo simulation, UVCE modeling

Abstract

Facility layout for the process industry plants is an important element in the basic engineering stage, which is fundamental in the design of the process, influencing significantly the future safety performance of the operation and its related costs. A control room serves as an operations centre where the facility can be monitored and controlled, therefore, its location and design is a priority for the safety of the process unit operation. Given that the control room is the 'core' of control and monitoring of the process, any significant damage to it resulting from a process incident might jeopardize the whole process and generate subsequent incidents. This work presents a novel methodology intended to serve as an additional tool for aiding the decision-making process regarding the location of control rooms, taking into account the risk of unconfined vapor cloud explosion (UVCE) and several passive barriers against it. The methodology is based on a mathematical model that bonds the design and localization of a control room into a computational tool, which has been developed taking into account key characteristics of a process or plant layout for explosion scenarios. The tool combines stochastic estimation of atmospheric parameters through raw Monte Carlo simulations with a mixed integer nonlinear programing (MINLP) optimization to determine the control room location. This process also helps to choose the best construction strategy for the control room among a portfolio of strategies fed to the model by the user. The final results of the methodology are the optimal location for a control room in an existing facility and the minimum costs related to the risk level to which it is exposed.

1. Introduction

A process incident such as a fire, explosion or dispersion can cause injuries and/or fatalities, but also a disruption of operations and undesired consequences, business-continuity wise. This makes determining an optimal facility layout and the location of the control room, one of the priorities for safe operations within a production plant. Given that facility layout is one of the elements determined in the early plant design stages, it also represents one of the first safety barriers in a hazard-incident representation such the one presented in Figure 1.

Figure 1: Hazard-Incident representation and location of facility layout as a safety barrier (FLB) and additional safety barriers (SB) [1].

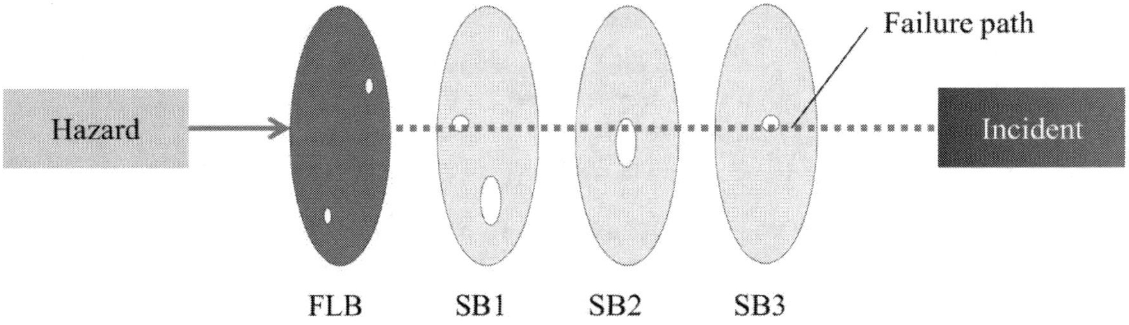

Figure 2 represents the desired function of facility layout as a safety barrier, including the adequate location of the control room. However, this desired representation is not always taken into account by the design philosophy implemented, which leads to process incidents with low probability of occurrence and high consequences, such as Flixborough (1974), Texas City (2005), Buncefield (2005), T2 Labs (2007), Puerto Rico (2009) and Amuay (2012) [2]. Although these incidents involve facility layout issues, Flixborough and Texas City involve serious damage to control rooms and its consequences has made them examples of scenarios and failures that should always be taken into account when determining the final facility layout.

Traditionally the facility layout problem has been approached with heuristic and technical recommendation published by several organizations such as the Center for Chemical Process Safety (CCPS) [3]. However, the integration of such approaches with current edge-cutting engineering tools such as non-linear optimization algorithms and solving platforms allow presenting novel approaches according to characteristics of the problem. In general, three types of problem can be identified: formulation, space treatment, unknown parameters estimation. For each type of problem a different approach is needed and this is summarized in Table 1.

Table 1. Approaches used depending on type of problem

Type of problem	Solution approach	Advantages	Disadvantages
Formulation	Linear [4]	The solution of the problem is usually simple and assures a global optimum under linearity assumptions	Does not allow the modeling of non-linear physic-chemical behavior and/or requires simplifications and strong assumptions
	Non-linear [3][5-7]	Applicable approach to complex systems with a high degree of non-linearity	Obtaining a solution requires more iterations than with a linear approach
Space treatment	Continuous [3][5-7]	Allows determining the exact location of buildings and equipment. Use of different shapes to represent footprints.	Complicated mathematical formulation. Obtaining a solution requires high computational times.
	Discrete [4]	Simple formulation and the solution	Exact locations are not possible. Shapes for

Type of problem	Solution approach	Advantages	Disadvantages
		can be obtained after a few iterations.	footprints are limited to squares o rectangles.
Unknown parameters estimation	Stochastic [3][5-7]	Better representation of real behavior.	Complex procedure that demands large amounts of input data.
	Deterministic [6]	Simple procedure based on worst-case scenario approach.	Derived distribution show few applications.

This work defines describes a methodology which identifies the sets of parameters, variables, functions and restrictions that represent the control room's location problem. Furthermore, the methodology implements a strategy that allows reducing the atmospheric conditions uncertainty when modeling UVCE and dispersion scenarios. The output of the modeling (given by overpressure) allows determining risk values, with consequences on people and infrastructure. Finally a case study is presented and the results analyzed.

2. Methodology

Although no exact and unique solution exists for a highly complex engineering problem such as facility layout, the proposed methodology (Figure 2) intends to support the decision making process in which final layouts are determined and the protection strategies are selected. The methodology is configured as a non-linear optimization of discrete control room's location, having an underlying process of stochastic parameters (atmospheric, risk) estimation. This configuration of the problem requires the following assumptions:

- A square process plant area is selected
- The process plant has only one hazardous process unit, hence only one hazards source
- The process plant area is divided into a uniform computational grid
- A control room can occupy one and only one node of the computational grid
- Historic atmospheric data must be available

The methodology begins by characterizing the hazardous process unit in terms of the possible Loss of Primary Containment (LOPC) scenarios, allowing an estimation of the liberation quantities of hazardous materials (HazMat). From the LOPC, failure trees can be depicted, allowing a classical calculation of the UVCE occurrence probability. Here the liberation of HazMat stops driving the analysis and focus is shifted towards the next driven force: atmospheric conditions. In order to take into account real atmospheric data of the determined location of the plant, it is necessary to acquire the historical atmospheric data, process it, organize it and finally fit it to probability distributions. These distributions are the input for the next step, the Monte Carlo simulation. Since the final aim is to obtain the resulting consequences of the UVCE, both dispersion and an explosion models are implemented within the Monte Carlo simulation that has a fixed number of iterations, *N*. The output of the Monte Carlo simulation is a probability distribution of overpressure-impulse measurements for each node of the grid. This constitutes a risk map, which combined with the calculation of the damages that a determined level of overpressure can cause to humans or infrastructure (i.e. control room) at a specific node, leads to the consequences calculation. The previous steps and costs information on materials, construction and installation of the control room, feed the non-linear optimization algorithm with the required data to run it and obtain an optimum control room's location.

Figure 2: Detailed steps of the proposed methodology

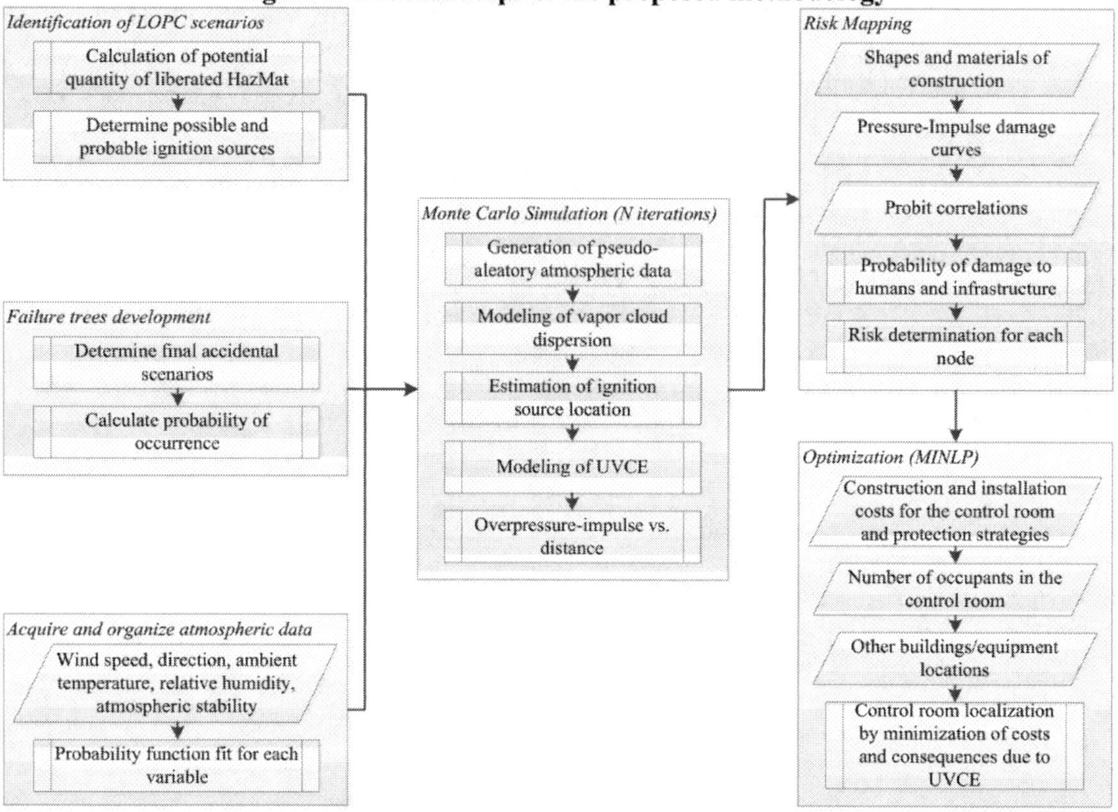

Following the general description of the methodology, it is implemented on a case study and the results are analyzed to establish advantages, disadvantages and improvement opportunities to be implemented in future work.

3. Case study: Hexane-Heptane separation unit

In order to implement the proposed methodology, a facility at Corpus Christi, TX, USA is selected as case study [13]. The hazardous process unit consists of a distillation unit and the following assumptions apply to the subsequent analysis of the system:

- The HazMat consists of 28 tons of flammable material
- The operating temperature of the distillation tower is 130°C and 160°C for top and bottoms, respectively
- The operating pressure is 4 barg
- The vapor transport pipeline has a 0.5 internal diameter and a 25 m length
- The liquid transport pipeline has a 0.15 internal diameter and a 25 m length
- The process plant area is 100 m in length and 100 m wide

- The grid consists of uniform nodes of 10 m by 10 m (see Figure 3)

The previous information and assumptions allow implementing the methodology in 7 sequential steps. The following is a summary of the most important developments in each step.

Figure 3: Representation of the grid and the nodes (labeled) in the process plant area

k01	k02	k03	k04	k05	k06	k07	k08	k09	k10
k11	k12	k13	k14	k15	k16	k17	k18	k19	k20
k21	k22	k23	k24	k25	k26	k27	k28	k29	k30
k31	k32	k33	k34	k35	k36	k37	k38	k39	k40
k41	k42	k43	k44	Hazardous process unit		k47	k48	k49	k50
k51	k52	k53	k54			k57	k58	k59	k60
k61	k62	k63	k64	k65	k66	k67	k68	k69	k70
k71	k72	k73	k74	k75	k76	k77	k78	k79	k80
k81	k82	k83	k84	k85	k86	k87	k88	k89	k90
k91	k92	k93	k94	k95	k96	k97	k98	k99	k100

3.1 Hazards identification

This step aims at identifying a representative group of scenarios that can result in a UVCE accidental event due to the LOPC of the involved HazMat. The found applicable scenarios are catastrophic failure of a vessel within the unit (column, condenser, reboiler), catastrophic failure of the vapor transport pipeline and catastrophic failure of the liquid transport pipeline.

The use of proper source models allow calculating that for both pipeline failures, the total HazMat available would be released in less than 120 seconds in vapor state at averaged atmospheric conditions. Therefore, it is assumed that any of the three mentioned scenarios has a leak/liberation time of 120 seconds. After the characterization of the LOPC, a event tree is developed to obtain the frequency of occurrence of the possible accidental events, using annual occurrence frequencies and event probabilities within the tree, obtaining the results presented in Table 2 [13].

Table 2. Annual occurrence frequencies for the identified accidental events [11]

Scenario	Probability	Frequency (year^{-1})
UVCE	0.34	6.5×10^{-6}
Flash Fire	0.34	5.1×10^{-6}
BLEVE	0.25	5.1×10^{-6}

3.2 Atmospheric data

A 10 year period (2002-2012) of atmospheric data with entries by the hour for the Corpus Christi area was obtained from the National Climatic Data Center (NCDC-NOAA) [19]. Wind speed & direction, ambient temperature, cloud coverage and relative humidity were extracted from the whole data set. This filtered data set allowed determining the Pasquill stability class each entry [8], using the relationship between solar elevation angle (determined an NCDC tool [17]) and cloud coverage as shown in Table 3. The final results of this step allow obtaining 109000 data registries, input for the next step.

Table 3. Relationship between solar elevation angle and cloud coverage

Cloud coverage	Solar elevation angle α		
	60°< α	35°< α<60°	15°< α<35°
Less than 4/8	Strong	Moderate	Minor
Medium altitude clouds	Moderate	Minor	Minor
Low altitude clouds	Minor	Minor	Minor

3.3 Dispersion and explosion models

The SLAB model negative floatability gases, is selected to model the dispersions due to its fit to the case study conditions. Such conditions involve a HazMat (hexane-heptane mixture) with a density higher than that of the air by a 2.5 factor. This model calculates the atmospheric dispersion solving a variant for mass, momentum, energy and species spatial and temporal averages [22]. Additionally, the SLAB model is considered a 'top hat' model, presenting the great advantage of its repetitive use within the Monte Carlo simulation. Once the model is applied, a concentration-distance profile is generated for each data set being evaluated and the ignition point is located as shown in Figure 4. An analysis of the model and the results obtained show that the accuracy of the SLAB model in the near-field is poor [22].
The TNT model is selected for the UVCE intensities calculation (overpressure and impulse) for each node of the computational grid imposed on the process plant area. Even though this model can be considered as the simplest one, its easy implementation in line with the Monte Carlo simulation and its well-known limitations [23] make it an attractive option. This model relies on an explosion efficiency factor (η) that is usually assumed to be between 1% and 10%, however a fixed 1% value is assumed since higher values presented unrealistic results in the near field when compared to typical UVCE overpressure values [4][8][25] .

Figure 4: Assumptions regarding ignition source location

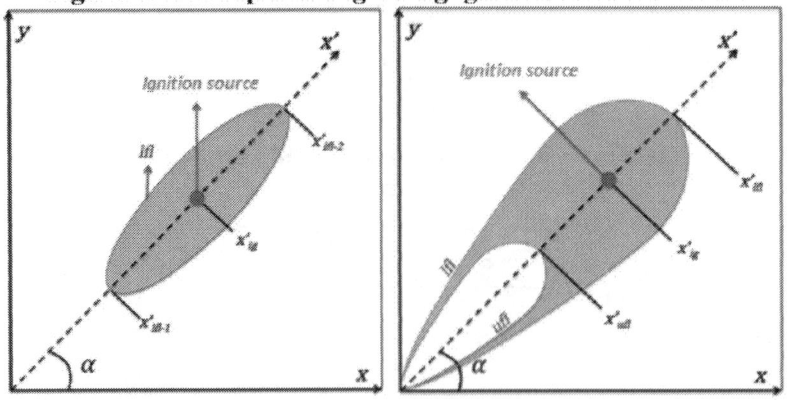

Figure 5: Example of a concentration-distance profile with its ignition source located

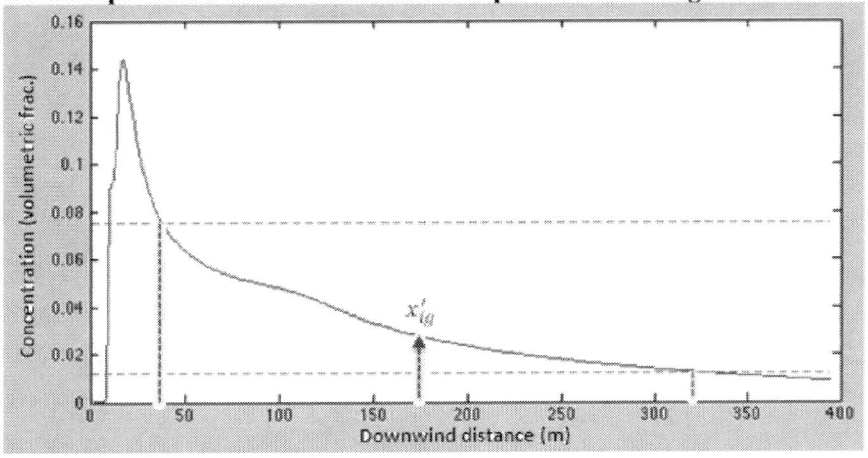

3.4 Monte Carlo simulation and risk mapping

Steps 3.1 and 3.2 are needed inputs for the Monte Carlo simulation and its output is given by the probability distributions of the fitted results of the N iterations performed over the dispersion and explosion models. For this case study the number of iterations was taken as 1000 as an initial guess, later proving to be a reasonable number of iterations when comparing computational time against the results' mean behavior. The results (profiles of concentration and overpressure against distance) are organized in form of histograms. From these results the expected values of overpressure (Figure 5) and impulse can be calculated for each node, necessary input to develop the risk maps. It must be noted that the expected value of the wind direction coincides with the non-symmetrical trend in the intensities maps such as Figure 5.

Risk mapping consist on assigning risk values or scores to each node of the grid taking into account structural damage to infrastructures and human affectation. The risk values for infrastructure damage depend on the type of protection strategy selected for the control room, having four different strategies as follows:

1. Steel reinforcement, walls (panels), roof and frame
2. No reinforcement, concrete reinforced walls, concrete roof and steel frame
3. Prefabricated concrete coating, concrete reinforced roof and walls, and steel frame
4. Reinforced masonry walls, light steel roof and no frame

For each construction configuration for the control room, there is an associated risk map calculated from Pressure-Impulse curves (PI) and Probit approach for human affectations. Once this risk maps are established assuming a Risk = Probability x Consequence relationship, the only step left previous to the optimization is the determination of additional parameters such as related costs of installation, construction and the location of buildings and equipment in the process plant area[4][5][12][18].

Figure 5: Overpressure map on the process plant area

35.01	47.41	29.74	29.24	28.53	26.52	24.48	22.61	21.11	20.09
30.49	30.84	30.17	31.36	32.44	31.44	28.44	25.06	22.39	20.70
28.73	29.89	33.48	40.49	48.05	47.33	38.31	29.73	24.43	21.57
28.45	31.39	40.55	64.92	109.75	109.16	63.14	37.52	26.97	22.47
28.60	32.77	48.20	109.78	Hazardous process unit		108.27	45.59	28.86	22.93
28.77	32.17	47.52	109.14			107.84	45.21	28.51	22.59
28.02	29.42	38.43	62.97	108.03	107.67	61.84	36.37	25.94	21.48
27.98	26.05	29.62	37.07	45.07	44.77	36.12	27.84	22.77	20.04
25.63	22.86	23.86	26.08	27.93	27.70	25.36	22.47	20.20	18.80
21.74	20.23	20.29	20.92	21.42	21.29	20.50	19.47	18.66	18.41

3.5 Optimization of the control room's location

The output of the 3.4 step along with some additional parameters, represent the needed input for to carry out the control room's location optimization. This step is performed through a MINLP formulation [24] of the optimization problem and a Simple Branch and Bound (SBB) solving algorithm, implemented on the GAMS software platform [25]. The optimization problem is structured as follows:

$$\begin{aligned} \min: \quad & z = f(x,y) \\ \text{s.t.} \quad & gj(x,y) \leq 0 \qquad j \in J \\ & x \in X \subseteq R^n \\ & y \in \{0,1\}^m \end{aligned}$$

The formulation can be divided into: parameters, sets, variables, constraints and objective function. A brief description of each one is presented as follows:

1. Parameters: the required parameters to solve the optimization problem are comprised of the protection strategies, PI curves, Probit parameters for human affectation, human occupancy, risk maps, construction/interconnection/installation costs, minimum separation distances, existing buildings/equipment & related location and the coordinates of each node within the computational grid (x_k, y_k).

2. Sets: the problem is divided in three sets. The first one (F) represents the existing buildings/equipment, the second one (K) the node set and the third one (R) the protection strategies.
3. Variables: the optimization problem solves two different kind of variables, integer and continuous. The former are given by binary variables that define the location of the control room at a specific node (k) and the active protection strategy (r). The continuous variables are given by the coordinates of the control room's location (x_c, y_c), but this is merely a formulation requirement since the possible value of these variables is given by the values of the node coordinates (x_k, y_k) specified as parameters.
4. Constraints: the following constraints assure that the obtained solution in the optimization is within physical and logical bounds:
    a. Control room's coordinates: the sum of the node-coordinate by binary variable of location product, of each node allows to determine the location of the control room.
    b. Unique location: the sum of the binary variable of location for each node must equal one.
    c. Protection strategy: the sum of all the binary variables of active protection strategy must equal one.
    d. Separation distances: the control room must be located taking into account the minimum separation distances established. This is a Pythagoras distances' formulation and finds its origins in [26].
5. Objective function: this function comprises the total cost related to the control room's obtained location and the protection strategy selected. Being a cost function, the optimization problem seeks to minimize it.

Once the elements of the optimization problem are defined, it is executed. The results show that the optimum control room's location for the case study is the *k87* node (using the 3rd protection strategy). It is important to note that this location does not correspond to the node with the lowest risk score in the grid, on the contrary, its risk score is 'medium'. The selection of this node balances the cost components involved in the control room's construction, installation and other key parameters evaluated in the objective function.

Regarding the optimum protection strategy, it is found to be the one that resist the highest levels of overpressure even when it also represents the most expensive one. This variable (protection strategy) and its influence demonstrates that it directly affects the decision making process and has a direct relationship with the plant's safety and business continuity.

4. Conclusions

This work proposes a methodology that integrates economic and safety analysis in order to determine the optimum location of a control room through the use of risk scores (risk mapping) and MINLP formulation of an optimization problem. This methodology seeks to aid the decision making process of facility layout.

The proposed methodology was applied to a previously documented case study and the results of it showed that it can effectively locate the control room within a discretized process plant area, while taking into account different protection strategies related to the control room's construction.

Nonetheless, the proposed methodology has diverse and important improvement opportunities in the treatment of the process plant space and the selection of dispersion and explosion models. These models do not allow the modeling of chemical species mixtures and do not take into account the interaction with objects, opposed to a Computational Fluid Dynamics (CFD) approach. The assumption that a building occupies a whole node tends to overestimate the real size of the facilities, inducing uncertainty to the results.

5. Perspectives

Future work should be focused in incorporating risk acceptability within the optimization formulation, as well as the interaction of the dispersion and the blast wave with the facility's buildings and equipment, e.g. through a CFD approach.

A spatially continuous approach should be developed in order to capture in the highest detail the consequences to humans due to structural damages, hence increasing the models sensibility to the number of occupants inside the control room.

Finally, it would be a considerable step forward to include dynamic risk scores related to the location of diverse hazardous process units and the risk represented by these (even if considered fixed).

6. References

[1] Sklet, S. 2006. "Safety barriers: Definition, classification, and performance". Journal of Loss Prevention in the Process Industries, 19, 494-506, http://dx.doi.org/10.1016/j.jlp.2005.12.004.

[2] Vyas, K. & Minaya, E. 2012. "Explosion at Amuay refinery in Venezuela kills 41; fire still burning". Hydrocarbon Processing. Punto Fijo, Venezuela.

[3] C. O. Diaz, R. Vázquez-Román, S. Jung, and M. S. Mannan, "A Comparison of Deterministic and Stochastic Approaches to Solve the Facility Layout Problem with Toxic Releases," in 19th European Symposium on Computer Aided Process Engineering, vol. Volume 26, Elsevier, 2009, pp. 93–98.

[4] S. Jung, D. Ng, C. D. Laird, and M. S. Mannan, "A new approach for facility siting using mapping risks on a plant grid area and optimization," Journal of Loss Prevention in the Process Industries, vol. 23, no. 6, pp. 824–830, Nov. 2010.

[5] S. Jung, D. Ng, J.-H. Lee, R. Vazquez-Roman, and M. S. Mannan, "An approach for risk reduction (methodology) based on optimizing the facility layout and siting in toxic gas release scenarios," Journal of Loss Prevention in the Process Industries, vol. 23, no. 1, pp. 139–148, Jan. 2010.

[6] C. Díaz-Ovalle, R. Vázquez-Román, and M. Sam Mannan, "An approach to solve the facility layout problem based on the worst-case scenario," Journal of Loss Prevention in the Process Industries, vol. 23, no. 3, pp. 385–392, May 2010.

[7] R. Vázquez-Román, J.-H. Lee, S. Jung, and M. S. Mannan, "Optimal facility layout under toxic release in process facilities: A stochastic approach," Computers & Chemical Engineering, vol. 34, no. 1, pp. 122–133, Jan. 2010.

[8] D. A. Crowl and J. F. Louvar, Chemical Process Safety: Fundamentals with Applications, 2nd ed. Prentice Hall, 2001.

[9] C. Cannalire, R. Clarke, B. Mazzarotta, and R. Nicoletti, "Optimizing safety of control rooms in case of flammable releases," Journal of Loss Prevention in the Process Industries, vol. 6, no. 5, pp. 313–317, Oct. 1993.

[10] Q. Baker, Explosion Risk ans Structural Damage Assestment Code (ERASDAC). Arlington, VA: 30th DoD Explosive Safety Seminar, Department of Defense Explosive Safety Board, 2002.

[11] DDESB (Department of Defense Explosive Safety Board), "Technical Paper 14." Approved Methods and Algorithms for DoD Risk-Based Explosives Siting, 2009.

[12] API, "Risk Based Inspection." Base Resource Document. API Publication 581, 2000.

[13] Center for Chemical Process Safety (CCPS), Guidelines for Chemical Process Quantitative Risk Analysis, 2nd ed. Wiley-AIChE, 1999.

[14] S. Mannan, Ed., Lees' Loss Prevention in the Process Industries, Third Edition: Hazard Identification, Assessment and Control, 3rd ed. Butterworth-Heinemann, 2005.

[15] F. Díaz Alonso, E. González Ferradás, J. Francisco Sánchez Pérez, A. Miñana Aznar, J. Ruiz Gimeno, and J. Martínez Alonso, "Characteristic overpressure–impulse–distance curves for the detonation of explosives, pyrotechnics or unstable substances," Journal of Loss Prevention in the Process Industries, vol. 19, no. 6, pp. 724–728, Nov. 2006.

[16] M. Swisdak, Simplified Kingery Airblast Calculations. New Hampshire: Naval Surface Warfare Center, 1980.

[17] Q. Baker and C. Oswald, "Vulnerability model for occupants of Blast Damaged Buildings." 34th Annual Loss Prevention Symposium, Session 3, 2000

[18] Center for Chemical Process Safety (CCPS), Guidelines for Facility Siting and Layout, 1st ed. Wiley-AIChE, 2003.

[19] National Oceanic and Atmospheric Administration, "Climate data online: map and application search," National Climatic Data Center, 24-Oct-2012. [Online]. Available: http://www.ncdc.noaa.gov/cdo-web/. [Accessed: 24-Oct-2012].

[20] National Oceanic and Atmospheric Administration, "Solar Calculation Details," National Climatic Data Center, 2009. [Online]. Available: http://www.esrl.noaa.gov/gmd/grad/solcalc/calcdetails.html. [Accessed: 02-Oct-2012].

[21] Center for Chemical Process Safety (CCPS), Guidelines for Use of Vapor Cloud Dispersion Models, 2nd ed. Wiley-AIChE, 1996.

[22] D. Enmak, User's Manual for SLAB: An Atmospheric Dispersion Model for Denser-than-Air Releases. California: University of California, 1990.

[23] Van der Bosch, C. J. H., Weterings, R. A. P. M. & Stoffen, N. C. P. v. R. d. G. 1997. Methods for the Calculation of Physical Effects: Due to Releases of Hazardous Materials (liquids and Gases): Yellow Book, Directorate-General for Social Affairs and Employment, Committee for the Prevention of Disasters.

[24] J.A. Caballero, I.E. Grossmann. "Una revisión del estado del arte en optimización, Revista Iberoamericana de Automática e Informática Industrial". RIAI, 4 (2007) 5-23.

[25] A. Brooke, D. Kendrick, A. Meeraus, R. Raman. "GAMS User's Guide". Washington DC, USA, 1998.

[26] J.E. Cadena, M.A. Ramos, J.M. Gómez, F. Munoz, "Facility layout optimization detailed research through MILP and MINLP", in AIChE - 2012 AIChE Spring Meeting and 8th Global Congress on Process Safety, Conference Proceedings , 2012.

9TH GLOBAL CONGRESS ON PROCESS SAFETY

Risk analysis for Overpressure from Pressure Vessel Burst

Seungho Jung
Air Products and Chemicals, Inc.
7201 Hamilton Blvd.
Allentown, PA 18195
Jungs2@airproducts.com

Elizabeth M. Lutostansky (lutostem@airproducts.com)
Joan M. Schork (shorkjm@airproducts.com)
Air Products and Chemicals, Inc

Prepared for Presentation at
American Institute of Chemical Engineers
2013 Spring Meeting
9th Global Congress on Process Safety
San Antonio, Texas
April 28 – May 1, 2013

UNPUBLISHED

Risk analysis on overpressure of Pressure Vessel Burst

Seungho Jung
Air Products and Chemicals, Inc.
Allentown, PA 18195
Jungs2@airproducts.com

Elizabeth M. Lutostansky, Joan M. Schork
Air Products and Chemicals, Inc

Keywords: Pressure Vessel Burst (PVB), explosion, Blast curves, geometry factor

Abstract

A pressure vessel burst (PVB) is a type of explosion that involves the burst of a pressure vessel containing gas at elevated pressure. Any vessel that is pressured or has the capacity to build high pressure before bursting can be a candidate for generating a PVB. At burst, a blast wave propagates outward from the source by the sudden expansion of a compressed gas. Fragment hazards are also present when the shell of the vessel along with attached external accessories is thrown. PVBs are possible with vessels containing any gas or mixture including inerts and flammables.

Historically, there have been incidents related to PVBs. Work that has been done to predict the overpressure impacts is covered in Guidelines for Vapor Cloud Explosion, Pressure Vessel Burst, BLEVE and Flash Fire Hazards, 2nd Edition by CCPS [1]. Moreover, as Facility Siting has received significant attention recently, Air Products has included PVBs in Facility Siting studies to safely locate occupied building.

This paper focuses on how to select an appropriate PVB model and correctly use it in calculating vulnerability inside an occupied building. It provides an overview of a simplified method for estimating the characteristics of pressure vessel bursting for practicing engineers.

1. Introduction

Explosions can be classified into two types, oxidations such as Vapor Cloud Explosions (VCE), dust explosions and condensed phase explosion and sudden expansions such as Pressure Vessel Bursting (PVB), Boiling liquid expanding vapor explosions (BLEVE), and rapid phase transitions (RPT). Pressure vessel burst (PVB) is involves a rupture of a vessel containing gas at elevated pressure. The source of the energy released in the blast is the compressed gas. A BLEVE is caused by the rupture of a vessel containing a pressurized liquid above its normal boiling point.

With a BLEVE, the liquid in the vessel undergoes a phase transition from liquid to gas. This flashing contributes to the energy released. Many engineers become confused by the difference between PVBs and BLEVEs because of their similarities. Both liquid and gas may be present in vessels undergoing PVB or BLEVE and blast effect calculations look similar. It is important to remember that in the case of PVB, liquids present in the vessel do not undergo phase transition while for BLEVEs, the primary source of energy is the phase transition. .

The scope of this paper is limited to PVBs. It presents a simplified method for obtaining PVB overpressure results for a building siting study.

2. Methods for predicting blast effects from pressure vessel bursts

A blast curve method is highly recommended because it is more easily implemented than doing numerical simulations. There is an outstanding set of blast curves based on a systematic one-dimensional numerical study by Tang et al. [2]. The Baker-Tang curves assume all of the energy is employed to make a blast effect rather than to be used into the fragments as kinetic energy. These curves are for spherical, free air burst which do not include ground reflection.

Figure 1. Positive overpressure curves for various vessel pressures (courtesy copied from Guidelines for Vapor Cloud Explosion, Pressure Vessel Burst, BLEVE and Flash Fire Hazards, 2nd Edition by CCPS [1])

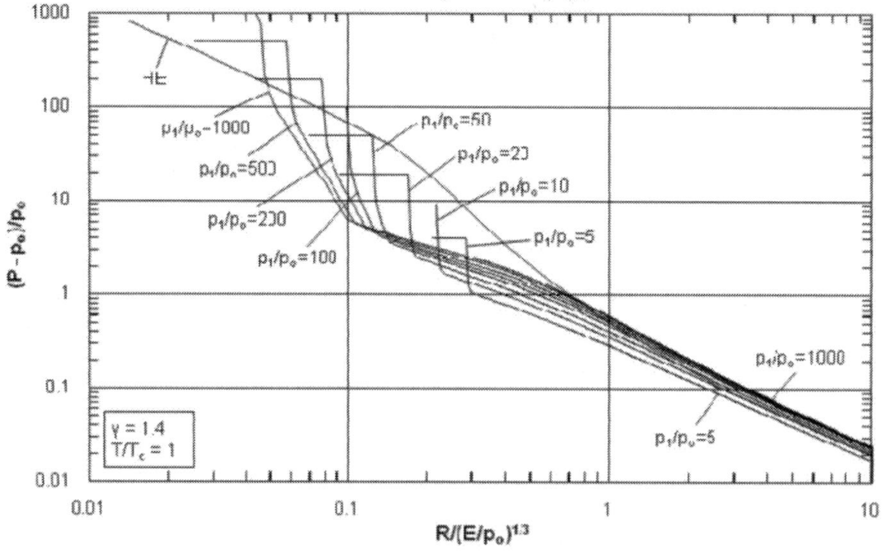

There were not much experimental data for PV burst but Table 1 shows the set of experimental data at relatively lower operating pressure. After early 1980s, there appeared many experimental studies that reported PV burst overpressure measurements but they cover different vessel shapes in different locations such as ground and air [3]. Therefore only data points with spherical free air burst are chosen to compare and validate against the Baker-Tang curves.

Table 1. Summary of literature data

Author	Year	Shape of vessel	Type of burst	Substances	Operating pressure ratio to atmospheric pressure (p1/p0)	Overpressure range
BAKER	1977	Spherical	Free air burst	Air, Argon	6	2.7~50 kPa
BAKER	1977	Spherical	Free air Burst	Freon 12	18 ~ 22	0.73~7.3 kPa

Figure 2 shows good validations against two experimental data sets qualitatively. As the peak overpressure line with the ratio, p1/p0=20 is higher than that with the ratio, p1/p0=5, the experiment data points with Air, Argon are distributed well above the data point with Freon 12.

Figure 2. Validation of Tang's curves on lower pressure cases against experimental data

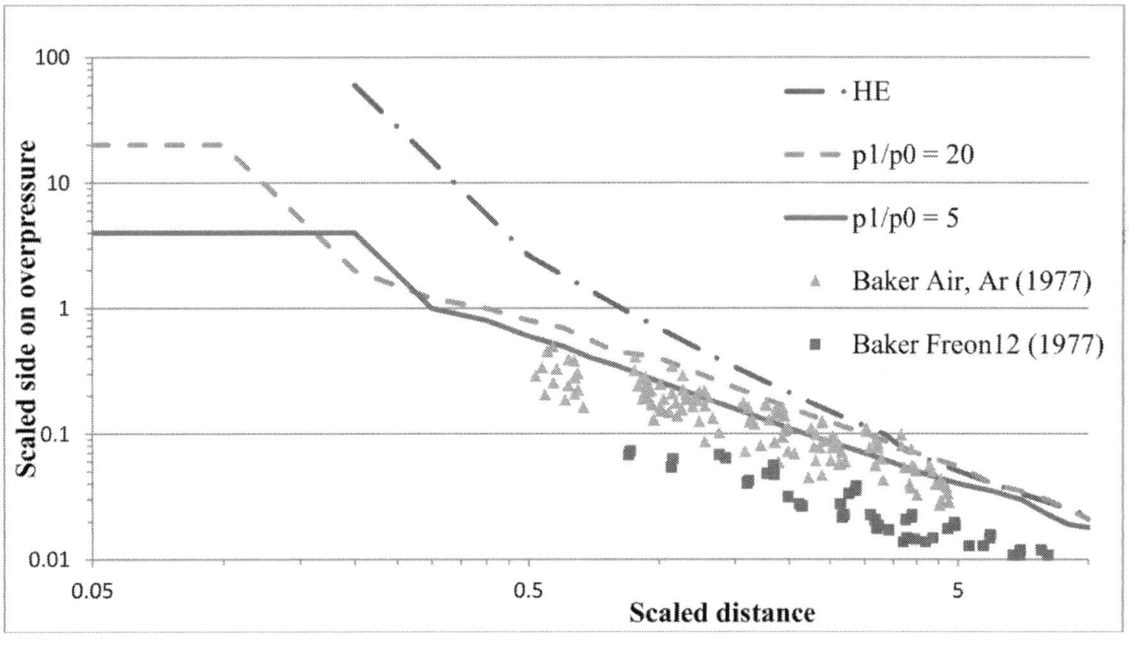

In the far-field, the HE line is close enough to all the other pressure ratio curves as seen in Figure 1. Though a recommendation was made to find the scaled side on overpressure based on p1/p0 ratio, it is quite helpful and faster to use the HE for the far-field further than scaled distance ($\bar{R}$) > 1. 6th order polynomial equation can better fit the HE curve in Figure 1 and suggested as

$$log\bar{P_s} = c_0 + c_1 log\bar{R} + c_2(log\bar{R})^2 + c_3(log\bar{R})^3 + c_4(log\bar{R})^4 + c_5(log\bar{R})^5 + c_6(log\bar{R})^6 \quad \text{[Eq. 1]}$$

where:

$\{c_0, c_1, c_2, c_3, c_4, c_5, c_6\} = \{-0.4102, -1.8282, 0.3661, 0.1122, -0.0830, -0.0042, 0.0046\}$

$\bar{P_s}$ is the dimensionless overpressure, or the overpressure scaled by ambient pressure (Multiply by 14.696 to convert to psig).

3. General calculating procedure for Pressure Vessel Bursts

3.1 Calculate PV energy

The internal energy U of a system is the sum of the kinetic, potential and intermolecular energies of all the molecules in the system. For an ideal gas it is

$$U = \frac{pV}{\gamma - 1} \quad \text{[Eq. 2]}$$

where p is the absolute pressure of the gas, V is the volume of the gas, and γ is the ratio of specific heats of the gas in the system, i.e. C_p/C_v. Therefore, the Brode's PV burst energy is

$$E_{ex,brode} = \frac{(p_1 - p_0)V_1}{\gamma_1 - 1} \quad \text{[Eq. 3]}$$

where the subscript 1 refers to initial state and the subscript 0 refers to the ambient conditions. For ground burst, the effective energy is doubled to account for the ground reflection as

$$E_{eff} = \frac{2(p_1 - p_0)V_1}{\gamma_1 - 1} \quad \text{[Eq. 4]}$$

3.2 Calculate scaled distance, $\bar{R}$

Once the explosion energy of a PV burst is estimated, the scaled distance, $\bar{R}$ can be obtained from

$$\bar{R} = R\left(\frac{P_0}{E_{eff}}\right)^{1/3} \quad \text{[Eq. 5]}$$

where R is the distance from the blast at which blast parameters are to be determined.

3.3 Determine scaled positive side on overpressure, $\overline{P_s}$

To use this graphical method, select the overpressure curve in Figure 1 that corresponds to the vessel burst pressure ratio p1/p0. Read the $\overline{P_s}$ = (p1-p0)/p0 value. For simplicity, if $\bar{R}$ is larger than 1, the equation correlated with the Highly Explosive (HE) line can be used

3.4 Adjust scaled positive side on overpressure, $\overline{P_s}$

There are two adjustment factors; one is for temperature, the other is for geometry. Higher temperature increases the overpressure near the vessel but the effect decreases with distance away from the vessel. If the target is far enough from the vessel, this adjustment effect is negligible and the use of ambient temperature in Figure 1 can be adequate.
Geometry adjustment factors are used with of cylindrical vessels and slightly elevated spherical vessels. The multiplier factors shown in Table 2, for$\overline{P_s}$ initially suggested by Baker et al. [4], are widely used.

Table 2. Factors for geometrical effect

Multiplication factors on $\overline{P_s}$ for slightly elevated spherical vessel	
$\bar{R} < 1$	2
$\bar{R} > 1$	1.1
Multiplication factors on $\overline{P_s}$ for cylindrical vessel	
$\bar{R} < 0.3$	4
$0.3 < \bar{R} < 1.6$	1.6
$1.6 < \bar{R} < 3.5$	1.6
$3.5 < \bar{R}$	1.4

However, these factors were made based upon high explosives experiments, and it was found by Geng et al. and written in the CCPS book [1] as the factors are not directly applicable to PVBs. It was suggested using a group of adjustment factor specifically for PVBs with bursting pressure ranges from 10 to 100 for both the cylindrical and elevated spherical PVBs. Note the factors vary continuously with scaled distance, $\bar{R}$. For simplicity, it is recommended that a factor of 1.8 be used for cylindrical vessels and a factor of 2.0 for slightly elevated spherical vessels as these are the maximum values given by Geng.

3.5 Calculate Vulnerability

Given the adjusted $\overline{P_s}$ the vulnerability of people inside a building is next determined. A vulnerability curve can be generated as Figure 3 using information from Guidance for the Location and Design of Occupied Buildings on Chemical Manufacturing Sites, published by the UK's Chemical Industries Association (CIA). The overpressure-vulnerability relationship for type 3, typical domestic building, is presented in Figure 3.

Figure 3. Vulnerability curve against overpressure (courtesy coped from Marx, et al. [5])

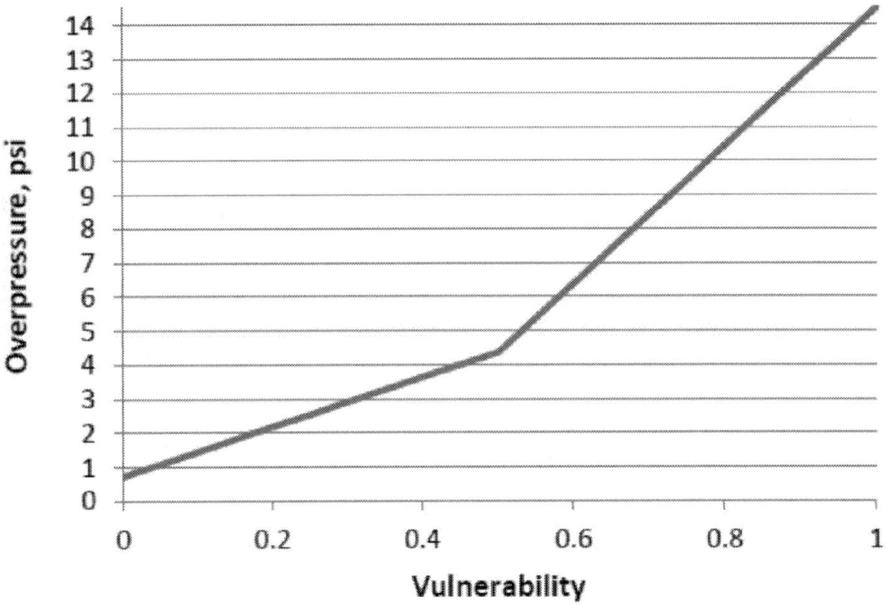

4. Example calculation case study

A $100m^3$-cylindrical pressure vessel at grade was under pressure test with nitrogen gas at 20 barg and it ruptured. There was a control room located 50 me from the vessel. What would be the side on overpressure at the control room? And what would be the vulnerability inside the control room if the building type is CIA type 3.
Table 3 presents the details of PVB calculation and recommendations.

Table 3. PVB calculation flowchart

Procedure	Details of calculation	Note
Collect data	Internal pressure is 21 bara The ratio of specific heats of nitrogen is 1.40 Ground burst, ambient pressure is 1 bara	
Calculate PV burst energy	$E_{ex,brode} = \frac{(p_1 - p_0)V_1}{\gamma_1 - 1} = \frac{(21-1)100}{1.4-1} = 500\ MJ$	
Calculate effective energy	$E_{eff} = \frac{2(p_1 - p_0)V_1}{\gamma_1 - 1} = 1\ GJ$	Multiply 2 if it is ground level explosion
Calculate scaled distance, $\bar{R}$	$\bar{R} = R\left(\frac{P_0}{E_{eff}}\right)^{1/3} = 50\left(\frac{10^5\ Pa}{10^9\ J}\right)^{1/3} = 2.32$	R = 50 m

Read or calculate scaled positive side on overpressure, $\overline{P_s}$	$\overline{P_s} = 0.13$	$\bar{R}$ is larger than 1, equation (1) is proper to use
Adjust scaled positive side on overpressure, $\overline{P_s}$	Multiply 1.8 for a cylindrical vessel geometry factor $\overline{P_s} = 1.8 \times 0.13 = 0.23\ barg$	
Vulnerability inside building	40 % from Figure 3.	CIA type 3 building
Recommendation	If the consequences are judged to be unacceptable, the blast-pressure effects shall be mitigated by the location of the equipment within suitable shields, walls, or barriers. Or the personnel in the control room should be evacuated when the test is ongoing.	

5. Conclusion

Young engineers are often told that BLEVEs are one of the highest impact phenomena in the process industry, even more so than PVBs. PVBs tend to have lower energy releases involved compared to BLEVEs since PVBs do not involve energy from liquid flashing. However, PVBs cannot be ignored since there can be many pressured gas vessels and pressure tests on a gas industry site. This paper provides an overview of method for estimating the characteristics of pressure vessel bursting for practicing engineers. It has been presented to include advanced modeling technology on PVBs and how to properly apply a simplification in adapting blast curves or equation and geometry effects.

6. References

[1] "Guidelines for Vapor Cloud Explosion, Pressure Vessel Burst, BLEVE and Flash Fire Hazards", 2nd ed., CCPS

[2] Tang, M. J., Cao, C. Y., Baker, Q. A., "Calculation of Blast Effects from Bursting Vessels", 1996 AIChE National Meeting

[3] Chen, Fei, Lutostansky, Elizabeth M., Schork, Joan M., "Improvement of BLEVE Overpressure Method Using Historical Data", 2011 CCPS conference

[4] W.E. Baker, J.J. Kulesz, R.E. Richer, R.L. Bessey, P.S. Westine, V.B. Parr, G.A. Oldham, "Workbook for predicting pressure wave and fragment effects of exploding propellant tanks and gas storage vessels", NASA CR-134906, NASA Scientific and Technical Information Office, Washington, 1975 and 1977

[5] Marx, Jeffery D., "API RP 752 Building Siting Using F-N curves", 2013 MKOPSC international symposium

9TH GLOBAL CONGRESS ON PROCESS SAFETY

Effects of Process Variables on Peak Relief Rates Estimated by Dynamic Simulation for a Multiple Distillation Column System

John Wilkins
Staff Process Consultant
Smith & Burgess LLC
5700 Northwest Central Drive Ste. 301
Houston, Texas 77092
John.Wilkins@smithburgess.com

Dustin Smith
Vice President of Operations
Smith & Burgess LLC
Dustin.Smith@smithburgess.com

Prepared for Presentation at
American Institute of Chemical Engineers
2013 Spring Meeting
9th Global Congress on Process Safety
San Antonio, Texas
April 28 – May 1, 2013

UNPUBLISHED

Effects of Process Variables on Peak Relief Rates Estimated by Dynamic Simulation for a Multiple Distillation Column System

John Wilkins
Staff Process Consultant
Smith & Burgess LLC
5700 Northwest Central Drive Ste. 301
Houston, Texas 77092
John.Wilkins@smithburgess.com

Dustin Smith
Vice President of Operations
Smith & Burgess LLC
Dustin.Smith@smithburgess.com

Keywords: dynamic simulation, VMG Sim, distillation columns, sensitivity analysis, pressure relief valves

Abstract

Process engineers and designers have recently turned to using dynamic analysis as a more realistic method for sizing relief systems. The primary goal of using dynamic analysis is to ensure that modifications are truly required instead of making unnecessary and costly upgrades of relief systems based on overly conservative calculations. In work previously presented by the authors, it has been shown that, for certain overpressure scenarios, the dynamic relief rates predicted by dynamic simulation are significantly affected by certain operating conditions [1].

This paper shows the effects of process variables on the relief loads estimated by dynamic simulation for multiple columns. This paper will also show how these effects would impact the flare load for a unit with multiple columns. The multiple column system is based on a system the authors have encountered at a refinery. The sensitivity analysis shows that changes in the column liquid levels significantly affect the initial relief times, relief loads, and relief durations.

1. Introduction

To stay in compliance with industry, regulatory, and typically corporate guidelines, it is important for a plant to conduct and document analyses of their relief systems. It also allows for plant engineers and designers to see the effects of their process on existing or proposed relief systems. Relief system analyses traditionally use steady-state equations to determine the relief rate used to size relief devices. For complex systems, these equations are based on conservative simplifying assumptions. This usually results in a larger relief rate that requires a larger relief device. The drawback to using traditional methods, when analyzing an existing system, is that they can overstate the required relief rate. Sometimes this results in the recommendation of costly relief systems upgrades where only minor modifications are needed.

Dynamic analysis of complex systems involves creating a simulation that calculates changes to the system over time. This analysis can be used to better estimate the relief requirements to protect from overpressure for a specific relief scenario. The primary goal of dynamic relief system analysis is to better understand how the system will respond to the upset scenario. An additional benefit is that the detailed analysis can be used to support making limited changes in lieu of more extensive modifications. The fundamental differences between these two methods are important to understand in order to make sure they are both being used to determine relief rates that are conservative.

Equation 1 shows a simplistic steady-state equation that can be used to determine the relief rate through a relief valve installed on top of a distillation column that has lost cooling, feed, and reflux, but has continued heat input (a typical total power failure scenario).

$$\dot{M}_{relief} = \frac{\dot{Q}_{reb}}{\Delta H_{vap}} \quad \text{[Eq. 1]}$$

In order to use the above equation, the right hand side of the equation would be evaluated at relief conditions, which may vary from normal operating conditions [2]. Assumptions such as the composition of the fluid on the process side of the reboiler can have effects such as a lower heat of vaporization or a larger reboiler duty that will increase the predicted relief rate.

Equation 2 is an example of how the above equation would be represented in a dynamic environment. It shows the same equation as functions of time as well as initial process conditions such as pressure, temperature, etc.

$$\dot{M}_{relief}(t, P_i, T_i, etc) = \frac{\dot{Q}_{reb}(t, P_i, T_i, etc)}{\Delta H_{vap}(t, P_i, T_i, etc)} \quad \text{[Eq. 2]}$$

Dynamic simulations calculate the changes in the system based on incremental time steps from an initial starting point. Therefore, the selection of initial conditions is important and will affect the results [1]. These initial conditions are typically based on operating conditions that represent the normal operations of the facility. Operating conditions are controlled in ranges, and depending on the specific item, there may be no such thing as a normal condition. Liquid levels, temperatures, pressures, and reflux rates for distillation columns can vary during operation and can change due to upsets elsewhere in the plant. This is important to consider when global relief scenarios like power failure and cooling failure affect the entire plant, as these variables are all essentially assumptions.

API 521 Section 5.22[3] states that if dynamic simulations are used for column relief system design, the model must be conservative with respect to the maximum relief load. It goes on to state that the assumptions used for the simulation *shall* be checked by sensitivity analyses to assess the impact on the column relief rate. A sensitivity analysis is the study of how the uncertainty of the input variables affects the output of a mathematical model. Therefore, the input variables of a dynamic simulation must be checked in order for the designer to ensure that the simulation is conservative.

2. Methodology

2.1 Setting up the dynamics file

A steady-state model containing 2 debutanizers, a depropanizer, a gasoline splitter, and 12 PSVs was created using VMG Sim. The flow diagram of this system is displayed in Figure 1.

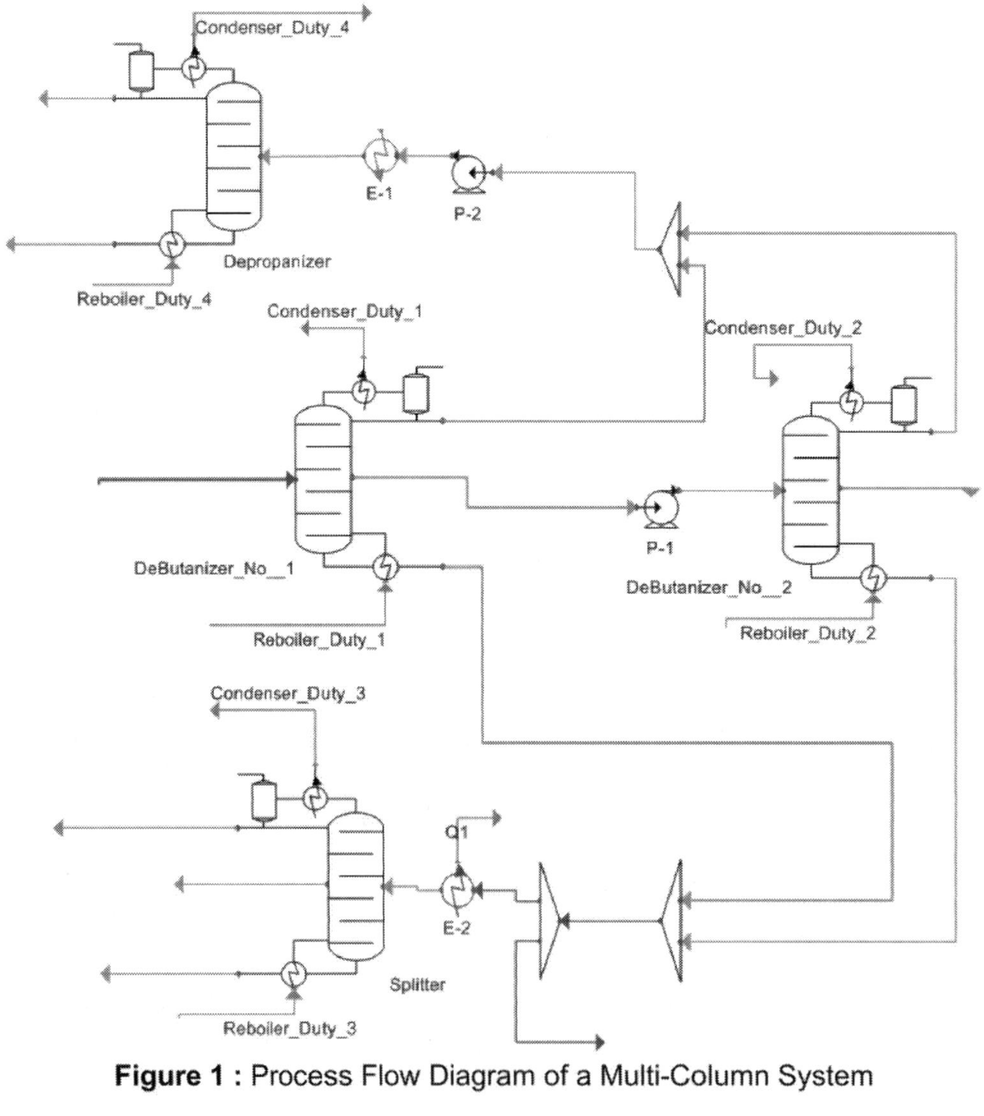

Figure 1 : Process Flow Diagram of a Multi-Column System

The steady-state model was used to create a dynamic model that uses 21 controllers set to match the operating parameters of the equipment. Controllers are used in dynamic simulations to achieve specified flowrates, pressures, temperature, and other physical conditions. The controller attempts to achieve the user's specification by opening or closing a valve located on a material or energy stream. This is different from a steady-state model where flowrates, duties, and physical properties are typically specified or calculated. The action of the controller is based on the tuning parameters provided by the user.

Twelve PSVs were added to the dynamic model to simulate relief. Table 1 displays information on these relief valves.

Table 1: Summary of PSV Information

Name	Location	PRV Size	Set Pressure (psig)
PSV-1	DeButanizer No. 1 vapor overhead	6Q8	184
PSV-2	DeButanizer No. 1 vapor overhead	6Q8	190
PSV-3	DeButanizer No. 1 accumulator	3L4	180
PSV-4	DeButanizer No. 2 vapor overhead	6Q8	190
PSV-5	DeButanizer No. 2 accumulator	1.5H3	190
PSV-6	Splitter vapor overhead	6R8	30
PSV-7	Splitter vapor overhead	6R8	30
PSV-8	Splitter vapor overhead	6R8	30
PSV-9	Splitter vapor overhead	6R8	30
PSV-10	DePropanizer vapor overhead	6Q8	290
PSV-11	DePropanizer accumulator	3K4	274
PSV-12	DePropanizer accumulator	3K4	274

In order to perform the sensitivity analysis, different column liquid levels were used as operating conditions. Setpoints of the respective level controllers for the columns were adjusted to achieve the desired liquid levels. Three simulations were created for a low liquid level of 20%, a medium liquid level of 40%, and a high liquid level of 60%. These percentages represent the fraction of liquid height in the lower section of the column that does not contain trays.

2.2 *Initiating relief*

Cooling water failure was chosen as the overpressure scenario to review as it affects all of the four columns simultaneously. Total loss of cooling water is expected to result in the loss of overhead condenser duties, which may lead to overpressure in the column due to excess vaporization. This was accomplished by specifying duties of zero for all condensers. It was assumed that all of the condensers lost cooling at the same time.

All controllers in the dynamics model were turned off (set to manual) to avoid taking credit for positive controller action. This was done in accordance with API 521 Section 4.2.4[3]. Changing the controller mode from "automatic" to "off" causes the controlled valve to hold its position during relief.

For each liquid level, the integrator (essentially an electronic strip chart recorder in VMG Sim) was turned on and recorded the pressure in the columns as it slowly increased. The overhead PSVs opened at their respective set pressures and flow data was collected. The sum of the mass flow rates relieved by the PSVs was recorded in order to view the total load into the flare.

3. Results

Graphs of dynamic relief for the three liquid levels are displayed below. PSV-5 was not graphed as it does not contribute a significant relief load. The initial time (t = 0), is the time at which the cooling to the tower was stopped and the overpressure scenario began. Note that the change in liquid levels shifts the magnitude of the relief rate and the peak point in time for some of the towers.

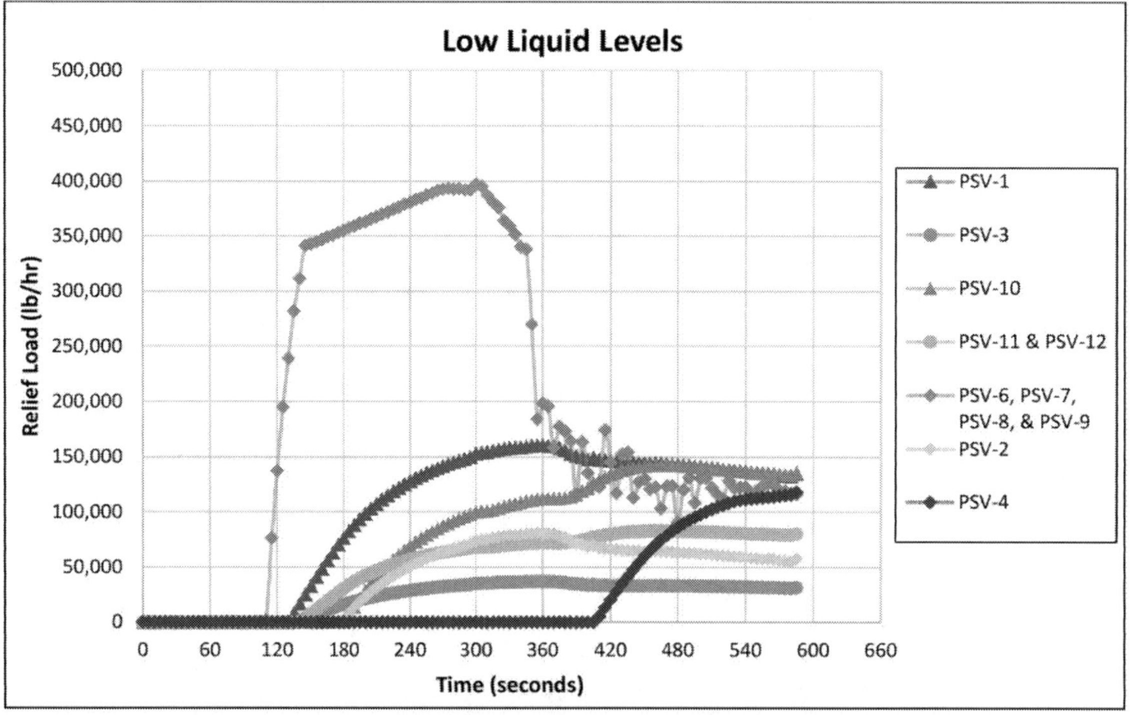

Figure 2: Cooling failure relief rates with low liquid levels

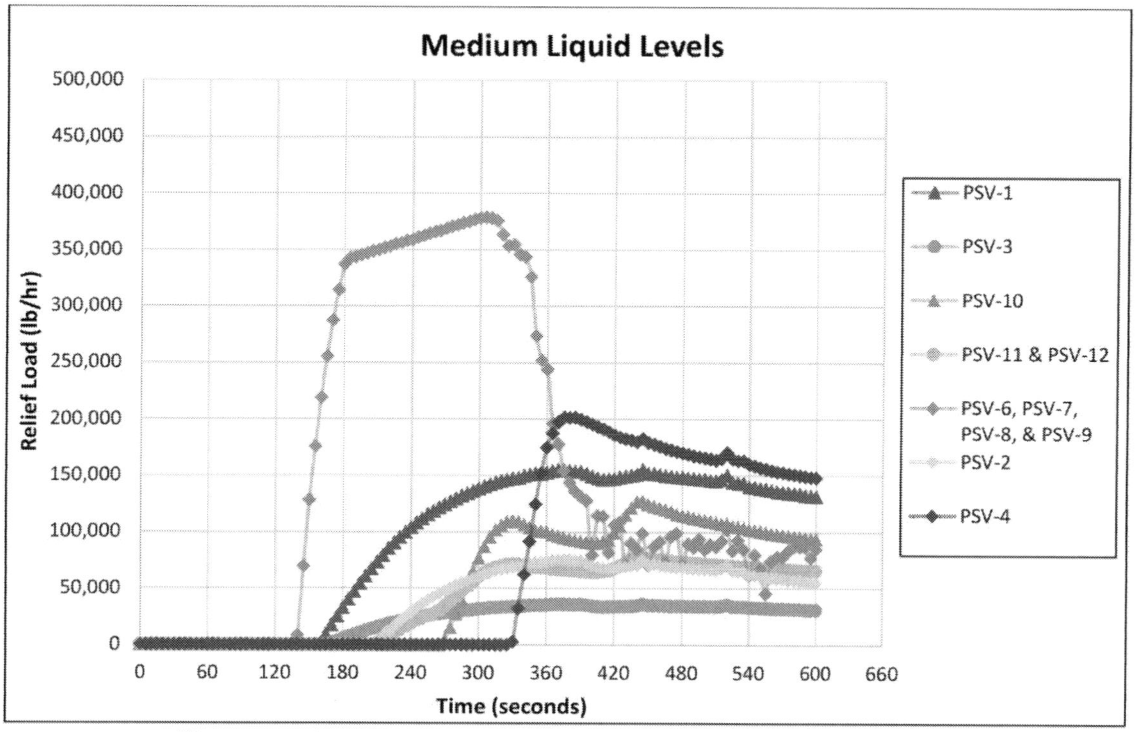

Figure 3: Cooling failure relief rates with medium liquid levels

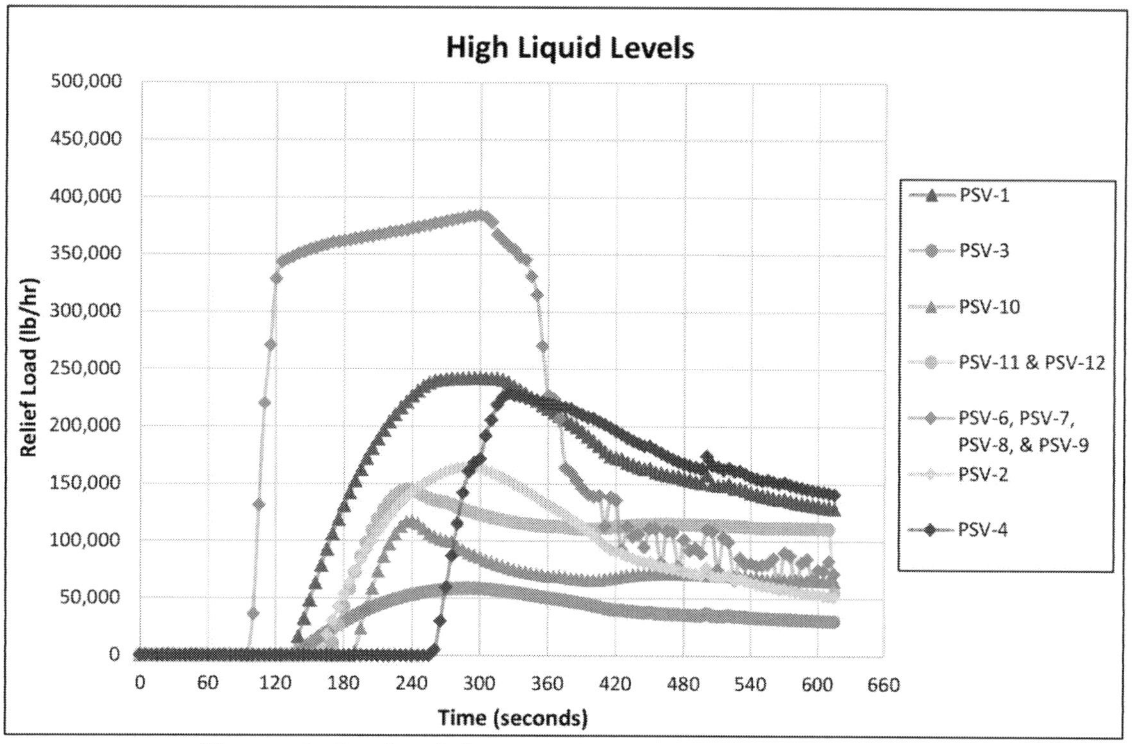

Figure 4: Cooling failure relief rates with high liquid levels

Figure 5 shows the combined flare loads for the three liquid levels. The curves were made by adding the relief loads from the previous graphs over the same time frame. The combined load represents the load that would enter the flare header. The horizontal line represents the steady-state relief load for the four columns estimated by traditional methods.

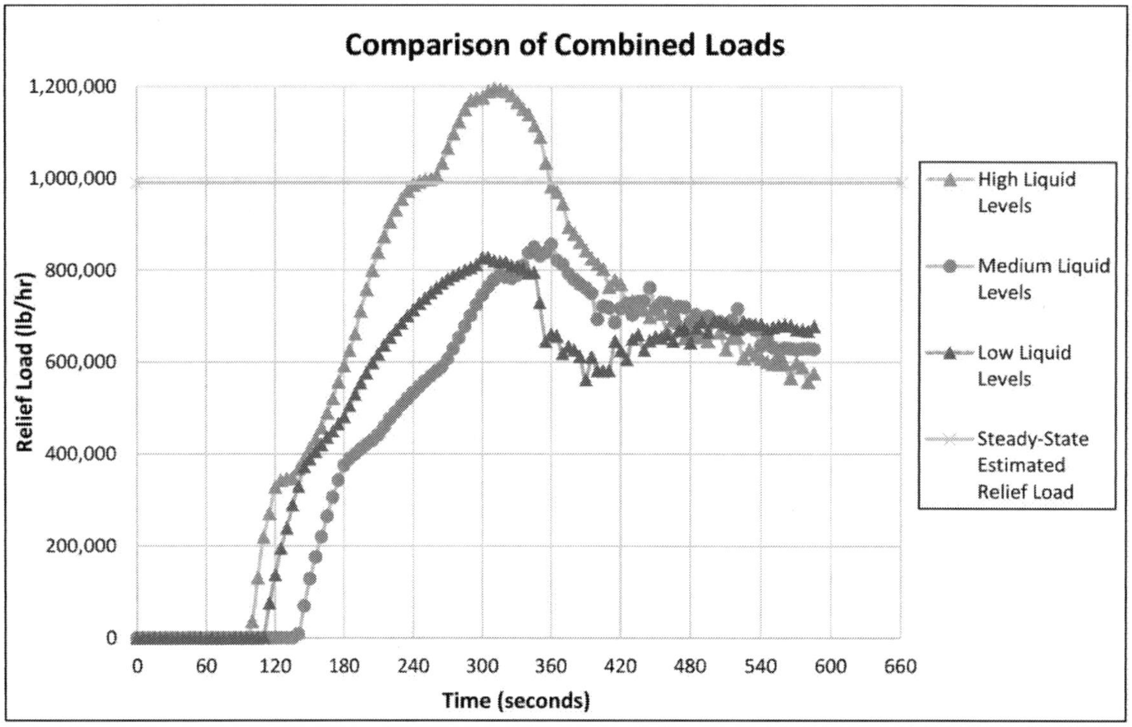

Figure 5: Combined cooling failure loads for different column liquid levels

The steady-state estimation of the combined relief load of the columns assumes that the reboilers have a reduced capacity due to the reduction in the log mean temperature difference (see Reference 1 for an additional explanation). The dynamic model was based on a constant heat input, and thus may over-predict the duties of the reboilers under upset conditions. A further improvement could be made by modeling the reboilers based on their individual UA characteristics.

The three curves in Figure 5 have roughly the same shape as well as time of peak flow. The noticeable difference is the peak relief load for the high liquid level curve. There is a 43% increase in the peak of the combined relief load for all PSVs when the liquid levels are increased by 50%. Lowering the liquid levels by 50% had little effect on the combined peak load.

The increase in the combined peak load appears to be caused by two factors. The peak flows of PSV-1 and PSV-2 located on the 1st debutanizer increased by 57% from normal liquid levels to high liquid levels. The peak time for PSV-4 located on the 2nd debutanizer lowered when the liquid level increased, which resulted in PSV-4 contributing significantly to the combined load peak. Changing the liquid levels in the columns affected the time to initial relief, the time to reach peak load, and the height of the peak for these valves. This behavior is present to a lesser extent in the other relief valves.

4. Conclusions

Dynamics simulation can be a useful tool to show how a system reacts to sudden changes and how the system uses control valves to stabilize itself. Dynamic analysis is very different from hand-calculated methods which are traditionally used to size relief systems. Although dynamic analysis can be used to size relief systems, it must be done knowing how process variations affect the relief rates.

The results show that variations in the liquid levels affect relief times and peaks for a system of multiple columns. Increasing the liquid levels of four columns in a system by 50% increased the combined peak load into the flare by 43%. This was due to increases in the peak flowrate of two PSVs and the shifting of a large peak towards the combined load peak.

This result shows that changes in process variables can have a large impact on the relief rate used to size a flare system. Consideration must be given to the effect process variables can have on the time of initial relief, the time to reach the peak load, and the magnitude of the peak flow.

Because there are multiple variables that can impact the peak relief rate for a flare, a full sensitivity analysis of a system with multiple columns would be time consuming and costly. However, if dynamic analysis is used to size a relief system, a sensitivity analysis must be performed to ensure that the analysis will lead to a safe design.

5. References

[1] Wilkins, John & Dustin Smith, P.E. *Effects of Process Variables on Dynamic Relief Load Estimates for a Depropanizer and Debutanizer*. Oct. 2012. Electronic poster presented at the AIChE Regional Process Technology Conference, Texas City, TX.

[2] Cristea, Nicholas and Dustin Smith, P.E. *Making Relief Load Estimates Match Reality*. 2013

[3] ANSI/API Standard 521 5th Edition, ISO 23251, May 2008

9TH GLOBAL CONGRESS ON PROCESS SAFETY

A Hybrid Model for Characterizing the Source of Hazardous Material Release

Shahryar Khajehnajafi, PhD
Anjana Meel, PhD
SAFER Systems LLC
5284 Adolfo Road, Suite 100
Camarillo, CA 93012
shah@safersystem.com

Prepared for Presentation at
American Institute of Chemical Engineers
2013 Spring Meeting
9th Global Congress on Process Safety
San Antonio, Texas
April 28 – May 1, 2013

UNPUBLISHED

A Hybrid Model for Characterizing the Source of Hazardous Material Release

Shahryar Khajehnajafi, PhD
Anjana Meel, PhD
SAFER Systems LLC
5284 Adolfo Road, Suite 100
Camarillo, CA 93012
shah@safersystem.com

Keywords: hazardous chemicals, chemical spill, source of release, source of hazardous material, plume model, Source Area Locator (SAL), source characterization, Gaussian plume model, Bayesian method, Bayesian inference, hybrid model, odor source, source location estimation, burro

Abstract

On Sept 10, 2012, Los Angeles Fire department and 911 operators received many complaints about a strong foul smell. The complaint covered an area of 10,000 square miles. There were many speculations for the source of the release; sewage, chemical spill, pipeline carrying natural gas. However, after many hours of investigation by South Coast Air Quality Management it was discovered that the source of the foul odor was dead fish at the Salton Sea churned up by a thunderstorm that had occurred that evening. The Salton Sea is a 376-square-mile saltwater lake about 150 miles southwest of Los Angeles. The odor remained strong for up to 150 miles downwind of the source, and quite unusual. Fortunately the above case did not pose any health effects.

It is common for industrial sites carrying hazardous chemicals to receive calls from surrounding communities complaining of a chemical presence in the ambient air. Following a complaint, industrial sites must work promptly check and verify if the odor source is related to a chemical release on their property, especially when there are other industrial sites in the vicinity, for mitigating and reducing the extent of the possible hazard. Locating the source and amount of release for predicting the geographical extent of the exposure in a timely manner is, therefore, essential to take an effective response action for evacuation or shelter-in-place.

With only basic knowledge of meteorological information; wind speed, wind direction, and stability along with two or more gas concentration measurements, it is possible to quickly frame the zone of chemical release. This is an inverse problem to the usual forward dispersion calculation in which the source location and release rate is known and the concentrations at sensors positions are predicted. There are many publications in the area of source characterization; location, and amount of chemical being released. The two most cited methods are: Bayesian inference using Markov Chain Monte Carlo (MCMC) sampling methodology, and adjoint advection- diffusion equation. The Bayesian methodology fits well for application to atmospheric dispersion where ambient and gas sensor data are stochastic in nature, however, it takes a long time to converge.

In this paper, we briefly describe a simple least-square based method for source characterization using Gaussian plume model. We apply the Bayesian method to address the same problem and compare the results of the two methodologies for few selected release scenarios as well as a set of field data. Subsequently, we will present a hybrid model using Bayesian inference which utilizes the result of the simple least- square model as a seed for narrowing the search domain and reducing the time for locating source of chemical release. We present the improvement in the run time using the hybrid model which makes it suitable for emergency response and provides a reasonably accurate prediction of source characterization.

1 Introduction

In this paper, the focus is on estimating the source area location of a hazardous release using the sensor measurements obtained from sensors located throughout a facility. If parameters of a system are known, the outcomes of measurements can be predicted. That is known as the 'forward problem'. On the contrary, estimation of the values of the parameters of the system using the results of the measurements is called a 'backward problem'. Identifying source parameters for location, quantity of release using sensor measurements is a backward problem. Tarantola (2005) discussed the inverse problem theory and methods for model parameter estimation in great detail. Usually a forward problem has a unique solution, but the inverse problem can have multiple solutions. Therefore, in an inverse problem, any apriori information should be utilized explicitly and uncertainties in the data should be handled carefully. When nonlinearities are available, it is important to describe uncertainty and apriori information using probability distribution. Bayesian theory based methods can be used to describe uncertainty. When systems are not nonlinear, least-square methods are considered very efficient. Both least-square methods and Bayesian Theory based methods have their own advantages and disadvantages. Least-square method usually minimizes the error between measurements and prediction to calculate the system unknowns while Bayesian theory based methods updates the prior probability distribution of unknown parameters using the sensor measurement data to obtain a posterior distribution of the unknown parameters, which provides the updated information of the system.

Thus the source area location and release rate is indeed a very challenging and nonlinear problem, even for identifying the source area from a single sensor concentration measurement. Even for a single sensor measurement, there can be multiple solutions for release location all of which may lie on a line/curve with varying distance and release rates. For example, a large release rate at a further distance from a sensor may result in the same concentration values observed by a fixed sensor as a small release rate at a location closer to the sensor. When more than one sensor is involved, the problem becomes even more complicated.

The Bayesian theory has been used by several researchers for estimating the source area location. Yee et al. (2007, 2008) developed a Bayesian probabilistic inferential solution for source reconstruction (both location and emission rate) from a limited noise concentrated data and the methods were tested for JU2003 field study in Oklahoma City (urban dispersion) and the ETEX (long-range dispersion on continental scales). Kim et al. (2007) developed a source reconstruction algorithm using dynamic Bayesian inference and applied it on Yonggwang atmospheric tracer experiment in Korea to show that the posterior modes of the release point and the released source rate for this experiment converged to their true values. Keats (2009) developed Bayesian inference technique for source determination in atmospheric environment. Adjoint (backward) dispersion models were used to reduce the computational effort required from calculating one [forward] plume per possible source configuration to calculating one [backward] plume per detector. Markov chain Monte Carlo (MCMC) is used to efficiently sample from the posterior distribution for the source parameters; both the Metropolis- Hastings and hybrid Hamiltonian algorithms are used. These algorithms were applied on a full-scale field experiment (Joint Urban 2003) in Oklahoma City and the results from the algorithms were also compared with the US Environmental Protection Agency's standard CMB model using a test case involving PM data from Fresno, California. Neuman et al. (2005) used an urban puff model [the Defense Science Technology Laboratory's (Dstl) Urban Dispersion Model (UDM)] for prediction plume evolution in complex urban geometries that enabled rapid urban dispersion simulations by combining traditional Gaussian puff modeling with empirically deduced mixing and entrainment approximations. In their work, a preliminary reconstruction of an atmospheric release event was demonstrated using stochastic sampling algorithms and Bayesian inference together with the rapid UDM urban puff model based on point measurements of concentration. This model was applied to a prototype isolated building and for observations and flow conditions taken during the Joint URBAN 2003 field campaign at Oklahoma City. Ideally, the urban puff model is used to approximate the source location and strength. The more accurate CFD model can then be used to refine the solution. Zheng et al. (2011) reviewed inverse calculation approaches for source determination and classified them into probability modeling methods and optimization modeling methods. According to them, probability modeling methods are demonstrated to be insufficient during emergency responses due to their lacking of enough prior information of unknown parameters, while optimization modeling methods are efficient and become a new trend in source determination. Sharan et al. (2011) applied least square data assimilation for identification of point source emissions from limited number of atmospheric concentration measurements. Monache et al. (2008) applied Bayesian inference and Markov Chain Monte Carlo sampling to reconstruct a contaminant source on a continental scale. By running the Bayesian MCMC algorithm on a large parallel cluster the inversion results could be obtained in few hours as required for emergency response to continental-scale releases. Senocak (2008) developed a method for stochastic event reconstruction of atmospheric contaminant dispersion using Bayesian Inference. Cowles and Carlin (1996) wrote a review article describing the markov chain monte carlo convergence diagnostics. Convergence criteria to the right solution and optimum step size for chain movement are important topics in the application of Markov Chain Monte Carlo method. Haario et al. (2001) described the importance of choice of proposal distribution for MCMC methods for convergence of the algorithm. An adaptive metropolis algorithm with Gaussian proposal distribution was introduced.

As shown in the above references, a number of researchers have worked in source location estimation using Bayesian probabilities methods. Since probability based methods alone cannot be sufficient during emergency response due to long convergence time periods, in the approach proposed in this paper, first an initial solution is quickly estimated using a least-square based method based on backward wind corridor. Subsequently, to obtain a more rigorous solution using Bayesian probabilistic method (referred as MCMC), the initial solution based on least square method is provided as a seed solution along with a search area/boundary. Thus, this two-stage method decreases the overall convergence time for probabilistic method by providing an intelligent apriori information. Using this approach, the emergency responders can quickly start responding based on the initial solution and more accurate information using a probabilistic method can be provided to them in due course of time. Thus, a combined least square-Bayesian approach ensures that even a probabilistic method provides a solution in much smaller time frame.

2 Approach

In this section, our hybrid approach for estimation of source area location is described. At first, source area location is estimated using a simple and intelligent technique developed using backward wind corridor and least-square based method (Khajehnajafi and Burla, 2008). The least-square method is an optimization based approach and several key features of this approach are listed in Table. 1. This method is very efficient and gives a good estimate for source location for most of the wind stability conditions, except when the wind is too stable because the slope of wind corridor is very narrow.

Another approach used for source location estimation is a probability based method using Bayesian theory where a probabilistic distribution is assumed for source area location and the sensor measurements and markov chain monte carlo (MCMC) simulations are used to estimate the source area location. Since this is a statistical approach, it takes significantly more time to converge to a value. Key features of this approach are listed in Table 1.

A Hybrid approach is a combination of the two approaches as listed in Table 1. Deterministic approach assumes that the parameters of the system are known completely without any uncertainty while probabilistic approach accounts for uncertainty in the model at the expense of longer time for convergence. Therefore, having a method that can narrow down the initial search zone for the Bayesian approach could be very useful in decreasing the convergence time. Thus, a hybrid approach is suggested in this paper where at first a Source Area Locator (SAL) model is used to narrow down the area and then Bayesian approach (referred as MCMC) is used to zero down on the actual source area location. The idea is that the combination of these two is a great method that accounts for weaknesses of both the models and makes it a very efficient model to narrow down upon a location very close to the actual source for all wind and stability conditions. Each approach on its own can provide reasonable results with some limitation either in terms of a further away solution for certain atmospheric conditions, or really long convergence time. Thus, the hybrid approach is a very valuable approach to estimate the source area location. Next, specific details of each of the methods are provided.

Table 1: Deterministic, Stochastic, and Hybrid Approach to Source Area Location

Deterministic (Least-square based)	Stochastic (Probabilistic)
• Assumes that all the system parameters are known perfectly • Usually faster runtime • Good for emergency response situations • Referred as SAL – Deterministic approach based on least-square based method to narrow down the search area	• Assumes uncertainty in system parameters • Usually longer run-time • Probabilistic approach • Based on Bayesian - Markov Chain Monte Carlo method (MCMC)
Hybrid Approach	
• Initially solved using deterministic approach based on least –square method SAL • Solution of deterministic approach is used to narrow down the search area for Bayesian - Markov Chain Monte Carlo method (MCMC). Thus, reducing the run time and obtaining a more accurate prediction of source parameters.	

2.1 SAL (Least Square Based Model)

The least square based method is referred to as SAL (Source Area Locator) in this paper. In this technique, it was demonstrated that with only the basic knowledge of meteorological information; wind speed, direction, and stability along with two or more gas concentration measurements, it is possible to quickly frame a region for the source of chemical (Khajehnajafi, and Burla 2008). A technique called 'reverse corridor' or 'back trajectory' is used to narrow the search domain for release location as seen in Figure 1 (a). This simple methodology uses the horizontal dispersion obtained from surface data at receptor locations. The construction of the corridor is key to reduce search area, therefore, reducing the computational cost. In Figure 1(a) the yellow region is the search domain if only sensor 1 and 2 were impacted. Introducing sensor 3 value, the search area is further reduced to the meshed zone. Adding additional sensors may narrow the search area; however, it will significantly increase the simulation time and increase the nonlinearity of the problem.

For the simple example shown in Figure 1(a) initially, sensitivity of the model to main parameters for dispersion, e.g. wind meander, sensor concentration fluctuations, and atmospheric stability was studied. A Gaussian model was used for that study. The result of SAL model for different stability classes A through F (1 through 6) showed that the accuracy of the model prediction for source area decreased with the increasing in stability class especially for higher stability classes 5 and 6 and at very low release rates. The results for the source location and the release rates, even when the model was simulated with the noisy sensor data, viz. 40% and 80%, matched very well with the true data when very fine mesh size was use (except for highly stable conditions at low release rates) indicating that the back trajectory method coupled with a simple Gaussian plume model is in fact a very simple yet plausible technique for locating source of pollutant.

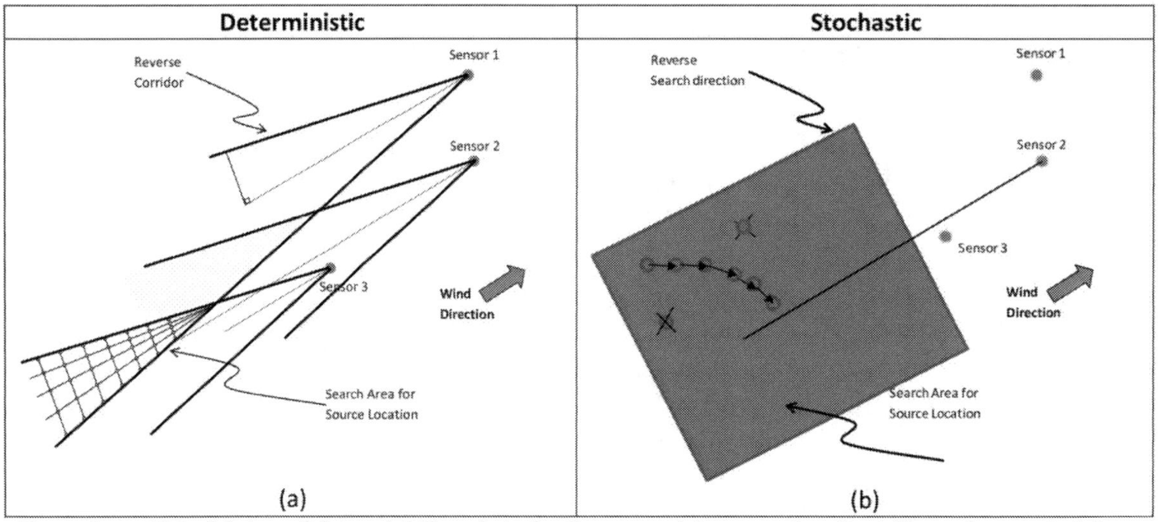

Figure 1: **Pictorial view of deterministic and stochastic approach to source location of hazardous material release**

2.2 Bayesian Model (Markov Chain Monte Carlo)

As mentioned in the introduction section, MCMC alone takes a lot of time to converge. Herein, based on the solution from the SAL model, an area was outlined as a narrowed search domain for the Bayesian probabilistic method. If the initial guess is chosen at random, the outcome from MCMC may take much longer time to converge and may even fails to converge sometimes because of broader search zone. In Bayesian method, a prior distribution is defined for parameters of interest (For example the location and release rate in this case). The likelihood function is estimated using the predicted concentration and measured sensor concentration. Then both prior distribution and likelihood distribution are used to estimate the posterior distribution. The posterior distribution gives us the latest knowledge of the original parameters of interest. The Bayes's theorem for the source reconstruction can be written as (Monache et al. 2008).

$$p(x,y,q|\,Sensor\ Data) = \frac{p(Sensor\ Data|x,y,q)p(x,y,q)}{p(Sensor\ Data)} \propto p(Sensor\ Data|x,y,q)p(x,y,q) \tag{1}$$

Where p( ) is the probability distribution. The x, y, q are the source coordinates and the release rate. *Sensor data* is the measurements from the different sensors. Bayes theorem relates the posterior distribution, $p(x,y,q|\,Sensor\ Data)$, to the product of the probability of measurements given the source parameters, $p(Sensor\ Data|x,y,q)$, also called the likelihood function, and the probability of the source parameters prior to any knowledge of the measurements, $p(x,y,q)$, also called the prior. For this study, x, y, q are assumed to be unknown parameters. One can include other parameters in the unknown list.

2.2.1 Prior Distribution

Let *m* be the model parameters i.e. all the parameters for which prior need to be defined. x, y, q are the prior parameters, x coordinate of the release location, y coordinate of the release location and the release rate, respectively. The prior distribution can be written as:

$$p(m) = p(x)p(y)p(q) \tag{2}$$

2.2.2 Likelihood Function and Acceptance Condition

The likelihood distribution is function of CP, predicted concentration for a sensor, and CM, measured concentration by an actual sensor. Here, the likelihood function is assumed to be of the following form (Monache et al. 2008):

$$\ln[(CM|S)] = \ln[L(\sigma, CP, CM)] = -\frac{\sum_i^N [log_{10}(CP_i) - log_{10}(CM_i)]^2}{2\sigma^2} \tag{3}$$

Where L is the likelihood function, CP_i is the predicted concentration values at the sensor location *i*, CM_i is the sensor measurements for sensor *i*, σ is an error parameter chosen accordingly to expected errors in the observations and predictions. *N* is the number of sensors. *S* is the state vector formed by all the unknown parameters (*x*, *y*, *q*). If the measured concentration values spread over several order of magnitudes, taking the logarithm of both predicted and measured values prevents large concentrations from dominating in the likelihood computation. The above likelihood is relatively simple, where the *N* prediction errors are assumed independent and expected to be proportional in size to the predicted concentration (as a result of log transforming both the observed and predicted concentration). If large errors are expected in the prediction/or measurements, large values of sigma should be assumed resulting in a broader-flatter posterior distribution. Here a sigma value of 0.8 is assumed. For a more complex scenario, it can also be treated as an unknown parameter and assigned a prior distribution.

The posterior distribution is sampled with an MCMC procedure via a Metropolis-Hasting algorithm.

2.2.3 Concentration Prediction Model

The concentration prediction model is based on a Gaussian plume dispersion model as shown below (Kim et al. 2011).

$$CP_i(x, y, z) = \frac{q}{2\pi U \sigma_y \sigma_z} \exp\left(-\frac{y^2}{2\sigma_y^2}\right) \tag{4}$$

Where, $CP_i(x,y,z)$ is predicted concentration at a particular location (x, y, z). q is the emission rate, U is mean wind speed, H is height of the release, x is distance along the wind, y is distance along the horizontal crosswind direction, z is the distance along the vertical axis, σ_y, σ_z are the standard deviations in the horizontal crosswind and vertical directions that are also called as Gaussian plume dispersion parameters and are defined empirically for different stability conditions. σ_y, σ_z are defined below:

$$\sigma_y = \zeta_1 x\,(1 + 0.0004x)^{-0.5}, \qquad \sigma_z = \zeta_2\, x \tag{5}$$

2.2.4 Sampling Procedure and Algorithm

Various algorithms exist for the sampling with MCMC. We have assumed uniform prior for the prior parameters. For the MCMC method, x is in the range -10,000 to 10,000 and y is the range 10,000 to 10,000 with (0,0) as the release location. q is uniform prior between 0.1 to 10 kg/s. U is the value given for the experiment of interest. H is assumed at the ground level. Ideally, parameter ranges can be assumed for wind direction, wind speed, ζ_1, ζ_2. However, for this study we have kept these constant. Of course, ζ_1 and ζ_2 will vary based on the stability class.
For the hybrid approach, the release location is first determined using the SAL approach. Limits for x, y are assumed to be ($-1000 + x_{sal}$, $1000 + x_{sal}$) and ($-1000 + y_{sal}$, $1000+ y_{sal}$) where x_{sal} and y_{sal} are release location coordinates obtained using the least-square method (SAL). Range for q is 0.001 to 10 Kg/s. The new value of x, y, q is obtained using adaptive sampling based on the current value and a step size.

2.2.5 Algorithm

An initial value for each parameter is sampled from its prior distribution. System parameters are assumed to be uniform prior with range as described above. The prior distributions are not completely uncertain as a good initial range is provided using the SAL method. Using the likelihood function, these sampled values from the prior distribution are selected to be part of the posterior distribution based on some criteria. A number of iterations involving source term sampling and dispersion simulation are executed until convergence to the posterior distribution is reached. Accepted parameter values collected during all the iterations constitute a Markov Chain which is defined as a set of consecutive values of parameter in which the probability of a value depends only upon the previous sampled value. Each iteration of the solution that becomes part of Markov Chain consists of following steps (Monache et al. 2008):

1. New value for source term is proposed (x_{prop}) according to $x_{prop} = x + dx$. Where x is the current value and dx is displacement from the current value modeled as a random walk sampled from a Gaussian distribution with zero mean and a standard deviation specified from the current step size. The Prior distribution $p\ (m)$ is used as a target distribution to estimate a prior likelihood of x_{prop}. If the prior likelihood of x_{prop} exceeds that of x, the proposed value x_{prop} replaces x. If not, a random (Bernoulli) "coin flip" determines whether the new proposal, even with its lower prior likelihood will be accepted. This ensures that sampling doesn't get stuck at a random point.

2. Repeat step 1 for the y and q parameters considered in the prior distribution. The assumption is that independent sampling of each parameter ensures that they reflect their likelihood better than sampling all of them at once and testing their likelihood
3. The predicted concentration for sensors is simulated using the current values of all the prior parameters using equation 4.
4. Likelihood of the current values of all the parameters which forms the proposed state is calculated using the predicted concentration and observed concentration by sensors. If the proposed likelihood is higher than the likelihood of previous state, it is accepted. If not, a random (Bernoulli) coin flip is used to ensure that the search reconstruction procedures continues to search the complete space of proposed source term parameter states, preventing the procedure from remaining indefinitely within the neighborhood of a local extreme.

After the proposed state is calculated, it is accepted if

$$\ln[L_{new}] - \ln[L] \geq \ln[rnd(0,1)] \quad (6)$$

Where L_{new} is the likelihood value of the proposed state and L is previous likelihood value and rnd(0,1) is a random number generated from a uniform distribution in the internal [0,1].

At a particular time instant, likelihood test for proposed state using measured data from all sensors at a particular time instant determines the acceptance or rejection of this state. Steps 1-4 are repeated to create a new proposed state with respect to all the parameters until the Markov Chain converges for a particular time instant. Note that in case of time varying information; step 1 to 4 can be further repeated for each time instant.

3 Case Studies: Comparison with the Field Data

3.1 BURRO EXPERIMENTS

BURRO experiments from the 1980 LNG (Liquefied Natural Gas) spill tests conducted by Lawrence Livermore National Laboratory (LLNL) and Naval Weapons Center (NWC) were used to compare the predictions from the model with the experiments. BURRO series experiments are one of the most comprehensive experimental analyses available in the literature for sensor measurement data of a chemical release (Koopman et al. 1980, Goldwire et al. 1980). In total nine experiments were conducted. Five experiments from the 1980 LNG (Liquefied Natural Gas) spill tests were used for this analysis. It is very difficult to get access to experimental data in this field and BURRO data are indeed very useful to compare modeling predictions with the experimental data for such a complex set of sensors.

This particular experimental setup employs close to 20-30 sensors. Not all sensors will see concentration values all the time so only the sensor with non-zero concentration reading are analyzed in this situation. This in itself is debatable as non-zero sensors may provide some information about the location which otherwise can be missed in case of uncertainty. Ideally, non-zero sensors should be included as well but for simplicity reasons we have not included those here and can be part of a further study. The next level of complication arises from the dynamic nature of the measured sensor values due to plume meander caused by horizontal turbulent fluctuation in the atmosphere. In ideal situation one expects the sensors to measure concentrations reliably based on the release rates. However, the uncertainty in wind direction, wind speed, measurement errors due to hardware and software and variability in atmospheric stability condition can contribute to a lot of uncertainty in the sensor data. For example, see the BURRO experiments report for concentration values measured by sensors for few experiments. Because of these uncertainties in the values and the simplified nature of our modeling approach, careful and intelligent decisions need to be made about the measured sensor readings that are used for this analysis as only a single measured value is used for each sensor in this analysis. Nonetheless, a more complicated model can be created which can include few/all aforementioned uncertainties at the expense of higher simulation time and resources. These may be included in the future studies of the project after carefully analyzing the tradeoffs involved in various approaches.

In our approach, the spill scenarios conditions and measured sensors values were used to obtain solution from SAL and MCMC approaches and then predictions for source area locations from the two models were compared against the actual experiment release location. In particular, five different spill scenarios from BURRO experimental series i.e. BURRO 2, BURRO 3, BURRO 7, BURRO 8, and BURRO 9 covering a range of atmosphere stability conditions are reported in this study (Table 2).

The release location was denoted as (0, 0) position in a new coordinate system, which is about 45 degree from the true north. The wind conditions were measured along the true north conditions. Therefore, wind directions for the new coordinate system are calculated by adding 45 degrees to wind direction values reported in the BURRO experiments. A plot in the new coordinate system containing actual release location and sensor locations is shown in Figure 2. The sensors were mainly distributed in four different arches as shown in Figure 2. Detailed location of these sensors and other towers can be found in the BURRO report part 1. The wind fields for different experiments are reported in the BURRO report part 2. To run SAL, a constant wind field (i.e. constant wind speed and wind direction) is assumed. For these short duration experiments, wind field does not change a lot so this is a reasonable assumption. Undoubtedly including variability in the wind field would result in a better prediction at the expense of more computational time.

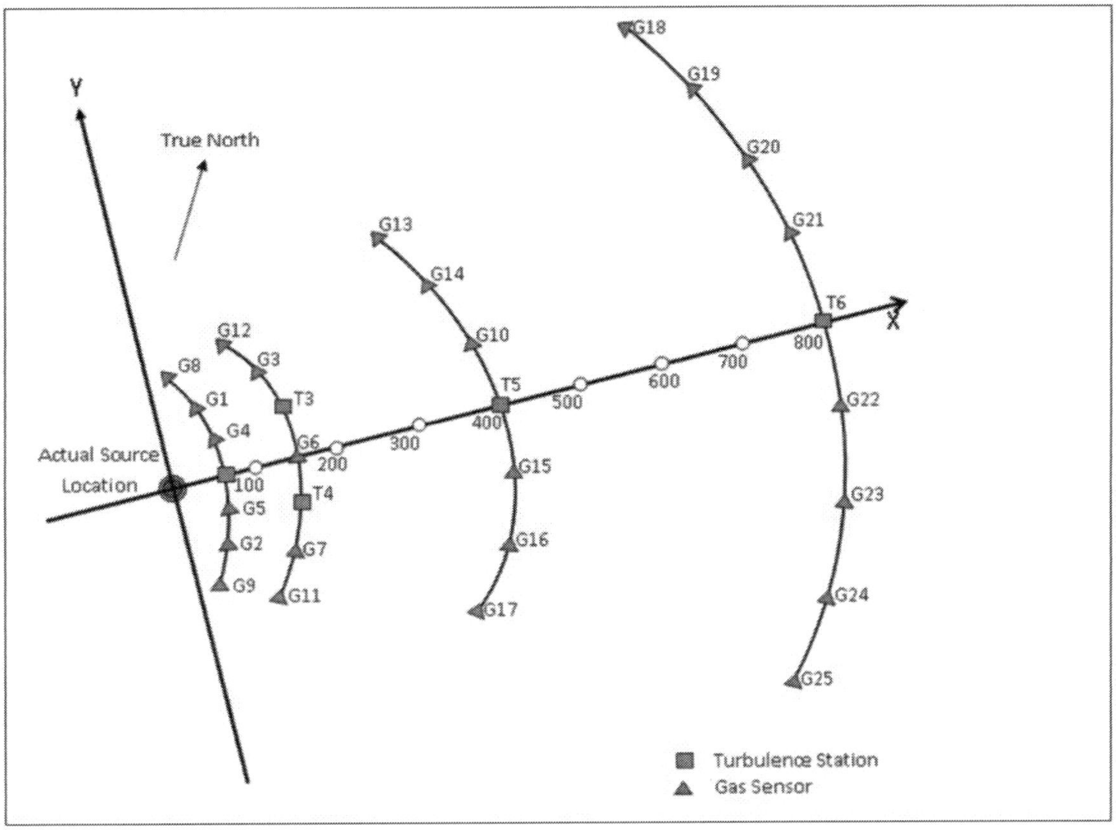

Figure 2: **Burro experiment setup of sensors.**

Table 2: Spill conditions for BURRO 3, BURRO 7, BURRO 8, and BURRO 9

Experiment Name	Chemical	Wind direction	Stability	Wind speed (m/s)	Spill Rate (m3/min)	Spill duration
BURRO 2	LNG	266	2	5.4	11.9	173
BURRO 3	LNG	270	2	5.4	12.2	166
BURRO 7	LNG	253	3	8.4	13.6	174
BURRO 8	LNG	280	5	1.8	16	107
BURRO 9	LNG	277	4	5.7	18.4	79

3.2 SAL Sensitivity to Different Cost Functions

A reasonably good approximation for source area location using SAL is our first target. To accomplish that goal we had to identify the best cost function. Three cost functions were identified; A simple concentration difference objective function **abs**(c_m-c_p), referred as CF 1, Normalized concentration based objective function [**abs**(c_m-c_p)/c_m], referred as CF 2; Distance and concentration normalization based objective function [abs (c_m-c_p)/c_m *$(x_{dist}^2+y_{dist}^2)^{0.5}$], referred as CF 3. Where c_m is the measured concentration and c_p is the predicted concentration.

Only a single measurement value per sensor is used for this analysis. Since it is a dynamic release scenario, various values can be selected from the sensor measurements. A sample of sensors output for Burro 3 is shown in Figure 3. Two set of data was used for SAL study. One was based on the single best sensor value for the entire time zone which reflects the best/average sensor values and the other is based on the maximum value for the entire time. All the sensors with non-zero sensor values were included in the analysis. For SAL simulation, thirteen different values of release rates ranging from 0.01 to 10 are used for source location. A local source point was identified which was average of these thirteen values and was used as an initial point for the next level search. First a coarse grid is used to narrow the search zone. Then a finer mesh grid search was carried out around that zone for various ranges of release rates to identify the source area location.

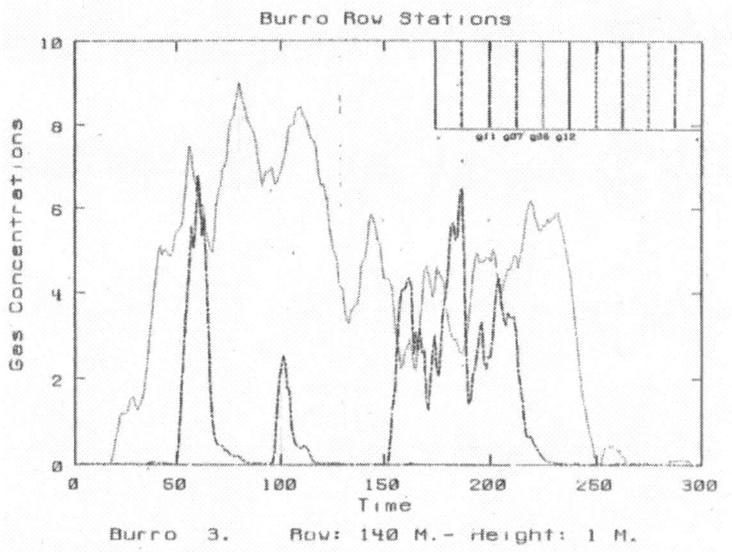

Figure 3: Gas sensors concentration measurement at 140 m arc.

The following tables represent the single sensor value obtained from each sensor positioned in the experimental grids for this study along with the source location predicted by SAL using three different cost functions.

3.2.1 Burro 2

Table 3: Sensor's value from BURRO 2 experiment

Sensors	G2	G5	G4	G1	G8	G3	G12	G13
ppm	8	5	1.2	1	2	5	4	0.02

Table 4: Source area location for BURRO 2 using sensor values

WD=266	CF1	CF2	CF3
Stability	2	2	2
X location	-65.1	-57.5	-56.6
Y location	9.2	7.5	8.8

3.2.2 Burro 3

Table 5: Sensor's value from BURRO 3 experiment

Sensors	G2	G5	G4	G1	G7	G6
ppm	9.0	4.0	17	15	4	5

Table 6: Source area location for BURRO 3 using sensor values

WD=270	CF 1	CF 2	CF 3
Stability	2	2	2
X location	-73.4	-117.2	-42.5
Y location	-14.8	40.2	16.3

3.2.3 Burro 7

Table 7: Sensor's value from BURRO 7 experiment

Sensors	G12	G3	T3	G13	G14	G18	G19
ppm	9.0	0.78	0.34	2.8	2.6	0.4	0.05

Table 8: Source area location for BURRO 7 using sensor values

WD = 253	CF1	CF2	CF3
Stability	3	3	3
X location	-111.6	69.5	191.6
Y location	-49.6	59.0	148.6

3.2.4 Burro 8

Table 9: Sensor's value from BURRO 8 experiment

Sensors	G9	G2	T2	G4	G1	G8	G11	G7	T4	G6	T3	G3
ppm	15	6	8	11	5	8	17	12.5	3	6	5	4.8
Sensors	G12	G10	G16	G15	T5	G17	T6					
ppm	2	0.02	0.7	0.7	2	3.8	0.25					

Table 10: Source area location for BURRO 8 using sensor values

WD = 280	CF1	CF2	CF3
Stability	5	5	5
X location	-633.9	-285.4	-355.2
Y location	119.8	70.6	77.9

3.2.5 Burro 9

Table 11: Sensor's value from BURRO 9 experiment

Sensors	G2	T2	G4	G7	T4	G6	T5	G16	T3	T6
ppm	6	1	7	1.2	2.2	6	0.25	0.2	4.5	0.2

Table 12: Source area location for BURRO 9 using sensor values

WD=277	CF1	CF2	CF3
Stability	4	4	4
X location	-70	63.5	66.3
Y location	8.8	-3.9	-5.3

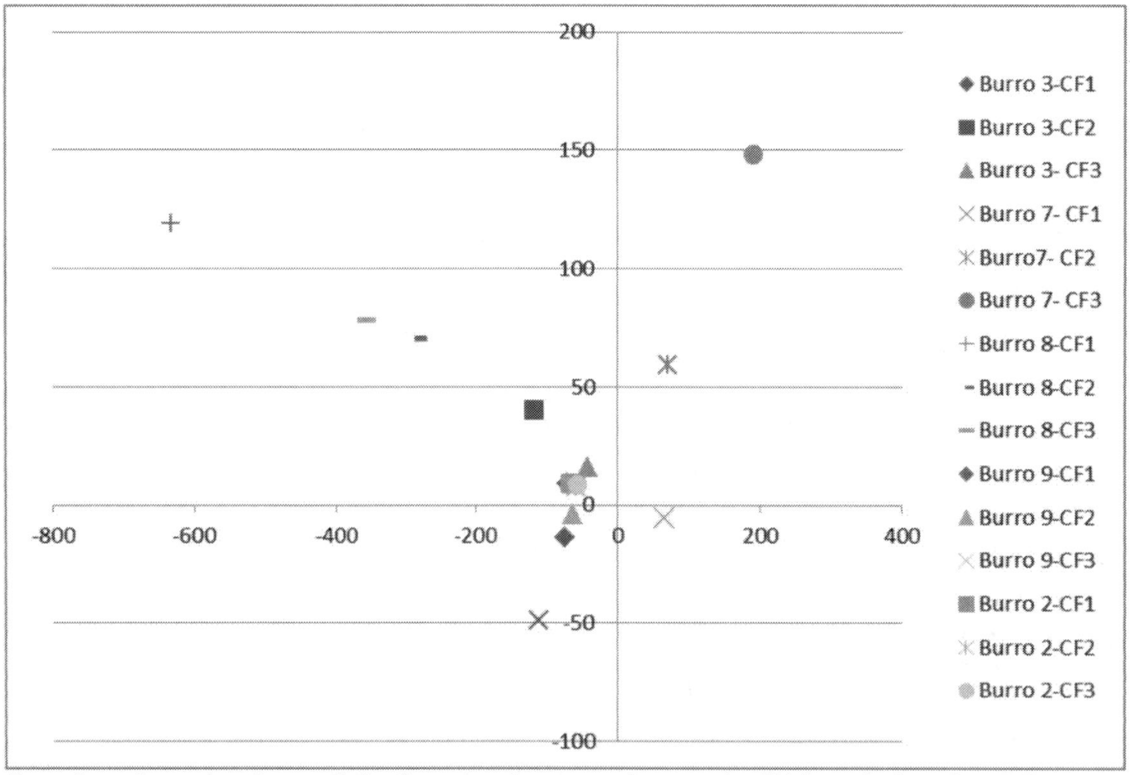

Figure 4: Source location prediction for various Burro experiments using SAL (actual release is at 0,0).

From this analysis we can see that the CF2 give better results out of all three objective functions. It was found that other than stability case 5, in most cases the source area was located within (+/- 70, +/- 70) range of the actual release location. SAL took only few seconds to run for all cases.

3.3 Hybrid Model

Subsequently, the solution from the hybrid approach with an initial guess from SAL to narrow down the search domain using Bayesian theory was obtained. BURRO MCMC refer to the results obtained using the hybrid approach in Figure 5. BURRO SAL refers to results obtained using the least-square based methodology. As evident from Figure 5, BURRO MCMC provides good solution for all stability conditions. Improvements in source location were seen for almost all the cases using the hybrid approach. These simulations were done for a wide variety of stability ranges including unstable to stable atmospheric conditions. Improvements seen in unstable cases using hybrid MCMC approach are more profound than stable cases as shown later in this report. The hybrid model improved the run time by a factor of two to three times.

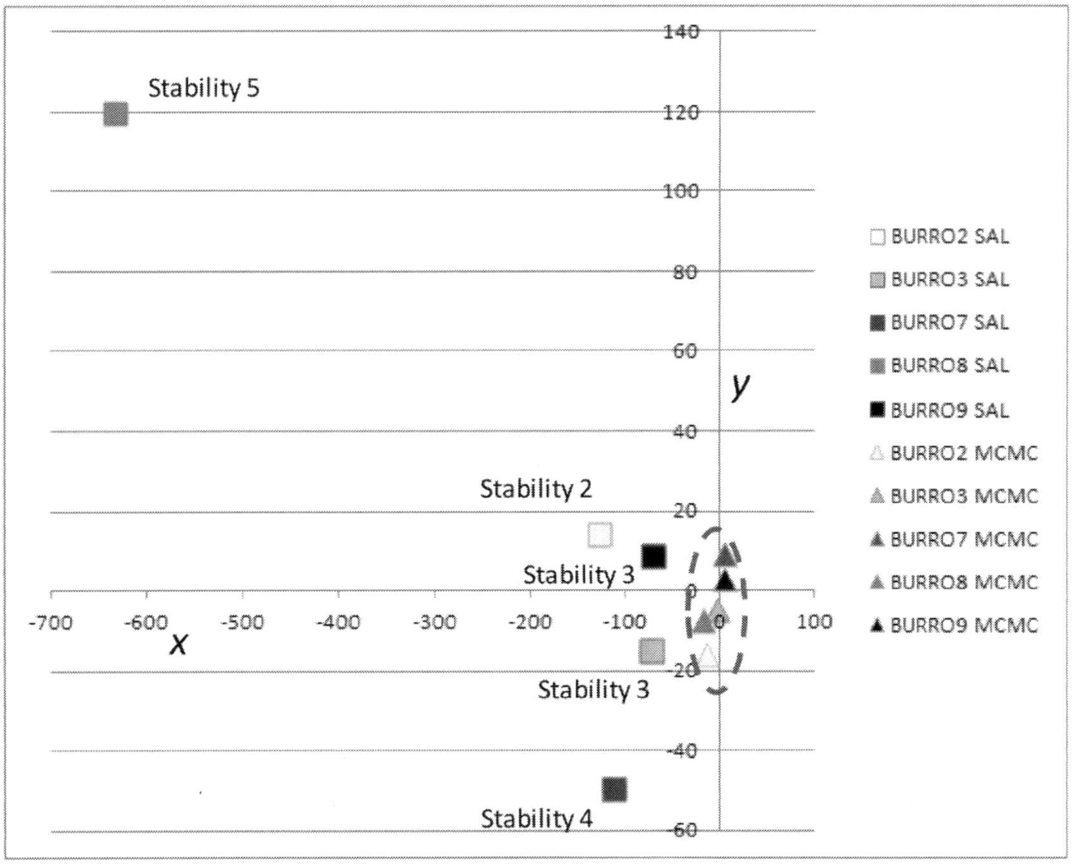

Figure 5: Burro experiment result from SAL and hybrid MCMC approach

Time-varying wind field, uncertainty in wind direction and speed, dynamic sensor readings, release duration, release time may be included in prior parameters for the hybrid approach where the main focus will be on including uncertainties and then identifying the location of the release.

If sensor data is available in dynamic fashion, these hybrid models can be extended to accommodate the dynamic Bayesian model (Johannesson et al. 2004) techniques.

4 Conclusion

The focus of this paper is mainly on estimating release location using the sensor measurements and providing an estimate for the release rate. An important method to estimate source area location using a hybrid approach based on least-square method (SAL) and bayeisan theory (MCMC) was provided. It was shown that the method provides a good solution for BURRO experiment data series for all stability conditions.

5 References

[1] Tarantola A., " Inverse Problem Theory and methods for model parameter estimation", Society for industrial and applied mathematics 2005, Philadelphia.

[2] Johannesson G., Hanley B., Nitao J., Dynamic Bayesian Models via Monte Carlo – An Introduction with Examples -, UCRL-TR-207173, Lawrence Livermore National Laboratory, October 2004.

[3] Cowles M. K. and Carlin B.P. ,Markov Chain Monte Carlo Convergence Diagnostics: A Comparative Review, Journal of American statistical association, vol 91 (434), 1996, pp. 883.

[4] Yee E., Lien F-S, Keats A., D'Amoursc R. , Bayesian inversion of concentration data: Source reconstruction in the adjoint representation of atmospheric diffusion, Journal of Wind Engineering and Industrial Aerodynamics, Vol 96, 2008, pp. 1805–1816.

[5] Yee E., Bayesian Probabilistic Approach for Inverse Source Determination From Limited and Noisy Chemical or Biological Sensor Concentration Measurements, Chemical and Biological Sensing VIII, Proc. of SPIE, Vol. 6554, 65540W, 2007.

[6] Kim J.Y., Jang H.K., and Lee J.K, Source Reconstruction of Unknown Model Parameters in Atmospheric Dispersion Using Dynamic Bayesian Inference, Progress in Nuclear Science and Technology, Vol. 1, 2011, pp. 460-463.

[7] Keats W.A. , Bayesian inference for source determination in the atmospheric environment, PhD thesis, University of Waterloo, Waterloo, Ontario, Canada, 2009.

[8] Neuman S., Glascoe L., Kosovi´c B., Dyer K., Hanley W., Nitao J., Event Reconstruction for atmospheric releases employing urban puff model UDM with stochastic inversion methodology, American Meteorological Society Conference, 2005, Atlanta GA.

[9] Zheng X., Chen Z., Inverse calculation approaches for source determination in hazardous chemical releases, Journal of Loss Prevention in the Process Industries 24 (2011) 293-301.

[10] Sharan M., Singh S. K., and Issartel J.P, Least Square Data Assimilation for Identification of the Point Source Emissions, Pure and Applied Geophysics, August 2011.

[11] Monache L.D. et al., Bayesian Inference and Markov Chain Monte Carlo Sampling to Reconstruct a Contaminant Source on a Continental Scale, Journal of Applied Meteorology and Climatology, pp 2600-2613, vol 47, 2008.

[12] Senocak I., Hengartner N. W., Short M.B., Daniel W.B., Stochastic Event Reconstruction of Atmospheric Contaminant Dispersion using Bayesian Inference, Atmospheric environment, Vol 42 (33), 2008, pp7718-7727.

[13] Koopman R. P., Baker J., Cederwall R. T., Goldwire H. C., Hogan W. J. LLNL/NWC (Lawrence Livermore National Laboratory/Naval Weapons Center) 1980 LNG (Liquefied Natural Gas) Spill Tests. Burro Series Data Report.

[14] Goldiwre H. C., Baker J., Koopman R. P., Cederwall R. T., Hogan W. J. LLNL/NWC (Lawrence Livermore National Laboratory/Naval Weapons Center) 1980 LNG (Liquefied Natural Gas) Spill Tests. Burro Series Data Report: The Appendices.

[15] S. Khajehnajafi, S. Burla, Prediction of Source Location and Rate of a Chemical Release for Emergency Response: Back Trajectory Approach, AIChE Spring Meeting, 2008

[16] Hogan W.R., Cooper G.F., Wallstrom G.L., Wagner M. M. and Depinay J-M, The Bayesian aerosol release detector: An algorithm for detecting and characterizing outbreaks caused by an atmospheric release of *Bacillus anthracis, Statist. Med. 2007;* **26**:5225–5252.

[17] Haario H., Saksman E., Tamminen J., An adaptive metropolis algorithm, Bernoulli, Vol 7 (2), 2011, pp. 223-242.

(43ak) Guidelines for Developing Site-Specific Human Error Rates and Human IPLs for LOPA Using a Safety Climate Approach

Farheen Khan, Human Factors Practice, Exponent, Inc., Chicago, IL, Sunil D. Lakhiani, Human Factors, Exponent, Failure Analysis Associates, Wood Dale, IL and Delmar Morrison, Thermal Sciences, Exponent, Inc., Warrenville, IL

Layer of Protection Analysis (LOPA) is a simple order of magnitude quantitative risk analysis technique commonly applied in the process industries. This powerful tool is especially useful for refining specific hazard scenarios during a process hazard analysis or a risk analysis. In LOPA, a hazard scenario is decomposed into a series of events, which may include the frequency or probability of equipment failures, personnel actions, and physical processes. By one estimate, 80 percent of incidents involve human error; thus, most risk analyses consider the likelihood and effect of human factors on potential scenarios. For LOPA, human error includes events initiated by the human in the work system and events where the human element failed to act as an independent layer of protection against an adverse event.

Although the LOPA method is widely used, there are few guidance sources for refining the human error component. The currently available standard LOPA estimates for the occurrence of human error assume optimal human factors conditions and that work systems are static environments unchanging over time. Work systems are, in fact, dynamic environments, which evolve over time through planned changes, as well as unplanned adaptations. Thus, no work system is completely free of error at any given time. Current LOPA estimates, therefore, fail to account for the dynamic nature of work systems where change from original conditions is not only possible but even necessary from time to time. This paper will examine the deficiencies in the LOPA human error probability estimates and propose a strategy that utilizes a safety climate approach to understanding error causation and incorporating it as a fundamental building block to human error estimation.

Benchmarking – An Important Milestone in the Journey towards PSM Excellence

Prerna Jain

Reliance Industries Limited

India

prerna.jain@ril.com

[1]CindiePridy, [2]Ravi Ramaswamy, [3]Robert Cravens

[1,3]Phillip Townsend Associates, Texas

[2]Reliance Industries Limited, India

cpridy@ptai.com

ravi.ramaswamy@ril.com

rcravens@ptai.com

Prepared for Presentation at

American Institute of Chemical Engineers

2013 Spring Meeting

9th Global Congress on Process Safety

San Antonio, Texas

April 28 – May 1, 2013

UNPUBLISHED

Benchmarking – An Important Milestone in the Journey towards PSM Excellence

Prerna Jain

Reliance Industries Limited

India

prerna.jain@ril.com

[1]CindiePridy, [2]Ravi Ramaswamy, [3]Robert Cravens

[1,3]Phillip Townsend Associates

[2]Reliance Industries Limited

Keywords: benchmarking, gap analysis, metric, performance gap, process safety management system, risk based process safety element, survey

Abstract

Reliance Industries Limited (RIL) has implemented a Process Safety Management (PSM) system through a structured and systematic approach over the last 6 years. In its PSM journey, RIL realized that there is a continuing need to improve and sustain PSM performance and to develop an understanding of the current status of RIL's existing PSM system, a benchmarking process of PSM practices, with reference to industry peers, is required. This was essential in order to identify and address the gaps to best performers within the peer group, define specific action items to obtain significant performance improvements, and ultimately achieve the best systems in the world in field of Process Safety Management.

To accomplish this goal, RIL conducted a PSM Benchmark Program Study designed by Phillip Townsend Associates (PTAI) and CCPS® at its six largest manufacturing locations in India. By helping to understand the relationship between process safety outcomes and management practices; this benchmarking program provided several key approaches for improving overall PSM performance. The program covered benchmarking of six elements related to a world class process safety management system. They are: process safety culture; compliance with standards, codes, regulations, and laws; hazard identification and risk analysis; asset integrity and reliability; management of change; and measurement and metrics.

The results of the study provided element specific quartile performance of each of the six manufacturing sites and Reliance as a whole in comparison to the industry peers who participated in the PTAI study. The gap analysis report presented by PTAI to the six sites

PSM team and senior management of RIL upon completion of the benchmarking study identified key areas of improvement for implementing and evolving effective Process Safety Management System at RIL.

This paper, presented by RIL and PTAI, discusses the benchmark process, methodology followed for the program, the value proposition of Process Safety Benchmarking, benefits of the program, key applications and lessons learned by RIL and new directions for improving PSM. The paper specifically focuses on the journey to success by realizing the gaps to top performers in the peer group, enhancing PSM knowledge and spreading PSM best practices.

1. Introduction

Chemical production and the associated processing industries dealing with hazardous and toxic materials have a significant commercial and societal responsibility to eliminate and/or mitigate the loss of primary containment or fires related to process materials while also ensuring that adequate safety systems are in place. Process Safety Management systems differ from personnel safety management in that process safety focuses directly on the processes, equipment and technology involved in causing an incident. The broader application of process safety management decision making and leadership behavior needs to address loss of containment events of greater and lesser consequences, safety system challenges and failures that could have led to failure, and operating discipline and management behaviors that ensure adequate layers of protection are being maintained. The focus shifts from the lagging indicators that track incidents to leading metrics that are used in a proactive manner to prevent incidents and continuously improve as shown in Figure 1: Process Safety Indicator Pyramid (API RP 754, 2010).

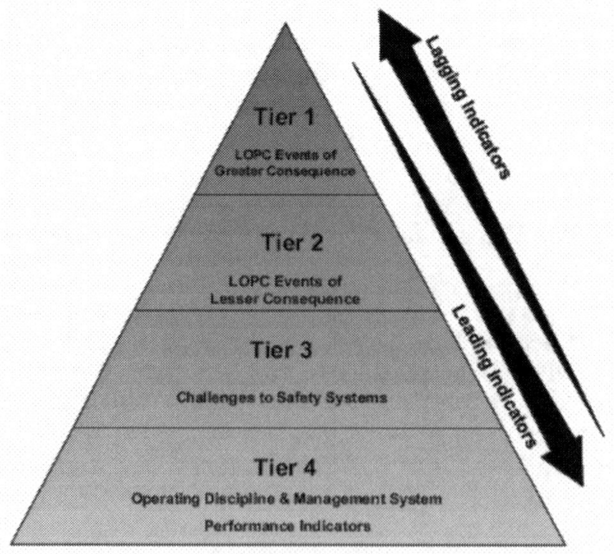

Figure 1: Process Safety Indicator Pyramid

The vision and recommendation of industry-wide process safety originated with the setting up of Centre for Chemical Process Safety (CCPS) by AIChE as a result of the December 3, 1984 Bhopal incident killing more than 3000 people due to toxic Methyl Iso Cyanate gas release. The need for continued focus on process safety got reinforced with the 2005 explosion at the BP refinery in Texas City, Texas. Since then, the Center for Chemical Process Safety (CCPS®) has identified 20 process safety elements to be used as guidelines for a risk based safety process (CCPS®, 2007). In 2009, the CCPS® and Phillip Townsend Associates, Inc. (PTAI) entered into an agreement to develop a benchmarking program for Process Safety Management Systems (PSM).

Viewed from a risk based perspective, the process safety elements can be grouped into four categories: Commitment to Process Safety, Understanding Hazards and Risk, Managing Risk, and Systematic Learning from Experience. PTAI and CCPS® designed the PSM benchmark selecting six of the 20 elements: Process Safety Culture, Compliance with Standards, Hazard Identification and Risk Analysis, Asset Integrity and Reliability, Management of Change, and Measurement and Metrics (CCPS®, 2007). The key goal was to seek an understanding of the relationship between process safety outcomes and management practices.

This paper discusses the details of the PSM benchmark, the value proposition to be gained, and specifically the experience of Reliance Industries Limited (RIL) in developing an understanding of the current status of RIL's existing PSM system, and the practices that will improve and sustain PSM performance. The objective of RIL is to take PSM systems and practices to the top quartile performance through a benchmarking process of industry peers relative to PSM practices. Specifically, the paper focuses on the most important decisions and behavior changes that leadership can make to spread process safety awareness throughout the organization, close the gap to the top performers, develop central and site specific action plans for improvement, and raise the internal level of all organizational units to the that of the top internal performer.

2. Methodology

Benchmarking identifies how well a company is doing compared to its peers in the industry, what the industry top performance is, how large the gap is to the top performers within the industry, and create a stimulus for continuous improvement. Benchmarking programs seek to understand the relationship between outcomes and management practices. The PSM benchmarking program is a systematic study of process safety between companies that will allow sites/companies to compare their management systems with the goal of improving the design and implementation of management practices that impact process safety.

2.1 Value Proposition of Process Safety Benchmarking

PSM benchmarking can elevate a company's process safety performance through four distinct steps:

2.1.1 Identify performance gaps

The identification of a company's performance in relation to Performance Quartiles will highlight key drivers for performance enhancement versus its peers. These quantifiable results replace perceptions with a data-based reality check. In addition, these results identify if the enterprise policies, practices, and procedures are operating as intended and are achieving the desired results.

2.1.2 Create performance enhancement goals

Data driven information provides a measure of the enterprise effectiveness – emphasizing strengths, weaknesses and potential for improvement over time. This helps focus attention on the critical aspects of an enterprise that create risk. It also underscores what improvements should be made – providing quantifiable performance enhancement goals.

2.1.3 Measure the effectiveness of improvement programs

PSM benchmarking can be a starting point for goal setting and performance initiatives aimed at boosting step change improvements in process safety. The clear outcome is that this will add significant shareholder value through reduced risks of catastrophic incidents.

2.1.4 Maintain continuous improvement

The PSM benchmark helps maintain a stimulus for continuous improvement. It provides the necessary motivation to ensure consistent top performance results.

2.2 Benchmarking Process

A peer group is established by PTAI of sites/companies agreeing to compare their process safety programs and performance in an effort to identify best practices and drive improvement. Ensuring consistency and protection of each participant's interests are critical to a successful benchmark. A confidentiality agreement was completed between PTAI and RIL along with execution of a participation agreement. The Data Collection Instrument (DCI), providing clear and consistent definitions for all participating sites/companies was made available. The use of a secure site over the internet ensured security of data and ease of use by participants as they are able to see all the survey questions and select their answers. A kickoff discussion to align key stakeholders at each of the sites is critical. In RIL case, this was done through a series of webinars between PTAI consultants, RIL's Centre for HSE Excellence and stakeholders at each of the participating sites.

RIL chose to benchmark PSM in six of its large Refining and Petrochemical manufacturing sites in India and a team of site focal points, corporate coordinator and PTAI worked in a seamless manner to plan and execute the benchmark program. Following submission of the data, PTAI validates the data on a site and company basis and provides analysis of the results. Validation was enhanced with RIL by individual site visits to each of the six sites by PTAI. This provided the opportunity for discussion and verification of the results reported to the on-line survey. Participation by key RIL personnel during this process added to the effectiveness and accuracy of data reported and provided a seamless voice to the sites between RIL's Center for HSE Excellence and PTAI.

In particular, statistical analysis of the data was performed by PTAI to evaluate various drivers of process safety using a Gap Analysis comparing RIL's performance to the top quartile of the peer group. The quantitative and qualitative results were distributed to participants with numerical scores for each element along with an overall site/company specific score. Strengths, as well as the main opportunities for improvement to close the gap towards the top performers were identified. A discussion of the study results and gap opportunities was held with the executive management in an executive management summary with specific individual gap analysis presentations to the top Line and HSE management of each of the six RIL sites benchmarked.

2.3Benefits of PSM Benchmarking

Management review is an integral part of moving from lessons learned to systematic improvement of PSM. Reviews of the performance gaps can identify potential problems and address them before an incident occurs; as well as, identify opportunities for improvement to move from current levels to the level of top performers. PSM benchmarking helps an organization prioritize the allocation of process safety resources, both financial and human, to focus on the highest identified risk areas.

Among the benefits of PSM benchmarking, there are certain ones that can have a big impact on an organization. There should be a marked increase in staff awareness of process safety issues. This awareness will expand the issues from the Health & Safety group to Engineering to Operations. Based on the organizational level of results discussion, it will potentially increase awareness from site management to impact those that are responsible for setting corporate strategy. A key learning for RIL was the differences between the individual site scores resulting in a changed perception by the Center for HSE Excellence and executive management. PSM benchmarking creates a common understanding and language on process safety matters. Benchmarking can also reduce risks by potentially providing an incident early warning signs.

Sometimes the opportunities for improvement can be very different from the perceived gaps in an organization. Identifying and gaining insight on perceived gaps versus actual gaps is greatly enhanced by an independent benchmark which effectively creates consistency across multiple sites within a company and with the others in the peer group.

3. Key Applications and Lessons Learned

Traditional process safety surveys have identified a list of both quantitative and qualitative issues. Translating the qualitative issues into measurable, actionable steps is a key issue for senior management decision making and allocation of resources. Benchmarking results must be measurable and use a scoring methodology that will not only identify the performance gaps, but also quantify the impact of selected actions. The

PSM benchmarking process designed by CCPS® and PTAI includes a scoring methodology which enables participants to have measurable action plans. Benchmarking is a systematic way to make site-to-site company comparisons and view a company's performance level in relation to that of top performers.

The application of the benchmarking results to identify performance gaps provides meaningful insight for organizations as they develop goals and the associated improvement plans to reach those goals. Repeated benchmarking monitors the rate of improvement. One example of a lesson learned that applies to numerous organizations relates to Process Safety Culture. The findings from the Process Safety Culture section indicate the largest range of results compared to the other sections. The underlying cause of the low scores was a direct outcome of not monitoring the process safety culture through the use of an employee survey. Without a way to assess their work force's values, attitudes and behaviors pertaining to process safety, an organization will have no meaningful basis for making and measuring improvements.

The global process safety incidents have produced tremendous need for the use of PSM benchmark assessment to measure and improve process safety performance. Initially focused on fatality rates, manufacturers quickly learned that having strong PSM system and effective layers of protection were important to prevent any loss of containment scenarios.

At RIL, PSM Benchmark assessment and results helped in identifying following three main interrelated but distinct benefits or observations: (a) Internal Benchmark of various RIL sites, (b) learning through process of understanding the benchmark questionnaire, and (d) need for a detailed analysis of existing systems implementation. We briefly discuss and illustrate examples of each benefit or observation, highlighting how the results could be used.

(a) Internal Benchmark of RIL multiple sites:

PSM Benchmark program conducted for six RIL sites brought to the surface best practices followed at different sites like: Number of indicators tracked and analysis of metrics trend, Quality control reviews of Process Hazard Analysis reports, action plan clearly defined from the Operational Discipline survey results, employee involvement through various PSM campaigns and some other actions related to operational discipline. These practices were beyond the existing standard and system requirements and became good learning for other sites. Also, these best practices will be additional inputs to the group level company documents on PSM. For example, the delta between the site with the highest score and the site with the lowest score in the Compliance with Standards section was 17%; offering a significant opportunity to analyze internal best practices and improve the level of performance to the highest score.

(b) Learning through process of understanding the questionnaire

The benchmarking questionnaire itself generated opportunities for learning new knowledge. Understanding the "why" part of questions called for discussions among various functions at each of the manufacturing sites and reinforced the understanding of PSM. A self-assessment process got started on the compliance to PSM requirement. Following table gives certain examples of learning and new knowledge for the questionnaire itself:

Table 1. Examples of learning and new knowledge from Benchmark questionnaire

Sr. no.	PSM Element	Question	Learning
1	Hazard Identification and Risk Analysis	How are overdue HIRA action items escalated in your organization?	Tracking system to be further strengthened for overdue HIRA action items.
2	Asset integrity and reliability	How is the metal thickness data from the periodic inspections analyzed for trends?	Thickness measurement trends should be analyzed, remaining life projections should be calculated, and adjustments to be made to inspection intervals.
3	Process safety culture	If your organization has utilized a survey tool to assess Process Safety Culture, how many questions were on the survey?	PSM Culture Survey to be established for the organization.

The validation visits from PTAI enhanced the quality of responses and the participation of various functions in the PSM benchmarking program.

(c) Need for a detailed analysis of existing systems implementation

The PSM benchmark results identified areas for further analysis to enhance the systems implementation for improved performance. At RIL, the trend of Process Safety incidents is varying for the last three years and there is no clear trend of reduction in loss of containment events from the metrics, inspite of implementing PSM systems for the last four years. For this, RIL has initiated a group level six sigma project on reduction in loss of containment events with support of an inter-site team. The project follows Define, Measure, Analyze, Improve and Control

(DMAIC) methodology of six sigma. Following is a brief of activities initiated for the above:

- A multisite team has been formed with a six sigma Black Belt guide.
- Literature on Process Safety and Six sigma circulated to team members.
- Project charter was prepared and circulated to team members and other stakeholders for inputs/suggestions and firmed up.
- Cause-Effect Matrix Format developed and Process mapping completed for PSM elements.
- Audit Report Analysis.
- Incident investigation report analysis.

4. New Directions and Next Steps

The PSM benchmark program results gave new insights to the RIL team on the status of PSM within RIL and in relation to the other participants. The gap analysis was converted to actionable recommendations at Group Level and individual site level. Next steps in the PSM excellence journey for RIL have been identified and these are

- Develop a formal system of Process Safety Culture Survey.
- Review of existing PSM metrics and make necessary amendments.
- Review Human Reliability/Human Factors/ Human-Machine Interface requirements, develop best practice requirements and make necessary amendments in Standards.
- Develop program for assessment of compliance with standards/codes/practices for existing equipment.
- Establish system to manage deviations from the industry consensus standards.
- Roll out the Management of Change –Personnel process formally.
- Actual implementation of Standards/Procedures/Practices in field:
    - Meeting target inspection/testing intervals, backlog of overdue inspections
    - Closure of recommendations of PHA, QRA studies
    - HIRA team members to meet qualification requirements
    - Closure of Audit recommendations
    - Implementation of temporary Management of Change
- Review RIL Group PSM Standards and Procedures for gaps and make necessary amendments.

A key benefit of the program for RIL is an Internal Benchmarking among its major manufacturing sites and sharing the best practices within the group. All the sites have developed action plans to close their own gaps in PSM practices.

The collaboration between PTAI and RIL gave new directions for RIL as well as for the Benchmark program itself. A development of two distinct Plan Do Check Act Assess (PDCAA) cycles, one for RIL and the other for PTAI has been the result of this collaboration.

4.1 Overview of PDCAA cycle

PDCAA is a 5-step methodology for process improvement. The stages are shown in Table 2.

Table 2. PDCAA stages

Plan	Establish objectives and how they will be achieved
Do	Put the plan into effect
Check	Verify that the process achieved the desired results
Act	Analyze any differences and their causes. Take action to improve things.
Assess	Analyze and reflect on evidence to make informed and consistent judgments on an ongoing basis for continual improvement.

The PDCAA Cycle for RIL is addressed through the following:

- Planning upgrade of RIL PSM standards to industry best levels and establishing systems to manage deviations, and developing appropriate leading and lagging metrics.
- The do part of the cycle will address the actual implementation of standards and practices in the field at different sites through a process of sharing the best practices by institutionalizing the inter-site network teams for the PSM elements.
- Check part will address strengthen the current PSM metrics tracking and review systems, IT enablement of dashboards for Process Safety at different levels of organization, and reflecting the updated requirements of compliance in the PSM audit protocols.
- A robust system of closure of recommendation and actions arising out of various studies and audits through a system of IT enabled universal work list and linking it to individual performance management system is under development which form the Act step of the cycle.

- Periodic reviews and assessment of PSM will complete the whole cycle of improvement.

4.2 Plan-Do-Check-Act-Assess- RIL Example: (Hazard Identification and Risk Assessment)

Plan

In this step, process needs are analyzed by examining a range of data available from the Benchmark assessment gap analysis relating questions in program to performance results. Data can be analyzed at the individual sites level and the company level. In this process, aspects on HIRA management system, including HIRA methodologies, team composition, criteria for HIRA team leaders/facilitators, level of training, techniques, risk criteria, recommendations and their status were covered.

Do

This step asks what group standards require and how they will be assessed. Forward action plan is made by looking at what is lagging in actual implementation in field ; such as, the timely closure of HIRA recommendations and minimum criteria/qualification for team members. Sites develop individual goals to improve their performance.

Check

In the "check" step, formal and informal assessments take place continually, to strengthen the performance of Metrics on HIRA elements. Sites can make use of comparative data to identify trends to study; for example, overdue HIRA action items.

Act

In this example the "act" step calls for standardization or having a robust system to ensure closure of overdue action items. Results from this cycle become input for the "assess" phase.

Assess

In this step, process of periodic reviews and monitoring of HIRA action items is done and tracked to ensure continual improvement.

4.3 Plan-Do-Check-Act-Assess- PTAI Example: (The benchmarking process and Process Safety Culture)

The PDCAA Cycle has also provided the opportunity for continuous improvement for PTAI and the PMS benchmarking survey.

Plan

Through the validation process the iterative nature of conference calls and site visits proved to be of significant value in providing accurate and consistent understanding of the survey questions. This resulted in comparable and meaningful results among the sites and RIL total scores.

Do

Reviewing the current benchmarking process and seeking feedback from RIL identified that enhanced validation and the creation of a more iterative process could be incorporated into the benchmarking process. A key area of needed improvement supporting the benchmarking process in the Process Safety Culture section is the measurement of Process Safety Culture. Both RIL and PTAI determined the need for a formal system for a Process Safety Culture Survey that provides more consistent measurement and comparable results of the safety culture that can be utilized for RIL internal site to site comparisons and RIL's sites to the external peer group. RIL and PTAI have discussed future collaboration on this project and the possible implementation to all of the RIL sites for internal benchmarking and review.

Check

Informal and formal process assessments are the key to continuous improvement of the PSM Benchmarking Survey and process.

Act

The act step calls for standardization of a formal process assessment tool to measure the benchmarking efforts required by participants and the actual benchmarking survey tool.

Feedback from participants needs to be evaluated and appropriate improvements to the program parameters and the benchmarking process implemented.

Assess

This is the process of continuous improvement and an environment of continuous learning. PTAI will periodically review the process and enhancements with CCPS to ensure an evolving process that grows in value-added elements.

5. Key success factors

The success of the PSM benchmarking survey hinged on four key factors; management support of a systematic approach, the strength of the validation process, the ability to report results with a detailed gap analysis that could be translated into specific action plans at a central and site level, and the seamless voice of RIL's Center for HSE Excellence and PTAI through strong communication and a systematic effort.

The management presentation of results was opened with comments from RIL's President, HSE, emphasizing the strategic importance and focus of PSM as a systematic process and not simply a process safety program. His comments were directed and focused on the areas he felt most important to taking RIL to the next level. This sent a very clear message of the strategic focus placed on PSM at RIL.

The continued dialogue and collaboration between RIL and PTAI has enhanced the double loop, PDCAA cycle. As a critical success factor this creates ongoing and new opportunities for both.

6. Conclusion

Process safety is a mature, but continuously evolving concept with practices and procedures varying widely among the refining, chemical and chemical processing industry. Moving from a compliance model to a risk-based model requires a comprehensive approach to managing process safety. Substantial improvement opportunities exist for even the current top performers. Specifically, no site has made a perfect score on any of the risk-based elements of the PSM benchmark.

The PSM benchmarking program is a unique methodology to allow companies to identify how large the gap is to top performers and to "learn" how to prioritize resource allocations based on the greatest impact opportunity to improve their process safety management system. This shifts the understanding of current practices to the lessons learned that are measurable and actionable. The benchmarking process identifies top quartile management practices while highlighting not only a company's strengths, but also potential development opportunities. The program's scoring system provides a marker to track improvements and measure the gaps within the company sites and against external top performers.

Using a systematic Process Safety Management System benchmark drives lessons learned to strategic and organizational learning.RIL considers periodical assessment of its PSM through repeat benchmarking at regular intervals as a good value proposition to monitor trends of performance and to achieve pacesetter performance compared to industry peers.

7. References

[1] American Petroleum Institute, API RP 754, "Process Safety Performance Indicators for the Refining and Petrochemical Industries, First Edition," 2010.

[2] Center for Chemical Process Safety CCPS®, "Guidelines for Risk Based Process

Safety," John Wiley & Sons, Inc., Hoboken, NJ, ISBN 978-0-470-16569-0: liii-liv, 2007.

9TH GLOBAL CONGRESS ON PROCESS SAFETY

"Process Safety Management - A Practical View"

German Luna-Mejias, P. Eng.
ACM Facility Safety
Suite 700 940 6th Ave SW, Calgary, AB, Canada
gluna@acm.ca
guatire2@gmail.com

Prepared for Presentation at
American Institute of Chemical Engineers
2013 Spring Meeting
9th Global Congress on Process Safety
San Antonio, Texas
April 28 – May 1, 2013

UNPUBLISHED

"Process Safety Management - A Practical View"

German Luna-Mejias, P. Eng
ACM Facility Safety
Suite 700 940 6th Ave SW, Calgary, AB, Canada
gluna@acm.ca
guatire2@gmail.com

Keywords: Process, Safety, Management, Practical

Abstract

Process Safety Management (PSM) was developed over 20 years ago to reduce hazards and risk for the Oil Industry. In the last 3 years serious accidents have happened in the Oil Industry that requires us to re-evaluate PSM programs. It has been observed in many HAZOP studies, that the HAZOP (part of a PSM program) does not reflect a real commitment to safety, but the "need" to meet regulations. This paper's initiative is to call organizations/companies to think and be safe instead of just fulfilling project or government rules. Full involvement of the organization, starting with managers, is required for the PSM program to work. In this paper, practical examples will be drawn from the successful implementation of a PSM program from several years ago in a refinery. Additionally, the reasons for this implementation to be considered successful will be discussed at length.

1. Introduction

1.1 History

On June 1st, 1974, a Cyclo-hexane vapor cloud explosion killed 28 workers and injured 53 others; 1800 houses were damaged in the nearby area in Flixborough, England. [1]

On Dec 3, 1984, an uncontrolled release of methyl-isocyanate (MIC) gas from a pesticide plant in Bhopal, India, resulted in more than 3,800 fatalities and approximately 11,000 injuries with different degrees of severity. [2]

In 1985, the American Institute of Chemical Engineers (AIChE) formed the CCPS (Chemical Center for Process Safety) quickly set out to publish its first process safety guideline book: Guidelines for Hazard Evaluation Procedures.

On July 6, 1988, the offshore platform Piper Alpha, located in the North Sea and owned by Occidental Petroleum, suffered a series of explosions and fires that resulted in the deaths of 165 crew members and the total loss of the platform. [3]

On Oct 23, 1989, an explosion in the Phillips refinery in Pasadena, Texas killed 23 people and injured 314 more. Impact of the explosion was registered as a 3.5 on the Richter scale. [4].

In 1990, the American Petroleum Institute issues the recommended practice 750: "Management of Process Hazards" which can be considered the first step onto the Process Safety Management programs. [5].

In 1992, OSHA formally recognized the vital importance for managing highly hazardous chemical processes within the US. A company that is involved in highly hazardous chemical processes or activities is mandated to implement a process safety management system. [9]

In 1994, CCPS modified the elements of the PSM program recommended in API RP-750 [5] to issue a new "standard" for PSM that includes more elements: "Guidelines for implementing Process Safety Management System"

After a period of "relative calm", big accidents in the oil and gas industry occurred again:

- BP Texas city, 2005 [6]
- BP Macondo, Gulf of Mexico 2010 [7]
- San Bruno, California 2011 [8]

Many disasters have impacted the process industry, from Flixborough to Macondo and elsewhere many more. Through the ashes of these tragedies, strong change has been advocated by the public and their government's regulators to hold companies accountable for their negligence.

1.2 Process Safety Management

Process Safety Management (PSM) is a methodology that was developed after the lessons from several major accidents showed that such events can be difficult to prevent using traditional occupational health and safety approach, which tends to focus on the interface between the individual operator and the equipment or process.

The 3 keywords of this philosophy are:

- PROCESS: related to process issues. i.e. fires, explosions, etc
- SAFETY: Safety regulation, safety concern
- MANAGEMENT: anyone is a manager in some degree, requires employment participation.

PSM is the application of management principles and systems to the identification, understanding and control of process hazards to prevent the release of highly hazardous chemicals such as toxic, reactive, flammable and/or explosive substances (process-related events), that may lead to injuries and incidents.

The goal is to minimize process incidents by evaluating the whole process. The PSM standard (OSHA 29- Part 1910.119) [9] is the regulatory framework for process industries in the US and is quickly becoming the industry "best practices" standard for responsible companies worldwide.

The PSM system originally suggested by the CCPS (Guidelines for implementing Process Safety Management System) consists of 12 elements:

1. Accountability: objectives and goals
2. Process knowledge and documentation
3. Capital project review and design procedures
4. Process risk management
5. Management of change
6. Process and equipment integrity
7. Human factors
8. Training and performance
9. Incident investigation
10. Company standards, codes and regulations
11. Audits and corrective actions
12. Enhancement of process safety knowledge.

The intent of this paper is to present some of the elements of PSM that were considered the most important for a specific company; not to debate the number of elements or which one is more important than the other.

Regulations should not be the sole driving force for the process industry to "think safe". Applying the PSM program accurately and effectively requires more than just the requirement to fulfill government regulation but a whole company culture to be involved in making the workplace safe. This paper's initiative is to encourage organizations/companies to think and be safe instead of simply fulfilling a project or complying with government regulations.

The examples given in this paper correspond to a company where PSM was not mandatory by law, but it was part of the company's standard as a way to make sure the company complied with the International O&G industry standards.

2. The practical Approach

In this section, examples of the approach taken by an O&G company to implement PSM are given for what was considered the most relevant part of the program. Not all elements are discussed as the key elements described below will support all other elements. The practical approach also includes unfortunate situations that happened before the implementation of the program to emphasize the need for it.

Elements analyzed in this document are:

- Management commitment (Element 1)

- Process Safety Information (Element 2)
- Process Hazard Analysis (Elements 4 & 5)
- Operating Procedures (Elements 6,7 & 8)
- Hot work Permit (Element 6)
- Training (Element 8)
- Pre-start up Safety Review (Element 9)
- Emergency Planning and Response (Elements 4 & 12)
- Program Audits (Element 11).

2.1 Management Commitment

Management commitment is fundamental to the PSM program. Exercising and eating well is something that people need to do for their own good and not because it is a mandate or a command to obey. Exactly the same happens with the PSM program: if people are not convinced of the importance of safety, the PSM program won't last. The first safety best practice identified, and one which is essential to any good safety program, is a management commitment to safety.

No company can have a successful PSM program if the management of that company is not committed to the concept of safety. Every company has an obligation to ensure, as much as possible, the safety of its employees. Not only does the company have a moral obligation to ensure the safety of their workforce, it is also good business to do so. Keeping the number of incidents down will also help the company to maintain few lost time days, low work injury indexes and increased profitability.

In a particular refinery, the General Manager wanted to develop a PSM program and designated a top manager to be in charge of the PSM program. This designated manager's first action was to convince all top managers of the importance of the PSM program. Convincing the other managers was not based on: "I said so, and I have been appointed by the GM" but with education and conviction.

The training of managers and cascading this knowledge to all personnel with supervisory responsibilities started with the video forum about the Piper-Alpha accident (one of the objectives of the forum was to recognize that this scenario may happen to that facility). The Piper-Alpha video, in conjunction with API 750 video, was presented to all personnel.

"Safety oriented Management behavior" was a key, as the employees know who the managers are, and they observed their commitment to safety. All personnel understood that they were "living by example" instead of "giving lip service to their own safety rules". The exposure of management risk and demonstration of Safety Concept by action prevented the scenario of: "Well, management doesn't need to operate safely, neither do we."

Additionally, financial incentives to personnel were also created during the earlier phases of the program to promote participation within it. After a period of time, people became conscious about safety and they started to behave safely even when "nobody is looking"; not because of a reward but they were convinced of PSM as a way of life.

Part of the management commitment to safety was the implementation of safety related training programs with the same level of importance as technical training. The program was successful because there was support from the entire management team, and management was involved in the formulation, communication and enforcement of its safety program. All levels of management were visibly demonstrating their commitment to safety.

2.2 Process Safety Information.

Process Safety information (PSI) includes design information, mechanical design and working knowledge. These 3 elements are linked to good quality P&IDS, which was the driving reason the company decided to move in the direction to update the P&IDs as the first step under the PSI element.

This was critical to the facility as it was a 60 year old refinery. In several cases the installation of jump-overs, minor changes and similar modifications were done in the field without any analysis or documentation on the P&IDs. Some minor incidents (spills, spray, etc.) were found to be a consequence of inadequate information on P&IDs.

As a result, a team composed with very experienced operators and young engineers was formed to update all the P&IDs of the facility, with two main purposes: pair young engineers with very experienced personnel and teach them how to update P&IDs and to comply with PSM.

A few years later, when most of the 30+ year operators were retired, these updated P&IDs were successfully used to train the new group of operators in the facility.

Another group, supported by Technical Services, began work to update and maintain the MSDS (Material Safety Data Sheets) for the facility. This action proved to be very effective for the firefighters as they were properly trained in the materials they may have to deal with in case of emergency. Before that they were relying on an operator telling them what to do during an emergency scenario.

Nowadays, this situation (outdated PSI) is a concern for many facilities: In a study performed in 2010 by OSHA IN 2010 [10]; 22.2% of the deviations (not compliance with PSM) were associated to non-updated PSI.

2.3 Process Hazard Analysis (PHA)

The purpose of this analysis is to minimize the likelihood of the occurrence and the consequences of a dangerous substance release by identifying, evaluating, and controlling the events that could lead to releases.

The management team and all personnel were asked to answer the following questions:

- Is my facility safe to operate?

- In the case of an event happening, am I protected?
- Is my facility protected?

This is how the HAZOP concept was introduced to the facility. A group of people were trained in HAZOP facilitation and they were in charge of implementing the PHA program.

Implementation of PHA, as with any other type of new procedure, was a challenge. At that time only the HAZOP was considered for implementation; later, additional PHA methods were added to the portfolio. Manual recording of PHA scenarios was a challenge, but the use of PHAs (manual or automated) continues to be vital to gauge whether a facility is protected or not.

In the earlier stages of the implementation, PHA-HAZOP was used as part of Management of Change element, but later was included in the evaluation cycle of all projects.

2.4 Operating Procedures

In the old facility, procedures were transmitted by oral tradition and there were very few written procedures, but even those written were not clear for non-experienced individuals. They were not developed with consideration for new personnel in that corresponding area.

In several opportunities, lack of written procedures led to accidents when one of the "guru's" was not there to execute a particular procedure or when a step was missed in a complicated procedure.

To prevent this from happening repeatedly, a team was built consisting of very experienced operators with experts in writing procedures and process engineers to start writing the procedures for each unit. Managers defined the most critical and most urgent ones to be done first. It was a medium-term effort that produced great results when a few years later the operations crews were modified to include people with technical background but almost zero operational experience.

2.5 Hot Work Permit

In traditional operation, the Operations Supervisor is also in charge of supervising the maintenance activities in his area. About 25 years ago, a request for a work permit was made to the Operations Supervisor; he erroneously determined that the risk was not too high, as he was busy with the day-to-day operation. He did not check the actual situation in the field. Unfortunately, the line where the work was going to be performed was not checked before and was filled with light naphtha (not properly drained) and this led to a fatality when welding.

After this accident and as a part of PSM program, a new work permit procedure was established:

- A dedicated Operations Supervisor for maintenance activities, being the only authority to sign-off work permits.
- A procedure was developed to include documentation and proper sign-off
- Training was developed for all involved parties about how to perform using the new Hot Working Permit, including contractors.

2.6 Training

The training of all personnel is a key element of Process Safety. The training must be on a continuous basis and at all levels of the organization. In this particular company, the kick-off of the PSM was to present the Piper-Alpha accident video to all levels of the organization from the top Management to the bottom of the organization. Also the API-750 Video was presented to all personnel, the objective: understand the risk, study the failures and realize that this situation was very likely to happen to that facility.

The importance of training is that it must to be continuous, refreshed on regular basis. Training must be developed for new employees and all contractors, with a rule of "nobody in" without safety training. The safety training included an exam; and those who do not pass the exam cannot get into the facility. To make sure PSM works, training is a key element and dedicated personnel to fulfill this target is recommended.

2.7 Pre start-up Safety Review

According with API RP 750 [5] the PSSR (Pre Start-up Safety Review) should be performed for new and modified facilities that are subject to this recommended practice to confirm that:

- Construction is according with specifications
- Procedures are in place (including safety procedures)
- PHA recommendations are closed out.
- Training for all operators is provided

It is fairly common to see confusion between PSSR for commissioning and PSSR during normal S/D for maintenance; as API 750 indicates, it is needed under both scenarios.

Before the implementation of PSSR, the responsibility of checking the facility prior to start up was only for operations. Normally this was done by an Operations Supervisor with other duties in the area and sometimes relying on another operator or on a Maintenance Supervisor. In this facility the need for PSSR was identified in an unfortunate incident:

An Operations Supervisor on his last night shift was doing the plant checking for start-up after he was informed by maintenance about completion of work. He was the only person in charge of checking the facility and provided the team with the "go ahead" for start-up 30 minutes before the end of his shift. The heating of the plant started and about 2 hours later a hot naphtha leak spilled over a hot heat exchanger producing a large fire that required 2 months of additional plant shutdown to repair damages. A good example of what "not to do".

A PSSR exercise takes a longer time to perform and needs a multidisciplinary team but will prevent incidents or accidents.

2.8 Emergency Planning & Response

Firefighter training is very important. In 1984, 100 municipal firefighters died when they were called to fight a fire in a heavy diesel tank in another location [11]. They handled it as a regular fire and high quantities of water were directed to the roof of the tank. No effort was made to cool off the walls of the tank and they were not aware of the presence of water in the bottom of the tank. When the diesel in the bottom of the tank reached the appropriated temperature, a boil over situation occurred.

One of the key elements in an Emergency Response plan is the knowledge of the hazards of a system that may create an emergency. As part of a PSM program, an intensive training on Process for non-technical personnel was given to firefighters (permanent and volunteers). Prior to the implementation of the PSM program, in the event of an emergency, the Firefighting Supervisor communicated directly with the Operations Supervisor on site, to obtain information about the potential hazards to people and facility. The Firefighting Supervisor had only a very limited knowledge of the facilities and associated hazards.

Emergency dry-runs were also implemented weekly according with a program developed on the most critical areas. This practice involved regular firefighters and volunteer firefighters.

Another step that was taken as a part of keeping the firefighters engaged was the annual recognition to all firefighters (rewards, celebrations, etc.).

2.9 Program Auditing

All the PSM elements should be audited on a regular basis to ensure the program effectiveness and performance. Programs like PSM require a lot of involvement from the Auditing team, the PSM leading team will define which parts of the program are most critical and it will be the start-up point for the auditing.

Audit process will always be needed, even when every person in the organization is "in the safety mindset".

3. Conclusions

- Implementation of the PSM program is a time consuming effort.
- Dedicated personnel are required for leading the change.
- Management actively involved in the PSM program is a must.

3.1 Tangible benefits from the implementation of PSM

- In a facility with a record of 4 fires per year, they were able to reduce the number of fires down to 1/year.
- Training of operators improved the operations of the facility reaching production records and reliability records.
- The Safety concept was built-in (be safe when nobody is looking)
- Ongoing, non-prescriptive, performance-based, participation at all levels.
- Facility received a Safety Award as "Safest in class" facility.

4. Recommendations

- PSM is a MUST for all O&G facilities
- It is a way of living! I encourage all O&G companies to follow this "way of living".

5. References

[1] Health and Safety Executive, 'The Flixborough Disaster: Report of the Court of Inquiry', HMSO, ISBN 0113610750, (1975)

[2] Chouhan TR "The Unfolding of Bhopal Disaster". Journal of Loss Prevention in the process industry **18** (4–6): 205–208. (2005).

[3] PatC-Cornell, M. E. "Learning from the Piper Alpha Accident: A Postmortem Analysis of Technical and Organizational Factors". Risk Analysis, Vol. 13, No. 2, 1993

[4] Bethea, R.M. "Explosion and Fire at the Phillips Company – Houston Chemical Complex, Pasadena, TX" Chemical Engineering Department, Texas Tech University, Lubbock, TX 79409-3121 (2007)

[5] API Recommended Practice API RP-750. "Management of Process Hazards". American Petroleum Institute, 1990.

[6] CSB. Investigation Report: Refinery Explosion and Fire (15 killed, 180 injured). Report # 2005-04-I-TX; Chemical Safety Board, Washington, DC. 2007

[7] British Petroleum Oil Company (BP). "Deepwater Horizon Accident Investigation Report" (2010)

[8] NTSB "Pacific Gas And Electric Company Natural Gas Transmission Pipeline Rupture and Fire" San Bruno, California. NTSB Report # PAR-11-01 (2011).

[9] OSHA. "Process safety management of highly hazardous chemicals". OSHA 29- Part 1910.119 (1992)

[10] Jordan Barab., "Learning from Industry Mistakes" presented at NPRA National Safety Conference, May 2010

[11] MARAVEN S.A. "Investigacion del Accidente de Tacoa" (Tacoa's Accident Investigation, In Spanish). Caracas, 1985.

(43an) Auditing the Process Hazard Analysis - Key Characteristics to Scrutinize in a Good-Quality HAZOP Study and Supporting Effective HAZOP Study Updates

Steven T. Maher[1], Michael A. Nour[1] and Kristin D. Norton[2], (1)Process Safety, Risk Management Professionals, Irvine, CA, (2)Process Safety, Risk Management Professionals, Inc., Irvine, CA

Although Process Safety Management (PSM), requiring good-quality Process Hazard Analyses, was promulgated in the United States in 1992, and HAZOP Study techniques date back to the mid-1970's, the HAZOP Study technique is still not uniformly applied in a high quality fashion. The objective of this paper is to provide experience-based tips on Process Hazard Analysis auditing, and some key characteristics that the auditor can scrutinize to assure quality implementation. Analogously, a similar set of characteristics that can facilitate future HAZOP Study updates can also contribute to a quality HAZOP Study that supports the spirit of Safety Management System objectives for the identification of hazards potentially impacting plant personnel, the community, and the environment. This paper will also review tips for structuring the HAZOP Study to facilitate later updates and revalidation.

Vapor Cloud Explosion Test of Air-Inflated Fabric Structure

Brad Horn
Baker Engineering and Risk Consultants, Inc.
3330 Oakwell Court, Suite 100
San Antonio, TX 78218-3024
BHorn@BakerRisk.com

Matthew T. Edel, P.E.
Baker Engineering and Risk Consultants, Inc.
3330 Oakwell Court, Suite 100
San Antonio, TX 78218-3024
MEdel@BakerRisk.com

Prepared for Presentation at
American Institute of Chemical Engineers
2013 Spring Meeting
9th Global Congress on Process Safety
San Antonio, Texas
April 28 – May 1, 2013

UNPUBLISHED

Vapor Cloud Explosion Test of Air-Inflated Fabric Structure

Brad Horn
Baker Engineering and Risk Consultants, Inc.
3330 Oakwell Court, Suite 100
San Antonio, TX 78218-3024
BHorn@BakerRisk.com

Matthew T. Edel, P.E.
Baker Engineering and Risk Consultants, Inc.
3330 Oakwell Court, Suite 100
San Antonio, TX 78218-3024
MEdel@BakerRisk.com

Keywords: vapor cloud explosion, explosion test, air-inflated structure, fabric structure, blast load, thermal load, structural response

Abstract

Portable buildings are gaining widespread use at refining and chemical processing facilities, particularly for facility siting requirements. Although trailers and blast resistant modules (BRMs) can be used for many modular applications, in certain situations fabric structures may provide a viable alternative. However, the blast response of fabric structures is neither clearly defined nor well understood throughout the blast design community.

One such modular fabric structure is an air-inflated structure. A manufacturer of this type of structure recently funded the testing of one of their specimens for the effects of a vapor cloud explosion (VCE). The VCE for the test was initiated using propane and an internal matrix of congestion inside a three-walled cubicle, referred to as the Deflagration Load Generator (DLG) test rig. The test involved a deflagration event that imparted a thermal flux and a maximum free-field pressure of 2.9 psi with an average duration of 31 ms at the front wall of the air-inflated structure.

The reduced blast loads and thermal effects inside the structure were compared to those outside. In addition, the attenuation of blast loads behind the structure was also examined, as well as shielding effects. The structural response of the test specimen was captured with high-speed video, and observations on the motion of the structure are provided.

1. Introduction

Baker Engineering and Risk Consultants, Inc. (BakerRisk) has developed a Deflagration Load Generator (DLG) test rig that can supply VCE blast loads to a variety of structures in order to provide structural response test data to manufacturers (i.e., product proof testing). The DLG rig produces a VCE blast load with a blast wave shape, pressure, and duration similar to the blast loads that are generated by accidental VCEs at chemical processing facilities. Dynamic Air Shelters, Ltd. (DAS) contracted a test of its manufactured air-inflated shelter.

The test involved imparting a VCE blast load on a DAS air-inflated shelter. The shelter was erected to the manufacturer's specifications and then subjected to a VCE blast load using the DLG test rig. The target blast loading for this test was a centrally peaked blast wave with a peak free-field blast pressure of 3.5 psi and 50 ms duration at the shelter wall closest to the DLG rig. Data recorded include pressure histories, thermocouple and heat flux measurements, and video.

2. Test Setup

This section describes BakerRisk's DLG test rig, the DAS air beam shelter, and the instrumentation used to measure and record data during the test.

2.1 DLG Test Rig

The DLG test rig is shown in Figure 1. The test rig is a steel box with one open side that is initially covered with a thin plastic sheet to allow a fuel-air mixture to be developed within the test rig. During testing, the test rig is filled with a propane-air mixture at a desired concentration, which is then ignited at the middle of the back wall. An obstacle array within the test rig causes the flame to accelerate and generate a blast wave. The resulting blast wave propagates out toward the target structure.

2.2 Air Beam Shelter

DAS constructed the air-inflated shelter shown in Figure 1, which was located directly in front of the DLG test rig during the test. The shelter system consisted of a ballistic-resistant wall and an air-inflated shelter with overall dimensions of approximately 46 ft. (14 m) wide by 56 ft. (17 m) long and 16 ft. (4.9 m) tall. The blast wall was made of PVC trapezoidal bags that were filled with material, in this case sand, and then hung between supporting wall frames spaced at approximately 6.7 ft. (2.0 m).

Figure 1. Air-Inflated Shelter (on left), Blast Wall, and DLG Test Rig (on right)

The air-inflated members were continuously pressurized and created the structural support for the shelter. Doors were located on two walls parallel to the direction of the blast propagation. The interior air beams are shown in Figure 2. Except for the instrumentation, the interior of the shelter was vacant during the test.

Figure 2. Air-Beam Columns

A schematic showing the location of the test rig, shelter and instrumentation is provided in Figure 3. The standoff distance between the test rig and the shelter was 21 ft. (6.4 m).

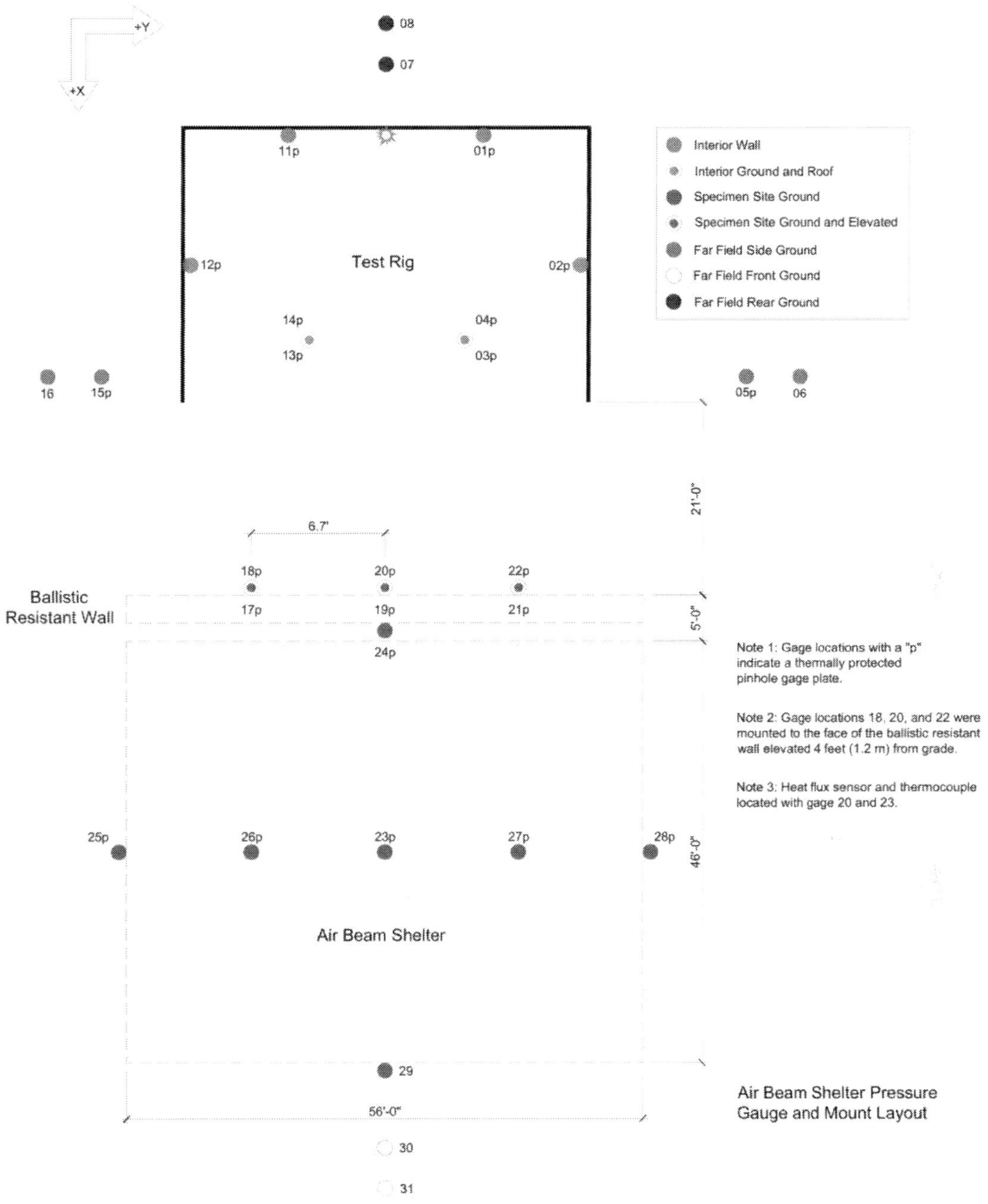

Figure 3. Instrumentation Layout Map

2.3 Instrumentation

During the test, the pressure, heat flux, and temperature histories were measured at locations of interest. Both regular and high speed video of the shelter structural response were captured.
A total of 29 dynamic pressure gauges sampled at 10,000 samples per second were used to measure the blast loads. Pressure gauge locations are shown in Figure 3; the gauge locations relative to the front of the test rig are provided in Figure 5. A total of 13 pressure gauges were positioned to measure the blast load at points of interest on the shelter and its components. Six gauges measured blast loads at the ballistic resistant wall, with three on the ground at the base of the wall and three mounted on the wall 4 ft. (1.2 m) above the ground gauges in order to measure the reflected load. One pressure gauge was located on the ground between the ballistic resistant wall and the shelter wall, and one pressure gauge was located at each shelter doorway. Three pressure gauges were located within the shelter interior, and one pressure gauge was located just outside the rear wall of the shelter.

Thermocouple and heat flux sensors were placed at the ballistic resistant wall and inside the shelter, coincident with pressure gauge locations 20 and 23. Each sensor was elevated 4 ft. (1.2 m) above grade and oriented toward the DLG test rig. Each thermocouple was sampled at 20,000 samples per second.

Video recording devices were used to document the test and record the shelter structural response to the blast loading. The high speed video (HSV) camera was positioned in the same plane as the leading edge of the ballistic resistant wall, while two high definition (HD) video cameras, two exterior spy cameras, and one interior spy camera recorded other angles.

3. Test Results

Figure 4 shows the structural response recorded by the high definition video camera, capturing at 30 frames per second.

Figure 4. VCE Loading Applied to Shelter

3.1 Blast Loads

A peak pressure measurement map with pressure values listed in green font is shown in Figure 5. Table 1 provides the blast load summary which includes the pressure gauge location, standoff distance from the test rig, measured peak pressure, positive impulse, and both the measured and calculated durations. The calculated duration assumes a perfectly triangular blast pressure history (i.e., duration is twice the impulse divided by the peak pressure). Table 2 provides a summary of the peak reflected blast pressures measured at the ballistic resistant wall and the corresponding calculated peak free-field (i.e., side-on, incident) blast pressures.

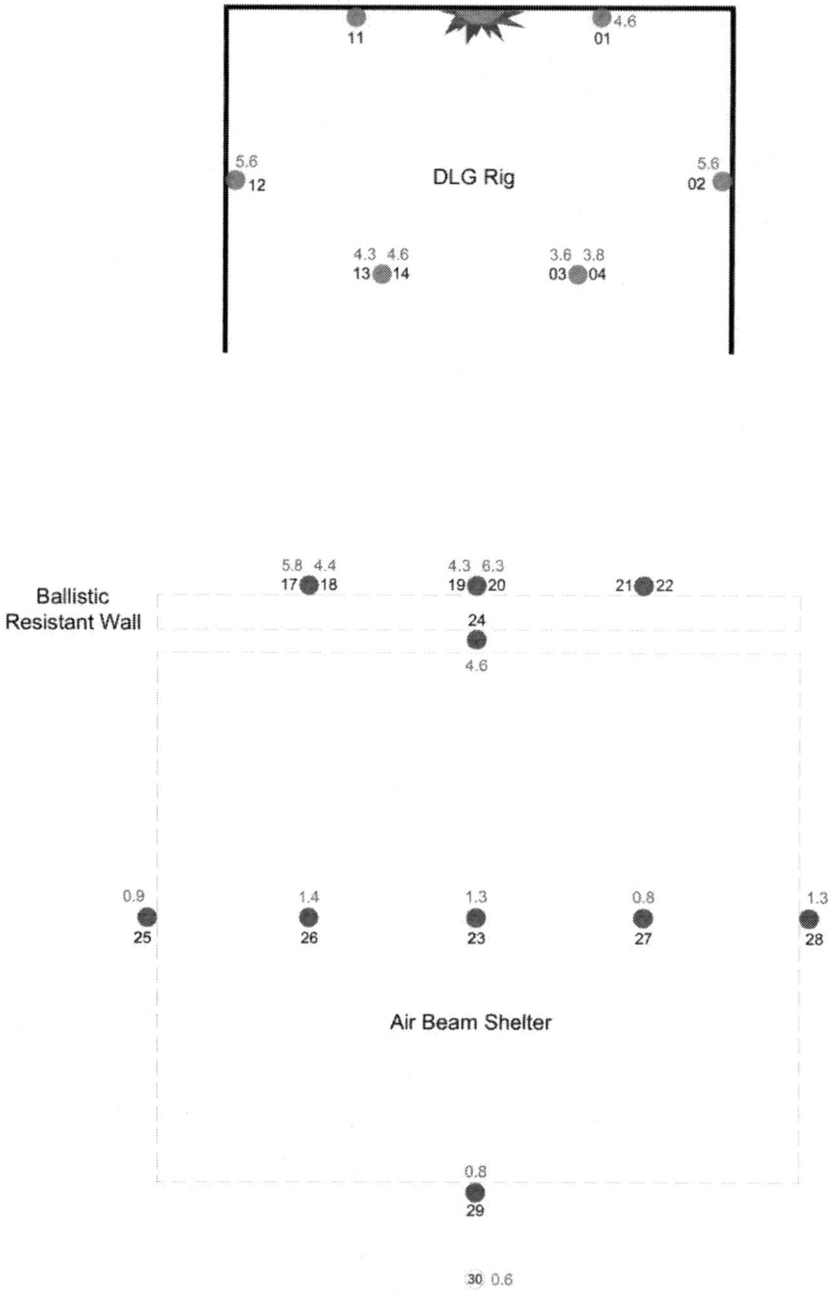

Note: Pressure values directly in front and behind the ballistic-resistant wall are reported as the adjusted peak values.

Figure 5. Peak Pressure Measurement Map (psig)

Table 1. Blast Load Summary

Location	Gauge	Distance From Test Rig (ft.)	Measured Peak Pressure (psig)	Measured Positive Impulse[1] (psi-ms)	Measured Duration (ms)	Calculated Duration (ms)
Rig Interior	2	0	5.6	206	120	74
Ballistic Resistant Wall[1]	17	21	5.4	90	48	31
	18	21	3.9	67	42	31
	19	21	3.9	70	51	32
	20	21	5.8	93	48	30
Front Shelter	24	24	4.6	95	57	41
Shelter Doorway	25	48	0.9	19	57	41
	28	48	1.3	28	57	41
Shelter Interior	23	48	1.3	29	53	45
	26	48	1.4	29	53	41
	27	48	0.8	16	53	41
Rear Shelter	29	71	0.8	16	45	41
Far Field	30	76	0.6	10	30	33

Table 2. Ballistic Resistant Wall Blast Pressure Summary

Location	Gauge	Distance From Test Rig (ft.)	Measured Peak Pressure (psig)	Adjusted Peak Pressure[1] (psig)	Calculated Free Field Pressure[2] (psig)
Ballistic Resistant Wall[1]	17	21	5.4	5.8	2.7
	18	21	3.9	4.4	2.0
	19	21	3.9	4.3	2.0
	20	21	5.8	6.3	2.9

[1] *Pressure values adjusted for pre-event signal ringing*
[2] *Calculated from reflection factor*

The maximum peak pressure in the DLG test rig was 5.6 psig (0.39 bar). Gauges at the ballistic resistant wall were positioned to measure the reflected load on the face of the wall or shelter.

The peak pressures and impulse at the wall were adjusted to account for the signal ringing prior to the arrival of the blast wave and are shown in Figure 7. The maximum adjusted peak pressure was 6.3 psig (0.43 bar). The pressure signals along the centerline of the shelter in the direction of the blast are shown in Figure 6. Two of the six gauges were invalid and were therefore discarded.

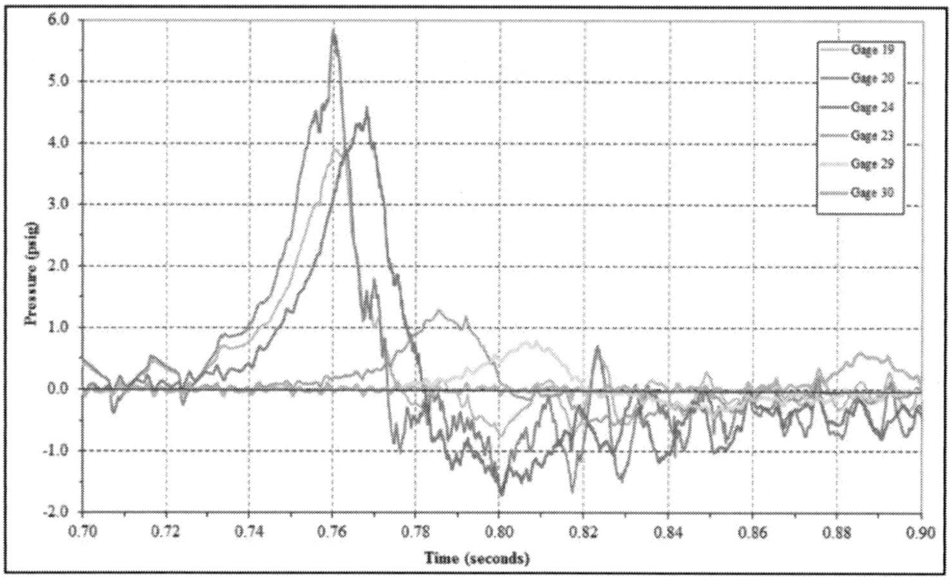

Figure 6. Pressure Histories along the Front Center Gauge Lane

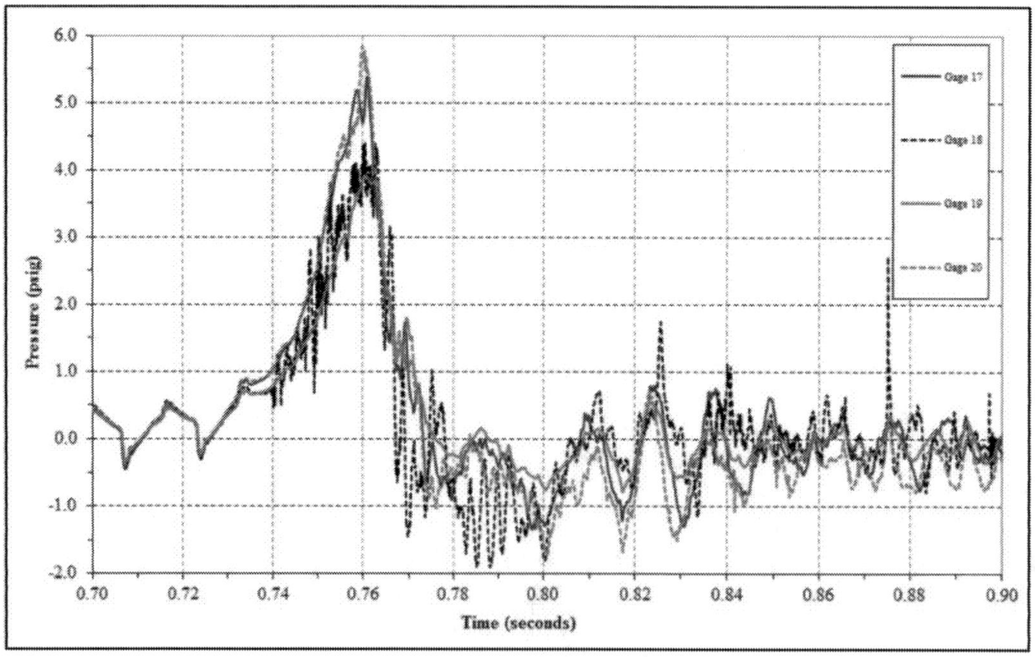

Figure 7. Pressure Histories at the Ballistic-Resistant Wall

The measured peak pressure and impulse between the ballistic resistant wall and air shelter were 4.6 psig (1.4 bar) and 95 psi-ms (6.6 bar-ms), yielding a calculated duration of 41 ms. Pressure histories were measured along the mid-span of shelter. The pressure histories at these mid-span locations are shown in Figure 8. For the doorway exterior locations, the largest measured peak pressure was 1.3 psig (0.09 bar) with an impulse of 28 psi-ms (1.9 bar-ms) and a calculated duration of 41 ms. The largest measured peak pressure among the interior gauges was 1.4 psig (0.1 bar) with an impulse of 29 psi-ms (2.0 bar-ms) and a calculated duration of 41 ms. Behind the shelter, the measured peak pressure was 0.8 psig (0.06 bar) with an impulse of 16 psi-ms (1.1 bar-ms) and a calculated duration of 41 ms.

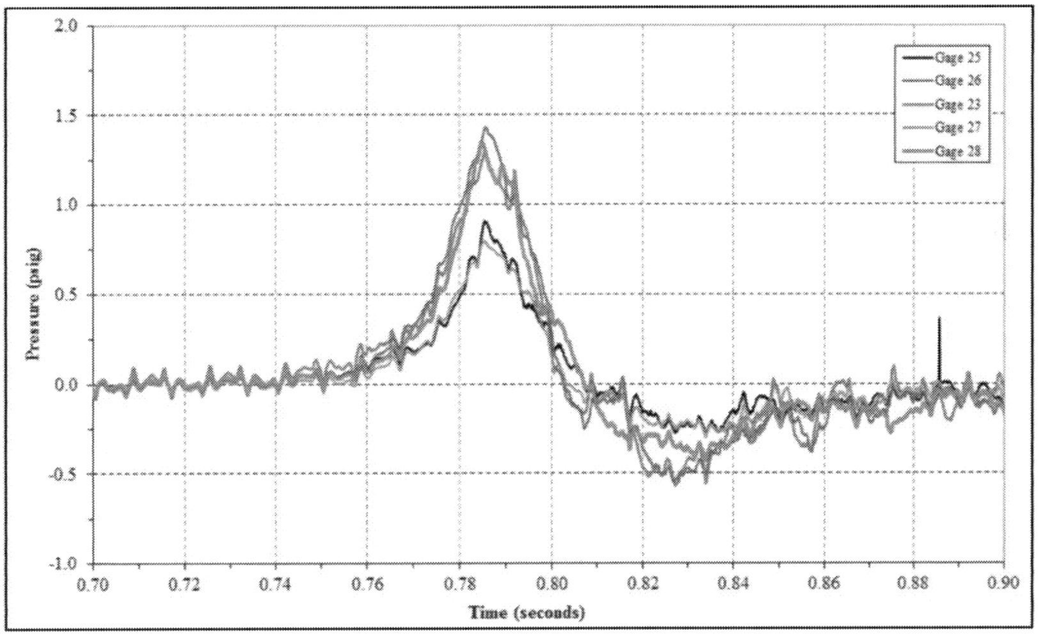

Figure 8. Pressure Histories at Shelter Mid-Span Locations

3.2 Thermal Loads

Figure 9 shows the measured temperature just past the exterior face of the ballistic resistant wall. This peak signal exceeded the range of the signal conditioner and was "clipped" at approximately 1100 °F (590 °C). The estimated actual maximum temperature at this location was just under 1500 °F (815 °C) based on an extrapolation of the signal. The measured signal inside the shelter indicates that the temperature had increased less than 5 °F (3 °C), as shown in Figure 10. That is, while the temperature just outside the shelter increased dramatically due to exposure to the fireball, the temperature within the shelter barely rose.

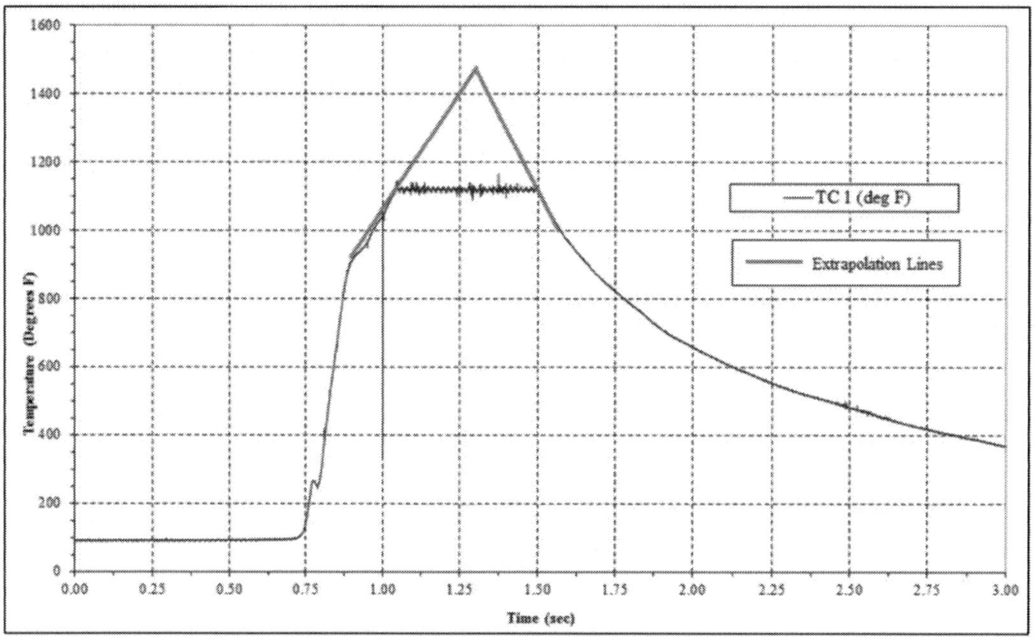

Figure 9. Exterior Temperature History

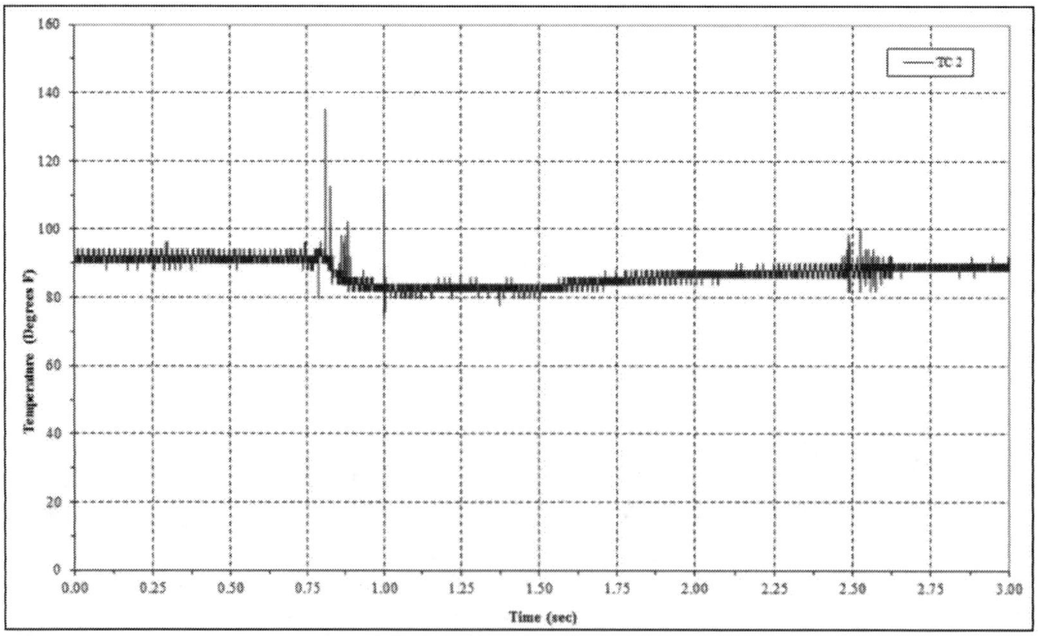

Figure 10. Interior Temperature History

The exterior heat flux signal is shown in Figure 11. This signal was filtered to provide a smooth curve with a maximum heat flux of 6 kW/m^2. The interior heat flux signal was invalid due to the low signal to noise ratio. The virtually zero interior heat flux is consistent with the measured temperature at this location.

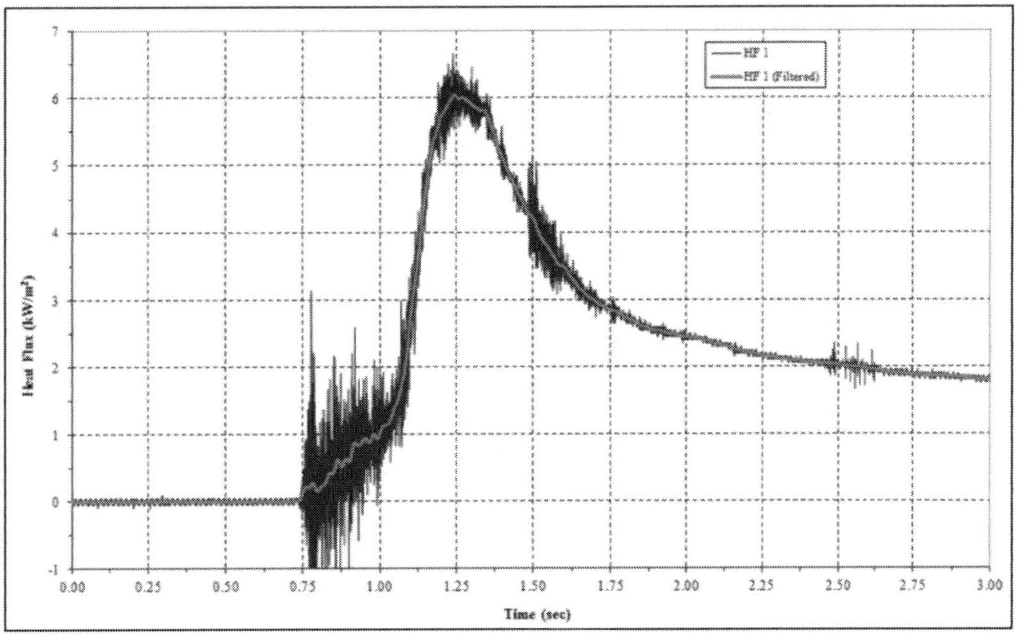

Figure 11. Exterior Heat Flux History

3.3 Structural Response

The air-inflated shelter sustained only minor damage from the VCE blast loading. The air-beams that create the arch section of the shelter experienced rapid temporary deflection as the blast wave propagated across the shelter. The ballistic resistant wall remained functional following the test, allowing safe egress from the shelter after the VCE. Figure 12 shows the wall deflection of about 1 ft. (0.3m) at the top. The yellow grid lines are approximately 1 ft. (0.3 m) apart. The yellow and blue markers indicate the roof deflected approximately 1.5 ft. (0.5 m) downward during the test and returned to the initial position within 2 seconds.

Figure 12. Shelter Deflection Comparison

4. Conclusions

A Dynamic Air Shelters, Ltd. air-inflated structure was subjected to a VCE having a blast load with a peak free-field pressure of 2.9 psi and a 31 ms duration at the wall nearest the DLG rig. The test resulted in a deflagration-type load with a centrally-peaked blast shape. The blast loads inside the structure were reduced from the levels recorded at the exterior ballistic resistant wall by approximately 75% but were comparable to the exterior pressures measured outside of the two doors. Shielding effects of the blast loads were anticipated directly behind the structure, but none were observed at this location relative to the loads recorded closer and further from the rig.

Significant thermal loads were observed on the front face of the structure due to its close proximity to the fireball emanating from the test rig. Yet even at nearly 1500 °F (815 °C) and 6 kW/m^2 at the exterior front face, the interior temperature increased less than 5 °F (3 °C) with virtually no heat flux. The air-inflated structure provided almost complete shielding from the thermal loads with negligible observable thermal damage.

The high speed and high definition video show a maximum of 1 ft. deflection of the ballistic resistant wall and less than 2 ft. deflection of the air-inflated shelter. The structure returned to its initial position within 2 seconds and had no permanent deflection. Internally, the structure had no debris hazards and remained inflated, allowing safe egress from the air beam structure.

9TH GLOBAL CONGRESS ON PROCESS SAFETY

Revisiting the Conditions for a CO2 Tank Explosion

Yu Zhang
Air Products and Chemicals Inc.
7201 Hamilton Blvd, Allentown, PA 18195-1501
zhangy15@airproducts.com

Joan Schork
Keith Ludwig
Air Products and Chemicals Inc.
7201 Hamilton Blvd, Allentown, PA 18195-1501
schorkjm@airproducts.com
ludwigka@airproducts.com

Prepared for Presentation at
American Institute of Chemical Engineers
2013 Spring Meeting
9th Global Congress on Process Safety
San Antonio, Texas
April 28-May 1, 2013

UNPUBLISHED

Revisiting the Conditions for a CO2 Tank Explosion

Yu Zhang
Air Products and Chemicals, Inc.
7201 Hamilton Blvd, Allentown, PA 18195-1501
zhangy15@airproducts.com

Joan Schork
Keith Ludwig
Air Products and Chemicals, Inc.
7201 Hamilton Blvd, Allentown, PA 18195-1501
schorkjm@airproducts.com
ludwigka@airproducts.com

Keywords: Boiling Liquid Expanding Vapor Explosion (BLEVE), Carbon dioxide, Overpressure

Abstract

Carbon dioxide is broadly used in a number of applications including food freezing, beverage production, enhanced oil recovery, supercritical extraction and welding. It is commonly stored as a pressurized liquid (e.g.: -18 °C/0 °F and 2 MPa/300 psia) in carbon steel tanks.

A number of CO2 tank explosions have been reported in the literature over the past 50 years. In order to design appropriate safety systems for liquid CO2 storage tanks, it is important to understand the worst case scenario. Should the energy release be modeled as a rapid vapor phase expansion (PV) or a Boiling Liquid Expanding Vapor Explosion (BLEVE)? The question arises because of a common definition of a BLEVE as an event resulting from the sudden loss of containment of a liquid that is at or above its atmospheric superheat limit temperature. Under the conditions listed above, CO2 is above its atmospheric boiling point but below its superheat limit. In this temperature range, will the vaporization rate be rapid enough to contribute to a blast overpressure in the event of a tank failure?

To answer this question, the authors made a critical review of information available in the literature including laboratory data and incident reports. This paper will describe the analysis conducted and present the results indicating that the potential for rapid vaporization should be included in evaluating the consequences of a CO2 tank rupture below the superheat limit.

1. Introduction

Carbon dioxide (CO2) is broadly used in a number of applications including food freezing, beverage production, enhanced oil recovery, supercritical extraction and welding. Since CO2 at its bubble point becomes a solid at a pressure below 5 bars, CO2 is commonly stored as a pressurized liquid at an elevated pressure of 20 bars absolute in carbon steel tanks.

When a pressurized liquid in an equilibrium state is depressurized instantaneously, it enters a superheated state (a meta-stable liquid above its boiling point) from which the liquid could vaporize in an explosive manner resulting in a blast wave and risk of flying fragments of the vessel. This is called a Boiling Liquid Expanding Vapor Explosion (BLEVE) [1,2]. When the degree of superheat of the liquid is small, the vaporization rate is not large enough to develop a significant blast effect. According to the Superheat Limit Theory developed by Kim and Reid [3], an explosive evaporation occurs when the depressurized liquid enters the spinodal decomposition temperature-pressure range. Based on thermodynamics principles and equation of state of fluids, Kim and Reid [3] predict a so-called BLEVE envelope (Figure. 1) which indicates the range of initial temperatures and pressures which will lead to a BLEVE following instantaneous depressurization. The superheat limit theory is cited in some industrial hazard evaluations as a criterion to determine whether or not a BLEVE should be considered [4,5].

The mechanism of BLEVE based on Kim and Reid's theory is briefly described here. When the boiling point of a liquid is reached, vapor bubbles start to form at active sites such as vessel walls. Nucleation at these sites is called *heterogeneous nucleation*. If bubble nuclei are large (approaching millimeter sizes), a small degree of superheat (defined as heating beyond the boiling point) is needed to make the bubble grow. If the bubble nuclei are small, for example when the vessel is very clean and smooth, they need a large superheat to grow. When there are not many active sites, the bubble nucleation must take place in the bulk of the liquid. Boiling in this case generally takes place at submicron nucleation sites at suspended particles, impurities, or ions. This requires a large degree of superheat to make the bubble grow. This means the liquid normal boiling point can be exceeded without boiling, and the liquid can be significantly superheated. There is, however, a temperature limit at a given pressure above which a phase transition from liquid to vapor has to happen, even though the bubble nulei are small. This superheat limit temperature (SLT) and pressure correspond to the conditions at which a spinodal decomposition of liquid happens. This can produce a very violent phase transition with boiling caused by *homogeneous nucleation* in the bulk of liquid. The superheated liquid state could also be reached by sudden pressure drop. For CO2, the experimental value of SLT at atmospheric pressure is 267 K (-6.2°C) [6].

Although the theory is commonly accepted, some authors question its applicability to a real case. For example, Birk and co-works [7] report that BLEVEs have been observed with propane when the pressure-liquefied gas was at ambient temperature (20 °C), well below its atmospheric superheat limit temperature of 53 °C. In the 2010 edition of the Center for Chemical Process Safety (CCPS) Guidelines to evaluate BLEVE [2], it says "There is no experimental evidence with a real pressure vessel where the phase change was by homogeneous nucleation. Homogeneous nucleation is extremely unlikely in industrial use due to temperature gradients, impurities, surface roughness and other factors that cause localized nucleation."

2. CO2 BLEVE

The end state of liquid CO_2 depressurized to 1 atmosphere is a solid and vapor mixture, which makes its depressurization process unique. CO_2 may evaporate explosively even when its temperature is below the STL, because CO_2 crystallizes at ambient pressure, thus presenting enough nucleation sites for liquid to flash [1].

Several fundamental questions are raised when considering CO_2 system. How does the formation of dry ice affect the blast energy? During the decompression process, does the vapor expand isothermally or isentropically? This initial report will only touch on two questions based on various incident reports on large scale CO_2 tank energy releases:

1. *Could a CO2 BLEVE happen when the storage conditions (temperature and pressure) are outside the so-called BLEVE envelope predicted by the superheat limit theory?*
2. *If it goes BLEVE, could we use the algorithm presented in the CCPS guide to perform the consequence analysis?*

3. Historical experiences

Table 1 summarizes major CO_2 tank incidents we found with documentation in the open literature. The Worms, Haltern, and Repcelak incidents are summarized in CCPS's guide to BLEVE [1-2]. Information about the incident at Yuhang is collected from news reports in Chinese [8-9], and information about Fukushima's incident is translated from reports in Japanese [10-12].

Table 1. Some major CO_2 incidents

Date	Location	Cause	Initial Pressure [bar]	Initial Temperature [°C]
13 Nov. 2008	**Yuhang, China***	Brittle fracture	23	-15
21 Nov. 1988	Worms, Germany	Overpressure	20	-20
2 Sept. 1976	Haltern, Germany	Brittle fracture	23	-15
1 Mar. 1969	Fukushima, Japan	Overpressure	20	-20
2 Jan. 1969	**Repcelak, Hungary***	Overfilling	15	-30

* Overpressure tags are extracted from these incidents

Out of the five incidents, only two overpressure damage data points (tags) are obtained, which are used to validate the algorithm employed in the CCPS guidelines. There were many tags on projectiles, but we find these less reliable than the overpressure tags. One data point is from the Repcelak incident: a warehouse about 15-30 meters away from the explosion center was torn down due to heavy damage. The other data point is from the Yuhang incident: Window glass shattered within a 400 meter radius of the explosion center. Criteria on overpressure damage to buildings are obtained from CCPS's guide to BLEVE [1].

3.1 Large scale CO2 tank energy release

Yuhang, Hangzhou, China, 13 November 2008

A transport ship carrying 130 cu. meter (95% full) of CO_2 exploded on November 13, 2008 in Yuhang, Zhejiang, China [8-9]. The tank's content was stored under a pressure of 23 bar at a temperature of -15 °C. At the moment of explosion, the ship was in the dock belonging to Rongsheng Chemicals. It was not in operation and there were few workers around as it was before regular business hours. The explosion completely destroyed the CO2 ship and sank H2SO4 and H2O2 carrying ships nearby. Two workers on the CO2 ship lost their lives instantly and 3 were injured due to projectiles.

Pieces of the ship were scattered over a 500 meter area around the explosion. The weight of these fragments varied from a few grams to several hundred kilograms. The shrapnel caused damage in several adjacent residential buildings. All the windows of a metal machining factory located around 500 meters away shattered. In the Yujiadouzhen, a village 500 meters away from the explosion, windows shattered in many residential buildings.

The cause of this accident is believed to be overloading and brittle failure of the CO2 tank. The tank was designed for on-shore use, but was modified by Rongsheng Chemicals for ship transportation. To lower the transportation cost, the company modified the level indicator, locked the relief valve, and overloaded the tank (95% filling level). The tank material, 16 MnR steel, has a working temperature from -20 °C to 475 °C. It was not suited for use under low temperature conditions.

Worms, Germany, 21 November 1988
This is one of the most catastrophic incidents involving a CO2 tank explosion. The detailed description of Clayton and Griffin is summarized here [13]. The liquid CO2 was stored in a 30 tons capacity tank at 20bar and -20°C. The vessel was shattered into a number of pieces in the explosion. Most of the tank was propelled into the Rhein River about 300 meters away from the epicenter. Only 20% of the tank was recovered originally. The Worms's investigation concluded that the vessel failed catastrophically, most likely due to overpressure from overheating with a frozen relief valve. The failure pressure was estimated to be between 31 and 51 bara. It was speculated that the failure caused a CO2 BLEVE [2].

Haltern, Germany, 2 September 1976
A rail car carrying 231,000 kg of CO2 exploded in Haltern, Germany [1]. The tank's contents were at 7 bar and -15°C. At the moment of the explosion, the car was passing through a railroad shunting yard. As it passed checkpoints, the car was observed to be releasing plumes of CO2 from the safety valve. It then exploded 15 m in front of a group of other rail cars. Parts of the tank were thrown up to 360 m. One person was killed in the explosion. Analysis of a recovered piece of the tank car showed that failure was due to brittle fracture. Brittle metal fracture is consistent with the near instantaneous overpressure created by a BLEVE event.

Fukushima, Japan, 1 March 1969
A CO2 storage tank in a steel mill exploded [10-12]. A slate-roofed factory within a 50 m radius was completely destroyed, and only its pillars remained. Windows of houses located within a 500 m radius were damaged. Before the explosion, the main safety valves were closed during repairs. Liquid remained in the tank (50% filling level) and the heat source of the external evaporator was left on. The vapor phase temperature and pressure in the tank kept rising, and eventually the tank exploded when the pressure exceeded the burst pressure of the tank. Based on heating element power, CO2 usage, and heating time, it is estimated that almost all liquid CO2 was turned into vapor before the explosion. Thus, this incident is determined to be a PV release.

Repcelak, Hungary, 2 January 1969

Two vessels containing liquid CO2 (15 bar, -30°C) in a CO2 production and filling plant exploded in rapid succession [1, 14]. The explosion completely destroyed the tank yard consisting of four liquid CO2 storage vessels and the equipment of the technological laboratory. Large fragments were scattered in a circle of approximately 400 m radius. A 2800 kg shell landed at a distance of 150 m and a fragment weighing 1000 kg landed 250 m away. During filling, the first vessel exploded. Some minutes later, another close-by vessel exploded. The second vessel probably failed as a result of impact from a fragment from the first vessel. The explosions tore the two vessels into pieces, and tore the third vessel off its foundation bolts. The third vessel was shot into the process laboratory like a rocket due to liquid CO2 rapid expansion through a hole on the bottom. The probable cause of the accident was overfilling the first tank due to a level indicator failure. The explosion of the second vessel was speculated to be caused by a CO2 BLEVE.

3.2 Small scale experimental investigation

Most experimental work on BLEVE has focused on flammable substances like LPG and propane [1,2]. Recently, there were several small scale lab tests performed to study CO2 BLEVE [15-18].

Kim and Reid, 1978

An important reason for considering the possible occurrence of BLEVE with CO2 is this paper by Kim and Reid which presents experimental and modeling work to investigate the phenomena of rapid depressurization of hot, high pressure liquids or supercritical fluids (CO2) [3].

The authors applied a thermodynamic model based on the spinodal fluid state, which represents the limit-of-stability for the liquid phase during expansion. For a BLEVE to occur, the substance has to find itself on the spinodal curve between 1 bar and the critical point where the curve ends. When a catastrophic failure occurs there is not time for heat to pass into the system and so the path during failure is adiabatic. In thermodynamic terms the entropy of the system is constant during failure and the conditions for BLEVE to occur means that the initial entropy has to be higher that the entropy of the intersection of the spinodal curve and the 1 bar isobar. As a result, a region of initial conditions that could give a spinodal state can be identified (Figure 1). Kim and Reid performed some simple experiments on depressurization of saturated liquid and supercritical CO2. None led to explosions regardless of the initial conditions relative to the envelope in Figure 1. The authors suggested the reasons for disagreement between theory and experiments were associated with factors like vapor bubbles on the wall of the vessel (sites for heterogeneous nucleation) and shock waves or disturbances from the mechanism that ruptured the burst disk of the test vessel.

Pettersen, 2002

Pettersen performed an experimental study aimed at clarifying the possible occurrence of BLEVE in CO2 systems [17]. A test vessel was built and instrumented in order to measure transient pressures during initial depressurization, and to record possible subsequent overpressure peak(s). The study concluded that none of the experiments, which were conducted over a wide range of initial conditions (3.5, 4.5, 6.0 and 10.0 MPa initial pressures with varying liquid fill level and discharge opening), showed signs of BLEVE, which would be characterized by a sudden repressurisation to a level significantly above the initial pressure.

Wei, 2009
This master thesis presents experimental demonstration of CO2 BLEVE which happened below the superheat limit temperature (outside the BLEVE envelop in Figure 1) [18].

The list below highlights (taken from Wei's thesis) the most significant findings:

- The 'Superheat Limit Temperature' theory for predicting occurrence of a BLEVE was not supported with experiments in this work. A CO2 BLEVE can also occur when the initial temperature is below the superheat limit temperature of CO2 (-13.8 °C). Nevertheless, considering influencing factors including CO2 filling level, potential gas leaking and CO2 mixture with different phase compositions, the theory may still be acceptable.
- Two possible approaches for a safer CO2 storage include using an initial storage pressure of around 2 MPa and developing a safety valve that can further slow down pressure drop when an unexpected vessel opening and depressurization occurs.

3.3 Consequence analysis

An established procedure for calculation of the consequence of a BLEVE event can be found in CCPS's guide to BLEVEs [1,2].

1. Calculate the explosion energy. Ongoing debates on the calculation of the explosion energy from a BLEVE event involve discussions about the BLEVE mechanism and are outside the scope of this study. In this work, we consider three options where the explosion energy is
    a. Internal energy difference between the initial and final states of an isentropic expansion process [1],
    b. Internal energy difference between the initial and final states of an irreversible adiabatic expansion [19],
    c. The PV energy (energy from the vapor phase, excluding that from flashing liquid).

   Please note that the explosion energy will need to be multiplied by a factor of 2 if the burst is a ground level, to obtain the effective working explosion energy.
2. Once the explosion energy of a BLEVE is estimated, overpressure can be determined by employing the overpressure- scaled distance curve by Baker.
3. Once overpressure is calculated, a multiplier has to be applied to account for vessel shape effect.

According to CCPS, the method presented above is based on the similarity of the blast waves of pressure vessel bursts and high explosives. This similarity holds only at some distance from the explosion. In the near field (scaled distance < 2), the peak overpressure and impulse from a pressure vessel burst differ greatly from those of a detonation of a high explosive, except when the pressure vessel is filled with a very hot high-pressure gas. Baker et al. developed a method which is used to solely predict blast pressures in the near field. However, it is based on results of numerical simulations and involves solving an implicit equation. The details of this refinement can be found in CCPS's Guidelines [1]. Note that this refinement for the near field cannot be made for non-ideal gases.

4. Results

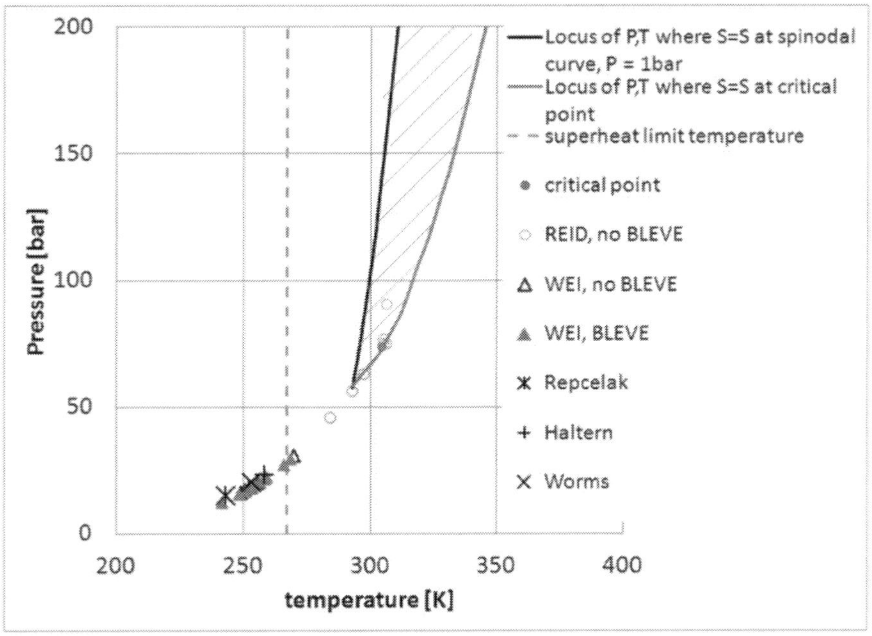

Figure 1. Summary of conditions involving CO2 BLEVE. Empty symbols are experiments with no BLEVE, and solid ones and crosses are experiments/events with BLEVE.

We summarize the literature data discussed in Section 3 in Figure 1. From Figure 1, we see that there is no clear dividing line between a BLEVE and a non-BLEVE event based on the superheat limit temperature, as indicated by experimental results and historical experiences. This might be due to the various processes involved in such an event, including nucleation of bubbles at the vessel wall. It is conclusive that the superheat limit temperature has no special significance to the BLEVE event observed in either past accidents or controlled experiments. Furthermore, Kim and Reid's experimental data along with others failed to produce a BLEVE in the laboratory when initial conditions were within the envelope predicted by the superheat limit theory.

We obtained two overpressure tags, one from Repcelak and the other fromYuhang:

1. The Repcelak: "light structure warehouse was so seriously damaged that it had to be pulled down." [14] The warehouse was 15 meters from the epicenter, and the overpressure was about 0.15-0.25 bar[1];
2. Yuhang: The windows of a metal machining factory located around 500 meters away, all shattered. In the Yujiadouzhen, a village 500 meters away from the explosion, windows shattered in many residential buildings [9]. The overpressure was about 0.035-0.07 bar [1].

Three different explosion energies, as discussed in Section 3.3, for each event are summarized in Table 2. The actual overpressure data *vs.* calculated overpressures using the CCPS's recommended method is shown in Figure 2. If the correct energy is chosen, the points should fall on or near the $y = x$ curve. Clearly, these results suggest that the explosion energy released from the incidents is more consistent with BLEVE type of explosion-energy where energy from the liquid phase contributes significantly.

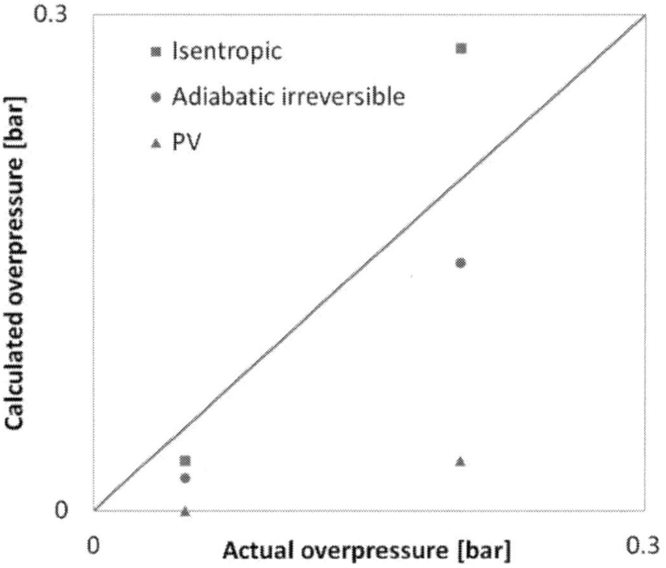

Figure 2. Actual overpressure versus calculated overpressure using different explosion energies for two events.

Table 2. Explosion energy, calculated overpressure (COP), and actual overpressure (AOP) tag for Repcelak and Yuhang incidents

Location	Initial filling level	Isentropic expansion [J]/COP[bar]	Adiabatic irreversible [J]/COP[bar]	PV [J]/COP[bar]	OP tag distance to epicenter R_tag [m]	AOP at R_tag [bar]
Repcelak	80%	3.5E9/0.28	9.9E8/0.15	3.0E7/0.03	15±5	0.20±0.05
Yuhang	95%	1.3E10/0.03	4.9E9/0.02	3.6E7/0.003	400±100	0.05±0.02

5. Conclusion

Through a critical review and analysis of historical and experimental reports of CO2 tank explosion, we conclude that a CO_2 tank rupture could lead to a BLEVE event even when the initial storage temperature is below the superheat temperature limit of CO_2. The consequence of a BLEVE event should be considered when assessing the risk associated with CO2 storage.

6. References

[1] "Guidelines for Evaluating the Characteristics of Vapor Cloud Explosions, Flash Fires, and BLEVEs, Center for Chemical Process Safety", *American Institute of Chemical Engineers*, New York, 1994

[2] "Guidelines for Vapor Cloud Explosions, Pressure Vessel Burst, BLEVE and Flash Fire Hazards, Center for Chemical Process Safety", *American Institute of Chemical Engineers*, New York, 2010

[3] Kim, M.E., Reid, R.C., "The rapid depressurization of hot, high pressure liquids or supercritical fluids", *Chemical Engineering at Supercritical Fluid Conditions*, Ann Arbor Science, 1983

[4] Calculation of BLEVE conditions for carbon dioxide, www.criticalprocesses.com/BLEVE.htm (accessed Jan. 10, 2013)

[5] "Final public report safety study for liquid logistics shipping concept (DNV)", http://cdn.globalccsinstitute.com/sites/default/files/publications/19006/final-public-report-safety-study-liquid-logistics-shipping-concept.pdf (accessed Jan. 10, 2013)

[6] Abbasi, T., Abbasi, S.A., "Accidental risk of superheated liquids and a framework for predicting the superheat limit," *Journal of Loss Prevention in the Process Industries,* 20 (2007) 165-181

[7] Birk, A.M., Cunningham, M.H., "The boiling liquid expanding vapor explosion", *J. Loss Prev. Process Ind.*, 7 (1994) 474–480

[8] Li, W., Di, G., Wang, R., "The cause analysis of a liquid CO2 tank explosion on a ship", *Materials Science Forum,* 689(2011)461-467

[9] Yuhang CO2 cargo ships explosion, http://zjnews.zjol.com.cn/05zjnews/system/2008/11/14/014973390.shtml (in Chinese, accessed Jan. 10, 2013)

[10] "High Pressure Gas Safety Inst. of Japan. Rupture of liquefied carbon dioxide in a low-temperature storage tank", *Accident investigation report (outline),* 1969 (in Japanese)

[11] "High Pressure Gas Safety Inst. of Japan", *Examples of accidents, Collection of examples of high-pressure gas accidents,* 1982 (in Japanese)

[12] Kitagawa, T., "Vapor explosion in liquid carbon dioxide tank", *Analysis of explosion hazard.* 1980 (in Japanese)

[13] Clayton, W.E., Griffin, M.L., "Catastrophic failure of a liquid carbon dioxide storage vessel", *Process Safety Progress* 13 (1994) 202-209.

[14] Voros, M., Honti, Gy, "Explosion of a liquid CO2 storage vessel in a carbon dioxide plant", *Loss Prevention and Safety Promotion in the Process Industries*, 1974

[15] van der Voort, M.M, van den Berg, A.C., Roekaerts, D.J.E.M., Xie, M., de Bruijin, P.C.J., "Blast from explosive evaporation of carbon dioxide: experiment, modeling and physics", *Shock Waves* 22 (2012) 129-140

[16] Bjerketvedt, D., Egeberg, K., Ke, W., Gaathaug, A., Vaagsaether, K., Nilsen, S.H., "Boiling liquid expanding vapour explosion in CO2 small scale experiments", *Energy Procedia* 4 (2011) 2285-2292

[17] Pettersen, J, "Experimental study on boiling liquid expansion in a CO2 vessel", 5th IIR Gustav Lorentzen Conference on Natural Working Fluids, Guangzhou, China, September 2002

[18] Ke, W., "CO2 BLEVE (Boiling Liquid Expanding Vapor Explosion)", Master Thesis, Telemark University College, Porsgrunn, Norway, 2009

[19] Chen, F, Schork, J.M., "Improvement of BLEVE Overpressure Method Using Historical Data. Conference Proceedings", 7th Global Congress on Process Safety. Chicago, IL, 2011.

New Ignition Source "Exploding Wire" for the Determination of Explosion Characteristics of Combustible Dusts in the 20-L-Sphere

Marc Scheid
Federal Institute for Materials Research and Testing (BAM),
Department 2 "Chemical Safety Engineering"
D-12200 Berlin, Germany
marc.scheid@bam.de

Alexander Klippel, Rico Tschirschwitz, Volkmar Schröder, Stefan Zirker, Christian Kusche
Federal Institute for Materials Research and Testing (BAM), Department 2 "Chemical Safety Engineering"
D-12200 Berlin, Germany

Prepared for Presentation at
American Institute of Chemical Engineers
2013 Spring Meeting
9th Global Congress on Process Safety
San Antonio, Texas
April 28 – May 1, 2013

UNPUBLISHED

New Ignition Source "Exploding Wire" for the Determination of Explosion Characteristics of Combustible Dusts in the 20-L-Sphere

Marc Scheid
Federal Institute for Materials Research and Testing (BAM),
Department 2 "Chemical Safety Engineering"
D-12200 Berlin, Germany
marc.scheid@bam.de

Alexander Klippel, Rico Tschirschwitz, Volkmar Schröder, Stefan Zirker, Christian Kusche
Federal Institute for Materials Research and Testing (BAM),
Department 2 "Chemical Safety Engineering"
D-12200 Berlin, Germany

Keywords: dust explosions, dust explosion testing, exploding wire, explosion characteristics, fuse wire, ignition source

Abstract

According to international standards several safety characteristics of dusts are determined in the 20-L-sphere (also known as SIWEK-Chamber). Dust cloud ignition is carried out using pyrotechnical igniters. Due to various disadvantages of pyrotechnical igniters the need for alternative ignition sources arises again and again. An alternative ignition source could be the so-called "exploding wire" or "fuse wire". The paper presents test results of a comparative study between both ignition sources for the determination of the safety characteristics Maximum Explosion Pressure and Maximum Rate of Explosion Pressure Rise of selected dusts in the 20-L-sphere. In addition to that the flame/arc propagation of both ignition sources was analysed by high speed and IR camera recordings. Tests in a windowed autoclave were performed to get information on the influence of turbulence generated due to dispersion of the dust on the shape of the generated flame and its propagation. Turbulence measurements in the windowed autoclave as well as in the 20-L-sphere with a LDA system allowed investigation if the igniter led to turbulence increase at the moment of ignition in comparison to tests without ignition.

1. Introduction

Safety characteristics are essential for the determination of explosion hazards during handling of combustible dusts and for the design of safety measures. The safety characteristics Maximum Explosion Pressure p_{max}, Maximum Rate of Explosion Pressure Rise $(dp/dt)_{max}$ and Lower Explosion Limit LEL are determined in closed vessels such as the 20-L-sphere (also known as SIWEK-Chamber). Tests are preformed according to international standards, for example EN 14034 part 1 and part 2 [2,3] or ASTM E1226 [1]. In order to determine the safety characteristics the dust samples are dispersed in air and ignited using two pyrotechnical igniters with energy contents of 1000 J (LEL) or 5000 J (p_{max}, $(dp/dt)_{max}$). Since the igniters could have an influence on the test results they are defined in the standards.

Due to various disadvantages of pyrotechnical igniters such as high costs, legal requirements concerning its storage and use as well as high energy input in comparison to most ignition sources in practice, the need for alternative ignition sources arises again and again; see also Krietsch [6]. Such an ignition source should be less expensive, readily available and the operator should be able to use it without a certificate of competence. In addition to that, the ignition energy should be adjustable over a wide range and if possible allow energy amounts less than 1000 J. Compared to the ignition energies described in the standards such energy amounts are more realistic for ignition sources found in practice, such as electrostatic discharges, mechanical sparks or hot surfaces.

An alternative ignition source which fulfils these requirements is the so-called "exploding wire" or "fuse wire". This type of ignition source is used for the determination of explosion limits of gases and is described in the standard EN 1839 [4].

This paper presents test results of a comparative study in order to find out, if the exploding wire is suitable as an alternative ignition source for the determination of explosion characteristics of dusts. For the study Maximum Explosion Pressure and Maximum Rate of Explosion Pressure Rise values from different dusts were determined with both ignition sources in the range of 100 J to 2000 J. The dusts were selected such that different combustion mechanisms were considered. High speed and IR camera recordings allowed comparison of flame front and electrical arc generated from the igniters. Calorimetric and electric measurements obtained information on the ignition energy of the igniters delivered to the mixture. With additional tests in a windowed autoclave it was possible to get information on the influence of pre-ignition turbulence on the shape of the generated flame/arc and its propagation as well as information on the turbulence generated from the ignition source. Turbulence measurements in the windowed autoclave were realised with a LDA system. The same system was used for tests in the 20-L-sphere to get information on the turbulence at the moment of ignition.

2. Test assembly and test settings

2.1 20-L-sphere

Main component of the test assembly was the 20-L-sphere (Cesana and Siwek [5]; EN 14034 1-2 [2,3]). In order to reach almost homogeneous dust dispersion into the test chamber the so called "rebound nozzle" was used.

Tests with pyrotechnical igniters as an ignition source were performed using the standard control unit of the 20-L-sphere. For tests with exploding wires an external ignition unit was necessary. Depending on the ignition energy different systems with isolating transformer, or capacitors were used. An isolation amplifier allowed the determination of the ignition energy by measuring current and voltage.

2.2 Windowed autoclave

A windowed autoclave with cylindrical shape was used for testing in order to get information on flame/arc propagation in a closed vessel as well as on the shape of flame/arc generated from both ignition sources. In this autoclave it was also possible to determine the influence of turbulence on the generated flames. The volume of the autoclave was approximately 11 L. Its diameter was 220 mm. Optical accessibility via the window allowed high speed recordings at different positions and turbulence measurements. A fan which was placed on the bottom of the autoclave enabled average turbulent velocity fluctuations (RMS-value) up to 1 m/s. In addition to that the system allowed measurements of turbulence generated from the ignition source.

2.3 Ignition sources

One of the fundamental differences of the two ignition sources pyrotechnical igniter and exploding wire is the ignition mechanism. While for the exploding wire an electrical arc and plasma is generated the pyrotechnical igniter emits flames and burning solids.

The ignition sources were placed in the middle of the vessel and connected to the lid of the vessel, such as it is described in the test standards. According to the European standards for the determination of the explosion characteristics Maximum Explosion Pressure and Maximum Rate of Explosion Pressure Rise two pyrotechnical igniters with energy of 5000 J shall be used, firing horizontally in opposite directions (EN 14034 1-2).

Instead of the ignition energy described in the standards tests were performed with ignition energies of 100 J, 500 J, 1000 J and 2000 J. In the first tests only a single ignition source with ignition energies of 100 J, 500 J and 1000 J was used. Further tests with a modified lid allowed tests with two igniters with ignition energies of 500 J and 1000 J each. For all tests with one ignition source the igniter was mounted such that it produced a flame or an electrical arc downwards. Electrodes and connected igniter for tests with a single ignition source are shown in Figure 1. In tests with two igniters the ignition source was installed such that the flame propagated in nearly horizontal direction such as it is described in the standard.

Pyrotechnical igniter

All pyrotechnical igniters used for this work have the same composition as the standard igniters used for determination of explosion characteristics in the 20-L-sphere. They contained 40 % by weight zirconium metal, 30 % by weight barium nitrate and 30 % by weight barium peroxide.

Exploding wire

The principle of the exploding wire is the evaporation and ionization of metal particles from a wire due to a sufficient current flow within milliseconds. During that process a metal vapour is created in which an electrical arc is generated between the electrodes because of the electrical conductivity of the plasma.

For all tests a nickeline wire with a diameter of 0.12 mm was used. In tests with a single ignition source the wire was tensed between the ends of two auxiliary electrodes which were connected two the main electrodes of the 20-L-sphere, see Figure 1. For tests with two exploding wires the lid of the 20-L-sphere had to be modified. It was designed such that it contained four electrodes.

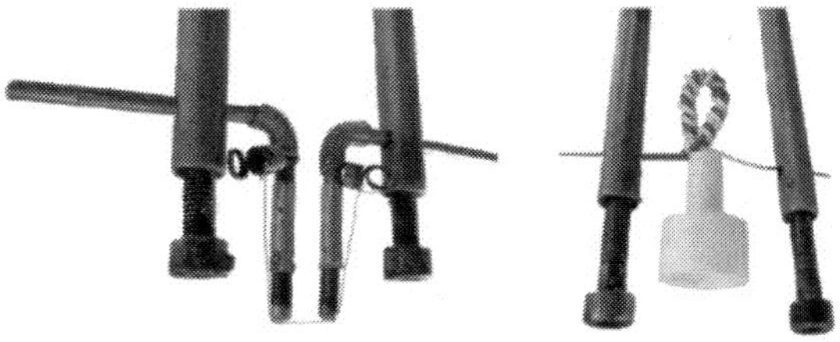

Figure 1. Ignition source for tests with a single igniter: Exploding wire with electrodes (left side) and pyrotechnical igniter (right side)

Ignition unit with isolating transformer

First tests with the exploding wire were performed with ignition energy of 100 J. In these tests the electrical power which is needed for melting the wire and generating the arc was supplied by an isolating transformer similar to that used for the determination of explosion characteristics of gases and vapors according to EN 1839 [4]. This system and its disadvantages such as not reproducible ignition delay are described by Scheid [7].

Ignition unit with capacitors

In tests with ignition energies of 500 J, 1000 J and 2000 J an ignition unit with capacitors was used instead of the isolating transformer. The main advantages of this new type of ignition unit are the higher ignition energies which the unit can produce and that it is able to ensure an exact ignition delay. A choke enabled elongation of the electrical arc during each test.

2.4 Recordings of flame front and electrical arc

In order to record the propagation of flame/arc of both igniters a high speed camera system was used. It allowed recordings of 5000 frames per second. So it was possible to determine maximum flame volume and length as well as the time delay for activating the ignition source and its burning time. In addition to that a fast acting IR camera system was used to estimate the maximum temperature level generated from the igniters. The IR camera allowed recordings with 850 frames per second.

2.5 Measurements of ignition energy (delivered to the mixture)

A combustion calorimeter working with the dry method allowed measurements of the ignition energy of both ignition sources. Dry method means, that the brass bomb of the calorimeter is covered with an insulating styrofoam and not with a water bath. This allows a more sensitive direct measurement of the temperature changes in the metal. Additionally an isolation amplifier was used to measure the burning current in the exploding wire as well as the voltage and enabled the determination of ignition energy generated in the electric arc during each test.

2.6 Turbulence measurements

A one-dimensional laser-Doppler-system (LDA) was used for turbulence measurements in the 20-L-sphere as well as in the windowed autoclave. Due to the big window of the autoclave the turbulence was measured at three different positions around the ignition source. Since the optical accessibility in the 20-L-sphere could only be realized via a windowed flange with a diameter of 40 mm it was only possible to measure the turbulence at one position next to the ignition source.

2.7 Tested dusts

Dust explosions generally show complex reaction mechanism, which depend e.g. on the chemical composition of the dust, the dust concentrations and the flow conditions. Van der Wel [8] differs between different reaction mechanisms depending on whether the reaction takes place in the gas phase due to evaporation or at the solid surface in form of gaseous products, solid or liquid material. The dusts used in this work were chosen such that they represent the different chemical classes and mechanisms as well as the dust explosion classes found in practice. As a result dusts from lignite, maize starch, niacin and anthraquinone were chosen. In addition to that a steel dust which was generated during a shot blasting process was tested.

3. Test results

3.1. Ignition energy measurements

Measurements with the isolation amplifier lead to different test results depending on the boundary conditions of the measurements. Table 1 shows the test results for measurements of ignition energy outside the test vessel without dust and flow, inside the test vessel with flow but without dust and inside the test vessel with flow and with a dust concentration of approximately 1000 g/m³ - 1250 g/m³ maize starch. Ignition unit was the isolating transformer.

Table 1. Ignition energy of igniter determined with the isolation amplifier; ignition source isolating transformer; ignition energy 100 J

Experiment-No.	Determined ignition energies		
	Outside the apparatus [J]	Inside the apparatus without dust [J]	Inside the apparatus with dust [J]
1	93.61	98.42	98.88
2	86.84	94.19	84.14
3	90.92	89.27	88.24
4	89.80	101.59	48.85
5	91.62	82.96	94.13
6	84.46	98.58	83.89
7	90.40	94.74	100.63
8	94.41	91.37	74.36
Ø	**90.26**	**93.89**	**84.14**
SD	**3.30**	**5.97**	**16.67**

Table 1 shows that the ignition energy values for tests without dust (inside and outside the test vessel) were almost similar. As a trend the values with flow (tests inside the vessel) and their standard deviation seem to be slightly higher. Tests with dust led to slightly lower ignition energies and a significant higher standard deviation of the values. Most of the tests led to ignition energies of more than 90 % of the adjusted value (set point adjustment). However, some tests with dust led to lower values. In one test less than 50 % of the adjusted value was measured. As a result the dust particles seem to influence the test results in some tests.

Figure 2 shows the arithmetic mean of the determined ignition energies as a function of dust cloud concentration for maize starch and lignite for adjusted ignition energy of 100 J. From the course of the mean values follows, that there is only a minor influence of the dust concentration on the ignition energy up to concentrations of approximately 1250 g/m³. For higher concentrations the influence is significantly higher.

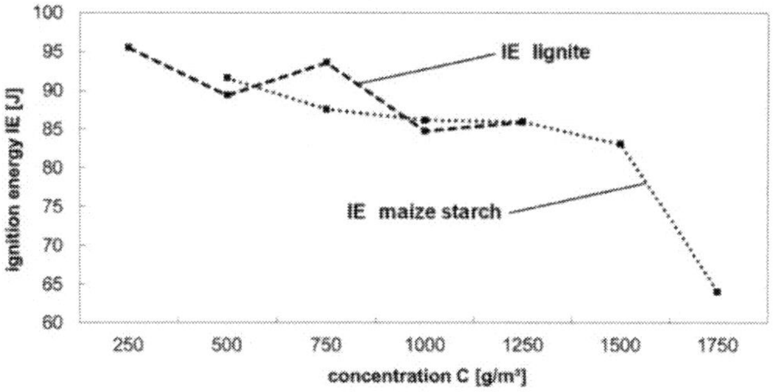

Figure 2. Arithmetic mean of ignition energy of igniter as function of dust cloud concentration for maize starch and lignite; adjusted ignition energy 100 J

Comparable test results were determined for ignition energies of 500 J, 1000 J and 2000 J using an ignition unit with capacitors. But as a trend the influence of the dust cloud on the ignition energy seems to be lower. For all tests the determined ignition energy was at least 80 % of the set point adjustment. Energy measurements with the combustion calorimeter confirmed the test results.

3.2 Propagation of flame/arc of ignition source

Tests in the windowed autoclave with maximum fan speed were performed in order to get information if pre-ignition turbulence has an influence on the flame/arc propagation of a 1000 J pyrotechnical igniter or an exploding wire with the same energy. However, for both ignition sources no visible influence on flame/arc propagation and maximum flame/arc volume was found.
Figure 3 shows the flame propagation for a 1000 J pyrotechnical igniter in comparison to an exploding wire with the same energy for tests in the windowed autoclave (no reaction of dust). The flame recordings show, that the flame of the pyrotechnical igniter propagates 3-4 times faster than the flame of the exploding wire and reaches a much higher maximum volume.

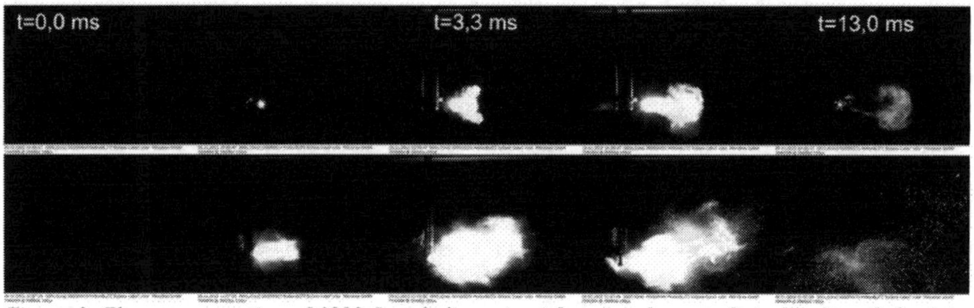

Figure 3. Flame propagation of 1000 J exploding wire (above) and pyrotechnical igniter (below) in a windowed autoclave with a volume of 11 L recorded with high speed camera

Recordings with a fast IR camera allowed an estimation of the maximum temperature level in the generated flame/arc and the maximum volume of flame or arc. Figure 4 shows four sequences of the flame/arc propagation of exploding wire (first and third sequence) and pyrotechnical igniter (second and fourth sequence) recorded with the IR camera. The sequences show a period of approx. 5 ms within the first 15 ms after triggering the igniter. Temperature range in the first two sequences was adjusted such that temperatures between 200 °C and 650 °C were visualized.

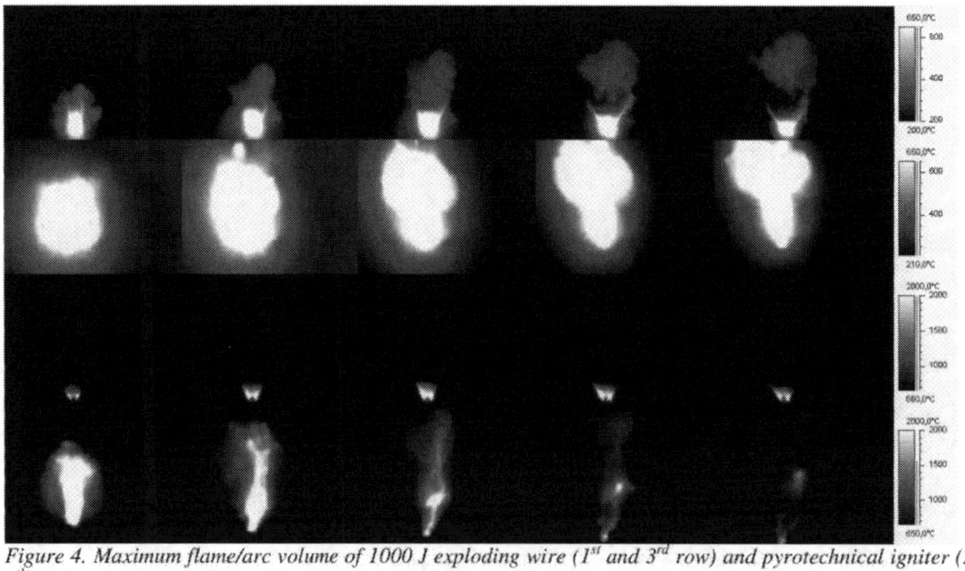

Figure 4. Maximum flame/arc volume of 1000 J exploding wire (1st and 3rd row) and pyrotechnical igniter (2nd and 4th row) recorded with a fast IR camera system

A comparison between sequence one and sequence two shows that the volume with the highest temperatures generated from the pyrotechnical igniter is much larger than the volume with such temperatures generated from the exploding wire. While for the exploding wire the volume with the highest temperatures does not increase with the propagation of the flame, it increases for the pyrotechnical igniter.
Sequence three and four visualize the temperature range between 650 °C and 2000 °C. For the exploding wire the highest temperatures were determined close to the wire. The volume in which such temperatures were determined stays almost constant in the whole sequence. In contrast to that the pyrotechnical igniter seems to generate a large volume with such high temperatures in from of a jet immediately after it was triggered. However, the volume with the highest temperatures seems to decrease significantly within 2 ms.

3.3 Turbulence measurements in 20-L-sphere and windowed autoclave

In tests in the windowed autoclave with operating fan (generated RMS-value of approx. 1 m/s) no differences in the measured RMS-value could be determined between tests without ignition and with a 1000 J exploding wire. Tests with the pyrotechnical igniter resulted in RMS-values of 2.5 m/s - 4.5 m/s depending on the measuring position. In tests in which the fan was turned off the exploding wire led to RMS-values of 0.5 m/s while the pyrotechnical igniter led to values of 2 m/s - 3 m/s. As a result turbulence measurements in the windowed autoclave showed that the turbulence generated from a 1000 J pyrotechnical igniter is stronger than the value generated from an exploding wire with the same ignition energy.
LDA-measurements in the 20-L-sphere showed, that the dispersion of dust with the rebound nozzle generated a flow with a maximum velocity of more than 40 m/s. This flow velocity decreased rapidly with decreasing pressure in the dust storage chamber and with closing the inlet valve. At the moment of ignition the flow velocity reached a value of approximately 10 m/s. The measured RMS-value at the moment of ignition was approximately 15 m/s.
In comparison to the turbulence generated in the windowed autoclave the value determined in the 20-L-sphere is much higher. Hence the test results from the windowed autoclave could not be transferred to the 20-L-sphere. As a result LDA-measurements in the 20-L-sphere were necessary. These measurements showed for tests with 1000 J ignition sources that at the moment of ignition for none of the igniters an increase of the RMS-value could be measured. However, the pyrotechnical igniter showed a not negligible velocity increase in one direction in comparison to the velocity generated due to the dispersion of dust.

3.4 Explosion behavior

Figure 5 presents the explosion pressure of maize starch as a function of the dust cloud concentration determined with both ignition sources with ignition energy of two times 1000 J. Both curves show a comparable maximum of approximately 8.2 bar for the exploding wire and 8.4 bar for the pyrotechnical igniter. For both igniters the maximum was determined for a concentration of 1000 g/m³. Deviations in explosion pressure between the different series were very low for dust explosion testing.
The determined rate of explosion pressure rise values as a function of the dust concentration are found in Figure 6. All test results show a good conformity concerning maximum rate of explosion pressure rise values which were almost similar. However, the reproducibility was not as good as for the explosion pressure. As a result the test values of the different test series differed more for both ignition sources. But this is very often found in dust explosion testing. Comparable test results were found for all tested dusts and ignition energies.

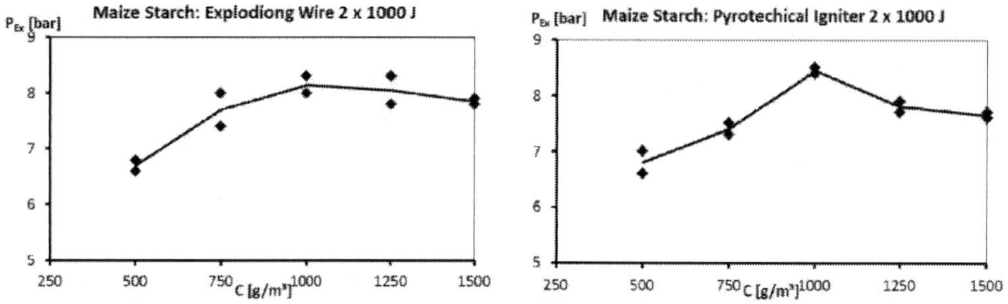

Figure 5. Explosion pressure of maize starch as function of dust cloud concentration determined with pyrotechnical igniters and exploding wires with ignition energies of 2 x 1000 J

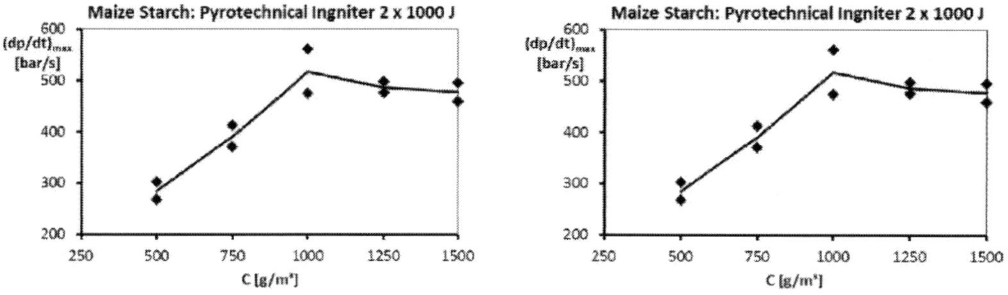

Figure 6. Rate of Explosion pressure rise of maize starch as function of dust cloud concentration determined with pyrotechnical igniters and exploding wires with ignition energies of 2 x 1000 J

Table 2 gives an overview of all determined p_{max} and $(dp/dt)_{max}$ values as well as the K_{St} values which were calculated from the $(dp/dt)_{max}$ values.

A comparison of the results determined with two 500 J ignition sources with one 1000 J ignition source in order to find out if the use of two ignition sources leads to different test results did not come to a clear result. While for lignite tests with two ignition sources led to slightly higher values for the Maximum Explosion Pressure, for maize starch one ignition source led to slightly higher values. The determined rate of explosion pressure rise values were for both dusts slightly higher in tests with two ignition sources. However, for both tested dusts the deviations in explosion pressure and explosion pressure rise between the different test series seem to be slightly lower for tests with two ignition sources.

Table 2 also includes test results which have been determined with two 5000 J igniters, which is the ignition energy described in the test standard for determining p_{max} and K_{St}. The tests showed that the Maximum Explosion Pressures values determined with ignition energies of 100 J, 500 J, 1000 J and 2000 J are less than 10 % lower than those with 10000 J. For steel dust and niacin the values determined with 10000 J were slightly lower than the values determined with the other ignition energies. However, the determined $(dp/dt)_{max}$ values and therefore the K_{St} values differed a lot from the values determined according to the standard procedure, especially for tests with comparatively low ignition energies. Tests with 10000 J led to values which were up to 30 % higher than the values determined with low ignition energies. Only tests with 2000 J ignition sources led to reasonable test results.

Table 2. p_{max}, $(dp/dt)_{max}$ and K_{St} values for all tested dusts determined with exploding wires and pyrotechnical igniters with ignition energies of 1 x100 J, 1 x 500 J, 2 x 500 J, 1x 1000 J 2 x 1000 J and 2 x 5000 J.

Dust		1 x 100 J		1 x 500 J		2 x 500 J		1 x 1000 J		2 x 1000 J		2 x 5 kJ
		Ex. Wire	Pyro. Igniter	Ex. Wire	Pyro. Igniter	Ex. Wire	Pyro. Igniter	Ex. Wire	Pyro. Igniter	Ex. Wire	Pyro. Igniter	Pyro. Igniter
Lignite median 37 µm	p_{max}	8.4	8.1	8.1	7.6	8.1	8.5	7.9	7.7	8.3	7.9	8.4
	$(dp/dt)_{max}$	601	545	681	630	711	619	671	608	723	705	784
	K_{St}	163	148	185	171	193	168	182	165	196	191	213
Maize starch median 14 µm	p_{max}	8.4	8.5	8.4	8.6	8.0	8.2	8.2	8.4	8.2	8.4	8.7
	$(dp/dt)_{max}$	408	470	528	480	500	527	487	468	504	530	614
	K_{St}	111	127	143	130	136	143	132	127	137	144	167
Niacin median 14 µm	p_{max}	-	-	8.3	8.2	-	-	8.2	8.0	-	-	7.9
	$(dp/dt)_{max}$	-	-	953	942	-	-	920	874	-	-	1051
	K_{St}	-	-	259	256	-	-	250	237	-	-	285
Anthraquinone median 3 µm	p_{max}	-	-	-	-	-	-	8.1	8.0	-	-	8.4
	$(dp/dt)_{max}$	-	-	-	-	-	-	1113	999	-	-	1379
	K_{St}	-	-	-	-	-	-	307	271	-	-	374
Steel dust median 24 µm	p_{max}	-	-	-	-	-	-	4.1	4.1	-	-	3.8
	$(dp/dt)_{max}$	-	-	-	-	-	-	398	394	-	-	432
	K_{St}	-	-	-	-	-	-	108	107	-	-	117

4. Conclusion

Measurements of ignition energy delivered to the mixture of the ignition source exploding wire using two different methods showed that the ignition energy generated with the exploding wire was reproducible. For ignition energies of 500 J and higher, the dust cloud did not influence the ignition energy significantly. However, some tests with ignition energy of 100 J and dust concentrations with more than 1000 g/m³ showed that the dust cloud was able to influence the ignition process such that the electric arc extinguished earlier. This resulted in lower ignition energy.

High speed recordings showed that the flame generated from the pyrotechnical igniter propagated faster than the flame/arc generated from the exploding wire and reached a larger volume. Recordings with a fast IR camera system confirmed these test results. In addition to that the recordings showed that as well the volume with a temperature level comparable to the Minimum Ignition Energy of most dusts as the volume in which very high temperatures were measured was much larger for the pyrotechnical igniter. In contrast to that in tests with the pyrotechnical igniter the temperature decreased earlier as in tests with the exploding wire. As a result of high speed and IR recordings as well as energy measurements for the 100 J exploding wire the igniter should have a significant influence on the determined p_{max} and $(dp/dt)_{max}$ values. However, such an effect was not determined during the measurements of p_{max} and $(dp/dt)_{max}$ of five different dusts. All tests resulted in comparable values of p_{max} and $(dp/dt)_{max}$ for both ignition sources.

A comparison of p_{max} values determined with two 500 J ignition sources with one 1000 J ignition source in order to find out, whether the use of two ignition sources will lead to different test results, did not come to a clear result. The determined pressure rise values were for both dusts slightly higher for tests with two ignition sources. However, for both tested dusts deviations in explosion pressure and rate of explosion pressure rise seem to be slightly lower for tests with two ignitions sources.

The p_{max} values determined with ignition energies of 100 J, 500 J, 1000 J and 2000 J were less than 10 % lower than values determined according to the test standard with two igniters of 5000 J each. As a result the influence of the ignition energy of the igniter on the Maximum Explosion Pressure seems to be almost negligible.

In contrast to that $(dp/dt)_{max}$ values determined according to the test standard led to 30 % higher values, especially in tests with ignition energies of 100 J and 500 J. Only tests with 2000 J ignition sources led to reasonable test results. The reason for that could not be solved totally. It is assumed that a combination of different effects has to be considered such as turbulence generated from the ignition source and the flame/arc volume of the ignition source. However, the effect of the flame/arc volume should result in different test results for exploding wire and pyrotechnical igniter. However, this could not be determined. Probably the turbulence generated from the igniters has a strong influence especially on the $(dp/dt)_{max}$ values. Tests in the 20-L-sphere showed for all ignition sources with energies up to 2000 J that the turbulence generated from the igniters seems not to have a significant influence on the turbulence field generated due to dispersion of the dust (pre-ignition turbulence). This is probably different for higher ignition energies.

In a next step tests with ignition energies up to 10000 J with two pyrotechnical igniters and two exploding wires will be performed. In addition to that a validation of the exploding wire with a greater number of dusts is planned.

5. References

[1] ASTM E1226-12, "Standard Test Method for Explosibility of Dust Clouds," ASTM International, US, 2012

[2] EN 14034–1, "Determination of explosion characteristics of dust clouds - Part 1: Determination of the maximum explosion pressure p_{max} of dust clouds," Beuth-Verlag, Berlin, Germany, 2011.

[3] EN 14034–2, "Determination of explosion characteristics of dust clouds - Part 2: Determination of the maximum rate of explosion pressure rise $(dp/dt)_{max}$ of dust clouds," Beuth-Verlag, Berlin, Germany, 2011.

[4] EN 1839, "Determination of explosion limits of gases and vapours," Beuth-Verlag, Berlin, Germany, 2003.

[5] Cesana C., Siwek R., "Manual KSEP 7.0," Kühner AG, <kuhnersafety.com/tl_files/kuhner/product/safety/PDF/B000_071.pdf>, Accessed 2013-01-04

[6] Krietsch A., Scheid M., Lasry M., Suter G., Teske H.-J., Stahmer K. W., Arnold A., Nemcak O., Cesana C., Siwek R., "Validation of alternative igniters for the determination of explosion characteristics of dust clouds in 20-l-sphere and 1m³-vessel," Proceedings 13th Int. Symp. on Loss Prevention and Safety Promotion in the Process Industries 6th – 9th June 2010, Brugge, 2010

[7] Scheid M., Kusche C., Schröder V., Barth U., "Tests on Suitability of the Ignition Source "Exploding Wire" for the Determination of Explosion Characteristics of Combustible Dusts in the 20-L-Sphere," Chemical Engineering Transactions, vol. 31, 2013 (in print)

[8] Van der Wel P. G. J., "Ignition and Propagation of Dust Explosions," Dissertation, Delft University Press, the Netherlands, 1993

Operational Discipline

Improving Process Safety Management Implementation through Behavior and Cultural Change

Charles A. Soczek
Principal Consultant
DuPont Sustainable Solutions
E.I. DuPont De Nemours and Company

Abstract

In 2007, US OSHA initiated a Petroleum Refinery Process Safety Management National Emphasis Program. This was followed in 2009 with a similar program focused on Chemical operations. These programs developed a series of findings, many of which were results of failure to implement OSHA 1910.119 PSM regulatory requirements. Many facilities had well documented programs, but there was poor implementation. In DuPont it was recognized that implementation is an Operational Discipline issue.

Operational Discipline (OD) as defined by DuPont is "the deeply rooted dedication and commitment by every member of an organization to carry out each task, the right way, and each time". OD is recognized as the key to successful implementation of a Process Safety Management program, turning a paper program into reality. Typically, companies have addressed OD improvements through behavioral modifications, but while the gains from programs to modify behavior can be substantial, they are difficult to sustain. Cultural change, while far more difficult to achieve, can produce more sustainable results. When done together, substantial improvements in PSM performance can be both rapid and sustained. This presentation will discuss the key attributes of operational discipline as well as provide an outline of the process to address behavior and culture change is provided.

Introduction

Operational Discipline (OD) as defined by DuPont is "the deeply rooted dedication and commitment by every member of an organization to carry out each task, the right way, and each time". Per the Business Dictionary, OD is defined as assurance that specific activities necessary for successful daily operation of a company's operations are running as they should be. This can include specific goals, missions, and guidelines. Whichever definition is chosen, OD is important to success of all aspects of an operating facility.

Process Safety Management has been in place for over twenty years. Many companies have good PSM systems, but based on the CSM Special Emphasis studies, implementation is poor. It is being increasingly recognized that Operating Discipline is a key factor that drives PSM implementation. Operating Discipline helps turn a paper PSM program into a reality

Improving Operating Discipline

Advances in Social/Psychological theory are showing that operating discipline improvements are influences by two factors, Behavior Change and Culture change. The Dejoy model helps explain this.

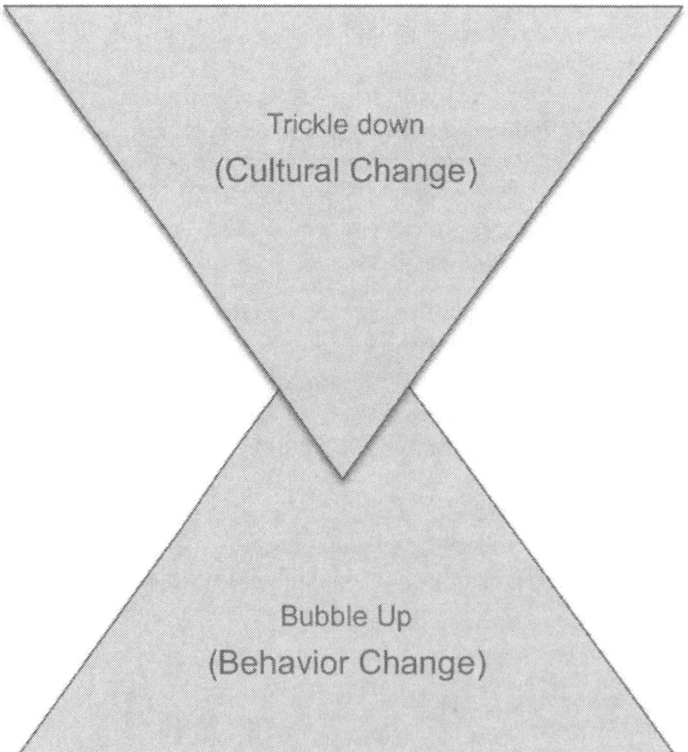

The basis for the Dejoy[1] model is that behavior change can be made quickly, but improvements are not sustainable without a large continuing effort, and that even with this, results will plateau (Bradley Curve). Safety culture change takes much longer, but once in place, superior performance can be both achieved and sustained when coupled with behavior improvement. In the Dejoy model, cultural change is driven by leadership whereas behavior change is driven by supervisors. This is an important distinction and helps focus the role of leadership and the role of supervision.

In a study of some DuPont sites, it has been found that multi-faceted Operating Discipline programs are in place. These programs address both behavior change and cultural change. Programs vary, and can consist of consists of a large number of small efforts or fewer larger programs involving all members of the organization. Either can be successful.

Dejoy[1] did a comparison of a cultural based program and a behavior based program.

Characteristic	Behavior -Based changes approach to safety	Cultural Based changes approach to safety
Background/origin	Operant psychology/ behavior modification	Organizational behavior/ anthropology
Key aspects	Bottom up approach	Top-down
	Analytic/data driven	Intuitive/ethnographic
	Setting Specific	Setting Specific
	continuous process	Self-sustaining
Typical Implementation	Identify and define critical behaviors	Assess aspects of culture (values, beliefs, assumptions, etc.)
	Set performance goals	Devise alternative vision
	Observe/sample behavior	Work with leadership and employees to implement change
	Provide contingent feedback and reinforcement	
Principal Strengths	Specific technology	Emphasizes organizational change
	Objective/empirical	Focuses on basic causes
	Shop floor focus	participatory (often)
	Participatory (usually)	Comprehensive
	Positive	
Principal Weaknesses	Victim-blaming	Diffuse technology
	Minimizes environment	Subjective/intuitive
	Focuses on immediate cause	Indirect

The above chart has highlighted a few of the key differences between the behavioral and cultural approach. In the bottom up, the emphasis is on the operators, maintenance workers, and contractors to follow procedures and rules. This requires good processes and standards and includes items as good communications, housekeeping, documentation, following procedures, and adequate resources. The top-down cultural improvement process is driven by leadership and is reflected in efforts such as leadership by example, building pride in the organization, expression of values, risk sensitivity, and being a learning organization that is innovative, agile, and continuously improves..

Some Definitions

- Operant conditioning psychology (sometimes referred to as instrumental conditioning) is a method of learning that occurs through rewards and punishments for behavior
- Psychological anthropology is an interdisciplinary subfield of anthropology that studies the interaction of cultural and mental processes. The subfield tends to focus on ways in which humans' development and enculturation within a particular cultural group—with its own history, language, practices, and conceptual categories—shape processes of human cognition, emotion, perception, motivation, and mental health. It also examines how the understanding of cognition, emotion, motivation, and similar psychological processes inform or constrain our models of cultural and social processes. Each school within psychological anthropology has its own approach
- Ethnography is a qualitative method aimed to learn and understand cultural phenomena which reflect the knowledge and system of meanings guiding the life of a cultural group

In behavior based approach, performance goals are set, whereas in a cultural based approach leadership works to build a vision of a better place (zero incidents).

Behavior programs are focused on the shop floor, whereas cultural change is focused on everyone in the organization.

Operating Discipline Characteristics

Organizations that have a high level of operating discipline exhibit an identifiable set of characteristics. DuPont has defined these characteristics as follows:

OD Characteristics

» Leadership by example
» Sufficient & capable resources
» Employee involvement
» Active lines of communication
» Strong teamwork
» Shared values
» Up-to-date documentation
» Practice consistent with procedures
» Absence of shortcuts
» Excellent housekeeping
» Pride in the organization

» Risk sensitivity
» Innovation, agility, and continuous improvement
» Justice and fairness

It is recognized that behavior change of people and cultural change are very complex. It is not proposed that the set of characteristics above is comprehensive, and it is expected that over time additional characteristics will be recognized and added to this list.

Implementation of an Operating Discipline Improvement Program

The process developed to implement an operating discipline program starts with an assessment based on 6 Sigma methodology. In the assessment, each characteristic is defined by a set of CTQs (Critical to Quality) and CTXs that define the CTQ.

These steps are

1.Pre-work

- Define scope of the assessment
- Form client OD team
- Develop assessment schedule
  - Capture input from 50% or more of the population
  - Capture current relevant data (see appendix A)
- Assessment team orientation
- Develop current state hypothesis

2. Field data collection (reach 50% of the workforce)

- Focus Groups – operators, mechanics, first line (2 hours)
- One-on-One- managers and leaders (60 minutes)
- Field Tours- employees actively performing work

3. Analysis

- Review and summarize key findings
- Apply a rating to each CTQ per rating guidance

4. Recommendations

- Each low rate CTQ has a recommendation(s) based on best practices. These best practices come from DuPont sites.
- The completed Assessment Model is used as the report to management

5. Action Plan

- Owned by line management based on findings

6. Continuous Improvement

    - Conduct reviews of the assessment and periodically update the action plan with additional "best practices"
    - Conduct re-assessments no more frequently than once per year

Examples of the Assessment Questions

Below is an example of the assessment tool developed that is used to understand each of the fourteen OD characteristics.

OD Characteristic: Leadership by Example

Leaders are demonstrating unrelenting passion for SHE and continuously model the behavior they expect of others

CTQ-1 Visible leadership presence and involvement across the organization

CTX – Build value in leaders that this is a critical part of their role and hold them accountable with specific expectations

- Is visible leadership evaluated as part of the overall personnel review process for line leaders, and if so, how ? Do Leaders at all levels have COTs relating to Operational Discipline?
- Do line leaders consistently take specific corrective action steps for non-conformance to SHE requirements?
- Line management is actively involved with planning, leading safety meetings & communications, and emphasizing broad participation
- Is visible leadership included in new supervisory/manager training and if so, is it effective?
- Is there consistency in approach and focus over time as leaders change?
- Does leadership review action plans to assure alignment with goals and objectives?
- Does leadership allocate adequate resources to support the safety program?
- Does leadership provide a personal example of following standards and procedures? (Wearing required personal protective equipment when in the field; using handrails, etc.)

CTX – Define and communicate what is "Visible Leadership" throughout the organization

- Does the site have formal/informal guidelines for leadership time spent in the field? How does the site measure this and is it effective
- Has the site clearly defined opportunities for visible leadership and personal leadership participation at all levels?
- Is leadership at all levels routinely in work areas conducting Safety Observations audits or other field evaluations?
- Has site leadership clearly articulated a vision and goals for Operational Discipline?
- Does leadership conduct periodic reviews of activities with all safety related teams?
- Does leadership critique audit and reports and follow-up to assure implementation of recommendations?

This characteristic has three CTQs and six CTXs. This is similar to the other characteristics.

The assessment process generates a large volume of information. The assessment team takes this information and develops action items. These items must be prioritized. Below are some examples of action items for leaders taken from an operating facility that could be generated from the process described above.

1. Increase field time and double safety audits
2. Devote 50% of PM meeting to safety
3. Off hours audits and safety coaching
4. Continue small things matter theme
5. Establish designated time in the field for leadership on safety
6. Revamp weekly safety audit meeting
7. Focus team on elimination of slips trips and falls
8. Review all letters of commendations

Summary

Sustainable improvements in Operating Discipline take time, and require facilities to continuously work on improving the safety culture and the behavior of people. This continuing focus on Operating Discipline is a common characteristic of high performing facilities.

References

(1) "Behavior change versus culture change: Diverging approached to managing workplace safety" David M. Dejoy Safety Science 43 2005.

*** Formatting Issue - Best Available Paper/Graphic ***

(43g) Risk Analysis for Pressure Relief Devices – Managing Plant Safety and Reliability

Jose R. Aguilar-Otero Sr., Technology Application, COMIMSA, Saltillo, Mexico

Risk Analysis for Pressure relief devices – Managing Plant Safety and Reliability

Pressure relief devices are one of the most critical assets for plant risk and safety, therefore it is essential to be able to ensure PSV's operation, reliability and availability through the implementation of risk mitigation or prevention actions, either, corrective, preventive or predictive; these actions will allow plant owners to manage the risk associated with the device operation (or failure). This paper is intended to show a risk assessment using Weibull analysis focused on the identification for risk reduction opportunities and the selection for the best chance and dates to implement risk mitigation actions. Risk assessment methodology is based on API RP 581, Risk Based Technology.

Historically, maintenance planning for a PSV was carried out, based on pressure tests and engineering judgment; a pressure (POP) test is usually conducted to evaluate device's availability and functional condition and its frequency is linked to the protected equipment's maintenance program, regardless whatever the failure mode. Nevertheless, now a days, more quantitative and rigorous methodologies are required not just for maintenance planning but for developing a risk management strategy.

The methodology is a risk assessment. We evaluate the probability of failure (POF) times the consequence of failure (COF), for both, a valve failure and a protected equipment failure. Two failure modes are evaluated: fail to open on demand (FOD) and leak (LEAK). POF is calculated analyzing three main issues:

1. Historical inspection records and time in service, using a two parameters Weibull distribution, as follows:

(1)

2. Probability of failure from the protected assets,

(2)

3. Probability of failure from process over pressure scenarios.

(3)

Consequences of failure (FOD, LEAK) are evaluated as a function of consequence area and as a function of its economic impact, for a valve failure and for a protected equipment failure.

FOD consequences are evaluated for four leak scenarios (1/4", 1", 4" and catastrophic rupture) for each protected equipment and using an overpressure as the discharge pressure. This overpressure is a higher pressure than the normal operating conditions. In this approach impact zones are used to evaluate fire, explosion and toxicity effects, in the same way we do it for a traditional quantitative consequence analysis. Meanwhile, for the leak failure mode, consequences are evaluated using a financial approach, no impact area is considered. Consequences categories for the leak mode are as follows:

- Loss of production costs
- Maintenance cost (repair costs)

***** Formatting Issue - Best Available Paper/Graphic *****

- Product costs

The leak mode includes both, a small leak and a moderate leak. Moderate leak means valve stuck open, while small leak means leakage past valve (No seal).

Therefore, the global risk is calculated using the two risks approaches,

(4)

Risk calculation involves a straightforward review of the process operating conditions using a layer of protection analysis (LOPA), it also requires analyzing the process conditions and initiating events than can challenge or demand the PSV and the layers affecting the initiating event frequency. The methodology also involves a process for evaluate as many overpressure scenarios as possible; this scenarios will also challenge the pressure relief device.

This risk analysis methodology has the following benefits:

a) Perform a What if...? Scenarios analysis, and evaluate risk mitigation actions

b) Update analysis modeling, using new data

c) Understanding risk drivers

d) Understanding event sequence for risk contributors

e) Set up a risk management strategy

The risk assessment was performed for 100 PSV, located on several offshore facilities. Results show risk contribution for every failure mode, risk ranking for the whole set of PSVs, new maintenance dates as well as reliability and availability estimations. The use of this approach is very useful, because we are able to deal with large data samples and also fits well with small data samples. Results can be expressed using failure mode information or time to failure information.

(43h) Enterprise Risk Management--Adopting Risk-Based Decision Making in a Chemical Enterprise From the Board Room On Down Through All Functions and All Levels

Jerry L. Jones, PE, Consultant, Menlo Park, CA

Enterprise risk management (ERM) practices have been in place for decades in many businesses. However, following a number of failures of large financial, energy and telecommunications ventures, the interest in broader adoption of effective ERM systems has increased dramatically. During the last decade in response to recommendations and guidance by leading professional societies concerned with financial accounting and corporate governance practices, many more publicly owned non-manufacturing and manufacturing businesses have begun implementing very structured and rigorous ERM systems.

While risk has historically been defined in relation to hazards (including natural perils) as a function of the probability and consequences (harm and financial loss) of an undesired exposure event, under various recent ERM guidelines risk has been more broadly defined in relation to business objectives:

o **Risk isthe effect of uncertainty on meeting objectives** (ISO 31000-2009: Risk Management Principles and Guidelines adopted in the US as ANSI/ASSE Z690.2-2011)

o **Risk isthe possibility that an event will occur and adversely affect the achievement of objectives** (COSO Framework for Integrated Risk Management, (Committee of Sponsoring Organizations for the Treadway Commission, 2004)

A key argument made for adopting ERM principles by both public and private companies and using risk-based decision making processes throughout the enterprise is the opportunity the approach offers to more effectively and efficiently protect current value (preventing losses, the downside of risk) and to create value (the upside of some risk exposures) for owners and other stakeholders. A further incentive is the potential to establish long-term strategic advantage by being able to more effectively and efficiently manage risks than competitors.

Both of the ERM system guidelines cited above recommend rigorous and comprehensive oversight of risk management practices by the Board of Directors and active engagement of the executive leadership team in managing the key risks. Such practices are exactly what have been recommended in relation to chemical process safety by a number of investigation panels or boards in the US and abroad following major chemical process industry accidents during the last decade.

In this presentation the wide range of risk exposures that must typically be managed by a chemical enterprise (showing that process safety is one of many requiring leadership attention) will be briefly reviewed with some specific examples provided for relevant consensus standards or guidelines for managing those risks. The following list of objectives that an ERM should achieve will be discussed along with some the challenges they present.

The ERM system should assure that:

o The company's significant existing risk exposures are identified, assessed and managed in a consistent manner by the various functional organizations and the results of such risk management activities regularly reviewed by the executive leadership team and the Board of Directors,

o The "right" risks are being taken based on consideration of risk/reward tradeoffs and the "risk appetite" of the organization, and resources for risk management are allocated commensurate with the risk exposures and risk control needs,

o Individual at all levels are actively engaged in risk-based decision making and risk management so as to leverage off the knowledge and experience of the entire organization,

o Individual responsibilities and accountabilities for ongoing risk control and monitoring are clearly defined in all functional organizations at all levels, including risk management related reporting requirements with metrics (both leading and lagging indicators) that go to the executive leadership team and to the Board of Directors,

o Internal auditing and self-assessment processes are in place to regularly evaluate the effectiveness of risk management processes and controls so as to keep risk exposures below the agreed upon risk tolerance limits.

A four-step business process for risk identification, analysis, risk-based decision making, and risk controls monitoring/reporting will be described that is consistent with ERM consensus guidelines and others including those for risk-based process safety. This generic business process can be readily adopted by various functional organizations within companies of all sizes to manage a wide range of types of risk exposures and used to support other business processes requiring risk-based decision making (e.g. process hazard analysis, product safety analysis, product quality assurance, management of change, incident investigations, supply chain security, mechanical reliability analysis, and business continuity analysis).

(43i) The Chemical Safety Board and the Prevention of Combustible Dust Hazards

Manuel R. Gomez, Chemical Safety Board, Washington , DC and Christina Morgan, Office of Recommendations, US Chemical Safety Board, Washington, DC

THE CHEMICAL SAFETY BOARD AND THE PREVENTION OF COMBUSTIBLE DUST HAZARDS

This presentation will:

- Briefly summarize the CSB's rationale for recommending an OSHA general industry combustible dust standard;
- Review OSHA actions up to date;
- Comment on the difficulties OSHA may be facing in the conduct of combustible dust rulemaking; and,
- Describe the CSB's ongoing and planned efforts to advocate successful rulemaking.

Performing a HAZOP on Start-up/Shutdown Procedures

Fareed Ebrahim
ACM Automation Inc.
#700, 940 – 6th Avenue SW
Calgary, Alberta
febrahim@acm.ca

German Luna-Mejias, P. Eng
ACM Automation Inc.
gluna@acm.ca

Prepared for Presentation at
American Institute of Chemical Engineers
2013 Spring Meeting
9th Global Congress on Process Safety
San Antonio, Texas
April 28 – May 1, 2013

UNPUBLISHED

Performing a HAZOP on Start-up/Shutdown Procedures

Fareed Ebrahim
ACM Automation Inc.
#700, 940 – 6th Avenue SW
Calgary, Alberta
febrahim@acm.ca

German Luna-Mejias, P. Eng
ACM Automation Inc.

Keywords: Procedure HAZOP, Start-up/Shutdown

Abstract

Statistics show that approximately 65% of process safety related incidents occur during the start-up/shutdown operations of the plant. Yet, the bulk of the risk assessments performed today focus on the normal operation of the plant. A major contributor to the lack of start-up/shutdown risk assessments is the lack of a strongly documented methodology. One method of performing this assessment is through the HAZOP of start-up/shutdown procedures. This HAZOP involves a step by step review of the procedure with consideration given to the sequencing, delay, or missing of steps involved in the procedure. By performing this review, required permissives are identified, contingencies are developed and essential safeguards are identified. This will help ensure the operations team is confident that they are performing the start-up/shutdown in a safe manner. This paper focuses on the framework for guidelines, procedures and templates that are necessary to perform this HAZOP. A practical approach with examples is presented.

1. Introduction

Currently, it is common practice to reserve a deviation in a continuous process HAZOP for start-up and shutdown issues. Unfortunately, during analysis, this deviation if often not explored to its full potential. This abrupt analysis is often due to the lack of available information and expertise, lack specificity within the deviation, and time constraints. Yet, statistics provide a compelling argument to improve the rigour associated with start-up/shutdown analysis. Approximately 65% of process safety related incidents occur during these modes of operations1. To improve the rigour associated with the assessment of start-up/shutdown, the HAZOP technique must be further extended to the procedures for these modes of operation. Many sources are available which explore the fundamentals of performing such a HAZOP1,2, and therefore, this paper will focus on developing a methodology by expanding upon the commonly used HAZOP technique for normal operating modes.

2. Preparing for the HAZOP

In order to ensure that a complete and accurate assessment is performed, compiling the correct documents and team members for the HAZOP is of paramount importance. Just as there are key documents required for a process HAZOP, specific documents are also required for the start-up/shutdown procedure HAZOP.

2.1 Required Documentation

Firstly, the procedure for start-up/shutdown must be available. Just as the P&IDs are the central document during a process HAZOP, the procedure will be the main focus in the start-up/shutdown analysis and therefore, should be the most up-to-date revisions available. This procedure can then be split into "nodes" or subsystems for analysis. Nodes or subsystems may be defined as each individual step in the procedure.

Depending on the complexity of each step, it may be required to split the step into simplified actions, which in turn become the node for analysis. Conversely, procedure steps may be combined to develop a node based on timing requirements and the requirement for actions needing to be performed simultaneously. Timing concerns associated with procedures will be further discussed with the HAZOP deviations. As the implications of each procedure step may not be immediately clear to the HAZOP facilitator, it is of utmost importance that nodes are developed in conjunction with the operations personnel.

Once the nodes have been identified, and the complexity of the procedure understood, a decision must be made on the degree of analysis for the HAZOP to be performed. Two methods of analysis can be employed for the HAZOP – a "What-If?" approach, or a guideword method. The "What-If?" approach follows the same method as that used for a continuous process. If only an extremely high level analysis is required, the "What-If?" methodology may be applied to the entire procedure instead of applying it to nodes. Conversely, the application of the guideword method provides a more structured approach. Generally, as the complexity of the procedure increases, more guidewords should be employed in order to ensure a thorough analysis is completed. PHA software3 contains pre-packaged guidewords to be used for analysis. Explanation and application of the guidewords will be explored in Section 3.

Supporting documents for the procedure HAZOP include the following:

- As-Built P&IDs
- Isolation markup drawings - showing spec blind placement and position as per start-up/shutdown modes
- Shutdown keys - including permissives
- Alarm lists
- Alarm/Trip bypass procedures - including contingency plans

It is important to note that the shutdown keys and alarm lists may not outline which systems may be bypassed during a start-up and shutdown. Therefore, it is critical to obtain the bypass information, as well as the operating window for all instrumentation that will be referenced during the start-up/shutdown operation.

2.2 Required Personnel

In addition to compiling the correct documents for the session, it is of equal importance to ensure the correct personnel are in the room during the HAZOP. Similar to a process HAZOP, representation from the process team, instrumentation, and operation is required. It is suggested to have extra representation from the operations group to provide a more diverse perspective for the brainstorming process. Furthermore, if the HAZOP is being performed at the commissioning phase, it is recommended for the construction manager or equivalent representative to be present.

3. Performing the HAZOP

3.1 Introducing Guidewords

Risk assessment software[3] often includes preloaded libraries of guidewords for the HAZOP of procedures. Table 1 below lists seven guidewords that can be used for the HAZOP. These guidewords are listed in a suggested order for use in the HAZOP. Guidewords may be changed, added or deleted based on the HAZOP team experience, process being analyzed, and corporate guidelines. Note that the performance of a start-up/shutdown procedure HAZOP does not negate the requirement to perform a pre-start-up safety review.

Table 1. Suggested Guidewords and Description for Procedure HAZOP

Guideword[3]	Description
Less	Not enough is done to adequately execute procedure action.
More	Action is performed to a further extent than specified.
No	Action is fully omitted.
Reverse	Actions in procedure are performed out of specified sequence.
Other than[3]	Action is performed that does not align with what is specified.
As well as[3]	The operator does another action, in addition to the specified step being done correctly.[2]
Sooner/Later[3]	Timing of action is incorrect.

All seven guidewords may not be used for each start-up/shutdown procedure being analyzed. Guidewords should be applied based on the complexity of the procedure. An example of a simplified method is the two-guideword approach[2] is contained in Table 2 below.

Table 2. Two-Guideword approach for Procedure HAZOP[2]

Guideword	Description
Omit	Step is not done or part of the step is not done.
Incorrect	Step is not performed as intended.

3.1 Application of Guidewords

In order to further illustrate the application of each guideword, a subsection of a mock, simplified procedure (listed below) for a Reformer will be used. Guidewords from the two word approach will not be further explored as they can be considered a subset of the initial seven. Note that nodes will be assumed to be the individual procedure steps for the purpose of this illustration. Consequences are not to be considered complete for use in industry HAZOPs but for the purpose of example only.

Simplified Excerpt of Reformer Procedure:

1. Once pilots are lit, start lighting main burners (two per cell).
2. After completing the first ten pilots, light another ten but evenly distributed per cell.
3. Wait for 30 minutes or until the heater outlet temperature is stable
4. Continue lighting burners to increase the heater temperature at a rate of 50°C/hour until a temperature of 380°C is reached.
5. Once temperature is stable and maintains at 380°C, start injecting HP Steam and continue lighting burners as required to prevent the heater temperature from dropping below 350°C.

3.2.1 Less

Example Deviation from Step 3: Adequate time is not allotted by operator for heater outlet temperature to stabilize in an attempt to speed up start-up.

Consequence: Potential for a sudden temperature increase in heater leading to refractory damage or flame impingement.

3.2.2 More

Example Deviation from Step 1: More than two burners are lit per cell in order to speed up start-up of unit.

Consequence: Potential for a sudden temperature increase in heater leading to refractory damage or flame impingement.

3.2.3 No

Example Deviation from Step 4: Additional burners are not lit by operator.

Consequence: If steam is injected into the heater when temperature is below 350°C there will be deactivation of catalyst within heater tubes.

3.2.4 Reverse

Example: Deviation from Step 3: Additional burners are lit prior to stabilization of heater outlet temperature.

Consequence: Potential for a sudden temperature increase in heater leading to refractory damage or flame impingement.

3.2.5 As well as

Example: Deviation from Step 2: Steam is injected by operator as well as lighting the additional burners.

Consequence: If steam is injected into the heater when temperature is below 350°C there will be deactivation of catalyst within heater tubes.

3.2.6 Other than

Example: Deviation from Step 5: Natural Gas is injected into the heater instead of Steam.

Consequence: Overheating of tubes in reactor leading to tube damage and potential explosion.

3.2.7 Sooner/Later

Example: Deviation from Step 5: Steam injected to heater prior to temperature reaching 350°C.

Consequence: If steam is injected into the heater when temperature is below 350°C there will be deactivation of catalyst within heater tubes.

3.3 Applying a Risk Ranking

As HAZOPs have developed over time, there has been a continued emphasis to add more semi-quantitative aspects to the study. A major element of this semi-quantitative aspect is the application of a risk matrix. Unfortunately, risk matrices are designed using a per year time frame. This per year concept is of course not applicable to a procedure. Therefore, when performing the procedure HAZOP, two options are available. One, is to avoid the risk ranking and perform a fully qualitative analysis. Second, is to apply the risk matrix by avoiding the yearly references contained in the likelihood, and setting a default likelihood of making an error in the procedure to correspond with highest likelihood available (ie. frequent) on the matrix. This is to quantify the fact that there is a possibility to make a mistake in executing the procedure at any given opportunity for execution. For example, if a procedure step is incorrectly worded, there is an opportunity for error every time that step is executed. The consequence levels outlined in the risk matrix are not time constrained and therefore are easily applied to a start-up/shutdown procedure HAZOP. Safeguards can then be applied in a fashion similar to process HAZOPs by dropping the likelihood by one level.

3.4 Adding Safeguards

When applying safeguards to a procedure, it is of course of utmost importance to adhere to the guidelines set out in IEC-61511 if they will be used as independent protection layers. In addition to this, it is extremely important to ensure that the safeguard will not be bypassed during the start-up or shutdown. In some cases, safeguards through a safety instrumented system may be disabled during start-up or shutdown. It is extremely important to identify these bypasses and ensure safeguards are not credited that will not be active. If a safeguard is bypassed, refer to contingency planning documents in order to identify an alternative. Furthermore, it is of utmost importance to ensure that the safeguards will in fact be within their operating range during start-up and shutdown. One control related safeguard that is far more predominant in start-up/shutdown procedure HAZOPs when compared to process HAZOPs are permissives. It is therefore important to have these permissives identified, optimally on the shutdown key. In addition to instrumentation safeguards, a double check to the operator performing the specified action also acts as a valid safeguard.

4. Conclusion

The application of the HAZOP technique to a start-up/shutdown procedure is an effective way to improve the quality of analysis in these modes of operation. Just as in a normal operating mode HAZOP, specific documentation and team members are required. Furthermore, the level of rigour of the study may be adjusted by using a "What-If?" approach or by varying the number of guidewords to be used. A risk ranking may be applied during a procedure HAZOP, but caution must be applied in order to avoid diluting the risk of a situation by ensuring the procedure is analyzed on a per opportunity (instead of a per year) basis. Safeguards must also be applied with caution due to the potential for bypasses or conditions being out of the operating window of certain instruments during start-up/shutdown. It is hoped that the application of the HAZOP methodology to start-up and shutdown procedures will greatly reduce the number of incidents that occur during these modes of operation.

5. References

[1] Bridges, Bill. "Paper: How to Efficiently Perform the Hazard Evaluation Required for Non-Routine Modes of Operation (Start-up, Shutdown, Online Maintenance)." Proc. of 2011 Global Congress on Process Safety., 2011

[2] CCPS. "9. Extensions and Special Applications." *Guidelines for Hazard Evaluation Procedures.* New York: Center for Chemical Process Safety of the American Institute of Chemical Engineers, 1992. 257-67.

[3] Dyadem International Ltd. *PHA-Pro.* Computer software. Vers. 8.2.2.0.

(43p) Chasing the Protection of Intermediate Bulk Containers, Year 10

John A. Leblanc, Engineering Standards, FM Global, Norwood, MA

The challenges of protecting composite intermediate bulk containers (IBC) storing liquids that burn have been under investigation by a number of groups for more than 10 years. The path to resolving this challenge chosen by FM Global looked at defining the fire hazard created by the liquids and then understanding how the containers reacted to fire. This effort has resulted in a strong understanding of the fire hazards and protection options for several liquid-IBC combinations:

Rack storage of Oil and alcohol in IBCs

Silicone fluids in palletized IBCs

Very high flash point liquids in palletized IBCs

Propylene glycol in palletized IBCs

Storage of Empty IBCs

Defining an FM Approved IBC

This paper will provide an overview of the fire hazards associated with storing liquids that can burn in IBCs and present the new protection options and their associated test programs.

Process Safety Challenges in the Hazardous Waste Industry

Hope A. Luebeck
Veolia Environmental Services, L.L.C.
700 East Butterfield Road, Suite 201
Lombard, IL 60148
Hope.Luebeck@Veoliaes.com

Prepared for Presentation at
American Institute of Chemical Engineers
2013 Spring Meeting
9th Global Congress on Process Safety
San Antonio, Texas
April 28 – May 1, 2013

UNPUBLISHED

Process Safety Challenges in the Hazardous Waste Industry

Hope A. Luebeck
Veolia Environmental Services
700 East Butterfield Road, Suite 201
Lombard, IL 60148
Hope.Luebeck@Veoliaes.com

Keywords: Process Safety, Hazardous Waste Industry, Recyclers' Safety Council

Abstract

The OSHA Process Safety Management (PSM) Standard, 29 CFR 1910.119, was promulgated in 1992 to prevent the release of highly hazardous chemicals (HHCs) following a series of tragic industrial accidents. During the last 20 years, the emphasis of this standard has primarily been focused on refineries, chemical plants, and similar processes. However, companies in the Hazardous Waste Industry (HWI) also handle HHCs and face a special set of challenges as they strive to be PSM-compliant in their operations.

Hazardous Waste Industry facilities are often smaller than typical petrochemical, chemical and pharmaceutical plants. They have fewer resources dedicated to their operations and, as a result, generally make slower progress in advancing their PSM programs. Reasons for these tendencies will be discussed, as will initiatives which are being taken to strengthen their programs.

1. Introduction

History has taught us that the most influential changes result from tragedies. The Process Safety Management (PSM) Standard 29 CFR 1910.119 [1] was established by the Occupational Safety and Health Administration (OSHA) in 1992 after a series of deadly industrial accidents including the methyl isocyanate release at a Union Carbide pesticide plant in Bhopal, India, in 1984 [2] and the vapor cloud explosion at the Phillips 66 chemical complex in Pasadena, TX in 1989 [3]. The 14-element safety standard includes specific requirements for managing risks associated with processes which utilize Highly Hazardous Chemicals (HHCs) such as were involved in the noted catastrophes. More recent events such as the BP Texas City, TX disaster in 2005 [4] have motivated OSHA to renew its scrutiny of process safety practices of PSM-covered organizations through a series of National Emphasis Programs [5].

There have been serious accidents in the Hazardous Waste Industry (HWI) as well, though these have not had the significant impact or notoriety as those in the chemical or petrochemical industries. Such recent HWI events include The Environmental Quality Company fire in 2006 [6] and the Veolia Environmental Services explosion and fire in 2009 [7] (See Table 1). As these sorts of incidents continue in the HWI as well as in other industries, companies must continue to further develop their PSM programs and prepare for potential future NEP inspections.

Table 1. Some Noteworthy Industrial Process Safety-Related Incidents

Industry	Company	Location	Year	Deaths	Injuries
Petrochemical	Union Carbide	Bhopal, India	1984	2000+	Unknown
	Phillips 66	Pasadena, Texas	1989	24	132
	BP	Texas City, TX	2005	15	180
Hazardous Waste	The Environmental Quality Company	Apex, NC	2006	0	30
	Veolia Environmental Services	West Carrollton, OH	2009	0	4

Many of the same issues that challenge the petrochemical industry in their quest to be PSM-compliant are shared by the HWI. However, the size and nature of their operations are often quite different; thus, their approach to PSM compliance can vary significantly.

This paper explains some unique aspects of PSM which apply to companies in the Hazardous Waste Industry (HWI). When the acronym HWI is mentioned, this is intended to apply broadly to all companies having business operations involving hazardous waste. A subset of HWI companies is Hazardous Waste Treatment (HMT) facilities. The HMT acronym is used to emphasize operations related to the treatment (e.g. purification or neutralization), recycling, or destruction of hazardous wastes.

2. Operational Differences by Industry

Companies in the refinery, chemical or pharmaceutical industries typically utilize the same types and quantities of raw materials on a daily basis. They tend to manufacture or store a limited number of products or families of products. Their manufacturing operations occur according to cycles or routine production runs. A given plant has one or more primary technologies which are key to their particular business. These facilities often have recognized that PSM was applicable to their operations long ago and thus have reasonably mature PSM programs.

Companies in the HWI have more variable operations. They utilize an assortment of unit operations to handle, process, recycle or destroy the chemical and biological wastes they receive on a regular or periodic basis. They combine, decant, recycle, fuel blend, distill, incinerate or otherwise dispose of the materials they receive from their customers. These are shipped to their facilities in a variety of container types and material phases. PSM challenges shift on a daily basis since the nature of what they need to process changes frequently and can fluctuate widely. PSM programs in the HWI must be adept at effectively managing these inconsistent hazards.

3. General PSM Challenges for the HWI

Companies in the HWI have to be prepared for a wider variety of process hazards due to the nature and kind of substances they accept from their customers. They refer to waste shipments of a given type they receive as a 'stream'. Even trace amounts of certain chemicals can have a tragic effect if these contaminate incoming streams without the HWI recipients' knowledge. HMT facilities need close ties with their customers to truly understand the nature of the wastes they receive and know how repeat streams may vary in chemical concentrations of major components and may contain different kinds of constituents.

The number and volume of these streams are always changing, even from the same customer. Customers of HWI companies don't intentionally generate waste regularly and the amounts of each are often related to either their production volumes or the number of process upsets they have experienced. So companies in the HWI can't always anticipate how much of what kinds of streams they will process in a business cycle. The generation of customer waste is extremely unpredictable. HWI companies have to be prepared for almost anything and have equipment which is designed for just that. Since waste can be generated at any time, HMT facilities need to be nimble and be able to respond quickly to new business or business for a newly-generated stream of waste from an existing customer. Compliance concerns which may result from new business need to be promptly addressed.

HWI companies are usually comprised of small facilities (with less than 100 employees per site) which have a multitude of regulations which they need to meet, including RCRA, RMP, and numerous others in addition to PSM. Personnel in these more-compact organizations tend to wear many hats and are distracted by the assorted duties cast upon them during the normal workday. They many even include the owner/operator of such a facility. Expediency in addressing PSM compliance issues may be affected by the amount of resources available in such organizations.

HMT facilities tend to have smaller budgets than comparable size facilities in larger companies or those in other industries. As such, monies to upgrade facilities, build infrastructure, or address hazard reduction projects can be more limited. Business cycles have a huge impact on what funding is available to address all compliance needs. And with the HWI downstream of the manufacturing industry, the HWI often goes as manufacturing does.

Companies and operating sites in the HWI have had multiple owners over the years, perhaps more than those in the chemical or petrochemical industries. This has led to mixed safety cultures both at individual sites and across organizations. Dealing with PSM issues may be controversial, since senior personnel or management might have experienced one or more safety cultures and younger personnel others. Companies must apply extra effort to address identified deficiencies since these cultural viewpoints may lead to disagreement regarding the extent of work necessary to effectively resolve compliance issues.

Organizations in the HWI industry previously treated their activities as simply waste-related operations, even if they were recycling or creating renewed end-products through processes such as distilling or fuel blending. But they really are another kind of chemical plant, albeit a more unique sort. These facilities need to be just as focused on maintaining their waste processing equipment as though they are generating a virgin product. Operations can have deadly consequences from improperly processing or having uncontrolled releases of highly hazardous chemicals, which is the reason for the PSM regulation in the first place. HWI companies have improved their performance in this arena in recent years, especially as OSHA continues its Chemical NEP activities [5].

4. Specific Challenges by PSM Element

Employee Participation

Effective Employee Participation can be limited by the number of personnel available at HMT facilities to participate in, lead and drive improvements and PSM-related activities. Companies must allot time and resources for their employees to participate in training, hazard analyses, safety committees, operating procedure reviews, incident investigations, emergency drills, etc. required for their PSM program to be successful. Site management may need to be resourceful in giving personnel time to meet PSM program obligations.

Process Safety Information (PSI)

Efforts must be ongoing to keep PSI current. Personnel must be encouraged to review and update P&IDs, equipment files, chemical and process hazard information, consequences of deviation in procedures and similar documents so they remain accurate and useful. One of the hardest challenges seems to be keeping good records of PSM-covered equipment in easy to access files. Older facilities have limited equipment design or code information available, which can prompt re-inspection or re-certification of equipment to verify these critical details. Sometimes design literature may need to be requested from a manufacturer if it becomes damaged or misplaced. If changes of service (say of storage tanks) have occurred but were poorly documented, records may need to be re-created and verified before inclusion in PSI databases.

Some HMT facilities have difficulty describing their 'process chemistry' per the requirements of the PSM standard. This is a dilemma unique to the HWI. True process chemistry is often not intended in typical HWI operations. If this applies to a given location, the site's process safety documentation must clearly explain this circumstance.

Process Hazard Analysis

A comprehensive Process Hazard Analysis (PHA) program is critical to understanding any site's hazards and ensuring that appropriate controls and acceptable levels of risk are assumed. Many PSM-covered sites in the HWI are older facilities which suffer from process designs which did not adequately cover facility siting concerns at the time of construction. If rebuilding such sites is not possible, locations instead have to focus on administrative and engineering controls to reduce or mitigate risk to their personnel and property from their operations. And facilities with multiple sites operated independently can face inconsistencies in risk tolerances when coming under a corporate PSM umbrella.

Some HWI organizations heavily rely on consultants to lead their PHAs. This can create a double-edge sword: a site gets support for a needed activity, but if they don't define the requirements for the PHA adequately nor hire a suitable consultant, the PHA results yielded may be substandard. The PHAs may not cover enough hazardous aspects of a process if their consultant is not properly informed and knowledgeable. PSI details and reports for these PHAs may be limited in their scope and may omit critical information (background or otherwise) relevant to the study. Sites must be vigilant to ensure that their PHA activities be thoroughly completed as required.

Operating Procedures
HMT facilities typically have very good basic operating procedures due to the nature of their operations. There would either be more accidents or more hazardous spills if these facilities were not being run safely. But, as is sometimes the case in any industry, the written procedures which exist may not be adequately comprehensive in their descriptions of consequences of deviation or emergency actions to take, including explaining what conditions warrant the shutdown of a process. Some procedures may direct the operator to consult with supervision during atypical or emergency operations. While this is not necessarily incorrect, what does an operator do when the process goes haywire on a weekend or off-shift when less supervision or technical support is available? Authors of such procedures need to clearly spell out common out-of-control responses and/or define when emergency shutdown should take place to provide the operators with the best ready guidance at all times.

On the maintenance side of operations, many more procedures often need to be composed to describe inspection, testing and preventive maintenance (ITPM) activities and how repairs and replacement of equipment ought to take place. A significant number of procedures should exist to define ITPM for PSM-critical equipment systems and controls. When predictive and reliability-centered maintenance (RCM) programs become utilized by more advanced organizations, the number of maintenance procedures should increase proportionally to define and address these specialized activities.

Training
Training can be executed in a variety of ways. Some companies depend upon computer-based training for much routine or refresher training, since employees can determine when they have time available to meet the required obligations and can fit in training modules while they are present during normal work hours. When facilities do not have this option available, they may need to be resourceful with training solutions. A familiar pattern is to schedule hourly workers for day long training sessions when they have a day off. Even this becomes an issue for companies who do not have excess staff to backfill work slots for other employees undergoing training. Companies can also organize special 'training day or days' when operations are intentionally curtailed so that particular training or orientation programs may be held for the entire facility at one time. Possibly other creative solutions may be sought to balance staffing with training needs and requirements as situations dictate.

Contractors
Managing contractors can be tricky, depending on the number and type of crafts or specialty groups on which a facility may rely. Overseeing the screening, performance and insurance

requirements, among other factors, can require constant review. Staff dedicated to birddogging records and updates of same can help to keep these appropriately current. Ensuring that contractors have proper desired skills, certifications, and current site orientation may be another story. Some sites find it cumbersome to track down qualifications of contractors and have regular orientation sessions with them. That is why many smaller sites often partner with local/regional safety and health groups which specialize in OSHA and other common safety program training. In the greater Chicago area, one such group is the Three Rivers Manufacturing Association, which has a specialty Safety Center [8]. Similar such organizations exist in other parts of the country. For an annual membership fee, the safety groups can ensure their customers that a variety of contractors serving a region have certain basic safety and health skills, which then make it easier for member companies to quickly orient contractor personnel to their PSM-covered sites when jobs are required.

Sites in the HWI may also have their own employees serve as contractors for customer facilities when they work on special projects or are embedded at customer sites. These HWI personnel may need additional PSM-related training to be appropriately compliant both for their employer as well as for the customer. This is another unique aspect of the HWI business.

Pre-Start Up Safety Reviews (PSSR)
It is less common in the HWI for companies to debottleneck or re-design their processes. They don't generate as many cases where PSSRs are required. As a result, the PSSR process tends to be less well-known and less understood by site personnel. More time needs to be taken whenever PSSR opportunities exist to refresh personnel's memory of the purpose of the review and to ensure that a thorough and proper study is made of the new equipment or system before it is commissioned and HHCs are introduced to the process.

Mechanical Integrity
Maintaining the mechanical integrity of PSM-critical equipment is complicated for any organization. What makes this particularly difficult for smaller companies is establishing and staying on top of inspection, testing and preventive maintenance (ITPM) for what is often older equipment that had historically been run until it was broken. Emphasis is needed on implementing aspects of preventive and predictive maintenance to enhance the reliability and service life of their critical equipment, whether this is done in-house or via contract resources. Maintenance employees need appropriate training and/or certification according to the ITPM skills expected of them. Improving ITPM practices may even involve a type of 'culture change' which shifts focus from reactive to preventive/predictive maintenance practices. Management influence is essential to support this enhanced execution of maintenance activities.

Management of Change (MOC)
The MOC process is not only to be used for new equipment, capital projects, or changes in personnel with PSM responsibilities, but is also for new streams to be processed in an HMT facility. As waste compositions change from allowed norms, reviews need to occur to ensure that new streams or significant variations in existing ones will not result in any new hazards in the anticipated handling of the materials or the associated process operations. Changes in streams should likewise result in re-evaluating the materials of construction of the process equipment which may handle the new or different stream so that unanticipated failures of

equipment do not occur. Furthermore, since newer equipment is less common in the HWI, employees need to be diligent in understanding the difference between 'replacement in kind' (RIK) and non-RIK situations so they know when to apply MOC appropriately.

Hot Work

Hot Work operations in an HMT facility can be confusing to personnel. Some sites have difficulty defining hot work areas due to the fact that handling of waste streams may occur in many more locations around the site than at a standard manufacturing plant. Employees may need additional orientation and training to recognize Hot Work situations and zones where Hot Work operations may apply. As employees become more knowledgeable of their site's Hot Work requirements, proper preparation for and permitting of all Hot Work activities should occur as expected.

Incident Investigation

HMT facilities tend to have well-established incident investigation and emergency response planning. Many of their environmental permits already require these to be fairly robust. HWI organizations, like many other organizations, need to place more emphasis on leading indicators rather than lagging indicators. The new focus on standard API RP 754 *Process Safety Performance Indicators for the Refining and Petrochemical Industries* [9] is further emphasizing this trend. One leading indicator that is easy to track is near misses. Good PSM programs already have discipline in identifying incidents or spills to report. Now these need to identify events which may not have had an adverse result, but may have had the potential for the release of a highly hazardous chemical. Understanding the dangerous potential of these near miss situations can help sites to avoid tragic consequences from ignoring such warning signs.

Emergency Planning and Response

There needs to be diligence in conducting enough annual emergency response drills and in holding debriefings to help identify weaknesses in response programs and make necessary changes to improve. HWI organizations focus heavily on spills due to the nature of their environmental permits, but should equally consider drills on equipment failures or releases from pressure relief valves, for example. This 'balanced approach' to emergency drills should result in more comprehensive emergency response procedures and staffs which are better prepared to handle a greater range of adverse events.

Compliance Audits

Sites in the HWI are used to compliance audits from regulators on EPA and similar programs and tend to be well-versed in preparing for such audits. PSM auditing activities likewise need to be ongoing, including self-audits, to support PSM gap analyses which ought to be routinely occurring. Management should regularly review the results of these audits and put in place systems to successfully address identified deficiencies.

Trade Secrets

HMT sites do not tend to have many trade secrets, but, as with other organizations, they need to ensure that these are made available to personnel who need them to conduct PSM compliance activities. These personnel may need to enter into confidentiality agreements regarding access to this critical information to perform their normal duties.

5. Advancing PSM Programs at any PSM-Covered Facility

PSM programs are only as good as their corrective actions and improvement efforts. If you find problems in your systems but don't address them, it is as though you had no knowledge of them. Shortcomings identified in audits but which were not addressed have been contributing factors to major incidents, such as that which occurred in Texas City, TX in 2005 [4]. Management teams in every company need dedication to follow through on the recommendations from hazard analyses, incident investigations and compliance audits, among others, to address deficiencies and to correct issues. They must ensure continual improvement activities and projects exist and are led by appropriate champions. A variety of solutions may be needed to engage the work force to achieve better success with these aspects of PSM.

Sister plants can work together or company divisions can pool resources when there is an inadequate number of technical staff at a given location to lead PSM efforts. Today's internet-based applications and array of electronic gadgetry make these interactive communications easier than ever without excessive travel burdens.

Companies may consider utilizing process safety consultants to supplement their programs. This is a great way to get new ideas and access to industry best practices from trusted sources. Third-party auditors may also be employed to make unbiased suggestions for improvement. However, be sure your company actively partners with these groups for more successful PSM program advancement.

Organizations should get involved with trade organizations, professional societies and other groups that conduct seminars and training on process safety topics. Technical conferences are a great way to network with others in your own industry and learn about current PSM topics and newly identified best practices. The Global Congress on Process Safety (GCPS), which is a part of the Spring Meeting of the American Institute of Chemical Engineers (AIChE), is a fine example of an ongoing conference series dedicated to Process Safety education [10].

Companies in the HWI also have a new organization to supplement their PSM compliance efforts. The Recyclers' Safety Council (RSC) was created in 2011 by a consortium of industrial companies who recycle waste solvents and purify them for reuse. The intent of the RSC is to share best practices related to safety programs and compliance with the PSM standard. No similar group previously existed specifically for companies who recycle waste solvents. The RSC promotes collaboration among organizations who understand the unique set of challenges they face in achieving safety and process safety goals. My company, Veolia Environmental Services, is one of the founding members of this Council. Please contact me, the author of this article, for more information on participating in RSC activities, including an annual safety-themed conference.

6. Conclusions

HWI facilities, and HMTs in particular, require special diligence to effectively manage and progress their PSM programs. The size and nature of their operations require them to seek creative and possibly atypical measures to meet PSM compliance challenges. Though they are sometimes less advanced than their counterparts in other industries, they remain dedicated to addressing process safety gaps and continually improving their PSM systems. The recent formation of the RSC is an example of this commitment to process safety excellence.

PSM program success for any company is a function of the availability and dedication of the employees and the leadership of management. Limited resources, while a challenge, should not be an impediment to compliance. Be proactive and focused on your program and it will be possible to have a thriving process safety culture in your organization, regardless of size or the nature of your company's business.

7. References

[1] Process Safety Management of Highly Hazardous Chemicals, 29 CFR 1910.119, Occupational Health and Safety Administration , Occupational Safety and Health Standards.

[2] Jackson B. Browning, Jackson Browning Report, "Union Carbide Disaster at Bhopal (1993)" Available at www.bhopal.com/reports-and-studies/union-carbide-disaster-at-bhopal.

[3] "OSHA: Phillips 66 Company Houston Chemical Complex Explosion and Fire: A Report to the President, U.S. Department of Labor, Washington, D.C. (1990)".

[4] Baker, James et al 2007 "The Report of the BP U.S. Refineries Independent Safety Review Panel".

[5] OSHA National Emphasis Programs: Available at: http://www.osha.gov/SLTC/processsafetymanagement/index.html

[6] U.S. Chemical Safety Board Case Study 'Fire and Community Evacuation in Apex, N.C.', April 16, 2008. Available at: www.csb.gov/investigations .

[7] U.S. Chemical Safety Board Case Study 'Veolia Environmental Services Flammable Vapor Explosion and Fire', July 21, 2010. Available at: www.csb.gov/investigations .

[8] Three Rivers Manufacturing Association; Available at www.trma.org.

[9] API RP-754 Process Safety Performance Indicators for the Refining and Petrochemical Industries, 2010. Available at: www.api.org .

[10] Global Congress on Process Safety; Available at: www.aiche.org/resources/conferences.

A Novel Methodology for Maintaining Mechanical Integrity of High Pressure Relief Systems

Sandipan Laskar
Sr. Process Safety Manager
Clean Harbors Inc.
2027 Independence PKWY S
La Porte, TX 77571
laskar.sandipan@cleanharbors.com

Prepared for Presentation at
American Institute of Chemical Engineers
2013 Spring Meeting
9th Global Congress on Process Safety
San Antonio, Texas
April 28 – May 1, 2013

A Novel Methodology for Maintaining Mechanical Integrity of High Pressure Relief Systems

Sandipan Laskar
Sr. Process Safety Manager
Clean Harbors Inc.
2027 Independence PKWY S
La Porte, TX 77571
laskar.sandipan@cleanharbors.com

Keywords: Relief System, Mechanical Integrity (MI)

Abstract

Safe operation and maintenance of emergency systems are of primary importance to the process industries, including upstream oil and gas. The upstream production facilities are usually protected from overpressurization by high integrity pressure protection systems (HIPPS) which shuts in the well when local pressure levels exceed a certain threshold. The pressure relief system is often installed as a last safety barrier, when HIPPS fails. One of the critical elements of emergency system is proper design, documentation and to maintain high reliability of this system.

In particular, the pressure relief system is an integral part of upstream design and operation; prevention of an accidental release of toxic, flammable or noxious materials from a production facility requires correct design and construction, safe operation, and proper documentation of this system.

This paper proposes a basic change in the way relief systems are documented - a requirement in order to comply with OSHA 1910.119 and thereby describes a new methodology and a change in documentation to meet OSHA requirements and other industry standards (such as API 520/521, API-RP 75, 14J). This paper also describes the steps involved in a high pressure relief system evaluation and introduces concept of ever-greening to maintain a high reliability of this system. The paper illustrates how an up-to-date relief system leverages management of change (MOC) and helps in setting up the foundation for a strong overall process safety management (PSM) program. The paper will identify the drivers and the benefits of the proposed methodology for pressure relief documentation as well as explore the interdependency of the proposed program elements.

The paper will assist upstream facility personnel to utilize good engineering practice in design and documentation of safety relief systems, to meet both onshore and offshore process safety requirements, and to develop consistent company best practices for process safety management. In addition the paper will describe a comprehensive equipment based analysis protocol, preparation of documentation sufficient for design, operational, and audit purposes, and discuss maintenance both of the physical system and design data.

1. Introduction

Most oil and gas production, refining and petrochemical organizations have some kind safety management program that stems from the requirements outlined under OSHA's CFR 1910.119. Some have a very extensive program that covers most of the 14 elements of Process Safety Management (PSM) and is managed via an integrated platform while others follow an ad hoc approach that does not span all of the elements or is disjointed.

Maintaining accurate and current Process Safety Information (PSI) is a cumbersome and difficult task to accomplish due to the numerous sub-elements of Safety Management Program (SEMP). One of the key sub-elements of PSI maintenance is the management of the pressure relief documentation. Historically this is an element that has not been given the attention it deserves. However, recent incidents in the industry have elevated the importance of maintaining accurate and current pressure system documentation.

The ultimate goal is to allow all plant personnel to understand the basis of the current plant relief and flare header design and ensure that this current plant relief and flare header design provides adequate overpressure protection for all equipment. Thus, a new methodology for relief system documentation is created that is customized to fit the needs and resources of the operating company and had the ability to relate to the other Process Safety Information elements.

2. Pressure Relief Analysis

Pressure relief systems are applied in various installations in the oil and gas industry. The function of a pressure relieving system is to provide a positive means of protection for plant equipment from overpressure above limits permitted by the relevant codes and to dispose safely of any resultant relief of the process fluids.

The pressure relief analysis provides an efficient way to document the basis for relief system design basis. The major reasons for undertaking the pressure relief analysis by an operating company are listed below.

- Safety Initiatives
- Design/Capital Projects
- Audit/Compliance Regulations
- Corporate Standardization
- Acquisitions

Identification of the causes of overpressure, quantification of relieving rates required complying with codes, sizing of relief devices and the design of the relief disposal system is a complex task. Use of a systematic method for identification of contingencies, together with a set of project specific standards, provides the basis for sound relief system design. Quantification of contingencies using established calculation methods and process simulation is then used to size pressure relief devices and the associated relief header piping network. This approach provides an auditable safe design for a new plant and existing facilities. The different steps involved in a pressure relief analysis are illustrated below.

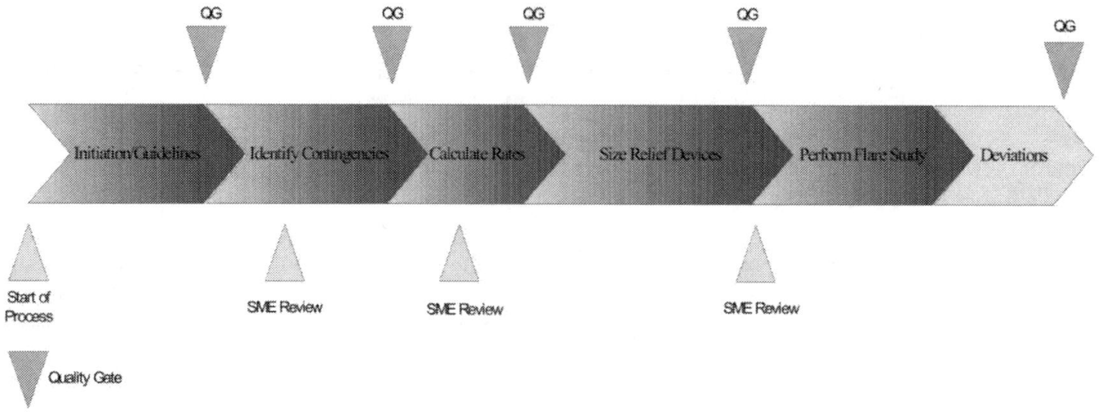

Figure 1 –Typical Phases of Pressure Relief Analysis

Mostly the pressure relief analysis is done as a one-time comprehensive project during the design phase or if there are any changes in the process. The major issues with one time comprehensive studies are listed below.

- Reactive instead of proactive approach
- Based on snap shot
- Underutilized valuable dormant data
- Missed training opportunities
- Out-of-sync with typical MOC programs
- Spend millions in 1-2 years for complete overhaul
- Increase risk of missing safety issues

3. Mechanical Integrity

Mechanical integrity is defined as the program or activities which ensure that equipment in a specified process is designed, constructed, installed and maintained to minimize the risk of accident release of high hazardous chemicals. Mechanical integrity can be achieved through the use of followings.

- Proper design to assure that equipment is suitable for its intended service
- Procedures to assure that we know how to properly operate, inspect and maintain equipment
- Training to assure everyone understands the procedures and the importance of doing things the right way
- Inspection to find flaws or deficiencies that may occur over a period of time or to confirm the status of equipment if nothing has changed
- Repair to maintain integrity
- Quality assurance to make sure that whenever something is done it is the done the right way.

Though there are many codes, standards and recommended practices that are available as a guide to ensure compliance with this element of PSM but still Mechanical Integrity (MI) presents the greatest implementation challenge. Mechanical integrity can be defined as an element of process safety management used to encompass activities that are designed to assure that mechanical equipment in an operating facility has the integrity to perform in the manner and service for which it was designed, locate, characterize, and rectify areas of deficiency that might lead to a condition that would be unsafe and endanger human life, health and/or property, assure that equipment is built and maintained in such a way as to perform it's desired function and ensure the ongoing durability and functionality of equipment achieved through a systematic approach. Although maintenance departments have successfully installed, repaired, and replaced plant equipment for decades, many of these same maintenance departments have struggled with OSHA's PSM requirements. One major challenge is prioritizing resources. Opportunities to improve will always exist; however, it is often difficult to effectively allocate money and manpower. Another challenge is simply getting organized. To meet the requirements of maintaining mechanical integrity, it is necessary to develop a program of testing safety devices and inspecting for both internal and external corrosion and erosion.

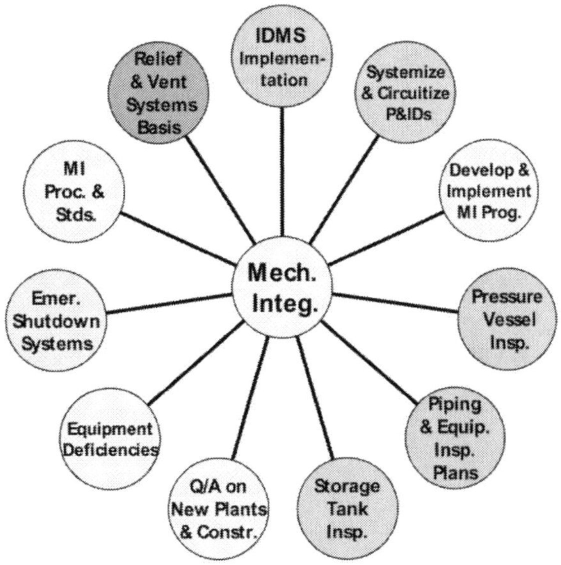

Figure 2 -Elements of Mechanical Integrity (OSHA's 29 CFR 1910.119, Para. j)

Mechanical integrity assures integrity and reliability of pressure equipment which in turn results in fewer safety incidents; reduction of mechanical failure (minimizes downtime and enables better planning), less process upsets and improves maintenance consistency and efficiency.

OSHA 1910.119 was written so that it would cover highly complex and hazardous processes handling a wide variety of hazardous chemicals. Production facilities are some of the simplest processes covered under this regulation. On the other hand, API RP 75 is written specifically for oil and gas production. It addresses offshore operations but much of the information is applicable to onshore operations as well. API helps in interpreting the application of OSHA's requirements to oil and gas facilities.

Many of the mechanical integrity activities are performed as part of a plant's maintenance program. Generally the OSHA guidelines rely on recognized and generally accepted good engineering practice (RAGAGEP), but allows companies to utilize alternate methods as long as they can prove that performance resulting from that method meets the standard. The specifications and techniques that form RAGAGEP are generally understood to mean recommended practices from the American Petroleum Institute (API), codes from the American Society of Mechanical Engineers (ASME), standards from the International Standards Organization (ISO), certification agency standards (DNV, Lloyd's, ABS), and other sources in regular use. The different elements of the PSM and SEMP program, is as shown below.

OSHA's PSM	SEMP
1. Employee Participation	1. Safety and Environmental Information
2. Process Safety Information	2. Hazard Analysis
3. Process Hazard Analysis	3. Operating Procedures
4. Operating Procedures	4. Training
5. Training	5. Pre-Startup Review
6. Contractors	6. Assurance of Quality and Mechanical Integrity of Equipment
7. Pre-startup Safety Review	7. Safe Work Practices
8. Mechanical Integrity	8. Management of Change
9. Hot Work	9. Investigation of Incidents
10. Management of Change	10. Emergency Response and Control
11. Incident Investigation	11. Audit of Safety and Environmental Management Program Elements
12. Emergency Planning and Response	12. Records and Documentation
13. Compliance Audits	
14. Trade Secrets	

The OSHA PSM guidelines are enforceable throughout the United States and specific US holdings for facilities having a threshold volume of hazardous chemicals, but are not applicable where another federal agency has occupational safety and health jurisdiction, including in federal waters controlled by Bureau of Ocean Management Regulation and Enforcement (BOEMRE, formerly Minerals Management Service (MMS)). Oil and gas drilling and servicing operations and normally unmanned remote facilities are specifically excluded from the standard, but typical facilities in the upstream industry such as production facilities, gas oil separation plants, and gas plants are covered if the threshold chemical volumes, typically hydrocarbons, are present.

In today's economic climate, we are operating with fewer people and demanding more from them than ever before. Therefore, it is imperative that the amount of documentation required be reduced to the minimum necessary to improve communication and increase safety. Excessive documentation places a time requirement that distracts people from their true and safe work practices. Documentation must be kept simple and concise so that it will be easy to use, with essential information readily at hand. The goal of the methodology presented in this paper is to improve safety without creating a large paperwork burden.

4. Methodology

The main goal of the proposed methodology is to develop a program that impacts the culture of the organization, thereby, drives and sustains accurate pressure relief information on a continual basis. The new program must leverage the overall, existing Safety Management Program with focus on the specific task at hand; keep perspective of risk in the current design. The new methodology eradicates the short comings of a one-time comprehensive pressure relief analysis and provides a standardized approach to ensure high reliability of the pressure relief system. The proposed methodology is further sub-divided into six different elements. These elements are Workflow, Resources and Processes, Tools, Training, Key Performance Indicators and Auditing. All these elements are interdependent for an overall success of the proposed methodology. The workflow process starts with a relief system review and generates the PRV Impact checklists, this checklist can be handed over to the MOC program. The actual workflow of the process is shown below.

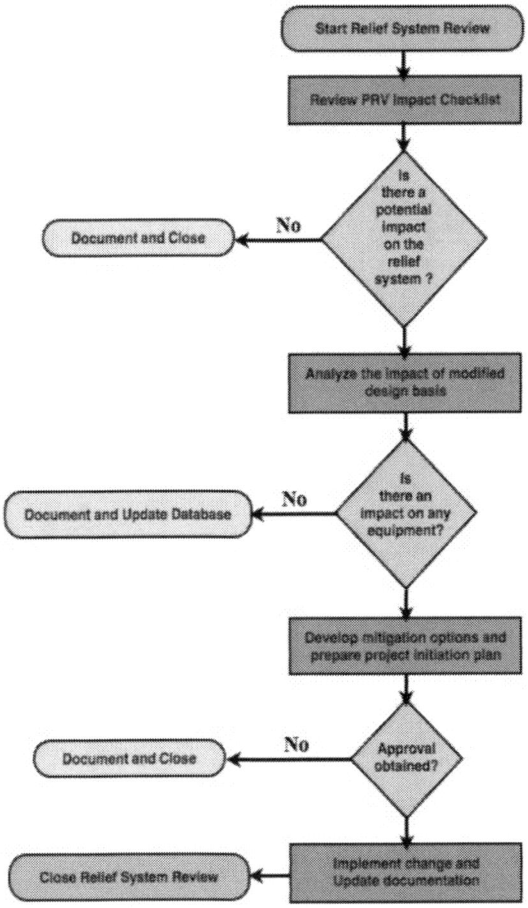

Figure 3 –Pressure Relief Valve Program Workflow

The steps involved in maintaining the data integrity and control of the proposed methodology are mentioned below.

- Develop a Relief System Impact checklist
- Incorporate checklist in the MOC program
- Control access to electronic databases
- Establish within MOC an exception check process for reconciling and validating data changes
- Develop a revalidation process for the plant-wide flare/relief system evaluation
- Complete a revalidation of each study within five years of the previous study
- Perform a complete redo of the study on a cyclical basis
- Manage the resolution of recommendations
- Establish a system of records for data
- Pool data into a central repository

The other elements of the program are to designate an owner for the program and appoint subject matter experts for engineering and information technology. An advisory group both internally at the site and externally at corporate level should be formed to support the program. There are readily available software in the market today to facilitate the actual calculations and sizing of pressure relief valves. Care should be taken to ensure that the tools used meet corporate standard, could be tailor-fitted to compliment internal company practices and also adapt to industry technology and technical practices.

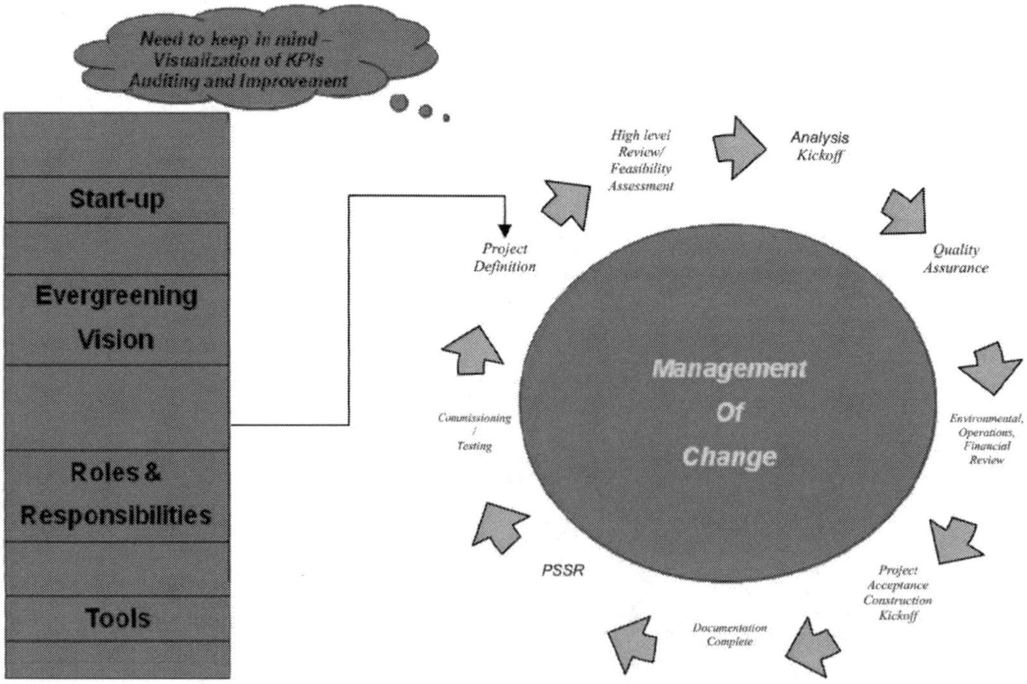

Figure 4 –Pressure Relief Valve Program Leverage Management of Change (MOC)

The above figure enunciates the importance of the proposed methodology in leveraging the existing management of change (MOC) program. The key performance indicators of the program are the data accuracy between different repositories, leverage existing information to measure performance and the program pull dormant data from relief system analysis to measure other PSM metrics. The major benefits of the proposed program are shown below.

- Standard approach
- Continual training
- Defined success factors
- Repeatable process
- Known roles and responsibilities
- Updated and accurate documentation

5. Conclusion

A good program work process starts from the projects to operations handover step on new construction or debottlenecks and runs all the way through the lifecycle of the plant encompassing modifications as a result of small and large projects. It provides a platform to communicate the PSI requirements through all disciplines at the operating facility in a concise manner and holds everyone at a high standard to meet these set requirements by continuously auditioning its effectiveness and providing training and improvement measures.

The cause of overpressure and the amount of fluid that is required to be removed from the system (vessel and piping) is unique for each installation. The proposed methodology ensures process safety without creating a large paperwork burden. The proposed methodology provides operating and maintenance staff, supervisors and foreman with the information, training and motivation they need to conduct their business in a safer manner. It is also intended that the information presented will help in proper documentation and thereby enable a faster response in case of an emergency.

The above mentioned methodology ensures most cost effective process for maintaining the mechanical integrity of high pressure relief systems while contributing up-to-date documentation of the emergency systems and thereby leveraging the existing MOC program. Particularly for high profile vertically integrated companies operating both onshore and offshore facilities, coordination of programs and practices provides an opportunity to streamline internal processes, comply with the new offshore regulatory environment, and operate more safely than ever before.

6. References

[1] "API RP 75, Recommended Practice for Development of a Safety and Environmental Management Program for Offshore Operations and Facilities", third edition. 2004. Washington, DC: API

[2] "Oil and Gas and Sulphur Operations in the Outer Continental Shelf--Safety and Environmental Management Systems", Bureau of Ocean Energy Management, Regulation and Enforcement Implementation (BOEMRE), *Final Rule*, 30 CFR Part 250, Volume 74, Number 199, (October 15, 2010)

[3] "Petroleum Refinery Process Safety Management National Emphasis Program", U.S. Occupational Safety and Health Administration (US OSHA), Directive CPL 03-00-004, (June 7, 2007)

[4] "PSM Covered Chemical Facilities National Emphasis Program", U.S. Occupational Safety and Health Administration (US OSHA), Directive CPL 09-06-002, (July 27, 2009)

[5] "Survey for API RP 75", Bureau of Ocean Energy Management, Regulation and Enforcement Implementation (BOEMRE), (September. 17, 2010)

[6] "API 520 Part 1, Standard for Sizing, Selection, and Installation of Pressure-Relieving Devices in Refineries, eighth edition. 2008. Washington, DC: API

A Framework for Managing Near-Miss Incidents

Presenter: GC Shah/Principal Author: GC Shah
Affiliation: Wood Group Mustang inc.
Address: 16401 Park Ten Place Blvd.; Houston, TX. 77084
email address: ghanshyam.shah@mustangeng.com

[

Prepared for Presentation at
American Institute of Chemical Engineers
2013 Spring Meeting
9th Global Congress on Process Safety
San Antonio, Texas
April 28 – May 1, 2013

UNPUBLISHED

A Framework for Managing Near-Miss Incidents

Presenter: GC Shah/Principal Author: GC Shah
Affiliation: Wood Group Mustang Inc.
Address: 16401 Park Ten Place Blvd.; Houston, TX. 77084
email address: ghanshyam.shah@mustangeng.com

Keywords:

ANSI/API RP-754-2010: Standard that creates "performance indicators" for process safety.
HAZID: Hazard Identification.
HAZOP: Hazard and Operability Study.
Lagging Indicators: Failures which have resulted in loss.
Leading Indicators: Indicators or evaluations that help predict the risk of unsafe events.
Near-miss incidents: Incidents that did not result in any loss, but have a potential for loss.
PHA: Process Hazard Analysis.

Abstract

Today, there is considerable emphasis being placed on learning from near-miss incidents and applying additional safeguards which would minimize possibility of escalating adverse impact of these events in the future. Not all near-miss incidents have potential for major adverse impact. It is important to focus on near-miss incidents which could result into incidents with major adverse impact. This paper applies a modified "Risk Assessment Matrix," that is commonly used for conducting PHA assessments such as HAZID and HAZOP, to estimate potential future risk indication conveyed by near-misses. In a strategic sense, near-miss management spans all work systems of an organization and its culture: IT system, operations, legal, and many others.

1. INTRODUCTION

In the hydrocarbon industry, as well as in many other industries for that matter, the "minor" events occur fairly often. While most of these events probably would end there and not result into major adverse impact, some of the events, if they recur perhaps under different circumstances, could have major impact on safety, environment, asset or a company's reputation. For instance, the event where a worker misses a rung on the ladder and falls and gets bruised may not develop into a major event. On the other hand, an event where there is a pipe leak which creates a cloud of flammable vapor that gets dissipated because of high wind, could under slightly different circumstances, result into a major event (potential fire and explosion). Even small leaks of flammable gases, if allowed to accumulate, could cause a fire and/ or an explosion.

OSHA defines near miss as "*incidents where no property was damaged and no personal injury sustained, but where, given a slight shift in time or position, damage and/or injury could have occurred.*" The definition seems to suggest that, the existing safeguards failed to work; but luckily the incident turned out to be a minor event. However, sooner or later luck may run out and a near miss can turn into a major adverse event. In terms of probabilities, one may say that statistically, there is a greater probability of occurrence of 'small impact" events and much smaller probability of high consequence or high impact events. From cost considerations, high impact events with high frequency simply can't be sustained financially.

Once near-miss events which have a potential to turn into "major" events have been identified, resources can be allocated to either strengthen the existing safeguards or add new safeguards in order to minimize the likelihood of occurrence of a major event.

2. IDENTIFYING POTENTIAL "TIME BOMBS."

Some experts suggest that near-miss is an accident waiting to get you into big trouble. The goal is to let it keep waiting!

How do we identify with a good degree of certainty that a near-miss will grow into a large-impact event? Experience with operations and maintenance could be a big help. You can develop checklists of incidents likely to develop into major events and incidents which are not likely to develop into large-scale events. Before we look at these lists, it is worth considering the meaning of the term "large-scale?" Stated broadly, the events that could cause one or more of the following may be regarded as the events that are likely to develop into "large-scale events."

OSHA recordables
Lost-time
Environmental RQ (Reportable Quantity) events
Environmental or safety violations
Events that could become a legal liability to a company
Public complaints

Some examples of specific events which could potentially turn into events with major adverse impact are:

Pin-hole leaks (e.g., pipes or flanges and fittings handling flammable, explosive or toxic materials; steam or condensate leaks on old boiler systems);
Leaking flanges, leaking mechanical seals;
Leaking tanks with flammables and/or toxic materials;
Spills under Reportable Quantity;
Lack of or poor lubrication of large pumps and/or compressors because pre-startup checks were not made or there was no checklist;
A worker narrowly escapes being asphyxiated in a confined space where nitrogen is being compressed;
Purge cycle on a furnace not working properly;

Welding sparks falling on hydrocarbon containing pipes;
Workers fail to understood a procedure (and therefore don't follow it) as it contained high-level technical jargon;
An isolation valve fails in open position;
Security infraction incidents (including cyber security);
Others.

It may be convenient to categorize the above events into the following groups:

- *Failure to follow procedures (OR poorly written or nonexistent procedures)*
- *Control system (and safety system) failures*
- *Old (older) equipment in challenging service*

Similar other systems of classification may be developed for your company.

Examples of events which will not likely develop into major events are:

Minor bruises or cuts;
Falls with a minor injury;
Potable water leaks;
A worker experiencing headache and taking non-prescription medication;
Other similar examples may be cited.

3. A FRAME WORK FOR NEAR-MISS MANAGEMENT

To be effective, the framework should be efficient. Workers should be encouraged to report ALL near-misses. Near-miss management start NOT after a near-miss, but way before such an event occurs. Guidance on leading indicators (as well as lagging indicators) is developed by API-754-2010. As Figure 1 shows near-miss is not a narrowly defined safety issue of simply reporting events; it spans through the entire organization and systems.

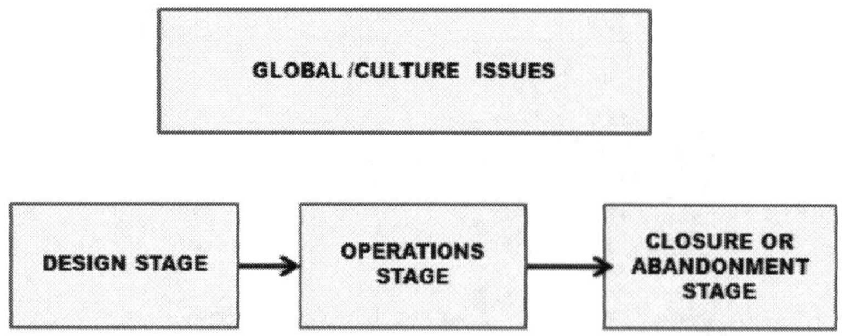

FIGURE 1: NEAR-MISS INCIDENT MANAGEMENT STAGES

3.1 Global and Cultural Issues

- Follow "good neighbors" policy. Establish good rapport with neighbors and regulators.
- Strategically important documents should receive thorough checks so as to avoid mis-interpretation.
- Encourage mutual trust between workers and management.
- Avoid bureaucratic systems in near-miss management.
- Encourage high-tech professionals (e.g., IT networks, advanced controls) to translate technical jargon into lay-person's language. The key is that persons, for whom the procedures (developed by technical professionals) are intended, should understand clearly.
- Ensure there are adequate resources to address near-miss incidents in reasonable time. Do not "dilute," safety.
- Acquisitions and divestitures also should (and typically they do) receive due diligence.

3.2 Design Stage:

- Ensure design is in accord with the applicable industrial standards and regulations.
- As far as practicable, follow the philosophy of fault-tolerant design. Stated differently, design should a mishap occur, there should be appropriate mitigative measures to constrain impact resulting from the mishap.
- Ensure documentation is easy to access and not ambiguous. (Workers referring to the documents in the future should get a clear understanding of the meaning and intent of the design).
- Perform periodic risk assessments; address findings thoroughly, and document.

3.3 Operations Stage:

- Encourage a collaborative rapport between your technical experts or engineers and operations and maintenance workers and contractors.
-
- Introduce New Technology only after a thorough risk assessment.
- Implement frequent training.
- Ensure the procedures and instructions are CLEAR to workers. Seek their input before implementing them.
- Conduct periodic audits of the facility. Some companies term these as "Loss Control Audits."
- Encourage workers to report ALL near-misses.
- Identify near-miss incidents which have potential to develop into major events. You may work out checklists workers can use to identify such incidents or consider the Risk Matrix shown later in this section. Of course checklists should be thorough so that workers won't miss out a near-miss that has a potential to grow into a major event.
- Provide resources to address near-misses within a reasonable time.

3.4 Closure and Abandonment

- Develop closure and abandonment policies. Apply Risk Management criteria to this phase of the organization.

3.5 Risk Matrix for Near Miss Incident Management

Figure 2 below shows a proposed Risk Matrix for near miss incident management. Risk is defined as a product of likelihood and consequence (severity of the outcome of an incident). Risk is grouped into three levels:

- High
- Medium
- Low

All near-miss incidents must be looked into. However, incidents identified as those having a potential for developing into "major" events require further investigation and possibly adding new safeguards or making existing safeguards more reliable or robust. Near miss management should be efficient. A small team of experienced workers can apply Risk Matrix (or checklists).

Figure 2: Risk Matrix for Near Miss Incident Management

		Risk Matrix for "Near-Miss" Incidents		
Consequence	3	3	6	9
	2	2	4	6
	1	1	2	3
		1	2	3
		Likelihood		

		Risk Matrix for "Near-Miss" Incidents		
		Likelihood		
		1	Relatively low likelihood of occurrence once in 10 yrs.	
		2	Relatively moderate likelihood of occurrence once every yr.	
		3	Relatively high likelihood of occurrence several times a year.	
		Consequence		
		1	Relatively "low" or "potentially low" level of impact: This events are NOT likely to develop into recordable or lost time or RQ events or higher.	
		2	Moderate level of impact or potential for medium level of impact such as OSHA recordable or lost time incident or RQ events.	
		3	High impact or potential for "high" impact: For example, large scale environmental or safety impact, or potential for damage to reputation.	
		Suggested Corrective Action		
		BLUE	Document the event. Further resources for corrective action may not be warranted.	
		YELLOW	Investigate and implement corrective action in "reasonable "time.	
		RED	Investigate. Should receive top priority and urgency in implementing corrective action.	

4. CONCLUSION

Near miss incident management starts way before a near miss occurs. In a strategic sense, the near miss management pervades all layers of an organization and its interactions with external systems. Strong safety culture and trusting relationships are important components of an effective near miss management framework.

9TH GLOBAL CONGRESS ON PROCESS SAFETY

HAZOP Analysis and Debottleneck for Laboratory Operation in the Semiconductor Industry

Tianxing Cai
Dan F. Smith Department of Chemical Engineering
Lamar University, Beaumont, TX 77710, USA
tcai@lamar.edu

Qiang Xu
Dan F. Smith Department of Chemical Engineering
Lamar University, Beaumont, TX 77710, USA
qiang.xu@lamar.edu

Prepared for Presentation at
American Institute of Chemical Engineers
2013 Spring Meeting
9th Global Congress on Process Safety
San Antonio, Texas
April 28 – May 1, 2013

UNPUBLISHED

HAZOP Analysis and Debottleneck for Laboratory Operation in the Semiconductor Industry †

Tianxing Cai
Dan F. Smith Department of Chemical Engineering
Lamar University, Beaumont, TX 77710, USA
tcai@lamar.edu

Qiang Xu*
Dan F. Smith Department of Chemical Engineering
Lamar University, Beaumont, TX 77710, USA
qiang.xu@lamar.edu

Keywords: HAZOP Analysis, Semiconductor Industry, Laboratory Hazards, Operation Debottleneck

Abstract

Unlike the semiconductor manufacturing process with fully automatic equipment, the laboratory operation in the semiconductor industry will still involve multiple types of manual handling for chemical analysis, reliability test, material characterization and failure analysis. The test operators or lab engineers have different levels of exposure to a lot of chemicals and physical hazards in multiple phases. This paper will describe the straightforward HAZOP (Hazard and Operability Study) process to identify the hazards and assess the risks of laboratory operations. The analysis will be deployed in the ways of manpower, machine, materials, method and environment. The analysis output can be regarded as the aid to safer laboratory design, operator training and real-time fault diagnosis. This will help to debottleneck and achieve operation safety and occupational health in the semiconductor industry.

---

† For presentation at AICHE 2013 Spring Meeting ,9th Global Congress on Process Safety.
* All correspondence should be addressed to Prof. Qiang Xu (Phone: 409-880-7818; Fax: 409-880-2197; E-mail: Qiang.xu@lamar.edu).

1. Introduction

Semiconductor industry revenue in 2012 is expected to reach $323.2 billion, up a slight 3.3 percent from last year's revenue of $312.8 billion, according to an IHS iSuppli Global Manufacturing Market Tracker report from information and analysis provider IHS.
Based on the professional estimation, the growth in the semiconductor industry from 2013 to 2015 will average between a more encouraging 6.6 to 7.9 percent, as shown in the figure below, with total semiconductor revenue by 2015 rising to some $397.7 billion. In the coming several years, the strong semiconductor growth can be seen.

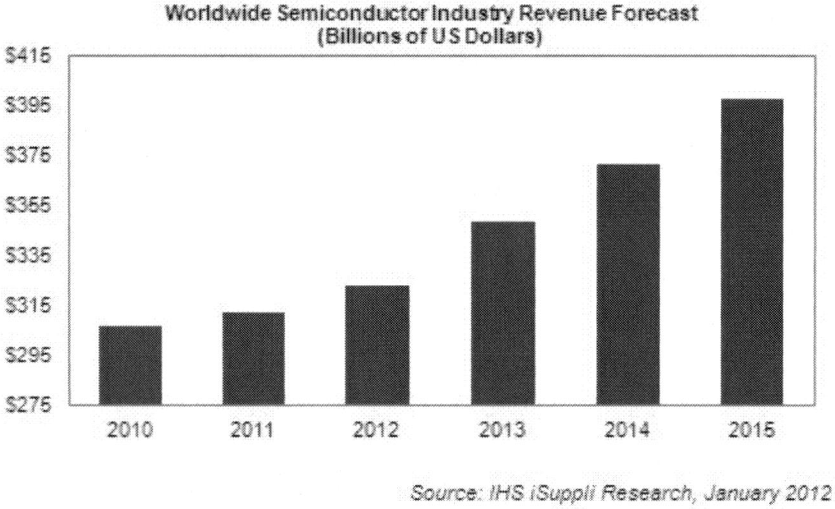

Figure 1: Worldwide Semiconductor Industry Revenue Forecast

Coupled with the fast development, the semiconductor employment has globally stretched. In USA, recently on Dec 11th, 2012, the Semiconductor Industry Association (SIA) announced that U.S. semiconductor industry jobs stretch across almost every region of the country and into the majority of states, according to an analysis of government data. California leads all states with 47,100 semiconductor jobs, followed by Texas (28,800), Oregon (23,400), Arizona (18,800), Massachusetts (10,100) and New York (7,600). According to BLS, the U.S. semiconductor industry's workforce grew by 3.7 percent over the previous year. In comparison, jobs throughout the broader U.S. economy increased by 1.2 percent over the same time period.

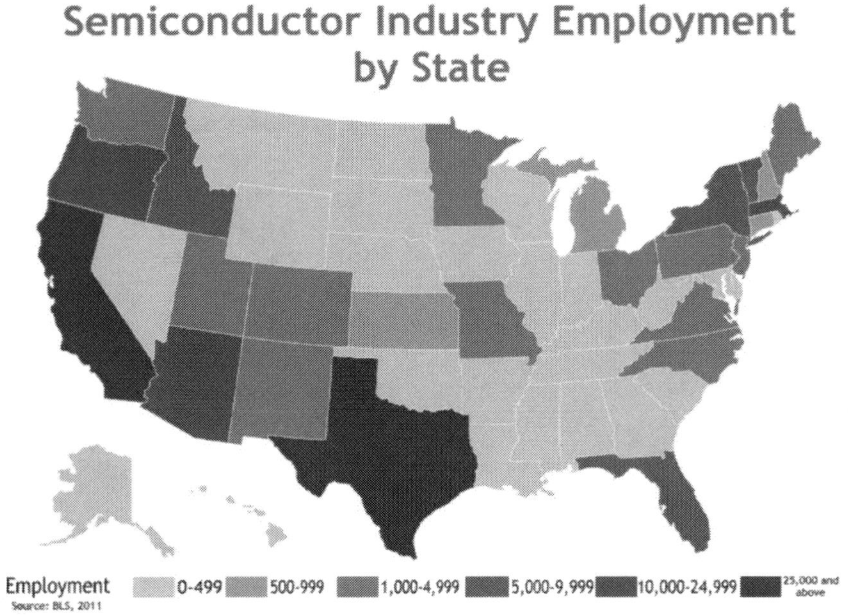

Figure 2: Semiconductor Industry Employment by State

Ever since the invention of transistor by the Bell Telephone Laboratories in 1948, the semiconductors industry has expanded greatly. The continuous fast development and innovations have expedited the increase of the semiconductor industry employment, which has further invoked an unavoidable concern of occupational safety to the semiconductor manufacturing operation.

Unlike the semiconductor manufacturing process with fully automatic equipment, the laboratory operation in the semiconductor industry will still involve multiple types of manual handling for chemical analysis, reliability test, material characterization and failure analysis. These operations will involve the handling of advanced and complex equipment or instruments such as Gas Chromatography (GC), Gas Chromatography-Mass Spectrometry (GC-MS), Fourier Transformed Infra Red Spectroscopy (FT-IR), Ultraviolet Visible Spectroscopy (UV/Visible), Liquid Chromatography (LC), Thermal and Elemental Analysis (TEA). Furthermore, due to rapid changes in this industry, manufacturing processes and their associated hazards may change completely every few years. These changes make hazard assessments more difficult to complete and require that they be conducted more often. Common hazards may include exposure to solvents, acid and caustic solutions, toxic metals, and radiation. Below Table 1 has listed the typical hazardous chemicals with their roles in the manufacturing process and potential health problems linked to the exposure. The test operators or lab engineers have different levels exposure to a lot of chemicals and physical hazards in multiple phases.

Table 1:Hazardous Chemicals in the Semiconductor Industry

Chemical name	Role in manufacturing process	Health problems linked to exposure
Acetone	Chemical-mechanical polishing of silicon wafers	Nose, throat, lung, and eye irritation, damage to the skin, confusion, unconsciousness, possible coma
Arsenic	Increases conductivity of semiconductor material	Nausea, delirium, vomiting, dyspepsia, diarrhea, decrease in erythrocyte and leukocyte production, abnormal heart rhythm, blood vessel damage, extensive tissue damage to nerves, stomach, intestine, and skin, known human carcinogen for lung cancer
Arsine	Chemical vapor deposition	Headache, malaise, weakness, vertigo, dyspnea, nausea, abdominal and back pain, jaundice, peripheral neuropathy, anemia
Benzene	Photoelectrochemical etching	Damage to bone marrow, anemia, excessive bleeding, immune system effects, increased chance of infection, reproductive effects, known human carcinogen for leukemia
Cadmium	Creates "holes" in silicon lattice to create effect of positive charge	Damage to lungs, renal dysfunction, immediate hepatic injury, bone defects, hypertension, reproductive toxicity, teratogenicity, known human carcinogen for lung and prostate cancer
Hydrochloric acid	Photoelectrochemical etching	Highly corrosive, severe eye and skin burns, conjunctivitis, dermatitis, respiratory irritation
Lead	Electroplated soldering	Damage to renal, reproductive, and immune systems, spontaneous abortion, premature birth, low birth weight, learning deficits in children, anemia, memory effects, dementia, decreased reaction time, decreased mental ability
Methyl chloroform	Washing	Headache, central nervous system depression, poor equilibrium, eye, nose, throat, and skin irritation, cardiac arrhythmia
Toluene	Chemical vapor deposition	Weakness, confusion, memory loss, nausea, permanent damage to brain, speech, vision, and hearing problems, loss of muscle control, poor balance, neurological problems and retardation of growth in children, suspected human carcinogen for lung and liver cancer
Trichloroethylene	Washing	Irritation of skin, eyes, and respiratory tract, dizziness, drowsiness, speech and hearing impairment, kidney disease, blood disorders, stroke, diabetes, suspected human carcinogen for renal cancer

In order to mitigate the potential risk during the chemical process operation, HAZOP study is always conducted. A hazard and operability study (HAZOP) is a structured and systematic examination of a planned or existing process or operation in order to identify and evaluate problems that may represent risks to personnel or equipment, or prevent efficient operation. The

HAZOP technique was initially developed to analyze chemical process systems, but has later been extended to other types of systems and also to complex operations and to software systems. A HAZOP is a qualitative technique based on guide-words and is carried out by a multi-disciplinary team (HAZOP team) during a set of meetings.
This paper will describe the straightforward HAZOP process to identify the hazards and assess the risks of laboratory operations. The analysis will be deployed in the ways of manpower, machine, materials, method and environment. The analysis output can be regarded as the aid to safer laboratory design, operator training and real-time fault diagnosis. This will help to debottleneck and achieve operation safety and occupational health in the semiconductor industry.

2. General Methodology

Several tools and methodologies are adopted to deploy the HAZOP Analysis and debottleneck for laboratory operation in the semiconductor industry. This has included Laboratory Operation Flow Diagram (LOFD), HAZOP Meeting, Parameters and Guide Words

2.1 Laboratory Operation Flow Diagram (LOFD)

The method applies to existing or planned laboratory operation for which design information is available. Similar with the commonly used process flow diagram, this commonly includes a laboratory operation flow diagram, which is examined in small sections, such as individual items of laboratory test instruments, glassware or chemicals between them. For each of these, a operation design intention is specified. For example, in a failure analysis laboratory, a standard solution of aqua regia (the combination of HNO_3 and HCl with the volume ratio of 1:3) may have been used. The intention to remove gold layer in the integrated circuit so that the further visual inspection can be conducted to check the failure mode. The HAZOP team then determines what are the possible significant deviations from each intention, feasible causes and likely consequences. It can then be decided whether existing, designed safeguards are sufficient, or whether additional actions are necessary to reduce risk to an acceptable level. Besides the operation safety, HAZOP can also give the detailed instruction on the opinion whether the potential deviation will affect the quality of experiment or test analysis result.

2.2 HAZOP Meeting

When HAZOP meetings were proceeded by the involved laboratory engineers, technicians and operators, they were generally scheduled for one to two hours per day. For a medium-sized laboratory where the total number of test items to be considered is 600 (items of equipment and pipes or other transfers between them), about 20 such meetings would be needed.

2.3 Parameters and Guide Words

The key feature is to select appropriate parameters which will apply to the design intention. The general words used in the laboratory operation are usually solution concentration, solution temperature, sampling volume, and instrument detection limit. In the above example, it can be seen that variations in these parameters could constitute deviations from the design intention of experiments. In order to identify deviations, a set of guide words is applied to each step of the experiment process. The current standard Guide Words are as follows:

Table 2: Guide Words for HAZOP Study of Laboratory Operation in the Semiconductor Industry

Guide Word	Meaning
NO OR NOT	Complete negation of the design intent
MORE	Quantitative increase
LESS	Quantitative decrease
AS WELL AS	Qualitative modification/increase
PART OF	Qualitative modification/decrease
REVERSE	Logical opposite of the design intent
OTHER THAN	Complete substitution
EARLY	Relative to the clock time
LATE	Relative to the clock time
BEFORE	Relating to order or sequence
AFTER	Relating to order or sequence

2.4 HAZOP Software

Various software programs are now available to assist in HAZOP meetings. Even though they are always used in the chemical industry, the principles will be quite applicable to laboratory operation in the semiconductor industry.

2.5 HAZOP Team

HAZOP is normally carried out by a team of people with roles as follows

Table 3: Team Members for HAZOP Study of Laboratory Operation in the Semiconductor Industry

Name	Position	Role
HAZOP Group Leader	Lab Manager	someone experienced in HAZOP and laboratory operation, to ensure that the method is followed carefully
Recorder	Lab Engineer 1	someone to ensure that problems are documented and recommendations passed on
Designer	Lab Engineer 2	someone to explain any design details or provide further information
User	Lab Operator	someone to consider it in use and question its operability, and the effect of deviations
Specialist	Lab Engineer 3	someone with relevant technical knowledge
Maintainer	Lab Technician	someone concerned with maintenance of the process.

3. HAZOP Management System

For the purpose of effective implementation of the HAZOP management policy and solid application of introduced general methodology, the laboratory HAZOP system should be set up. The system has coupled general management policy (operation specification and work instructions), environment control (temperature and humidity), operation safety (chemical storage, hazards assessment, chemical identification and MSDS), equipment management (PM, calibration, MSA), chemical control (data analysis and shelf life time control), training and certification (orientation, OJT, regular qualification). The whole system has covered multiple layers of analysis factors. The first layer will include manpower, machine, materials, method and environment.

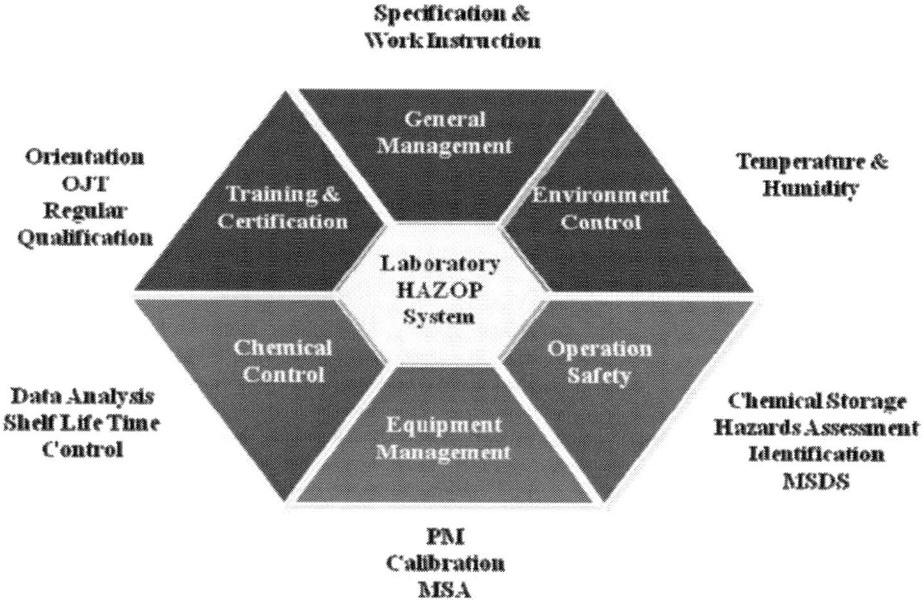

Figure 3: HAZOP Management System for Laboratory Operation in the Semiconductor Industry

These factors can be expanded with below fishbone diagrams so that the second layer of factors will be included into HAZOP matrix system: for example, sample collection and sample preparation will be considered in the material part; training, parallax, reproducibility, practice and ergonomics will be considered in the manpower part; samples, test methods, workmanship and standards will be considered in the method part; discriminations, repeatability, bias, linearity and calibration will be considered in the instrument part; vibration, lighting, temperature and humidity will be considered in the method part.

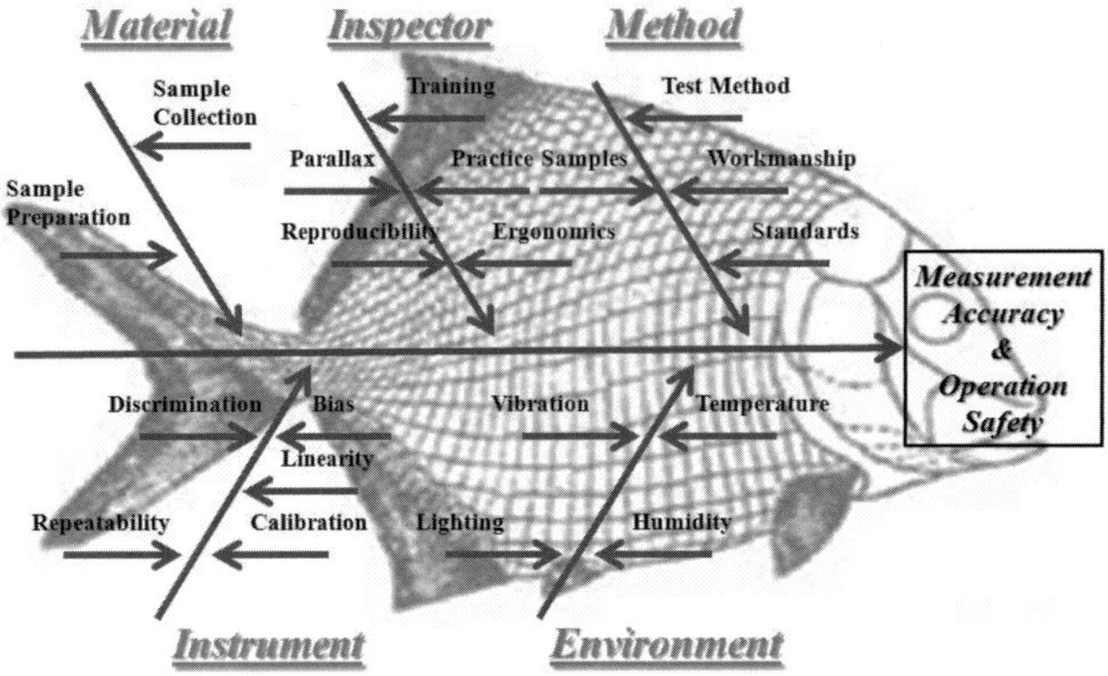

Figure 4: Fish Bone Diagram for HAZAP Study for Laboratory Operation in the Semiconductor Industry

Next, these practices can form a closed PDCA (Plan, Do, Check and Act) loop for continuous improvement of HAZOP activity. The close loop include four phases: plan, do, check and act. In the first phase of planning, the procedures and schedules to conduct the study should be identified and the input data should be collected. Then, the deviation should be identified by the application of key words and the causes and consequences should also be determined in the second phase. Next, the improvement actions need to be followed up. It is better to restudy the necessity of current HAZOP system. After the verification of the above items, the study report needs to be generated. Then the activity of HAZOP study should be evaluated to confirm whether the current scope, objective and responsibility need to be updated. Furthermore, the team should decide whether they need more team members to cover a larger scope.

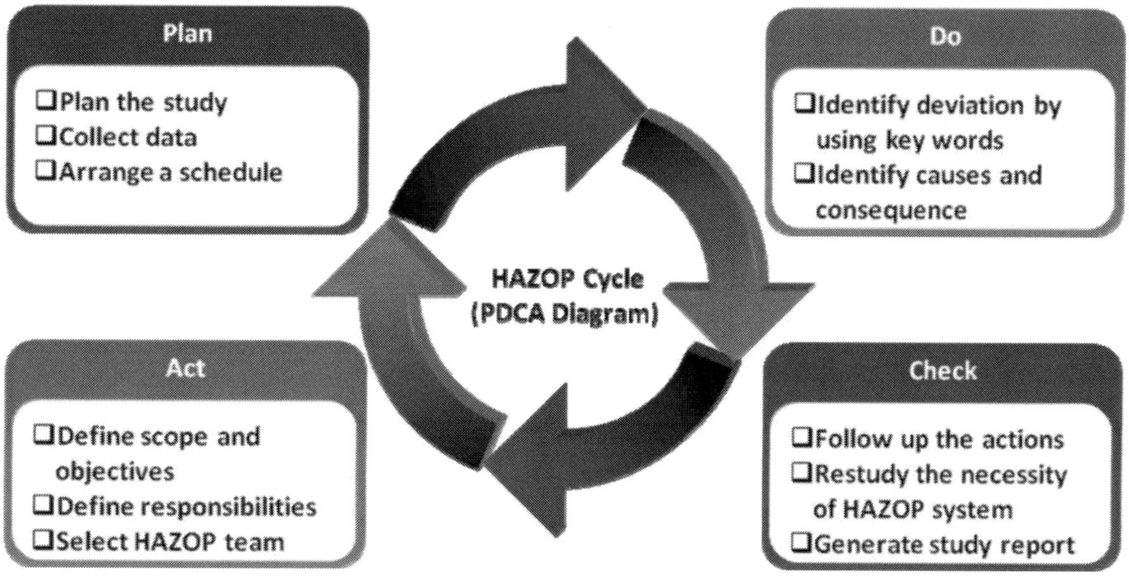

Figure 5: PDCA (Plan, Do, Check and Act) Diagram for HAZAOP Study for Laboratory Operation in the Semiconductor Industry

The analysis output can be regarded as the aid to safer laboratory design, operator training and real-time fault diagnosis. This will help to debottleneck and achieve operation safety and occupational health in the semiconductor industry.

4. Application Illustration

To demonstrate the efficacy of the developed systematic methodology, one case study has been conducted as the illustration of the methodology application in the real practice of plating bath monitoring of wafer bumping in the semiconductor industry. Wafer bumping is an advanced packaging technique that attaches solder spheres or gold bumps to the bond pads on a selected die. Gold bumping is generally used for Tape Automated Bond (TAB) assembly or Anisotropic Conductive Film (ACF) flip-chip assembly. Solder bumping is primarily used for reflow flip-chip assembly which unites the components face down, directly with the substrate or board through conductive bumps on the bond pads. These bumps deliver the electrical, mechanical and thermal interconnection which provides direct contact between the chip package and the device. Another process is called stud-bumping. Unlike gold and solder bumping, this process can be done at the die level. Wafer bumping also known as flip chip has traditionally been used for high-end niche applications. However in the last few years it has gone mainstream to include a wider range of consumer applications geared towards both the performance driven market and the form factor driven market.

Wafer bumping provides the electrical and mechanical connection between a die and substrate, which is one of the most critical elements of any flip chip package structure. Predominantly lead-tin and lead-free solders at the present time, these connections - or bumps - must exhibit superior adhesion to the die, minimal resistance, and result in high assembly yields. Solder bumps are formed by using either thin film metal deposition or ball loading techniques. Currently the advanced semiconductor assembly and test company such as Amkor offers state-of-the-art capability in electroplated solder technologies including the electroplating capability to support fine pitch bumping and copper pillar structures.

Wafer bumping utilizes thin film deposition, photolithography and plating techniques to place solder or gold onto IC (Integrated Circuit) lead pads. The wafer solder bump process is to produce the solder bumping of the wafer pad. Then it is going to melt the bumping by heat in assembly. This technology can substantially reduce IC size and have the advantages such as high density, low sensitivity, low cost and effective heat dissipation. This bumping process is suitable for flip chip assembly for LCD, memories, microprocessors, and microwave RF ICs applications. The process includes sputter UBM (under Bump Metallurgy), Photolithography, Plating and Etching etc., This bumping is suitable to apply on TAB, Flip Chip. The detailed process flow chart has been illustrated in below flow chart.

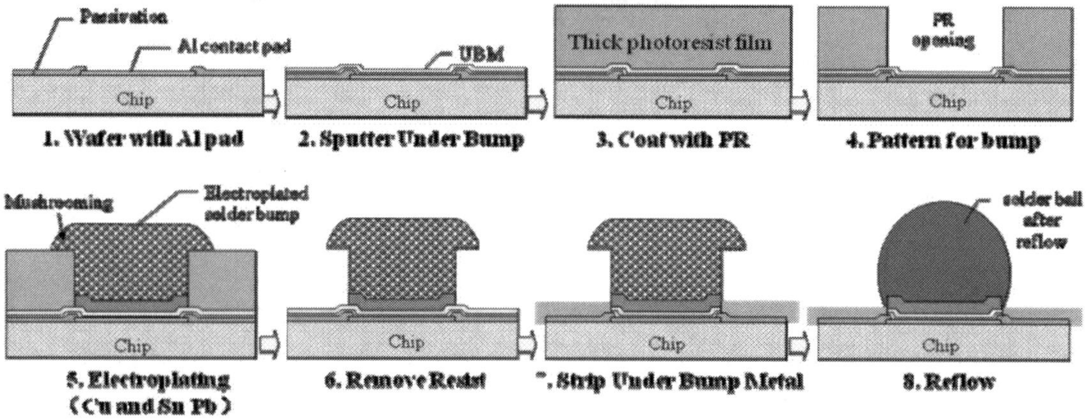

Figure 6: Process Flow Chart for Wafer Bumping Process

One of the key test item for wafer bumping operation is the electroplating bath monitoring and composition control of solder materials and additives. Plating solutions must be routinely analyzed in order to maintain the recommended bath formulation and to preempt the occurrence of problems related to improper levels of bath constituents. Contaminant levels in the solutions must also be monitored. Manufacturers of plating systems establish optimum specifications to ensure maximum solution efficiency and uniformity of deposits. The various factors that cause the concentrations of bath constituents to deviate from their optimum values are as follows: 1. drag-out; 2. solution evaporation; 3. chemical decomposition; and 4. unequal anode and cathode efficiencies. Regular analysis of plating solution becomes the responsibility of process engineering and quality control group.

The techniques employed for the quantitative analysis of plating solutions are classified as volumetric (titrimetric), gravimetric, and instrumental. Volumetric and gravimetric methods are also known as "wet" methods. The analyst must select the method that is best suited and most cost effective for a particular application. The wet methods outlined here are simple, accurate, and rapid enough for practically all plating process control. They require only the common analytical equipment found in the laboratory, and the instructions are sufficiently detailed for an average technician to follow without any difficulty. The determination of small amounts of

impurities and uncommon metals should be referred to a competent laboratory, as a high degree of skill and chemical knowledge are required for the determination of these constituents. Instrumental methods differ from wet methods in that they measure a physical property related to the composition of a substance, whereas wet methods rely on chemical reactions. The selection of an instrument for the analysis of plating solutions is a difficult task. Analysts must decide if the cost is justified and if the analytical instrument is capable of analyzing for the required substances with a high degree of accuracy and precision. Instruments coupled to computers can automatically sample, analyze, and record results. Mathematical errors are minimized and sample measurements are more reproducible than with wet methods. Instrumental methods are also extremely rapid when compared with wet methods. Analytical instruments frequently used in the analysis of plating solutions can be categorized as spectroscopic, photometric, chromatographic, and electro-analytical. Spectroscopic methods (flame photometry, emission spectrometry, X-ray fluorescence, mass spectrometry, and inductively coupled plasma) are based on the emission of light. Photometric methods (spectrophotometry, colorimetry, and atomic absorption) are based on the absorption of light. Chromatographic methods (ion chromatography) involve the separation of substances for subsequent identification. Electro-analytical methods (potentiometry, conductometry, polarography, amperometry, and electrogravimetry) involve an electric current in the course of the analysis. The instrumental methods, comprehensively reviewed below, are most applicable to plating environments.
Here, the manual titration procedures to determine copper concentration in the sample plating bath is used as illustration because the manual titration will have more potential safety concerns compared with the instrument analysis methods. Below is the procedure of the test.

PROCEDURE:

1. Preparation of EDTA (Ethylenediaminetetraacetic Acid) (0.01M). Dry EDTA for 2 hours at 80°C. GMW = 372.2., or check the reagent bottle. Dissolve needed grams in a 500 ml vol. flask. Store in a conditioned polyethylene container. EDTA solutions attack glass and deteriorate after 1 or 2 days.
2. Preparation of known copper solution. Place 0.2 g to 0.25 g of pure Cu. wire (or shot) (weighed on an analytical balance) in a 100 ml beaker. Add 15 ml conc. HNO3 and 1 ml conc. H2SO4. Place a watch glass over the beaker. Boil until dense white fumes of SO3 cease. Transfer carefully to a 200 ml vol. flask. Remember to rinse the watch glass. Dilute to the mark.
3. Preparation of unknown copper solution. Weigh 0.5 to 0.6 grams of unknown into a 200 ml vol. flask. Add 30 ml of 1:5 nitric acid. Heat until all particles are dissolved. Dilute to the mark.
4. Titration with EDTA - murexide indicator. Do this step for 4 samples of known and 4 samples of unknown.

KNOWN: Place 10 ml of Cu known solution (step 2) in 250 ml Erlenmeyer flask. Add pH buffer drop-wise until the solution turns completely blue and the precipitate first formed dissolves. Add about 100 ml. Add more buffer if solution becomes cloudy. Add 1 tablet of murexide indicator and titrate with EDTA until color changes from yellow (green) to deep violet. If the volume of EDTA is less than 10 ml, repeat using 25 ml liquid of Soln. 2.

UNKNOWN: Place 25 ml of Cu unknown solution (step 3) in 250 ml Erlenmeyer flask. Add pH buffer drop-wise until the solution turns completely blue and the precipitate first formed dissolves. Add 1 tablet of murexide indicator and titrate with EDTA until color changes from yellow (green) to deep violet. If the vol. of EDTA solution is less than 10 ml. use a larger liquid of Soln. 3.
Remarks: The copper-murexide complex is pure yellow, but the color at the starting point of the titration is greenish depending on the amount of copper present. The green is the mixture of yellow (indicator complex) and blue (copper-tetramine complex). As the titration proceeds the tetramine complex is transformed to the less intensely colored EDTA complex and the yellow of the indicator complex becomes permanent.
CALCULATION: Calculate the molarity of the EDTA. Calculate percent copper in the sample using the value of the Molarity of the EDTA found. Include the spread sheet for the calculation.

Based on the available procedures to determine the copper concentration in the sample solution of plating bath, HAZOP factors can be expanded so that the second layer of factors will be included into HAZOP matrix system: for example, sample collection and sample preparation will be considered in the material part; training, parallax, reproducibility, practice and ergonomics will be considered in the manpower part; samples, test methods, workmanship and standards will be considered in the method part; discriminations, repeatability, bias, linearity and calibration will be considered in the instrument part; lighting, temperature and humidity will be considered in the method part. These factors are listed as the column. Then, the HAZOP guide words will be listed as the row. The related information of the deviation stored in LOD(Laboratory Operation Database) and HPD(HAZOP Project Database) can be selected into the blank spaces from the drop-down list box in the above expanded table. Next, the data record of regular monitor results will be proceeded into data analysis to validate the impact of the above deviation . The data analysis result will couple with the expanded HAZOP table to quantitatively identify the potential deviation of operation safety and test accuracy.

5. Conclusion

The laboratory operation in the semiconductor industry will have more possibilities of manual handling for chemical analysis, reliability test, material characterization and failure analysis. Therefore higher attention should be paid to the laboratory operations. As illustrated in this paper, the methodology of HAZOP study procedure can be applied to identify the potential deviations in both experiment accuracy and operation hazards. Multiple layers of factor consideration by the application of 4M1E (Man, Machine, Material, Method and Environment) will help the brainstorm to give the aid to safer laboratory design, operator training and real-time fault diagnosis. This will help to debottleneck and achieve operation safety and occupational health in the laboratory operation semiconductor industry.

6. Reference

[1] Chepesiuk, R., "Where the Chips Fall: Environmental Health in the Semiconductor Industry." Environmental Health Perspectives, Volume 107, pp.452-457, 1999

[2] Swann, C. D., & Preston, M. L., "Twenty-five years of HAZOPs", Journal of Loss Prevention in the Process Industries, Volume 8, No 6, pp.349-353, 1995

[3] Amkor Technology Wafer Bumping Service Introduction, http://www.amkor.com /go/turnkey-services/wafer-bumping

(accessed Jan. 5, 2013)

[4] Nolan, D.P., "Application of HAZOP and What-If Safety Reviews to the Petroleum, Petrochemical and Chemical Industries". William Andrew Publishing/Noyes, 1994

[5] Lawley, H. G., "Operability studies and hazard analysis", Chemical Engineering Progress, Volume 70, No 4, pp.45, AIChE,1974

[6] Chemical Industries Association., "A Guide to Hazard and Operability Studies", 1977

[7] IHS iSuppli Research., "Semiconductor Industry Revenue to Ensure Slow Growth in 2012",http://www.isuppli.com/Semiconductor-Value-Chain/News/Pages/Semiconductor-Industry-Revenue-to-Endure-Slow-Growth-in-2012.aspx

(accessed Jan. 5, 2013)

[8] U.S. Bureau of Labor Statistics (BLS) Data Report,2011

[9] Semiconductor Industry Association, http://www.sia-online.org/

(accessed Jan. 5, 2013)

[10] Bolmen, Richard A., "Semiconductor Safety Handbook: Safety and Health in the Semiconductor Industry", Noyes Publications, 1998

9TH GLOBAL CONGRESS ON PROCESS SAFETY

Best Practice Management of Change & PSSR Process

Michael Bearrow
Optimized Systems and Solutions Inc.
8303 McHard Road, Suite 200, Houston, TX 77053
Michael.E.Bearrow@o-sys.com

Prepared for Presentation at
American Institute of Chemical Engineers
2013 Spring Meeting
9th Global Congress on Process Safety
San Antonio, Texas
April 28 – May 1, 2013

UNPUBLISHED

Best Practice Management of Change & PSSR Process

Michael Bearrow
Optimized Systems and Solutions Inc.
8303 McHard Road, Suite 200, Houston, TX 77053
Michael.E.Bearrow@o-sys.com

Keywords: MOC, process safety code, recommendation,

Management of Change (MOC) and Pre Start-up Safety Review (PSSR) are still the most challenging elements of OSHA's Process Safety Management (PSM) standard, the EPA's Risk Management Program (RMP) rule, and now, the US Department of Interior's Bureau of Safety and Environmental Enforcement's Safety and Environmental Management Systems (SEMS). Even though the PSM standard has been around since 1992 and the industry has been managing change for several decades, we still can get it wrong, sometimes with disastrous results. The mindful or diligent efforts of many working in concert are necessary to ensure that change is identified, analyzed and executed in a quality way. Just when we get it right on paper and get the workforce up-to-speed, we have employee turnover, neglect and sometimes regulatory change. New actors and a constantly changing script make it hard to manage change efficiently and effectively. This paper discusses how the chemical process industry has defined the MOC and PSSR best management practices and how they should be automated.

Management of Change (MOC) and Pre Start-up Safety Review (PSSR) are still the most challenging elements of OSHA's Process Safety Management (PSM) standard, the EPA's Risk Management Program (RMP) rule, and now, the US Department of Interior's Bureau of Safety and Environmental Enforcement's Safety and Environmental Management Systems (SEMS). Even though the PSM standard has been around since 1992 and the industry has been managing change for several decades, we still can get it wrong, sometimes with disastrous results. The mindful or diligent efforts of many working in concert are necessary to ensure that change is identified, analyzed and executed in a quality way. Just when we get it right on paper and get the workforce up-to-speed, we have employee turnover, neglect and sometimes regulatory change. New actors and a constantly changing script make it hard to manage change efficiently and effectively.

Many high-profile companies have addressed this challenge by creating and automating standard processes across their manufacturing or production facilities. MOC and PSSR are two of these major processes that have been successfully automated. Technology has been used to automate each step of the MOC workflow, enabling the collection of change ideas, formalizing the analysis of the ideas, and documenting the approval and execution of the change. Reports and dashboards can provide visibility of the health or current state of your MOC and PSSR processes. The use of software can enable the standardization of best practices. This paper discusses how the chemical process industry has defined the MOC and PSSR best management practices and how they should be automated.

What is a best practice?

"Best Practice" is an overused buzzword, like so many, and they are hard to define. Similar terms include Industry Standard, Best of Breed, Lesson Learned, Core competency, and Customer-centric, Downsizing, Drinking the Kool-Aid, Eating your own dogfood, Granular, Herding cats, Holistic, Low Hanging Fruit, Mindshare, Mission Critical, Management Visibility, Pain point, Return on Investment, Seamless, Touch point, Value-add and Visibility. Fortunately, there is a working definition of a best practice.

A **best practice** is a method, process or technique that consistently shows results superior to those achieved by other means. It is hard to prove that you have developed a standard process--like MOC--that enables your company, facility or organization to achieve repeatable superior performance, but it has been done.

Some other terms synonymous with best practices are good operating practice, good agricultural practices, good manufacturing practice, good laboratory practice, good clinical practice, good distribution practice, and **R**ecognized **A**nd **G**enerally **A**ccepted **G**ood **E**ngineering **P**ractices (REGAGEP). In the chemical process industry, the PSM standard relies on REGAGEP to ensure facilities adopt best practices. REGAGEP is specified in federal statutes from OSHA and EPA, and is also mentioned in the American Chemistry Council (ACC) Responsible Care® Process Safety Code.

How do you identify a Best Practice?

Everyone would like to be known as the best practice company. Even if you did not create the best practice, identifying them and using them can achieve the same results. If you can tick the box next to these five requirements, then you may have a best practice:

1. Gives a company or facility a tangible competitive advantage.
2. Takes advantage of technology.
3. Improves EHS performance, product quality, and also lowers cost.
4. Gives management more visibility, control and influence over outcomes.
5. Must be a working system; i.e. not a theoretical best practice

The benefits of having such a best practice might separate your company from your competition or keep you from costly business disruption, fines and/or law suits. Many of the best practices have common themes, including: ensuring consistency; standardization; simplifying processes; zero redundancies; leveraging technology and web-based software solutions; and single source for all data (one source of truth). It may be difficult for you to prove you have adopted/created a best practice. However, it is easy to identify a best practice when a confluence of large and small companies adopt the same technique or process and are successful.

In the context of this paper, we are focused on best practice MOC and PSSR automation. Now that we know what a best practice is, let's define change and discuss how to manage it.

What is Change & Why Do We Manage it?

"Change" means to transform or cause to be different. In the chemical process industry, change is omnipresent and required to stay competitive. William Edwards Deming, a famous American statistician, professor, author, lecturer and consultant, once said, "It is not necessary to change. Survival is not mandatory." In order to survive, a company must learn, change and continuously improve to survive. Successful change requires you to learn from your experience and the experience of your peers and make changes that will improve your competitiveness.

Change is a normal part of business, but uncontrolled or poorly managed change has resulted in some of the worst disasters in the chemical process, oil and gas and refining businesses. According to Ian Sutton, a noted PSM expert, *"The root cause of all accidents is uncontrolled change. Leaving aside sabotage and other malicious acts, all industrial facilities are designed and operated to be safe, clean and profitable - yet incidents continue to occur. In every case, the fundamental cause of the incident is that someone, somewhere lost control of the operation, i.e., they allowed operating conditions to deviate beyond their safe range. Hence, the proper management of change is the foundation of all safety and accident prevention programs; an effective Management of Change (MOC) creates an atmosphere of "no surprises".*

Whether it is Bhopal, Piper Alpha or Deepwater Horizon, time after time the difficulty of managing change confounds some of the most proven processes of the most successful companies in the world. As a consequence of these disasters and hundreds more, we have rules, standards, regulations and laws to help ensure that companies manage change better. Guidance documents on managing change include the following sources:

- *US EPA's Risk Management Programs for Chemical Accident Prevention (40 CFR PART 68) or EPA RMP*
- *US OSHA's Process Safety Management of Highly Hazardous Chemicals standard (29 CFR OSHA 1910.119) or PSM,*
- *US Department of Interior's Bureau or Safety and Environmental Enforcement (BSSE) - 30 CFR Part 250 Subpart S - Safety and Environmental Management Systems (SEMS),*
- *OHSAS 18001, 2007 revision - ANSI/AIHA Z10-2005.*
- *European Union Law Seveso II Directive (see Annex III of EU Council Directive 96/82/EC).*
- *API-691 - Risk Based Machinery Management (RBMM) (in development)*

What kinds of change are there?

All ideas for improvement require change. No one ever suggests that we should do exactly what we are doing today, tomorrow. This fact is lost on a segment of the chemical process industry who do not think of recommendations as changes. Recommendations should be treated as "in-kind" changes because they are. They do not have to follow a formal and rigorous MOC and PSSR process, but they are certainly recommendation for change.

All change can be separated into three categories: corrective actions; preventive actions; and continuous improvements. Corrective actions are meant to fix something that is broken, or at least to improve efficiency and functionality. These ideas for change originate from accident investigations, incidents and negative operating experiences. Preventive actions can also come from incidents, but are more commonly created from near misses, safety observations, audits, hazard reviews, inspections, emergency response critiques, and/or general surveillance or observations. These ideas for change help ensure that we address upstream issues or causes before a negative outcome (incident, fine, loss) results. The final kind of change is continuous improvement. These are generally ideas for increasing production, reducing manufacturing costs or increasing product quality.

No matter what kinds of changes you have, they all should be managed. No matter how good an idea is, it must be analyzed for impact (positive and negative) and executed expertly. Ideas can also be rejected or denied if the risk or cost of implementation exceeds the benefits or value of the change. If the change affects the design and/or operation of a facility, it is likely going to require formal change management or MOC. The diagram below illustrates three kinds of change in the idea universe.

Figure 1.0

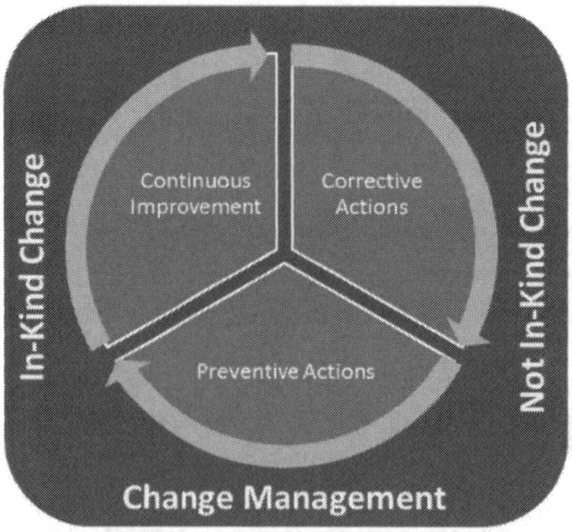

OSHA'S Two Categories of Change

The common way to refer to an idea that requires formal change management as defined by OSHA PSM or EPA RMP is an MOC or Management of Change (Not in Kind). MOCs are managed for changes to equipment, materials, technology, personnel, processes, procedures or anything that deviates from the original design, processes, operating procedures, or maintenance program.

In the chemical process industry, we refer to two categories of change. These are "Replacement In-Kind" and "Replacement Not In-Kind". These odd-sounding categories can have ambiguous meanings, but are explained by Ian Sutton, a noted process management expert, as follows:

"If an equipment item is to be replaced with one that is functionally identical, i.e., if the new item is built to the same specification as the old one, then the change is "in-kind". Otherwise it is "not-in-kind", and the MOC process has to be followed before the change can be implemented."

... The most challenging aspect of managing change is identifying that the proposed modification is in fact a change.

1. *Management of Change is needed if the change is Not-In-Kind.*
2. *A Not-In-Kind change is one where Management of Change is needed.*

In other words, the terms "Management of Change" and "Not-in-Kind" tend to be defined in terms of one another.

The second difficulty to do with the In-Kind/Not-In-Kind decision, noted above, is that all changes are, when analyzed deeply enough, not-in-kind."

What is the role of PSSR in the MOC process?

Pre-Startup Safety Review, or PSSR, is not actually part of the MOC process per se. It is intended to be a redundant check or review to ensure that an MOC has been implemented in a quality and safe manner. The PSM standard states:

The employer shall perform a pre-startup safety review for new facilities and for modified facilities when the modification is significant enough to require a change in the process safety information. Process safety information includes all information about the process. That covers about everything. The pre-startup safety review shall confirm that prior to the introduction of highly hazardous chemicals to a process:

- *Construction and equipment is in accordance with design specifications;*
- *Safety, operating, maintenance, and emergency procedures are in place and are adequate;*

- *For new facilities, a process hazard analysis has been performed and recommendations have been resolved or implemented before startup; and modified facilities meet the requirements contained in management of change, paragraph (l);*
- *Training of each employee involved in operating a process has been completed.*

When is Management of Change (MOC) Required?

OSHA requires the employer to establish and implement written procedures to manage changes (except for "replacements in kind") to process chemicals, technology, equipment, and procedures, as well as changes to facilities that affect a covered process. The MOC procedures shall assure that the following considerations are addressed prior to any change:

- The technical basis for the proposed change;
- Impact of change on safety and health;
- Modifications to operating procedures;
- Necessary time period for the change;
- Authorization requirements for the proposed change;
- Employees involved in operating a process and maintenance and contract employees whose job tasks will be affected by a change in the process shall be informed of, and trained in, the change prior to start-up of the process or affected part of the process;
- Updates to process safety information and operating procedures.

The industry has been practicing process safety for over 20 years. The definition of MOC and PSSR are well understood. Some companies have developed great procedures for addressing both of these elements in an integrated fashion using Internet-based web applications and powerful database engines. These software systems simplify the complex MOC process and enable standardization, ensuring a higher level rigor which is difficult to impose with paper-based execution. While the best practices for MOC and PSSR work practices continue to evolve, the following sections reveal what is currently accepted as best practice.

Best Practice MOC Part 1 – Collecting Change Ideas from the Idea Generators

As established above, continuous improvement is impossible without implementing change ideas that can come from anywhere. The problem is how can organizations empower their workforce to initiate ideas, capture and evaluate them then act on the ones that will improve the business? Can you imagine how successful your business would be if you could leverage all of the grey matter at your disposal to solve production problems?

In order to efficiently and effectively capture these ideas, we need to leverage technology. Using Internet–based software can make it easy for ideas to be collected at the source. The easier we make it for staff members to pass on their ideas, the more of these ideas we will receive with time to act on them. One best practice that has surfaced to address this need is to design an electronic suggestion or idea box. Web-based quick entry screens can be accessed using a browser on any company computer to easily suggest a change or improvement. Once these ideas are entered, they are automatically communicated to idea reviewers (supervisors) for vetting, categorization (in-kind or not in-kind) and prioritization. No matter what happens to these ideas, the reviewer must provide feedback to the idea initiators on what happened to their ideas. If an idea results in tangible savings or competitive advantage, a tangible reward will serve to prime the idea generator for the future.

You can visualize the best practice MOC process using a series of funnels (below). Changes begin as ideas and they can be entered by anyone in the company into a single quick entry screen and collected in the main funnel. The supervisor is automatically notified that a new idea has been entered for their unit or area. Each idea is meticulously reviewed and categorized as a recommendation or a MOC. Using electronic submission, this centralized collection funneling system is infinitely more efficient and effective than word-of-mouth, a paper system and/or multiple systems that collect and manage ideas. One common system for collecting, vetting, prioritizing and categorizing ideas is a best practice.

Figure 2.0

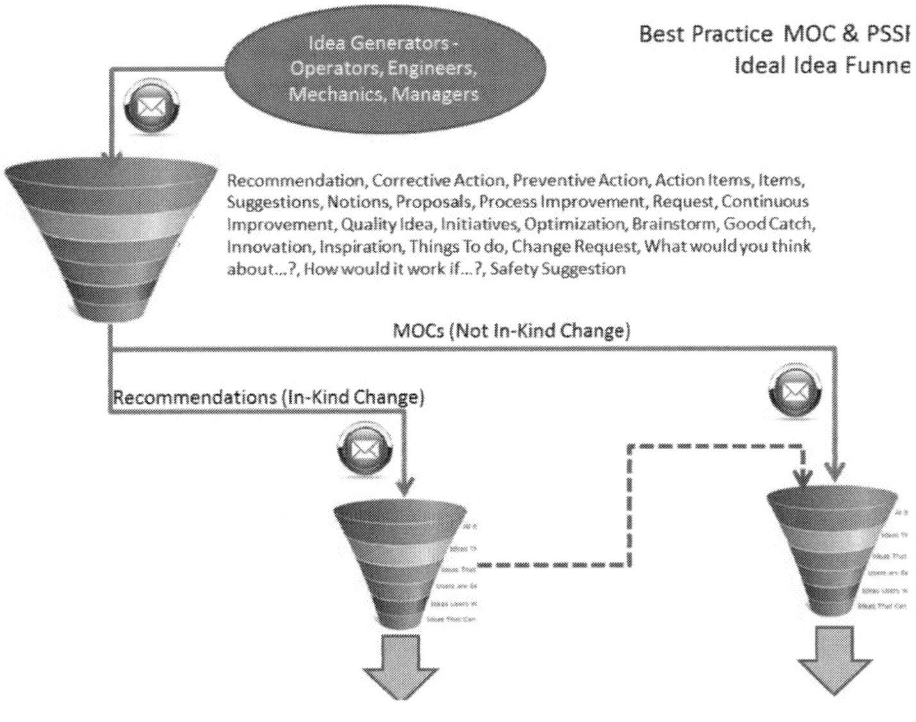

Best Practice MOC Part 2 – Managing the Recommendations (In-Kind Change)

Recommendations can come from audits, incidents, risk assessments, process hazard analysis, emergency response critiques, inspections, reviews, and surveillances. These recommendations are generally ideas for continuous improvement aimed at improving safety, environmental performance, quality or efficiency. Automating and standardizing the recommendation management process makes it repeatable, measurable and more readily understood. Utilizing software that illustrates the correct process and visually progresses users through the workflow is key to the acceptance and the eventual success of the MOC program.

Once the idea has been reviewed, assigned and categorized as a **recommendation**, or in-kind change, a simple workflow can be leveraged to ensure the idea is addressed in an efficient manner. A simple database can be used to document the steps in this process, but a best practice is to use software that illustrates the work process on screen as it is executed. The use of a visual workflow to navigate from step to step, showing progress with check marks or color-coding and enabling steps only when precursors (predecessor steps) are satisfied is a best practice. In addition, email notifications should be used to alert persons assigned to manage, approve and execute actions along the workflow. The more proactive the communication and visual the workflow is, the more intuitive it is to the end-user. The workflow presented below is used for hundreds of thousands of recommendations per year. The entire work process is automated from beginning to end.

Figure 3.0

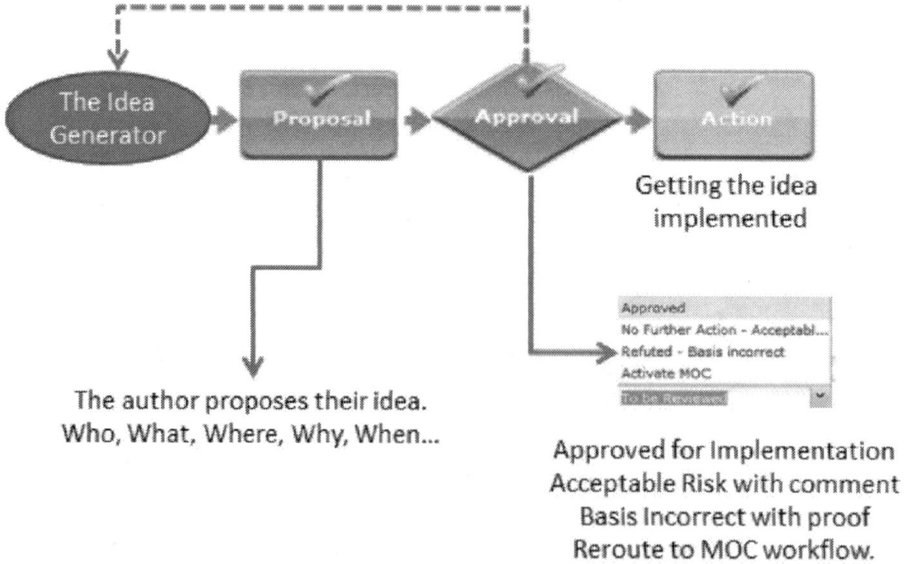

The workflow begins in the proposal step or in the "idea generator's" mind. The recommendation proposal must contain a title, a description (what needs to be done), a justification (why it is important) and a deadline (when it needs to be done) for getting this idea accomplished and a responsible person (RP). The reviewers of the proposal will ensure the accuracy and the resources are available to address the idea. The next step is to accept the idea into the system and automatically notify the RP. The acceptance is illustrated by a check or tick mark in the proposal icon.

After a review of the recommendation, the RP moves to the Approval step. It is in this step that they must decide whether to approve the idea, refute the idea, close the idea as an acceptable risk or decide it needs to be treated as an MOC. Approval of the idea will enable the creation of action items. Refuting a recommendation or addressing it as an acceptable risk should require proof or explanation. These comments must be documented for future reporting and analysis and automatically communicated to the idea generator.

Best Practice MOC Part 3 – Managing MOCs (Not In-Kind Change)

If the idea needs to be managed as a "Not in-Kind" change or MOC, a more complex workflow is required. Similar to the "In-Kind" change, it still requires a proposal, acceptance and assignment to a RP, but additionally will need a formal hazard review and planning. In almost all cases, a MOC also requires the pre-startup safety review (PSSR) step, a Startup Approval step and some post startup steps. Because of its complexity and the need to enforce a rigorous process, a "live" workflow showing what steps are necessary, which ones have been accomplished and which ones are left to do is essential. The workflow below is considered the best practice MOC and PSSR. There are no steps left out and no superfluous steps either.

Figure 4.0

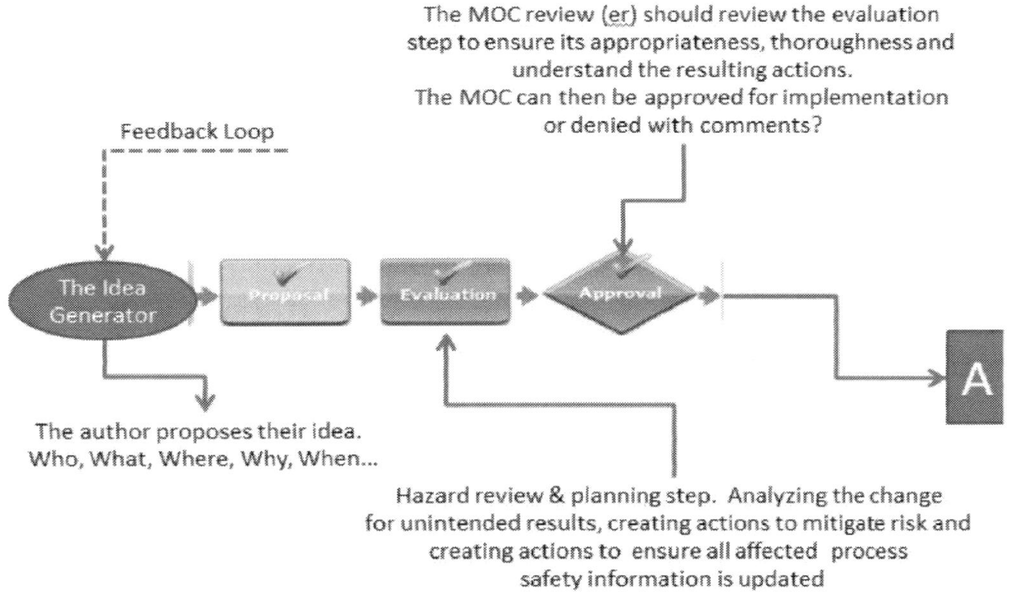

After the acceptance of the idea into the system, the RP receives a notification that they have been assigned a MOC to manage. The next step in the workflow is to "Evaluate" the MOC. This evaluation process is meant to ensure that the hazards of the change are reviewed, understood and mitigated, or at least a plan for mitigation is created. It is also meant to reveal the process safety information that will be affected and the training that will be necessary. Of vital importance is the creation of an action plan to make sure hazards are addressed and that appropriate training is done and all process safety information is completed. Just identifying the need falls very short of hitting the target. In fact, if you identify the need for action and then do not follow through, you have created the proverbial "smoking gun" and a red flag for auditors.

A smart checklist of questions should be presented to the RP in this step which are based on the category of the change. This eliminates the need to answer questions that have nothing to do with the change. Automation gives you the ability to tailor the evaluation process so that it is more effective and efficient to use. Asking evaluation questions unrelated to the change is very inefficient and can also make finding the right questions difficult. Even more automation can be leveraged to pre-assign reviewers to questions that only they should be answering and configuring automatic/predictable actions with associated questions. Here in this step we can make the comments and closure of all questions mandatory for moving to the next step. Until all questions are closed, we cannot move to the next step, the Approval step.

The Approval (to implement) step can be done by a person or groups of people. Their job is to look back at the MOC proposal, review the evaluation that was undertaken and judge whether it was done effectively by the right people. They should also take notice of the mitigating actions that resulted from that review. If more evaluation is required, the reviewer can loop back in the process and create more evaluation questions to satisfy. They should also ensure that actions have been tagged appropriately as pre- or post-startup. If the resulting actions or time have otherwise made this MOC a bad idea to execute, the MOC can be denied at this point with closure comments. Those comments are sent to the idea generator and should be documented in the system. If they are satisfied, then approval can be selected and we can begin implementing the MOC.

The MOC Action Step

Now the real work begins--updating procedures, marking up drawings, making sure all affected personnel are trained, writing work orders, getting an environmental permit, and all of the other actions required for completion to do a quality job. Once all actions (pre-startup) are completed, then you get a check mark over the action icon and we move on to the PSSR step.

Figure 5.0

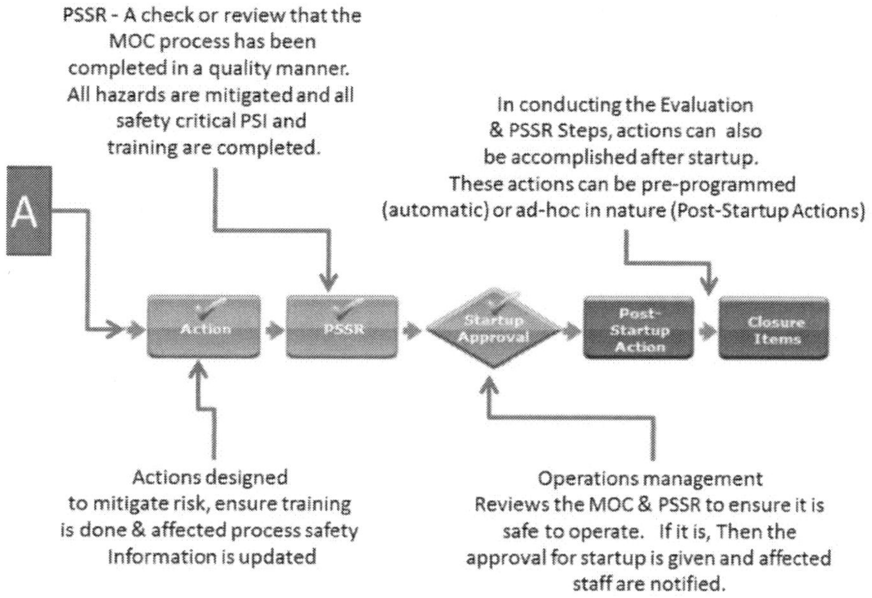

The PSSR Step

As previously stated, the PSSR is a redundant check to ensure the MOC has been properly executed. Like the Evaluation step, once we have chosen a category of change, automation can be used to ensure that the appropriate PSSR checklists/questions are presented to the RP. This increases the efficiency and effectiveness of this step as well. Once the PSSR is completed, we receive our check mark in that workflow step, and an automatic email is sent to the appropriate manager asking for permission to Startup (Startup Approval).

Startup approval is the formal acceptance of the change and the acceptance of the risk associated with it by operations or production. The person who completes this step should review all of the previous work steps in the workflow as well as physically review the work in the field.

The last two steps in the MOC workflow are closure items and post-startup actions. Although these are not referred to in any of the governing regulations, there will be non-safety related activities that are not required to be completed by startup and still need to be managed. This best practice will ensure these actions are accomplished and documented. Past Due email notifications are sent appropriately if the MOC becomes past due.

There are some distinct advantages to adopting the best practice MOC and PSRRs workflows discussed herein. As the best practice evolves, so must the users of the technology. Refresher training, upgrades and constant tuning and tweaking of the workflow are all part of continuous improvement of an MOC and PSSR program. This minimal investment is a small price to pay to avoid poorly managed changes or the opportunity cost of not collecting ideas from your idea generators.

How does Automation help?

Automation of the MOC and PSSR process helps in several ways. First, we can collect ideas at the source of the risk, whether it is the plant or on the platform. This is where the risk is and where continuous improvement is accomplished. Furthermore, it is also where most of the practical/achievable ideas for continuous improvement come from. The best practices must be useful and used where the money is made. Technology makes that possible.

Today's technology is perfectly suited for collecting, analyzing and sharing ideas. Accomplishing these actions without automation is nearly impossible. The days of paper-based MOCs are almost non-existent because automation makes these tasks more efficient and effective as well as less likely to be overlooked.

Reporting metrics and automatic alerts when MOC and PSSR actions are assigned, closed and past due ensure diligence and accountability. Technology can spread the news quickly to those who can create and execute a plan of action and do something with new ideas. Morse code, paper reports and snail mail do not get the job done like email communication and automatic reminders.

A central corrective and preventive action tracking software can help ensure that identified issues are dealt with in a safe and timely manner. Idea or recommendation management and action tracking can be accomplished with one database engine worldwide. That means one system to learn, maintain and continuously improve. It also means you have one global change management system to get things done, communicated and for generating reports. Managers can get the visibility they need to measure performance without bothering the people making the money.

An automated best practice MOC and PSSR system helps ensure that once you get it "right", you keep it "right". It also makes sure that change is appropriately identified, evaluated, approved and executed. It also makes poor change management easy to identify and correct through management surveillance.

Automation is not the goal; best practice is. Like reporting and management visibility and lots of other byproducts of this best practice, the goal is managing change efficiently and effectively.

Words of Caution

A cardinal sin of MOC & PSSR process automation is using a finance or maintenance technology that you already have to automate your best practice, because you already have it. This will sub-optimize a best practice faster than anything you can do, and it will likely defeat the purpose of the automation. Remember that 95% of your users are engineers, operators and mangers--not accountants, IT "geeks" or maintenance planners.

And finally, you can develop your own best practice MOC and PSSR, but to be successful as best practice software, you must have a full-time MOC best practice expert leading the development. Otherwise there will be several months of trial and error. Make no mistake; software development is difficult enough with unlimited resources, expert developers and patient end users. You cannot automate a business process you do not understand. Remember that managing change is a complicated management system for operations or production personnel. You cannot write a requirements document so thoroughly and clearly that a software development team can automate your MOC best practice workflow without constant oversight from the MOC experts.

9TH GLOBAL CONGRESS ON PROCESS SAFETY

Unifying Consequence Models for Natural Gas Pipelines in Petrobras Transpetro

Ana Paula Rodrigues Bahiense
PETROBRAS TRANSPORTE - TRANSPETRO
Av Presidente Vargas 328
Centro CEP 20091060
Natural Gas Division
Rio de Janeiro, Brazil
anapaulabahiense@petrobras.com.br

Maria Beatriz Winter Caputo
PETROBRAS TRANSPORTE - TRANSPETRO
Av Presidente Vargas 328
Centro CEP 20091060
Natural Gas Division
Rio de Janeiro, Brazil
Email: beatriz.caputo@petrobras.com.br

Prepared for Presentation at
American Institute of Chemical Engineers
2013 Spring Meeting
9th Global Congress on Process Safety
San Antonio, Texas
April 28 – May 1, 2013

UNPUBLISHED

Unifying Consequence Models for Natural Gas Pipelines in Petrobras Transpetro

Ana Paula Rodrigues Bahiense
PETROBRAS TRANSPORTE - TRANSPETRO
Av Presidente Vargas 328
Centro CEP 20091060
Natural Gas Division
Rio de Janeiro, Brazil
anapaulabahiense@petrobras.com.br

Maria Beatriz Winter Caputo
PETROBRAS TRANSPORTE - TRANSPETRO
Av Presidente Vargas 328
Centro CEP 20091060
Natural Gas Division
Rio de Janeiro, Brazil
Email: beatriz.caputo@petrobras.com.br

Keywords: pipeline, natural gas, consequence model

Abstract

Petrobras Transpetro has been contracting quantitative risk analysis for natural gas pipelines and their associated installations as a legal and internal requirement. These studies have been carried out by a series of consultants that estimate the magnitude of possible damage using different software available in the market. This fact led to a variety of results based on the premises adopted by consultants (most of them depending on the risk analyst) and differences regarding the peculiarities of mathematical models.

Petrobras Transpetro team, mainly field personnel, must use these data to carry out preventive actions inside vulnerable areas, and also to be aware of these ranges in case of a real emergency, thus providing evacuation and escaping actions, as necessary.

Based on this scenario, Petrobras Transpetro's Natural Gas Division had invested in purchasing a consequence-analysis software in order to model the effects of high-consequence loss events as an effort to unify consequence models for natural gas transportation pipelines at Petrobras Transpetro.

As per the above description, this paper aims to present results and benefits of the practical usage of unified consequence modeling information in risk management at Petrobras Transpetro's Natural Gas Division.

1. Introduction

Transpetro, a Petrobras wholly owned subsidiary, is the Brazilian leading logistics company dealing with transportation and storage of oil and byproducts, biofuels and natural gas. It operates 48 terminals (land & water), a fleet of 60 cargo ships, and a network of 15,000 kilometers of oil and gas pipelines interconnecting these facilities.) In addition, Petrobras Transpetro provides services to several distributing companies and to the petrochemical industry.
Although the contribution of natural gas to the country's energy matrix is still comparatively low, it is increasing in accelerated rate. On the other hand, the extent of the gas transmission pipeline network has doubled in the last 4 years. Nowadays, Petrobras Transpetro's Natural Gas Division is responsible for operating and maintaining 7.327km of onshore natural gas pipelines, 12 Compression Stations and 129 City Gates, all of them remotely operated by the National Operation Control Center (CNCO), located in Rio de Janeiro.
In Brazil, State & Federal Environmental Protection Agencies require every installation to carry out a Quantitative Risk Analysis including consequence and vulnerability analysis, frequency estimation and societal and individual risk calculation. These Agencies establish risk tolerability criteria and have, therefore, the responsibility of issuing operational licenses based on the risk results.
Beyond that, ANP (Agência Nacional do Petróleo, Gás Natural e Biocombustíveis), the Federal Agency responsible for regulating the oil & gas industry in Brazil, also requires risk assessments to be implemented for onshore pipelines, which may be either qualitative or quantitative, depending on the installation's complexity.
Based on governmental requirements and on internal procedures, Petrobras Transpetro has carried out Quantitative Risk Analysis (QRA) of natural gas pipelines, conducted by different Consultants, in different periods of time. The first QRA is often carried out when the Company requests an Environmental License, and then it's updated based on a regular basis. The review cycle for these studies depends on internal procedures and the License validity.
The results of consequence and vulnerability analysis, which are part of QRA studies, are the basis for pipelines' Emergency Plans. These results are used basically to assess whether there is population inside the vulnerable area and if it's necessary to evacuate people.
In case of onshore pipelines, that cross several communities and points of interest, such as schools, churches, hospitals and others, the information related to vulnerable areas is vital to adopt more adequate emergency response procedures.

2. Methodology

2.1 Software

Petrobras Transpetro uses PHAST - Process Hazard Analysis Software Tools, by DNV Risk Management Software (DNV RMS), to perform consequence and vulnerability analysis. PHAST contains discharge, dispersion and effects models to assess major hazards, foreseeing complex consequences from possible releases of hazardous materials.
The algorithm for evaluation of jet fire scenarios considers product characteristics (liquid, gas or two-phase), release rate and duration, height of release, orifice diameter and angle of the formed

jet (0° for horizontal jets and 90° for vertical jets). The software estimates flame dimensions and intensity.
For fireball scenarios, the model calculates its radius and duration. The critical parameters to be defined for a fireball model are: product mass involved in the fireball; strength of the thermal radiation source; and the fraction of the total heat that is released as thermal radiation.
The model simulates flash fires due to ignition of a vapor cloud, on an integrated basis, to the algorithm of atmospheric dispersion. The flammable cloud's size is used to model a transient front of flames that travels through the cloud. Thermal radiation in the receptor is calculated as a transient function due to variation of the flames' sizes inside the cloud.
This algorithm considers the speed, width and height of the flame, and concentration of the flammable vapors of the cloud during calculation time. The distance which corresponds to the Lower Flammable Limit (LFL) is assumed as the flame length. Result is calculated in terms of the distance reached by the flash fire.

2.2 *Premises*

2.2.1 Accidental Scenarios
Failure of a high-pressure natural gas pipeline can lead to various outcomes, some of which can pose a significant threat to people and property in the immediate vicinity of the failure location. For a given pipeline, the type of hazard, and the damage or injury potential associated with the hazard, will depend on the mode of line failure (i. e., leak vs. rupture), nature of the gas discharge (i.e., vertical vs. inclined jet, obstructed vs. unobstructed jet) and the time to ignition (i.e., immediate vs. delayed). The various possible outcomes are summarized in Figure 1. In mathematical simulations, methane was used as a representative for natural gas.

Figure 1: Possible outcomes from failure of a high-pressure natural gas pipeline

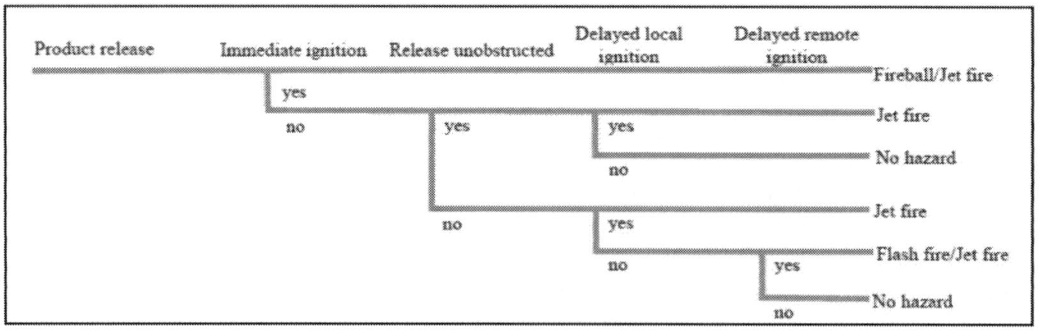

Although there is a great variety of outcomes that can result from failure of a high-pressure natural gas pipeline, the present study aims to analyze thermal effects due to jet fire.

2.2.2 Typical Pipelines
A set of 11 typical natural gas transportation pipelines was established to estimate physical effects due to natural gas possible scenarios. Typical pipelines considered in this study are listed on Table 1.

Table 1. Typical pipelines considered in this study

Pipeline	Pipeline Diameter	Operational Pressure
Pipe 1	8 in	70 bar
Pipe 2	10 in	100 bar
Pipe 3	12 in	100 bar
Pipe 4	14 in	100 bar
Pipe 5	18 in	100 bar
Pipe 6	20 in	120 bar
Pipe 7	22 in	74 bar
Pipe 8	26 in	100 bar
Pipe 9	28 in	100 bar
Pipe 10	38 in	100 bar

2.2.3 Meteorological data

The meteorological data considered in the mathematical simulations are presented in Table 2.

Table 2. Meteorological Data

Parameter	Day Condition	Night Condition
Ambient Temperature	25°C	20°C
Wind Speed	3 m/s	2 m/s
Atmospheric Stability classification*	C	E
Humidity	80%	80%
Soil Temperature	30°C	20°C

* Pasquill-Gifford Stability Categories

2.2.4 Physical Effects Considered

The Eisenberg Vulnerability Model, developed by the United States Coast Guard, was used for calculation of the exposed community vulnerability. This model uses probabilistic functions of the PROBIT type. It is a widely used approach to relate the phenomena intensity (like fire heat radiation) to their consequences.

Probit is a number related to the probability of damages, injuries or fatalities, expressed through the percentage of the population or physical installations affected.

It is relevant to emphasize that, conservatively, the vulnerable area corresponds to the area that could be affected in case of an accident scenario. The area predominantly affected will be influenced by local climatic and geographical conditions at the instant of the leak, such as interference by wind direction and speed, and may vary from person to person and equipment to equipment.

The following levels of interest were considered in this study:

1. Thermal radiation level of $4kW/m^2$ is frequently suggested in specialized literature as the pain limit after exposure of 30 seconds;
2. Thermal radiation level of $12.5kW/m^2$, which represents the limit to cause deaths - 1% fatality probability for an exposure time of 30 seconds;

The effects of thermal radiation arising from eventual jet fires were estimated for a height of one meter, understanding that this is a representative height for evaluating physical effects on people.

2.2.5 CETESB Requirements

CETESB is a State Environmental Protection Agency, from São Paulo State, publisher of the Quantitative Risk Analysis Manual: "Accidental risk from technological origin. Method for decision and reference terms", extensively used in Brazil, which establishes most of the premises adopted in this case study.

Although CETESB is a "local" Agency, it's considered as a benchmark all over the country. So, most of State Environmental Protection Agencies base their rules on CETESB requirements, including IBAMA, the National Environmental Protection Agency.

Premises based on CETESB Manual and used in this study are listed below:

- Simulations were carried out considering only pipeline buried stretches, calculating the effects of angular (45°) jet fires;
- Simulations also considered pipe rupture, considering the leak hole as 100% of pipe diameter.

3. Results

Results are presented in Table 3.

Table 3. Reaches (m) for Jet Fires Scenarios

Pipeline	Pipeline Diameter	Operational Pressure	Jet Fire Reaches (m)			
			Daily (kW/m^2)		Nightly (kW/m^2)	
			12,5	4	12,5	4
Pipe 1	8 in	70 bar	154	245	151	252
Pipe 2	10 in	100 bar	228	360	227	369
Pipe 3	12 in	100 bar	271	425	271	436
Pipe 4	14 in	100 bar	312	490	315	503
Pipe 5	18 in	100 bar	395	616	400	634
Pipe 6	20 in	120 bar	433	675	440	695
Pipe 7	22 in	74 bar	477	743	486	766
Pipe 8	26 in	100 bar	554	860	565	887
Pipe 9	28 in	100 bar	592	919	606	949
Pipe 10	38 in	100 bar	782	1209	803	1251

4. Conclusion

The results presented in item 3 show that the higher operational pressure and hole diameter, the higher would be the reaches of an accidental scenario of jet fire. The reaches shown in Table 3 could not be considered as an absolute value, instead it should be considered as a reference for the decision making process.

All models, including consequence models, present uncertainties. Input parameters that are modeled using the consequence analysis software may, and probably will, be different in an eventual emergency real case. Therefore, the results presented should be used as guidance to the emergency response team, so that adequate response is provided considering possible effects of

a jet fire on the pipe neighborhood, such as community evacuation, roads blocking, among others.
This study may be completed by simulating other possible outcomes, such as flash fires and fireballs, and by calculating thermal radiation levels.
As originally intended, the results present unified consequence modeling for jet fire scenarios involving natural gas pipelines, providing a quick reference to response preparedness and field personnel use in decision making process, in case of eventual emergencies.

5. References

[1] Manual de Orientação para a Elaboração de Estudos de Análise de Riscos, P4.261, CETESB, May 2003.

Additional references not cited:

[2] 8th Report of the European Gas pipeline Incident data Group - EGIG, December 2011.

[3] AMERICAN INSTITUTE OF CHEMICAL ENGINEERS (AICHE), 2000, "Guidelines for Chemical Process Quantitative Risk Analysis", New York, USA, 2nd Edition.

[4] "A Model for Sizing High Consequence Areas Associated with Natural Gas Pipelines", Gas Research Institute (GRI), 2000.

9TH GLOBAL CONGRESS ON PROCESS SAFETY

Automation Strategies: Avoiding Accidents During the Process of Draining Water From LPG Spheres

Carlos André Vaz Junior
Departamento de Engenharia Química, Escola de Química, Universidade Federal do Rio de Janeiro
Centro de Tecnologia, Bloco E, Sala E-209, Ilha do Fundão
cavazjunior@eq.ufrj.br

Maurício Bezerra de Souza Junior
Departamento de Engenharia Química, Escola de Química, Universidade Federal do Rio de Janeiro
Centro de Tecnologia, Bloco E, Sala E-209, Ilha do Fundão
mbsj@eq.ufrj.br

Diego da Silva Lione
Escola de Química, Universidade Federal do Rio de Janeiro
Centro de Tecnologia, Bloco E, Sala E-209, Ilha do Fundão
diegolione@ig.com.br

João Marcio Almeida Villaça
Escola de Química, Universidade Federal do Rio de Janeiro
Centro de Tecnologia, Bloco E, Sala E-209, Ilha do Fundão
joaovillaca@hotmail.com

Leonardo Salim Damian Aguiar
Escola de Química, Universidade Federal do Rio de Janeiro
Centro de Tecnologia, Bloco E, Sala E-209, Ilha do Fundão
leonardosdaguiar@gmail.com

Prepared for Presentation at
American Institute of Chemical Engineers
2013 Spring Meeting
9th Global Congress on Process Safety
San Antonio, Texas
April 28 – May 1, 2013

UNPUBLISHED

Automation Strategies: Avoiding Accidents During the Process of Draining Water From LPG Spheres

Carlos André Vaz Junior
Departamento de Engenharia Química, Escola de Química, Universidade Federal do Rio de Janeiro
Centro de Tecnologia, Bloco E, Sala E-209, Ilha do Fundão
cavazjunior@eq.ufrj.br

Maurício Bezerra de Souza Junior
Departamento de Engenharia Química, Escola de Química, UFRJ

Diego da Silva Lione
Escola de Química, UFRJ

João Marcio Almeida Villaça
Escola de Química, UFRJ

Leonardo Salim Damian Aguiar
Escola de Química, UFRJ

Keywords: LPG spheres, automation strategies, process of draining water

Abstract

The present paper has as its main objective to evaluate different strategies of automation, focusing on the reduction of risks on the water drainage procedures in spheres of liquefied petroleum gas (LPG). The condensation phenomenon of the fraction of the water found in gases during the storing process is largely observed in chemical plants. The condensed water will be accumulated in the bottom of the storing sphere and needs to be removed through a drainage procedure. Although being considered an operational routine process, such event was the main cause of at least two severe accidents: in 1966, in France, and in 1972, at the Refinaria Duque de Caxias (Duque de Caxias Refinery), in Rio de Janeiro, Brazil. One of the most critical decisions viewing to avoid that a new accident was to interrupt the drainage before the water could get completely drained. When the drainage continuous after all the water has been removed, the LPG begins to leak. The gas leaking phenomenon promotes the formation of a toxic and inflammable cloud, *par et passu* with a strong reduction of the temperature. In order to analyze the system, mathematical simulations of the drainage were carried out using the UniSim@ Design (Honeywell) simulator. Variations of the different properties (parameters) of the system were studied along the drainage. Understanding the dynamic of the process it is possible to develop automation systems which can preview and anticipate the best moment for the interruption of the water draining process. Traditional control architectures were suggested based on the PID control architecture. The "monitoring processes", that had been proposed by Seborg et al. [2], were tested as well. The monitoring system was considered more efficient, avoiding that natural fluctuations would anticipate and/or retard the action of the controller.

1. Introduction

The present study has as its objective to propose and to evaluate different controlling strategies for water draining systems in LPG (Liquefied Petroleum Gas) spheres. This analysis obtained its stimulus from the study of the accident that occurred in the Refinaria Duque de Caxias/Duque de Caxias Refinery, in Rio de Janeiro, in 1972. Such accident caused the explosion of a LPG sphere, resulting in several fatal victims and a large number of wounded people. The sphere's water draining system consisted, at the time, of a manual opening of the valve localized in the lower part of the reservoir. The employee in charge of such operation should be waiting for the estimated period of time so that practically all the water would have been removed. After this time, the employee should close the valve, interrupting the drainage and avoiding the gas leak. During the whole time of the drainage, the employee should stay in the area, following the process. In the accident of 1972, the employee left the place, not returning for the closing of the valve, causing the LPG leak which formed an inflammable cloud and the freezing of the equipment nearby. The low temperature generated by the liquefied gas leak promoted the freezing of the drainage valve, not allowing its closing. After the accident, the sphere's drainage system received several improvements, such as the inclusion of the blockage double system and safety instruments, even though the drainage procedures continued to be carried out essentially in a manual way.

Proposing to offer a more efficient and safe solutions for the drainage procedures of the water, this study used dynamic simulation. Through the simulator were made several drainage simulations, evaluating the behavior of several parameters of the process. Therefore, the automation strategy was based on the fact that the dynamics of the variables of the process can anticipate under a safe and exact way the ideal moment in which the end of the drainage should happen.

2. The Accident

The description made here is based on statements of the professionals that were working in the location at the time of the accident. The accident occurred in 1972, in the park of the LPG spheres of the Duque de Caxias Refinery, in Rio de Janeiro, Brazil. The storage of the LPG occurs under elevated pressure, allowing the maintenance of the gas in its liquefied state. Nevertheless, the elevated pressure also generates the formation of a condensed water layer in the bottom of the sphere. The valve and pipes localized in the lower part of the sphere are used for the water drainage. To remove the water is fundamental to ensure that the final product will meet the required specifications when in use. At the time of the accident, the space where the drainage pipes sat had a sole manual valve.

The conventional procedure to be carried out in 1972 by the employee on duty was to connect the hose to the drainage pipes. This hose was then positioned in a nearby ditch. The server would return to the place in order to open the valve. The drained water would be routed this way to the ditch. As the amount of water formed during the compression is known, as well as the flow of drained water, it is possible to estimate the necessary time for the water to be removed from the sphere.

According to instructions being followed at the time, it was expected that the employee on duty would stay in the location, following the process. Nevertheless, it is said that on the day of the accident the operator was absent at the exact moment of the end of the drainage, therefore not closing the valve as expected. Due to this fault, not only the water, but the gas as well, was sent to the draining pipes. Besides the formation of an inflammable cloud, the quick decompression of the liquefied gas promoted a strong reduction of the temperature of the line, causing the freezing of the valve. From this moment on, the operator became incapable to stop the leak. The emergency brigade of the company was called, but did not succeed in stopping the gas leak.

It was reported that a dense and large cloud of LPG was formed over the whole park of the spheres of the Refinery. The presence of an ignition source, never identified, detonated the cloud. This detonation, although causing light burning wounds in a few employees, was the main cause of the improvements that followed the accident. The detonation created a jet fire, thus forming a flame that touched the sphere, heating it.

Due to the seriousness of the accident and its duration, the fire was battled not only by the emergency brigade of the refinery, but also with the help of the employees of the companies of the neighborhood. Moreover, the first emergency teams were helped by the next shift of workers. This fact contributed to the increase of the number of victims of the accident.

After almost two hours of fire, the sphere that was leaking exploded in a phenomenon known as BLEVE. The explosion sent fragments over the other spheres, thus propagating the destruction. A sequence of explosions totally destructed the park of spheres of the refinery. Pieces of the reservoirs were found 800m away from their original place.

2.1 Implemented improvements

Similar to the accident reported, a few others occurred in other industrial plants in the same period, as the one in Feyzin, in France, in 1966 [1]. Although the task to remove the water formed in storage spheres of LPG continues to exist until today, several improvements were implemented. The ground of the spheres is now concreted including a slope, avoiding the accumulation of LPG under the sphere in case of a leak (pool fire). Other important changes were implemented in the drainage piping, which now present a larger number of valves, as shown in Figure 1. The opening-and-closing of the valve nearest to the sphere became remote, driven from the control room. The functioning from a distance has as its goal to allow that, even in case of fire, the valve will be controlled. The valve counts on a thermo-sensitive fuse, which promotes its closing in case the external temperature increases. Finally, the length of the drainage piping was also extended. The physical distance between the drainage point and the valves is set so that–even in a leaking scenario– the low temperatures would allow the freezing of the equipment. A larger number of valves in the system also contribute for a safer process. A project of "closed drainage" is being tested in Brazil, nowadays. This system should eliminate the gas release to the atmosphere, in case a leak happened. If a gas leak occurred during the drainage process, the system would send the gas to flare.

Figure 1: Drainage piping with a larger number of valves

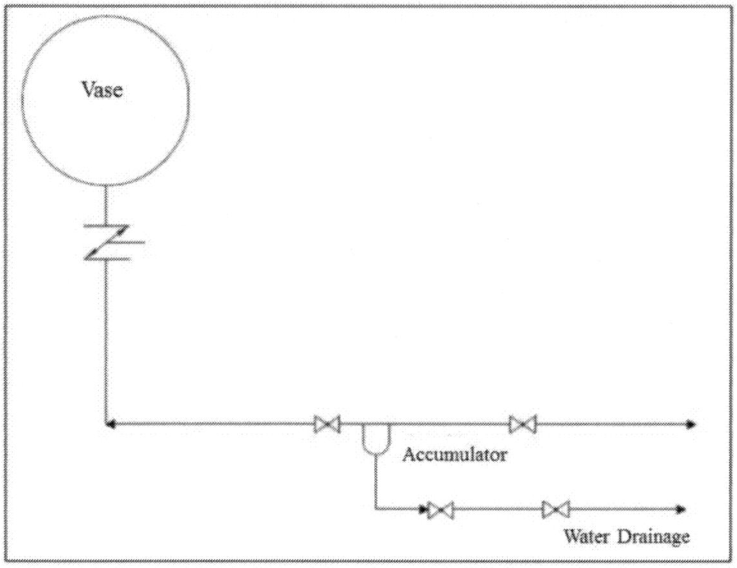

3. Simulator

Through the use of a dynamic simulator, it was possible to carry out several simulations of the drainage system. These results permitted the evaluation of the dynamic behavior of different variables of the process. From the behavior of the process parameters, different control strategies were proposed. The automation would allow the interruption of the drainage process in its ideal moment, thus avoiding the LPG leak.

In this paper we applied the software UniSim® Design. The simulation considers: the storage sphere, the valve directly connected to the bottom of the sphere, and the last blocking valve of the drainage line, as seen in Figure 2. For the simulation it was applied the Peng-Robinson's state equation. The LPG composition was presumed as the mixture formed by 70% of propane and 30% of butane. The amount of the condensed water in the bottom of the sphere, before the beginning of the drainage, was estimated in approximately 5% of the total volume. The diameter of the simulated sphere was 18.25 m. The diameter of the valve nearest to the sphere (VLV-100) was estimated to be 10 in, while the diameter of the most remote valve (VLV-101) was 1 in. The operating pressure of the sphere was presumed to be 8.5 kgf/cm^2 and its initial temperature, 29°C (environmental temperature). In order to simulate the drainage, the outlet flow (flow 5) pressure was adjusted for 1atm (atmospheric pressure). Finally, at the start of the drainage process the initial opening of both valves was presumed to be 100%.

Figure 2: System without PID control

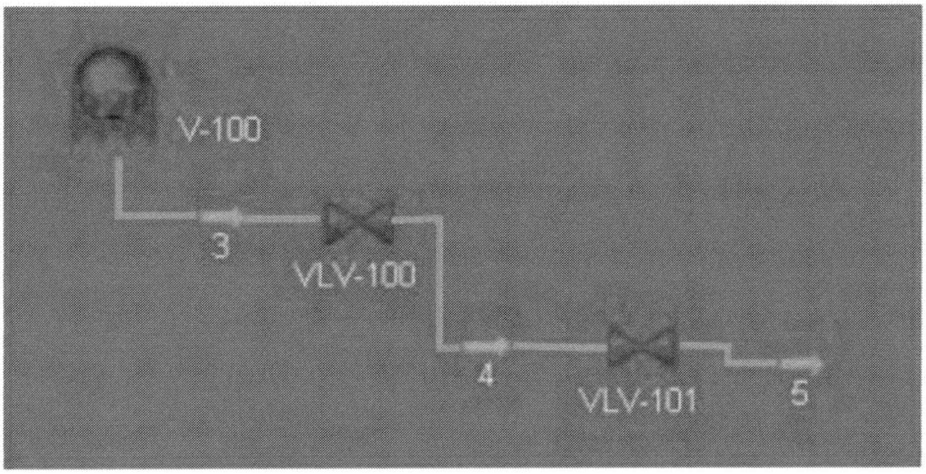

The system represented in Figure 2 does not have the PID controller to interrupt the process. This way, once both valves open, all the condensed water will be removed. After that, the gas release will occur in a situation similar to that observed in the 1972 accident.

After the simulations of the system represented in Figure 2, a PID controller was implemented (Figure 3). The control system sees, as its main goal, to allow that the drainage procedure be carried out safely and efficiently. To achieve this, it is necessary that the controller interrupts automatically the drainage before a leak of the LPG occurs.

Figure 3: System with PID control

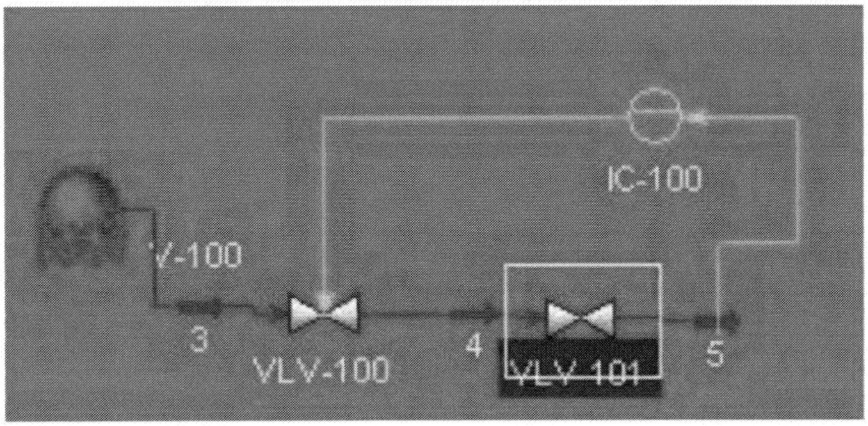

From the simulations study of the flow-diagram without PID controller (Figure 2), the controlled and manipulated variables were defined. The out-flow temperature (flow 5) was defined as being the controlled variable. Such choice was due to the fact that such process parameter presented high sensitivity to the beginning of the LPG leak, varying abruptly. The manipulated variable was defined as being the out-flow drainage of the tank, through the opening control of the valve (VLV-100).

4. Results

4.1 Process simulation without PID controller

First, the simulation of the system was done without the controller (Figure 2). In this simulation, the two valves were maintained opened during all the simulated time (1h40m). This time window was sufficient to simulate not only the water drainage, but the LPG leak as well.

The fractions of the phase water and LPG at flow 3 (closest to the sphere) are illustrated in Figure 4. One may notice that initially there is only a way-out for the water from the spherical tank. This occurs because the water is denser than the LPG, being accumulating so far at the bottom of the sphere. At instant t=1852 seconds, i.e., approximately 30 min. after the beginning of the simulation, the water fraction begins to decrease, initiating the way-out of the liquid LPG from the sphere. From this moment on, the water fraction of flow 3 decreases.

Figure 4: Fraction of the Phase x Time

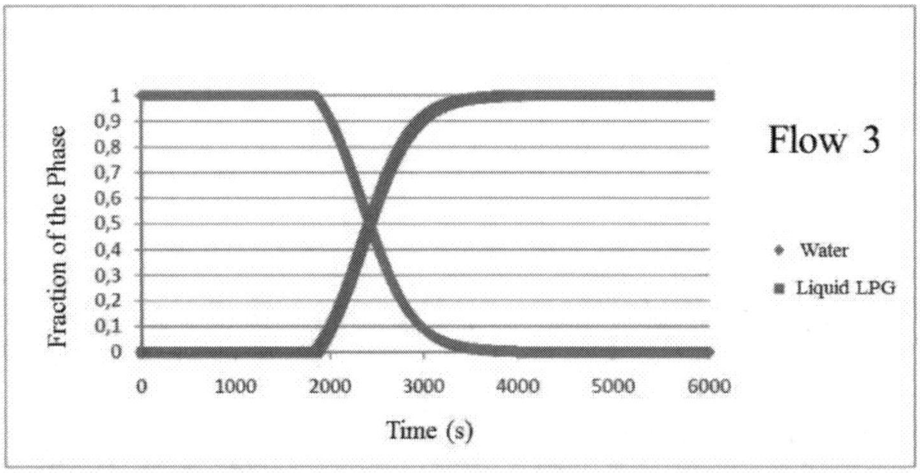

On the other hand, the fractions of flow 5 are represented in Figure 5. In a similar way to flow 3, one may notice that during the first moments of the simulation, the water fraction in the flow equals 1. From instant t=1855 seconds, the water fraction decreases, while the LPG fraction under a vapor state increases. The LPG presence under a vapor state occurs because this flow is being discharged on the environment, with an atmospheric pressure and an initial temperature of 29°C. Under such conditions, the LPG is found in a vapor phase. Therefore, the liquid LPG originated from the tank will vaporize after valve VLV-101.

As observed in the 1972 accident, the vaporization promotes a strong reduction of the line temperature, reaching -30°. C, as we can see in Figure 6. From instant t= 2170 seconds, the flow 5 temperature is already registering -23°C, and the LPG starts to leak in its liquid form. We shall highlight that, in a real scenario, this condition should be sufficient to promote the freezing of the nearest valve.

Figure 5: Fraction of the Phase x Time

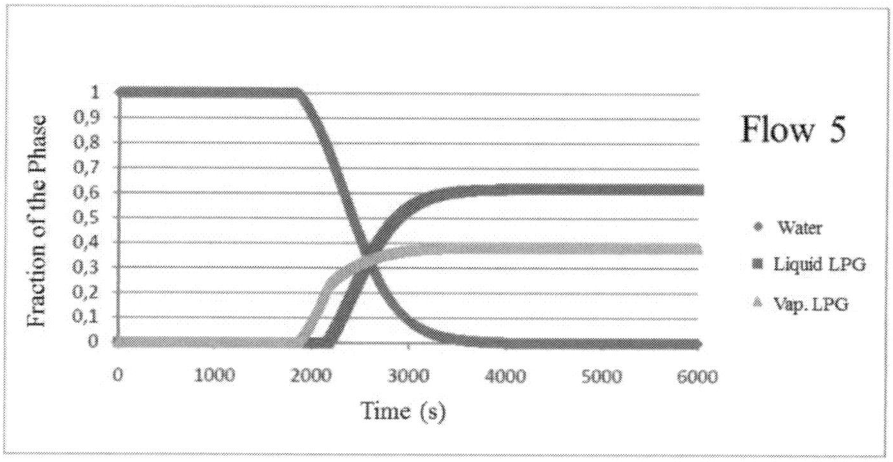

Figure 6: Temperature x Time

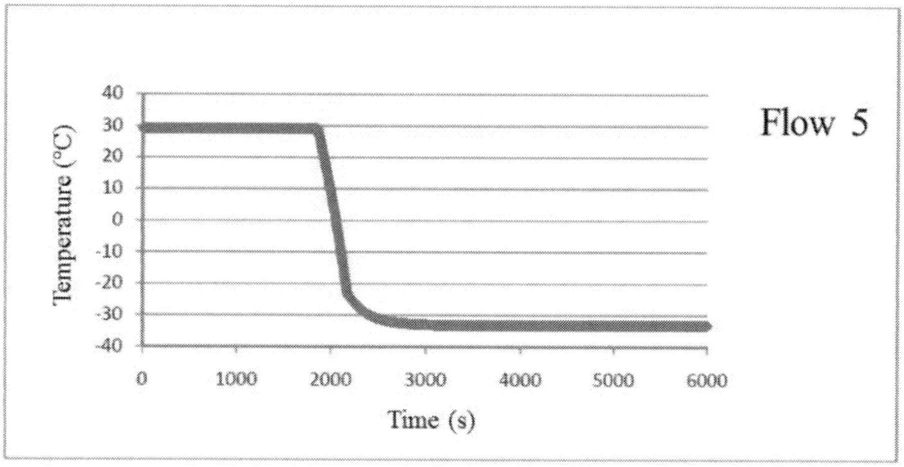

Through Figure 6 we can observe that the out-flow temperature (flow 5) is stable, approximately at the environmental temperature of 29° C, during the first 1855 seconds. In this period of time, only the water of the reservoir leaves it (Figure 5); from then on, the temperature decreases abruptly. At 1915 seconds, i.e., 1 minute after the beginning of the LPG leaking, the temperature was 20°.C, a fall rate of 10 degrees per minute. After 10 minutes, the temperature of the flow reached -30°C. This strong change of the temperature occurs due to the fast change of the phase suffered by the LPG.

From the simulation it was possible to observe that within the variables analyzed, the most sensitive to the beginning of the LPG leaking was the flow 5' temperature. This way, as mentioned, the control system was based on the temperature measures of flow 5.

4.2 Simulation of the process with different tunings of the PID controller

Intending to propose and test different automation strategies for the sphere's water drainage, the use of the PID methodology was initially investigated. Viewing to maintain the controller as simple as possible in a first approach, we opted for the use of a controller known as "P" (proportional). The tuning of the controller had, as its main goal, to maximize the withdrawing of the water from the spheres, thus avoiding the occurrence of a significant LPG leak. The chosen *modus operandi* for the controller was "direct": if the temperature starts to decrease the controller should be activated in order to reduce the flow by closing the valve.

Several simulations were made varying the gain (Kc) of the proportional controller. The set-point was adjusted for 19°.C, 10 degrees lower than the environmental temperature (29°C). In each simulation the following variables were evaluated: opening of the valve, the water fraction at flow 5 and the temperature in flow 5. The results are shown in Figure 7.

Figure 7: Water fraction in the flow 5 x Time

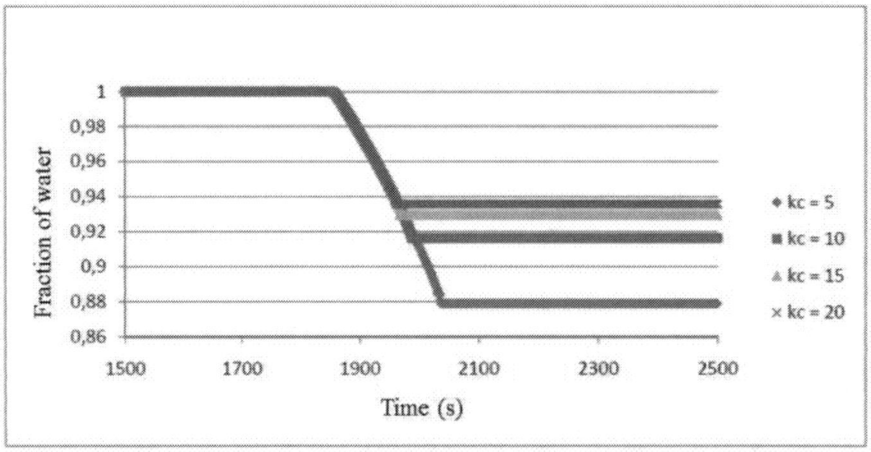

Figure 7 illustrates the behavior of the water fraction in flow 5 as time goes by, for different tunings of the control network. First of all, one may observe that only the water is being drained. From then on, a fall of this fraction occurs, indicating the beginning of the LPG leaking. As the water fraction decreases, the amount of LPG in the flow increases. As a consequence, the temperature decreases severely, according to Figure 8. It is possible to notice that the behavior of the temperature in the outflow follows the same profile of the water fraction in this flow.

Figure 8: Temperature of flow 5 x Time

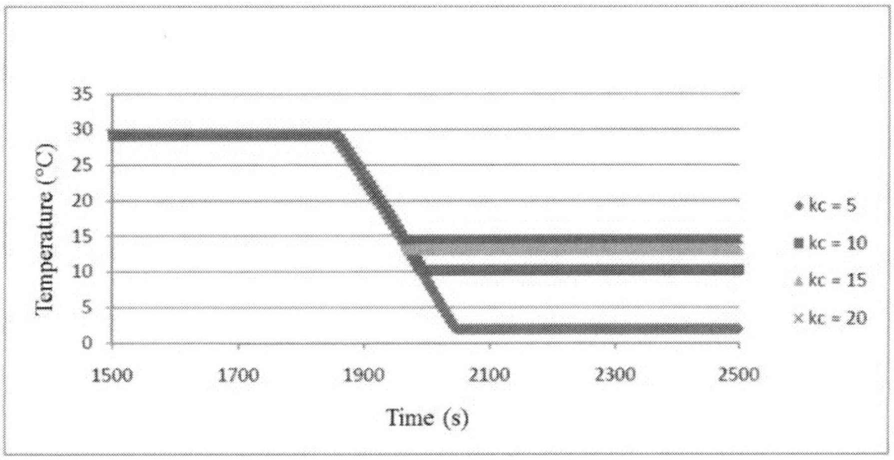

Figure 9 shows the percentage of the valve VLV-100 opening, as a function of time. Initially, while only the water is being removed from the sphere, the valve is kept in a totally open position (100%). As the simulation is carried out, such percentage decreases at the same time that the water content and the temperature decrease. The valve movement proceeds until the valve is totally closed (0% opening). The closing of the valve indicates that the controller is acting expecting the decrease of the leak, since temperature of flow 5 is coming down.

Figure 9: Opening of the valve x Time

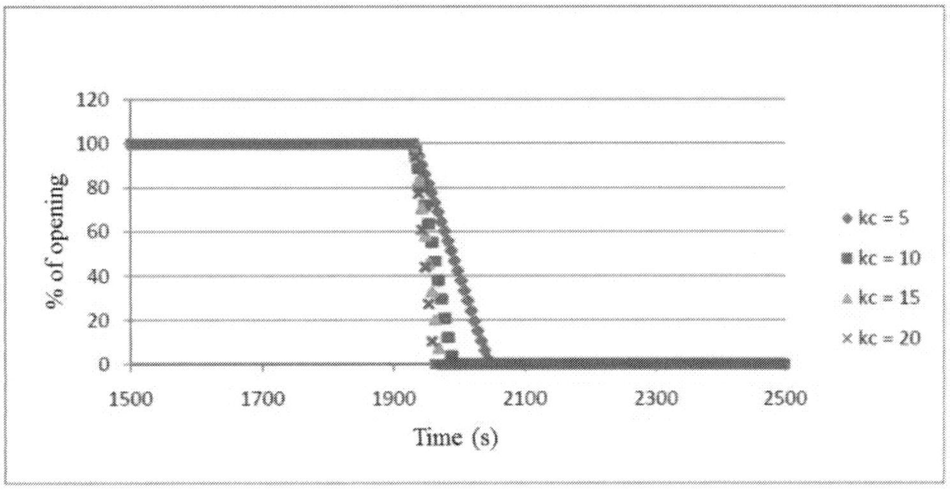

Comparing the behavior of the analyzed variables in relation to the values of the proportional controller gains (Kc), one may observe that the larger the gain, the quicker the valve closes and, therefore, the temperature and the water fraction are higher in flow 5 (Table 1). High gains minimize the amount of the LPG leaked, reducing the risks of frozen valves. Nevertheless, it is worth to highlight the fact that the closing time of the valve is still a function of the valve dynamics, which was not considered in this paper.

Therefore, for a given set-point, the increase of the gain of the controller promotes a quicker closing of the valve and drainage with a smaller LPG leak. It is interesting to notice that for a determined value of gain an increase on the value of the set-point would cause the same results.

Finally, we shall say that very elevated gains of the controller take the valve to act practically as a valve on-off. In this case, there are only two possible positions for the valve: totally closed or totally opened. Other types of control action (integral - I or derivative - D) were not tested, once the results based on the PID methodology took us to choose the on-off action.

Table 1. Dynamic behavior of the parameters of the process

Kc	Final Temperature - Flow 5 (°C)	Final Water Fraction - Flow 5
5	1.95	0.879
10	10.21	0.916
15	13.09	0.929
20	14.49	0.938
100	18.04	0.952
1000	18.74	0.955

4.3 Processes monitoring

In industrial processes, the maintenance of the operational conditions in general uses strategies of feedback or feed forward controller. A different approach was proposed by Seborg et al [2] and denominated “processes monitoring”. This approach can be employed with the purpose of following the dynamical behavior of the variables. The monitoring, when associated to tools of supervision and control, guarantees that the variables stay within specific limits and rapidly detect abnormal conditions of operation, implementing corrective actions, etc. This way, the monitoring contributes to the reduction of the variability of the process. In given scenarios, this strategy may present advantages on the use of traditional approaches of control.

In the present study, the variable measured is the temperature of the out flow. The methodology of “process monitoring” is based on the realization of a series of measures of the temperature of the out flow carried out in the beginning of the draining. During the initial phase, the parameter stays practically stable, its average value being used as reference. In a real situation this initial value would correspond approximately to the environmental temperature. As the drainage proceeds and there is a space for a small fraction of LPG, the temperature decreases. The values measured are then compared in real time with the reference value and the differences are then added. When the sum overcomes a specific value, the control action will occur; in this case, the closing of the valve. The strategy of processes monitoring is known as “Cumulative Sum” [2].

It is important to highlight the fact that the temperature of flow 5 – before and at the beginning of the draining process – is subjected to natural fluctuations, such as alterations of the environmental temperature. This normal variation can reduce the efficiency of the traditional approach of control based in a fixed set-point.

In this study, through a simulation the average of the temperature of the out flow was measured during the first 10 minutes of drainage, thus obtaining the average value of 29.17ºC. Through previous simulations it was observed that the decrease of the temperature, caused by the beginning of the leak, was of approximately 0.5º.C This was the maximum value of the cumulative sum adopted to determine the closing of the valve.

Using the cumulative sum, two seconds after the beginning of the LPG leak, the cumulative term became larger than the value stipulated as maximum, this way determining the closing of the valve. The quick closing of the valve, immediately after the beginning of the LPG leak, prevents the freezing of the valves. This approach seems to be much more stable and efficient than the proportional controller.

So, it is important to highlight the fact that in real operations the existence of fluctuations noises (interference) and fluctuations of temperature, or the composition of the LPG yet, could well difficult these parameters. The adoption of higher values as the maximum allowed at cumulative sum could appear adequate.

5. Conclusion

The large accidents of the past reveal the catastrophic potential of the events involving large LPG storage spheres. The draining procedure of the water accumulated in the bottom of the spheres was the main objective of this paper. Focusing on the automation of the procedure, we tried to simulate such process, based on dimensions, parameters and typical industrial conditions. We searched for the identification of a variable that could be indicative of gas leaking, being able to be applied as reference to indicate the best time to interrupt the draining. The temperature of the out flow seemed to be the most adequate parameter.

Different tunings of the proportional controller were tested. It was noticed that the larger the gain value (Kc), the smaller was the amount the LPG liberated at the end of the drainage. This way, the results converged to an action of control of the type on-off. This approach seemed to be strongly influenced by noise (interference) on the variable measured and by the environmental temperature observed.

The use of the strategy of process monitoring was also tested, based on the cumulative sum method. The results obtained were higher than those found for the proportional controller. The drainage was immediately interrupted after the occurrence of a small leak of LPG through the out flow. A "cumulative sum" of only 0.5º C, was adopted as the best. A quick interruption of the drainage process would have reduced the risks of freezing in the line and the emission of LPG to the atmosphere. Therefore, it seems to be an interesting option to increase the safety of the procedure.

6. References

[1] Kletz, Trevor. *Learning from Accidents*. 3rd edition, Gulf Professional Publishing, Woburn, MA, 2001.

[2] Seborg, Dale; Edgar, Thomas; Mellichamp, Duncan, *Process Dynamics and Control*, New York, 2nd edition, Wiley, New York, 2003.

(44aa) Complex BMS Made Simple

Dave Sheppard, Emerson Process Management, Hazelwood, MO and Nikki Bishop, Industry Solutions Group, Emerson Process Management, Roswell, GA

Complex BMS systems with multiple fuels and multiple burners have long presented a struggle for engineers and operators alike. This presentation will cover the methods used to simplify the implementation of a complex BMS in DeltaV SIS. The approach to design and use of standardized sequence modules and graphics greatly increases operator awareness of the operating conditions of the BMS. Included in this presentation will be a description of the design philosophy, the configuration standards, and the approach to operator graphics.

A three step approach is implemented. Break the BMS sequences down in to incremental parts that are easy to understand. Develop standard modules that implement these parts individually. Use a graphical interface the represents these individual parts in an easy to understand way for the operator. This approach is highly scalable to meet various complex BMS implementations.

With this approach it is easy to take a complex BMS and break it down in to small and simple to understand pieces. This allows greater alignment and understanding between the safety engineer, configurator, and operator. Ease of use and troubleshooting for the operator is also greatly increased through a standard set of graphics that show actions needed and first outs for trips.

Through this approach it is possible to take the design of how a complex BMS needs to function and actually implement that design cleanly and simply. An accurate system to capture the first out for a trip or loss of a permissive and also indication of the current state and actions needed to progress forward is needed to increase operator knowledge of what is going on in the BMS at all times.

"Using ESD Valves as Safeguards, Myth or Reality?"

German Luna-Mejias, P. Eng.
ACM Facility Safety
Suite 700 940 6th Ave SW, Calgary, AB, Canada
gluna@acm.ca
guatire2@gmail.com

Prepared for Presentation at
American Institute of Chemical Engineers
2013 Spring Meeting
9th Global Congress on Process Safety
San Antonio, Texas
April 28 – May 1, 2013

UNPUBLISHED

"Using ESD Valves as Safeguards, Myth or Reality?"

German Luna-Mejias, P. Eng.
ACM Facility Safety
Suite 700 940 6th Ave SW, Calgary, AB, Canada
gluna@acm.ca
guatire2@gmail.com

Keywords: Process, Safety, Emergency, Shutdown, valve, bypass

Abstract

An Emergency ShutDown (ESD) valve is a safety device to be actuated during an emergency situation to prevent hazardous consequences. The ESD valves are to be designed not to create additional hazards when actuated. Sometimes, due to operational constraints ESD valves are bypassed which contradicts best practices and creates an incremental risk for the facility, people and environment. Lack of regulations sometimes drives ESD valves' users to overlook or ignore best practices with the associated incremental risk. This paper presents experiences during HAZOPs/field observations where installed ESD valves are claimed as safeguards but are not (can be bypassed). The intent of this paper is to call the attention of designers, operators and managers of the potential hazardous consequences they may face if ESD valves are not used as per recommended practices.

1. Introduction

Our police force has to be right every time to prevent worst case scenarios, the "bad guys" only have to be right once for a major impact to occur. Exactly the same happens with safety; safety devices have to work when required; only one failure is enough to cause a major impact.

The use, absence or abuse of safety devices –e.g. emergency isolation valves- have been the cause of several accidents in the industry:

- A.P. Green Refractories Company – Explosion at Brick Kiln during start-up due to a main gas isolation valve bypass line leaking with 2 injured. Georgia, USA 1985. [1]
- Oil & Gas, Sonat Exploration Company - Explosion and fire due to misalignment of manual valves when feeding a low pressure system with HP Gas with 4 fatalities. Louisiana, USA. March 1998.[2]

- Ammonia Plant: Fire and explosion with substantial damage to a reformer-furnace due to bypass of control valve -with shutdown function- inadvertently open. No injuries, India, May 1998.[3]
- Esso-Australia Gas Plant Explosion due to the improper use of bypass valve around safety valve as the safety valve was not operable with 2 fatalities and substantial public welfare impact. Victoria, Australia. Sept 1998.[4]

As per these four examples, problems can happen anywhere in the world. In the last 3 years, when performing Process Hazards Analyses (PHA), the PHA teams indicated no concerns about bypasses around ESD valves, because there are no regulations that prevent this. Common responses were: "this is not regulated, why worry about it" or even worse: "recommended practices are recommendations, they are not mandatory". These practices are not only recommended practices but common sense. Additionally, as AIChE members (including professional denominations), the Code of Ethics indicates that welfare of the public and environment is paramount.

2. Definitions

Emergency isolation valves, emergency block valves, Emergency Shutdown valves or Automated Shutdown valve are different names for the same type of function valve.

A paper presented to GCPS in 2012 [5] indicates that the Emergency Block Valves (EBV) are namely different as per different codes, regulations and companies.

2.1 Additional definitions:

Emergency Block valves are used to control a hazardous incident. These are valves for emergency isolation and are designed to stop the uncontrolled release of flammable or toxic materials [6]

The ASV (Automated Shutdown Valve) is most commonly use in the Pipeline Safety Act 2011 [7]. The concept of the ASV is to mitigate the release of flammable or toxic materials, not the prevention of it. Pipeline Safety concern is H&SE not Process Safety.

2.1 ESD valve:

A valve with the following characteristics:

- Actuated valve
- Actuate upon a detection of a dangerous event defined by designers/operators and confirmed during PHA study.
- Provide protection against possible harm to people, assets or environment.
- Not used for control

An Emergency Shutdown valve –for the scope of this paper- is a valve that will prevent a hazardous situation from occurring (i.e. closing to prevent the overpressure of the vessel downstream of the valve) or mitigating the impact after the situation occurred.

2.2 Bypass valve:

Definition of bypass:

- go around something
- channel carrying a fluid around a part and back to the main stream

In the oil and gas industry is sometimes common to find bypasses around control valves; this situation is commonly used when it is expected that the valve may need maintenance during the operation of the plant. The plant can continue to run for a short period of time with a manual controlled function using a globe valve. This is a common accepted practice.

In this paper, the word "bypass" is a key element of the study, and follows the second definition: "channel carrying a fluid around a part and back to the main stream"; more specific: "around the ESD valve".

2.3 Norms and Standards

Depending on the system under review, there are several norms, recommended practices, regulations, etc. associated with the operation and installation of ESD valves. Nevertheless, none of these documents indicates or dictates providing a bypass around an ESD valve. A list of these documents is provided in the appendix A.

3. Installation and Use of ESD valves

In Viviano's work [5], the use of the "5W" (What, Who, When, Where and Why) helps to identify the need for emergency isolation valves. In this paper, it is considered that all 5W are being taken in consideration and the evaluation is moving onto the next questions: Is it correctly installed? It is a valid layer of protection? Does it protect against a hazardous event? Does it mitigate the impact?

Is my safeguard device reliable? In the O&G business a spurious trip is also a problem, but does not affect public welfare. Once the ESD valve is included in the detail design, how do we ensure it is going to work in the different scenarios that led to the installation of the ESD valve?

The first phase of the analysis it to perform the risk evaluation or risk assessment.

3.1 Risk Analysis

During a PHA, analysis will be done to ensure the different ESD valves are valid safeguards to prevent the hazardous event from happening. In this phase of the study a qualitative analysis is performed and team members will agree, or disagree, on the protection being used for prevention. A good facilitator should challenge the team about the effectiveness of the safeguard (e.g. is there enough Process Safety Time, Is the actuation of the ESD valve creating another hazardous condition(s)?).

On the other hand, since the PHA is a qualitative analysis, each safeguard is considered to be of the same "weight" or of the "same importance". If required, an additional safeguards evaluation may be carried out: LOPA.

LOPA (Layers of Protection Analysis) is a semi-quantitative analysis to determine the PFD (probability of failure on-demand) of the safeguards to meet the tolerable frequency requirements of each company.

Based on LOPA an Emergency Shutdown Valve may require being a SIL rated device to achieve the tolerable frequency for a specified event. SIL rated means that the associated elements of the safety device has to be designed, constructed and maintain to a minimum level of "probability of failure when demanded to operate" [8]

Previously it was mentioned that ESD valves should not be used to perform a control function as the reliability or SIL rating of the ESD valve may be compromised. The PFD of a typical BPCS (Basic Process Control System) is 0.1 (1 failure every ten times it is demanded) and the ESD valve may require a higher reliability (e.g. SIL 2).

3.2 Use of ESD for Accident/Incident Prevention

When referring to Emergency Block Valves or Emergency Shutdown valves, two functions (or purposes) should be studied: 1. Use of ESD valves for preventing the hazardous situation from happening (preventive barrier). 2. Mitigating the consequence (acting after the hazardous event happens).

3.2.1 Accident/Incident Prevention

Valves in this category are mainly used in industries (e.g. Oil & Gas) to prevent the hazardous event from happening.

In several scenarios briefly described in the introductory section of this paper, ESD valves were installed to prevent a hazardous incident from occurring as per the original plant design concept.

3.2.2 Accident/Incident Mitigation

Valves in this category are mainly used to prevent public exposure to a hazardous event, after the initiating event occurs. These scenarios are largely described and analyzed in documents such as Pipeline Safety Act 2011 [7].

The concerns of the regulator are focused more on spurious trips that may create a public impact due to the effect of cutting the gas supply to communities. In this case, the regulator may recommend a manually activated Isolation Valve. For this scenario, a LOPA is recommended to ensure that the Isolation valve will have a low PFD (probability of failure on demand).

4. ESD Valve Configuration

Sometimes for "ease the operation" the bypassing of ESD valves are permitted as there are no regulations or norms against it but common sense. These "ease operation" scenarios may remove the protection intent of the ESD valve.

4.1 Creating additional hazards.

Installing an ESD valve to reduce the risk is a good idea unless this addition creates additional hazards. If the ESD valve closes, it will isolate two systems: one at high pressure and other at lower pressure. What happen if the high pressure system cannot withstand the pressure? This analysis should be part of the PHA analysis.

Sometimes, a risk analysis is improperly done just to complete a "check mark" in a paper that means the activity was performed. Therefore a proper evaluation is not accomplished and the operator may believe the facility is safe to operate and it is not.

Unfortunately in many projects the PHA is just a "check mark" to achieve the project approval and the PSM is not strong enough to ensure a proper PHA is accomplished. In an evaluation done by OSHA [9] it has been found that improper or inaccurate hazard assessment is one of the main deviations from the safety programs.

In many PHAs studies it has been found that as no regulations "mandate" the owner to avoid bypassing the ESD valves they are frequently bypassed as it makes the start-up much easier and faster.

If the ESD valve is designed to "Provide protection against possible harm to people, assets or environment" or "actuated upon a detection of a dangerous event" it should not be bypassed. A general rule of Safety indicates that: "the important thing to do is to be safe when nobody is watching"

4.2 Bypassing ESD valve for start-up.

After the PHA and LOPA team agrees to install an ESD valve with specific SIL rating and confirms that no added risk will be created if valve is actuated, what is next?

Even if no regulation "mandates it" common sense indicates that if a bypass is required around an ESD valve the bypass has to be equipped with similar protection or equivalent to prevent the inadvertent operation of bypass or minimize the possibility of leaking.

What are the problems associated with improper bypassing of ESD valves?

- Operator error may remove protection
- Valve leakage may remove protection
- Any of these scenarios will invalidate the PHA previously performed and "due diligence" may be difficult to justify if an accident happens.

Filling the lines with products before start-up is a common practice, occasionally done by opening the bypass around the ESD valve. The filling of lines is not a problem but the increased pressure on the downstream side of the ESD valve may be a problem. That is what happened in the Esso accident (Victoria, Australia) as a valve misalignment created an overpressure of the receiving vessel.

Use of CSC (car seal closed) valves in the bypass around ESD valves is a partial protection as the isolation valve may leak overtime without the operator knowing it. Pressure equalization is a bigger issue than flow.

In the incident that occurred in the Brick Kiln the "bypass valve" around the main isolation valve was "somehow inadvertently open"; it means that to start-up the facility the protection was removed.

5. Conclusions

Potential damages to public and environment can be serious, including fatalities, as it has been shown in several accidents around the world.

In the Pipeline Safety Act 2011 [7], the major concern is the amount of gas release and the impact on public after the incident happens. It does not direct the users to provide preventive safeguards.

As there are no specific regulations or codes, the bypassing of safety devices is a concern that is very likely to be overlooked.

6. Recommendations

6.1 General

- Accident prevention requires ACTIVE leadership by management of safety issues.
- Effective implementation of the right controls to manage, mitigate or eliminate hazards and reduce risk.
- Management systems and attitudes towards safety go hand-in-hand in creating robust defenses.

This paper is written with the intention that the reader perform "due diligence" to make sure the area under his/her responsibility is safe.

6.2 Safe Design

- As a general rule, ESD valves should be installed based on PHA/LOPA
- The actuation of ESD valves must not create additional hazards.
- The required reliability of an ESD valve (PFD) must be evaluated by a LOPA or similar analysis.
- Bypass around ESD valve should only be permitted if an additional ESD valve with the same functionality is installed in the bypass.

6.3 Operating practices

In the accidents reviewed at the beginning of this paper, in most of the cases the inadvertent opening or closing of a valve was a key element to the problem. It is highly recommended to implement a PSSR (Pre Start-up Safety Review) before any start-up.

7. Final Remarks

- Bypass of a safety device may save time for operator, but may cost lives!
- Follow PSM or equivalent because you want to be safe, not because a regulation requires it!
- A general rule of Safety indicates that: "the important thing to do is to be safe when nobody is watching"

8. References

[1] OSHA. Accident Report: Explosion in Brick Kiln. Accident # 14542070, report ID 0418100. Washington, DC. 1985

[2] CSB. Investigation Report: Catastrophic Vessel Overpressurization (4 deaths). Report # 1998-002-I-LA; Chemical Safety Board, Washington, DC. 2000

[3] R.M. Chopde & K.M. Patel ., "Ammonia Plants: Past accidents and lessons learnt", Krishak Bharati Cooperative LTD, India. Presented at International Fertilizer Industry Association (IFA) Technical Conference, Louisiana, 2000.

[4] A. Hopkins, PhD ., "Lessons from Esso's Gas Plant Explosion at Longford"; Australian National University, 2000.

[5] J. Viviano ., "Emergency Block Valves"; presented at VIII Global Congress on Process Safety, 2012.

[6] " API Recommended Practice API RP-553:"Refinery valves and accessories for Control and Safety Instrumented Systems", Chapter 7. American Petroleum Institute, 1998.

[7] "Pipeline Safety Act 2011" U.S. Congress, Law 112-90 – Jan 3, 2012.

[8] CCPS., "Layer of Protection Analysis: Simplified Process Risk Assessment", Center for Chemical Process Safety. New York, NY. 2001.

[9] Jordan Barab., "Learning from Industry Mistakes" presented at NPRA National Safety Conference, May 2010

APPENDIX A

Regulations, codes and recommended practices for the installation of emergency block valves.

- API RP-553 "Refinery valves and accessories for Control and Safety Instrumented Systems" American Petroleum Institute, 2007.
- NFPA 30 "Flammable and Combustible liquid Code", National Fire Protection Association, 2012
- NFPA 497 "Recommended Practice for the Classification of Flammable Liquids, Gases, or Vapors and of Hazardous (Classified) Locations for Electrical Installations in Chemical Process Areas" National Fire Protection Association, 2012
- API RP-500 "Electrical Area Classification" American Petroleum Institute, 2000
- NFPA-58 "Liquefied Petroleum Gas Code" National Fire Protection Association, 2011
- API RP-2510 "Design and Construction of LP-Gas Installations at Marine and Pipeline Terminals, Natural Gas Processing Plants, Refineries, Petrochemical Plants, and Tank Farms", 4th Edition, 1978. American Petroleum Institute; recommended practice.
- NFPA 85 "Boiler and Combustion System Hazards Code", National Fire Protection Association, 2011
- API RP-560 "Fired heaters for General Refinery Services" American Petroleum Institute, 2001
- API RP-556 "Instrumentation, Control and Protective Systems for Gas Fired Heaters", American Petroleum Institute, 2000.
- API RP-520 "Part I: "Sizing, selection, and installation of pressure-relieving devices in refineries". Part II – "Installation" American Petroleum Institute, 2000
- API RP-521 "Guide for pressure-relieving and de-pressuring systems", American Petroleum Institute, 1998
- Pipeline Safety Act 2011: U.S. Congress, Law 112-90 – Jan 3, 2012.

(44ac) Contours When Pipelines Meets Facilities

Karthick Deivasagayam, Loss Prevention, SNC Lavalin Inc, Calgary, AB, Canada

Pipelines involved in oil & gas sector carries risk of fires and explosion along with the actual pipelines. This risk of fire and explosion are vulnerable in the places it could affect and become even more vulnerable in places of proximity of facilities where the risk could turn out into initiating of a major catastrophic events.

This should be addressed in the way of doing things.

QRA done for the most of the projects are specific to their scope alone. Boundary limits of the QRA Study is the same as that as the contractual scope of work. Whereas the risk and hazard boundaries are not contractual but as long as it could affect.

This paper will discuss about few pipeline accidents their consequences and the probable effect if they occurred in a different location (close to the facilities) and from discussing about the potential of what a pipeline accident can do and the focus will shift to the way these risks are approached the way what we do.

This paper will also discuss the way the risks of pipeline projects are captured and also drawbacks and grey areas of the existing approach. This paper also recommends the proposed approach that could be followed for the future risk assessments.

(44ae) Ozone Flammability and Its Affect On the Lower Flammable Limit of Simple Hydrocarbons

Paul Osterberg, Thomas Johnson and Thomas Frawley, Fauske & Associates, Burr Ridge, IL

As a result of ozone's strong oxidizing capabilities and ease of generation on an as-needed basis, ozone applications in process operations are gaining momentum. Previous work has demonstrated that ozone is flammable at high concentrations with a lower flammable limit of 8.4 vol.% in oxygen. However, there is limited data regarding the effect of ozone on the lower flammable limit of other gases known to be flammable in air. This work investigated the lower flammable limit of methane, ethane, and propane in various concentrations of ozone at atmospheric pressure and temperature. The tests were conducted in a 2-L vessel using a fuse wire ignition source at 1 percent ozone intervals spanning from no ozone to the LFL of ozone in the vessel. It is anticipated that the addition of ozone will decrease the LFL of each flammable gas as the concentration of ozone increases.

Evaluate Plant-Wide Safety Of Your Interlock System

Principal Author: Mohammed Naved Khan
Ingenero Technologies India Pvt Ltd.
Mumbai, India
nkhan@ingenero.com

Presenter: Swapnil Pathak
Ingenero Technologies India Pvt Ltd.
Mumbai, India
spathak@ingenero.com

Co-author: Jim Brigman
Ingenero Technologies Inc.
Houston, Texas, USA
jbrigman@ingenero.com

Prepared for Presentation at
American Institute of Chemical Engineers
2013 Spring Meeting
9th Global Congress on Process Safety
San Antonio, Texas
April 28 – May 2, 2013

UNPUBLISHED

Evaluate Plant-Wide Safety Of Your Interlock System

Principal Author: Mohammed Naved Khan
Ingenero Technologies India Pvt Ltd.
Mumbai, India
nkhan@ingenero.com

Presenter: Swapnil Pathak
Ingenero Technologies India Pvt Ltd.
Mumbai, India
spathak@ingenero.com

Co-author: Jim Brigman
Ingenero Technologies Inc.
Houston, Texas, USA
jbrigman@ingenero.com

Keywords: IRM, interlock relationship, interlock-process interference, holistic, isolated information, PHA, mitigation

Abstract

Interlocks serve as important safety systems in industrial settings, where they protect equipment from damage and employees from toxic and harmful releases from that equipment arising out of unsafe conditions. The same safety systems can lead to damage of process systems, if not analyzed properly. In some of the cases we encountered, initiation of interlocks in equipment caused severe damage to other equipment (due to change in composition) or initiation of interlock caused damage to the same equipment it was intended to protect (due to incorrect sequencing of interlocks).

The information available with the interlock design documents (like cause and effect diagrams) is limited to the vicinity of that particular interlock. It does not give any information about the relationship of the interlock, under study, with other process equipment or even with other interlocks (except for initiation). The relationship between safety interlocks and process system arises out of the behavior of the process system (e.g. change in composition) to the actions taken by the interlock. This paper proposes how a holistic view for these behavioral responses can be obtained by developing the IRM (Interlock Relationship Matrix). IRM is a single process document (excluding instrumentation and logic details) that provides information on the effects of all interlocks on each element of process system. Developing IRM is crucial in evaluating the interlock-process interference (behavioral responses) and is beneficial in managerial decision making to assess the likely damage to equipment and provide mitigation thereof.

1. Introduction

Interlocks form the most critical and independent layer of safety in industrial settings. They are primarily designed for:

- Equipment - protect the equipment from damage due to unsafe conditions (eg. over speed) and the resulting economic loss arising out of replacement and downtime.
- Personnel - protect the personnel from exposure to harmful and toxic materials that may get released from equipment damage (e.g. Seal leaks from rotating equipment).
- Plant - maintain the integrity of the plant (e.g. flare mitigation methods employed to limit the flaring during global utility failures and avoid exposure of un-burnt hydrocarbons to the general public).

The design intention of the approaches employed for interlock systems are two-fold; first is the isolated view to protect the equipment at hand and second is the global view to maintain integrity of the plant. Although the approach seems to be looking fair, they (even the global interlock system) lack in considering the interlock-process systems relationships in designing interlock systems.

The very definition of an independent protection layer provided by interlock systems (whether a single or global interlock) has an inherent element of relationship between interlock actions and the behavior of process systems i.e. the actions of the interlock system are taken independent of the responses of process system. This relationship may or may not be relevant as far as the functions of that particular interlock is concerned, but in no case it should cause upset in other process units that could jeopardize their integrity and cause irreparable damage. In the following sections, we will look into some of the incidences that lead to irrevocable equipment damage as a direct result of the interlock-process interference and we will suggest an approach as to how a holistic view for these behavioral responses of the process systems to interlock actions can be obtained for effective mitigation. Also the major points of differences between global interlock systems and holistic view of individual interlock systems will be highlighted.

2. Interlock Design Information

The design of interlock systems starts with the development of interlock diagrams or narratives. They provide key information as to the analysis and identification of scenarios that can cause damage to particular equipment (initiating causes) and the actions that are required to be taken for the protection of that equipment. They are of the following three types:

2.1 Narratives

Narratives provide descriptive statements for the various elements of the interlock system. Narratives are a short description stating the primary function of the interlock, events that can lead to unsafe conditions (initiating causes) and finally actions required by the interlock to mitigate these events. They are written in a simple but precise manner without making use of detail identification tags.

2.2 Cause and Effect (C&E) Diagram

C&E diagrams are the next level of detail information about the interlock. C&E diagrams are presented in a tabular format, detailing the initiating causes and the actions taken in the form of individual tag numbers, usually initiating causes are activation of suitable switches due to unsafe conditions and corresponding actions taken (opening or closing a valve). Actions taken are segregated for each initiating cause. They also provide information about time delays. C&E diagrams provide detail accounting of each element of the interlock.

2.3 Logic and Ladder Diagram

The information provided in logic and ladder diagrams is specific to the instrumentation part of interlock system. They provide the circuit diagrams of the relay logic hardware used for the interlock system under consideration.

3. Limitations of Design Information

Interlock design information is limited to the vicinity of the particular interlock under consideration. The actions taken (or required) are bound by the isolated view of the equipment to be protected. They are limited in their ability to evaluate and analyze the effect of actions taken by the interlock systems on other process systems.

The structure of the C&E and ladder-logic diagrams is incompatible with identifying and analyzing such kind of relationships. Interlock narratives can incorporate this information in a piece-meal fashion (similar to the way information is presented in a narrative - one interlock at a time), however the utility of such disintegrated information is questionable due to the following facts –

- Language - The information presented in narratives is written in easy to understand language and hence lacks a structured and objective approach.
- Lack of specific information - They are prepared usually at a stage where system specific detail information is not available and lot of re-work might be required to make them consistent and compatible to receiving detail information.

4. Need for Holistic View

We will discuss two incidences where interlock activation in an equipment led to severe damage in the same or other equipment.

4.1 Incident one – Activation of Interlock in Acetylene Hydrogenation Reactor of Ethylene Unit

Ethylene cracker unit was operating with an acetylene hydrogenation reactor located between fourth and fifth stages of cracked gas compressor after Deethanizer and before the chilling train that produces tail gas (front end configuration). Tail gas, usually, is a methane-rich stream that is used as fuel in cracking furnaces. Fuel gas system for crackers also has backup from imported LPG to handle contingencies.

For this case, interlock was designed for shutting down, boxing up and depressurizing acetylene reactor to purge out reactive hydrocarbons from the catalyst bed and prevent runaway reactions that might cause a meltdown of the reactor. However, shutting down the reactor also meant stopping of tail gas flow to furnaces. Thus LPG from backup system was lined up, which changed the calorific value of fuel to the furnaces. The cracking furnace temperature controller was working without duty correction to account for change in composition and, therefore, was sluggish in taking appropriate corrective actions and closing the valves. This time gap led to increased heat input to the furnaces which resulted in a spike in the tube metal temperatures of the coils and subsequent coil damage. Thus to protect acetylene reactor, interlock actions lead to damage of cracking furnace coils due to inappropriate handling of fuel gas composition change.

In this case, the behavioral response of process system to the actions of interlock was change in fuel gas composition to the cracking furnaces. This information was not apparently or directly visible from interlock design information.

4.2 Incident two - Activation of Total Shutdown interlock of Cracking furnace of Ethylene Unit

Cracking furnaces in Ethylene units usually have two levels of interlock shutdown in case of upset, first is steam standby with isolation of hydrocarbon feed and operating the firebox in hot condition and second is the total shutdown where firebox is cooled.

Usually, first interlock is designed for faster restart after a partial shutdown by sustaining close to operating conditions inside the firebox and maintaining minimum inventory of BFW in the steam drum. The second interlock is designed to maintain integrity of the furnace and cool the firebox, which might be operating close to 1100 deg C, to a safe condition. Both interlocks have different functional requirements and are designed in a way so as to be independent of each other. However, from process point of view, initiation of first interlock also provides the necessary time lag for controlled rate heat release from the firebox, by maintaining minimum firing, before the cooling down of the firebox is initiated by total shutdown interlock.

For the case under consideration, in order to maintain minimum BFW inventory for first shutdown, the trigger was set at significantly low-level point. This resulted in activation of both interlocks with minimum time delay resulting in rapid cooling of firebox. The coke layer deposited inside the coils has very different coefficient of expansion than coil metallurgy leading to unequal contraction and subsequent mechanical failure of coils.

The incident gives the interlock interference or dependence that may be sometimes necessary due to process behavior. Design information of both interlocks, when observed in isolation, was sufficient to perform their designated tasks. However without a holistic view of looking at them, integrity of the equipment was jeopardized.

With the above-discussed instances, it is evident that sole reliance on interlock design information to provide the hindsight necessary to avoid damage to other equipment is insufficient. It requires a holistic view, encompassing all individual interlock systems, to study the interlock-process relationship for each element of the process system to better understand these relationships rather than piece-meal and isolated interlock design information.

One can argue that global interlock systems responsible for maintaining the integrity of the entire plant can take care of these relationships. We think that such a system is practically infeasible. The reasoning starts with the very definition of an interlock which states that it is the actuation of the initiating causes that triggers the interlock responses irrespective of the process. For global interlock systems to take care of interlock-process relationships, all the elements of the process system must be allowed to trigger them (along with the upsets in global utility networks).

Another approach could be to trigger specific interlock systems put in place to protect these relationships. The practicality of such an interlock system is bound to fail because of two reasons. First the investment required in putting up these extra interlock systems or triggers is huge as the permutations and combinations of initiating causes from all process elements is gigantic. Second it can lead to loss of earnings and operational prudence due to the higher cost of production for increased number of instances of unsteady plant operation due to spurious activations.

5. IRM

We propose to develop IRM as a tool to give the necessary holistic view for understanding the interlock-process relationship. IRM is an acronym for Interlock Relationship Matrix. It is the final outcome of the study for evaluation of interlock actions and is a single document that shows the effect of actions of each interlock systems on all elements of the process system.

It is prepared in a tabular format with interlock systems arranged in rows and process system elements arranged in columns. A typical IRM is shown in Table 1 for reference layout.

Table 1. Typical IRM Layout

Elements of Process System	Element 1	Element 2	Element 3	Element 4	...
Interlock Systems					
Interlock 1					
Interlock 2					
Interlock 3					
...					

The effect of each interlock action is documented in the corresponding column for process elements. For example, if the actions of Interlock 2 change the feed composition of Element 3, then "change in feed composition" is typed in cell corresponding to Interlock 2 and Element 3.

IRM should include only those effects of interlock actions that have the potential to cause significant damage to process elements. This will avoid ambiguity and redundancy of information. Populating IRM with all effects of interlock actions reduces the utility of the tool. Existing safeguards should also be considered for evaluating the potential damage to the equipment. If damage is possible after considering existing safeguards, only then IRM should be updated with the relevant effect. For earlier example, it can be analyzed whether change in feed composition can cause damage to the equipment if it is not corrected (by the absence of density or composition meter) for changed composition.

However, reference of existing safeguards, for effects that have potential for damage, can be made in IRM to ensure and document that the analysis of the interlock actions has been conducted.

6. Why IRM

There are generic approaches available for evaluating hazards for process system. They can be studied under two categories: Software Evaluation and PHA.

6.1 Software evaluation techniques

These techniques include Software Systems Hazards Analysis like Fault/Event Tree Analysis, Failure Mode and Effects Analysis, Software Failure Mode, Effects, and Criticality Analysis, etc. They are deemed to analyze and identify whether software system components' operation or failure (functioning or lack of functionality) could result in hazards for process systems. It begins when the components are designed sufficiently, and is updated as their design matures to take care of the potential hazards.

Functional difference between these analyses and IRM is that they are more concerned with the effects of software (interlock) systems on process elements in case the software goes rogue and malfunctions (lack of functionality). Moreover, the analysis of components' operation (functionality) on process systems is unable to understand interlock-process relationship as it is concerned only for segregating and avoiding intermixing interlock (safety actions) and process (control actions) systems.

Whereas, in the case of IRM, it analyzes the effects of interlock actions on process element considering that software (interlock) system has fulfilled its functionality and there is no sharing of functions from either side. It is not intended to find any design fault in software (interlock) systems but to evaluate and understand the probable responses of process system to software (interlock) system actions.

6.2 PHA

PHA techniques include operational and process hazards analysis like HAZOP, LOPA, HAZAN, etc. These techniques use guidewords in combination with process parameters to evaluate deviation scenarios from normal operation and verify the utility of existing safeguards. If the existing safeguards are not sufficient, new safeguards are suggested. Safeguards can include interlocks, process control, alarms, relief valves, operating procedures, etc.

Key difference between PHA and IRM is that PHA considers only the process parameters as the potential causes for deviation and subsequent consequences. To mitigate such consequences, PHA provides the scope for defining actions required by a safety system. However, it does not evaluate the actions of a safety system like an interlock as probable causes of deviation.

IRM fills this gap of analyses of actions of an interlock as potential causes of deviation for the process elements and gives insight to unearth latent harmful events of interlock actions.

7. Developing IRM

IRM can be developed in a structured way such that the aftereffects of interlock actions can be analyzed under the considerations of core parameters critical to the industrial setting. This approach will avoid ambiguity and subjectivity.

Simple IRM layout discussed in Table 1 can be modified to include critical parameters as shown in Table 2.

Table 2. Typical IRM Layout with critical parameters (cp)

Elements of Process System	Element 1	Element 2	Element 3	Element 4	...
Interlock Systems					
Interlock 1	*cp1:*	*cp1:*	*cp1:*	*cp1:*	...
	cp2:	*cp2:*	*cp2:*	*cp2:*	...
	cp3:	*cp3:*	*cp3:*	*cp3:*	...
	...	...	...	...	...
Interlock 2	*cp1:*	*cp1:*	*cp1:*	*cp1:*	...
	cp2:	*cp2:*	*cp2:*	*cp2:*	...
	cp3:	*cp3:*	*cp3:*	*cp3:*	...
	...	...	...	...	...
Interlock 3	*cp1:*	*cp1:*	*cp1:*	*cp1:*	...
	cp2:	*cp2:*	*cp2:*	*cp2:*	...
	cp3:	*cp3:*	*cp3:*	*cp3:*	...
	...	...	...	...	...
...	...	...	...	...	...

For the refining and petrochemical industry, critical parameters are pressure, temperature, flow and composition. Other parameters like chemical reaction, liquid level, etc. can be evaluated from these four parameters. Each action of an interlock system should be evaluated with respect to above four parameters for understanding the relationship with each process element.

We will prepare IRM for the two incidents discussed in Section 4 as case study (Table 3 and 4). These tables will form a subset of the overall IRM for the plant.

Table 3. IRM for interlocks and equipment involved in Incident 1

Elements of Process System	Cold box	Cracking Furnace
Interlock Systems		
Front End Acetylene Reactor shutdown interlock	*Pressure:* no change, cold box is boxed up	*Pressure:* no change, OSBL pressure controller is designed for backup
	Temperature: no effect, temperature will rise gradually due to ambient heating	*Temperature:* no effect
	Flow: no flow to cold box resulting in loss of tail gas and H2	*Flow:* flow of tail gas will be replaced by LPG

Elements of Process System	Cold box	Cracking Furnace
Interlock Systems		
	Composition: no effect on cold box streams composition	*Composition:* methane-rich tail gas will be replaced by propane-butane rich LPG increasingly the volumetric calorific value, no heat value correction available. Likely to increase firing duty due to sluggish temperature control of furnace coils.

Table 4. IRM for interlocks and equipment involved in Incident 2

Elements of Process System	Cracking Furnace
Interlock Systems	
First shutdown interlock of Cracking Furnace	*Pressure:* no change in firebox and furnace coil pressure
	Temperature: firebox temperature is maintained at lower level by minimum firing.
	Flow: less flue gas flow due to low firing
	Composition: NA
Second shutdown interlock of Cracking Furnace	*Pressure:* firebox pressure will rise to ambient due to unavailability of ID fan
	Temperature: firebox temperature will reduce rapidly to ambient due to fuel gas shutdown. Controlled rate heat release is required to avoid rapid cooling by maintaining minimum firing conditions (first shutdown interlock) for some duration or stopping flue gas instantly by closing the damper. However, neither the duration of firebox operation under first shutdown is sufficient, nor the damper will close to allow controlled rate heat release and avoid thermal shock. Likely damage to coils.
	Flow: no flue gas flow
	Composition: NA

8. Integration of IRM with PHA

A structured IRM described in Section 7 provides compatibility to integration with PHA techniques such as HAZOP, LOPA, etc. PHA technique like HAZOP analyzes each and every parametric deviation on process system that is applicable for an industrial setting. For refining and petrochemical industry, the parameters include pressure, temperature, level, flow, composition, chemical reaction, mixing, contaminants, special procedures, etc.

Development of IRM along with PHA study starts by identification of a subset of critical or primary parameters out of the PHA parametric set on which other parameters depend (like pressure, temperature, flow and composition). It then incorporates and analyzes actions of a safety system (interlock) as potential sources of deviation (in addition to the parametric deviations) for each process element (equipment or node) where the deviations of interlock actions are evaluated under the guidance of critical parameter subset.

9. Conclusion

Interlock design information, in the form of narratives, cause and effect diagrams or ladder/logic diagrams, is limited to the vicinity of the interlock system. It is unable to evaluate and analyze the effect of actions taken by the interlock systems on other process systems. This inability might result in irrevocable damage of equipment or potential personnel exposure to toxic materials as evident from the two case studies presented in Section 4. These damages can be avoided by understanding the interlock-process relationship to gauge probable process responses to interlock actions.

The interlock-process relationship is not visible explicitly in evaluation techniques like Software Systems Hazards Analysis (Fault/Event Tree Analysis, Failure Mode and Effects Analysis) and PHA (HAZOP, LOPA, HAZAN). These techniques intend to eliminate operational hazards caused by either malfunctioning of the software (interlock) system or deviations in process parameters. However, they do not take into consideration the probable causes that can occur in process elements after the actions of a software (interlock) system have successfully executed.

IRM (Interlock Relationship Matrix) can overcome the isolated nature of interlock design information by providing a holistic view of all interlock systems and process elements in the plant (in the form of a matrix) and evaluate critical interlock-process relationship under the guidance of critical parameters applicable for the industrial setting. It can also fill the gap in other evaluation techniques by analyzing actions of a safety system (interlock) as potential causes of deviation for the process elements. It provides efficient incorporation of new interlock systems or functions for existing process systems and vice-a-versa by giving a ready platform for analyzing the interlock actions on the process system. IRM is a managerial decision making instrument that provides valuable information to assess the likely damage to equipment and provide mitigation thereof.

Understanding interlock-process relationship is crucial to avoid jeopardizing mechanical integrity of the process system by the very interlock system installed to protect them and IRM makes it happen.

10. References

1. "Guidelines for Hazard Evaluation Procedures", 3rd edition, Center for Chemical Process Safety (CCPS), John Wiley & Sons, 2008.
2. "Lees' Loss Prevention in the Process Industries, Hazard Identification, Assessment and Control", Volume 1, 3rd edition, Sam Mannan, Elsevier Butterworth-Heinemann, 2005.

Additional references not cited:

1. "Instrumentation and Control systems", Princeton Plasma Physics Laboratory, ES&HD 5008 Section 2, Chapter 10, Revision 6, 2005.
2. "Code Walk-Through", 1984, Dunn, Robert, Software Defect Removal, McGraw-Hill, Inc.
3. "Event Trees and their Treatment on PC Computers", Limnious, N. and J.P. Jeannette, Reliability Engineering, Vol. 18, No. 3, 1987.
4. "A Guide to Hazard and Operability Studies", Chemical Industry Safety and Health Council of the Chemical Industries Association, Alembic House, London, UK.
5. "HAZOP and HAZAN", Klutz, T.V., Institution of Chemical Engineers, UK, 1986.
6. "Nuclear Surety Design Certification for Nuclear Weapon System Software and Firmware", Air Force Regulation 122-4, Department of the Air Force, 24 August 1987.
7. "Petri Net Theory and Modeling of Systems", Peterson, J.L., Prentice Hall, 1981.
8. "Procedures for Performing A Failure Mode and Effect Analysis", MIL-STD-1629A, Department of Defense, 24 Nov 1980.
9. "Generic Techniques in Reliability Assessment", Fussel, J., Noordhoff Publishing Co., Leyden, Holland, 1976.

Virtual Reality Applied to Training and Process Safety: A Semi-Industrial Boiler Case Study

Carlos André Vaz Junior
Departamento de Engenharia Química, Escola de Química, Universidade Federal do Rio de Janeiro
Centro de Tecnologia, Bloco E, Sala E-209, Ilha do Fundão
cavazjunior@eq.ufrj.br

Thiago Souza de Oliveira
Escola de Química, Universidade Federal do Rio de Janeiro
Centro de Tecnologia, Bloco E, Sala E-209, Ilha do Fundão
thiagowsousa@gmail.com

Maurício Bezerra de Souza Junior
Departamento de Engenharia Química, Escola de Química, Universidade Federal do Rio de Janeiro
Centro de Tecnologia, Bloco E, Sala E-209, Ilha do Fundão
mbsj@eq.ufrj.br

Prepared for Presentation at
American Institute of Chemical Engineers
2013 Spring Meeting
9th Global Congress on Process Safety
San Antonio, Texas
April 28 – May 1, 2013

UNPUBLISHED

Virtual Reality Applied to Training and Process Safety: A Semi-Industrial Boiler Case Study

Carlos André Vaz Junior
Departamento de Engenharia Química, Escola de Química, Universidade Federal do Rio de Janeiro
Centro de Tecnologia, Bloco E, Sala E-209, Ilha do Fundão
cavazjunior@eq.ufrj.br

Thiago Souza de Oliveira
Escola de Química, Universidade Federal do Rio de Janeiro

Maurício Bezerra de Souza Junior
Departamento de Engenharia Química, Escola de Química, Universidade Federal do Rio de Janeiro

Keywords: virtual reality, process simulation, training, 3D plant model

Abstract

The short-term time given to learn a particular task associated with the demands for higher levels of production to increase profits are potential reasons for the generation of accidents. Seeking to overcome this situation in training a large number of operators in a similar environment in which they will work, virtual applications were developed. A 3D reality, integrating phenomenological simulation developed in MATLAB, and Virtual Reality (VR) using Plant Design Management System© (PDMS), were applied in this work. Process simulators using virtual reality are applied in several areas, like flight simulators and firefighters' training. Additionally, the developed 3D plant model based tool allows different safety tasks, such as the optimum location of fire extinguishers, fire hoses and escape routes, among others. As a case-study, a semi-industrial boiler, located in Universidade Federal do Rio de Janeiro, Rio de Janeiro, Brazil, was chosen. The interaction between the user and 3D model was addressed by developing a process data interpreter, so that the model may been modified on the basis of actions taken by the operator and by the phenomenological simulation. As the developed tool improves the operator's understanding of the process, it may minimize rework, accidents and damage. Additionally, this methodology may be replicated in other industrial plants.

1. Introduction

Combine safety process, higher levels of production and environment cares is a permanent challenge to be overcome. Face this fact becomes much harder in the current days due to the increasing complexity of plant process, much ambitious quality range and the short-term time given to learn a particular task. In particular, the complexity of the chemical industry also affects in process safety.

Modern chemical processes use several electronic controls and layers of protection. However, even with all the complex components and systems, the human error still are the responsible by an expressive number of incidents in the industry [1]. It is understood as essential to the operator the full knowledge of the system in which he works. But with the growing demand for increased productivity, professionals have with this feature becomes increasingly rare.

For this scenario, industries are increasingly opting for the use of Virtual Reality (VR) training, where the VR can be defined as a computer-generated three-dimensional environment, updated in real time, allowing human interaction across multiple input-output devices [2]. Virtual reality is used in several studies that allow a greater understanding of safety and process control in many areas of engineering, such as analyzes of alarms allocation, analysis of electrical area classification, determination of routes of leakage, among others.

This work has as objective the development of a simulation tool of operational dynamics of industrial equipment, engaging in 3D graphical representation through an electronic model. To improve the study and the application, this tool was implemented to a plant containing a semi-industrial boiler, as a case study.

2. Material and Methods

In this work, the integration between process simulation and 3D graphical representation was done to provide the interaction with the user. Two specific tools were used: MATLAB and AVEVA Plant Design Management System (PDMS). The MATLAB was used to develop the phenomenological model of the boiler studied in the plant. The PDMS was used to develop the 3D model of the plant. PDMS also work as a bridge between data generated from process simulator (MATLAB) and the 3D model itself in PDMS. The system was used to train operators and safety projects.

2.1 virtual reality

The PDMS is a software engineering that has the capability of modeling three-dimensional CAD (Computer-Aided Design), allowing multiple concurrent users on a single model. The software can be used in several areas. He has a complete interactivity with the environment detailing dimensional, possessing style similar to that adopted by the software of the Windows environment.

The software has a number of benefits that among the most important, is the ability to extract a list of quantitative component containing its data of materials, weights, and other information. Importantly, such a list, depending on the size of the unit may have more than a thousand pages, which greatly increases the chance of inconsistencies, considered the possibility of creating this document manually. Thus, the software allows an important gain in terms of time, eliminating delays, the automated generation of various documents and reports, plans and two-dimensional views and isometric piping, and other documents. The user can check the information contained in the 3D model and compare them to equivalent 2D designs.

2.2 phenomenological simulation

The boiler operation was simulated in MATLAB software. To provide this, was used the model created by ÅSTRÖM & BELL [3], who presented a dynamic model for a boiler, based on basic principles of conservation of mass and energy. Justifies the adoption of this model here for its simplicity and because it has been validated with real data by the authors [3].

2.3 case study: boiler studied

The usability of the tool developed in this work has been tested and applied in a small boiler, model CV-500-VDM. This boiler is located in a pilot plant, which in turn is in the Utilities Central Laboratory of Chemical Engineering, Escola de Química, Universidade Federal do Rio de Janeiro, Rio de Janeiro, Brazil. In the Figure 1 will be shown a picture taken from the plant. All the related equipment was identified in it.

Figure 1: Studied pilot plant containing the boiler simulated

3. Results and discussion

3.1 process simulation

Using the MATLAB and the developed program was possible integrate "process simulation" and "virtual reality". To verify if the data is trustful was produced some graphics and analyze their behavior through steps given in some input variables. The Figure 2 represents the temperature behavior after some different steps given in heat flow, Q.

Figure 2: Temperature behavior after steps in heat, Q

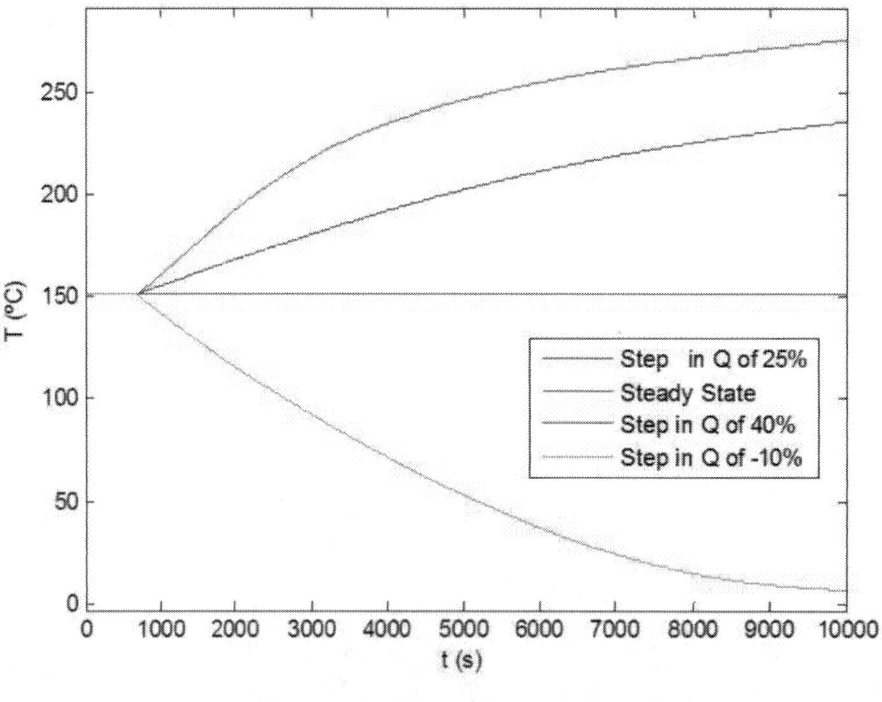

It is possible to see that if the heat is increased, the temperature consequently increase as well. By other side, if the heat is decreased, the temperature also decreases, following the dynamics of the boiler from energy and global mass balance. In the Figure 3, a similar result is obtained, where a step is given in Q and the pressure is observed after it.

Figure 3: Pressure behavior after steps in heat, Q

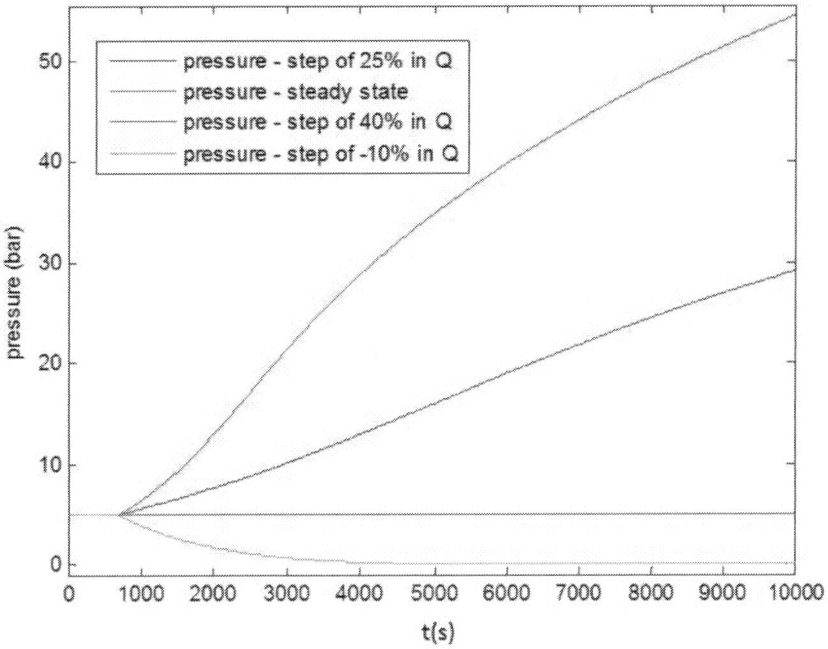

3.2 virtual reality tool

The developed tool has in summary three main functions as results. The first one is provide the integration between the simulations taken in MATLAB. The 3D model shows to the operator if the plant is operating in normal situation or if there is some equipment in alarm status. The model can shows data tables too. In the Figure 4, is possible to see some equipment analyzed by the integration system and showing if they are in normal operation (green), in maintenance (gray), in stand by (yellow) or in alarm (red). Beyond that, it is possible to see relevant data process, like pressure, temperature and level, directly in the model.

The second function is providing the training of the operator. In this work, the training was applied to "boiler starting". During the training the operator can see instructions and complementary texts. The 3D model change through the actions taken, interacting with the operator. The user has the free movement possibility during the training. In the Figure 5 is possible to see the virtual operator during the training. The water level and pump status are shown in the Figure 5 too.

Figure 4: Equipment analyzed by the integration tool

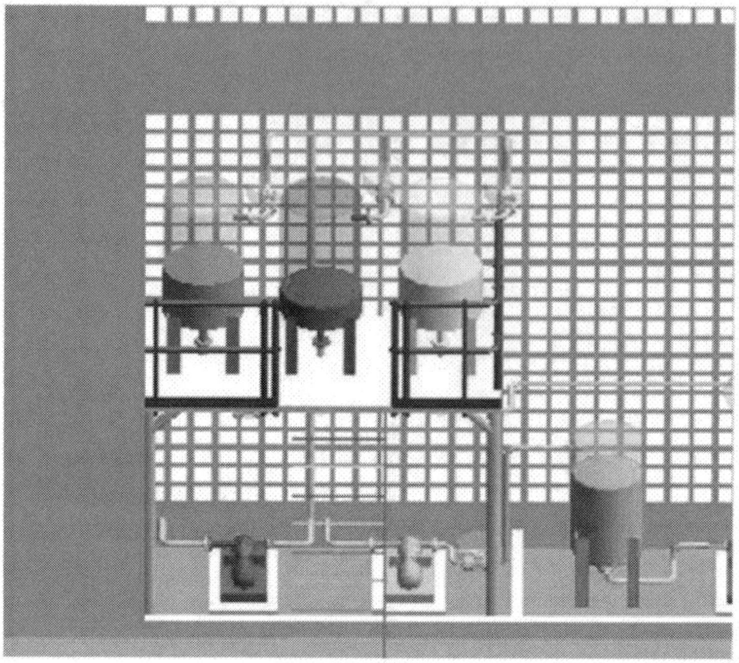

Figure 5: Model during the training

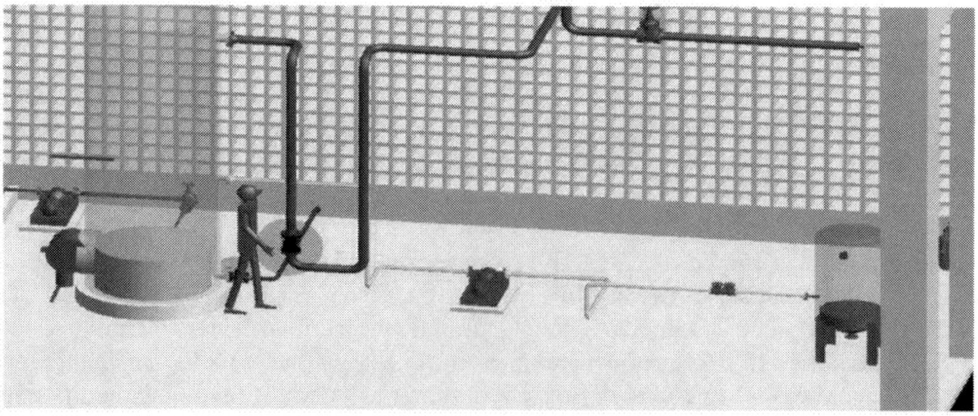

The third function is work as a "support tool", helping the safety analyzes in the plant. The safety analyzes can be applied during project of new process. The virtual reality can be used to design escape routes and emergency signaling. An example of the escape route chooses is showed in the Figure 6.

Figure 6: Virtual tool used to know the best exit to be taken in a case of emergency

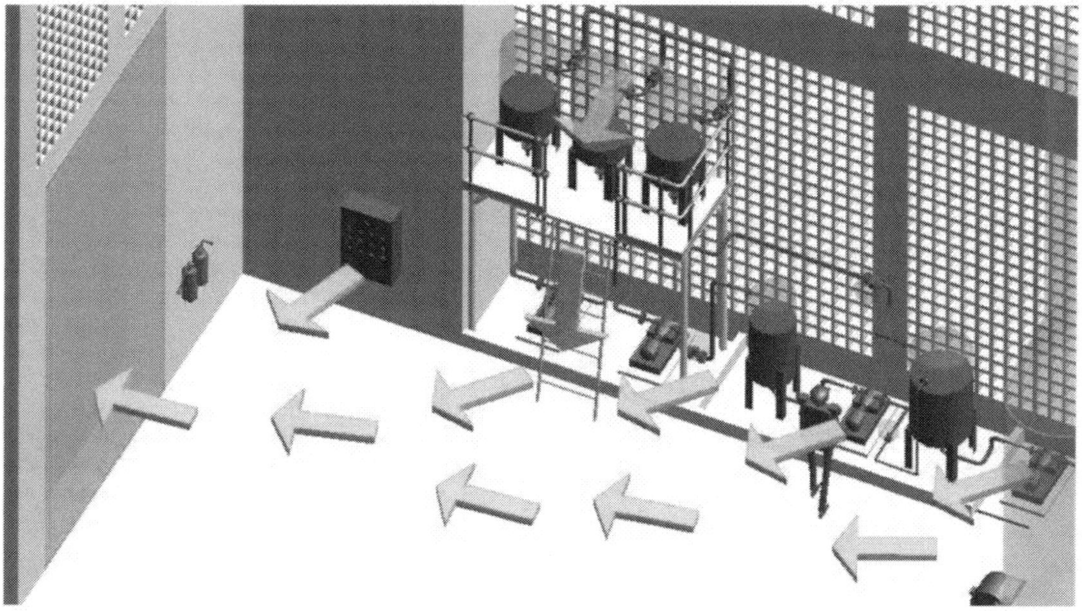

4. Conclusion

The integration of virtual reality and process simulation allow plant operators acquire knowledge about real process. This solution offers low cost and it's safer than use real equipment during operation training. In other words, performing training in the virtual reality tool developed, the user can get information about the plant without actually going to it.

About the development of a three-dimensional supervisory system prototype, it is concluded that it is valid and interesting for training purposes. With the training tool, it is possible to obtain many advantages, due to the operator interaction with the virtual three-dimensional model. So, it is concluded the development of a similar system for all industrial plants can facilitate the training of new operators, facilitating the determination of operations and/or maintenance actions in the plan. Through this, it is possible study the best way to access the site without the need to perform a complete analyzes of the activity at the site to be developed. Furthermore, the safety analyzes can be very useful for the operation team.

About study in general, it can be concluded the use of virtual reality, primarily for training purposes and in addition as a support tool to the operation unit, has much contributing to healthy plant operability. The implementation of virtual reality is becoming an increasingly viable option in view of the development of systems software and hardware capable of supporting the tools developed so that the cost of such equipment is not high as it would at the last decade of previous millennium.

5. References

[1] Lasala, K.P; Filgueiras, L.V.L; Gigley, H.; Ullman, R., *Designing Systems for Reliable Human Performance*, IEEE Tutorial Video, 1997.

[2] Pamplona Jr, A. R. S., *Hidrogeradora Virtual: Utilização de Técnicas de Realidade Virtual Desktop para o Estudo de uma Unidade Hidrelétrica de Energia*, Dissertação de Mestrado, Programa de Pós Graduação em Engenharia Elétrica, Centro Tecnológico, Universidade Federal do Pará, 2006.

[3] Åström, K. J.; Bell, R.D., *Drum Boiler Dynamics*. Automatica, v.36, n.3, p. 363-378, 2000.

Investigations on the influence of different inert gases on the laminar flame propagation in lycopodium/air mixtures

Hannes Kern
Montanuniversität Leoben
Franz-Josef Strasse 18, 8700 Leoben
hannes.kern@unileoben.ac.at

Bruno Wimmer, Harald Raupenstrauch
Montanuniversität Leoben
Franz-Josef Strasse 18, 8700 Leoben
bruno.wimmer@stud.unileoben.ac.at, harald.raupenstrauch@unileoben.ac.at

Prepared for Presentation at
American Institute of Chemical Engineers
2013 Spring Meeting
9th Global Congress on Process Safety
San Antonio, Texas
April 28 – May 1, 2013

UNPUBLISHED

Investigations on the influence of different inert gases on the laminar flame propagation in lycopodium/air mixtures

Hannes Kern
Montanuniversität Leoben
Franz-Josef Strasse 18, 8700 Leoben
hannes.kern@unileoben.ac.at

Bruno Wimmer, Harald Raupenstrauch
Montanuniversität Leoben
Franz-Josef Strasse 18, 8700 Leoben
bruno.wimmer@stud.unileoben.ac.at, harald.raupenstrauch@unileoben.ac.at

Keywords: dust explosion, flame velocity, limiting oxygen concentration

Abstract

The laminar flame velocity is a key parameter in the description of explosions. The determination of the flame velocity (laminar or turbulent) of dust/air flames is relatively complex due to the characteristics of dusts in general. In context of investigations on the limiting oxygen concentration of different powders, the influence of the oxygen concentration and different inert gases on the laminar flame propagation in lycopodium/air mixtures was investigated. In the course of this study, a new apparatus for the determination of the flame velocity was developed. The apparatus mainly consists of a vertical duct with a diameter of 140mm that is split in three main sections. The dust input system lies in the head section of the apparatus and is realized by a small conveyor screw. The bottom section of the apparatus houses the electrical spark ignition system and a gas inlet system. The gas inlet creates a steady flow of a chosen gas. The flame propagation is detected by IR photodiodes along the 2m long main section of the apparatus. The dust concentration is measured by light extinction. The determination of the flame velocity was carried out at dust concentrations of $100g/m^3$, $200g/m^3$ and $300g/m^3$ as well as around the lower explosion level of lycopodium. The oxygen concentration was varied in steps of 1Vol.-% from 21Vol.-% down to the limit where no ignition could be achieved. To investigate the influence of the inert gas on the flame propagation carbon dioxide and nitrogen were used as inert gases. The results of the recent studies are used for further studies on the limiting oxygen concentration of different dusts and the validation of a new CFD model for the simulation of dust explosions in large vessels.

1. Introduction

Dust explosion research mostly relies on sum parameters like the minimun ignition energy (MIE), the maximum explosion pressure or the limiting oxygen concentration (LOC) for the description of explosion processes and material characterization. For mathematical modeling these sum parameters show a lot of uncertainties due to the influence of experimental design. Therefore parameters like the laminar flame velocity are commonly used for the description of the combustion process itself. For combustible gases the determination of the flame velocity is relatively simple compared to combustible dusts. The biggest challenge concerning dusts is the creation of a homogeneous dust cloud with low turbulence.
In the course of this experimental study about the influence of the oxygen concentration on the flame velocity of lycopodium/air flames a new testing device was designed. With this apparatus it is possible to measure the flame speed within a 2m combustion tube, a new method for the calculation of the laminar flame speed from the results is yet to be developed. For current research we relied on present methods. Even if the laminar flame velocity can't yet be measured directly with this device, the influence of changing atmospheric conditions can be observed very well.

2. Experimental investigations

2.1 Method

Various apparatus for the measurement of the flame speed of dust/air mixtures are found in literature [1]. The testing apparatus used for current research consists of three major parts. As shown in Figure 1 the main sections of the apparatus are the top section, the main explosion tube and the bottom section. All elements are built out of a Plexiglas (PMMA) tube with a diameter of 140mm and a wall thickness of 5mm.
The head element houses the dust input system and is equipped with a flame arrestor (Figure 2). The dust input is carried out by a screw conveyor that can be regulated by the operating voltage. The input system allows the generation of dust concentrations from 50g/m^3 to 500g/m^3 (Lycopodium).
The bottom section contains a gas inlet system and the ignition source. The ignition source is an electric spark igniter (~200mJ), consisting of two 3mm steel electrodes and a high voltage generator. The gas inlet is designed as a circular tube with multiple nozzles to generate a homogeneous gas flow along the main section of the apparatus. The 2000mm long main section of the apparatus is equipped with 5 photodiodes to measure the flame propagation along the tube. The first diode is mounted 10cm above the ignition point, further diodes are placed every 40cm.

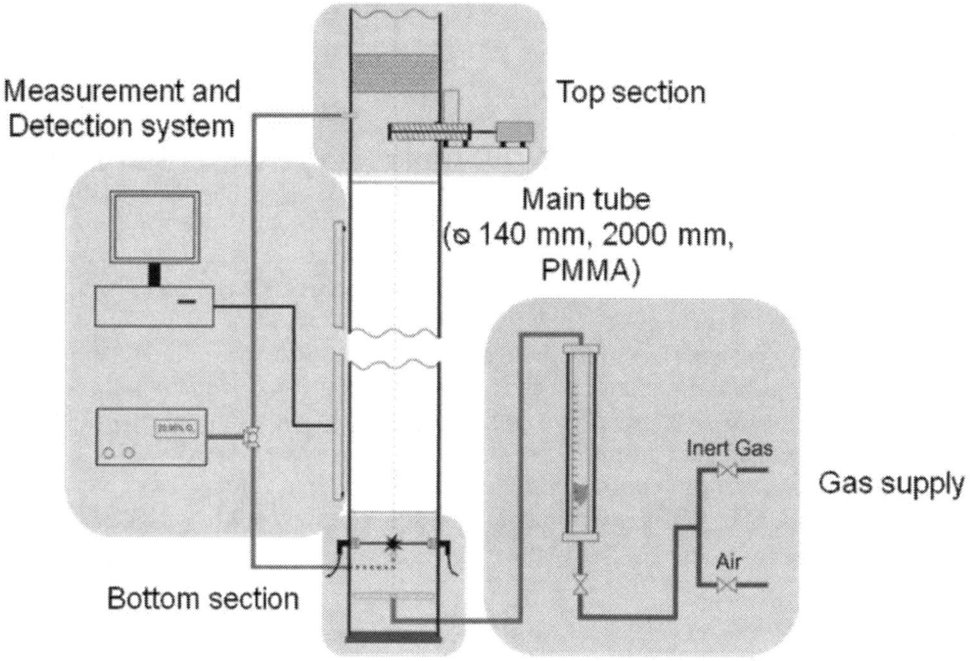

Figure 1: Experimental setup for the determination of the flame velocity

Two of the photodiodes are used to monitor the dust concentration in the tube before ignition by the extinction of IR light emitted by IR diodes. The emitter diodes are turned off shortly before the ignition by the electric spark.

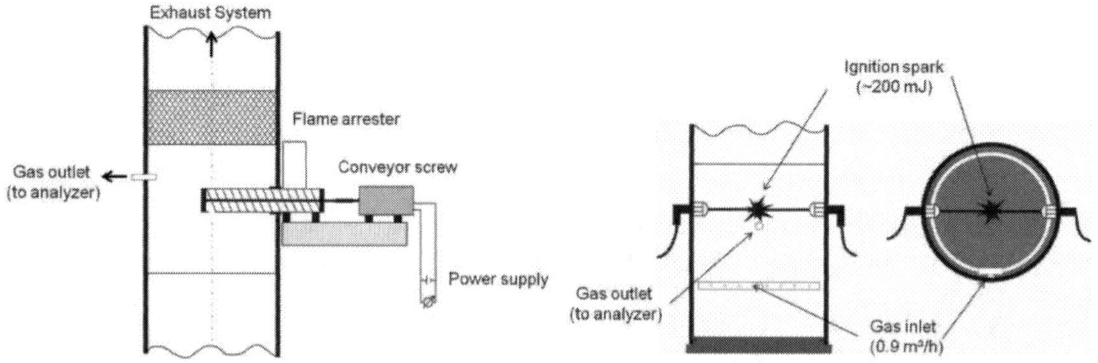

Figure 2: Top section (left) and bottom section (right) of the experimental setup

In the measuring mode a gas flow of 0,015m³/min is provided through the bottom to realize different atmospheric compositions. The dust concentration is observed throughout IR extinction. The oxygen concentration inside the tube is measured by an oxygen monitor at the head section of the apparatus. When the extinction reaches a constant level along the tube, the emitter diodes are switched off and the ignition source is activated.

2.2 *Analysis of flame propagation*

For dust explosion research in general, laminar flame velocity is a relevant parameter. Different methods for the determination of the laminar flame velocity are known, like using burners to create laminar dust flames or tube methods like described above. The data sets produced from all of these methods are more or less deviating from laminar conditions due to the experimental conditions. Figure 3 shows a sequence of pictures taken during the measurement of the flame velocity at a dust concentration of 300g/m³ and an oxygen content of about 15Vol.-% using nitrogen as inert gas. The picture shows a more or less turbulent combustion close to the ignition point and a relatively closed flame shape at around 800mm. At lower concentrations the flame tends to be more closed throughout the whole measuring distance. The flame therefore shows a more laminar character at lower concentrations. The turbulence effects may have occurred due to the falling particles themselves.

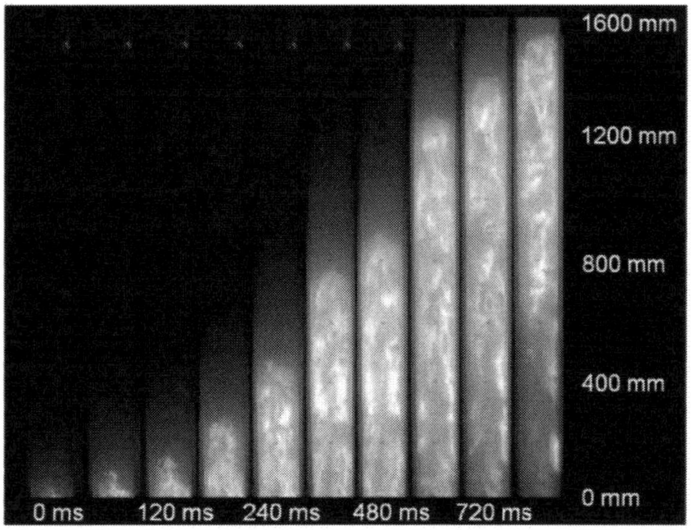

Figure 3: Flame propagation within the testing device

Andrews and Bradley [2] described a method for the calculation of the laminar flame velocity out of relations regarding shape and surface of the flame. Andrews and Bradley's approach (1) was taken into account for further investigations. A video analysis system was installed to determine the flame geometry.

$$S_L = \frac{A'}{A_{fl}} \cdot \left(S_{fl} - u\right) \qquad (1)$$

For a lycopodium concentration of 50g/m³, the flame surface (A_{fl}) and the area cross section (A') were interpolated and calculated by a Matlab code. The measured flame velocity (S_{fl}) was 0.62m/s. Calculations with Andrews and Brandly's relation resulted in a laminar flame velocity (S_L) of around 0.20m/s.

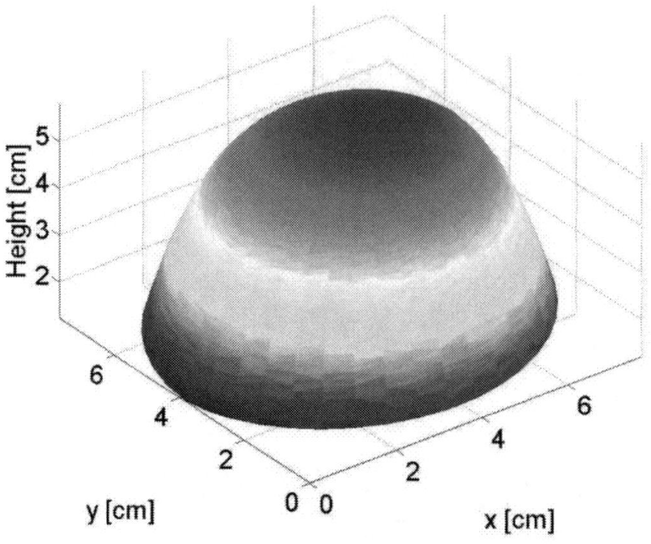

Figure 4: Flame shape at 50g/m³, surface area around 150cm²

Figure 4 shows an example for the flame shape of a 50g/m³ lycopodium/air flame. Compared to the results described by Eckhoff [1], the measured flame velocity is well in the range of velocities measured in other apparatus. For concentrations of around 50g/m³ the flame became very unstable in some measurements, so that higher concentrations were chosen for further research on the influence of the oxygen concentration. Due to the clear connection between actual and laminar burning velocity, only the actual flame speed was measured further on, to avoid deviations from the flame shape analysis.

2.3 Results

Measurements of the flame velocity (S_{fl}) have been carried out at three different dust concentrations. In order to keep the flame stable, concentrations of 100g/m³, 200g/m² and 300g/m³ were investigated. The oxygen concentration was lowered by nitrogen and carbon dioxide as inert gases. Measuring sequences started at ambient conditions (~21Vol.-% oxygen) and were lowered in steps of 1Vol.-% down to the point where no ignition could be reached.

Table 1: Flammability limits

	N_2	CO_2
Dust conc. [g/m³]	cO_2 [Vol.-%]	
100	12.0	15.0
200	12.0	15.0
300	13.0	15.5

As described in Table 1, no ignition could be achieved below 12Vol.-% during nitrogen inerting and 15Vol.-% when carbon dioxide was used. Bartknecht [3] described a relation between the oxygen concentration and the required ignition energy for lycopodium/air mixtures. Regarding the observed ignition limits during the experiments, the results show very good accordance with Bartknecht's values. For an oxygen content of around 12Vol.-% Bartknecht estimated an ignition energy between 150 and 250mJ which is well in the range of the ignition source used for the experiments. Figure 5 shows the development of the flame velocity for nitrogen inerting.

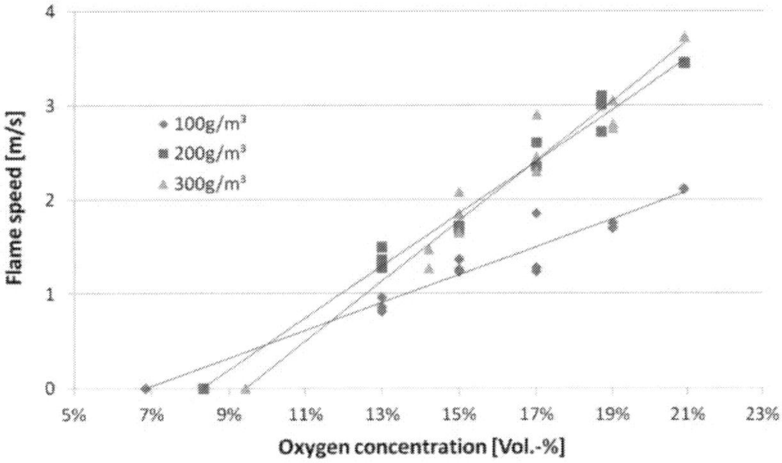

Figure 5: Development of flame speed during nitrogen inerting

At ambient conditions the flame velocity increases with the dust concentration. The increase of the flame velocity in the range from $100g/m^3$ to $200g/m^3$ is significantly bigger than from $200g/m^3$ to $300g/m^3$. This effect can be explained by stoichiometry [4].

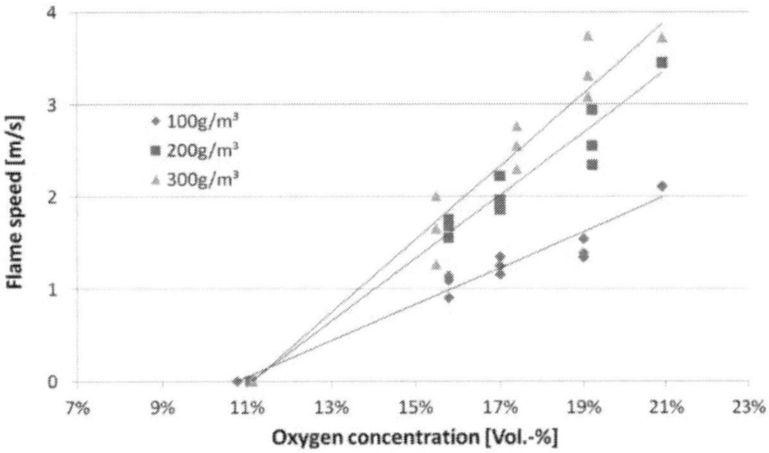

Figure 6: Development of flame speed during carbon dioxide inerting

At lower dust concentrations the combustible material is the limiting component for combustion while at higher dust concentrations oxygen becomes the limiting factor, which is why the flame velocity nearly stagnates. For all dust concentrations the flame velocity decreases with lower oxygen concentrations. The flame velocity drops faster at higher concentrations than at 100g/m³. That is also true for carbon dioxide inerting, as it can be seen in Figure 6. At Figure 5 the lines for 300g/m³ and 200g/m³ intersect at around 17Vol.-%. This is presumably due to problems with the gas measuring system and could not be observed at the carbon dioxide experiments. The values for zero flame speed are estimated.

As mentioned above values for zero flame speed were estimated to be able to compare possible results with the limiting oxygen concentrations of lycopodium. The extrapolated values are given in Table 2.

Table 2: Estimates limiting oxygen concentrations

	N_2	CO_2
Dust conc. [g/m³]	cO_2 [Vol.-%]	
100	6.8	10.8
200	8.4	11.1
300	9.4	11.1
average	8.2	11.0

Siwek [5] found a relation between the limiting oxygen concentrations (LOC) measured during nitrogen inerting and the LOC values for carbon dioxide inerting (2).

$$LOC_{CO2} = 1.36 \cdot LOC_{N2} \quad (2)$$

Due to the described problems with measurement attempts were made to recalculate the LOC for nitrogen inerting out of the measured values for carbon dioxide inerting. The estimation resulted in a value of 8Vol.-% for nitrogen inerting which is highly consistent with the 7.5Vol.-% published in BIA Report 12/97 [6] and the average value that was measured.

3. Conclusions

During the investigations on the flame speed of lycopodium/air mixtures a new apparatus for the determination of the flame velocity was developed. The results from the method regarding laminar flame speed showed good accordance with published values at ambient conditions. With the present experimental setup it was also possible to determine the influence of varying atmospheric conditions on the flame propagation in a vertical duct. Carbon dioxide and nitrogen were used as inert gases. Due to the relatively low ignition energy provided by the spark igniter, the flame propagation around the known limiting oxygen concentrations for lycopodium could not be reached. Extrapolations on the limiting oxygen concentrations based on the measured data have shown good accordance with values measured in the ISO 20L sphere. In addition to the the influence of inert gases on the flame propagation other deviations from atmospheric conditions will be investigated in future research. The results from the current experimental work are used for the validation of a new CFD model for the calculation of dust explosions in complex geometries.

4. References

[1] Eckhoff R. K. (3rd edition) (2003) Dust Explosions in the Process Industries. Gulf Professional Publishing

[2] G.E. Andrews, D. Bradley (1972), Determination of burning velocities: A critical review, Combustion and Flame, Vol. 18/1, Elsevier 1972

[3] Bartknecht W. (1993), Explosionsschutz – Grundlagen und Anwendung. Springer – Verlag

[4] Crowl D. A. (2003), Understanding Explosions, Center for Chemical Process Safety (CCPS) of the American Institute of Chemical Engineers, 1st edition, Wiley-AIChE, New York 2003

[5] Siwek R. (1989), Zuverlässige Bestimmung explosionsschutztechnischer Kenngrößen in der 20l-Laborapparatur, VDI Berichte 701, S215-262

[6] BIA – Report 12/97, Brenn- und Explosionskenngrößen von Stäuben, Hauptverband der gewerblichen Berufsgenossenschaften, November 1997

(44al) Fire Protection and Life Safety Issues in Semiconductor and Related Facilities

Jonathan Eisenberg, Rolf Jensen & Associates, Framingham, MA

Semiconductor and related process facilities contain numerous fire and explosion hazards that impact life safety and operations. This presentation discusses the major fire protection and life safety design aspects for such facilities, including specific requirements contained in the International Building & Fire Codes, as well as applicable NFPA codes and standards. Project examples will be used to illustrate how these aspects are addressed in facility and process design.

The presentation will include the following:

Location and design of Use Group H-5 occupancies – Major IBC Chapter 4, IFC Chapter 18, and NFPA 318 fire protection and life safety requirements, including vertical and horizontal location, occupancy separation, code approach for associated mechanical spaces including interstitial levels, electrical classification, ventilation, fire suppression (overhead and local), fire and gas/vapor detection, and emergency alarms.

Location and design of Use Group H-2, H-3 and H-4 storage areas – Building and fire code requirements for vertical and horizontal location, occupancy separation, electrical classification, ventilation, fire suppression, fire and gas/vapor detection, and spill control/secondary containment. Approaches and ideas on the number, location and function of flammable liquid, flammable and pyrophoric gas, and highly toxic gas storage areas, including piping and delivery philosophy.

Special suppression and detection – Fire protection system approaches for hazards found in a clean room/fab area, eg. wet bench chemistry involving flammable liquids, and processing that uses hydrogen and/or silane mixtures. Discussion of local suppression options, as well as specialty gas and flame detection. We will give our thoughts and experiences on how these systems interact with the rest of the building and the responding fire department.

Industrial Issues – Hydrogen detection; combustible dust hazards; NEC Article 500 electrical classification; explosion protection; and several examples from our industrial/chemical project experience

Licensing and permitting – Discussion of issues faced by owners of fab and clean room facilities that store and use hazardous materials, where the hazards are not clearly understood outside of the semi-conductor industry.

Lower Flammability Limits of Mixtures of Hydrogen and Methane at Atmospheric and Sub-atmospheric Pressures

Hai Le
Mary Kay O'Connor Process Safety Center, Artie McFerrin Department of Chemical Engineering, Texas A&M University
Room 200, Jack E. Brown Building, Texas A&M University, 3122 TAMU College Station, TX 77843-3122
haile@tamu.edu

Yi Liu, M. Sam Mannan
Mary Kay O'Connor Process Safety Center, Texas A&M University
Room 200, Jack E. Brown Building, Texas A&M University, 3122 TAMU
College Station, TX 77843-3122
mannan@tamu.edu, liuyi.chemical@mail.che.tamu.edu

Prepared for Presentation at
American Institute of Chemical Engineers
2013 Spring Meeting
9th Global Congress on Process Safety
San Antonio, Texas
April 28 – May 1, 2013

UNPUBLISHED

Lower Flammability Limits of Mixtures of Hydrogen and Methane at Atmospheric and Sub-atmospheric Pressures

Hai Le
Mary Kay O'Connor Process Safety Center, Artie McFerrin Department of Chemical Engineering, Texas A&M University
Room 200, Jack E. Brown Building, Texas A&M University, 3122 TAMU College Station, TX 77843-3122
haile@tamu.edu

Yi Liu, M. Sam Mannan
Mary Kay O'Connor Process Safety Center, Texas A&M University

Keywords: lower flammability limit, hydrogen, methane, pressure

Abstract

The knowledge of the LFLs of hydrogen mixtures with hydrocarbons is essential since hydrogen is present in various mixtures with hydrocarbons in many processes and applications. The lower flammability limits of various mixtures of hydrogen and methane in air were determined experimentally in a closed cylindrical stainless steel vessel (inner diameter 10.22 cm, length 100 cm) with upward flame propagation at room temperature (20 ^{o}C) and initial pressures of 1.0 atm, 0.5 atm, and 0.1 atm. It was found that the LFLs of mixtures of the same hydrogen fuel fractions did not change much when the initial pressure decreased from 1.0 atm to 0.1 atm. And pressure has more impact on the LFLs when there is more hydrogen in the fuel mixture. The application of Le Chatelier formula to predict the LFLs of the mixtures was also evaluated. Overall, Le Chatelier formula can predict the LFLs of mixtures of hydrogen and methane very well at all initial pressures.

1. Introduction

Lower flammability limit (LFL) is an important parameter used in the prevention of fire/explosion when handling combustible gas mixtures. By definition, LFL is the minimum concentration of gas in air capable of propagating flame upon ignition;[1] in other words, if the concentration of a gas is smaller than its LFL, it is no longer flammable. Therefore, LFL is widely used in the calculation of the level of inerting needed to prevent the formation of flammable mixtures.[2] In these calculations and other fire risk assessments, it is advised to select and use the proper value of LFL for the operating condition since LFL is affected by pressure, temperature, direction of flame propagation and other conditions.[1]

The literature has a vast number of LFLs of various gases determined with different apparatuses at both atmospheric and non-atmospheric conditions, e.g. high temperatures, pressures, etc. LFLs of pure gases are more readily available than LFLs of mixtures of gases. Whenever LFLs of mixtures of gases are not available, methods have been developed to predict them. One of the widely used methods is Le Chatelier formula. It is an empirical mixing rule to predict the LFL of a mixture of gases as below: [3]

$$LFL_{mix} = \frac{1}{\sum_i^n \frac{y_i}{LFL_i}}$$

where LFL_{mix} is the LFL of the mixture, y_i is the mole fraction and LFL_i is the LFL of the ith component. For mixtures at non-atmospheric condition (e.g. high pressure, temperature) the LFLs of the components at the same condition should be used to accurately estimate the LFLs of the mixtures.

The knowledge of the LFLs of hydrogen mixtures with hydrocarbons is important since hydrogen is present in various mixtures with hydrocarbons in many processes and applications. Examples of such mixtures are refinery gas streams which contain various percentages of hydrogen (up to 90%) and hydrocarbons (methane, ethane, C_3, C_{4+}.).[4] Hydrogen is also added to natural gas, consisting mostly methane, to reduce CO, CO_2 and NO_x emissions[5-7], improve flame stability,[6, 7] and extend the lean operating limit of spark ignited engines.[8] It is critical to understanding the LFLs of these hydrogen mixtures in order to handle them safely.

Reference to the literature shows that studies on the LFLs of mixtures of hydrogen and hydrocarbons at non-atmospheric conditions, particularly sub-atmospheric pressure conditions, are very limited. Therefore, to enhance our understanding about the LFLs of hydrogen mixtures, this study focuses on investigating the LFLs of various mixtures of hydrogen and methane at atmospheric and sub-atmospheric pressures. Specifically, the LFLs of mixtures of hydrogen and methane were determined experimentally at room temperature (20 °C) and initial pressures of 1.0 atm, 0.5 atm and 0.1 atm. All experiments were carried out in a closed cylindrical vessel with upward flame propagation. The application of Le Chatelier formula to predict the LFLs of mixtures of hydrogen and methane at atmospheric and sub-atmospheric pressures was also verified using the LFLs of pure hydrogen and methane generated by the same experimental method and conditions.

2. Experimental setup

The experimental apparatus and procedure is illustrated schematically in Figure 1 and briefly described below.

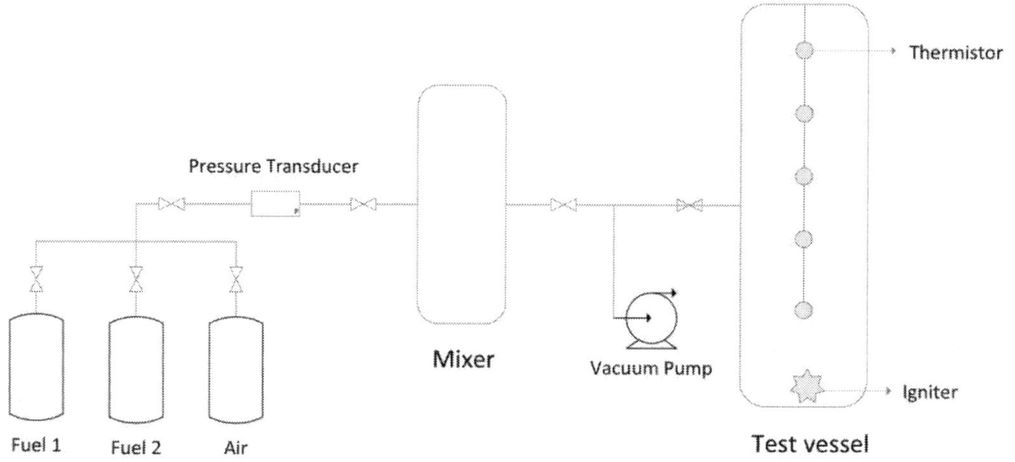

Figure 1: Experimental Setup

The LFLs of mixtures of hydrogen and methane were determined in a test vessel which was a closed stainless steel cylindrical vessel whose design was similar to that developed and used by the U.S. Bureau of Mines.[1] The vessel was 100 cm long with the inner diameter of 10.12 cm and outer diameter of 11.43cm. The igniter was installed at the bottom of the vessel and was capable of providing 10 J of energy consistently. The pressure inside the test vessel was measured by a pressure transducer placed at the top of the vessel. A set of 5 thermistors was positioned along the center axis of the test vessel to detect flame propagation. The distance between the top thermistor and the igniter was 75 cm. Further details about the configuration of the test vessel and its components can be found in our previous published work.[9]

High purity fuels and air from pressurized cylinders were loaded into the mixing vessel. The amount of the fuel and air were determined based on the partial pressure basis. The fuels and air were thoroughly mixed in a mixer which was a stainless steel vessel containing a Teflon block which slid along the vessel when it rotated lengthwise to enhance mixing. After the mixing, the test mixture was permitted to flow to the test vessel at the desired initial pressure measured by the pressure transducer. Then the mixture was ignited from the bottom, and the flame propagation was detected by the 5 thermistors inside the test vessel. Signals from the thermistors were obtained by a Keithley data acquisition card (Keithley® KPCI-3102, 225 signals per second at 0.05% accuracy) installed in a desktop computer, and the experimental process was controlled by a LabView® program. All experiments were carried out at room temperature (20 ^{o}C) and initial pressure of 1.0 atm, 0.5 atm and 0.1 atm. After each experiment, combustion products were purged from the test vessel which was then flushed with nitrogen and vacuumed to be ready for the next experiment.

A mixture is considered flammable if it can propagate flame to the top thermistor upon ignition;[9, 10] this behavior was defined as continuous flame propagation.[9, 10] For each mixture composition, 10 experiments were repeated and the probability of continuous flame propagation was recorded. Between two concentration values different by one experimental step size, one has the probability of continuous flame propagation equal or less than 50%, the other has the probability more than 50%. The former was chosen as the LFL, unless the former's probability is 0%, the later was chosen as the LFL. [9, 10]

3. Experimental Results and Discussion

3.1 Lower flammability limits (LFLs) of mixtures of hydrogen and methane at atmospheric and sub-atmospheric pressures

The experimentally obtained LFLs of various mixtures of hydrogen with methane at room temperature (20 °C) and initial pressure of 1.0 atm, 0.5 atm and 0.1 atm are shown in Figure 2. Overall, the LFLs of the mixtures of the same hydrogen fuel fractions did not change much when the initial pressure decreased from 1.0 atm to 0.1 atm. This is understandable since the LFLs of pure hydrogen and methane also changed very little with the initial pressure. When the fraction of hydrogen in the fuel mixture increased from 0% to 100%, the LFLs of the mixtures showed larger deviation from each other under the change of pressure from 1.0 atm to 0.1 atm. This suggests that pressure has more impact on mixtures of hydrogen and methane when there is more hydrogen in the mixture. However the impact is not significant since the degree of changes of the LFLs is small. It is worth to note that the LFLs of the mixtures was the lowest at 0.5 atm which means that there is an increased risk of fire/explosion when the initial pressure is reduced from 1.0 atm to 0.5 atm. Thus, more attention and precautions are recommended when handling mixtures of hydrogen and methane at pressures lower than 1.0 atm.

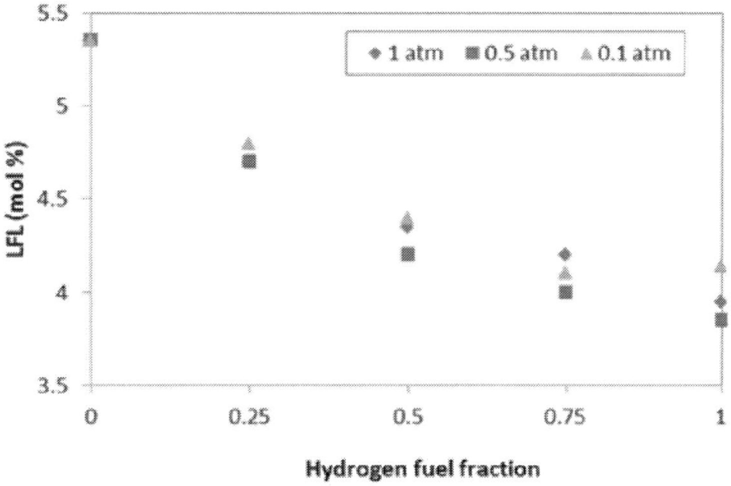

Figure 2: LFLs of mixtures of hydrogen and methane at atmospheric and sub-atmospheric pressures and room temperature (20 °C)

3.2 The application of Le Chatelier formula to the LFLs of mixtures of hydrogen and methane at atmospheric and sub-atmospheric pressures

The LFLs of mixtures of hydrogen and methane were calculated based on Le Chatelier formula using the LFLs of pure hydrogen and methane determined experimentally with the same experimental setup and conditions. Comparisons between the calculated LFLs and experimental LFLs are presented in Figure 3, Figure 4, and Figure 5 for 1.0 atm, 0.5 atm and 0.1 atm respectively. In general, Le Chatelier formula can predict the LFLs of the mixtures very well at all initial pressure conditions; in other words, the prediction of Le Chatelier formula for hydrogen and methane mixtures is not affected by changes in the pressure from 1.0 atm to 0.1 atm. The maximum deviation between experimental and calculated LFLs is 7.0 %. And the deviation is the largest when the fraction of hydrogen in the fuel mixture is 0.5.

The application of Le Chatelier formula to the upper flammability limits (UFLs) of the mixtures of hydrogen and methane will be evaluated in the future when we complete the determination of the UFLs of the mixtures experimentally.

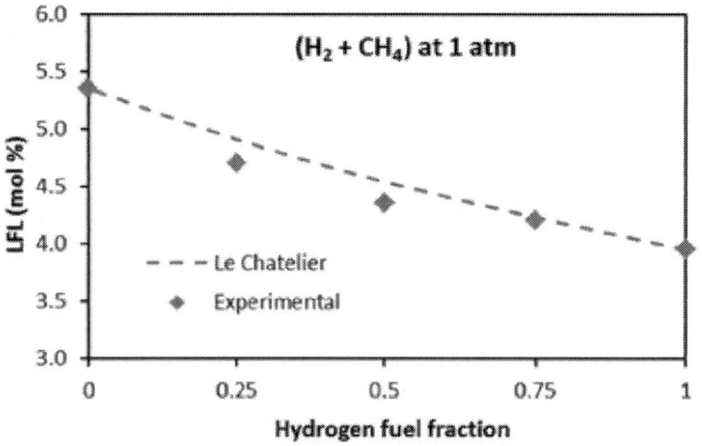

Figure 3: LFLs of mixtures of hydrogen and methane at room temperature and 1.0 atm (experimental and calculated by Le Chatelier formula)

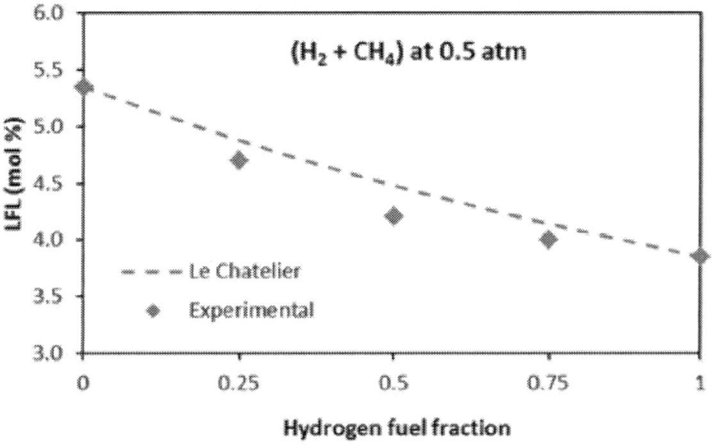

Figure 4: LFLs of mixtures of hydrogen and methane at room temperature and 0.5 atm (experimental and calculated by Le Chatelier formula)

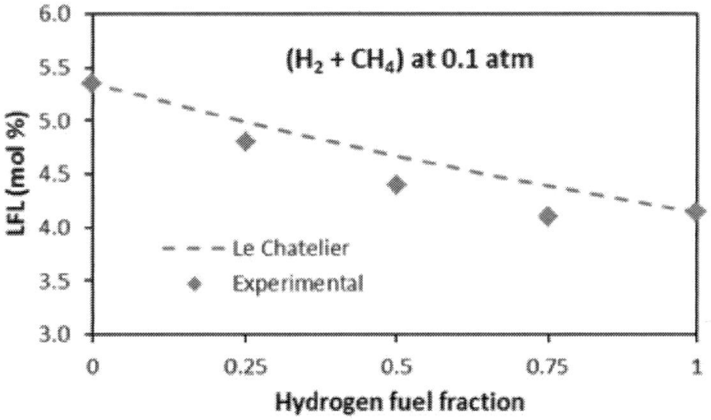

Figure 5: LFLs of mixtures of hydrogen and methane at room temperature and 0.1 atm (experimental and calculated by Le Chatelier formula)

4. Conclusion

The lower flammability limits (LFLs) of mixtures of hydrogen and methane in air were determined experimentally at room temperature and initial pressure of 1.0 atm, 0.5 atm and 0.1 atm. The application of Le Chatelier formula to predict the LFLs of the mixtures was also evaluated. It was found that:

- The LFLs of mixtures of the same hydrogen fuel fractions did not change much when the initial pressure decreased from 1.0 atm to 0.1 atm. Pressure has more impact on the LFLs of the mixtures when there is more hydrogen in the fuel mixture. The LFLs of the mixtures were the lowest at 0.5 atm which implies an increased risk of fire/explosion when the initial pressure decreases from 1.0 atm to 0.5 atm.

- Le Chatelier formula can predict the LFLs of mixtures of hydrogen and methane very well, and the prediction of Le Chatelier formula is not affected by changes in the pressure from 1.0 atm to 0.1 atm.

5. Acknowledgement

This research was sponsored by the Mary Kay O'Connor Process Safety Center, Artie McFerrin Department of Chemical Engineering, Texas A&M University.

6. References

[1] Coward, H.F. and G.W. Jones, *Limits of flammability of gases and vapors*, in *Bureau of Mines Bulletin*, 1952.

[2] Crowl, D.A. and J.F. Louvar, *Chemical Process Safety: Fundamentals with Applications, Second Edition*. 2001: Prentice Hall.

[3] Mashuga, C.V., *Determination of the combustion behavior for pure components and mixtures using a 20L sphere*, in *Chemical Engineering*, 1999: Michigan.

[4] Kaldis, S.P., G.C. Kapantaidakis, and G.P. Sakellaropoulos, *Simulation of multi-component gas separation in a hollow fiber membrane by orthogonal collocation — hydrogen recovery from refinery gases.* Journal of Membrane Science, 2000. **173**: p. 61-71.

[5] TerMaath, C.Y., et al., *Emissions reduction benefits from hydrogen addition to midsize gas turbine feedstocks.* Int J Hydrogen Energy, 2006. **31**(9): p. 1147-1158.

[6] Guo, H., et al., *The effect of hydrogen addition on flammability limit and NOx emission in ultra-lean counterflow CH4/air premixed flames.* Proc Combust Inst, 2005. **30**(1): p. 303-311.

[7] Hawkes, E.R. and J.H. Chen, *Direct numerical simulation of hydrogen-enriched lean premixed methane–air flames.* Combust Flame, 2004. **138**(3): p. 242-258.

[8] Bell, S.R. and M. Gupta, *Extension of the lean operating limit for natural gas fueling of a spark ignited engine using hydrogen blending.* Combust Sci Technol, 1997. **123**(1-6): p. 23-48.

[9] Le, H., S. Nayak, and M.S. Mannan, *Upper Flammability Limits of Hydrogen and Light Hydrocarbons In Air at Sub-atmospheric Pressures.* Ind. Eng. Chem. Res, 2012. **51**: p. 9396-9402.

[10] Wong, W.K., *Measurement of flammability in a closed cylindrical vessel with thermal criterion*, in *Artie McFerrin Department of Chemical Engineering, Texas A&M University,* 2006.

Informatics Rich Integrated Chemical Safety Risk Assessment (ICSRA) Platform

Muthukumarasamy Karthikeyan1, Renu Vyas2, HV Prasad3

1 Centre of Excellence in Scientific Computing and Digital Information Resource Centre, CSIR-National Chemical Laboratory, Pune, India, CSIR-National Chemical Laboratory, Pune, India
2 Chemical Engineering and Process Development, CSIR-National Chemical Laboratory, Pune, India, CSIR-National Chemical Laboratory, Pune, India
3 Chemical Engineering and Process Development, CSIR-Indian Institute of Chemical Technology, Hyderabad, Habsiguda, Hyderabad-500 607, India India

Abstract

Inherently Safer Practices (ISP) is a systematic innovative scientific methodology integrated with the process, design and operations for facilitating the improvement of process safety at all stages of process life cycle. ISP constitutes intensification (minimization), substitution, attenuation, limitation of effects and simplification as procedural options for implementation in process plants.

For this study we conducted literature data search focused on recent domestic and international incidents where the primary cause was related to chemical reactivity and runaway reactions; Several parameters were taken into consideration including chemical manufacturing (i.e., raw material storage, chemical processing, and product storage) and other industrial activities involving bulk chemicals, such as storage/distribution, waste processing, and petroleum refining.

For purposes of the incident search, only reactive incidents that caused serious consequences would be examined. Source of data: Four decades of data collection involve searching over 40 data sources, focusing on incidents where the primary cause was related to chemical reactivity. For the purposes of the investigation and analysis, an "incident" was defined as a sudden event involving an uncontrolled chemical reaction—with significant increases in temperature, pressure, and/or gas evolution etc.,

The proposed experimental and simulation platform will serve as a repository for plant-specific information, and facilitates the classification and ranking of Indian chemical industry - for the first time - through a Risk index.

Introduction

Computational and Simulation of Industrial chemical reactions to predict runaway reactions and conditions as safety measures. To our knowledge, no systematic studies on inherent safety are yet reported in India. The studies comprise important research areas e.g., process evaluation, estimation of reactivity, hazard identification and evaluation, assessment and prevention of thermal runaways, early fault diagnosis etc. The computational risk assessment platform intends to collect new chemical process safety and risk data, and plant-specific incident data from participating industries across the world for model building and simulation.

To integrate and include a dynamic interactive information platform, viz., chemical safety risk ranking and incident reporting system, by the application of High performance Computing (Cloud, Grid and Distributed systems) for simulation, modeling and response for use by Govt., Industry, Emergency Responders, Research and academic communities, and the public. Upgradation of Thermo chemical Laboratories to assess the proposed/existing industry from the context of implementing ISP, i.e., towards safer process chemistry including selection of solvents, safer process routes, safer process operating conditions for preventing Runaways, early detection of runaways, evaluation/selection of process and designs to handle lesser inventories etc.

Methodology

The methodology devised comprises recognition of the hazards and risks posed by the process, and efforts to reduce or control them to the lowest practical levels, keeping in view the principle objectives of the business. Here we studied data based on literature to identify and collect chemically significant industrially important data pertaining to physico-chemical, biological and toxicological profiles of chemicals (reactants, reagents, catalysts, solvents, products, side products) for their suitability and reusability in simulation and modelling. This includes development of suitable algorithms, chemical reaction component descriptors and methods to simulate experimental conditions and validation through experiments to improve the predictions of risk assessment and prevention of disaster in industrial conditions (Lab-scale, pilot plant and large scale manufacturing).

This process also includes mapping known literatures (Publications, patents and reports) to the existing knowledge for interactive risk assessment of chemicals and incidents in industrial environment. The computational approaches include classical methods, semi-empirical methods, and quantum methods to determine properties at the molecular level. We are developing a computational problem-solving platform and these tools, data and methods to build ICSRA a cloud-computing infrastructure for integrated chemical safety risk assessment of chemicals manufactured, imported and exported from world chemical sources. Proper documentation and accurate predictive models on chemicals of Indian origin would facilitate compliance with other global compliance on chemicals being exported especially in REACH, EPA, FDA and other global federal regulation context.

Development of QSPR (Quantitative Structure Property Relationships) models to relate macroscopic behaviour to molecular properties with detailed molecular computations to obtain thermo chemical data for reactive materials and to predict calorimetric data based on molecular properties. These open source based tools are integrated to this platform (ChemInfoCloud) includes identification and assessment of hazards due to chemical reactivity and risk assessment.

For this systematic study we considered major potential chemical group (CG) which are frequently encountered in industrial reagents like Organic peroxides, Organic Nitrites, Unsaturated Hydrocarbons, Epoxides, Hydrides and Hydrogen, Metal acetalides etc. Efforts to study and model major Chemical Reaction Types (CRT) include Decomposition, Polymerization, Pyrophoric, Peroxide Former, Water sensitive and Oxidation reactions. Molecular computing would help to predict heat of formation, electronic energy, core-core repulsion, dipole, no. of filled levels, ionization potential, molecular weight, topological, electronic and three dimensional descriptors for critical components of reactants, products, solvents.

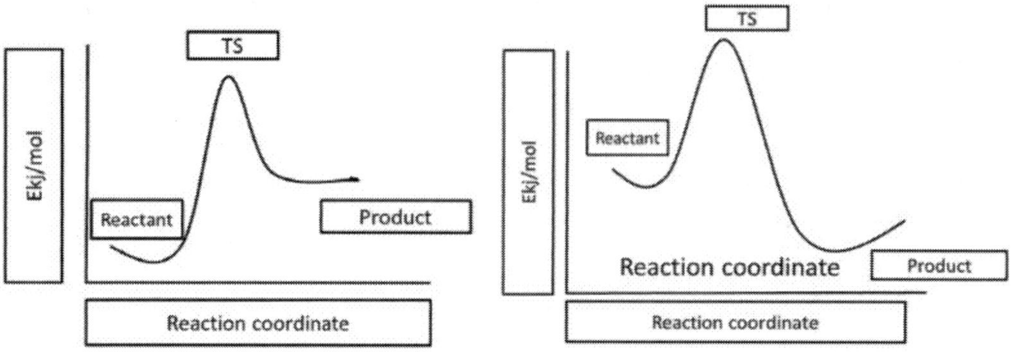

Figure-1: Computational Modeling generic types of chemical reactions along the reaction co-ordinates

Theoretical study of transition states of chemical reactions, computation of activation energies and experimental validation to predict that one reaction pathway is favoured over another reaction pathway. In order to predict the type of polymerization and runaway reactions enthalpy of reactions and kinetics was calculated using the semi-empirical quantum methods, molecular modelling and dynamics simulations of reactants under simulated diverse experimental conditions of Industrial Chemical Processes (ICP), including homogeneous and heterogeneous catalysis.

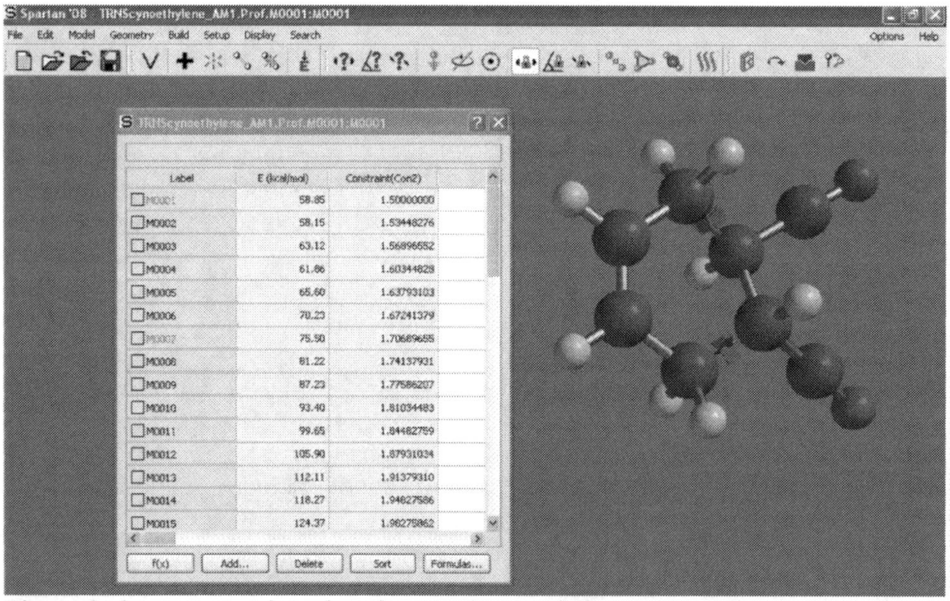

Figure 2: Spartan user interface for reaction modeling

Reactants **Transition State** **Product**

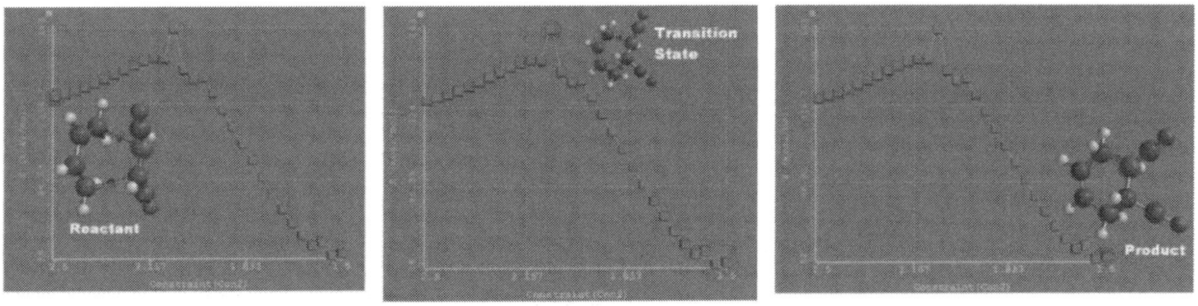

Figure: 3 Computational modelling of organic reactions (Exothermic)

Reactants	Transition State	Product
Heat of Formation: **489.651** kJ/mol Solvation Energy: SM5.4/A -9.308	Heat of Formation: **595.583** kJ/mol Solvation Energy : SM5.4/A -21.665	Heat of Formation: **246.229** kJ/mol

Table:1 Sample data of predicted Heat of formation for reactants, transition state and product

E (TS-Reactant)	E (Reactant-Product) ** heat generated!
E1=595.583-489.65 =105.933 kJ/mol =25.32 kcal/mol	E2=489.65-246.229 =243.421 kJ/mol =58.19 kcal/mol

Table:2 Computed energy difference

Digital mapping of incidental data concerning process safety and their availability to decision makers would help effective risk assessment and rapid response during unexpected events. During the presentation the detailed roadmap, current scenario and future direction of the chemical safety risk assessment platform will be discussed and the input from delegates for this meeting would be considered for integration into the tool if required.

References

1) Organic Reactions Series. John Wiley & Sons
2) http://downloads.wavefun.com/Spartan10Manual.pdf

IT Enablement For
Effective Implementation
And Monitoring of
Process Safety Management

Ajay Kumar/ Pawan Mittal

Reliance Industries Limited

Jamnagar

Ajay.kumar@ril.com/ Pawan.mittal@ril.com

IT Enablement for various PSM elements

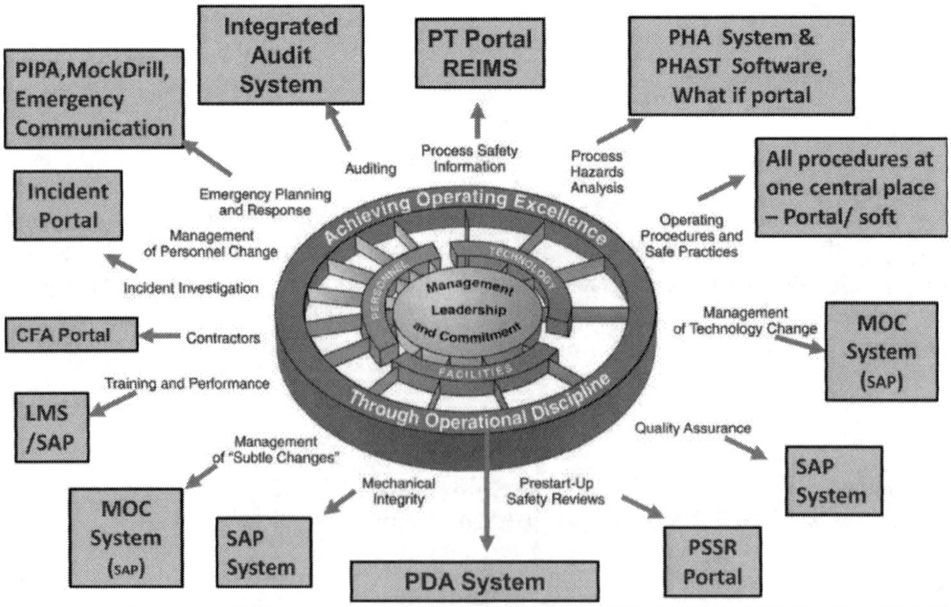

Governance and Control : Corrective Action Tracking System

Advantages of IT Enablement of PSM Elements

- **Process Technology (PT)**
  - ✓ Single repository for storage of all PT related documents
  - ✓ Easy access to PT documents to all concerned.
  - ✓ Easy updation of PT documents

- **Process Hazard Analysis (PHA)**
  - ✓ Automation of all 11 steps of PHA
  - ✓ Linkage of all relevant information required for PHA
  - ✓ Easy Tracking of PHA study progress
  - ✓ Consequence Modelling using PHAST software
  - ✓ Easy recordkeeping of PHA study reports & effective networking
  - ✓ Recommendation management in CAT system

Advantages of IT Enablement of PSM Elements

- **Management of Change Technology & Subtle (MOC – T & S)**
  - ✓ Single repository for recordkeeping of all changes
  - ✓ Controls provided for Hazard Analysis (Pre Screening PHA Checklist)
  - ✓ Tracking & recordkeeping of approval process
  - ✓ Tracking of PT document updating while MOC close out

- **Pre Start up Safety Review (PSSR)**
  - ✓ Automation of PSSR Checklist
  - ✓ Control provided for steps of PSSR e.g. Self Assessment , Level II & Level I
  - ✓ Pending checklist point tracking in CAT system
  - ✓ Permanent records of PSSR points & its close out status

Advantages of IT Enablement of PSM Elements

- **Learning Management System (LMS)**
  - ✓ Online nomination for the training program is available
  - ✓ Online training modules & validation sheets are available
  - ✓ Easy retrieval & tracking of employees' training records
  - ✓ Tracking of the pending training modules for planning training programs
- **Incident Reporting Investigation & Learning**
  - ✓ Reporting of incident in iRIL system.
  - ✓ Online team formation.
  - ✓ RCA tool interface
  - ✓ Uploading of investigation report in system.
  - ✓ Interface with CAT system for Recommendation till closure
  - ✓ LFI sharing to all concerned
  - ✓ Incident Investigation Quality
  - ✓ SAP based incident management system is under development.

Advantages of IT Enablement of PSM Elements

- **PIPA (Pre Incident Planning & Assessment)**
  - ✓ PIPA software used for management of emergency scenarios
  - ✓ Each plant selected credible scenarios & developed plan with the help of PIPA software
  - ✓ Rehearsal of PIPA scenarios & plans is ensured through emergency mock drills
- **Contractor Field Audits (CFA)**
  - ✓ Online scheduling & recordkeeping of field audits for various contractor agencies
  - ✓ Effective records of audit observations e.g. % Good Citizens observed during audit round, Score achieved by contractor agencies
  - ✓ Overall Tracking of Severity Index (Plant/Site wise)
  - ✓ Tracking of Contractor Field Audits e.g. Scope Vs. Completed

Advantages of IT Enablement of PSM Elements

- **Progressive Disciplinary Action System (PDA)**
  - ✓ Developed for recordkeeping of violation / negligence of safety rules & procedures
  - ✓ Easy retrieval of records
  - ✓ Uniformity in record maintenance of PDA at all RIL sites.
- **Corrective Action Tracking (CAT)**
  - ✓ Interface with iRIL, Audit Management, PSSR, Mock Drill
  - ✓ Tracking of Recommendation at one location
  - ✓ Mail Alert for Pending and Overdue Recommendation

Thanks

(44at) Sulphur Dust Explosions an Overview

Karthick Deivasagayam, Loss Prevention, SNC Lavalin Inc, Calgary, AB, Canada

This paper will focus on the sulphur dust, its explosion energy, available models and capabilities of modeling its explosion energy, the equipment susceptible for sulphur dust explosion, various prevention& mitigation strategies of sulphur dust explosion, and industrial applied technology to sulphur dust explosion mitigation. This paper will begin with increase of solid sulphur production from the refineries, its present forms of products and case studies.

The technologies and methodologies for sulphur dust explosion, frequency data, consequence modeling, prevention and mitigation of sulphur dust as applied in a present day sulphur plant.

Discussion for future developments will be discussed in this paper as well.

Simplified Uncertainty Analysis of LOPA Results

Raymond "Randy" Freeman
S&PP Consulting
12303 Lake Shore Ridge
Houston, TX 77041

rafree@yahoo.com

ABSTRACT

Layer of Protection Analysis (LOPA) is a semi-quantitative risk evaluation tool. The methodology is widely used in setting Safety Integrity Level (SIL) targets for Safety Instrumented Systems (SIS). The methodology is often used to evaluate the adequacy of existing protective systems against corporate risk targets. The LOPA method is based on the use of conservative estimates of the failure probability of protective systems.

Freeman previously presented a method for the quantification of uncertainty in a LOPA study based on the application of variance contribution analysis (VCA) techniques. The use of VCA to evaluate the uncertainty in a LOPA study requires the evaluation of the sensitivity of the LOPA equation to changes in each of the variables. Freeman used direct analytical computation of the partial derivatives of the LOPA equation for each of the uncertain variables. While mathematically correct, this use of a partial derivative analysis during an actual LOPA review is unlikely. Faster and easier methods are needed if LOPA uncertainty analysis is to be applied by practicing engineers.

This paper explores the use of approximations to compute the required uncertainty information without the use of the partial derivative computations. Numerical approximations (perturbation methods) are found to allow for the estimate of the partial derivatives with a maximum error of 16%. When evaluating simple safety systems designed per IEC 61511, the error is found to be no more than 10% for SIL-1 or SIL-2 capable systems. Compared to direct Monte Carlo simulation methods, the numerical approximation method is much more efficient. Monte Carlo simulation takes approximately 10,000 function evaluations to arrive at a result. For the base line problem studied, the numerical approximation method will take only 5 function evaluations.

BACKGROUND

Layer of Protection Analysis (LOPA) [Ref. 1] is a widely used method for the evaluation of the frequency of undesired events. LOPA is one of the methods proposed in the ANSI/ISA 84.00.01 standard [Ref. 2] for the determination of Safety Integrity Level (SIL) targets. The SIL target [Ref. 3] is used by process and instrument engineers for specification of equipment to reduce the frequency of undesired events such as overpressures, spills from tanks, releases to flare systems, fires, etc.

Freeman [Ref. 4, 5] developed a method for evaluating the uncertainty in a LOPA study using a mathematical method called Variance Contribution Analysis. The VCA method outlined in Freeman's paper require the evaluation of the partial derivative of the basic LOPA equation for each of the uncertain variables. There are several problems with the application of this method in an actual LOPA study, including:

1. Most engineers would not be comfortable evaluating a series of partial derivative functions while leading a LOPA study.
2. The LOPA team leader must develop the uncertainty parameters for each initiating event and each IPL during the analysis. A set of pre-calculated tabulations of the uncertainty parameters would be needed to speed the analysis.
3. The application of Monte Carlo simulation as used in Freeman's paper is not a topic familiar to most process engineers - nor is Monte Carlo simulation a simple methodology to implement for a real LOPA scenario.

The fundamental issue with the use of VCA in an actual LOPA study is the evaluation of the variance contribution from each uncertain parameter. This paper is directed at rectifying these deficiencies in the application of VCA to the evaluation of uncertainty in a LOPA study.

REVIEW OF LOPA METHODOLOGY

A LOPA [Ref. 1] review is a semi-quantitative risk evaluation of undesired events in a process. The LOPA begins with the definition of the set of undesired events (fires, explosions, toxic chemical releases, etc.). Next, potential safeguards (such as pressure relief valves, high/low level or pressure safety shutdown systems, emergency shutdown systems, etc) are reviewed to determine if there are independent protection layers (IPLs) that will prevent the undesired event from occurring. The frequency of the undesired event is then computed by combining the frequency of initiating event with the combined probability of failure on demand (PFD) of the IPLs as:

$$\text{Frequency of undesired event} = (\text{Initiating event freq}) \times \prod_{i=1}^{n} \text{PFD of the IPLs} \qquad \text{(Eqn. 1)}$$

Or in more compact form:

$$\text{Freq} = (\text{IE}) \times \prod_{i=1}^{n} P_i \qquad \text{(Eqn. 2)}$$

Where:

Freq = Frequency of the undesired event being studied by the LOPA method
IE = Initiating event frequency or the starting event frequency for the scenario under study

$P_i =$ Probability of Failure on Demand (PFD) of an Independent Protection Layer (IPL) counted as reducing the frequency of the undesired event.
$i =$ Subscript indicating the ith IPL
$n =$ Total number of IPLs counted in the LOPA study as satisfying the criteria for an IPL

Equation 1 is referred to as the basic LOPA equation. The basic LOPA equation of Eqn. 2 can be restated as:

$$Freq = (IE)\ P_1\ P_2\ P_3\ \ldots\ P_n \qquad \text{(Eqn. 3)}$$

All of the variables in Eqn. 3 are random variables with inherent uncertainty. The resulting calculated frequency of the LOPA scenario will therefore be uncertain.

APPLICATION OF VARIANCE CONTRIBUTION ANALYSIS TO THE BASIC LOPA EQUATION

Using variance contribution analysis, the mean and variance for the LOPA equation are found directly per the work of Freeman [Ref. 4, 5]. For the expression:

$$Y = F(x_i) \qquad \text{(Eqn. 4)}$$

Where:

Y = dependent random variable computed from x_i
$F(x_i)$ = function that converts (maps) random variable x_i into Y

The mean of Y can be estimated using the following approximation:

$$E(Y) = F[E(x_i)] \qquad \text{(Eqn. 5)}$$

Where:

$E(Y)$ = expected value of random variable Y = mean of Y
$E(x_i)$ = expected value of random variable x_i = mean of x_i

The variance of Y can likewise be estimated as:

$$V(Y) = \sum_{i=1}^{n} \left[\frac{\partial Y}{\partial x_i}\right]^2 V(x_i) \qquad \text{(Eqn. 6)}$$

Where:

$V(Y)$ = variance of random variable Y
$V(x_i)$ = variance of random variable x_i

The application of the VCA method to the evaluation of the uncertainty in a LOPA calculation requires the evaluation of the partial derivative term in Eqn. 6. Previously, Freeman computed the partial derivative directly using the methods of calculus and partial differential equations. However, the direct analytical evaluation of each of the terms in a LOPA expression is time-consuming and prone to error.

The basic LOPA equation of Eqn. 3 is restated as:

$$Freq = (IE)\ P_1\ P_2\ P_3\ \ldots\ P_n \qquad \text{(Eqn. 3)}$$

Note that Freq is first order with respect to each random variable. Therefore, the partial derivative of the Frequency (Freq) with respect to a random variable will become a function of the form:

$$\frac{\partial Freq}{\partial P_1} = (IE)\ P_2\ P_3\ \ldots\ P_n \qquad \text{(Eqn. 7)}$$

The partial derivative may be evaluated numerically rather than analytically. The numerical approximation to the partial derivative may be found by simple evaluation of the Freq near the mean values of each random variable. For all cases to be reviewed in this paper, a 10 percent perturbation about the expected value of the random variable will be used to complete the numerical approximation calculations for the partial derivative term in a VCA study.

For the simple case where all of the IPLs are independent and share no common inspection or repair times, the numerical evaluation of the partial derivative of Freq with respect to the variable P_1 in Eqn. 3 may be estimated as:

$$\frac{\partial Freq}{\partial P_1} \approx \frac{\Delta Freq}{\Delta P_1} = \{[E(IE)\ 1.1E(P_1)\ E\ (P_2)\ldots E(P_n)] - [E(IE)\ E(P_1)\ E(P_2)\ldots E(P_n)]\}/0.1E(P_1)$$

(Eqn. 8)

Where:
$E(P_i)$ = expected value of random variable P_i = mean of P_i

The numerical estimate reduces to:

$$\frac{\partial Freq}{\partial P_1} \approx \frac{\Delta Freq}{\Delta P_1} = \left(\frac{1.1-1}{0.1}\right)\ [E(IE)\ E(P_2)\ E(P_3)\ \ldots\ E(P_n)] \qquad \text{(Eqn. 9)}$$

$$\frac{\partial Freq}{\partial P_1} \approx \frac{\Delta Freq}{\Delta P_1} = [E(IE)\, E(P_2) \ldots E(P_n)] \qquad \text{(Eqn. 10)}$$

Equation 10 is exactly the same as the analytical expression presented in Eqn 7. Thus, for simple LOPA expressions involving multiple independent protection layers, the result obtained by a numerical estimate will be the same as that obtained by the analytic expression obtained from the partial differentiation of the scenario basic LOPA equation. No error will be introduced by using the numerical approximation method to evaluate the partial derivative.

There are occasions in a LOPA review where the resulting basic LOPA equation may become non-linear. For example, if two or more IPLs are implemented in a safety instrumented system, the resulting LOPA equation becomes non-linear with respect to inspection frequency. For example a SIL-2 interlock design might involve two redundant instrumentation systems installed in parallel to reduce the frequency of an undesired event.

Appendix A to this paper evaluates the cases up to, and including, a SIL-4 capable design with 4 safety interlocks all tested on the same test period. The results of the analysis of Appendix A are summarized in Table 1. For most practical LOPA problems, a numerical analysis of the partial derivative term (Eqn. 6) will produce a result within 16% of that obtained using analytical methods.

Table 1.
Error Introduced in LOPA Uncertainty Calculations by the Use of Numerical Estimates of the Partial Derivative Term

Case	Type of System	Error In LOPA Uncertainty* %
n = 1	SIL-1 Design or BPCS interlock	0
n = 2	SIL-2 Design or two SIL-1 interlocks	5
n = 3	SIL-3 Design or three SIL-1 interlocks	10
n = 4	SIL-4 Design or four SIL-1 interlocks	16

* Error = (Numerical Approximation Estimate – Analytical)/Analytical

The second problem with the application of the VCA method to LOPA uncertainty analysis is the need for pre-calculated tabulations of the mean and variance of the initiating event frequencies and IPL PFDs. Appendix B of this paper presents the method used as well as the results of calculating the needed uncertainty parameters using the data presented in the CCPS LOPA book. These tabulations may be used directly in a LOPA uncertainty study. The application of these simplifications is best illustrated using an example.

EXAMPLE PROBLEM

Previously, Freeman [Ref. 4, 5] presented an example problem to demonstrate the direct application of VCA to the uncertainty in the LOPA calculations. The same example is used for comparison in this work.

Consider the process system shown in Figure 1. The bottom sump level in separation tower (T-200) is controlled by a level controller (LIC-101) that regulates a control valve LCV-102. The tower is normally operating at a pressure of 500 psig and is designed for a maximum pressure of 750 psig. The liquid flow from T-200 goes to an intermediate vessel (Hold Drum D-205) prior to additional processing. D-205 is designed for a maximum pressure of 100 psig and normally operates at 75 psig.

Of concern is the failure of the level control system in T-200 to maintain the liquid seal at the bottom of the tower. If the liquid level in the T-200 sump is lost due to an instrumentation failure, the Hold Drum could be exposed to pressures as great as 500 psig. A pressure of 500 psig is a factor of 5 above the design pressure of the Hold Drum. Catastrophic failure of the Hold Drum is probable.

This problem is generically referred to as a high-low pressure interface problem or as a liquid blow-through case. Two protective systems are in place to prevent the Hold Drum from being over pressured.

1. The Separation Tower T-200 is provided with a SIL-1 low level interlock to prevent the loss of the liquid seal in the tower sump.
2. The Hold Drum D-205 is provided with a relief valve (PSV-206) sized to prevent overpressure if the tower liquid seal is lost.

LOPA EVALUATION OF THE EXAMPLE

For the example problem, the initiating event of concern is a basic process control system (BPCS) failure, resulting in loss of level control in the Separation Tower. From the CCPS LOPA book [Ref. 1], the initiating event frequency of a BPCS failure is 1 event in 10 years.

The low level protective interlock system was designed and installed in accordance with the ANSI/ISA 84.00.01 standard [Ref. 3] as a SIL 1 protective system. By definition, the probability of failure on demand (PFD) of a SIL-1 system is between 0.01 and 0.1. The LOPA book recommends using 0.1 as the PFD of a SIL-1 system.

The relief valve on the Hold Drum D-205 is designed for this hypothetical event. A value of 0.01 is suggested by the LOPA book as an estimate of the PFD for the relief valve of not opening on demand. However, for this analysis, a more conservative PFD value of 0.016 is used based on a review of the work by Bukowskia [Ref 6].

The LOPA formulation of the frequency of the Hold Drum being overpressured is:

$$\text{Freq} = \text{IE x SIS failure x PSV failure} \qquad \text{(Eqn 11)}$$

$$\text{Freq} = (0.1 \text{ event/yr}) \text{ x } (0.1) \text{ x } (0.016) = 1.6 \text{ x } 10^{-4} \text{ events/yr} \qquad \text{(Eqn 12)}$$

In a LOPA, the frequency of 1.6×10^{-4} events per year would typically be compared with applicable risk acceptance criteria to determine if additional safeguards are needed. Many LOPA teams might recommend the addition of a critical high level alarm outside of the basic process control system, with adequate time for operator intervention, to reduce the risk of this liquid blow-through scenario.

Due to conflicts in turnaround schedules, the site management is interested in extending the test interval on the SIS system by 3 months, from 2 years to2.25 years. The management asks the LOPA team:

> "Will extending the SIS test interval by 3 months require the provision of additional IPLs to protect the system?"

At this point in the analysis, the engineer would not be able to explain to management the magnitude of the underlying uncertainty in the computed frequency of the Hold Drum overpressure. The next section of this paper attempts to answer this, and related, questions.

LOPA UNCERTAINTY

The basis for the evaluation of the LOPA uncertainty comes from two sources:

1. The LOPA model presented in Equation 11
2. The uncertainty in the parameters of the LOPA model

Rewriting Equation 11 in shorter terms, we obtain:

$$\text{Freq} = (\text{IE})\,(\text{SIS})\,(\text{PSV}) \qquad \text{(Eqn 13)}$$

Where:

Freq is the frequency of the Hold Drum D-205 being over pressured to the point of failure
IE is the initiating event frequency of the basic process control system failing in a dangerous state
SIS is the PFD of the SIL-1 low level interlock on the Separation Tower T-200
PSV is the PFD of the relief valve (PSV-206) opening and relieving the overpressure event

The variables in Equation 13 are all random variables. The variable SIS is actually the result of a simple calculation involving the equipment failure rate and test interval. For a negative exponentially distributed failure rate for a non-repairable device, the PFD is approximately:

$$\text{SIS} = \lambda\, T/2 \qquad \text{(Eqn 14)}$$

Where:
λ is the failure rate of the SIL-1 low level interlock on the Separation Tower T-200
T is the inspection or test interval for validation that the interlock is working and is functional
Equation 14 is an approximation of the negative exponential distribution. For λ $T/2 < 0.2$, the approximation is very good (error of 10%) and will be used throughout this paper.

Substituting the SIS failure rate and test interval into Equation 13 yields the basic LOPA equation for this scenario as:

$$\text{Freq} = (\text{IE})\,(\lambda)\,(T/2)\,(\text{PSV}) \qquad \text{(Eqn 15)}$$

The initiating event frequency in the basic LOPA analysis was taken directly from the LOPA book as 0.1 events per year. Discussions with site operations group indicates that the IE frequency is uncertain and might be as low as 0.05 events per year or as high as 0.5 events per year.

Likewise, the failure rate of the SIS interlock (λ) has significant uncertainty. The original plant design targeted this interlock as SIL-1 with a risk reduction factor of 50 based on a two-year testing interval, or a SIS failure rate of 0.02 failures per year. However, the failure rate may be as low as 0.01 events/year or as high as 0.1 events/year. From Appendix B (Table B-6) for a SIL-1 interlock designed with an RRF=50, the mean PFD is 4.3E-2 and the variance is 4.1E-4.

The test interval (T) may be targeted as once every 2 years. However, it might be as short as once every 1.5 years if the site takes advantage of an unexpected shutdown to test the system. It might be as long as 2.5 years, if the production run of the system is extended.

Appendix B has a discussion of relief valve PFDs. The Bukowskia [Ref 6] Process Safety Valve (PSV) PFD of 0.016 is higher than the CCPS LOPA book value and was used in the above LOPA calculation of the frequency of overpressure of the Hold Drum. For this uncertainty analysis, a mean PSV PFD of 0.016 with a standard deviation of 4.7×10^{-3} was used as the best estimate of the probability of failure of the relief valve to open on demand. The PSV variance would be $[4.7 \times 10^{-3}]^2$ or 2.2×10^{-5}.

These data may be used to establish the probability distribution for each of the uncertain variables in Eqn. 15. Assuming a triangular probability distribution (Ref. 7), the mean and variance may be calculated as:

$$\text{Mean} = (A + B + C)/3 \qquad \text{(Eqn 16)}$$
$$\text{Variance} = (A^2 + B^2 + C^2 - AB - AC - BC)/18 \qquad \text{(Eqn 17)}$$

Where:
A = Minimum Value
B = Mode
C = Maximum Value

The mean and variance for the initiating event (IE), the failure rate of the SIS (λ) and the inspection interval are calculated and presented in Table 2.

Table 2
Triangular Probability Distribution Parameters

Model Variable	Units	Minimum A	Mode B	Maximum C	Mean by Eqn 16	Variance by Eqn 17
Initiating Event (IE)	Events per Year	0.05	0.1	0.5	2.167E-01	1.014E-02
λ of SIS (From Table B-6 with RRF=50)	Failure per Year	0.01	0.02	0.1	4.3E-02	4.1E-04
Inspection Interval (T)	Years	1.5	2	2.5	2.000E+00	4.167E-02

The variance contribution for each uncertain (random) variable may now be computed using the numerical approximation for the partial derivative term presented above and derived in detail in Appendix A to this paper. Restating Eqn 15 as the basic LOPA equation for this scenario:

$$\text{Freq} = (\text{IE})\ (\lambda)\ (T/2)\ (\text{PSV}) \qquad \text{(Eqn 15)}$$

Using a 10% perturbation about the means of each random variable, the partial derivative term may be estimated. For example the partial derivative portion of the contribution from the uncertainty in the initiating event may be estimated as:

$$\frac{\partial Freq}{\partial IE} \approx \frac{\Delta Freq}{\Delta IE} = \{[1.1\ E(IE)]\ E(\lambda)\ E(T/2)\ E(PSV) - E(IE)\ E(\lambda)\ E(T/2)\ E(PSV)\}/(0.1\ E(IE))) \qquad \text{(Eqn 16)}$$

$$\frac{\partial Freq}{\partial IE} \approx \frac{\Delta Freq}{\Delta IE} = [1.1]\ E(\lambda)\ E(T/2)\ E(PSV) - E(\lambda)\ E(T/2)\ E(PSV)/(0.1) \qquad \text{(Eqn 17)}$$

$$\frac{\partial Freq}{\partial IE} \approx \frac{\Delta Freq}{\Delta IE} = [(1.1 - 1)/0.1]\ E(\lambda)\ E(T/2)\ E(PSV) \qquad \text{(Eqn 18)}$$

$$\frac{\partial Freq}{\partial IE} \approx \frac{\Delta Freq}{\Delta IE} = E(\lambda)\ E(T/2)\ E(PSV) \qquad \text{(Eqn 19)}$$

Likewise the partial derivative contribution to the variance of the other random variables may be found as:

$$\frac{\partial Freq}{\partial \lambda} \approx \frac{\Delta Freq}{\Delta \lambda} = E(IE)\ E(T/2)\ E(PSV) \qquad \text{(Eqn 20)}$$

$$\frac{\partial Freq}{\partial T} \approx \frac{\Delta Freq}{\Delta T} == E(IE)\ E(\lambda)\ (1/2)\ E(PSV) \qquad \text{(Eqn 21)}$$

$$\frac{\partial Freq}{\partial PSV} \approx \frac{\Delta Freq}{\Delta PSV} = E(IE)\ E(\lambda)\ E(T/2) \qquad \text{(Eqn 22)}$$

It should be noted that these results are identical to those obtained by taking the partial derivative of the basic LOPA equation for this scenario (Eqn 15) with respect to each of the random variables. With these estimates for the partial derivatives, the variance contribution from each of the uncertain random variables may now be computed (see Table 3). Note that the results in Table 3 are identical with those previously obtained by Freeman [Ref. 4, 5] using more formal mathematical methods. The same result could be obtained by simple numerical substitution. Equations 19 through 22 and those in Table 3 are presented to allow the reader to understand why the numerical results of a numerical approximation are the same as obtained using analytical methods.

Table 3.
Contribution of Each Model Variable
to Frequency of Drum Overpressure Variance

Model Variable	**Units**	**Equation for Contribution**	**Variance Contribution**	**Percent of Total Variance**
Initiating Event (IE)	Events per Year	$[E(\lambda)\ E(T/2)\ E(PSV)]^2\ V(IE)$	4.874E-09	41%
λ of SIS	Failure per Year	$[E(IE)\ E(T/2)\ E(PSV)]^2\ V(\lambda)$	4.874E-09	41%
Inspection Interval (T)	Years	$[E(IE)\ E(\lambda)\ (1/2)\ E(PSV)]^2\ V(T)$	2.351E-10	2%
Relief Valve (PSV)	none	$[E(IE)\ E(\lambda)\ E(T/2)]^2\ V(PSV)$	1.947E-09	16%
		Totals	**1.193E-08**	**100%**

Since the variance contribution and total variance are the same whether computed via an analytical method or by a numerical approximation method, the accuracy of the resulting will be the same as previously reported by Freeman [Ref 4, 5]. Table 4 presents a comparison of the computed results with those previously obtained by Freeman using both an analytical and a Monte Carlo method for estimating the frequency of overpressure of the Hold Drum in the example problem.

Table 4
Comparison of Results via Numerical Approximation,
Using Theoretical and Direct Monte Carlo Simulation of Hold Drum Overpressure

Method	Mean event/yr	Standard Deviation event/yr	Value of 70% Upper Limit event/yr	Value of 95% Upper Limit event/yr
Theoretical	1.50E-04	1.09E-04	2.07E-4	3.29E-04
Numerical Approximation	1.50E-04	1.09E-04	2.07E-4	3.29E-04
Monte Carlo	1.51E-04	1.19E-04	1.78E-4	3.90E-04
Basic LOPA	1.6E-04	Not Computed	Not Computed	Not Computed

With the results of the uncertainty analysis now available, we can answer the managements question in the example problem on the impact of extending the inspection and test interval by 3 months. As seen in Table 3, the variance contribution of the inspection interval is only 2% of the total. Increasing the inspection interval by 3 months will not appreciably impact the computed 95% upper limit on the frequency of the Hold Drum overpressure.

SIMPLIFIED METHOD FOR QUANTIFYING LOPA UNCERTAINTY

Based on the results obtained using the numerical approximations outlined in this paper, a simplification of the LOPA uncertainty method outlined by Freeman [Ref. 4, 5] is possible. With the exception of the method used to compute the partial derivatives for the variance contribution, the revised procedure will be the same as that which was previously presented. The change to the previously presented method is indicated in *italic font* below.

1. Complete the LOPA study using the methods outlined in the CCPS LOPA book [Ref. 1].

2. Write the basic LOPA equation for the scenario of interest in the form of the product of initiating event frequency and the PFD of the IPLs. An example of the basic LOPA equation is given in Equation 15.

3. Define the uncertainty in the terms of the expanded LOPA equation. The uncertainty can be given as the upper and lower range of the possible values (uniform probability distribution), as the upper, lower and recommended values (triangular distribution), or as a mean and standard deviation (normal distribution). See the example above for guidance in the evaluation of safety instrumented system interlocks.

4. Compute the expected value of each variable in the LOPA equation. The equations for the mean and variance for the uniform, triangular and normal probability distributions are presented for various probability distributions in Vose [Ref. 7]. *For commonly used LOPA initiating events and IPLs, the mean and variance are tabulated in Appendix B to this paper and may be used directly.*

5. Compute the expected value or mean of the LOPA equation using the mean value of each of the variables in the basic LOPA equation.

6. *Compute the sensitivity of the result from the LOPA equation by estimating the partial derivative of the basic LOPA equation with respect to each of the variables using a 10% perturbation about the mean or expected value of each random variable. See the problem worked above as an example of this step. The theory for this simplification is presented in Appendix A to this paper*

7. Compute the variance of the scenario frequency of the basic LOPA equation by use of the variance contribution of Equation 6. This entails multiplying the variance of each of the uncertain variables in the basic LOPA equation by the square of its sensitivity (obtained in step 6), as evaluated at the mean. Sum the resulting terms to obtain the overall variance of the result of the basic LOPA term.

8. Compute the 95% upper confidence limit on the computed frequency by use of the standard normal factor, Z, [Ref. 8] as:

   $$Z = [x_i - E(x)]/\sigma \qquad \text{(Eqn 23)}$$

   Where:

   σ = standard deviation of the frequency of the scenario of interest from the basic LOPA equation obtained from step 7. Note that the variance of a random variable is the square of the standard deviation of the random variable.

   $E(x)$ = the expected value of the frequency of the scenario of interest from the basic LOPA equation obtained from step 5

   For the 95% upper limit, $Z = 1.64$. Rearranging Eqn. 10 allows for the direct calculation of the corresponding value of the 95% upper confidence limit as:

   $$X_{95\%} = 1.64\,\sigma + E(x) \qquad \text{(Eqn 24)}$$

   Where

   $X_{95\%}$ = is the upper 95% limit on the computed frequency of the scenario of interest from the basic LOPA equation.

9. Compare the 95% upper confidence limit on the frequency of the scenario of concern with that obtained using the basic LOPA method (step 1). If the difference is greater than one order of magnitude, revise the LOPA study accordingly to incorporate better information obtained in the uncertainty analysis. Improve the integrity of the LOPA IPLs or identify additional IPLs to drive the risk to a tolerable level.

RESULTS AND CONCLUSIONS

A numerical approximation can be used to estimate the variance contribution of each uncertain variable in a basic LOPA. Use of a 10% perturbation about the mean or expected value of each random variable was found to be adequate to define the variance. The numerical approximation is exact for cases where all of the IPLs are independent. In the case where 4 IPLs share a common test or maintenance frequency, the resulting error in the uncertainty estimate will be less than 16%.

A simplified method has been developed to evaluate the uncertainty in the basic LOPA equation. Using available data, the simplified numerical approximation method estimates the mean and variance of the basic LOPA equation and then computes the uncertainty range in the resulting frequency of the scenario of concern. For the example studied with 4 random variables, a total of 5 evaluations of the basic LOPA equation were all that were required to characterize the uncertainty. The results are found to be in excellent agreement with those obtained using analytical methods. When compared to a Monte Carlo simulation of 10,000 function evaluations, the calculation speed improvement is approximately a factor of 2,000.

A set of tables for the mean and variance of initiating events (IE) and independent protection layers (IPLs) are presented in Appendix B of this paper. These can be used directly by the analyst to estimate the uncertainty in an IE or IPL.

The methodology presented in this paper is simple and direct. It may be programmed into any standard LOPA software package to allow the users the ability to estimate the uncertainty in their calculated result.

ACKNOWLEDGMENTS
This paper is an expanded version of a paper entitled ''Uncertainty in LOPA; Part 2 – Simplified Methods Applied to Safety Systems'' presented at the 2012 Mary Kay O'Connor Process Safety Symposium, October 23-25, 2012, College Station, TX. The author wishes to thank Kathy Kas, Angela Summers, Robert Johnson, Dan Crowl and other individuals who gave comments and suggestions for improving this article.

LITERATURE CITED

1. CCPS, *Layer of Protection Analysis – Simplified Process Risk Assessment*, American Institute of Chemical Engineers, New York, 2001
2. Instrumentation, Systems and Automation Society, Functional Safety: Safety Instrumented Systems for the Process Sector—Parts 1, 2, and 3, Standard ANSI/ISA 84.00.01, 2004. USA implementation of IEC 61511.
3. R.A. Freeman, "Using Layer of Protection Analysis to Define Safety Integrity Level Requirements, pp 185 – 194, *Process Safety Progress*, Vol. 26, No. 3, 2007
4. R.A. Freeman, "Uncertainty in LOPA Studies," pages 623-634, Proceedings of Mary Kay O'Connor Process Safety Center International Symposium held October 25-27, 2011 in College Station, Texas
5. R.A. Freeman, "Quantifying LOPA Uncertainty," pp 240 – 247, *Process Safety Progress*, Vol. 31, No. 3, 2012. Published in Wiley Online Library (wileyonlinelibrary.com), DOI 10.1002/prs.11493
6. Julia V. Bukowskia and William M. Goble, "Analysis of Pressure Relief Valve Proof Test Data," pp 24-29, *Process Safety Progress*, Vol 28, No.1, 2009
7. David Vose, *Risk Analysis – A Quantitative Guide, 3rd Edition*, John Wiley, 2008
8. Paul L. Meyer, *Introductory Probability and Statistical Applications*, John Wiley, New York, 1972

Appendix A – Numerical Approximation to Partial Derivatives

The basic LOPA equation of Eqn. 2 can be restated as:

$$Freq = (IE)\ P_1\ P_2\ P_3\ \ldots\ P_n \qquad \text{(Eqn. A-1)}$$

Where:
Freq = Frequency of the undesired event being studied by the LOPA method
IE = Initiating event frequency or the starting event frequency for the scenario under study
P_i = Probability of Failure on Demand of the Independent Protection Layer (IPL) counted as reducing the frequency of the undesired event.
n = Total number of IPLs counted in the LOPA study as satisfying the criteria for an IPL

The probabilities of failure on demand of the IPLs may be restated in terms of a failure rate and a time between tests as:

$$Freq = (IE)\ (\lambda_1)\ (T_1/2)\ (\lambda_2)\ (T_2/2)\ \ldots\ (\lambda_n)\ (T_n/2) \qquad \text{(Eqn A-2)}$$

Where:

λ_1 is the failure rate of the safety instrumented system, i

T is the inspection or test interval for validation that the interlock is working and is functional

i is a subscript indicating the ith IPL

If all the IPLs are tested or inspected at the same test interval

$$T_1 = T_2 = T_3 = T_n \qquad \text{(Eqn A-3)}$$

Forming a new parameter K as:

$$K = (IE)\ (\lambda_1)\ (\lambda_2)\ \ldots\ (\lambda_n) \qquad \text{(Eqn A-4)}$$

The basic LOPA equation may now be re-written as:

$$Freq = K\ (T/2)^n \qquad \text{(Eqn A-5)}$$

When evaluated at the expected value of the random variables, the parameter K will not change and can be treated as a constant. How good is the numerical approximation for the partial derivative of Freq with respect to time as n increases from 1 to n?

The numerical approximation to a derivative requires the selection of a step size for the change in the variable of interest. In the limit as the step size goes to zero, the resulting answer will become the exact analytical expression for the derivative. However, from a practical point of view the selection of the numerical step size is constrained by two problems:

1. Too large and the function may change too much and the resulting numerical approximation will be very bad.
2. Too small and the numerical approximation will be limited by round-off error of the computer.

The step size of 0.1 T was chosen as the balance point between the two numerical problems. For the following cases (1 through 4), a step size of 0.1T was used as the basis for estimating the partial derivatives required for computations in the variance contribution analysis. Note that the step size for the evaluation of the derivative is arbitrary. A smaller step size could have been selected (say 0.01 T) and the resulting errors would decrease. However, the potential for numerical calculation errors would increase.

Case 1, n=1

For n = 1, restating the LOPA frequency becomes:

$$Freq = K\,(T/2) \qquad \text{(Eqn A-6)}$$

$$\frac{\partial Freq}{\partial T} = 0.5\,K \qquad \text{(Eqn A-7)}$$

The numerical approximation would become:

$$\frac{\Delta Freq}{\Delta T} = K\,\left(\frac{1.1^1 - 1}{0.1T}\right)(T/2)^1 \qquad \text{(Eqn A-8)}$$

$$\frac{\Delta Freq}{\Delta T} = K\,\left(\frac{0.1}{0.1}\right)1/2 \qquad \text{(Eqn A-9)}$$

$$\frac{\Delta Freq}{\Delta T} = 0.5\,K \qquad \text{(Eqn A-10)}$$

This numerical approximation is exactly the same as the analytical result. For n =1, there will be no error in the computation of the partial derivative.

Case 2, n = 2

For n = 2, restating the LOPA frequency becomes:

$$Freq = K\,(T/2)^2 \qquad \text{(Eqn A-11)}$$

From basic calculus, the analytical expression for the partial derivative of equation A-5 when n=2 is:

$$\frac{\partial Freq}{\partial T} = 0.5\,K\,T \qquad \text{(Eqn A-12)}$$

The numerical approximation would become:

$$\frac{\Delta Freq}{\Delta T} = (\frac{1.1^2 - 1}{0.1T})\ K\ (T/2)^2 \qquad \text{(Eqn A-13)}$$

$$\frac{\Delta Freq}{\Delta T} = (\frac{0.21}{0.1})\ (1/2)^2\ K\ T \qquad \text{(Eqn A-14)}$$

$$\frac{\Delta Freq}{\Delta T} = (\frac{2.1}{4})\ K\ T \qquad \text{(Eqn A-15)}$$

$$\frac{\Delta Freq}{\Delta T} = (0.525)\ K\ T \qquad \text{(Eqn A-16)}$$

The error in the numerical approximation is:

Error = (Estimate - Actual)/Actual x 100 (Eqn A-17)

Error = (0.525 -0.5)/0.5 x 100 = 5% (Eqn A-18)

Thus, for a SIL-2 type of design, we would expect that no more than a 5% error in the LOPA uncertainty estimate would be introduced by the use of the numerical approximation for the partial derivatives.

Case 3, n =3

For n = 3, restating the LOPA frequency becomes:

$$Freq = K\ (T/2)^3 \qquad \text{(Eqn A-19)}$$

From basic calculus, the analytical expression for the partial derivative of equation A-5 when n=3 is:

$$\frac{\partial Freq}{\partial T} = K\ (3/8)\ T^2 \qquad \text{(Eqn A-20)}$$

The numerical approximation would become:

$$\frac{\Delta Freq}{\Delta P_1} = (\frac{1.1^3 - 1}{0.1T})\ K\ (T/2)^3 \qquad \text{(Eqn A-21)}$$

$$\frac{\Delta Freq}{\Delta P_1} = (\frac{0.331}{0.1})\ (1/2)^3\ K\ T^2 \qquad \text{(Eqn A-22)}$$

$$\frac{\Delta Freq}{\Delta P_1} = (\frac{3.31}{8})\ K\ T^2 \qquad \text{(Eqn A-23)}$$

$$\frac{\Delta Freq}{\Delta P_1} = (0.3875)\ K\ T^2 \qquad \text{(Eqn A-24)}$$

The error in the numerical approximation is:

Error = (Estimate - Actual)/Actual x 100 (Eqn A-25)

Error = (3.31/8-3/8)/(3/8) x 100 (Eqn A-26)

Error = (3.31-3)/(3) x 100 = 10% (Eqn A-27)

Thus, for a SIL-3 type of design, we would expect no more than a 10% error in the LOPA uncertainty estimate to be introduced by the use of the numerical approximation for the partial derivatives.

Case 4, n =4

For n = 4, restating the LOPA frequency becomes:

$$Freq = K\ (T/2)^4 \qquad \text{(Eqn A-28)}$$

From basic calculus, the analytical expression for the partial derivative of equation A-5 when n=4 is:

$$\frac{\partial Freq}{\partial T} = (1/4)\ K\ T^3 \qquad \text{(Eqn A-29)}$$

The numerical approximation would become:

$$\frac{\Delta Freq}{\Delta P_1} = \left(\frac{1.1^4 - 1}{0.1T}\right) K\ (T/2)^4 \qquad \text{(Eqn A-30)}$$

$$\frac{\Delta Freq}{\Delta P_1} = \left(\frac{0.464}{0.1}\right) (1/2)^4\ K\ T^3 \qquad \text{(Eqn A-31)}$$

$$\frac{\Delta Freq}{\Delta P_1} = \left(\frac{4.64}{16}\right) K\ T^3 \qquad \text{(Eqn A-32)}$$

$$\frac{\Delta Freq}{\Delta P_1} = (0.290)\ K\ T^3 \qquad \text{(Eqn A-33)}$$

The error in the numerical approximation is:

Error = (Estimate - Actual)/Actual x 100 (Eqn A-34)

Error = (0.29 – 0.25)/(0.25) x 100 = 16% (Eqn A-35)

Thus, for a SIL-4 type of design, we would expect no more than a 16% error in the LOPA uncertainty estimate to be introduced by the use of the numerical approximation for the partial derivatives.

Table 1 presents a summary of the results of these calculations. In summary, the use of a numerical approximation for estimating uncertainty in a LOPA analysis will introduce no greater than a 16% error in the results. A 16% error in the LOPA uncertainty is well within the accuracy of the Variance Contribution Analysis method for estimating the LOPA uncertainty. This result eliminates the need for time-consuming evaluation of partial derivatives during a LOPA uncertainty analysis. Instead, a simple numerical analysis using a 10% perturbation about the mean of the inspection frequency is adequate to evaluate the partial derivatives needed in a LOPA uncertainty study. The results of this appendix are general and will apply to any basic LOPA equation formulated up to a 4th degree in one or more random variables.

Appendix B – Derivation of Tables for LOPA Uncertainty Analysis

The use of Variance Contribution Analysis to estimate the uncertainty in an estimate of the frequency of an accident requires a source for estimating the statistical parameters of the initiating events (IEs) and independent protection layers (IPLs). This appendix presents a methodology for estimating the required statistical parameters using available published data.

Initiating Events

The initiating event frequencies commonly used in a LOPA study are presented in Table 5.1 CCPS LOPA book [Ref. 1]. Included in Table 5.1 are ranges for the frequencies of each initiating event as determined from a literature search during development of the book. The uncertainty in the frequency data is expressed as a range (minimum to maximum), with a particular value assigned for use during a LOPA review. This type of uncertainty can be described by the triangular distribution [Ref. 7]. The triangular distribution has three parameters:

A = Minimum value
B = Mode
C = Maximum value

For a triangular distribution, the mean and variance are:

$$\text{Mean} = (A + B + C)/3 \qquad \text{(Eqn. B-1)}$$
$$\text{Variance} = A^2 + B^2 + C^2 - AB - BC - AC \qquad \text{(Eqn. B-2)}$$

For the data in Table 5.1, the minimum value becomes A, the maximum value becomes C, and the value used in LOPA becomes the mode, B. The mean and variance of the initiating event frequencies of the CCPS LOPA book Table 5.1 are calculated using Equations B-1 and B-2 and are presented in Table B-1 below. Note that the triangular distribution parameters (A, C and B) are defined in Table B-1 as the Min A, Max C, and Used in LOPA B respectively.

Table B-1. Mean and Variance of LOPA Initiating Events

Initiating Event	Units	Min A	Max C	Used in LOPA B	Mean per unit	Variance per unit^2
Pressure Vessel residual failure	per year	1E-07	1E-05	1E-06	3.7E-06	5.0E-12
Piping residual failure - 100 meter - full breach	per year	1E-06	1E-05	1E-05	7.0E-06	4.5E-12
Piping Leak - 10% section - 100 meter	per year	1E-04	1E-03	1E-03	7.0E-04	4.5E-08
Atmospheric Tank Failure	per year	1E-05	1E-03	1E-03	6.7E-04	5.4E-08
Gasket Packing Blowout	per year	1E-04	1E-02	1E-02	6.7E-03	5.4E-06
Turbine / diesel engine overspeed with casing breach	per year	1E-04	1E-03	1E-04	4.0E-04	4.5E-08
External impact by backhoe, vehicle, etc.	per year	1E-04	1E-02	1E-02	6.7E-03	5.4E-06
Crane Drop	per lift	1E-04	1E-03	1E-04	4.0E-04	4.5E-08
Lightning strike	per year	1E-04	1E-03	1E-03	7.0E-04	4.5E-08
Safety valve operates spuriously	per year	1E-04	1E-02	1E-02	6.7E-03	5.4E-06
Cooling water failure	per year	1.E+00	1E-02	1E-01	3.7E-01	5.0E-02
Pump seal failure	per year	1E-02	1E-01	1E-01	7.0E-02	4.5E-04
Unloading / loading hose failure	per year	1E-02	1E-01	1E-01	7.0E-02	4.5E-04
BPCS failure	per year	1E-02	1E-01	1E-01	7.0E-02	4.5E-04
Regulator Failure	per year	1E-01	1E+00	1E-01	4.0E-01	4.5E-02

The CCPS LOPA book Tables 6.3, 6.4, and 6.5 present the probability of failure on demand of passive, active mechanical and human operator IPLs. Using the same of data as for the initiating events allows for the uncertainty of IPLs to be calculated using Equations B-1 and B-2. The results of these calculations are summarized in Tables B-2, B-3, and B-4 below.

Table B-2. Mean and Variance of LOPA Passive IPLs

IPL Description	Min A	Max C	Used in LOPA B	Mean per unit	Variance per unit^2
Dike	1E-03	1E-02	1E-02	7.0E-03	4.5E-06
Underground Drainage System	1E-03	1E-02	1E-02	7.0E-03	4.5E-06
Open Vent (no valve)	1E-03	1E-02	1E-02	7.0E-03	4.5E-06
Fireproofing	1E-03	1E-02	1E-02	7.0E-03	4.5E-06
Blast Wall / Bunker	1E-03	1E-02	1E-03	4.0E-03	4.5E-06
Flame /Detonation Arrestors	1E-03	1E-01	1E-02	3.7E-02	5.0E-04

Table B-3. Mean and Variance of LOPA Active Mechanical IPLs

IPL Description	Min A	Max C	Used in LOPA B	Mean	Variance
Relief Valve*	1E-05	1E-01	1E-02	3.7E-02	5.1E-04
Rupture Disk	1E-05	1E-01	1E-02	3.7E-02	5.1E-04
BPCS	1E-02	1E-01	1E-01	7.0E-02	4.5E-04

Note the asterisk (*) in Table B-3 for relief valves The issue of relief valves not opening when needed is s topic that has been discussed extensively by Bukowskia [Ref. 6]. Bukowskia found that there is a significant probability that a relief device is not working when installed. A PFD value of 0.016 was suggested as an estimate for the PSV not opening on demand due to initial installation problems. Table 1 in the paper by Bukowski presents an estimate for the uncertainty in the PFD of the relief valve in terms of the standard deviation. For the data sets evaluated, the mean PFD of the PSV may be taken as 0.016 with a standard deviation of 4.7×10^{-3}. The corresponding PSV variance would be $[4.7 \times 10^{-3}]^2$ or 2.2×10^{-5}. The data of Bukowski indicate a PFD of about half that of the mean based on the range of the CCPS book Table 6.4 with a smaller variance. The LOPA value of 0.01 is lower than the value obtained using Bukowski's paper. The analyst completing a LOPA uncertainty study should carefully evaluate which data set to use for representing the uncertainty in the relief valve PFD.

Table B-4. Mean and Variance of LOPA Human Operator IPLs

IPL Description	Min A	Max C	Used in LOPA B	Mean	Variance
Human action with 10 minutes response time	1E-01	1E+00	1E-01	4.0E-01	4.5E-02
Human action with 40 minutes response time	1E-02	1E-01	1E-01	7.0E-02	4.5E-04

Safety Instrumented Systems
Safety instrumented systems are often provided in process design to control undesirable events such as overpressure, runaway reactions, or loss of level interface control. Note for the purposes of this paper the use of the phrase "safety interlock" means:

> "A Safety Instrumented Function (SIF) implemented in a Safety Instrumented System (SIS) to achieve a target Safety Integrity Level (SIL) designed and installed in accordance with the requirements of ANSI/ISA S84 – 2004 [Ref. 2]"

The S84 standard defines a range of acceptable performance operation of the interlocks in terms of a safety integrity level (SIL). The various SIL levels are defined by the probability of failure on demand (PFD) of the interlock, where the RRF is related to the PFD as:

$$RRF = 1/PFD \qquad \text{(Eqn. B-3)}$$

For example, an interlock designed for SIL-1 must have a PFD between 0.1 and 0.01, or by Equation B-3, an RRF between 10 and 100. Table B-5 presents the defined range of PFDs and RRFs for various SIL classes.

Table B-5. Probability of Interlock Failure on Demand of Various SIL Classes

IPL Description	Min PFD	Max PFD	Used in LOPA	Min RRF	Max RRF
SIL-1	1E-02	1E-01	1E-01	10	100
SIL-2	1E-03	1E-02	1E-02	100	1,000
SIL-3	1E-04	1E-03	1E-03	1,000	10,000
SIL-4	1E-05	1E-04	1E-04	10,000	100,000

There are two strategies available for the interlock design engineer to use:

1. Complete the design using rules of thumb, evaluate the RRF, and accept the design if the RRF falls in the desired range (as specified by the standard).
2. Complete the design to a desired risk reduction factor (RRF)

These two different design strategies will result in different probability distributions for the resulting uncertainty in the performance of the interlock. In the first case, any design that falls in the desired RRF range will be acceptable. The resulting uncertainty will be uniformly distributed from the minimum to maximum acceptable RRF. The uniform probability distribution may be characterized by two parameters, A and C:

A = Minimum value
C = Maximum value

The mean and variance of the uniform distribution are:

$$\text{Mean} = (A + C)/2 \qquad \text{(Eqn. B-4)}$$

$$\text{Variance} = ((C\text{-}A)^2)/12 \qquad \text{(Eqn. B-5)}$$

In the second case, the design is targeted to a particular value of the RRF. This RRF becomes the mode (parameter B) of the triangular probability distribution of outcomes. Equations 1 and 2 may be used to compute the mean and variance of the resulting distributions.

Since a range of RRFs are available for each selected interlock SIL, a set of three possible targets was selected to represent the SIL range. These are:

Target RRF = 25 percent of allowed range (Eqn. B-6)
Target RRF = 50 percent of allowed range (Eqn. B-7)
Target RRF = 75 percent of allowed range (Eqn. B-8)

The uncertainty in the interlock performance was calculated for both design strategies for SIL-1, SIL-2, SIL-3 and SIL-4 at each of the three targeted SIL levels (Eqns. B-6, 7, and 8). The results of these calculations are presented in Tables B-6, B-7, B-8, and B-8.

For design strategy 1, Table B-6 presents the uncertainty parameters for each interlock SIL class when no target RRF is assigned prior to design. For example, such a case could occur if the design engineer were told to simply provide a SIL 2 interlock with no specification of the corresponding RRF.

Table B-6. Uncertainty Parameters for Interlock PFD for Various SIL Classes
(No target RRF set prior to design)

IPL Description	Mean PFD	Variance of PFD
SIL-1	5.5E-02	6.8E-04
SIL-2	5.5E-03	6.8E-06
SIL-3	5.5E-04	6.8E-08
SIL-4	5.5E-05	6.8E-10

For design strategy 2, the interlock RRF is selected prior to the start of design. This selected RRF then determines the design target PFD, which becomes the mode (parameter C, Equations B-1 and 2) for the triangular distribution. Tables B-7, B-8 and B-9 present the results for the computation of the mean and variance of the PFD of interlocks designed for various RRFs.

Table B-7. Uncertainty Parameters for Interlock PFD for Various SIL Classes with Target RRF at 25% of Allowed Range

IPL Description	Min PFD A	Max PFD C	Used in LOPA	Design RRF	Mode B	Mean PFD	Variance of PFD
SIL-1	1E-02	1E-01	1E-01	25	4.0E-02	5.0E-02	3.5E-04
SIL-2	1E-03	1E-02	1E-02	250	4.0E-03	5.0E-03	3.5E-06
SIL-3	1E-04	1E-03	1E-03	2,500	4.0E-04	5.0E-04	3.5E-08
SIL-4	1E-05	1E-04	1E-04	25,000	4.0E-05	5.0E-05	3.5E-10

Table B-8. Uncertainty Parameters for Interlock PFD for Various SIL Classes with Target RRF at 50% of Allowed Range

IPL Description	Min PFD A	Max PFD C	Used in LOPA	Design RRF	Mode B	Mean PFD	Variance of PFD
SIL-1	1E-02	1E-01	1E-01	50	2.0E-02	4.3E-02	4.1E-04
SIL-2	1E-03	1E-02	1E-02	500	2.0E-03	4.3E-03	4.1E-06
SIL-3	1E-04	1E-03	1E-03	5,000	2.0E-04	4.3E-04	4.1E-08
SIL-4	1E-05	1E-04	1E-04	50,000	2.0E-05	4.3E-05	4.1E-10

Table B-9. Uncertainty Parameters for Interlock PFD for Various SIL Classes with Target RRF at 75% of Allowed Range

IPL Description	Min PFD A	Max PFD C	Used in LOPA	Design RRF	Mode B	Mean PFD	Variance of PFD
SIL-1	1E-02	1E-01	1E-01	75	1.3E-02	4.1E-02	4.3E-04
SIL-2	1E-03	1E-02	1E-02	750	1.3E-03	4.1E-03	4.3E-06
SIL-3	1E-04	1E-03	1E-03	7,500	1.3E-04	4.1E-04	4.3E-08
SIL-4	1E-05	1E-04	1E-04	75,000	1.3E-05	4.1E-05	4.3E-10

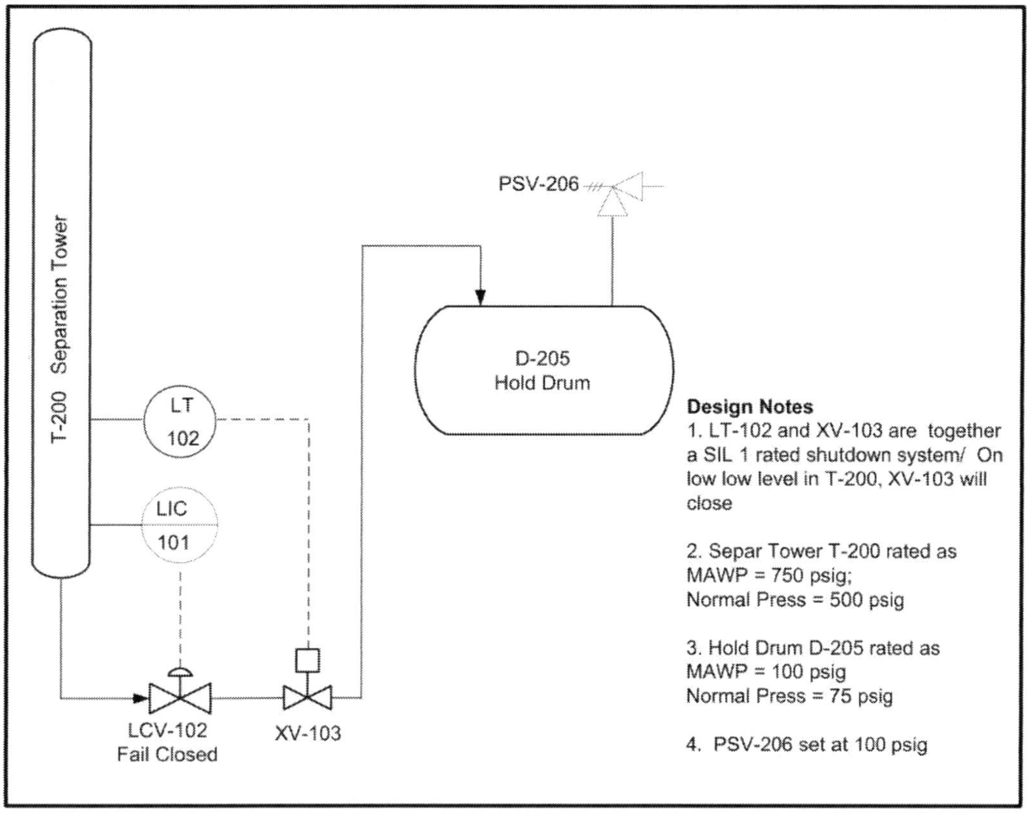

Figure 1. Piping and Instrument Diagram for Example

The Impact of Ozone on the Lower Flammable Limit of Hydrogen in Vessels Containing Savannah River Site High Level Waste

Paul Osterberg
Fauske & Associates, LLC
16W070 83rd Street
Burr Ridge, IL 60527
Osterberg@fauske.com

Carol Sherburne
Savannah River Remediation, LLC
Savannah River Site, Aiken, SC
Carol.Sherburne@srs.gov

Tom Johnson, Thomas Frawely
Fauske & Associates, LLC
16W070 83rd Street
Burr Ridge, IL 60527
Johnson@fauske.com, Frawely@fauske.com

Prepared for Presentation at
American Institute of Chemical Engineers
2013 Spring Meeting
9th Global Congress on Process Safety
San Antonio, Texas
April 28 – May 1, 2013

UNPUBLISHED

The Impact of Ozone on the Lower Flammable Limit of Hydrogen in Vessels Containing Savannah River Site High Level Waste

Paul Osterberg
Fauske & Associates, LLC
16W070 83rd Street
Burr Ridge, IL 60527
Osterberg@fauske.com

Carol Sherburne
Savannah River Remediation, LLC
Savannah River Site, Aiken, SC
Carol.Sherburne@srs.gov

Tom Johnson
Fauske & Associates, LLC
16W070 83rd Street
Burr Ridge, IL 60527
Johnson@fauske.com

Thomas Frawely
Fauske & Associates, LLC
16W070 83rd Street
Burr Ridge, IL 60527
Frawely@fauske.com

Keywords: hydrogen, ozone, flammability, lower flammability limit

Abstract

The Savannah River Site, in conjunction with AREVA Federal services, has designed a process to treat dissolved radioactive waste solids with ozone. It is known that in this radioactive waste process, radionuclides radiolytically break down water into gaseous hydrogen and oxygen, which presents a well defined flammability hazard. Flammability limits have been established for both ozone and hydrogen separately; however, there is little information on mixtures of hydrogen and ozone. Therefore, testing was designed to provide critical flammability information necessary to support safety related considerations for the development of ozone treatment and potential scale-up to the commercial level. Since information was lacking on flammability issues at low levels of hydrogen and ozone, a testing program was developed to focus on filling this portion of the information gap. A 2-L vessel was used to conduct flammability tests at atmospheric pressure and temperature using a fuse wire ignition source at 1 percent ozone intervals spanning from no ozone to the Lower Flammable Limit (LFL) of ozone in the vessel, determined as 8.4%(v/v) ozone. An ozone generator and ozone detector were used to generate and measure the ozone concentration within the vessel *in situ*, since ozone decomposes rapidly on standing. The lower flammability limit of hydrogen in an ozone-oxygen mixture was found to decrease from the LFL of hydrogen in air, determined as 4.2 % (v/v) in this vessel. From the results of this testing, Savannah River was able to develop safety procedures and operating parameters to effectively minimize the formation of a flammable atmosphere.

1. Introduction

Currently, in support of the Federal Facilities Agreement Act, tanks at the Savannah River Site (SRS) used to collect high level nuclear waste are being emptied, cleaned and grouted; and the waste permanently dispositioned in a series of specialized treatment processes. The majority of this waste removal is accomplished by a Bulk Waste Removal (BWR) campaign which empties the tank of easily accessible liquid supernatant and other liquid, followed by salt dissolution/removal and, finally, heel removal. At the end of the BWR campaign, fewer than 20,000 liters of sludge (less than 2 inches depth) remains in the tank. Currently, this final residual material is cleaned from the tank during a process called Bulk Oxalic Acid Cleaning (BOAC), whereby heated, 6 weight percent oxalic acid is added to the tank to a sufficient level to enable mixing. While BOAC has been effective in sludge dissolution and removal, the metal oxalates formed during oxalic acid dissolution cause downstream complications during evaporation (e.g. foaming) in addition to requiring excessive sludge washing (thereby greatly increasing the overall volume of waste to be dispositioned) to meet feed qualifications of the Defense Waste Processing Facility. This prompted research, testing and acceptance of a treatment process that utilized the proven effectiveness of oxalic acid for sludge dissolution and mobilization, yet removed the offending metal oxalates from the solution without adding liquid volume or contaminants. The Enhanced Chemical Cleaning (ECC) process was thus developed to replace BOAC stage of waste removal.

The Enhanced Chemical Cleaning (ECC) process uses ozone to promote the oxidation of metal oxalates produced during the dissolution of sludge in the Savannah River Site (SRS) waste tanks. The process is held at 70°C to improve decomposition kinetics and 10 psig to improve ozone solubility. A simplified depiction of the ECC process is shown in Figure 1, below.

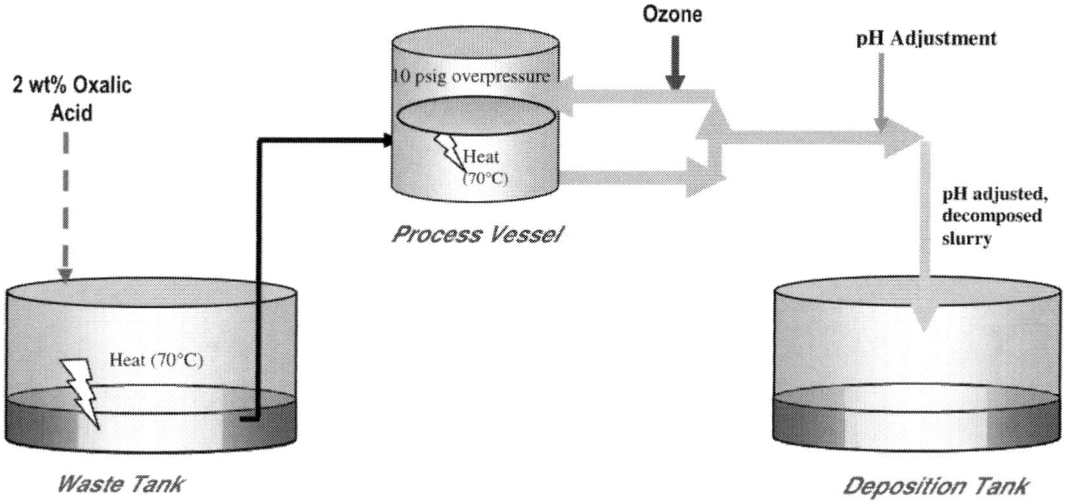

Figure 1: Schematic of the Enhanced Chemical Cleaning Process

In this process, the ozone reacts with the metal oxalates to form metal oxide and hydroxide precipitants, and the CO_2, O_2, H_2O and any unreacted O_3 gases are discharged into the vapor space. The major chemistry employed in the ECC process vessel is the reaction of ozone with the metal oxalates to form metal oxides, metal hydroxides, CO_2 and H_2O by the reactions shown below:

$$O_3 + H_2O \rightarrow O_2 + 2OH\bullet$$
$$Fe_2(C_2O_4)_3 + 6\ OH\bullet \rightarrow 6CO_2 + Fe_2O_3 + 3H_2O$$
$$H_2C_2O_4 + O_3 \rightarrow 2CO_2 + H_2O + O_2$$

While only the Iron oxalate reaction is listed above, formation of the powerfully reactive hydroxyl radical initiates many oxidizing reactions, effectively putting most waste constituents into their highest oxidation state.

In addition to the non-radioactive metals in the waste, however, the SRS radioactive waste also contains a variety of radionuclides, hence, hydrogen gas is also present in the vapor space of the ECC system. In order to prevent a buildup of hydrogen, and, hence, a flammability/explosion concern, waste tanks are purged and the hydrogen concentration monitored to ensure that the vapor space of each waste tank and all related processing equipment remains below the Lower Flammability Limit (LFL) of hydrogen. Furthermore, a comprehensive flammability control program is utilized to ensure that the LFL of the vapor space of a waste tank will not be attained for at least 7 days without forced ventilation. Because the use of ozone/oxygen in the processing equipment may also introduce ozone/oxygen into the vapor space of the waste tanks, the impact of this gas stream on the LFL of hydrogen must be understood for all possible operating scenarios of both normal and off-normal situations, with particular emphasis at the elevated temperatures and pressures of the typical process operating conditions. Oxygen is a known accelerant in combustion reactions, but while there are data associated with the behavior of hydrogen/oxygen environments, relevant studies addressing the effect of ozone on the flammability limit of hydrogen proved scarce. Further, discussions with industry experts verified the absence of data in this area and indicated that laboratory testing, specific to defined operating parameters was needed to comprehensively address the issue. Testing was thus designed and commissioned to provide the data necessary to support safety related considerations for the ozone treatment process.

2. Experimental

2.1 Flammability Apparatus

In order to ensure consistency between this testing and other documented flammability testing, the test methodology utilizes the guidelines established in ASTM E-918-83 (2005), "Standard Practices for Determining Limits of Flammability of Chemicals at Elevated Temperature and Pressure" [1]. Using this standard as the testing basis, a test setup was developed and is given in Figure 2.

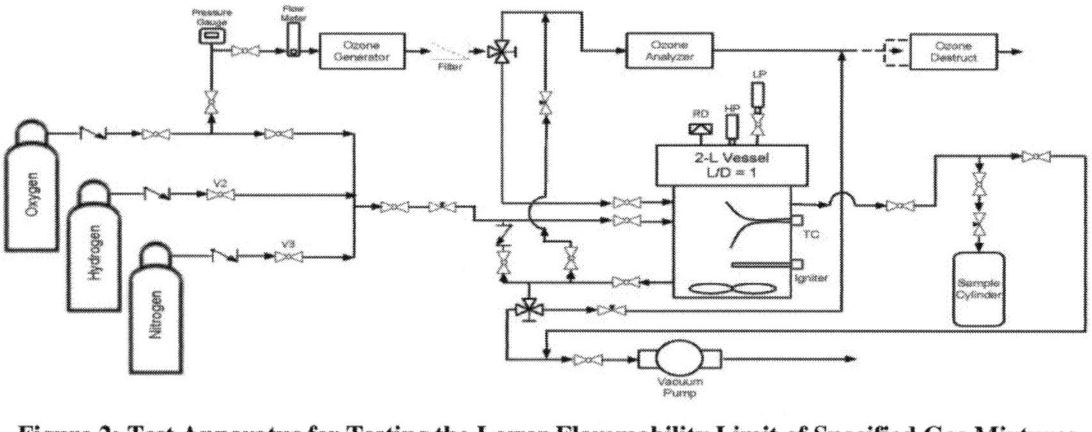

Figure 2: Test Apparatus for Testing the Lower Flammability Limit of Specified Gas Mixtures

The testing was conducted in a 2-L vessel rated to 3,000 psi at 500°C with an L/D ratio of one. There are two pressure transducers mounted on the vessel: a high accuracy (0.05% full scale) low range pressure transducer (0-25 psia) used for preparing the gas mixtures by partial pressure and a high accuracy (0.05% full scale) high range pressure transducer (0-200 psia) used to monitor the pressure after the mixture is ignited. High purity research grade gases were used for the preparation of the gas mixtures: nitrogen (99.9999%), oxygen (99.999%), and hydrogen (99.9999%). Two thermocouples were placed in the vessel to measure the temperature at the center and at the upper area of the vessel. A stir bar was placed at the bottom of the vessel to provide mixing of the gases. The gas mixture was ignited using an exploding fuse wire positioned a third of the way from the bottom. The energy provided by the fuse wire was approximately 10 joules. An ignition energy of 10 J was chosen because it produces very little wire spatter and a cylindrical shock wave is formed upon the vaporization of the wire because it is above the 5 J/cm threshold [2].

Ozone is generated on-site using a high ozone concentration generator. The ozone concentration of the ozone/oxygen stream is analyzed with an in-line UV absorption unit designed to enable accurate measurements of ozone from 0.01 wt. % ozone in oxygen to 20 weight percent.

In order to minimize undesired ozone decomposition, the materials used for this testing were carefully chosen. The test vessel, tubing, and fittings are all made of 316 stainless steel and passivated per ASTM A967 [3]. The packing material and/or seals for the valves were made of PTFE or fluorocarbon FKM.

2.2 Experimental Method

Prior to testing, the thermocouples, pressure transducers, ozone analyzer, and data acquisition instrumentation are checked to ensure they are functioning properly. The 2-L vessel is opened and wiped down with acetone. A 10mm 40 gauge tinned copper wire is wrapped around the igniter electrodes. The vessel is sealed and checked for leaks through pressurizing the vessel. This is followed by a vacuum and pressure purging procedure with a diluent gas (N_2 or O_2) depending on the test conditions. The gas manifold and piping are vacuumed and pressure purged with the respective gas prior to addition into the vessel. The purging helps minimize trace contamination of other gases as well as to remove residual moisture.

By assuming that these gases behave ideally, the gas concentration is determined based on partial pressure addition. Prior to gas filling, the vessel is evacuated down to approximately 0.2 psia. The magnetic stirrer is turned on to provide mixing during the gas filling step. Each gas is added slowly to the desired concentration based on the partial pressure calculation. During addition, there is a slight temperature rise due to the compression of the gases; therefore, after each gas addition, the mixture is allowed to reach equilibrium. The gas manifold and piping are purged with the subsequent gas prior to being added to the mixture.

For the ozone tests, oxygen is fed into a corona discharge ozone generator. A potentiometer on the unit controls the power to the elements. Thus, by adjusting the oxygen flowrate through the generator and tuning the potentiometer, the precise ozone/oxygen ratio specified in the test matrix can be produced. The ozone concentration is measured with an in-line ozone analyzer that is placed on the outlet line of the vessel.

Early scoping tests performed to optimize the test setup indicated that if ozone was added to the test vessel while the vessel was under a vacuum, the ozone decomposed very rapidly, making it difficult to establish the desired test conditions. By feeding the ozone/oxygen mixture through the process vessel at atmospheric pressure, then monitoring the ozone concentration at the discharge, an accurate determination of the ozone concentration in the test vessel, just prior to the addition of hydrogen, may be made. The test vessel can then be evacuated just enough to add the prescribed volume of hydrogen for the test, minimizing loss of ozone.

Once the mixture has been prepared, the magnetic stirrer is turned off. The mixture is allowed to relax prior to ignition so that a quiescent mixture is tested rather than a turbulent mixture. The low pressure transducer valve is closed and the data collection is started. The mixture is manually ignited using the fuse wire providing a nominal 10 joules. The data acquisition records the pressure and temperature time traces in the vessel. A mixture is considered to be flammable if the pressure rise after the ignition event is greater than or equal to 7% in accordance with ASTM E918. The procedure is repeated until there are a minimum of two demonstrations conducted at the lowest fuel concentration to be considered flammable and three at the highest fuel concentration considered non-flammable.

3. Results and Discussion

3.1 Hydrogen Lower Flammable Limit

The lower flammable limit for initially quiescent hydrogen-air mixtures was determined in order to validate the test equipment setup and the procedure. The LFL for hydrogen in air was determined as 4.2 volume percent based on the ASTM 7% pressure rise criteria. A comparison of this result to data obtained by other researchers is summarized in Table 1.

Table 1: Hydrogen LFL in Air, this testing compared to others

Source	Hydrogen Concentration (vol. %)	Comments
LFL using 5% pressure rise criteria		
SAFEKINEX [4]	4.1	2.8-L vessel
	3.9	6-L vessel
	4.1	14-L vessel
	4.7	40-L vessel
	5.8	1250-L vessel
	4.1	2000-L vessel
LFL using 7% pressure rise criteria		
Present Work	4.2 ± 0.1	2-L vessel (L/D = 1)
Crowl and Jo [5]	5.7	20-L spherical vessel
Cashdollar et al. [6]	5.0 ± 0.5	8-L chamber
	6.0 ± 0.5	20-L chamber
	6.5 ± 0.5	120-L spherical vessel
LFL using a visual criterion		
Coward and Jones [7]	4.0	Flammability Tube

The experimentally measured LFL is in close agreement with the accepted lower flammable limit of 4 vol. % which was obtained through visual criterion [7]. Many other researchers have shown that it is difficult to obtain this value using the 7% pressure rise criteria. This has been explained through hydrogen's high diffusivity as compared to oxygen and nitrogen [6].

The majority of tests using the pressure rise criteria are performed with the ignition source in the center of the vessel. Central ignition is used for the determination of the explosion severity of flammable mixtures because it gives the maximum rate of pressure rise and the maximum overpressure. However, central ignition may not be useful when determining the limits of flammability. It has been shown that ignition source location has an impact on the flammability limits of methane [8]. As the location of the ignition source is lowered, a larger percentage of the combustible mixture participates in the reaction with minimal heat loss to the vessel wall. This results in a higher explosion overpressure, thereby widening the flammable region. The present work uses an ignition source placed approximately one-third from the bottom of the vessel.

3.2 Ozone Lower Flammable Limit

Initially, tests were conducted to determine the extent of ozone decomposition over time inside the combustion vessel at ambient conditions. This was accomplished by preparing an ozone-oxygen mixture and holding the mixture in the combustion vessel for approximately 30 minutes while stirring. After the specified hold time the ozone concentration was measured. It was found that there were negligible amounts of ozone decomposition that occurred in the vessel.

The LFL of ozone in oxygen was determined as 8.4 volume percent. Previous studies on the flammability of ozone in oxygen are presented in Table 2.

Table 2: Ozone LFL in Oxygen, this testing compared to others

Source	Ozone Concentration (vol. %)	Comments
Present Work	8.40 ± 0.17	2-L vessel (L/D = 1)
Mizutani et al. [9]	10.5	1-L cylindrical vessel, Electric spark
Koike et al. [10]	10.0	1-L cylindrical vessel, 1 J electric spark
Cook et al. [11]	9.5 ± 0.5	10 cm i.d. glass bulb, 0.47 J electric spark

The present work reveals that the explosion limit of ozone in oxygen is consistent with previously reported values; however, it is lower than other literature data which can be attributed to the differences in the experimental method and procedure. The lower explosion limit obtained in this work is most likely a result of a combination of factors: the type of ignition source, location of the ignition source, and vessel size and geometry. In these tests, the ignition source energy was setup to give approximately 10J which is much greater than that used in previous work. Flammability limits are a function of ignition energy, thus, by increasing the strength of the ignition source the flammable limits tend to widen [12]. There also tends to be some variation in the flammable limits which is influenced by the type of ignition source: a fuse wire as compared to an electric spark [13]. This may be an effect of the energy density provided by each type of ignition source and the amount of energy actually being transferred to the mixture. As discussed previously, the ignition source location may be playing a part in the lower value obtained.

The differences in the size and dimensions of the explosion vessels could have affected the lower flammable limit [13]. As the vessel size is increased, the heat losses to the wall become negligible. Through minimizing heat losses to the vessel wall, more heat is transferred to the combustion reaction, therefore, promoting flame propagation. This will result in a widening of the flammable region and combustion can occur at lower temperatures.

3.3 Lower Flammable Limit of Hydrogen in Ozone/Oxygen Mixtures

The experimental results for the lower flammable limits of hydrogen in the ozone/oxygen are shown in Figure 3. The solid circles represent a flammable mixture and the open circles represent a non-flammable mixture.

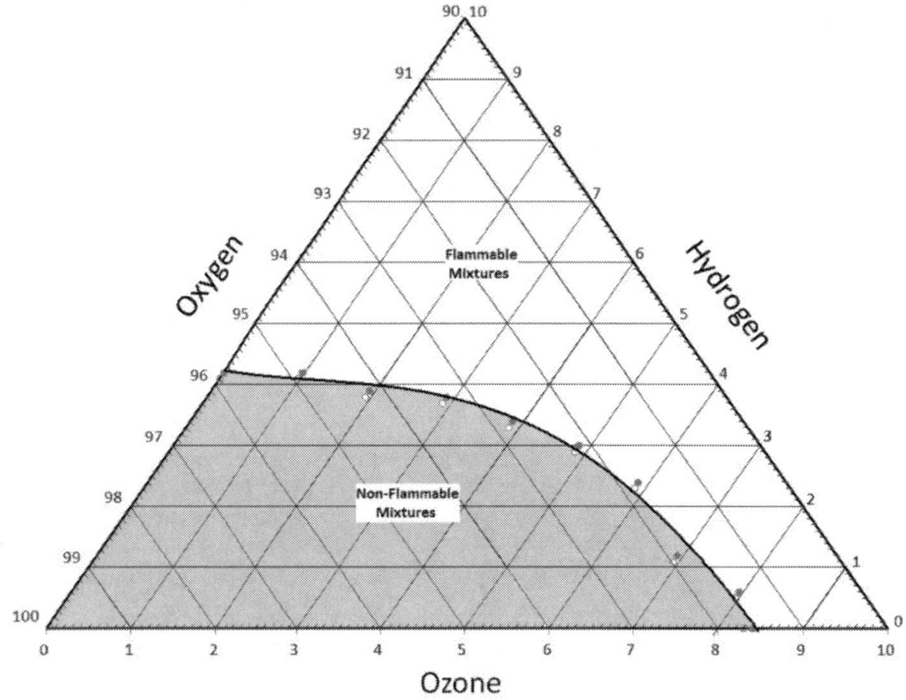

Figure 3: Flammability envelope for hydrogen/ozone/oxygen mixtures (all values are in volume percent).

Typically, it is of interest to estimate the flammability limits of multi-component mixtures. In the absence of experimental data, estimation methods can be useful if the flammable limits of pure components are known. Le Chatelier [14] developed an empirical equation for predicting the flammable limits of fuel mixtures,

$$LFL_{MIX} = \left(\sum_{i=1}^{n} \frac{y_i}{LFL_i} \right)^{-1} \qquad \text{Eq. [1]}$$

where y_i is the mole fraction of the ith component on a combustible basis and LFL_i is the lower flammable limit of the ith component in volume percent.

A comparison of the LFL predictions to the experimentally determined LFL of hydrogen in various ozone-oxygen mixtures is shown in Figure 4. It was found that the LFL of hydrogen decreases as the ozone concentration in the mixture increases. This result is consistent with the values approximated through Le Chatelier's rule; however, the experimental results are higher than the predicted calculations. Previous studies have shown that Le Chatelier's rule provides close approximations for many simple hydrocarbons; however, it is less accurate for mixtures with hydrogen and generally not applicable for substances that undergo decomposition [15]. However, from a safety stand point, Le Chatelier's rule provides a conservative estimate of the LFL.

The experimental data was further compared to flammability predictions from the calculated adiabatic flame temperature (CAFT) method. The flame temperature criteria used to assess whether a mixture is flammable or not was established by first calculating the adiabatic flame temperature at the determined LFL of hydrogen in air and ozone in oxygen. Three flame temperatures were used – the flame temperature of 8.4% ozone in oxygen, the flame temperature of 4.2% hydrogen in air and an average of these two flame temperatures. The LFL of hydrogen in various concentrations of ozone/oxygen was then calculated from these flame temperatures.

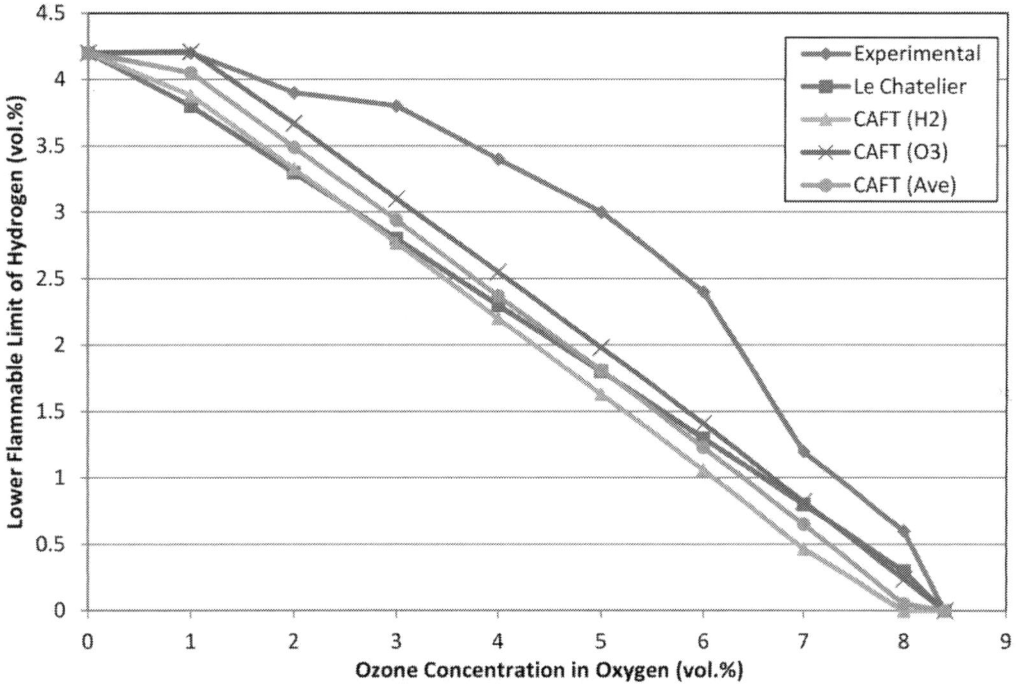

Figure 4: Comparison of lower flammability limit of hydrogen in ozone/oxygen to calculated adiabatic flame temperature predictions.

The calculated adiabatic flame temperature (CAFT) predictions are very similar to the predictions given by Le Chatelier's empirical equation. The deviations of the experimental data from the estimations in the flammability limits of a hydrogen/ozone/oxygen mixture lie within the unpredictability of ozone. This unpredictability comes from the instability of ozone, which tends to readily decompose to form oxygen. Furthermore, there are many different reactions that could be taking part in these experiments: H_2-O_2, O_3 decomposition, O_3-O_2, and H_2-O_3. The data seems to suggest that as the ozone concentration is increased, the reaction mechanism that takes precedence changes.

The deviation between the predicted and the experimental values may be two-fold. First, the reaction mechanism taking preference during the combustion reaction changes as the ozone concentration is increased. Secondly, the concentration of ozone participating in these reactions may be changing as well. Tests conducted on ozone decomposition in oxygen have shown that only 5.3% of the ozone decomposes for a gas mixture containing 6.7 mole % ozone and as high as 90% decomposition when the ozone concentration was 11.2% in the mixture [11]. In any case, additional work is needed to fully understand the nature of these experiments.

3.4 Temperature and Pressure Effect on the Lower Flammable Limit of Hydrogen-Ozone-Oxygen Mixtures

It has been reported previously that the ozone decomposition rate increases at the temperature of its environment increases [16]. Therefore, stability tests were performed again to determine the extent of ozone decomposition at the elevated test conditions. This time it was found that the ozone decomposition rate could not be considered negligible. With initial mixtures containing 5 vol.% ozone in oxygen at approximately 75°C, the ozone decomposition rate was approximately 0.1 vol.% ozone every 10 minutes.

The results from the tests performed at elevated temperatures and pressures are shown in Figure 5. It was found that as the temperature is increased the LFL decreases and the opposite was observed when the pressure was increased.

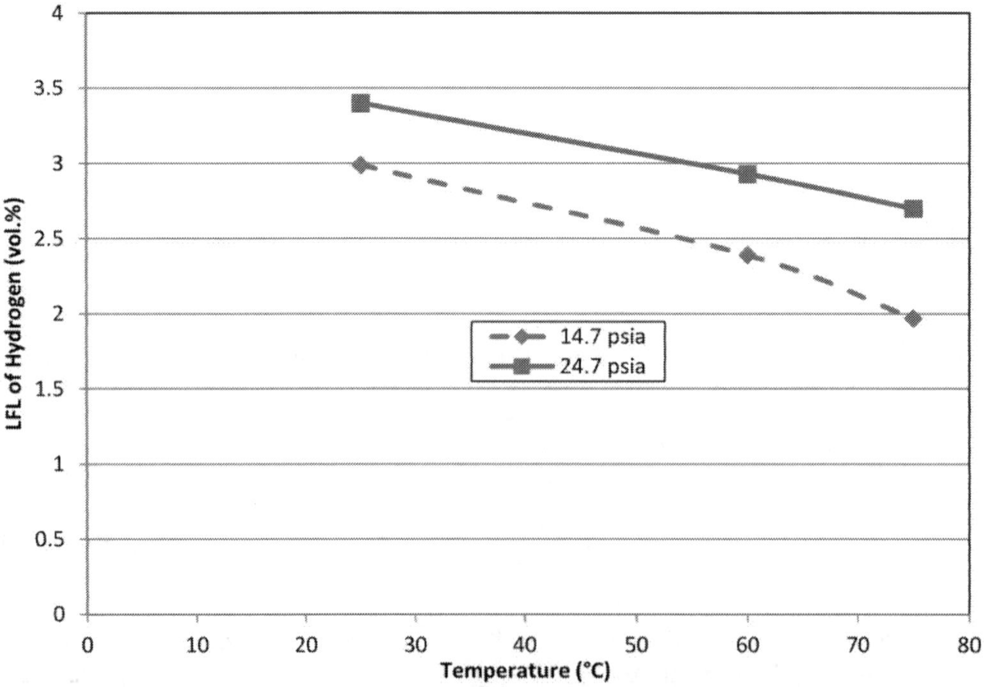

Figure 5: Effect of temperature and pressure on the flammability limits of hydrogen-ozone-oxygen mixtures. Tests were performed with 5 vol.% ozone in oxygen.

The increase in the LFL as a result of the increase in pressure is generally not characteristically observed for flammable gases; however, this is a characteristic of hydrogen flammability [17]. This result leads to the conclusion that the reaction mechanism responsible for the increased LFL as the pressure is increased for hydrogen is still the dominate mechanism when ozone is added. Furthermore, it is noticed that as the temperature is increased the decrease in the LFL is fairly linear throughout the range of temperatures tested. The LFL results from these tests show that the LFL is more susceptible to elevated temperatures when compared to results of hydrogen-air or hydrogen-oxygen. One explanation for this is that at the increase temperatures more of the ozone participates in the reactions that occur in the vessel. Previous research on ozone decomposition has found that only a part of the ozone decomposes when the ozone concentration is below its LFL. At the higher temperatures, the reaction rates increase as well as the instability in the ozone molecule.

3.5 Effect of Relative Humidity on the Lower Flammable Limit of Hydrogen-Ozone-Oxygen Mixtures at Elevated Temperatures and Pressure.

Tests conducted with the addition of water vapor have shown a slight increase in the LFL as shown in Figure 6. Since there is an additional component (water vapor) added to the mixture, the results should be compared on a total combustibles (hydrogen plus ozone) basis rather than just on the concentration of hydrogen in the mixture. This is because as more water vapor is added to the mixture, the concentration of ozone decreases; therefore, even if more hydrogen was needed to be added to the mixture, the total amount of combustibles may have remained constant to produce a flammable mixture.

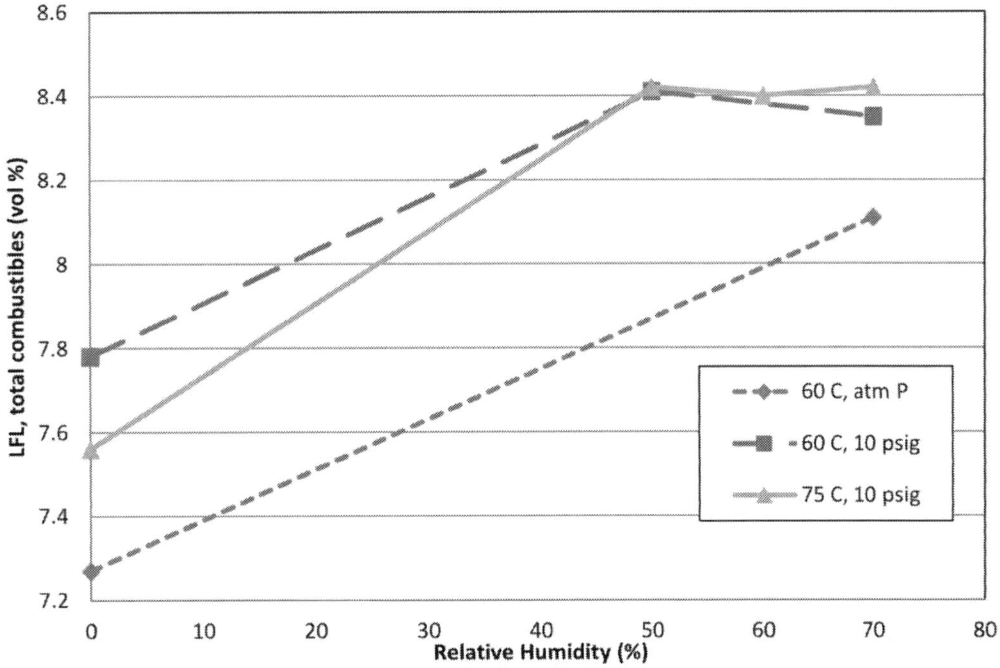

Figure 6: Effect of relative humidity on the LFL of hydrogen-ozone-oxygen mixtures. Tests were performed with 5 vol.% ozone in oxygen.

The slight increase in the LFL could be a result of either the addition of the water vapor or decrease in ozone concentration. The addition of water vapor to the mixture helps extract heat from the reactions occurring in the vessel (combustion or decomposition), thereby, quenching the flame propagation by lowering the reaction rate. Furthermore, the observed increase in the LFL could be a result of a decrease in the ozone concentration. It is known that both hydrogen and water act as a catalyst for ozone decomposition; therefore, the actual concentration inside the vessel prior to igniting the mixture is lower than the reported concentrations.

4. Impact on Operation

This testing was designed to provide data under the conditions considered most optimal to produce a deflagration. The geometry and materials of construction of the testing vessel; the location of the fuse wire; the magnitude of the supplied energy; the careful minimization of diluents and other contaminants; and meticulous procedural detail to maintain integrity of the ozone to the maximum extent practical, result in data that reflect not the expected process conditions, but those that enhance the possibility of flame propagation. For this reason, there is believed to be considerable conservatism in the indicated results. Per the vendor, the maximum possible ozone concentration producible by the planned ECC Ozone generator is 8 volume percent (typical maximum expected concentration is 6.8 vol%), less than the 8.4 minimum volume % concentration shown to be flammable in a pure O_2 environment at the optimally conservative conditions established in this testing. As both water vapor and nitrogen, together constituting approximately 6.5 % of the actual feed stream, are powerful diluents, it is believed not credible that deflagration can occur at this maximum feed concentration condition. In addition, once the ozone stream contacts the waste stream, the many simultaneous oxidizing reactions will rapidly decompose available ozone to well below flammable levels. Further, because the radiolytically generated hydrogen quantity is negligible compared to the supplied ozone/oxygen stream (0.0004 moles per minute H_2 vs 76 moles per minute ozone/oxygen), even the total H_2/O_3 mixture, without crediting decomposition reactions, does not approach flammable concentrations. Finally, even at the "end" of the ECC batch cycle, when most of the metal oxalates have been decomposed, testing has indicated that the ozone concentration in the vapor space of the ECC process vessel reaches a concentration of no more than 3 vol%, remaining well below concentrations of concern.

The major issue for the ECC operation established by this testing is the impact of the data when applied to off normal conditions. While it is possible to discontinue ozone addition to the reaction vessel at any time, the radiolytic hydrogen generation rate continues, varying slightly as ambient pressures and temperatures change. Relative to the data generated and analyzed in this testing, the ECC hazards analysis team will re-evaluate off normal conditions (e.g. those during which process exhaust ventilation is lost) such that issues involving mixtures of hydrogen and ozone in the vapor space can be appropriately controlled.

5. Conclusion

The LFL of hydrogen in air was determined and is in good agreement with the literature data. Ozone-oxygen mixtures were found to be flammable at concentrations above 8.3 vol.% based on the ASTM E918 7% pressure rise criteria for flame propagation. This result is lower than previously reported values which can be explained through the variations in the test setup and procedure. Tests performed with hydrogen in various concentrations of ozone in oxygen have shown that the LFL of hydrogen decreases as the concentration of ozone in the mixture increases.

The estimations of the LFL for various mixtures of hydrogen and ozone in oxygen was performed using Le Chatelier's rule as well as the CAFT method. It was found that both methods underestimate the LFL for mixtures of hydrogen and ozone.

The effects of temperature, pressure, and relative humidity were found for hydrogen-ozone mixtures. The trends observed through these tests were consistent with the general trends in reported results for hydrogen. Overall, the difficulty with some of these tests ultimately lies with the unpredictability of ozone.

6. References

[1] ASTM. (2005). ASTM E 918-83 Standard Practice for Determining Limits of Flammability of Chemicals at Elevated Temperature and Pressure. Philadelphia, PA: American Society for Testing and Materials.

[2] Mashuga, C. V. (1999). *Determination of the Combustion Behavior for Pure Components and Mixtures using a 20-Liter Sphere.* Phd Thesis. Michigan Technological University. Ann Arbor: UMI Company.

[3] ASTM. (2005). ASTM A967-05, Standard Specification for Chemical Passivation Treatments for Stainless Steel Parts. Philadelphia, Pa: American Society for Testing and Materials.

[4] Project SAFEKINEX, Contract No. EVG1-CT-2002-00072, Deliverable No.8, *Report on experimentally determined explosion limits, explosion pressures and rates of explosion pressure rise – Part 1: methane, hydrogen, and propene,* March 2006.

[5] Crowl, D. A., & Jo, Y.-d. (2009). A Method for Determining the Flammable Limits of Gases in a Spherical Vessel. *Process Safety Progress , 28* (3), 227-236.

[6] Cashdollar, K. L., Zlochower, I. A., Green, G. M., Thomas, R. A., & Hertzberg, M. (2000). Flammability of methane, propane, and hydrogen gases. *Journal of Loss Prevention in the Process Industries , 13*, 327-340.

[7] Coward, H. F., & Jones, G. W. (1952). *Limits of Flammability of Gases and Vapors.* BuMines, 503.

[8] Van den Schoor, F., Norman, F., & Verplaetsen, F. (2006). Influence of the ignition source location on the determination of the explosion pressure at elevated initial pressure. *Journal of Loss Prevention in the Process Industries* , 459-462.

[9] Mizutani, T., Matsui, H., Sanui, H., & Yonekura, M. (2001). Decomposing detonation and deflagration properties of ozone/oxygen mixtures. *Journal of Loss Prevention in the Process Industries , 14*, 559-565.

[10] Koike, K., Nifuku, M., Izumi, K., Nakamura, S., Fujiwara, S., & Horiguchi, S. (2005). Explosion properties of highly concentrated ozone gas. *Journal of Loss Prevention in the Process Industries, 18*, 465-468.

[11] Cook, G. A., Spadinger, E., Kiffer, A. D., & Klumpp, C. V. (1956). Explosion Limits of Ozone-Oxygen Mixtures. *Industrial and Engineering Chemistry , 48* (4), 736-741.

[12] Bartknecht, W. (1981). *Explosions, Course Prevention Protection.* New York: Springer-Verlag.

[13] Project SAFEKINEX, Contract No. EVG1-CT-2002-00072, Deliverable No. 2, *Report on the experimental factors influencing explosion indices determination*, September 2003.

[14] Le Chatelier, H. (1891). Note on the Determination of Firedamp by Flammability Limits. *Ann. Mines , 19*, 388-395.

[15] Babrauskas, V. (2003). *Ignition Handbook.* Issaquah: Fire Science Publishers.

[16] Koike, K., Inoue, G., Suzuki, M., Saida, J., & Komatsu, I. (1998). Decomposition Characteristics of Concentrated Ozone. *Journal of Chemical Engineering of Japan , 31* (2), 195-200.

[17] Schroeder, V., & Holtappels, K. (2005). Explosion Characteristics of Hydrogen-Air and Hydrogen-Oxygen Mixtures at Elevated Pressures. *International Conference on Hydrogen Safety.* Pisa.

9TH GLOBAL CONGRESS ON PROCESS SAFETY

Process Safety Considerations for New DOE Depleted Uranium Hexafluoride Conversion Plants

Peter M. Burban, Ph.D.
Department of Energy
Portsmouth Paducah Project Office
Peter.Burban@lex.doe.gov

Prepared for Presentation at
American Institute of Chemical Engineers
2013 Spring Meeting
9th Global Congress on Process Safety
San Antonio, Texas
April 28 – May 1, 2013

UNPUBLISHED

Process Safety Considerations for New DOE Depleted Uranium Hexafluoride Conversion Plants

Peter M. Burban, Ph.D.
Department of Energy
Portsmouth Paducah Project Office
Peter.Burban@lex.doe.gov

Keywords: depleted uranium hexafluoride, dry conversion, fluidized bed, uranium oxide, hydrofluoric acid, layers of protection, independent safety system, defense in depth, basic process control system, HF detectors, passive safety systems, Department of Energy

Abstract

The Department of Energy (DOE) OH & KY depleted uranium hexafluoride (DUF6) facilities are first of a kind operating plants designed to convert DUF6 into a safer form for reuse and/or disposal. Safety played a lead role in process design with HAZOP results guiding implementation of layers of protection to reduce risks associated with several materials including UF_6, H_2 and HF. A stand-alone independent safety control system functions to provide independent monitoring and control capability for those processes that have significant safety considerations.

1. Introduction

The Department of Energy Piketon, OH and Paducah, KY DUF6 facilities, Figure 1, are new fluid bed operating plants designed to convert approximately 700,000 metric tons of depleted uranium hexafluoride into a safer form for reuse and/or disposal. Depleted uranium hexafluoride has been generated in the United States since the mid-1940's as a byproduct of enriching uranium for both civilian and military applications at three gaseous diffusion plants located near Oak Ridge, Tennessee; Paducah, Kentucky; and Piketon, Ohio. This paper will highlight important process safety design considerations implemented that are critical to ongoing operations. Key aspects of the process hazards analysis and some lessons learned will be shared regarding chemical hazards associated with HF.

Figure 1. Portmouth (top) and Paducah DUF6 facilities

2. DUF6 Conversion Process and Safety Considerations

2.1 Process Description

The source DUF6 material, stored at both plants in approximately 63,000 large steel cylinders, is fed direct from the storage cylinders. A DUF6 feed cylinder, heated in a specially designed autoclave, vaporizes DUF6 which is fed with steam and hydrogen, blended with nitrogen, to a high temperature fluidized bed reactor for conversion to uranium oxide powder and hydrofluoric acid vapor. Some elements of the chemical process are described by Urza [1] and a general schematic of the process is given on the internet [2]. The uranium oxide powder transfer system loads powder into modified feed cylinders. The HF recovery system condenses the water and HF vapors in the process offgas, producing aqueous HF acid which is temporarily stored in tanks and later loaded for shipment to HF vendor facilities. The environmental green aspects of this approach include reuse of the feed DUF6 cylinders, regeneration of KOH used in scrubber operation and recovery of aqueous HF for commercial sale.

2.2 Overall Process Safety

A major safety advantage of the conversion process is low pressure reactor operation. The 'dry' conversion process uses superheated steam and nitrogen diluent to ensure safe reactor hydrogen concentrations. Administrative control of the feed cylinders ensures only depleted UF_6 is processed, eliminating any nuclear criticality event.

Hazard evaluations were performed to develop accident scenarios with the potential for exposure to HF, the greatest chemical hazard. Unmitigated consequences and risks were estimated for the credible events and compared to DOE risk guidelines. This paper focuses how safety significant engineered features and administrative controls are integrated in layers of protection to prevent or mitigate accidents associated with the major HF risks [3].

An Integrated Control System (ICS) controls and monitors the facility automated systems and associated components during normal operations. The system also mitigates accidents involving the conversion process and any of the systems or activities that have an interface with the process. The ICS consists of BPCS (Basic Process Control System) and ISS (Independent Safety System) subsystems. The purpose of the BPCS is to provide monitoring and control for normal operations from the Control Room and/or local operator workstations. The purpose of the ISS is to provide independent monitoring and control capability for those processes that have significant safety considerations. The ISS does not rely on the ISS human machine interface or BPCS to perform interlock actions.

3. Safety Features

3.1 Feed System Safety Features

A major component of the feed system is the autoclave, which serves to vaporize DUF6. The autoclave also serves as a pressure containment boundary (at several multiples of the supply pressure) to mitigate the consequence of an accidental release of DUF6 from a breached feed cylinder/piping in the autoclave. Since UF_6 reacts with atmospheric moisture to generate HF vapor, both HF and UF_6 releases have similar hazards classifications.

In the event of an over pressurized autoclave leading to activation of the pressure relief, the vent gases are directed to a controlled plenum. Automatic interlocks shut off DUF6 flow and/or electrical power to the cylinder heating system for an autoclave high pressure condition, as well as the following abnormal operating conditions: cylinder over-temperature, DUF6 header high-pressure, and an open autoclave door. The ISS monitors UF_6 and HF detectors located in the autoclave room which, upon alarm, terminate DUF6 flow to the downstream piping via an autoclave isolation valve. Additional administrative controls lead to shutdown of the reactor system. These layers of protection provide 'defense in depth' to reduce worker exposure from an abnormal UF_6 feed cylinder/system release.

3.2 Reaction System Safety Features

Each reactor serves as a primary containment boundary for reactants and products including DUF6 and HF. The ISS monitors UF_6 and HF detectors located in the reactor system room and adjoining HF condenser room which, upon alarm, terminate DUF6 flow to the downstream piping via the autoclave isolation valve. The BPCS in-line process oxygen monitors alarm on detection of oxygen in the process off gas. Additional administrative controls lead to shutdown of the reactor system.

Upon significant UF6 or HF leak detection, specific administrative controls instruct control room operations to terminate hydrogen flow via an automatic hydrogen feed isolation valve (a redundant manual isolation valve is also closed). Additional administrative controls lead to shutdown of the reactor system. These layers of protection provide 'defense in depth' to reduce worker exposure resulting from breach of the reactor system.

3.3 HF Recovery System Safety Features

Off gases from each reactor pass through individual partial condensers which produce an aqueous HF stream collected in individual primary receivers located in the HF recovery room. Off gases continue flowing to first a water and then to a KOH scrubber. All scrubber outlet streams merge and flow to a final backup KOH scrubber prior to venting out the facility stack. Multiple scrubbers reduce the likelihood of stack HF emissions. A secondary overflow HF receiver for each primary mitigates primary receiver level control problems. The HF recovery system units are skid mounted and reside in a common secondary containment dike mitigating the effect of HF leaks.

The ISS monitors several HF detectors located in the HF recovery room which, upon alarm, terminate DUF6 flow to the specific reactor via the autoclave isolation valve. The ISS also monitors HF detectors located in the HF recovery area for HF leaks in the HF transfer line which, upon alarm, terminates flow from the HF receiver tanks. Upon significant HF leak detection, specific administrative controls instruct control room operations to terminate DUF6 and hydrogen flow via an automatic hydrogen feed isolation valve (a redundant manual isolation valve is also closed). Additional administrative controls lead to shutdown of the reactor system. These layers of protection provide 'defense in depth' to reduce worker exposure resulting from breaches in the HF recovery system.

An important series of lessons learned include proper maintenance of HF process lines to eliminate chronic leaks. In addition, care was taken for stress relief of all piping flange connections, and modifications were made to address piping areas experiencing vibration. The use of HF vapor highly sensitive paint was helpful to identify the onset of leaks. Additional isolation valves were installed to improve the safety and ease of corrective maintenance.

3.4 HF Storage Tank Farm Safety Features

The HF storage area includes several passive, active and administrative safety controls. Each HF storage tank is situated in a secondary liquid dike that can hold the entire capacity of the 10,000 gallon tank in case of a leak. The tanker truck and rail loadout areas have a gravity drain equipped with a pumping system to transfer spilled HF to either a containment dike or a storage tank. This minimizes the spill area to reduce the potential on and off site consequences of an HF leak.

The ISS also monitors HF detectors located in the HF storage area which, upon alarm, terminates flow via the HF storage tank automated isolation valve. In addition, the ISS terminates HF flow from the HF receiver tanks. Specific administrative controls instruct an operator to be in required attendance for all transfers. This operator observes each HF storage tank to mobile tanker transfer and can terminate the transfer via an emergency button which stops HF transfer flow even before HF detection by the ISS. Upon alarm, following administrative control, the operators in the control room isolate the HF recovery system and then shutdown the reactor system.

These layers of protection provide 'defense in depth' to reduce worker exposure resulting from leak accidents at the HF storage tank system.

3.5 Additional Safety Features

Several fire suppression systems minimize the likelihood of a fire causing a cylinder breach and release of HF and/or DUF6. Strategically placed vehicle barriers add "defense in depth" to reduce consequences associated with collision with storage tanks and staged feed cylinders. Safety design elements were also incorporated for natural event accident initiators so that process buildings and equipment meet design criteria for forces such as seismic and high winds. Safety management requires the use of appropriate personal protection equipment for all work activity with potential HF exposure. These layers of protection provide 'defense in depth' to reduce worker exposure resulting from HF leak accidents in various systems described above.

4. Discussion

The process safety approach at the DUF6 facilities blends active, passive and specific administrative controls. Considering the other chemical and radiation hazards at the facility, the ability to maintain a safe operation depends on reliable machines and human performance of required operator actions. A rigorous preventative maintenance program is administratively controlled to maintain the reliability of all active safety systems at the facility in line with DOE safety requirements. The Integrated Safety Management Program at the DUF6 facility includes systematic approach to training for Operations. Training includes preparing each individual with the toolbox needed to address all specific administration controls, responses to alarms and recovery from abnormal operating conditions. Monitoring of human performance is an essential element for maintaining safe operations and continuous improvement.

Major facility start up challenges included accurate procedure development and updates, initial training and continuous training as the Operations knowledge base grew. Unique lessons were learned for operations in the four season climates of Ohio and Kentucky, necessitating adaptation of procedures, and emphasis on minor equipment changes to minimize the need for corrective maintenance. Both sites have had a good start to their multiple decade operations campaigns built on the cornerstone of a defense in depth process safety approach.

5. References

[1] Urza, Inaky J., "Conversion of uranium hexafluoride to uranium dioxide" US Patent 4,830,841

[2] "The DUF6 Process," www.bwconversionservices.com, accessed on Feb. 11, 2013

[3] DOE, Final Environmental Impact Statement for Construction and Operation of a Depleted Uranium Hexafluoride Conversion Facility at the Portsmouth, Ohio, Site, DOE/EIS-0360, prepared by Argonne National Laboratory for U.S. Department of Energy- Oak Ridge Operations, Office of Environmental Management, Oak Ridge, Tenn. June 2004. energy.gov/nepa/downloads/eis-0360-final-environmental-impact-statement

(44b) Minimum Ignition Energy Study of Flowing Heat Transfer Fluid Aerosols: Experimental and Theorectical Approach

Szu-ying Huang[1], Xinrui Li[1] and M. Sam Mannan[2], (1)Mary Kay O'Connor Process Safety Center, Artie McFerrin Department of Chemical Engineering, Texas A&M Univerisity, College Station, TX, (2)Mary Kay O'Connor Process Safety Center, Artie McFerrin Department of Chemical Engineering, Texas A&M University, College Station, TX

Liquid-gas aerosol has been a significant concern in process industries for its complexity and unpredictability of formation, ignition probability, combustion mechanisms, and flame propagation with impact on onshore/offshore facilities. Heat transfer fluid, specifically, has been proved to have the possibility to form aerosol from pressurized leaking, spraying, and potential liquid impingment from surfaces. Having more combustible matter and higher energy density, the high-flash point aerosol may induce more hazardous fires and explosions even operated or stored under the safety limits. Among the indicators describing flammability, minimum ignition energy (MIE) is one of the most widely used and studied ones. It is capable to reflect the internal chemical properties of the material, as well as to make connection to analyze the possible facility ignition sources. This paper demonstrates MIE experimental result of commercial heat transfer fluid aerosols, and develops COMSOL-Multiphysics models as theoretical approach on MIE prediction of similar chemical species. The influence of aerosol droplet sizes, initial spray velocity, fuel-air equivalence ratio, and fluid temperature are analyzed. The characteristic ignition temperature, which is the "flash point" of aerosol system, is also developed to compare with single-phase material. Results of the study may be used for industry liquid handling, working fluid selection, mitigation design, and ignition elimination to prevent fire and explosion of aerosols.

See more of this Session: Poster Session
See more of this Group/Topical: Global Congress on Process Safety

Domino Effect Model Applied to the Hertfordshire Oil Storage Terminal (Buncefield Oil Depot)

Manuscript ID: 295031

Esteban Bernechea
Centre for Technological Risk Studies (CERTEC). Department of Chemical Engineering.
ETSEIB. Universitat Politècnica de Catalunya.
Av. Diagonal, 647. 08028-Barcelona (Spain).
esteban.bernechea@estudiant.upc.edu

Xavier Seguí
Centre for Technological Risk Studies (CERTEC). Department of Chemical Engineering.
ETSEIB. Universitat Politècnica de Catalunya.
Av. Diagonal, 647. 08028-Barcelona (Spain).
xavier.segui@upc.edu

Josep Arnaldos
Centre for Technological Risk Studies (CERTEC). Department of Chemical Engineering.
ETSEIB. Universitat Politècnica de Catalunya.
Av. Diagonal, 647. 08028-Barcelona (Spain).
josep.arnaldos@upc.edu

1. Introduction

On December 11, 2005, a series of explosions occurred at the Buncefield Oil Storage Terminal, located in Hemel Hempstead, Hertfordshire, England. The explosions resulted in large fires which spread throughout the installation, causing more than 40 injuries, the destruction of most of the installation and the emission of large clouds of black smoke to the atmosphere; the accident also caused significant damage to residential and commercial properties in the vicinity of the terminal, the evacuation of population centers near the terminal, and an overall economic loss estimated at close to £ 1 billion [1].

The Buncefield disaster is a perfect example of domino effect in the oil industry; the initial explosion, caused due to the overflow of one of the tanks, caused fires that engulfed more than 20 storage tanks, and resulted in the destruction of a significant part of the terminal. The domino effect is defined as a cascade of events in which the consequences of previous accidents are increased both spatially and temporally by following ones, thus leading to a major accident [2]. This definition describes perfectly the situation that occurred in the Buncefield terminal.

One of the characteristics that made this accident the object of many researches was that the overpressure wave ensuing from the first explosion was much higher than would be expected from a Vapor Cloud Explosion (VCE). However, the aim of this manuscript is not to model the exact explosion, or to determine the reasons that caused the overpressure resulting from it to be higher than expected; much works have been dedicated to this endeavor, producing satisfactory conclusions to the phenomenon. The objective of this work is rather to use the Buncefield Terminal layout to test a domino effect model proposed by the authors [3]; to achieve this, possible accident sequences after an explosion will be simulated, to ascertain if they are similar to the chain of accidents that took place during the real event. Also, Quantitative Risk Assessments (QRAs) will be performed using the layout: one in a traditional fashion [4], without accounting for domino effect, and another taking domino effect into account through the application of the proposed model. This is done to analyze the change in the iso-risk curves obtained for both QRAs.

First, a more detailed description of the accident is presented, followed by an explanation of the proposed domino effect model. After this, the way in which the model has been applied to the layout of the installation is described and the results obtained are presented and discussed. Finally, the conclusions reached by the end of the study are shown.

2. Description of the accident

A brief abstract of the incident as described in Volume 1 of The Final Report of the Major Incident Investigation Board [1] is presented in this section.

On Saturday 10, December 2005, a delivery of unleaded petrol started to arrive at Tank 912 in bund A. At about 05:30 am on 11 December, the safety systems that prevented the overfilling of the tank failed to operate. Petrol cascaded down the side of the tank and was collected in Bund A. As the released continued, a vapor cloud mixture of fuel and vapor was formed and started dispersing westwards. Up to 300 tonnes of petrol were released, and about 10% turned to vapor and mixed with air, eventually reaching dangerous concentrations. At 6:01 am on Sunday 11, December 2005, the first, in a chain of explosions, occurred. The explosions caused a huge fire which engulfed more than 20 tanks and burned for 5 days, destroying great part of the terminal.

3. Domino effect model

The model used in this work [3] was developed to assess the impact that the domino effect could have on the frequencies of the accidents used to perform QRAs on storage installations; however, it can also be used to produce stand-alone accident sequences.

The model is an algorithm that has been programmed in MATLAB; it follows the steps used to perform a QRA in a cyclical manner to produce accident sequences which, in the end, will produce frequencies of accidents in all the tanks that are part of the installation.

The first step to obtain the final frequencies of accidents is to input some initial data that describes the installation; these are: the number of tanks in the installation, the substances stored and their properties, the mass contained in each tank, the sizes of the containment dikes, the types of tanks (atmospheric, pressurized), the initial frequencies of accidents per tank and the distances between tanks.

After the characteristics of the installation have been defined, the initial Loss of Containment events (LOCs) that will occur in each of the units, along with their frequencies have to be specified. Using these LOCs, and knowing the substance stored, event trees can be used to propose sets of possible accidents, together with their probability of occurrence; when these probabilities are multiplied by the frequency of the LOC, the initial frequency of accident is obtained. The initial LOCs and the event tree used in this work are those proposed in "The Purple Book" [4]. The LOCs for atmospheric storage tanks and their frequencies are presented in Table 1.

Table 1: Initial LOCs in atmospheric storage units

Code	Definition	Frequency (y^{-1})
G.1	Instantaneous release of the complete inventory.	$5 \text{ x} 10^{-6}$
G.2	Continuous release of the complete inventory in 10 min at a constant rate of release.	$5 \text{ x } 10^{-6}$
G.3	Continuous release from a hole with an effective diameter of 10 mm.	$1 \text{ x } 10^{-4}$

The steps followed up to this point are the same as for a regular QRA. To produce the accident sequences, one of the tanks will be selected to suffer the initial LOC, and the effects and consequences of the accidents in this unit will be calculated for all other tanks. Depending on the consequences, a level of damage, corresponding to a new LOC will be assigned to each of the affected units. This was done using the threshold values presented in Table 2.

Table 2: Threshold values applied in the model

Accident	Effect	Threshold Values		Consequence
		Pressurized Tanks	**Atmospheric Tanks**	
BLEVE	Thermal radiation (kW/m2)	≥ 100	≥ 100 [5]	G.3 type release
		< 100	< 100 [5]	No consequence
Explosion	Overpressure wave (kPa)	≥ 100	≥ 25 [6]	G.1 type release
		$53 < \Delta P < 100$ [7]	$20 < \Delta P < 25$ [7]	G.2 type release
		$30 \leq \Delta P < 53$ [7]	$14 \leq \Delta P < 20$ [7]	G.3 type release
		$\Delta P < 30$ [5]	$\Delta P < 14$ [8]	No consequence
Jet fire	Jet length (m) Heat Radiation (kW/m2)	$dj \geq d$ [5]	$dj \geq d$ [5]	G.1 type release
		$dj < d$ & $I \geq 40$ [5]	$dj < d$ & $I \geq 10$ [5]	G.3 type release
		$dj < d$ & $I < 40$ [5]	$dj < d$ & $I < 10$ [5]	No consequence
Flash fire	-	-	-	No consequence
Pool fire	Thermal radiation (kW/m2) Pool duration (min)	$dp \geq d$ [5]	$dp \geq d$ [5]	G.1 type release
		$60 \leq I$ & $tp \geq 10$ & $dj < d$ [9]	$15 \leq I$ & $tp \geq 10$ & $dj < d$ [9]	G.2 type release
		$I \geq 8$ & $dj < d$	$I \geq 8$ & $dj < d$	G.3 type release
		$I < 8$ & $dj < d$	$I < 8$ & $dj < d$	No consequence

dj = jet length; dp = pool diameter; d = distance between tanks; tj = duration of jet fire; tp = duration of pool fire

Once new LOCs resulting from an initial accident have been defined in the affected units, the effects and consequences of subsequent events on surrounding remaining units will be estimated; this will lead to other LOCs in other tanks, producing more and more branches of possible events. For each of the levels of propagation, the frequencies of accidents will be calculated, using the frequency of the previous accident as the initiator and the same event tree for the probabilities of possible accidents. Once one of the accidents in a branch reaches a frequency lower than $1 \text{x} 10^{-9}$, the sequence will stop, as it is deemed that this frequency is too low to be taken into account in the QRA.

Figure 1. Domino effect model algorithm

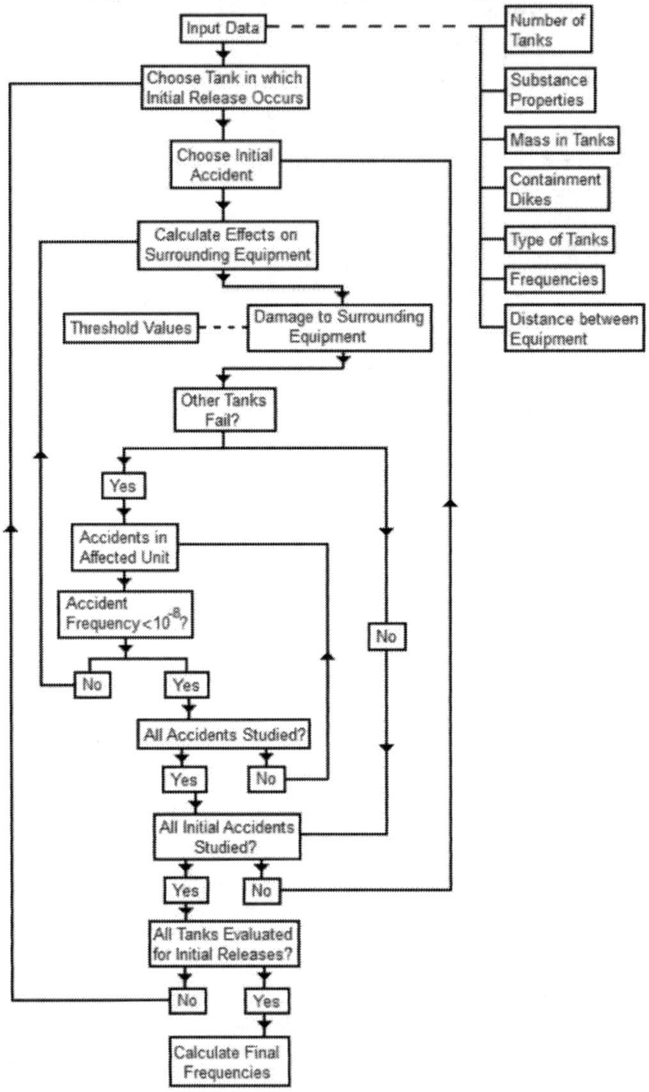

The process has to be repeated in loops, until all sequences have been explored, and all the tanks have been selected as initiators (suffering the initial LOC). In each step, the frequencies with which the different accidents occur in the different tanks are stored. Finally, these frequencies are added, by accident and by tanks, (for example, adding all the times that "accident 1" possibly occurs in "Tank 1"), whether due to an initial LOC event or due to the effect of an accident in another tank; these frequencies are referred to as final, and are the ones used in the QRA. A diagram of the algorithm that is followed in the methodology is presented in Figure 1.

The results obtained using the model can be compared against those obtained from a traditional QRA in order to assess the impact of domino effect on the risk associated to the installation. The model can also be used to produce a singular accident sequence, resulting from a specific accident in one of the units.

Figure 2. State of the terminal and proposed layout

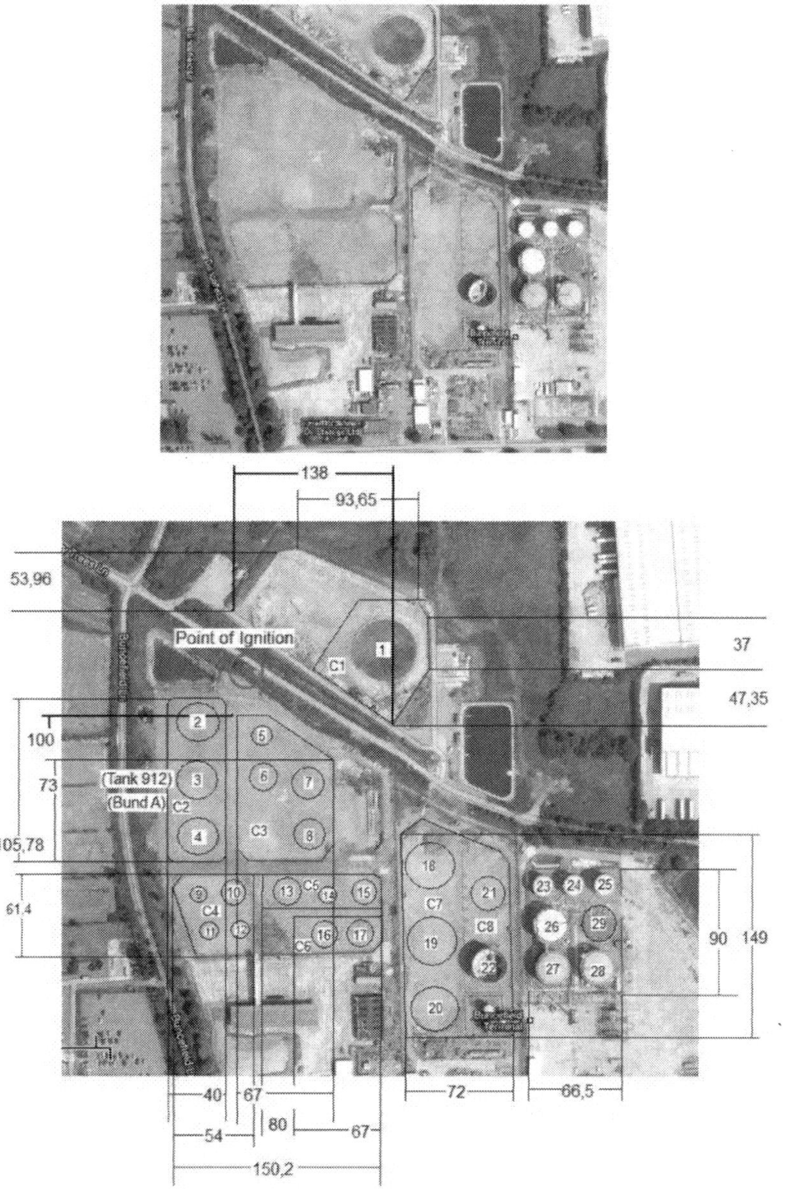

4. Applying the model to the Buncefield terminal

In this section the domino effect model will be applied to the terminal. First, to produce an accident sequence in a situation similar to the real accident and later, to produce iso-risk curves in two different QRAs.

4.1 Modeling the installation

To model the installation, the "Google Maps" tool was used to view the current state of the terminal, and obtain measures similar to those of the tanks that were destroyed and of the tanks that are still standing, as well as the distances between the tanks and the dimensions of the containment bunds. Figure 2 shows, on top, the latest image found of the state of the installation and on the bottom, the distances and dimensions used to model the installation. After estimating the dimensions of the installation, the tanks and bunds were renumbered for the application of the model; for example, Tank 912 becomes Tank 3, and Bund A is called C2.

4.2 Approaches to Accident Modeling

The vapor cloud that resulted in the explosion in the Buncefield Terminal is thought to have ignited at the emergency pump house located close to the containment bund A (in which Tank 912 was located), as shown in Figure 2. Since the domino effect model works by going through a fixed set of possible accidents in each of the tanks, the approach followed to produce the accident sequences was to define a fictional tank "Number 30" in the area in which the ignition of the flammable cloud is thought to have occurred; this tank is modeled containing a mass close to 150,000 kg of pentane.

This approach was followed to produce the accident sequence, but not to develop the QRAs, which were completed using the layout proposed in Figure 2, with all the tanks filled at full capacity.

4.3 Accident Sequence

Figure 3 shows the most probable incidents in the sequence produced when the domino effect method is applied to the case of an explosion derived from a G.2 type release in Tank 30 (the fictional tank created to model the initial explosion). The tanks are represented inside circles, while the possible accidents are shown inside diamonds (E stands for explosion, P for pool fire). The lines generated from the accident diamonds represent the level of damage caused by an event; these lines later divide, reaching different tanks, which means that the tank suffers the associated type of release after the previous incident. All the lines shown in the figure represent possible accidents identified by the methodology. The continuous lines represent the most probable accidents, which derive in the destruction or heavy damage of the affected unit, while the dotted lines represent accidents that do not cause catastrophic damages, or that are less probable; however, they are presented to show other possible scenarios produced when applying the algorithm. Once a tank is reached by a continuous line, it means that it has suffered a level of damage that will result in its destruction.

When an explosion (from a G.2 LOC) occurs in tank 30, the overpressure wave affects tanks 2, 3, 5, 6 and 7 in a catastrophic way; tank 1 suffers damages that lead to a G.2 release and tanks 4 and 8 are affected in a minor way. This is the first sequence of accidents, and it is understood that the failures and later releases occur almost simultaneously. After the mentioned tanks fail due to the explosion, other accidents will generate in each of them, specifically, explosions, flash fires or pool fires; since these last type of accident will be the most probable one (according to the event trees used to produce the sequences), it will be shown for each of the affected tanks.

Figure 3. Accident sequence developed using the domino effect model

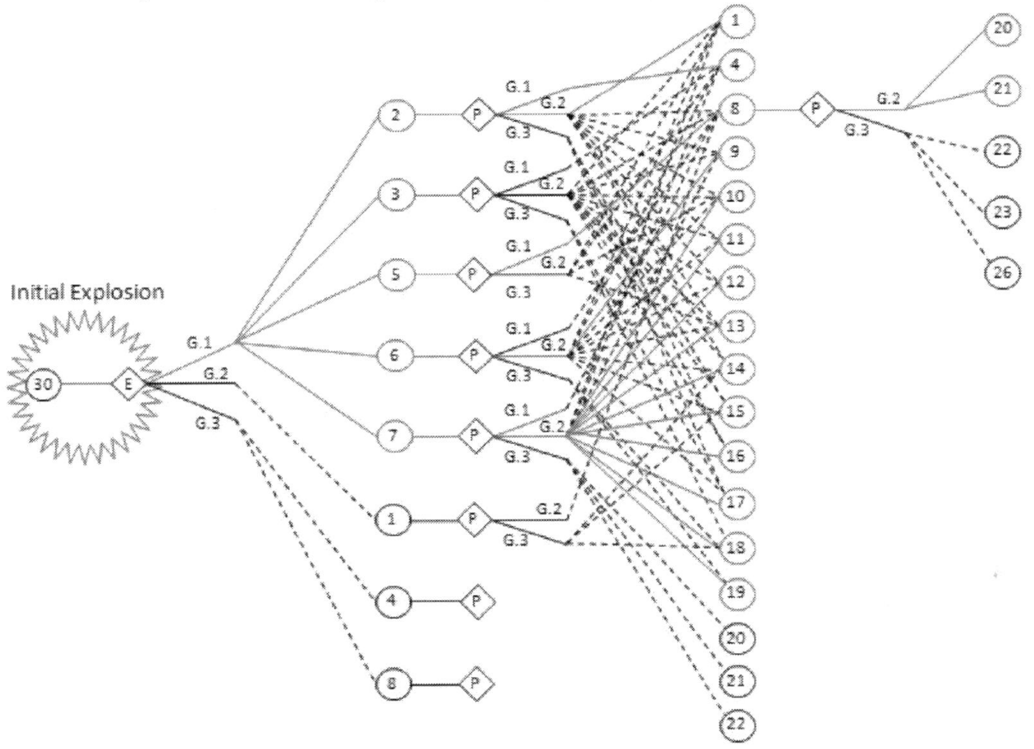

In the case that a pool fire occurs in Tank 2, Tank 4 will fail catastrophically due to the thermal radiation load it receives, while Tank 1 will also be affected in a major way, and also fail. There are other affected tanks, but these possible paths are not explored in detail here. It has to be noted that by the time Tank 2 fails, Tank 3 is also failing. A catastrophic failure in Tank 3 would also result in a pool fire that would further the chain of accidents, but none of these occurrences is explored. Pool fires in tanks 5, 6 and 7 would, according to some paths of the sequence obtained, result in the failure, and grievous damage, of tanks 8 through 19. If the sequence is furthered following a pool fire in Tank 8, tanks 20 and 21 would be damaged beyond repair, while tanks 22, 23 and 26 would suffer minor damages. It has to be noted that in the sequence explored, tanks 24, 25, 27, 28 and 29 do not suffer damages.

The sequence of accidents explored, which is the one that presented higher frequencies in each of its events, results in the destruction of 21 out of the 29 tanks that are part of the installation; this is in accordance to the level of damage in the real Buncefield terminal after the situations resulting from the explosion were controlled. It has to be said that this is one of many possible sequences produced by the algorithm, several of which are different from the real occurrence; however, branches of the possible sequences could be trimmed, in order to choose occurrences that are deemed to be more probable.

4.4 Application to QRA

Two QRAs were performed using the layout presented in Figure 2, the first using the normal procedure and the regular frequencies obtained from the original LOCs and the event trees; the second, using the frequencies of accident obtained from applying the domino effect model. The iso-risk curves obtained for the studies are presented in Figures 4 and 5.

Figure 4. Regular iso-risk curves

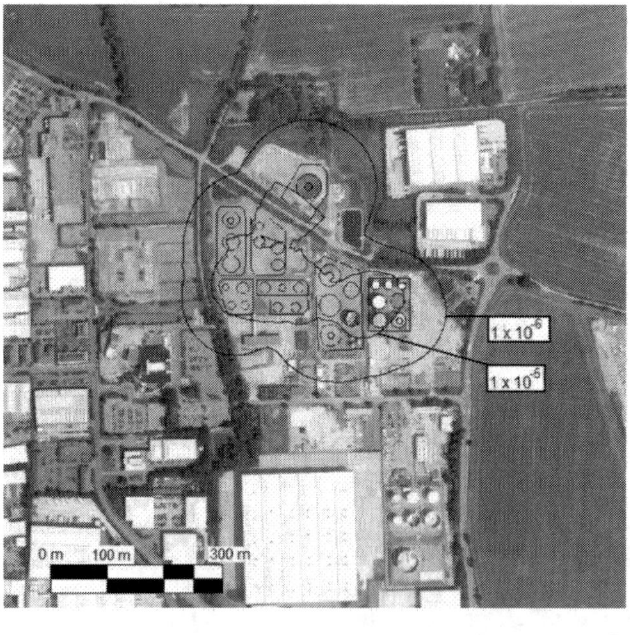

It is clear when observing the figures presented above that the individual risk associated to the installation increases when the domino effect methodology is applied. When regular frequencies are used, only 10^{-5} and 10^{-6} curves appear, while a 10^{-4} curve appears when using the modified frequencies; it encompasses the complete layout of the installation, and in some points, extends some 80 m from the limits of the terminal. The 10^{-4} increases it size so much because many lesser accidents with low frequencies appear in the accident sequences generated by the algorithm. The 10^{-5} curve also increases its area significantly, growing to the point of covering the complete evaluated installation, and reaching close to a 100 m from the limit of the terminal on its furthest point. The 10^{-6} curve is the one that presents a less appreciable change; to the north of the installation the curve suffers almost no alteration, to the south, it increases its length some 30 m and to the east and west, it increases some 20 m in each direction.

Figure 5. Iso-risk curves accounting for domino effect

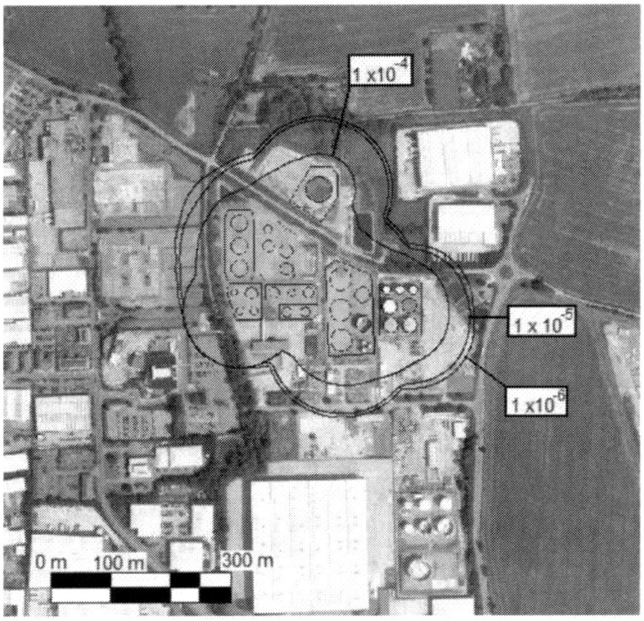

It is interesting to note that when performing a QRA of a storage installation composed of atmospheric tanks, the accidents with highest frequencies of occurrence are the pool fires, followed by flash fires and explosions. This means that the iso-risk curves related to the type of installation studied will not extend over great distances, as occurs when evaluating installations that store pressurized flammable liquids or toxic gases. Therefore, although the increase in the 10^{-6} curve, which, in many European countries is the most representative at the moment of evaluating external affectation, seems marginal, it is important, due to the type of installation that has been evaluated.

5. Conclusions

The domino effect model applied to a layout similar to that of the old Buncefield terminal produced many accident sequences; one of them is thought to be similar to the real accident that occurred in the facility in 2005. This means that the model is indeed suitable to be applied on real layouts, and that it can be used to help assess the risk related to a projected or existing installation, taking into account the possibility of domino effect occurrence.

The model can be applied with a minimum cost of time, and the frequencies obtained can be introduced in commercial software like "Riskcurves" without additional cost; QRAs could be performed taking domino effect into account, without the need to spend more time and money in the development of the study.

An increase in the risk associated to the installation was observed when the domino effect frequencies were used. The 10^{-6} curve does not suffer a significant increase in size due to the nature of the installation and its equipment.

Domino effect should be taken into account when performing QRA of industrial installations, as it can have an appreciable effect on the risk associated to the plant. This could lead to an impact on the Land Use Planning of the zone in which the facility is located.

References

[1] The Buncefield Incident 11 December 2005. The Final Report of the Major Incident Investigation Board. Volume 1, 2008. Buncefield Major Incident Investigation Board.

[2] Delvosalle, C, "Domino Effects Phenomena: Definition, Overview and Classification", First European Seminar on Domino Effects, Leuven, 1996.

[3] Bernechea, E.J., Vílchez, J.A., Arnaldos, J., 2012. A model for estimating the impact of the domino effect on accident frequencies in quantitative risk assessments of storage facilities. Process Safety and Environmental Protection, http://dx.doi.org/10.1016/j.psep.2012.09.004

[4] CPR 18E, 2005. Guidelines for Quantitative Risk Assessment (Purple Book). Directorate General for Social Affairs and Employment, The Hague, Netherlands.

[5] G. Antonioni, G. Spadoni, V. Cozzani, Application of domino effect quantitative risk assessment to an extended industrial area, Journal of Loss Prevention in the Process Industries, 22, 614-624, 2009.

[6] F. I. Khan, S. A. Abassi, Estimation of probabilities and likely consequences of a chain of accidents (domino effect) in Manali Industrial Complex, Journal of Cleaner Production, 9, 493-508, 2001.

[7] V. Cozzani, E. Salzano, The quantitative assessment of domino effects caused by overpressure: Part I. Probit models, Journal of Hazardous Materials, 107, 67-80, 2004.

[8] V. Cozzani, E. Salzano, Threshold Values for domino effects caused by blast wave interaction with process equipment, Journal of Loss Prevention in the Process Industries, 17, 437-447, 2004.

[9] G. Landucci, G. Gubinelli, G. Antonioni, V. Cozzani, The assessment of the damage probability of storage tanks in domino events triggered by fire, Accident Analysis & Prevention, 41, 1206-1215, 2009.

Proven Approach to Investigating Near Misses

Greg Smith, Senior Engineer
Process Improvement Institute, Inc. (PII)
7502 Chad Road
Harrison, TN 37341
Phone: (423) 413-0038
Fax: (865) 671-4944
e-mail: gsmith@piii.com

Gerald D. Burch, Senior Engineer
Process Improvement Institute, Inc. (PII)
e-mail: gburch@piii.com

William G. Bridges, President
Process Improvement Institute, Inc. (PII)
e-mail: wbridges@piii.com

Prepared for Presentation at
9th Global Congress on Process Safety
San Antonio, TX
April 28-May 1, 2013

UNPUBLISHED

Proven Approach to Investigating Near Misses

Steven (Greg) Smith, Senior Engineer
Process Improvement Institute, Inc. (PII)
7502 Chad Road
Harrison, TN 37341
Phone: (423) 413-0038
Fax: (865) 671-4944
e-mail: gsmith@piii.com

Gerald D. Burch, Senior Engineer
Process Improvement Institute, Inc. (PII)
e-mail: gburch@piii.com

William G. Bridges, President
Process Improvement Institute, Inc. (PII)
e-mail: wbridges@piii.com

Keywords: investigation, near miss, accident, incident, root cause, RCA

Abstract

Getting near misses reported is a major hurdle for most companies, one reason being is that the investigation system is not ready for the large influx of reports. Workers and management respond by being reluctant to potentially overload the investigation system (and overload themselves). This paper explains approaches to manage efficiency and effectiveness in near miss investigations that have been successful for handling near misses reporting within both large and small companies and to ensure high value from the investment of reporting and analysis. This paper builds on the updated paper presented in 2012 at GCPS on "Gains from Getting Near Misses Reported."

1. Introduction

Much attention has been given to the detection and correction of early warning signals as a means of improving safety performance.

Process safety events often have entirely different causal factors (human error, equipment malfunction, natural causes) and root causes (management system weaknesses) from those found in more traditional personal injuries.[1,2] Process safety incidents are more technical and complex in nature, usually involving multiple failures of protective barriers. The human actions and errors which cause them are often several steps removed from the incident itself. The human factors involved are not as obvious to investigators; the causes are more systemic than individual in their nature.

Even though personal injury rates have declined, significant injuries and financial losses have continued in a series of high profile, catastrophic (multiple injuries) process safety events. Because of inconsistencies in reporting standards and definitions, it is difficult to tell if these events are actually increasing in frequency or if their visibility has simply increased due to media attention and heightened public concerns. Either way, improved understanding and enhanced tools are required to improve our process safety performance.

In the process industries one such tool is near miss reporting, the early warning system equivalent to unsafe behaviors for process safety. Companies which have successfully developed the ability to recognize, report, investigate, and learn from near misses have shown dramatic reductions in loss events, losses, and injuries due to process incidents. In some cases, loss reductions of more than 90% have been achieved.[2]

2. Definitions

The following definitions are used throughout this article:

> *An incident is either an accident or a near miss.*
>
> *An accident is a sequence of unplanned events and conditions that result in harm to people, environment, process, product or image.*
>
> *A near miss is an unplanned sequence of events that could have caused harm if conditions were different or is allowed to progress, but did not in this instance.*
>
> *A causal factor is a human error (typically an error by the at-risk employee performing a task/job in the process) or a component fault/failure. Note that these human errors and component failures are probably caused by other humans making mistakes, and all errors are controlled by management systems. An incident typically has multiple causal factors. Natural phenomena can also be a causal factor.*
>
> *A root cause is a management system weakness that results in a causal factor. A casual factor typically has multiple root causes.*

The definitions above are the same as used in *Guidelines for Investigating Chemical Process Incidents, Second Edition* (CCPS, 2003).

3. Importance of reporting and learning from near misses

Investigating near misses is critical to preventing accidents, because near misses share the causes and root causes of accidents; they are one or two barriers away from the loss/accident. We are very likely preventing many apparently unrelated accidents when we prevent the ones that are obviously related. Figure 1 illustrates a hypothetical relationship between causal factors and root causes of accidents and those for near misses. From our experience, this relationship appears to be valid; see the case study section of this paper to see some of the proofs of this concept.

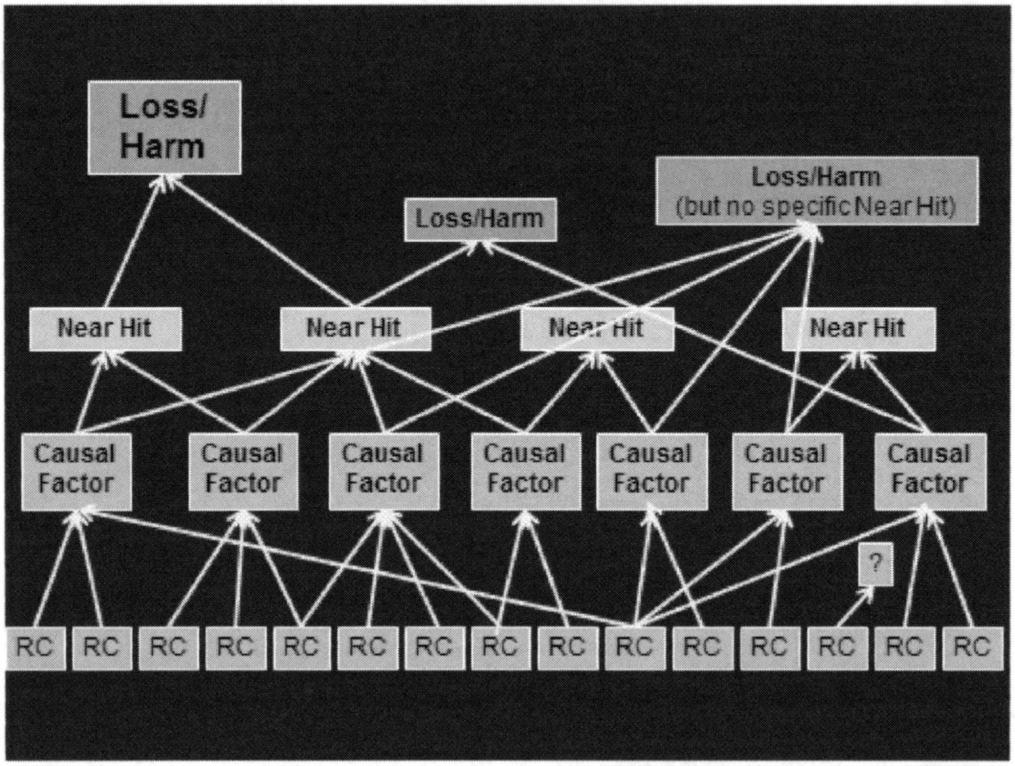

Figure 1 Interrelationship between the causes of Accidents (Losses) and the causes of Near Misses

As a brief explanation of Figure 1, root causes (management system weaknesses) make it more likely for a causal factor to occur. Combinations of causal factors (or in rare cases, perhaps a single causal factor) result in near misses. Additional causal factors result in losses (accidents) in increasing severity. A root cause can increase the likelihood of seemingly unrelated causal factors.

Formal and informal organizational surveys[3] detail the difficulties of implementing an effective near miss reporting system. In this paper, a number of barriers are outlined and potential solutions are proposed. The identified barriers are:

1. *Fear of disciplinary action*

2. *Fear of teasing by peers (embarrassment)*
3. *Lack of understanding of what constitutes a near miss versus a non-incident*
4. ***Lack of management commitment (no training provided on investigation techniques and procedures) and lack of follow-through once a near miss is reported (time is not allocated to investigate near misses, or corrective actions not implemented)***
5. ***An apparently high level of effort is required to report and to investigate near misses compared to low return on this investment***
6. ***There is No Way to investigate the thousands of near misses per month or year***
7. *Disincentives for reporting near misses (e.g., reporting near misses hurts the department's safety performance)*
8. ***Not knowing which accident investigation system to use (or confusing reporting system)***
9. *Company discourages near-miss reporting due to fear of legal liability if these are misused by outsiders*

4. Primary Barriers to Effective Investigation of Near Misses

Four of the barriers (4, 5, 6, and 8) tie directly to the organization's ability to effectively and efficiently investigate and resolve near miss incidents.

Of the four barriers related to getting near misses reported, the two that are most related to effective near miss investigations are **Barriers 6 and 8.**

Overcoming these barriers is described below by looking at the experience of organizations as they implement these concepts.

4.1 Barrier 6. There is No Way to investigate all of the thousands of near misses per month or year!

Normally, when the discussion of having a huge number of near misses reported, such as four per worker per month, the reaction will be shock and then a statement such as Barrier 6. This barrier is closely related to Barrier 5: *An apparently high level of effort is required to investigate near misses compared to the small gain perceived.* If a site has 500 staff as operators and maintenance craftsmen, then likely 25,000 near misses could be reported. At first glance, it can appear impossible to cope with, let alone investigate that number of near misses (incidents).

Maybe the most important solution to Barrier 5 and 6 is to first decide which near misses and losses/accidents need to be investigated. The best solution is normally stated as:

> ***Let front line foremen or supervisors decide if a near miss or accident needs to be investigated to root causes; the decision is made on the apparent Learning Value of the incident. How do you decide it a high or low learning value? The foremen or supervisors will easily decide based on their hands-on experience.*** *For instance, an additive pump failure may directly result in a*

minor and temporary impact to quality or production rates but does not appear to have the potential to cause a large loss. So the foreman of the shift decides to designate this as a low learning value incident and therefore not do an immediate RCA. But, then other similar pumps fail over a six month period. A query shows that after 12 months, there have been 3 failures of the same additive pump and 8 failures of 5 other similar pumps in different services. Based on the number of failures and cumulative losses, the company performs an RCA of the 11 failures, resulting in findings which when applied to all similar pumps and services, greatly reduces the production/quality impacts. The RCA of the grouping of incidents proved very cost effective.

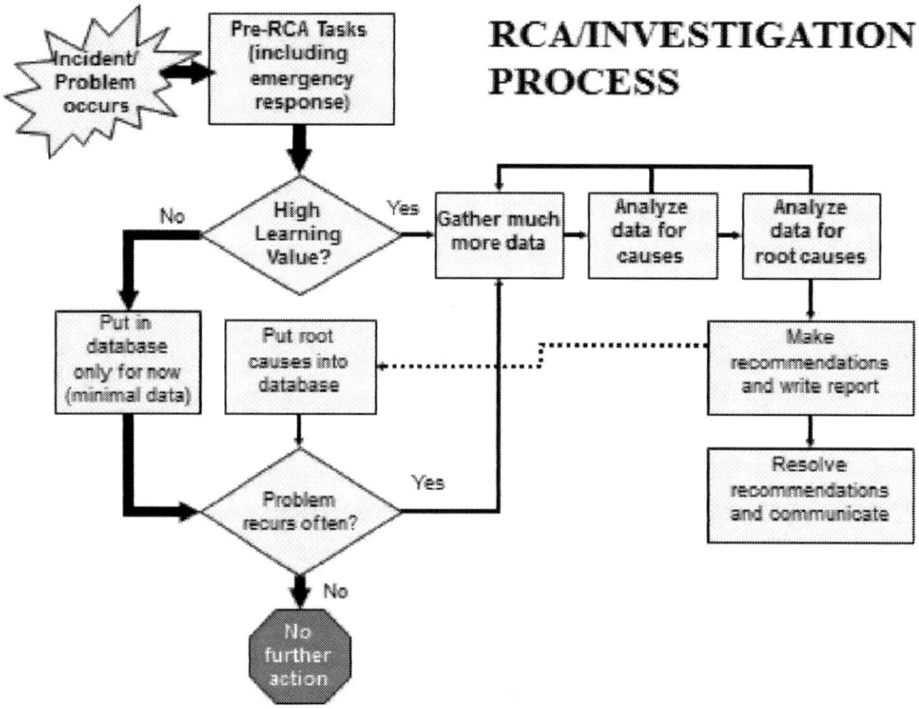

Figure 2: RCA/Investigation Flow Process

Figure 2 illustrates the process flow for an investigation system than can handle a large volume of near misses and losses/accidents. For this process to work:

- Be prepared for investigations by having enough staff trained in root cause analysis methods (or to help in the analysis, such as being able to interview peers).
- When the near miss, etc., is first noticed or reported by staff, let the frontline supervisor or foreman decide if it has high learning value.
- For high learning value incidents, investigate now. For low learning value incidents, put in the database now, along with the little data you have and any obvious causes. Do not investigate yet!

- Query the database every one to six months and perform Pareto or similar analysis to help decide which recurring events need to be analyzed in more detail.
- Take the root causes from investigations/RCAs and put in the database as well.
- Query the database every one to six month

4.2 Barrier 8. Not knowing which accident investigation system to use

One consideration that is not related to any of the barriers mentioned above, except marginally to Barrier 3 (lack of understanding of what a near miss is), is the scope of the investigation program. Some companies have one investigation system for occupational safety incidents, another one for process safety incidents, another for environmental releases, another for reliability issues, and yet another for quality and customer services issues. We have found that the same investigation approach and investigator training works well for incidents in any facet of a business. We believe there is merit in combining the systems and, in particular, in combining the incident databases. Combining the incident systems will require more work on defining near misses and in determining success in reporting near misses.

Consider having ONE incident reporting system with ONE approach for teaching employees the definition of a near miss and with ONE approach for doing incident investigations (including one approach for root cause analysis).

5. Effective and Efficient Incident Investigation and Root Cause Analysis

The incident investigation and root cause analysis (II/RCA) process should enhance organizational learning, reduce accidental losses, invest wisely in protective and mitigative systems, better protect people and equipment from harm, and continuously improve EHSS performance. An efficient and effective process ensures that scarce organizational resources are used appropriately in activities that matter.

The overall II/RCA process consists of the following steps:

- Data collection
- Causal factor analysis
- Root cause analysis
- Develop recommendations
- Reporting and follow through

5.1 Data collection

Every investigation begins by collecting, organizing, and preserving all relevant data as quickly as possible after the incident has been stabilized. Data comes in the form of personnel interviews, physical evidence, and documentation. All data is fragile and susceptible to loss or damage if it is not properly collected and preserved in a timely manner. All front line supervisors should be trained in the II/RCA process with particular emphasis on initiating the investigation, interviewing people, and collecting the physical evidence. Provide front line supervisors with an investigation tool kit containing a camera, barrier tape, collection bags and bottles, and appropriate forms and checklists.

5.2 Causal factor analysis

Once collected, data must be organized and analyzed to determine what happened – the chain of events (causal factors) that led to an undesirable occurrence. The most effective and proven methods for analyzing data are Event and Causal Factor Charting (ECFC) supplemented, if needed, with a Simplified Fault Tree Analysis for complex, chronic, or acute events. These are only effective when the rules are followed for their use – but with training and practice most people can be taught to identify causal factors using these basic tools for most of the incidents they will encounter.

Event and Causal Factor Charting: When an investigation team begins a root cause analysis (investigation), the first step is to prepare an Event and Causal Factors Chart (ECFC).

- Arranges event building blocks (sticky notes) to graphically depict cause-effect relationships between known events
- Allows for testing of "necessity" and "sufficiency"
- Best method when timing of events is important

Events make up the backbone of the ECFC. Event statements describe specific occurrences (e.g., "4-12 shift operator filled Tank 123," or "Control room operator acknowledged level alarm for Tank 123"). Conditions usually provide descriptive information (e.g., "Pressure was 1000 psig" or "The operator did not have hands-on training in this task") as opposed to stating action (e.g., "Operator placed Valve ABC into open position"). Primary events form the basic sequence in the diagram. Secondary events are actions that impact the primary events, but which are not directly involved in the situation. Figure 3[3] provides the steps for developing a casual factor chart.

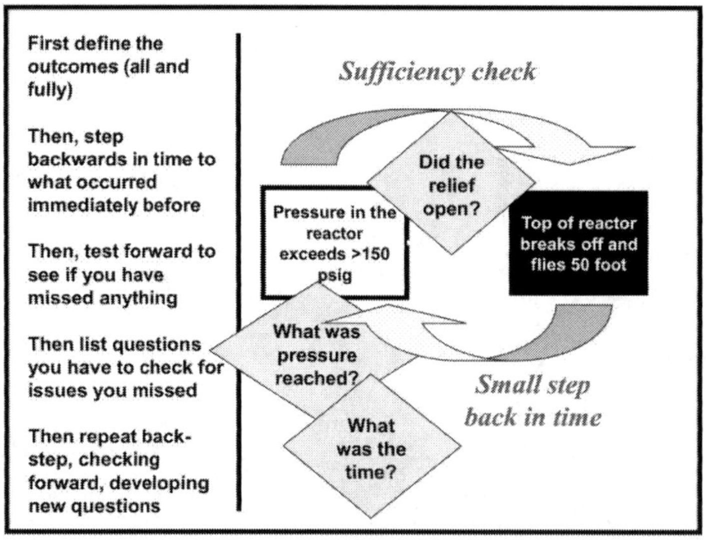

Figure 3: Steps for Casual Factor Charting

If applied correctly, ECFC will aid the investigators in generating a great many new questions. These will lead to more data gathering. Above are a couple of examples of how we show these questions on an ECFC. The answers to the questions sometimes become new events or conditions on the ECFC, as illustrated in Figure 4[3]:

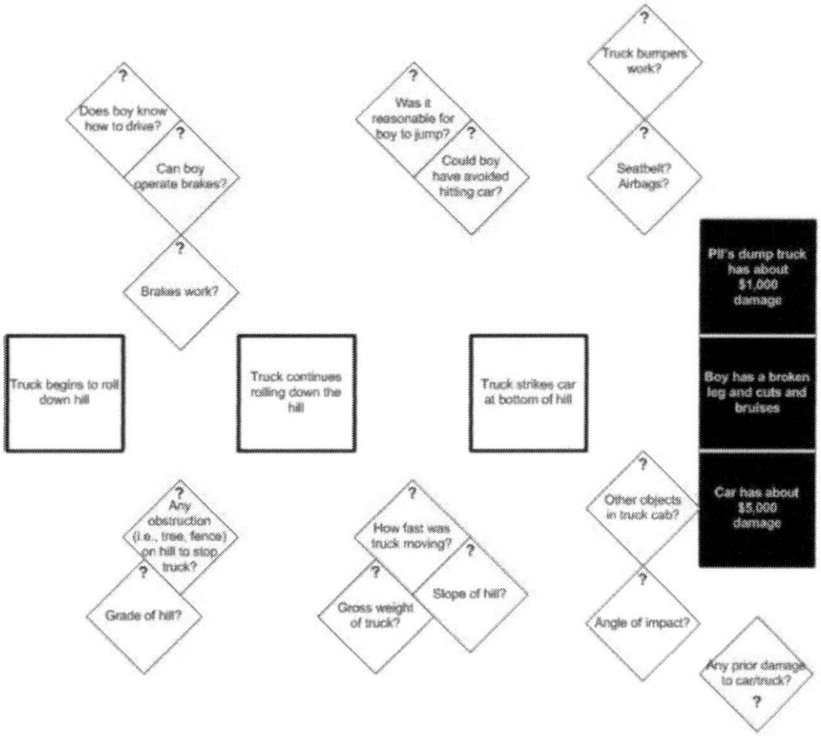

Figure 4: Example of a Partial Casual Factor Chart

The investigators will go back and forth between doing the ECFC, generating new questions, getting new data to answer the questions, and taking more steps backwards in time and sequence. This process continues until all of the data is exhausted and/or the investigators fully understand the sequence of the events and all of the mistakes and failures that led to the final ending condition or loss.

After the ECFC has been completed, the investigators are in a good position to identify the negative factors that influenced the course of events. These components are labeled causal factors. Causal factors, which may be in the form of events or conditions, are those items that are considered to be major contributors to the incident and as defined earlier, are either component failures or human errors or severe environmental problems.

The logic rules for constructing an ECFC are critical to success. These govern how small of step backwards the investigators take backwards in time and how many questions they generate as they test to ensure they understand each pair of building blocks.

5.3 Root cause analysis

When all the causal factors have been identified, the investigation team should begin asking "Why?" each causal factor occurred. Using a combination of the "5 Whys" and a Root Cause Chart™ (derived from the management oversight and risk tree process developed for US DOE's Savannah River Laboratory) shown in Figure 5, step down each of the root cause paths by:

- Select a causal factor from the causal factor chart or fault tree
- Work through the map for each causal factor
- Ask Why to help gathering the right data to get to the lowest level (root causes)
- Step down paths, identifying:
    - causal factor type (equipment, personnel, other)
    - root cause category
    - near root cause
    - root cause
- Record results on forms, at each step (see next page)

The Root Cause Chart™ structures the reasoning process for identifying root causes, identifies detailed root causes for each causal factor, facilitates consistency across all investigations, supports trending of "root causes" and "categories", and gives more thorough coverage of root causes by reminding investigator of all possible root causes.

Figure 5: Root Cause Chart™

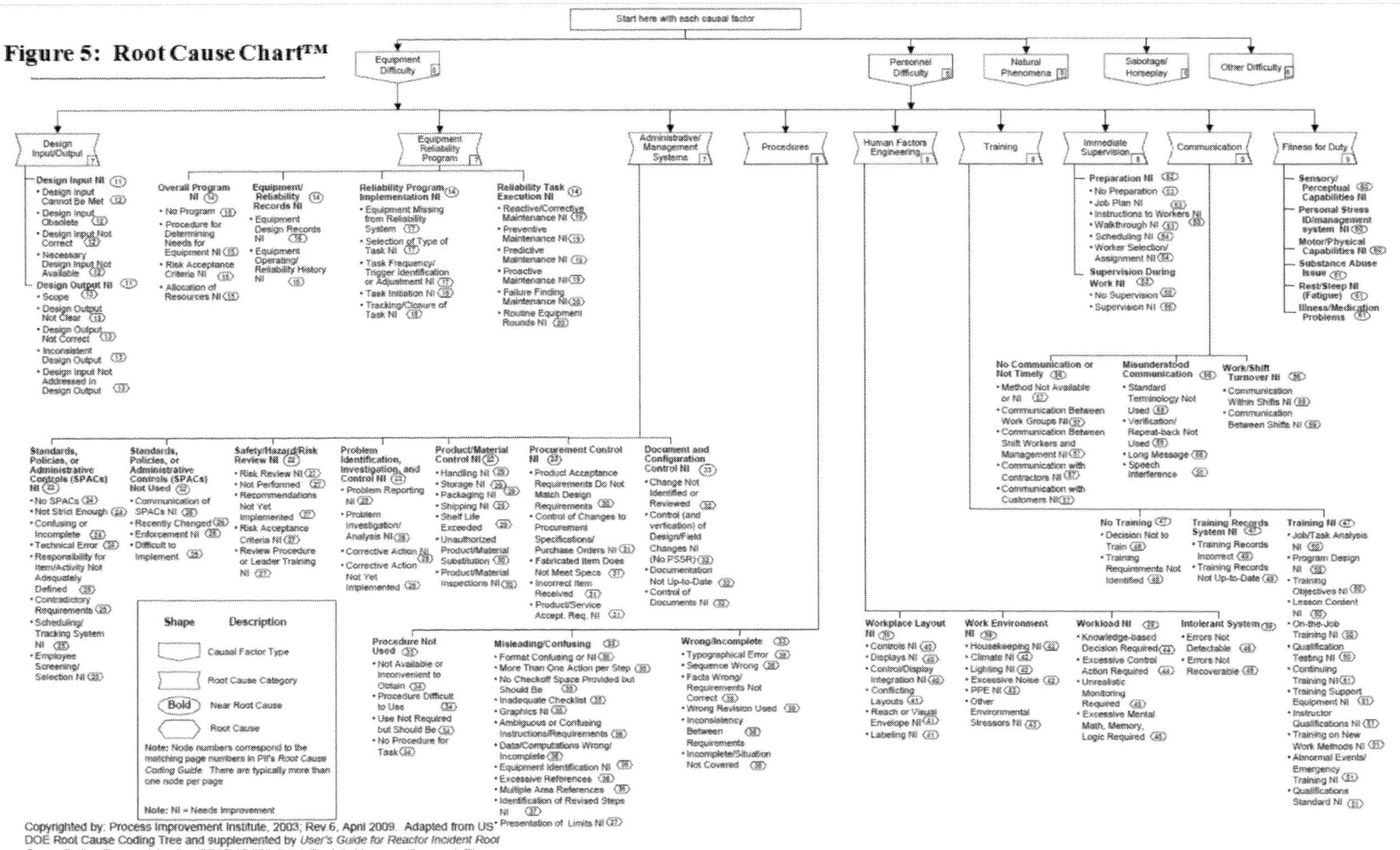

Copyrighted by: Process Improvement Institute, 2003; Rev.6, April 2009. Adapted from US DOE Root Cause Coding Tree and supplemented by *User's Guide for Reactor Incident Root Cause Coding Tree*, revision five, DPST-87-209, E.I. duPont de Nemours, Savannah River Laboratory, 1986. www.piii.com wbridges@piii.com

5.4 Develop recommendations

The investigation team should develop recommendations to remedy each root cause of each causal factor. Recommendations should describe the general objective to be accomplished and the consequences if not completed. Recommendations can also include possible solutions considered by the team and an example of how this could be accomplished.

5.5 Reporting and Follow Through

5.5.1 Simplify the reporting of the investigation/RCA results to the bare minimum needed

Part of the reason for the belief that it is impossible to investigate large numbers of near misses stems from the large reports currently required by the company for investigating incidents. Some companies insist on producing what they call "professional" reports of accidents, and these grow to 50 or 100 pages (half of the pages are attachments). Why produce such a large report? What is the use of that large of a report? What makes "size" of a report equivalent to "professionalism" of a report? Think about every aspect of the report and make sure it is needed. Normally, all that is needed is a:

- Cover sheet that includes the date, time, location, one or two sentence description of the near miss or incident, and a title that summarizes the incident at a glance. The cover sheet should also list the team members.
- Forms that have the causal factors filled in and the root causes filled in, with perhaps one or two sentences that explain the root cause. These forms can also contain the recommendations necessary to correct the root causes.
- That's It!

So, most near miss investigation results will be two pages or so and most loss/accident reports will be four pages or so.

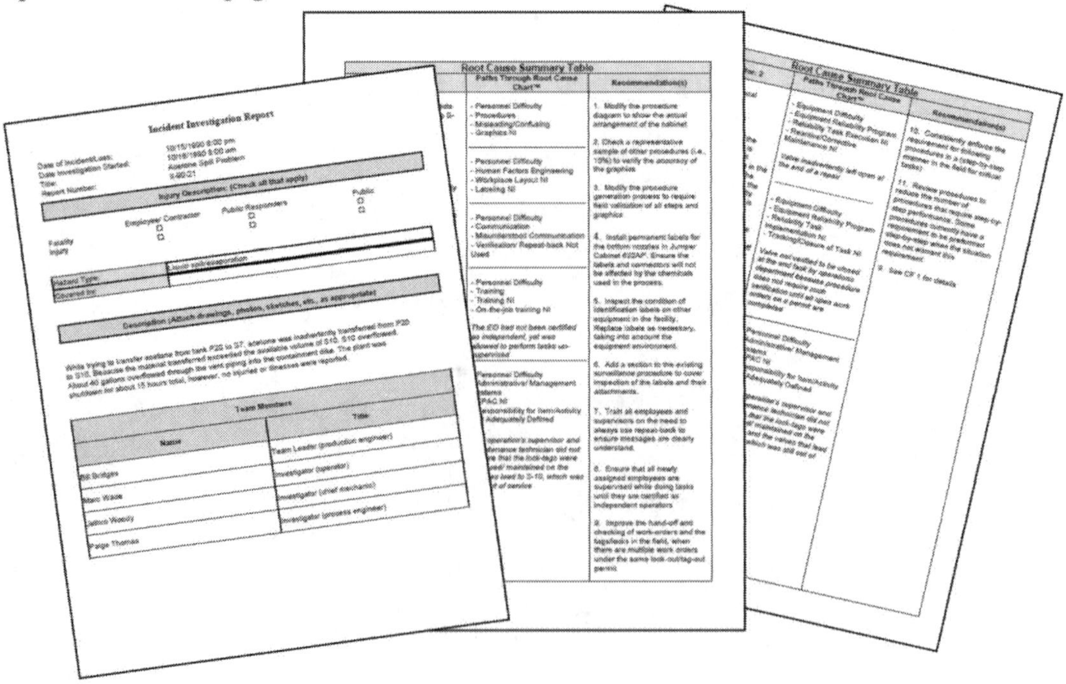

The complexity of results reporting has grown from the legacy of only investigating losses/accidents. When an organization gets a large ratio of near misses report and therefore a large number of investigations going on, the reports must shrink. This is a good thing. However, if you still have a major accident (which you won't have if you get a large number of near misses reported and investigated), then add more documentation to meet the needs related to litigation, regulatory interface, etc.

5.5.2 Deciding which incidents should be investigated

All near miss incidents should be reported and cataloged with basic information and apparent causes, but not all near miss incidents will warrant a full investigation. Only those incidents with either high potential impact or high learning value should be investigated using the II/RCA process. This is an excellent opportunity to engage staff and give them control over the process by granting front line supervisors the authority to decide which incidents have high learning value. For example, establish a shift target of investigating 15-20% of the near miss incidents reported and then let them decide which ones they want to investigate to meet their goal.

6. CASE STUDY: One Company's Experience with II/RCA

A large pharmaceutical company (more than 50,000 employees) initiated a drive to report and analyze large numbers of near misses. Before this initiative, they had about 3 near misses reported per loss event. Their loss categories included quality impacts, reliability issues, safety mishaps, and environmental issues. Also before the initiative, they had five incident reporting systems, investigation protocols, and databases. One of the first steps the company took towards the stated goal was to consolidate the five reporting and investigation systems into one system. They tried to use one of the existing database systems for this consolidation, but decided that all were deficient for this purpose. So, they modified the reliability incident database to make it suitable for all types of incidents. This took several months of effort (not full time though).

In parallel, they adopted a new investigation approach, which is the one described in this paper. This involved changing the database structure to provide pull-down menus for each branch on the Root Cause Chart. The chart was not modified. Next, a training course for lead investigators was customized to meet their new needs and to use their examples in the exercises and case studies. Then, this course was presented seven times in the USA and UK, thereby training an initial group of 200 investigators but more importantly training 7 instructors within the company. The eventual goal was to train 15% of the operating staff how to perform root cause analysis (RCA) of near misses and losses. The new database and consolidation of reporting and analysis systems (developed in parallel with the development of the customized training materials), was rolled out at the same time as the initial training courses. The reporting system was incorporated into their Lotus Notes e-mail system. The reporting forms were very simple and any worker could report a near miss or incident. Initial reports were routed automatically by the e-mail system to the worker's supervisor and viewable by any qualified investigator. A lead investigator on the same shift rotation and from the site (but not from the same group) was assigned to each incident; investigations typically began within one hour and on average were completed within the next hour.

These initial efforts were extremely successful and within about one year, the company achieved a ratio of more than 100 misses being reported per loss event. All near misses and loss events were captured in the database of incidents and about 20% were high value and immediately investigated. Typical reports were 2-3 pages. After the first year, the sites performed queries of the databases and based on Pareto analysis, performed RCA on the highest number of repeated "similar" incidents. In subsequent years, the queries were performed every three months.

This new system was much more productive in reducing losses than the previous collection of five systems. Overall, production and quality losses were reduced about 95% with little or no increase in time spent in RCAs (because the number and magnitude of losses decreased dramatically and so there was much less time spent on fire-fighting and major investigations).

7. Conclusions

Near miss reporting and investigation has been very effective in reducing losses due to process safety incidents. However, implementation has proven difficult for many companies due to several implementation barriers. The proven methods described will improve both the efficiency and the effectiveness of the II/RCA process. Consistently applying these methods will enable near miss reporting, providing early detection and correction of process safety warning signals, thereby reducing losses and improving process safety performance.

8. References

1. Busch, Carsten, *Heinrich's Common Cause Hypothesis: A tool for creating safety*, July 30, 2012, http://www.predictivesolutions.com/safetycary/heinrichs-common-cause-hypothesis-a-tool-for-creating-safety/
2. Wright, Linda and van der Schaaf, Tjerk, *Accident versus near miss causation: a critical review of the literature, an empirical test in the UK railway domain, and their implications for other sectors*, Journal of Hazardous Materials 111 (2004) 105–110.
3. Bridges, William, *Gains from Getting Near Misses Reported,* AIChE 8th Global Congress on Process Safety, 2012.
4. *Incident Investigation and Root Cause Analysis Course Materials*, Process Improvement Institute, Inc., 2003-2013.

9TH GLOBAL CONGRESS ON PROCESS SAFETY

Key Strategies for Implementing a Corporate Fatigue Risk Management System

David E. Cummings
E.I. DuPont de Nemours & Co.,Inc.
1007 Market Street, Wilmington DE 19898
david.e.cummings-1@dupont.com

Prepared for Presentation at
American Institute of Chemical Engineers
2013 Spring Meeting
9th Global Congress on Process Safety
San Antonio, Texas
April 28 – May 1, 2013

UNPUBLISHED

Key Strategies for Implementing a Corporate Fatigue Risk Management System

David E. Cummings
E.I. DuPont de Nemours & Co.,Inc.
1007 Market Street, Wilmington DE 19898
david.e.cummings-1@dupont.com

Keywords: fatigue, fatigue management, ANSI/API 755, human factors

Abstract

The First Edition of ANSI/API Recommended Practice 755 "Fatigue Risk Management Systems for Personnel in the Refining and Petrochemical Industries" was published in April 2010 as a result of the BP Texas City catastrophe of March 2005. In the last two years, many companies in the petroleum and chemical processing sectors have been working to implement the basic tenants of this new practice. An effective fatigue management program involves a multi-disciplined team approach to engage various functions of an organization, including PSM/EHS, HR, Medical, and Operations. This paper will explore the process, associated elements, and key issues integral to the design, development, and implementation of a formal fatigue management system across DuPont as a global manufacturing organization with a diversity of high hazard processes. This includes internal policies, training and development, standardized procedures and tools, and internal collaboration strategies.

1. Introduction

ANSI/API Recommended Practice 755 [1] was issued in 2011 based on a recommendation issued to the American Petroleum Institute by the U.S. Chemical Safety Board as an outcome of the BP Texas City 2005 incident investigation report. Prior to that time, there was little industry guidance or standards on the subject of fatigue management (FM). There is and has been substantial literature published and technical resources developed on the general issues and concerns with fatigue management in terms of circadian rhythms and the effects on human performance variables, and many industries such as aviation and commercial transportation have adopted specific standards with accompanying regulatory requirements.

E.I. du Pont de Nemours & Co. is a global science-based products and services company with over 200 manufacturing units across 70 countries involving nine business segments. It is engaged in technology solutions and innovation across a range of disciplines, including agriculture and industrial biotechnology, chemistry, biology, materials science and manufacturing, offering a wide range of products and services for markets including agriculture and food, building and construction, electronics and communications, general industrial, and transportation. Manufacturing technologies and site operations involves a myriad of process hazards, including flammable liquids and gases, highly toxic materials, reactives, combustible dusts, corrosives, asphyxiants, and thermal/mechanical energies. DuPont's philosophy and current practice is to apply process safety management systems and requirements to all manufacturing operations and locations using one global set of internal elements and technical requirements. Based on the company's current risk profile, process safety systems that help maintain a high degree of operational discipline and reduce the potential for human error, including those that could be related to employee fatigue, are key elements of an integrated risk approach.

DuPont's process safety management model and system has included a long standing element and requirements for "Training and Performance" where the performance portion of the element requires a specific focus on physical capabilities (fitness for duty) as well as diminished capacity (mental and psychological capabilities for process safety sensitive jobs, including the effects of fatigue) for operating and maintenance personnel working at all manufacturing locations. DuPont updated the "Training and Performance" element to specifically include an internal fatigue management policy as an integral part of the global process safety management program in January 2011. Initial implementation of the new technical requirements has been progressing since that time. The program embraces the following key elements of fatigue management as described in ANSI/API 755:

- Maximum hours of service
- Training for employees and line management
- Staff workload balance
- Work environment evaluations
- Medical Evaluations (personal risk factors)

Implementing a fatigue management system was integrated into corporate process safety management systems; however, active involvement of other key functions within the organization is important in this area. Within DuPont, the process safety competency serves as the corporate "owner" on FM, while other key functions such as Human Relations (HR), Integrated Health Services, and Occupational Health (EHS) all have played a meaningful and contributing role in developing and implementing portions of the FM process in the initial design stages, during implementation, and in an ongoing role as part of the "run and maintain" phase of this work. This collaboration has contributed greatly to the overall results.

2. Foundational Activities for FM Implementation

Building an internal fatigue management system based on the elements in the API standard requires a number of key strategic decisions and initiative planning prior to formal roll-out, as well as ongoing support, training, and tools during the active implementation phase. The following provides examples of many questions that were posed as part of DuPont's initial design strategy for fatigue management:

- Should the fatigue management program be a global or U.S. based only focus?
- What is the current state at all sites (does anyone already have an effective FM system)?
- What will be the major impacts on Operations at each site (resources, costs, tracking tools)?
- Who are the key stakeholders who should be involved in the design of the system?
- Which sites will be required to implement a local fatigue management system?
- How will fatigue management requirements be "codified" in corporate policies?
- What elements are consider mandatory versus suggested practices/guidance?
- What site roles and positions will be included in FM (and not included)?
- What is the required internal timing for implementation and how will it be tracked?
- When and how will the PSM auditing process incorporate FM elements during site reviews?
- How will interpretations and guidance on FM be captured and distributed?
- Will the program be required for contractors, and if so, which ones?
- Are there leveraged solutions and tools that can be developed to assist sites for effectiveness and efficiency?
- Who has primary responsibility to ensure compliance with Table 1 (employees or line managers)?

2.1 Site Surveys to Define Current State FM Practices

The first step that was taken to develop a fatigue management program across DuPont was to conduct a detailed survey across all manufacturing locations to accurately determine the present state relative to the main elements of ANSI/API 755. This included the following self-assessment parameters:

- Existence of site level fatigue management written programs and policies
- Existence of local policy limits on overtime and maximum consecutive work hours
- Past and present fatigue management training
- Existence of tools or tracking systems to measure work hours and overtime by work group and employees
- Local site systems and capabilities to integrate fatigue management into existing medical programs
- Forecasted requirements and needs with local labor organizations and contracts for fatigue management

Once the site survey was completed, the data was collated and utilized as inputs for the development of corporate policies, timing and implementation plans, management training, and development of internal FM training and guidance documents. The survey was collected and analyzed in a simple excel® tool similar to that depicted in Figure 1.

Site Fatigue Management Survey – Current State

What is your site's current level of implementation of Fatigue Management requirements? List all numbers that apply. 1. Leadership is aware. 2. The site is tracking hours. 3. Employees have received general awareness training. 4. The site is actively managing hours for Shutdowns/TurnArounds, Special Circumstances, or Normal Operation. 5. A Fatigue Management policy has been written. 6. An Exception Management policy has been developed.	If you listed No. 2 above, how is the site tracking hours? Is the tracking historical or real time? What system or tools are being used?			If you answered No. 4 above, when and how is the site managing hours?		Have you considered which employees would be included in a Fatigue Management program? If yes, how many at your site? Does this number include employees in LHT's? If yes, how many of the total are in LHT's?	How are your shifts set up, ie. 8 hour, 10 hour, 12 hour, rotating, 5 days, 7 days, etc.?	If your site has changed any policies, procedures, tracking, people's work times, etc. to move toward compliance, can you describe what you have done?	Site Changes
Level of Implementation	Tracking Hours	Tracking Hrs.?	How Tracked?	Managing Hours	Mgt. Hrs.?	Which Employ.	Shifts	Changes	Changed Schedule/ Pay?

Figure 1: Sample Site Survey Baseline Questions and Spreadsheet

2.2 Corporate PSM Standard Policies and Requirements

DuPont drafted and included a new fatigue management section in the global process safety management standard as part of the" Training and Development" element as depicted in Figure 2. Specific new content in the corporate standard was intended to incorporate the minimum mandatory requirements involving fatigue management elements described in ANSI/API 755, and not a be a re-statement of the guideline's full technical content. It was also defined with the standard scope that new internal fatigue management requirements would be applied to vital PSM roles within DuPont sites as well as to specific types of contractors, but only at higher hazard site operations (DuPont classifies all global manufacturing locations as higher hazard processes or lower hazard operations). The standard changes went through a broad internal peer review and was published in January 2011.

Figure 2: DuPont's Process Safety Management Model

2.3 Education and technical guidance to drive consistency

Internal webinars were developed and delivered which were targeted to several different key audiences, including site PSM leaders, operations leadership, site HR leaders, and contract administrators in order to develop foundational knowledge of the new FM program elements across the organization. The initial ANSI/API 755 standard as published is a performance based standard with specific prescriptive FM requirements such as Table 1 maximum hours of service limits, however, it became apparent early in the FM implementation process that numerous technical and interpretive questions existed both on the standard, as well as within DuPont's local site implementation efforts. In order to attain as much consistency as possible, a series of Frequently Asked Questions (FAQ's) were collected and formally documented with specific guidance. Once vetted and resolved, these FAQ's were distributed to sites and posted on the internal PSM website for retention and routine access. Additional questions and needs for interpretations have continued to surface from sites and functions and been added to the FAQ "evergreen" document even after almost two years of implementation progress. There is currently no effective platform within industry to collect and address these types of interpretations in a consistent manner.

As another step to implement consistency across all locations, DuPont developed and distributed a standard site based written fatigue management program or template which can be applied by each site as a base document and modified to add in local responsibilities, procedures, and line management practices.

There are also unique situations that may develop for plant operations that do not fit neatly into the existing definition and guidance in ANSI/API 755 involving "outages" as it applied to extended consecutive work shifts in Table 1. The existing definition for outages includes the following:

> *"Planned or unplanned interruption in the normal operations of a unit or plant, including mobilizing or de-mobilizing. Outages include but are not limited to such things as turnarounds, unit shutdowns, operational responses, etc."*

This definition does not cover all situations where an extended work set and shift schedules may be applicable and necessary to minimize fatigue, including commissioning new plants or substantive MOC projects where construction, PSSR's, start-ups, and lining out the process may be necessary involving operations, technical and R&D resources, response to natural disasters (e.g., hurricanes, earthquakes, etc.), mothballing or dismantlement activities, new product trials in pilot plants, etc. Each company should define when and how these types of situations should be managed in terms of unique conditions and the application of extended work hours and shift schedules with potential fatigue factors.

DuPont adopted an internal policy at the site level to develop a formal exception process for situations where the Hours of Service Limits (ANSI/API Table 1) may not be met due to specific circumstances. An exception process is discussed in the standard in Section 4.8.5 with general criteria, however; the specific procedures and details should be developed by each company. For DuPont, the exception approval process has been designed to be executed at the local level for a case-by-case exception by site or unit management. For longer term, permanent, or "standing" exceptions, off-site leadership approvals are required. In both cases, exceptions will document the situation, the rationale, and any relevant alternative controls or methods to be taken to mitigate the potential effects of fatigue. This includes use of a standardized form and line management approvals for each exception situation.

3. Shift Schedules and Maximum Hours of Service

API 755 includes specific requirements for maximum hours of service provided in Table 1 of the standard. The hours of service element in FM has received the most internal discussion, review and implementation work. A detailed analysis of existing shift rotation schedules at all U.S. site operations was conducted by DuPont HR to ascertain if any inherent conflicts existed in the current schedule structures that were not be fully compatible with Table 1. This included a specific emphasis on the number of consecutive shifts and the total hours of recuperative time for both day and night shift rotations. This review revealed that DuPont had over one hundred (100) different shift rotation schedules in use at all U.S sites for 8, 9, 10, and 12-hour operations and maintenance shifts. In a very limited number of cases, an inherent conflict in the site shift structure existed which did not fully meet the maximum "hours of service" in Table 1 (mostly involving the transition from day to night 12 hour shift schedules). Figure 3 describes the shift schedule analysis results at all US sites, and figure 4 provides an example of the detailed shift schedules which were generated and reviewed at a typical multi-unit site.

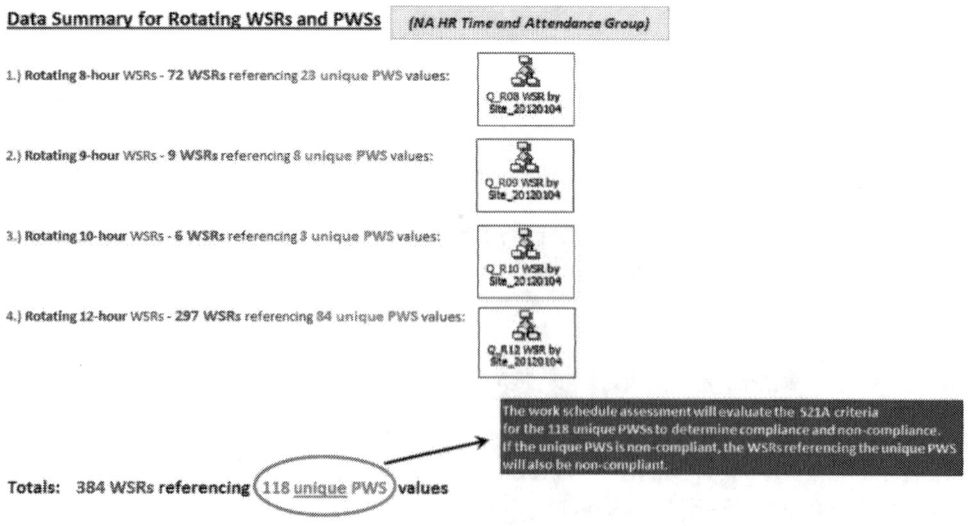

Figure 3: Data Summary for Plant Rotating Work Schedules

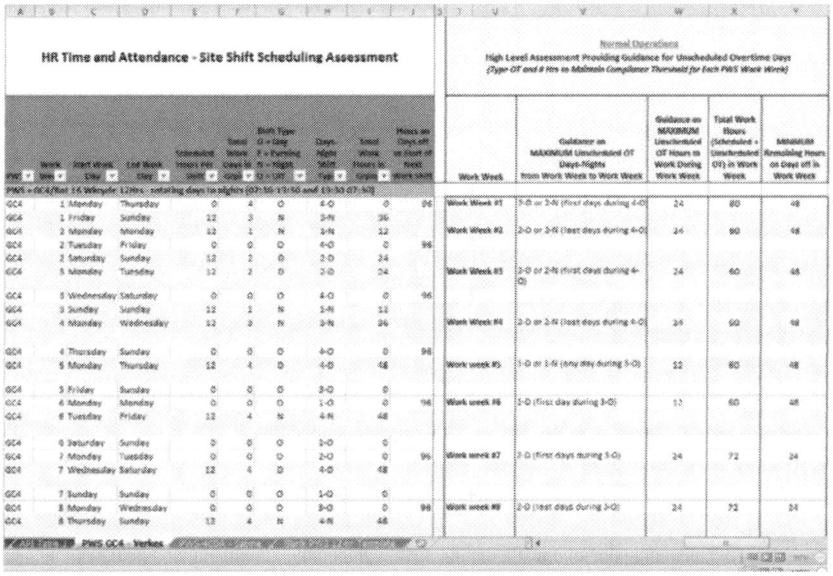

Figure 4: Sample Site Analysis – Time and Attendance Schedules

A significant part of DuPont's initial evaluation of existing shift schedules and site data survey questions focused on the types of tracking and software (scheduling) tools that were either already in use or could require development to track work hours, including scheduled and unscheduled overtime call-ins. In most cases, site systems tracking overtime rates for work groups but not specific employees other than annual salary statistics. Site level tracking systems which were cataloged that could be modified and used to monitor maximum hours of service and off site recuperation times included the use of an HR work hours and pay status, web based tools, security gate electronic logs, and access data bases. None of these existing systems as structured were deemed an acceptable and reliable means to proactively schedule, manage, and track individual employee work hours. A six sigma black belt resource and project was initiated in order to review each of the work hour and overtime tracking system attributes, and develop a standardized solution that could be used at all sites to help line management ensure that the maximum hours of work in ANSI/API 755 Table 1 maximum hours of work. As part of the six sigma project, the main attributes of an overtime tracking system were captured as illustrated in figure 5.

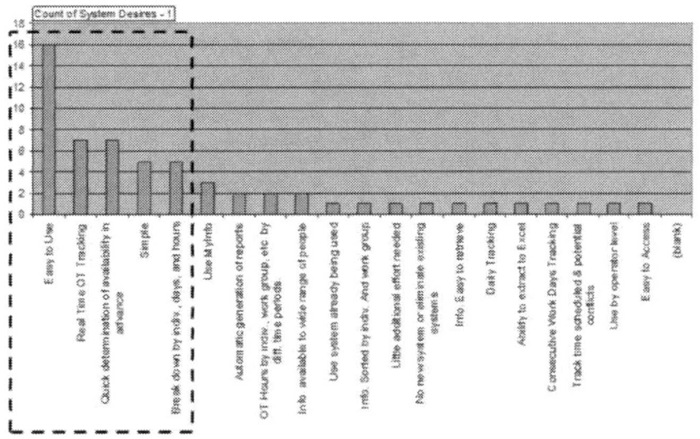

Figure 5 : Site Survey Data – Desirable Attributes for Electronic Scheduling and Tracking

4. Training

DuPont developed a standardized and leveraged training package which could be applied at each site in order to provide initial and refresher training on fatigue management. Two packages were developed ; one to target employees who were subject of FM (operators and mechanics), and one for line managers who supervise these employees. The guidance and criteria in ANSI/API 755 was used to create the technical content of each package. Initial training was designed to be one hour of information and cover the following topics:

- Basics of sleep and fatigue including causes, symptoms, and prevention strategies
- Understanding risk of on the job fatigue and impacts on safety
- Overview of the site fatigue management system including maximum hours of service criteria

DuPont occupational health resources have also previously partnered with external groups such as the Porcupine Health Unit and the Sudbury & District Health Unit in Canada to develop additional sources of fatigue management training such as the Shiftwork like Clockwork (SWLCW) training modules illustrated in Figures 6 and 7 . These materials provide significant details over nineteen (19) modules that can be used for both initial training of new employees and refresher training for existing employees. DuPont has adopted an internal FM refresher training frequency of three years.

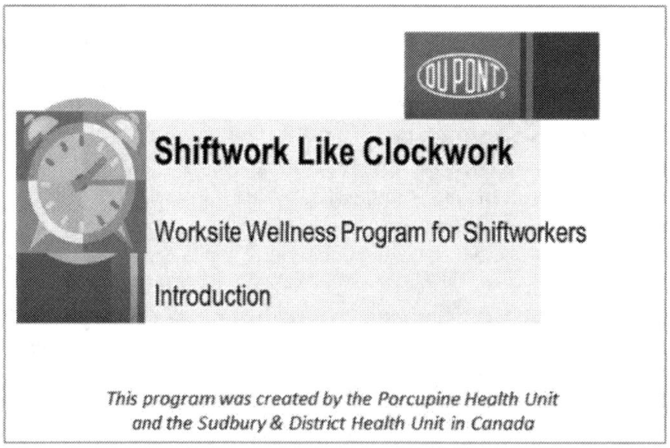

Figure 6: "Shiftwork Like Clockwork" Training Program

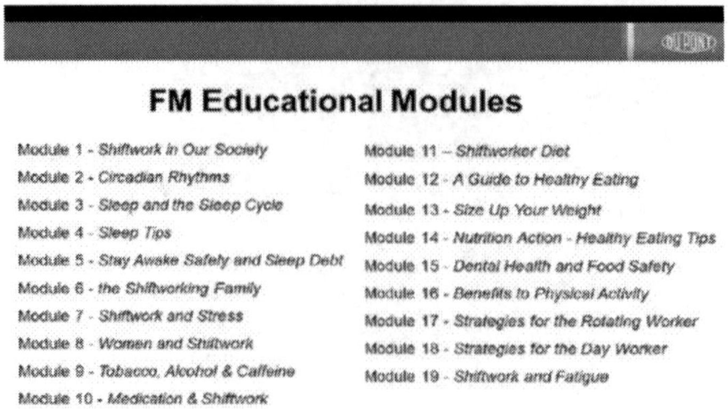

Figure 7: "Shiftwork Like Clockwork" Training Modules

5. Staff Workload Balance

Another key tenant of the ANSI/API 755 standard includes a focus to help ensure a staff workload balance is achieved and maintained within each operating area at each site. This should include an analysis of work activities on each shift, and a focus on considerations and impacts related to employee turnover, long term disability, extended time off due to personal reasons (vacation, leave of absence), staffing vacancies, and other sources of long term absenteeism. DuPont has adopted a policy that each site will conduct and document an annual review of staff workload balance within each higher hazard operating unit to periodically evaluate the status and performance in this area. A standardized set of evaluation criteria for the annual review was developed and distributed to all sites as a leveraged tool to help ensure consistency across all locations.

6. Work Environment

DuPont has utilized a series of PHA related tools and checklists to conduct baseline, cyclical and MOC related evaluations for over the past twenty years. A human factors checklist has been one of these historical tools. DuPont's strategy to address FM work environmental factors involved adding content to these existing tools in order to help ensure that fatigue management workplace factors were considered and evaluated during all cyclic and project related PHA's. Integrating this review into existing PHA processes has proven to be a more sustainable and effective practice rather than creating a "stand alone" fatigue management work environment assessment process.

7. Medical Evaluations (Personal Risk Factors)

DuPont Integrated Health Services, in coordination with the process safety competency, developed a fatigue management medical screening approach and screening form which could be used as a voluntary tool during periodic health exams for shift employees who may be subject to fatigue risks. The health services group also provided targeted FM training to site medical professionals (nurses, doctors, and NP's) in order to consistently apply FM medical aspects, including both the proactive (voluntary) approaches as well as "for cause" evaluations where potential fatigue has been observed by line management in the work place.

DuPont Medical also worked with the corporate HR, labor relations, and legal functions to develop and communicate consistent medical guidance and case management methods where employee fatigue concerns have been identified. This is an area where emerging issues, specific interpretations, and case law should be considered by each entity such as the following (examples only):

1) Where additional FM medical testing is prescribed based on an employee's physical status and personal risk factors (obesity, concerns about sleep apnea) which could relate to fatigue in the workplace, should the costs for further medical evaluations be borne by the employee as part of personal health care or as part of the occupational medical program for a given company?
2) How should an employee be managed if they refuse to participate in additional medical screening or specific diagnostic tests that focus on fatigue management?
3) Are certain medical conditions that can cause fatigue in the work place subject to the 1990 Americans with Disabilities Act (ADA) in the United States and if so, which ones, and what may be the expectations for reasonable accommodations?

8. Contractors

DuPont has also adopted internal process safety requirements for a fatigue management program applied to selective contractors who work in higher hazard process areas which typically is limited to resident maintenance service providers, test and inspection consultants, and specialized services resources who may be on-site only during major turnarounds.

Contractors are required to develop their own internal programs with the basic elements of fatigue management as described in ANSI/API 755. DuPont will typically audit their systems and performance at the site level on a periodic basis to verify the program is adequately implemented. Many major contractors have embraced an enhanced focus to manage fatigue risks. In one example, a resident maintenance contractor developed a modified approach for their employee work scheduling during an annual shutdown with a DuPont site in Protection Technologies in order to improve fatigue management practices. The shutdown typically lasts 15 days involving 12 hour shifts and the contractor reduced the shift duration to 11 hours and inserted a day off for each employee during the process. Feedback from the employees was very positive, and this modifications were managed so that there were no additional costs or delays in the shutdown schedule. This unique approach serves as a good example where the owner/operator (DuPont) worked in collaboration with contactor partners to achieve a common objective which also supported the business and operations.

9. Incident Investigation Processes

DuPont maintains a global internal incident investigation reporting, classification and investigation policy for both EHS and process safety incidents which addresses communications, investigation teams, root cause failure analysis (RCFA), recommendations tracking, and overall documentation. The existing policy does not specifically include or consider fatigue management components as they may relate to contributing factors and human error as part of the overall RCFA process. At some point in the future, the policy will be updated to include a focused consideration of fatigue during all formal investigations, particularly when human error has been identified. This may include a standardized review of employee work shift schedules and overtime for at least the previous two weeks prior to the incident date.

10. Conclusion

Based on the “lessons learned” from the BP Texas City incident in 2005 and the “new” ANSI/API 755 Recommended Practice, internal company initiatives to integrate fatigue management into existing risk management systems is a prudent and necessary step to help minimize the potential for human error and the associated episodic incidents in high hazard process operations. The process safety function should be the logical “process owner”, however, the initial systems design and implementation activities involving fatigue management necessitate a collaborative effort between other key company and site level resources and functions, including operations, human relations, health services and occupational health, in order to develop an integrated system and achieve maximum risk management benefits. Since the ANSI/API RP 755 standard includes primarily performance based content, implementation also may involve a substantial number of internal corporate or local policy decisions, and potential issues which require focused discussions and interpretations to promote consistency with internal applications at each site.

The initial FM implementation process may span two to three years in order to plan, educate, develop, implement and monitor site level organizations and accompanying results. Organizations that include fatigue management requirements in existing PSM policies should recognize that implementation progress may vary at each site depending on size, complexity, shift schedules, the existence of labor unions with collective bargaining agreements, and other factors. These potential variations on the rate of progress require a simplified process to track implementation across all applicable operating sites (see figure for an example of tracking based on the six main program elements). Integration of fatigue management evaluation criteria into existing PSM audit protocols and systems is typically the last step in a long term FM implementation and performance measurement process.

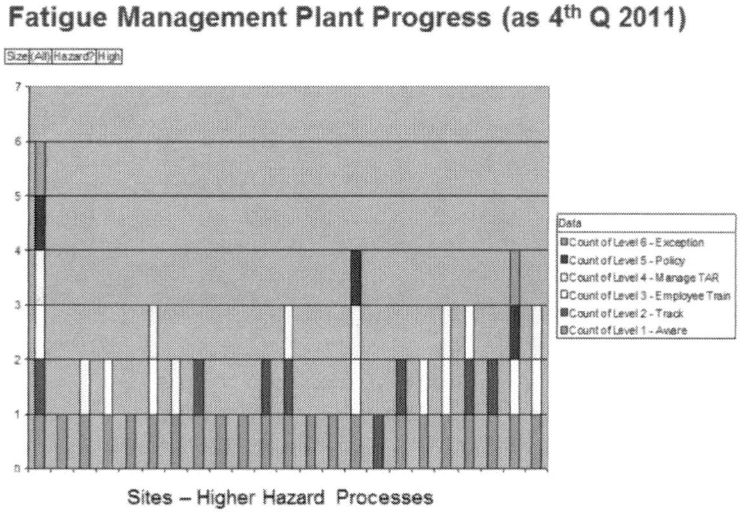

Figure 8: Example of Visual Management Board – Site Progress

11. References

(1) American National Standards Institute (ANSI)/American Petroleum Institute (API) Fatigue Risk Management Systems for Personnel in the Refining and Petrochemical Industries - RP 755 (First Edition, April 2010)

9TH GLOBAL CONGRESS ON PROCESS SAFETY

Consequences calculation for boilover and storage tank fires

Jaime E. Cadena
Chemical Engineering Department, Universidad de los Andes
Cra 1 Este No 19A – 40, Bogotá, Colombia
je.cadena21@uniandes.edu.co

Alejandra Cano[1], Juan Carlos Díaz[2], Alexis Vignes[3], Felipe Muñoz Giraldo[4]
[1,2,4]Chemical Engineering Department, Universidad de los Andes
Cra 1 Este No 19A – 40, Bogotá, Colombia
[3]INERIS
Technologique ALATA, 60550, Verneuil-en-Halatte, France
[1]na.cano744@uniandes.edu.co, [2]jc.diaz111@uniandes.edu.co, [3]Alexis.VIGNES@ineris.fr,
[4]fmunoz@uniandes.edu.co

Prepared for Presentation at
American Institute of Chemical Engineers
2013 Spring Meeting
9th Global Congress on Process Safety
San Antonio, Texas
April 28 – May 1, 2013

UNPUBLISHED

Consequences calculation for boilover and storage tank fires

Jaime E. Cadena
Chemical Engineering Department, Universidad de los Andes
Cra 1 Este No 19A – 40, Bogotá, Colombia
je.cadena21@uniandes.edu.co

Alejandra Cano[1], Juan Carlos Díaz[1], Alexis Vignes[2], Felipe Muñoz Giraldo[1]
[1]Chemical Engineering Department, Universidad de los Andes
Cra 1 Este No 19A – 40, Bogotá, Colombia
[2]INERIS
Technologique ALATA, 60550, Verneuil-en-Halatte, France

Keywords: Storage tank fires, boilover modeling, process safety, emergency response

Abstract

During the past decade, high loss incidents have been recorded due to fires in hydrocarbon storage facilities worldwide. An example is the storage tanks fire at a refinery in Amuay, Venezuela (August 25 2012, under investigation), which led to the loss of 650000 barrels of naphtha, around 41 fatalities and 80 injured. These incidents have high magnitude consequences, therefore an a-priori consequences analysis of these events become a key element for the decision-making processes to prevent, control and mitigate fire risks. The risk evaluation and mitigation for fire scenarios usually follows a prescriptive approach, therefore this work presents a tool for obtaining quick and conservative calculation of the thermal risks due to fire scenarios of hydrocarbons storage tanks facilities. These intensities are translated into consequences, emergency response requirements and key information for Contingency Plans development. To complement the fire scenarios, the tool also includes the calculation of effected distances for boilover phenomena. The results obtained by the fire tool were within an acceptable margin of error but it still conservative in comparison with real data. For the boilover tool different models were implemented among them the only one conservative for all the validation cases was the one proposed by Casal et. al. [1] for the feedback heat and the one proposed in the book Risk analysis in industrial facilities [2] for the boilover time.

1. Introduction

Storage tanks are a key element of any Oil & Gas operation since they constitute intermediate steps in the upstream operation e.g. in a refinery or a pipeline operation, and also in the downstream operations in which they are the source of final products to be delivered to the clients and customers. Nevertheless, even when hydrocarbons are stored without physico-chemical changes, risks are still present such as fire and boilover risks [3]. A fire in a storage tank is usually assumed as a pool fire in the surface of its contents, while a boilover is the ejection of the stored HazMat due to the fire and posterior boiling of a liquid with lower boiling point and higher density than the HazMat [2]. Storage tank fires and boilover scenarios represent low probability but high consequences events, making it necessary to apply available knowledge and strategies to analyze, assess, control and mitigate the related risks in order to avoid high capital and human loss as illustrated in Table 1 [4], which highlight that regardless of location or date, such events continue to occur. This makes it a key topic of study to avoid major hazard events and irreparable loss, especially for a region with a high industrial development such as Latin America [5, 6], in which the deployment of new Oil & Gas operations not only present emerging risks but also modify the existing ones [7].

Table 1. Storage tank fires and/or boilovers and related loss

Date	Location	Event	Related loss
13/12/1982	Bogotá, Colombia	Esso oil storage facility fire [8]	• 8000 people evacuated • Loss of 3 storage tanks
19/12/1982	Caracas, Venezuela	Explosion and subsequent oil tank fire [9]	• 153 fatalities • 7 missing • Severe damage to the facility
24/02/1986	Thessaloniki, Greece	Fire at an oil storage tank facility [10]	• 10 out of 12 tanks destroyed • US$330 M
11/12/2005	Buncefield, UK	UVCE and fire at hydrocarbon storage facility [11, 12]	• 43 injured • Over 2000 people evacuated • Low and moderate damage to surrounding buildings • Total destruction of most of the complex • Possible Detonation to Deflagration Transition (DDT) [13]
26/10/2009	Bayamon, PR, USA	Caribbean Petroleum Company fire (currently under investigation) [14]	• 1 injured • Costs around US$6.4 million • Around 5.6 million of product loss • Around 600 people evacuated • Damage to buildings in a radius over 1 mile (1.6 km)
25/08/2012	Amuay, Venezuela	PDVSA Amuay refinery explosion and fire (currently under invetsigation) [15]	• 48 fatalities (latest information) • Over 80 injured • Stop of the refinery with 640000 bpd capacity • Damage to at least two contiguous storage tanks, generating fire • 18 national guard and 15 civilian fatalities • Damage to around 200 residential buildings and 11 businesses • 33 families displaced • Damage to a military base

Although the subject has been addressed through different approaches [16, 17], in the field of process safety it is fundamental to understand the phenomena underlying the events of interest. For tank fires and boilover events this has been carried out for several decades, and a wide variety of empirical, semi-empirical and CFD models have become available to model such events [1, 18-23]. This work aims to develop a methodology to apply a set of the available models to calculate the consequences of the events of interest through its implementation in a user-friendly computational tool that allows obtaining quick and conservative results for either a preliminary risk analysis stage in which critical scenarios are identified or during actual emergencies in which knowledge of the possible outcomes is needed to establish mitigation strategies.

2. Methodology

In order to achieve the objectives stated above, a methodology (Figure 1) was proposed for the development and validation of the tool. First step is defining the scenario to evaluate, the tool use variables that influence tank fire and a boilover like some specific hydrocarbon properties such as viscosity, initial and final boiling point temperatures and combustion heat as input variables, next the tool implements models for the calculation of heat released by the fire the feedback heat and boilover and duration of fire times. Based on the last calculations the tool use models to determine the safe distances for both phenomena (boilover and fire) and in the case of the fire the tool also returns as output variables some mitigation strategies like refrigeration rings.

Figure 1. Methodology

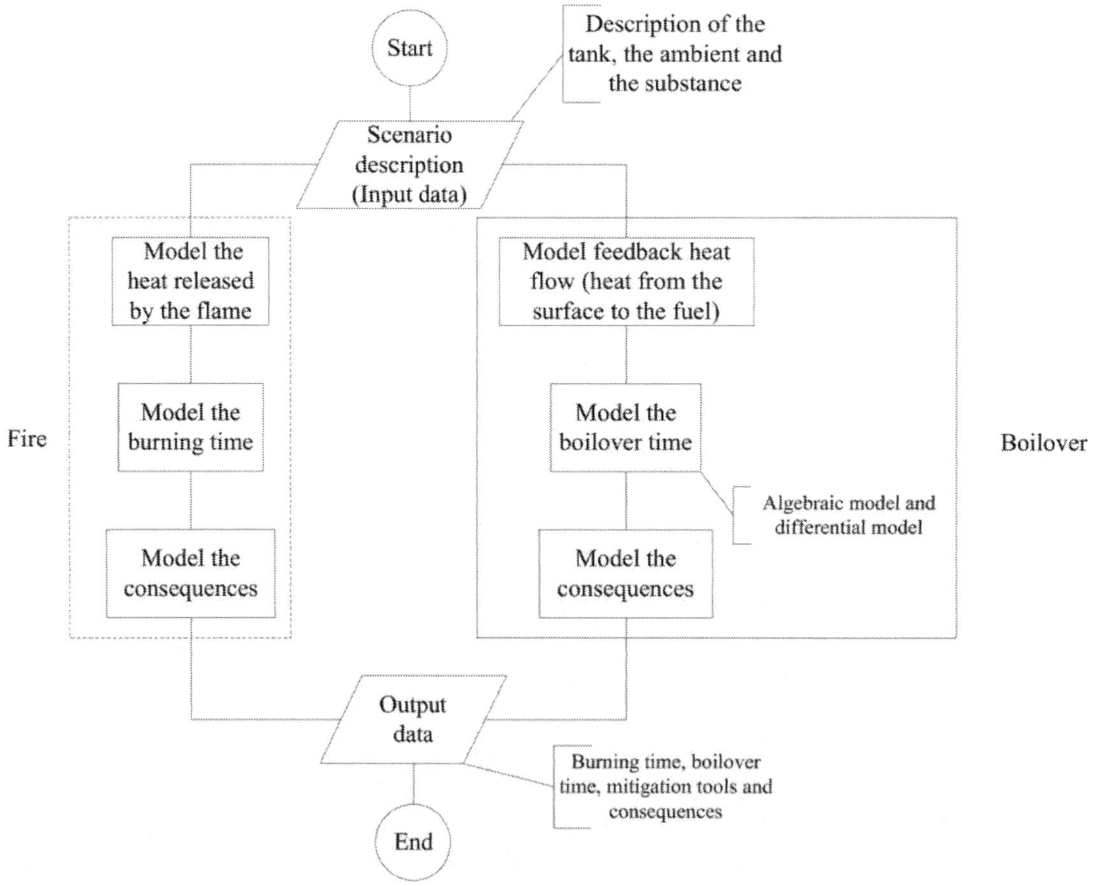

3. Tool implementation

The tank fire tool was developed in excel (visual basic) using macros for both, the heat released by the fire and the burning time. For the boilover there are multiple approximations, for the feedback heat there is an average value reported [2] and models for calculating it using a fraction of the total heat released by the fire, the difference among the last ones is the calculation of the total heat released that can be done using models with combustion heat and experimental corrections depending on the fuel [24] or heat balances between the tank and the surroundings [1], for the boilover time there is an algebraic approximation derived from a heat balance [2] and differential approximations; varying among them the initial and boundary conditions [1] . Some of them were implemented in excel and the ones involving differential approximations were implemented in matlab as this software allows multiples ways for solving differential equations in a short and easy way. Both Excel and Matlab have a user-friendly interface. The models selected for the two tools are presented on Table 2.

Table 2. Models used for the tool implementation.

		Model	Assumptions	Ref
Fire	Time calculation	$t_b = 4V/\pi D^2 v$ $v = m''/\rho$	-	[24]
	Heat calculation	$Q = m''\Delta H_{c,eff}(1-e^{-\kappa\beta D})A_{dike}$ $D = (4\ A_{dike}/\pi)^{1/2}$	- Experimental correction depending on the fuel	[24]
Boilover	Time calculation	1) $t_{boilover} = \rho\ c_p\ h_{HC}(T_{hw} - T_a)/(Q_f - m''(H_v + c_p(T_{0av} - T_a)))$ $T_{0av} = (T_{0min}\ T_{0max})^{1/2}$		[2]
		2) $\delta^2 T/\delta y^2 = (1/\alpha)\delta T/\delta t$	-Unidimentional	[21]
	Heat calculation	1) $Q_f \approx 60$ kW	-	[2]
		2) $Q_f = (4\chi/\pi)\ \rho_\infty\ c_{p\infty}(T_\infty\ g\ (T_f - T_\infty)\ D)^{1/2}$ $\chi = H_v/H_c$	- A heat conduction mechanism is assumed, no conduction in the fuel layer -The radiation is fully absorbed at the surface	[21]

The values for T_f in the heat calculation for the boilover, change depending on the assumptions used for the estimation, some authors use the adiabatic flame temperature as an approximation and others recommend approximated values like 1100 K [1] or values ranging from 1600 to 1900 K depending on the substance.

To solve the differential equation for time of boilover the following boundary conditions were established.

- Initial condition. [1]

$$\mathbf{T(y,0) = T_\infty} \quad \text{[Eq. 1]}$$

- Boundary condition 1.[1]

$$T(y(t),t) = T_s \quad [Eq. 2]$$

Where **y(t)**is defined by the following equation.

$$\delta y(t)/\delta t = m'' / \rho_f \quad [Eq. 3]$$

- Boundary condition 2.

$$-k_f * \delta T / \delta y |_{fuel} = -k_w * \delta T / \delta y |_{water} \quad [Eq. 4]$$

To achieve boundary condition 2, the modeling of the water layer is needed and it can be solved by the same equation for the fuel reported on Table 2. The first boundary condition for the water layer is stated on Equation 4 and the second boundary condition stated in Equation 5. This boundary condition implies assuming adiabatic conditions in the limit of the water layer with the bottom of the tank.

$$\delta T / \delta t |_{y=0} = 0 \quad [Eq. 5]$$

3.1 Input and output variables

For the fire tank tool is needed from the tank the diameter, the height, the type of roof and the capacity; the tool have all the characteristics for the fuel, it only need to be specified with the volume present in the tank. In addition if the studied facility has more storage tanks near to the one in fire (storage farm tanks) the distances from the one in fire to the others must be known. The output variables are the duration of the fire, the area involved on the fire, the zone classification (hot, warm, cold) and the mitigation and prevention actions [24] like if the refrigeration ring should be activated.
For the boilover the input values are the diameter of the tank the ambient conditions, specifying the fuel involved in the fire, the initial height of the fuel in the tank before the ignition the temperature of the heat wave when the boilover occurs and for one of the models the flame temperature. The output data consists of the boilover time and the distances at which lethally and irreversible consequences occur.

3.2 Consequence estimation

The consequences for both tank fire and boilover were calculated using Excel and both have dependence from the heat irradiated from the fire in the case of the tank fire and for the resulting fireball in the case of the boilover.
The consequence of the fire depends on the heat radiation determining the hot, warm and cold zone. In the hot zone the equipment will fail and the heat release will be lethal for human beings. The warm zone can be accessed with protection equipment, the firemen attack the fire there. Finally the cold zone is the one where the equipment is displayed, there is no need of protection equipment in this zone. The hot zone is determined by values where the heat is over 10 kW/m^2, the warm zone is the one with heat values between 2,5 kW/m^2 and 10 kW/m^2 and the cold zone is the one with values lower than 2,5 kW/m^2 [24]. The consequences of a boilover event are fireballs and in the ejection of ignited fuel around the tank. According to the radiation heat

release of the fire there can be determined lethal and irreversible (serious burns) effect distances corresponding to radiation heats of 1000 kW $m^{-2}$and 600 kW m^{-2} each one, they are estimated using the conservative assumption that the entire fuel in the tank participates to the fireball. The models used for the safe distances and zones are reported on Table 3 [2].

Table 3. Models used for consequence calculations.

	Consequence	Model	Ref
Fire	Irradiated heat	$d_{hotzone}=(QXr/4\pi\, q''_{hotzone})^{1/2}$ $d_{hotzone}=(QXr/4\pi\, q''_{warmzone})^{1/2}$	[24]
Boilover	Fireball	$d_{lethality}=k_{lethality}W^{0.45}$ $d_{irreversible}=k_{irreversible}W^{0.45}$	[2]

4. Results and analysis

To validate the tool it was compared with selected cases for each event (fire and boilover), the case selection was made using two criteria, the first one is searching for cases that are not affected by other events after the fire starts and the second one is searching for cases with complete information about the outputs and inputs for the tool mentioned above. The cases and the comparison with the tool results are described in this section with the comparison among the boilover time and the burning time.

4.1 Tank fire tool

For the tank fire tool the case selected was an accident that occurred at ConocoPhillips Company's Glenpool South tank farm, Glenpool, Oklahoma on April 7 2003. The event involved the tank 11 of the facility during the transport of Diesel from another tank [25] . The results obtained by the tool are reported on Table 4 and Table 5.

Table 4. Results related to the tank and the fire variables

Burning Tank Name:	11
Type of Roof:	CONE
Product:	Diesel
Nominal Volume (m3):	8420,00
Contents (m3):	1338,67
Height (m):	14,63
Diameter (m):	33,22
Area of Fire (m2):	1526,84
Time of Fire (hr):	8,39
Time of Fire (day):	0,35
Refrigeration Range (m):	71,89

Table 5. Results related to the consequences of the fire

	From (m)	To (m)
Hot Zone	0	64,298
Warm Zone	64,298	128,596
Cold Zone	128,596	N/A

The variable to compare among the case and the tool was the burning time because that is the variable reported in the few cases that also have all the input parameters. The percentage of error among the real case and the results reported by the tool is lower than 5%. In the case of the burning time the percentage of error reported may be due to the model, because the variables involved in its calculation (volume of the fuel and dimensions of the tank) don't require any assumptions and are directly specified by the case to evaluate. The possible source of error for this validation case may be directly related to it, as the tank was being filled at the moment of the fire the initial volume of fuel is not an accurate value.

4.2 Boilover tool

For the boilover tool different models were used, a comparison among them is also included in this section.

Case 1:

For the boilover, a large scale experiment was used for the study of boilover in large tanks using crude oil [26]. The results obtained by the tool with different models are presented on Table 6 and the distance effects are reported on Table 7. They are independent from the model used for heat and time of boilover so there is just one value for $d_{lethality}$ and one for $d_{irreversible}$ reported for the case.

Table 6. Case 1 results for the boilover time with the models proposed on Table 2.

	Tflame (K)		**tbo (s)**		**treal (s)**	**%error**	
Heat model 1	NA	NA	4067,67		4728	13,96637056	
Heat model 2	1100	1900	1127,07357	692,031935	4728	76,1617264	85,3631147
Time model 2	NA	NA	60604		4728	NA	

Table 7. Safe distances for the boilover consequences for case 1

dletal (m)	14,7314678
dirrev (m)	20,0281754

Case 2:

A fire on tank 3 containing crude at the Czechowice-Dziedzice Refinery on June 26, 1971 at 19:50 occurred due to a lightning hit [27]. For this case the results obtained by the tool are reported on Table 8.

Table 8. Case 2 results for the boilover time with the models proposed on Table 2.

	Tllama (K)		**tbo (hr)**		**treal (hr)**	**%error**	
Heat model 1	NA	NA	29,77614629		5,3	461,8140809	
Heat model 2	1100	1900	2,4275687	1,64560344	5,3	54,196817	68,9508784
Time model 2	NA	NA	1068,095		5,3	NA	

Table 7. Safe distances for the boilover consequences for case 2

dletal (m)	350,936098
dirrev (m)	477,115369

For case 1 the most accurate value reported by the tool was the one calculated by using an approximated value of 60 kWm^{-2} for the feedback heat, while the most accurate values in case 2 were the ones calculated by using the heat model 2, for this case the value using the heat model 1 is not conservative. This can be explained by the fact that the heat proposed by literature is very small compared with the one calculated by the second heat model, it worked well for a tank with an small amount of fuel but with great amounts of fuel that heat is not enough and will take more time to increase the temperatures in fuel layer and to take the water to its boiling point, making the boilover time bigger. So the second model may be better for modeling large tanks with big amounts of fuel, keeping the time in the same order of magnitude and conservative compared to the real time.

In the case of the differential model mentioned before for both cases the boilover time calculated with the tool is four orders of magnitude or more above the real time, this can be explained by the assumption around the model, first the model is one dimensional assuming that all the heat is transferred from the top to the bottom of the tank when it exist also a radial heat transference, this can also be seen by comparing the results from case 1 with the results from case 2 where the difference among the real time and the calculated time is proportional to the amount of fuel and the tank dimensions. Also the thermal diffusivities along with the rest of the fluid properties were assumed constant during the fire, despite the changes of temperatures in the fuel and water layers. One of the assumptions of the model implies assuming only a heat conduction mechanism, no convection inside the layer occurs, this last phenomena implies the change of density in different heights of the fuel layer and this difference in densities implies that the fuel in the top of the tank will eventually move to the bottom and the fuel in the bottom will eventually move to the top making the heat transference quicker and the boilover time smaller.

4.3 Comparison among boilover time and burning time

The case 2 described for the validation of the boilover tool have the complete information to use also the tank fire tool and can be used to compare the time of burning if the boilover doesn't occur with the time at which boilover occurs. Giving a little notion of what fraction of the total burning time takes the boilover to occur.

The results obtained by tool for the fire are presented on Table 8 and the consequences on Table 9.

Table 8. Results related to the tank and the fire variables for case 2

Burning tank name:	1
Type of roof:	cone
Product:	Crude Oil

Nominal volume (m3):	80000,00
Content (m3):	10142,89
Height (m):	14,7
Diameter (m):	33,00
Area of fire (m2):	1523,99
Time of fire (hr):	93,33
Time of fire (day):	3,89
Refrigeration rate (m):	57,11

Table 9. Results related to the consequences of the fire for case 2

	From (m)	To(m)
Hot zone	0,000	51,083
Warm zone	51,083	102,167
Cold zone	102,167	N/A

The boilover results are reported in the above section, the comparison was made using the real value reported in literature for the case and the only conservative value reported by the tool that was calculated using the heat model 2. The percentages reported on Table 10 are the fraction that takes the boilover to occur with respect to the total time of fire.

Table 10. Percentage boilover time/ burning time

	%
Real Value	5,67877424
Heat model 2	2,59294975

The boilover time is a little fraction of the total burning time, showing how important is to estimate the first one, modeling only the fire will give more time for mitigation actions but if a boilover occurs this time is reduced on a 95 % and being conservative on a 97%, the consequences are also greater for the boilover where the safe distances are superior than 477,11 m while the cold zone for the fire is above 102,167 m, about a fifth times less the boilover distance. Making the boilover a more dangerous event than the fire itself and a matter to be worried about.

4. Conclusions

The fire tool's validation results are within an acceptable error margin but still conservative, which implies that the used model describes the event and can be used for real calculations and being also simple to manipulate for the user because there is no need of complicated interfaces or calculations. In the case of the boilover the recommended model based on the results reported on this work is the one reported by Casal et. al. [1] for the feedback heat and the algebraic approximation for time because with them there were obtained for both cases used for validation a conservative result. This implies that some of the proposed models can actually be implemented on tools that can be used in real life cases helping to generate and design mitigation strategies.

5. Perspectives

For future work the differential solution for time will be studied calculating parameters like the thermal diffusivities depending on the temperature profiles and solving a two-dimensional model. Also validating the tool against more cases including cases that involve alternative fuels.

Nomenclature

Q: Pool fire heat release rate (kW)

m": Mass burning rate of fuel per unit surface area (kg $m^{-2}sec^{-1}$)

$\Delta H_{c,}$: Effective heat of combustion of fuel (kJ/kg)

A_f = A_{dike} :Surface area of pool fire (area involved in vaporization)

κβ: Empirical constant (m^{-1})

D: Diameter of the pool fire (m^2)

t_b: Burning duration of pool fire (sec)

V: Volume for liquid (m^3)

D: Pool diameter (m)

v: Regression rate (m sec^{-1})

ρ:Liquid fuel density (kg m^{-3})

c_p: Specific heat of the fuel at T_a (kJ $kg^{-1}K^{-1}$)

h_{HC} : Height of the fuel in the tank after ignition (m)

ΔH_v : Vaporization rate atT_{flame} (kJ kg^{-1})

T_a : Ambient temperature (K)

T_{hw} : Heat wave temperature when the boilover occurs (K)

T_{0av} : Average boiling temperature of the fuel (K)

Q_f: Thermal flow entering the fuel from its surface (kW m^{-2})

T_{0min}:

T_{0max}:

α: Crude thermal diffusivity

ρ_∞ : Air density (kg m^{-3})

$c_{p\infty}$: Specific heat of the air (kJ $kg^{-1}K^{-1}$)

T_∞: Ambient temperature (K)

T_f : Average flame temperature (K)

Xr:

$q''_{hotzone}$: Heat per unit area that humans and equipment can't tolerate (10 kW/m^2)

$q''_{warmzone}$: Heat per unit area that can be tolerated with protection equipment (2,5 kW/m^2)

W: Mass of the fuel at the beginning of the fire.

$k_{lethality}$: Lethality constant depending on the fuel.

$k_{irreversible}$: Irreversible consequences constant, depends on the fuel.

N. References

[1] J. Hristov, E. Planas-Cuchi, J. Arnaldos, and J. Casal, "Accidental burning of a fuel layer on a waterbed: a scale analysis of the models predicting the pre-boilover time and tests to published data," *International Journal of Thermal Sciences,* vol. 43, pp. 221-239, 2004.

[2] J. Casal, *Análisis del riesgo en instalaciones industriales*: Edicions UPC, 2001.

[3] K. Trevor, "Chapter 5 - Storage Tanks," in *What Went Wrong? (Fifth Edition)*, ed Boston: Butterworth-Heinemann, 2009, pp. 97-121.

[4] B. Zheng and G.-h. Chen, "Storage tank fire accidents," *Process Safety Progress,* vol. 30, pp. 291-293, 2011.

[5] E. Corportation, "2012 The Outlook for Energy: A view to 2040," 2012.
[6] CIA. (2011, 11/01/2012). *The World Factbook :: Oil Production*. Available: https://www.cia.gov/library/publications/the-world-factbook/fields/2173.html
[7] G. Gómez, S. Rodríguez, J. Cadena, and F. Muñoz, "Kletz's legacy for developing countries: Simple systems based on inherently safer design," *Journal of Loss Prevention in the Process Industries,* vol. 25, pp. 843-847, 2012.
[8] AP, "8000 flee Colombia gas fire," in *The Telegraph*, ed, 1982, p. 1.
[9] S. Mannan, F. P. Lees, and Knovel (Firm). (2005). *Lee's loss prevention in the process industries hazard identification, assessment, and control (3rd ed.)*. Available: http://www.knovel.com/knovel2/Toc.jsp?BookID=1470
[10] J. I. Chang and C.-C. Lin, "A study of storage tank accidents," *Journal of Loss Prevention in the Process Industries,* vol. 19, pp. 51-59, 2006.
[11] D. Bradley, G. A. Chamberlain, and D. D. Drysdale, "Large vapour cloud explosions, with particular reference to that at Buncefield," *Philosophical Transactions of the Royal Society A: Mathematical, Physical and Engineering Sciences,* vol. 370, pp. 544-566, February 13, 2012 2012.
[12] B. Investigation Board, "The Buncefield Incident," Health and Safety Commision, HSC2005.
[13] B. J. Lowesmith, G. Hankinson, and D. M. Johnson, "Vapour cloud explosions in a long congested region involving methane/hydrogen mixtures," *Process Safety and Environmental Protection,* vol. 89, pp. 234-247, 2011.
[14] U. S. C. S. a. H. I. B. CSB, "CSB Conducting Full Investigation of Massive Tank Fire at Caribbean Petroleum Refining; Investigative Team Plans to Thoroughly Examine Facility Safety Practices," in *News Room* vol. 2012, ed. Washington DC, USA, 2009.
[15] K. Vyas and E. Minaya. (2012) Explosion at Amuay refinery in Venezuela kills 41; fire still burning. *Hydrocarbon Processing*. Available: http://www.hydrocarbonprocessing.com/Article/3080380/Environment-and-Safety/Explosion-at-Amuay-refinery-in-Venezuela-kills-41-fire-still-burning.html
[16] C. Crippa, L. Fiorentini, V. Rossini, R. Stefanelli, S. Tafaro, and M. Marchi, "Fire risk management system for safe operation of large atmospheric storage tanks," *Journal of Loss Prevention in the Process Industries,* vol. 22, pp. 574-581, 2009.
[17] C. D. Argyropoulos, M. N. Christolis, Z. Nivolianitou, and N. C. Markatos, "A hazards assessment methodology for large liquid hydrocarbon fuel tanks," *Journal of Loss Prevention in the Process Industries,* vol. 25, pp. 329-335, 2012.
[18] C.-I. Lin and H.-F. Wang, "Research on Safety and Security Distance of Flammable Liquid Storage Tank," *Procedia Engineering,* vol. 11, pp. 51-60, 2011.
[19] M. Arai, K. Saito, and R. A. Altenkirch, "A Study of Boilover in Liquid Pool Fires Supported on Water Part I: Effects of a Water Sublayer on Pool Fires," *Combustion Science and Technology,* vol. 71, pp. 25-40, 1990/05/01 1990.
[20] T. Inamura, K. Saito, and K. A. Tagavi, "A Study of Boilover in Liquid Pool Fires Supported on Water. Part II: Effects of In-depth Radiation Absorption," *Combustion Science and Technology,* vol. 86, pp. 105-119, 1992/11/01 1992.
[21] I. M. Shaluf and S. A. Abdullah, "Floating roof storage tank boilover," *Journal of Loss Prevention in the Process Industries,* vol. 24, pp. 1-7, 2011.
[22] M. Muñoz, J. Arnaldos, J. Casal, and E. Planas, "Analysis of the geometric and radiative characteristics of hydrocarbon pool fires," *Combustion and Flame,* vol. 139, pp. 263-277, 2004.
[23] M. Muñoz, E. Planas, F. Ferrero, and J. Casal, "Predicting the emissive power of hydrocarbon pool fires," *Journal of Hazardous Materials,* vol. 144, pp. 725-729, 2007.
[24] NFPA, *SFPE handbook of fire protection*, 3rth ed., 2002.

[25] "Storage Tank Explosion and Fire in Glenpool, Oklahoma," National Transportation Safety Board, Washington DC2003.

[26] H. Koseki, Y. Natsume, Y. Iwata, T. Takahashi, and T. Hirano, "Large-scale boilover experiments using crude oil," *Fire Safety Journal,* vol. 41, pp. 529-535, 2006.

[27] H. Persson and A. Lönnermark, "Tank Fires Review of fire incidents 1951–2003," SP Swedish National Testing and Research Institute.

A practical view of risk reduction management of hardware and software for safety critical applications

Luis Duran
Product Marketing Manager Safety Systems
ABB
3700 West Sam Houston Parkway South suite 600
luis.m.duran@us.abb.com

Mike Krywonos
Safety Execution Center Manager
ABB
3700 West Sam Houston Parkway South suite 600
mike.krywonos@us.abb.com

Prepared for Presentation at
American Institute of Chemical Engineers
2013 Spring Meeting
9th Global Congress on Process Safety
San Antonio, Texas
April 28 – May 1, 2013

UNPUBLISHED

A practical view of risk reduction management of hardware and software for safety critical applications

Luis Duran
ABB
3700 West Sam Houston Parkway South suite 600
luis.m.duran@us.abb.com

Mike Krywonos
ABB
3700 West Sam Houston Parkway South suite 600
mike.krywonos@us.abb.com

Keywords: Safety, Risk Reduction, Safety Systems, Independent Protection Layers, Common Cause Failures, Functional Safety Management System, IEC61508, IEC61511, TÜV, third party certification, Human Factors

Abstract

Best in Class companies (both National and Global) continue to invest in safety systems; this in most cases is because safety is at the core of their production process and among their key performance metrics, closely linked to their productivity.

Many unfortunate industrial accidents such as the recent explosion of the Deep Water Horizon with the largest marine oil spill in history are evidence of the negative consequences they have on people, environment, production assets and corporations.

This paper discusses the accepted definition of safety and risk reduction in the industry and will discuss the best practices applied by engineers in the design of products for mission critical functions and the implementation of those products in projects.

Both in product design and implementation, this session will dedicate special attention to methodologies or work processes used to reduce potential common cause failures that might impair the operation of the safety system and ultimately expose the facility to a higher risk; and the human factor considerations recommended to ensure proper operation upon a critical condition or an emergency in the industrial process.

1. Introduction

Best in Class companies worldwide, defined as such for their high percentage of Overall Equipment Effectiveness (OEE) and low injury frequency rate; continue to show strong initiatives intended to expand the safety of their operation. From executive sponsored programs at the corporate level to the definition of proactive risk management strategies, these companies are investing in safety systems and processes; simply because in most cases, safety is identified among the core values of those companies, at the center of their production process and among their key performance metrics, closely linked to their productivity.

There are numerous publications and records of Industrial Accidents, a large number of them with terrible consequences, including not only loss of human life or environmental impact but also affecting the production assets in the site. As an example, in researching for this paper, the authors found record of at least 20 "notable" oil and gas offshore blowouts between 1980 and 2010.

As an example for this paper, the authors compared two serious incidents in the Oil & Gas sector; Piper Alpha operated by Occidental Petroleum and Deep Water Horizon leased to BP. Although both incidents are 22 years apart from each other, occurred in different geographies and have many differences in causes and consequences. They illustrate the risk and the potentially catastrophic dimension they can take in this industry and the impact on people, environment and corporations, which might cease to exist after these incidents.

2. Defining Safety and Risk

Safety is defined in the industry as a reduction of existing risk to a tolerable or manageable level while risk is a combination of the probability of a harmful incident and magnitude of the harm. Traditional design practices allocate the risk reduction across different and independent protection layers, the rationale behind it is simple "any system that can fail will fail" so the engineering best practice is to distribute the risk reduction tasks across multiple independent functions or systems one of these systems is a Safety Instrumented System.

2.1 Safety Instrumented Systems one of many Independent Protection Layers to Reduce Risk

A Safety Instrumented Systems (SIS) is a mission critical system designed following international design practices such as IEC61508 [1] Functional Safety Standard to reduce risk to the people in and around the production environment, the environment, the production asset and the business. In many cases the Safety Instrumented Systems are the last resource to prevent disaster.

SIS performance is measured by Safety Integrity Level (SIL) SIL 1 low, SIL 3 high o Risk Reduction Factor.

As mentioned earlier, the Safety Instrumented System is one of many functionally independent systems each intended to perform a task, as shown on Table 1 typically referred to as Independent Protection layers [2]

Table 1: Expected Functionality of Independent Protection Layers

Layer of Protection	Expected Functionality
BPCS:	Keeping the process under control
Alarms:	Alerting the operator of abnormal conditions and providing guidelines for appropriate operator response
SIS:	Automatically taking the process to a safe condition in case the abnormal condition goes out of control and the operators can't take any corrective action in time
Other layers:	Intended to mitigate the consequence of the hazard

The concept of Independent protection layers (IPL) can also be illustrated in Figure 1

3. Common Cause Failures

The idea of functional independent systems is an attempt to avoid Common Cause Failures (CCF).

Common Cause Failures [3] are failures that might affect more than one of these protection layers at the same time. As engineers, the tendency will be to focus on elements such as Heat, Humidity, Shock, and radio interference among others. However this paper also considers with special interest the human elements related to the design of safety critical systems which can contribute to additional failures.

As with other systems, some SIS problems are related to the Commercial Off the Shelf (COTS) products designed for a specific function (i.e. hardware limitations or poor documentation) while other problems are related to the use of those COTS product such as misapplication, user application programming or poor maintenance practices. Both areas can be addressed by implementing appropriate design best practices to reduce risk.

The industry has conceived these "best practices" as a series of steps that must be performed before putting one of these systems in operation (design, installation and commissioning). This paper will describe the two areas:

- Product Design and Implementation of the Commercial Off-the-Shelf (COTS) Product
- Application Design

Figure 1: Independent Protection Layers

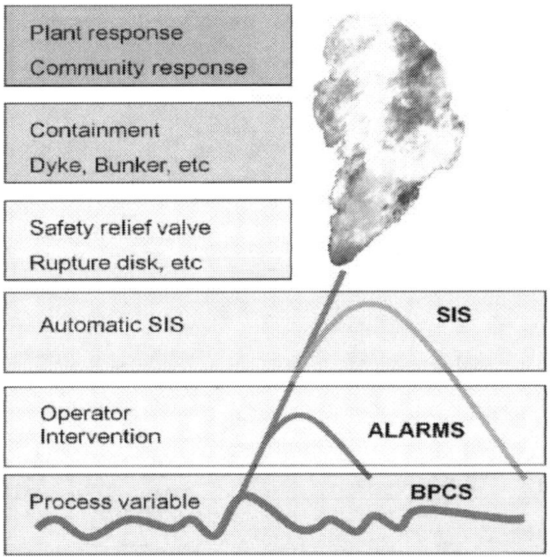

3.1 Reducing Common Cause Failures by Design

Traditionally or historically, Systems designed for SIS functions have relied on physical separation and redundancy [4] to reduce common cause failures. The authors identified technology changes across multiple generations of SIS over a period of 30 years. However over the same time frame the industry applied lessons learned from incidents to develop best practices in design and implementation of systems, the international industry standards used more often are IEC61508 and IEC61511 [5] (Figure 2)

3.2. Hardware Fault Tolerance (1st Generation SIS)

1st Generation systems rely on Hardware Fault Tolerance or redundancy to achieve reliability and availability as required for these applications, this practice is rooted in the late 1970s and early 1980s technology. As the majority of these systems were designed prior to the release of the Functional Safety standards, they do not follow the best practices found with regards to design, documentation and testing found in those standards.

Figure 2: Evolution of National Prescriptive Standard to International Performance Base Standards

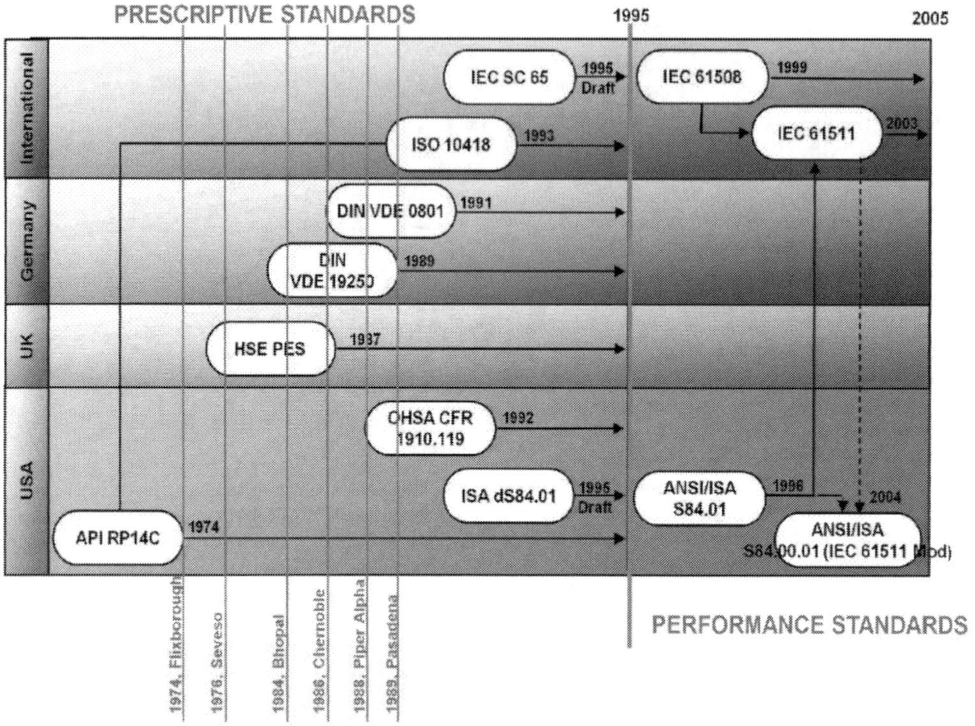

3.3. 2nd Generation Systems

As the standards became available (mid 1990s), vendors started to apply the design best practices and to pursue certification to those standards. This generation of systems was produced having third party assessment and certification but continued to rely on Hardware Fault Tolerance to satisfy the performance requirements. The authors found that a large portion of the systems applied as SIS in the market today, although designed to satisfy Functional safety standards haven't change dramatically in their use of hardware fault tolerance and basic software and hardware diagnostics.

3.4. Diverse Architecture and Implementation a 3rd Generation Systems

A more recent and different type of system, uses diverse redundancy, diverse implementation and active diagnostics [6] and can deliver not only reliability and availability required for the application while minimizing common cause failures found by the use of the same technology but also introduce additional protection to systematic failures normally related to human factors

3.4.1 Hardware Design – Diverse Hardware

This newer generation of systems use diverse processing hardware (multiple technologies) such as diverse operating systems and diverse hardware and diverse redundancy, both enhanced with diverse implementation (different implementation teams).

3.4.2 Application Execution in a SIL Compliant Environment

The use of diverse operating systems (use of different technologies) including the use of third party certified COTS and different execution path with different compiler rules extends the risk reduction found in the hardware design to the firmware/software environment in the system, including the use of limited variability software functions and provision of systematic capabilities as specified in the Functional Safety standards for critical applications.

3.4.3. Systematic Capabilities and Human Error

The concept of systematic capabilities and its associated performance measurement was introduced in the latest revision of IEC61508 standard (didn't exist as such in 1st Generation systems and was emerging in 2nd generation systems).

3.5. Safety and Network Security

1st and 2nd generation Safety Systems were not designed to reside on a networked plant, for the most part these systems allowed some communication interface or gateway but were not intended for integration, therefore in general are not equipped to counter security threats as those the industry has experienced in recent years.

A brief analysis of the definitions will show a clear differences Safety issues are random in nature and statistic analysis is possible security intentional in nature and statistical analysis will not measure performance.

The authors concur with industry standards and experts in the field. Network security may affect system performance and the safety of the installation. A system can't be safe without it being secure.

3.6. Design Best Practices and V-Model

As indicated earlier, industry standards in Functional Safety introduced the concept of Functional Safety Management System (FSMS) which includes a series of steps in the design, documentation and testing of the system, and should include Network Security as well.

Figure 3: Product Development V-Model

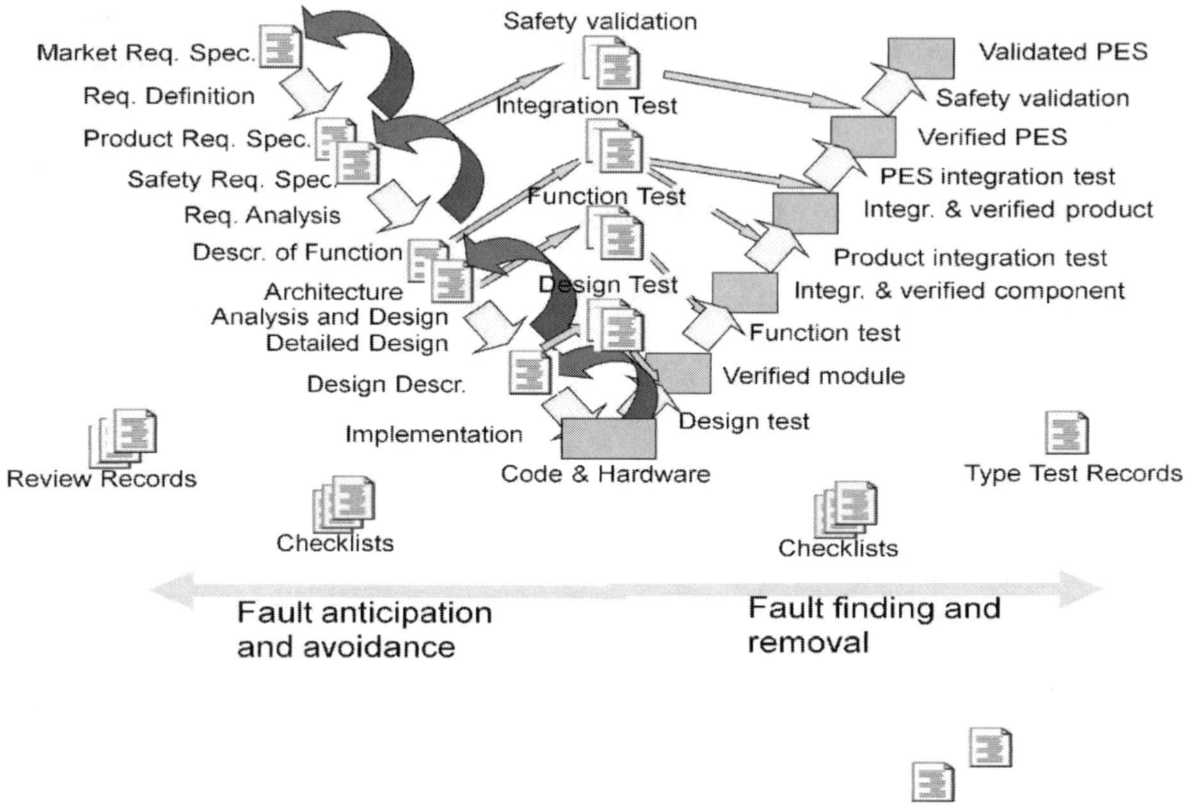

Today product development or design organizations responsible for releasing COTS products (hardware and software) intended for safety applications use design best practices as the V-Model (Figure 3) and are evaluated by a third party organization (i.e. TUV) and their FSMS is approved prior to the assessment of their products.

4. Engineering best practices to reduce risk in application design

As indicated previously, best practices in design apply both in the design of a Commercial Off the Shelf (COTS) Product and the Application Design or the application of a COTS tailored to reduce a particular risk in an industrial installation.

Best Practices in Application Design will apply a model similar to the V-Model used in the development of a product (Figure 3) and reinforced by a structured Quality Management system appropriate for Functional Safety applications.

Application Design starts with Hardware Design, which starts with a design that follows the COTS Product's Safety Manual and considers the particularities of the application as described on the Application Safety Requirements Specification to design a system that is less susceptible to product failures by considering aspects such as

- Power feeds
- Power Supplies
- CPUs
- I/O Modules Hardware
- Application Software
- Networks

4.1 Hardware

It's clear that hardware is not perfect, it's subject to fails, and the application design must consider fail recovery modes or additional risk reduction methodologies or options including diverse methods of performing shutdowns as:

- Hardwired wired pushbutton,
- Hardwired logic systems (with appropriate SIL rating)
- Solid State relays
- Etc.

4.2 Software

One critical aspect that has gained visibility in recent years is the software design of the application. Efforts in this area include the development of a well documented Software Functional Design, appropriate selection of software tools and libraries that meets the Safety Integrity Level (SIL) required for the application, Code Review practices including Independent Validation and Verification depending on the SIL, competence of the programmer and other elements described later in section 5.1. The features available in the COTS product to address the systematic capabilities requirements are a valuable asset for the programmer and contribute to reducing potential errors that will later affect the safety of the installation.

4.3 Functional Testing and Periodic Proof Testing

Another critical aspect is Functional Testing and Periodic Proof Testing, particularly because Not ALL System Faults are Self-Revealing and Covert Faults that may inhibit SIS action on Demand can only be detected by testing the entire system, this requires not only a Full Functional Test prior to commissioning, typically conducted as an integration test in the vendor or System Integrator facility and later in the final instrumentation room in the plant; but Periodical Functional Tests using a documented procedure to Detect Covert Faults and covering the entire SIS.

Functional Testing should Record and Analyze activation of SIS functions, and spurious activation of an Emergency Shutdown Valve due to a Process Shut Down, but this does not test the Entire Function of the same valve during an ESD action.

5. Automation can't check for human intervention

Once the design is complete, all of these systems will have different degrees of interaction with users, and different type of users with different levels of competence depending to their role (i.e. operation, maintenance personnel and engineers)

The most sophisticated automation can't prevent human error, industrial plants are designed with the highest accuracy in mind, with several separate safety loops often checking the integrity of process systems. This apparatus however can't check for human intervention and a small human error could cause an enormous catastrophe. 70% of reported incidents in the oil and gas industry worldwide are attributable to human error and account for in excess of 90% of the financial loss to the industry.

Results from research conducted by the Health and Safety Executive and published in the book Out of control: Why control systems go wrong and how to prevent failure [7] show the impact of the human element in industrial incidents throughout project lifecycle. This is shown on Figure 4.

International Functional Safety Standards (IEC 61508 and IEC61511) and previously ISA 84 introduced the concept of the Safety Lifecycle, describing the phases that should take place from concept to design to implementation and operation of a Safety Instrumented System. The Safety Lifecycle is a step in the direction to reduce the impact of human factors by establishing the proper design best practices, documentation reviews and validation and verification steps in the execution of a safety project. Additionally the standards introduce two important elements: Competence of Personnel mentioned briefly in Section 4.2 and Functional Safety Management System (FSMS).

Recent changes to International Functional Safety Standard IEC61508 have turned the requirements on FSMS and Competence of Personnel into normative clauses of mandatory compliance instead of a recommendation as was presented in the previous version of the same standard.

5.1 Functional Safety Management System (FSMS) and Competence Requirements

A Functional Safety Management System (FSMS) consists of proven and validated procedures, methodologies, templates and report outlines, covering those essential elements of hazard and risk assessment; SIL determination; SIS design and SIL capable hardware, SIS installation, SIS commissioning and validation and SIS operations and maintenance. A FSMS is used in the design of a COTS product and during the Application Design. It's typically an extension of the existing Quality Assurance/Quality Management process and lives from the conception to decommission of the SIS and beyond to become part of the organization best engineering practices and work processes.

Figure 4: Root Cause of Control System Failures

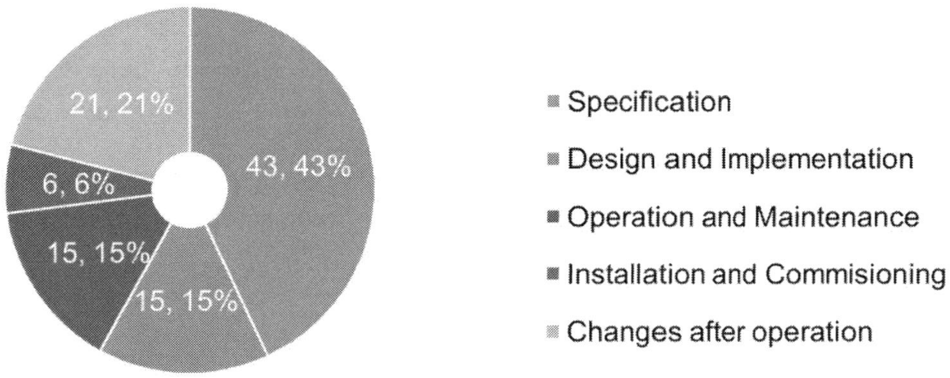

As indicated, the safety standards call for competence of personnel. This requirement shouldn't be interpreted as a need for safety experts in every aspect of the design, commissioning or operation and maintenance of a SIS, but instead as the need for an evaluation of the appropriate expertise and experience required to perform a particular job function in any phase of the safety lifecycle. In practical term those involved with activities related to a safety system should have demonstrated "competence" either by formal training, certification or accreditation or on-the-job relevant experience.

The reader might conclude that the standards today require competent and independent review groups because system designers don't trust product designers or other application designers and therefore we would rather be safe than sorry.

6. Summary

In summary

- Best in Class companies link safety to their success and invest in programs and systems to reduce their risk
- Engineers have attempted to reduce risk by minimizing the potential for common cause failures

- Common cause failures can occur in products (hardware or software) or the implementation of the application
- The industry has conceived best practices (i.e. FSMS) to minimize the impact of human error
- Human factors can't be ignored in the design, particularly in the application design
- Enforcement of these best practices via formal FSMS has proven to be a way to reduce the risk introduced by engineers… it's vital!

References

[1] IEC 61508 Functional safety of electrical/electronic/programmable electronic safety-related systems International Standard 2010

[2]Integrating Control and Safety: Where to draw the line.
Robin McCrea-Steele, TÜV FSExpert, Invensys-Premier Consulting Services

[3] Common Cause and Common Sense, Designing Failure Out of Your SIS
Angela E. Summers, Ph.D. and Glenn Raney, Oct 2000

[4] Integrated but separate, Advances in integrated and safety control,
Roger Prew, ABB , June 2009

[5] IEC 61511 Functional safety - Safety instrumented systems for the process industry sector International Standard 2003

[6] Providing Independent Layers of Protection with Integrated Safety Systems
Luis M. Duran, ABB and Ron Johnson Dow Chemical, Oct 2009

[7] Out of control: Why control systems go wrong and how to prevent failure

HSE Books ISBN 0-7176-2192-8

9TH GLOBAL CONGRESS ON PROCESS SAFETY

A practical view of risk reduction management of hardware and software for safety critical applications

Luis Duran
Product Marketing Manager Safety Systems
ABB
3700 West Sam Houston Parkway South suite 600
luis.m.duran@us.abb.com

Mike Krywonos
Safety Execution Center Manager
ABB
3700 West Sam Houston Parkway South suite 600
mike.krywonos@us.abb.com

Prepared for Presentation at
American Institute of Chemical Engineers
2013 Spring Meeting
9th Global Congress on Process Safety
San Antonio, Texas
April 28 – May 1, 2013

UNPUBLISHED

A practical view of risk reduction management of hardware and software for safety critical applications

Luis Duran
ABB
3700 West Sam Houston Parkway South suite 600
luis.m.duran@us.abb.com

Mike Krywonos
ABB
3700 West Sam Houston Parkway South suite 600
mike.krywonos@us.abb.com

Keywords: Safety, Risk Reduction, Safety Systems, Independent Protection Layers, Common Cause Failures, Functional Safety Management System, IEC61508, IEC61511, TÜV, third party certification, Human Factors

Abstract

Best in Class companies (both National and Global) continue to invest in safety systems; this in most cases is because safety is at the core of their production process and among their key performance metrics, closely linked to their productivity.

Many unfortunate industrial accidents such as the recent explosion of the Deep Water Horizon with the largest marine oil spill in history are evidence of the negative consequences they have on people, environment, production assets and corporations.

This paper discusses the accepted definition of safety and risk reduction in the industry and will discuss the best practices applied by engineers in the design of products for mission critical functions and the implementation of those products in projects.

Both in product design and implementation, this session will dedicate special attention to methodologies or work processes used to reduce potential common cause failures that might impair the operation of the safety system and ultimately expose the facility to a higher risk; and the human factor considerations recommended to ensure proper operation upon a critical condition or an emergency in the industrial process.

1. Introduction

Best in Class companies worldwide, defined as such for their high percentage of Overall Equipment Effectiveness (OEE) and low injury frequency rate; continue to show strong initiatives intended to expand the safety of their operation. From executive sponsored programs at the corporate level to the definition of proactive risk management strategies, these companies are investing in safety systems and processes; simply because in most cases, safety is identified among the core values of those companies, at the center of their production process and among their key performance metrics, closely linked to their productivity.

There are numerous publications and records of Industrial Accidents, a large number of them with terrible consequences, including not only loss of human life or environmental impact but also affecting the production assets in the site. As an example, in researching for this paper, the authors found record of at least 20 "notable" oil and gas offshore blowouts between 1980 and 2010.

As an example for this paper, the authors compared two serious incidents in the Oil & Gas sector; Piper Alpha operated by Occidental Petroleum and Deep Water Horizon leased to BP. Although both incidents are 22 years apart from each other, occurred in different geographies and have many differences in causes and consequences. They illustrate the risk and the potentially catastrophic dimension they can take in this industry and the impact on people, environment and corporations, which might cease to exist after these incidents.

2. Defining Safety and Risk

Safety is defined in the industry as a reduction of existing risk to a tolerable or manageable level while risk is a combination of the probability of a harmful incident and magnitude of the harm. Traditional design practices allocate the risk reduction across different and independent protection layers, the rationale behind it is simple "any system that can fail will fail" so the engineering best practice is to distribute the risk reduction tasks across multiple independent functions or systems one of these systems is a Safety Instrumented System.

2.1 Safety Instrumented Systems one of many Independent Protection Layers to Reduce Risk

A Safety Instrumented Systems (SIS) is a mission critical system designed following international design practices such as IEC61508 [1] Functional Safety Standard to reduce risk to the people in and around the production environment, the environment, the production asset and the business. In many cases the Safety Instrumented Systems are the last resource to prevent disaster.

SIS performance is measured by Safety Integrity Level (SIL) SIL 1 low, SIL 3 high o Risk Reduction Factor.

As mentioned earlier, the Safety Instrumented System is one of many functionally independent systems each intended to perform a task, as shown on Table 1 typically referred to as Independent Protection layers [2]

Table 1: Expected Functionality of Independent Protection Layers

Layer of Protection	Expected Functionality
BPCS:	Keeping the process under control
Alarms:	Alerting the operator of abnormal conditions and providing guidelines for appropriate operator response
SIS:	Automatically taking the process to a safe condition in case the abnormal condition goes out of control and the operators can't take any corrective action in time
Other layers:	Intended to mitigate the consequence of the hazard

The concept of Independent protection layers (IPL) can also be illustrated in Figure 1

3. Common Cause Failures

The idea of functional independent systems is an attempt to avoid Common Cause Failures (CCF).

Common Cause Failures [3] are failures that might affect more than one of these protection layers at the same time. As engineers, the tendency will be to focus on elements such as Heat, Humidity, Shock, and radio interference among others. However this paper also considers with special interest the human elements related to the design of safety critical systems which can contribute to additional failures.

As with other systems, some SIS problems are related to the Commercial Off the Shelf (COTS) products designed for a specific function (i.e. hardware limitations or poor documentation) while other problems are related to the use of those COTS product such as misapplication, user application programming or poor maintenance practices. Both areas can be addressed by implementing appropriate design best practices to reduce risk.

The industry has conceived these "best practices" as a series of steps that must be performed before putting one of these systems in operation (design, installation and commissioning). This paper will describe the two areas:

- Product Design and Implementation of the Commercial Off-the-Shelf (COTS) Product
- Application Design

Figure 1: Independent Protection Layers

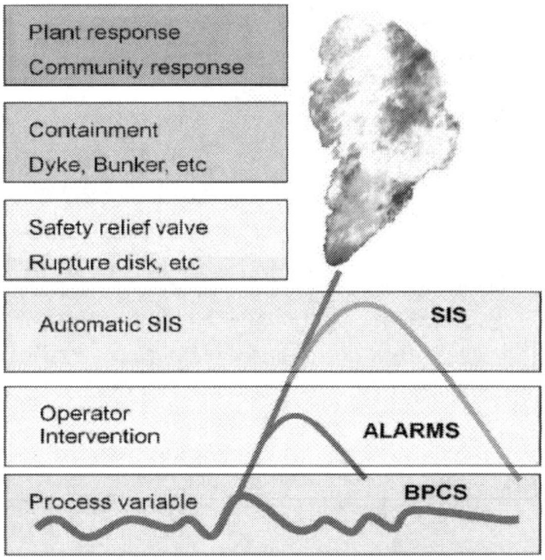

3.1 Reducing Common Cause Failures by Design

Traditionally or historically, Systems designed for SIS functions have relied on physical separation and redundancy [4] to reduce common cause failures. The authors identified technology changes across multiple generations of SIS over a period of 30 years. However over the same time frame the industry applied lessons learned from incidents to develop best practices in design and implementation of systems, the international industry standards used more often are IEC61508 and IEC61511 [5] (Figure 2)

3.2. Hardware Fault Tolerance (1st Generation SIS)

1st Generation systems rely on Hardware Fault Tolerance or redundancy to achieve reliability and availability as required for these applications, this practice is rooted in the late 1970s and early 1980s technology. As the majority of these systems were designed prior to the release of the Functional Safety standards, they do not follow the best practices found with regards to design, documentation and testing found in those standards.

Figure 2: Evolution of National Prescriptive Standard to International Performance Base Standards

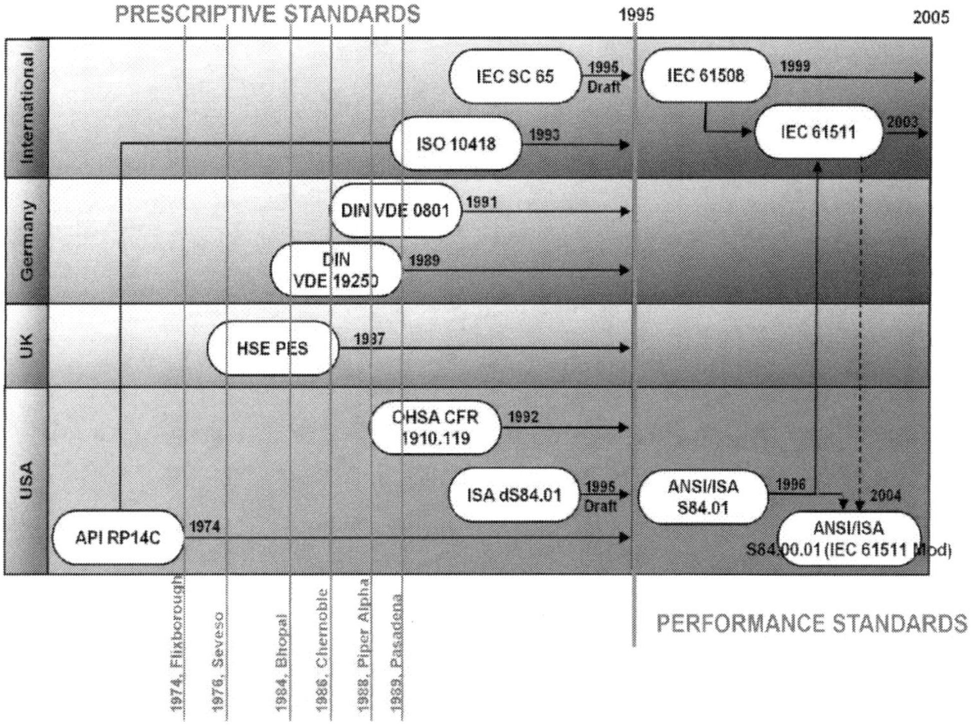

3.3. 2nd Generation Systems

As the standards became available (mid 1990s), vendors started to apply the design best practices and to pursue certification to those standards. This generation of systems was produced having third party assessment and certification but continued to rely on Hardware Fault Tolerance to satisfy the performance requirements. The authors found that a large portion of the systems applied as SIS in the market today, although designed to satisfy Functional safety standards haven't change dramatically in their use of hardware fault tolerance and basic software and hardware diagnostics.

3.4. Diverse Architecture and Implementation a 3rd Generation Systems

A more recent and different type of system, uses diverse redundancy, diverse implementation and active diagnostics [6] and can deliver not only reliability and availability required for the application while minimizing common cause failures found by the use of the same technology but also introduce additional protection to systematic failures normally related to human factors

3.4.1 Hardware Design – Diverse Hardware

This newer generation of systems use diverse processing hardware (multiple technologies) such as diverse operating systems and diverse hardware and diverse redundancy, both enhanced with diverse implementation (different implementation teams).

3.4.2 Application Execution in a SIL Compliant Environment

The use of diverse operating systems (use of different technologies) including the use of third party certified COTS and different execution path with different compiler rules extends the risk reduction found in the hardware design to the firmware/software environment in the system, including the use of limited variability software functions and provision of systematic capabilities as specified in the Functional Safety standards for critical applications.

3.4.3. Systematic Capabilities and Human Error

The concept of systematic capabilities and its associated performance measurement was introduced in the latest revision of IEC61508 standard (didn't exist as such in 1st Generation systems and was emerging in 2nd generation systems).

3.5. Safety and Network Security

1st and 2nd generation Safety Systems were not designed to reside on a networked plant, for the most part these systems allowed some communication interface or gateway but were not intended for integration, therefore in general are not equipped to counter security threats as those the industry has experienced in recent years.

A brief analysis of the definitions will show a clear differences Safety issues are random in nature and statistic analysis is possible security intentional in nature and statistical analysis will not measure performance.

The authors concur with industry standards and experts in the field. Network security may affect system performance and the safety of the installation. A system can't be safe without it being secure.

3.6. Design Best Practices and V-Model

As indicated earlier, industry standards in Functional Safety introduced the concept of Functional Safety Management System (FSMS) which includes a series of steps in the design, documentation and testing of the system, and should include Network Security as well.

Figure 3: Product Development V-Model

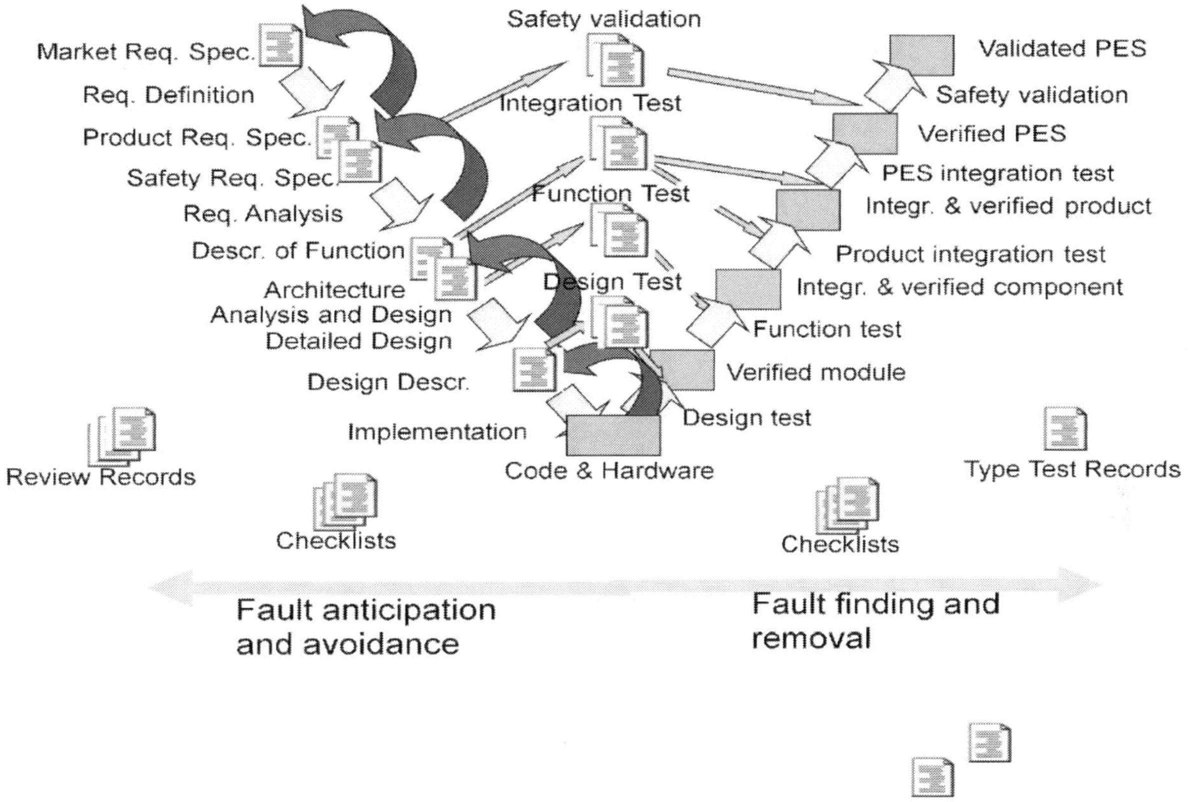

Today product development or design organizations responsible for releasing COTS products (hardware and software) intended for safety applications use design best practices as the V-Model (Figure 3) and are evaluated by a third party organization (i.e. TUV) and their FSMS is approved prior to the assessment of their products.

4. Engineering best practices to reduce risk in application design

As indicated previously, best practices in design apply both in the design of a Commercial Off the Shelf (COTS) Product and the Application Design or the application of a COTS tailored to reduce a particular risk in an industrial installation.

Best Practices in Application Design will apply a model similar to the V-Model used in the development of a product (Figure 3) and reinforced by a structured Quality Management system appropriate for Functional Safety applications.

Application Design starts with Hardware Design, which starts with a design that follows the COTS Product's Safety Manual and considers the particularities of the application as described on the Application Safety Requirements Specification to design a system that is less susceptible to product failures by considering aspects such as

- Power feeds
- Power Supplies
- CPUs
- I/O Modules Hardware
- Application Software
- Networks

4.1 Hardware

It's clear that hardware is not perfect, it's subject to fails, and the application design must consider fail recovery modes or additional risk reduction methodologies or options including diverse methods of performing shutdowns as:

- Hardwired wired pushbutton,
- Hardwired logic systems (with appropriate SIL rating)
- Solid State relays
- Etc.

4.2 Software

One critical aspect that has gained visibility in recent years is the software design of the application. Efforts in this area include the development of a well documented Software Functional Design, appropriate selection of software tools and libraries that meets the Safety Integrity Level (SIL) required for the application, Code Review practices including Independent Validation and Verification depending on the SIL, competence of the programmer and other elements described later in section 5.1. The features available in the COTS product to address the systematic capabilities requirements are a valuable asset for the programmer and contribute to reducing potential errors that will later affect the safety of the installation.

4.3 Functional Testing and Periodic Proof Testing

Another critical aspect is Functional Testing and Periodic Proof Testing, particularly because Not ALL System Faults are Self-Revealing and Covert Faults that may inhibit SIS action on Demand can only be detected by testing the entire system, this requires not only a Full Functional Test prior to commissioning, typically conducted as an integration test in the vendor or System Integrator facility and later in the final instrumentation room in the plant; but Periodical Functional Tests using a documented procedure to Detect Covert Faults and covering the entire SIS.

Functional Testing should Record and Analyze activation of SIS functions, and spurious activation of an Emergency Shutdown Valve due to a Process Shut Down, but this does not test the Entire Function of the same valve during an ESD action.

5. Automation can't check for human intervention

Once the design is complete, all of these systems will have different degrees of interaction with users, and different type of users with different levels of competence depending to their role (i.e. operation, maintenance personnel and engineers)

The most sophisticated automation can't prevent human error, industrial plants are designed with the highest accuracy in mind, with several separate safety loops often checking the integrity of process systems. This apparatus however can't check for human intervention and a small human error could cause an enormous catastrophe. 70% of reported incidents in the oil and gas industry worldwide are attributable to human error and account for in excess of 90% of the financial loss to the industry.

Results from research conducted by the Health and Safety Executive and published in the book Out of control: Why control systems go wrong and how to prevent failure [7] show the impact of the human element in industrial incidents throughout project lifecycle. This is shown on Figure 4.

International Functional Safety Standards (IEC 61508 and IEC61511) and previously ISA 84 introduced the concept of the Safety Lifecycle, describing the phases that should take place from concept to design to implementation and operation of a Safety Instrumented System. The Safety Lifecycle is a step in the direction to reduce the impact of human factors by establishing the proper design best practices, documentation reviews and validation and verification steps in the execution of a safety project. Additionally the standards introduce two important elements: Competence of Personnel mentioned briefly in Section 4.2 and Functional Safety Management System (FSMS).

Recent changes to International Functional Safety Standard IEC61508 have turned the requirements on FSMS and Competence of Personnel into normative clauses of mandatory compliance instead of a recommendation as was presented in the previous version of the same standard.

5.1 Functional Safety Management System (FSMS) and Competence Requirements

A Functional Safety Management System (FSMS) consists of proven and validated procedures, methodologies, templates and report outlines, covering those essential elements of hazard and risk assessment; SIL determination; SIS design and SIL capable hardware, SIS installation, SIS commissioning and validation and SIS operations and maintenance. A FSMS is used in the design of a COTS product and during the Application Design. It's typically an extension of the existing Quality Assurance/Quality Management process and lives from the conception to decommission of the SIS and beyond to become part of the organization best engineering practices and work processes.

Figure 4: Root Cause of Control System Failures

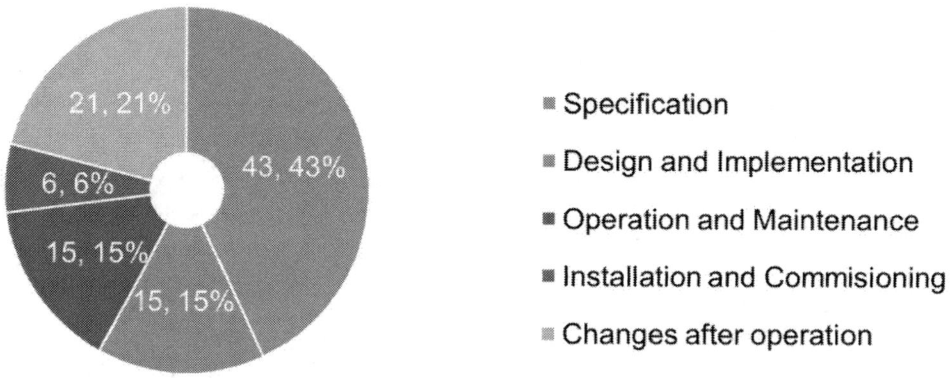

As indicated, the safety standards call for competence of personnel. This requirement shouldn't be interpreted as a need for safety experts in every aspect of the design, commissioning or operation and maintenance of a SIS, but instead as the need for an evaluation of the appropriate expertise and experience required to perform a particular job function in any phase of the safety lifecycle. In practical term those involved with activities related to a safety system should have demonstrated "competence" either by formal training, certification or accreditation or on-the-job relevant experience.

The reader might conclude that the standards today require competent and independent review groups because system designers don't trust product designers or other application designers and therefore we would rather be safe than sorry.

6. Summary

In summary

- Best in Class companies link safety to their success and invest in programs and systems to reduce their risk
- Engineers have attempted to reduce risk by minimizing the potential for common cause failures

- Common cause failures can occur in products (hardware or software) or the implementation of the application
- The industry has conceived best practices (i.e. FSMS) to minimize the impact of human error
- Human factors can't be ignored in the design, particularly in the application design
- Enforcement of these best practices via formal FSMS has proven to be a way to reduce the risk introduced by engineers… it's vital!

References

[1] IEC 61508 Functional safety of electrical/electronic/programmable electronic safety-related systems
International Standard 2010

[2]Integrating Control and Safety: Where to draw the line.
Robin McCrea-Steele, TÜV FSExpert, Invensys-Premier Consulting Services

[3] Common Cause and Common Sense, Designing Failure Out of Your SIS
Angela E. Summers, Ph.D. and Glenn Raney, Oct 2000

[4] Integrated but separate, Advances in integrated and safety control,
Roger Prew, ABB , June 2009

[5] IEC 61511 Functional safety - Safety instrumented systems for the process industry sector
International Standard 2003

[6] Providing Independent Layers of Protection with Integrated Safety Systems
Luis M. Duran, ABB and Ron Johnson Dow Chemical, Oct 2009

[7] Out of control: Why control systems go wrong and how to prevent failure

HSE Books ISBN 0-7176-2192-8

(44bj) Establishing a 3 Credit Course in PSM for Undergraduate Engineers

Deborah L Grubbe, Executive Mgmt, Operations and Safety Solutions, LLC, Chadds Ford, PA

Considering academic direction and industrial needs, the Purdue Chemical Engineering faculty and an industrial alumni team jointly sponsored, developed, executed and audited a 3 credit course in Process Safety Management for undergraduates in Engineering. This "team taught" course is now a required in the Purdue ChE curriculum, and has received enthusiastically high marks from the students. Originally conceived in 2010, the course went from concept to required in less than 24 months.

One of the alumni team members, with over 30 years in the process industries, will describe each step of the process: creating the support, drafting the syllabus, designing the course, executing the lectures, planning the ancillary support activities, and auditing the results of the 15 week, 45 lecture series. The final exam was the simulation of an emergency situation, which was modeled after a real life process safety incident. Actual student feedback will be shared. An eBook text has been developed for this course, and the speaker will conduct a short demonstration of this product as part of the overall presentation.

Investigations on the self heating potential of garage waste as basis for the hazard evaluation of bulk storage facilities

Hannes Kern
Montanuniversität Leoben
Franz-Josef Strasse 18, 8700 Leoben
hannes.kern@unileoben.ac.at

Christoph Spijker, Harald Raupenstrauch
Montanuniversität Leoben
Franz-Josef Strasse 18, 8700 Leoben
christoph.spijker@unileoben.ac.at, harald.raupenstrauch@unileoben.ac.at

Prepared for Presentation at
American Institute of Chemical Engineers
2013 Spring Meeting
9th Global Congress on Process Safety
San Antonio, Texas
April 28 – May 1, 2013

UNPUBLISHED

Investigations on the self heating potential of garage waste as basis for the hazard evaluation of bulk storage facilities

Hannes Kern
Montanuniversität Leoben
Franz-Josef Strasse 18, 8700 Leoben
hannes.kern@unileoben.ac.at

Christoph Spijker, Harald Raupenstrauch
Montanuniversität Leoben
Franz-Josef Strasse 18, 8700 Leoben
christoph.spijker@unileoben.ac.at, harald.raupenstrauch@unileoben.ac.at

Keywords: self heating, garage waste, differential reactor

Abstract

The self heating of bulk material is a major point to be considered in hazard analysis especially for waste management and waste treatment companies. The self heating potential for bulk material mainly depends on the composition of the material itself and on systemic influences like the geometry of the storage area, weather conditions or the storage temperature. Besides microbiological activity, chemical oxidation at low temperatures is the main reason for the self heating and auto ignition of bulk material. Data from the measurement of the chemical and biological oxidation kinetics can be used as basis for the simulation of storage areas by computational fluid dynamics. The measurement of the low temperature oxidation kinetics is carried out in a differential reactor, which is characterized by a small amount of sample material, isothermal conditions and very low turnover rates. Due to the fact that chemical oxidations reactions at low temperatures show very low turnover rates themselves, the experimental setup is laid out as a cycle process. The oxidation kinetics are measured by the analysis of the carbon monoxide (CO) and carbon dioxide (CO2) production rate at different temperatures. Usually a temperature range from 40°C up to 120°C is investigated. Waste materials from workshops contain significant amounts of grease and oil as well as residuals of metal dusts or chips dispersed on the surface of paper or cloth material used for cleaning purposes. The large surface combined with the fine dispersion of grease and oil accelerates the chemical oxidation significantly. To determine the influence of the oil and grease content on the oxidation kinetics, measurements of paper and cloth samples with different contents of oil were carried out in a differential cycle reactor. For the investigations biological cutting oil and oil with a petrochemical origin was investigated. Furthermore the influence of metal dusts on the oxidation characteristics of the different materials was investigated. The results of the investigations were compared with the oxidation kinetics of garage waste material with a mean diameter of 0- 40mm.

1. Introduction

The self heating of bulk material is a major point to be considered in hazard analysis especially for waste management and waste treatment companies. Besides safety issues the auto ignition of bulk storage areas is also a serious economical threat and occurs quite frequently. In industrial branches where material with an inhomogeneous composition (like waste treatment facilities) are processed the problem of self heating is even more present. Especially at residual derived fuel storage areas fires are also a serious ecological threat, due to composition of the materials themselves. The measurement of the low temperature oxidation kinetics combined with CFD [1] modeling used for the prediction of the self heating potential of bulk storage areas is an important element of hazard analysis and loss prevention. In the present study the self heating potential of garage waste was investigated. Garage waste shows a high tendency of self heating and therefore auto ignition is a major hazard. In the course of this study different influencing parameters on the self heating potential of garage waste were investigated, to be able to evaluate the risk of auto ignition. Self heating in general is a complex process that mainly depends on the parameters presented below [2]:

Material specific parameters

- composition
- chemical kinetics
- particle size distribution
- bed porosity
- heat conductivity

Apparatus specific boundary conditions

- geometry of the storage area
- size surroundings

Operating conditions

- temperature
- gas velocity

The chemical oxidation at low temperatures is besides microbiological activity the main reason for the self heating and auto ignition of bulk material. Basicly two main oxidation reactions occur at low temperatures:

$$a_{Fuel(s)} + b_{O_2(g)} \rightarrow c_{CO(g)} + d_{\mathrm{Rest}(s)} \qquad (1)$$

$$e_{Fuel(s)} + f_{O_2(g)} \rightarrow g_{CO_2(g)} + h_{\mathrm{Rest}(s)} \qquad (2)$$

The reaction of the solid fuel with oxygen gives carbon monoxide (1) and carbon dioxide (2). In a simple kinetic approach it is assumed that the solid fuel can be seen as bulk with a constant concentration compared to the oxygen concentration. After some simplifications the relations (3) and (4) can be derived from the Arrhenius relation for chemical reaction kinetics:

$$r_{CO} = k_0 \cdot \exp\left(\frac{-E_A}{R \cdot T}\right) \cdot y_{O_2}^n \quad (3)$$

$$r_{CO} = k_0 \cdot \exp\left(\frac{-E_A}{R \cdot T}\right) \cdot y_{O_2}^n \quad (4)$$

For further research it is assumed that the postulated reactions undergo only chemical limitation, because chemical limitation exceeds transport limitation by several orders of magnitude. The determination of the kinetic parameters is carried out by the observation of the CO and CO_2 production rates.

2. Experimental investigations

2.1 Sample Material

Waste materials from workshops contain significant amounts of grease and oil as well as residuals of metal dusts or chips dispersed on the surface of paper or cloth material. The large surface combined with the fine dispersion of grease and oil accelerates the chemical oxidation significantly [3]. To determine the influence of the oil and grease content on the oxidation kinetics, paper and cotton cloth samples with different contents of oil were chosen as model substances for further investigations. The calorific values as well as the chemical composition of the investigated materials can be seen in Table1. The oil content of the garage waste sample material was determined by Soxhlet extraction with cyclohexane as solvent. The sample had an oil content of around 15% and a particle size of up to 30mm.

Table 1: Composition of the sample material

	Carbon [%]	Hydrogen [%]	Nitrogen [%]	Calorific Value [kJ/kg]
Paper	40.41	6.70	0.61	16277
Cotton cloth	41.81	6.80	0.35	16484
Oil (arecal)	78.7	12.1	0.01	35911
Oil (WD 40)	85.7	13.3	0.04	42258

The product "arecal" is a cutting oil based on rape oil while "WD40" is a cutting oil derived from fossil hydrocarbons. Pre tests in the differential reactor described in 2.2 showed that arecal is far more reactive than WD40. That may be explained by the high content of unsaturated carbon bonds in rape oil, which are more reactive in terms of oxidation processes. Further investigations on the oxidation behavior have been carried out only with the more reactive arecal oil.

2.2 Method

The measurement of the low temperature oxidation kinetics is performed in a differential reactor (Figure 1), which is characterized by a small amount of sample material, nearly isothermal conditions and very low turnover rates. Due to the fact that chemical oxidations reactions at low temperatures show very low turnover rates themselves, the experimental setup is laid out as a cycle process. The oxidation kinetics is measured by the analysis of the carbon monoxide (CO) and carbon dioxide (CO_2) production rate at different temperatures. Usually a temperature range from 40°C up to 120°C is investigated.

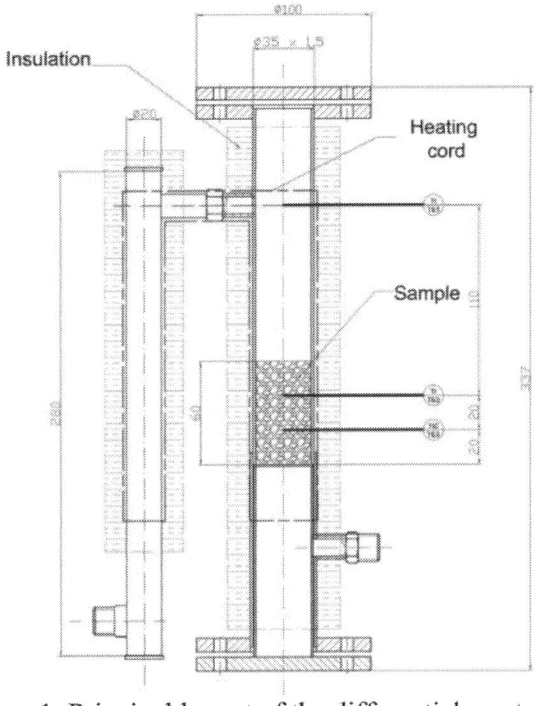

Figure 1: Principal layout of the differential reactor [2]

Figure 2 shows a typical of temperature development and gas production (e.g. carbon dioxide). At the beginning of the measurements the sample material inside the reactor is heated up above the highest planned measuring temperature to desorb carbon dioxide and carbon monoxide that may be adsorbed at the surface of the sample material. After the desorption process the sample material is cooled to the first measuring temperature and constantly purged with synthetic air (80%N_2, 20%O_2). When the sample reaches a constant temperature value the measurement process is started and the measuring cycle is closed. The areas of constant CO_2 increase in the concentration graph of Figure 2 represent these measurement cycles at different temperatures. At higher temperatures an increase of carbon dioxide production can be observed, which is directly linked to oxidation kinetics. The calculation of the kinetic parameters is carried out by the Arrhenius relation after the measuring procedure. A measuring cycle normally takes about 30min for one temperature.

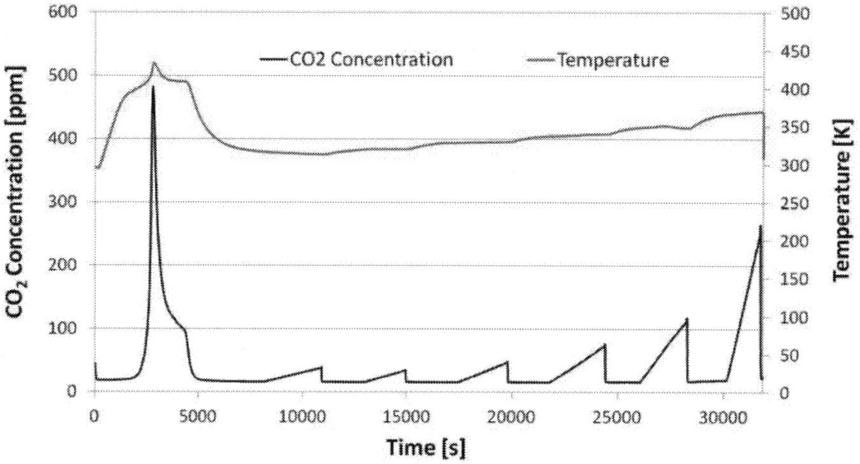

Figure 2: Principal layout of the differential reactor

2.3. Paper

Figure 3 below shows the CO production rates of paper with different oil contents. The "0% Oil content" line represents paper material not contaminated with arecal oil. The CO production rate increases with the oil content of the sample. The progress of the CO production rate with an oil content of 6.9% is widely comparable to the results of garage waste (~15% oil content).

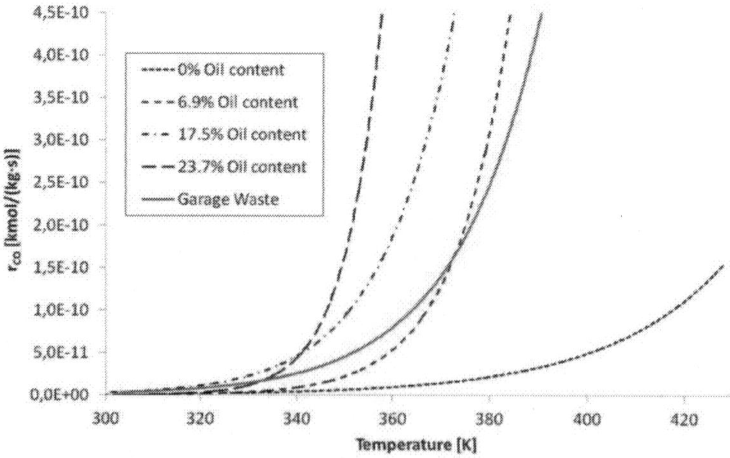

Figure 3: CO production rates of paper with different oil contents

For CO_2 production, as shown in Figure 4, the results were a bit different. The CO_2 production rate of garage waste was higher than all other rates measured. Interestingly the CO_2 production rate for the paper with an oil content of 17.5% was higher than for 23.7%. That may be explained by kinetic effects and is maybe linked to the significant amounts of CO that were produced by this paper sample.

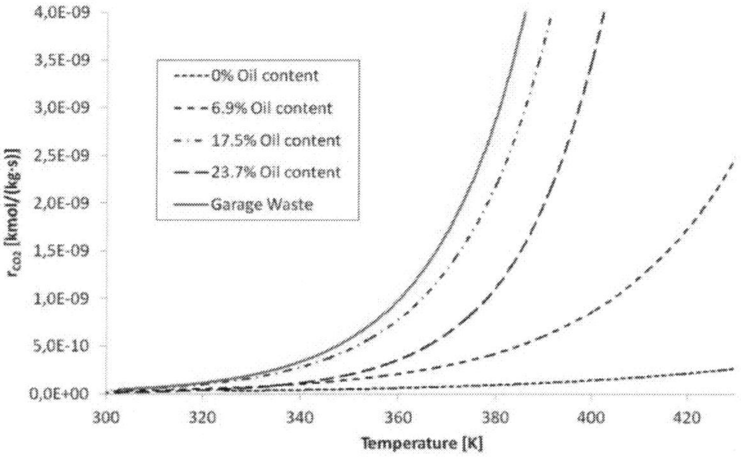

Figure 4: CO_2 production rates of paper with different oil contents

The kinetic data resulting from the calculations based on the Arrhenius relation are presented in Table 2.

Table 2: Kinetic data of the paper sample material

	CO		CO_2	
	E_a [kJ/mol]	K_0 Kmol/kg*s	E_a [kJ/mol]	K_0 [Kmol/kg*s]
Paper 0.0% Oil cont.	4.89E+04	3.25E-09	2.52E+04	9.74E-12
Paper 6.9% Oil cont.	8.70E+04	7.45E-03	3.45E+04	6.63E-10
Paper 17.5% Oil cont.	6.69E+04	3.17E-05	4.94E+04	3.84E-07
Paper 23.7% Oil cont.	1.02E+05	1.91E-00	5.48E+04	1.04E-06
Garage waste	5.59E+04	3.35E-07	5.22E+04	1.14E-06

2.4 Cotton cloth

During the preparation of the cotton sample material oil contents that were similar to the paper samples have been tried to be reached. Although there are smaller deviations in the oil content, the results are comparable.

Figure 5 shows the results for the investigation of the cotton cloth material. The results do not show good accordance with the garage waste material. But again it can be observed, that the sample materials with a high oil contents also show higher reactivity than the sample with lower contents.

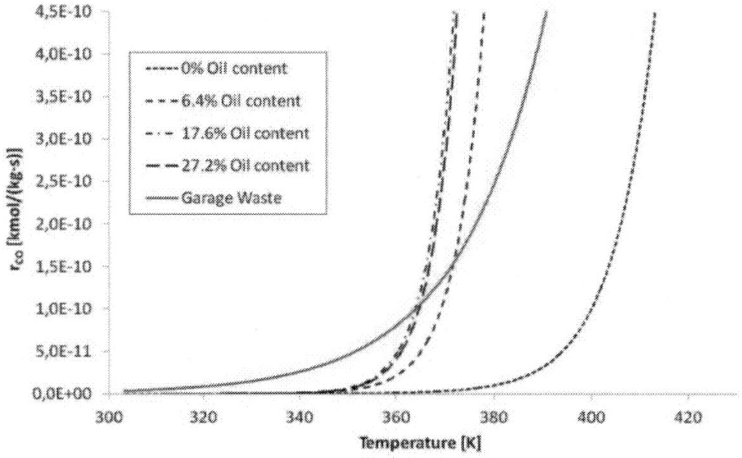

Figure 5: CO production rates of cotton cloth with different oil contents

The production rates from Figure 6 do not show a big deviation regarding the oil content, but the samples contaminated with oil clearly show a higher reactivity than the cotton material itself. In general all CO_2 production rates are comparable with the original garage waste sample material, which again indicates a strong link between oil content and reactivity.

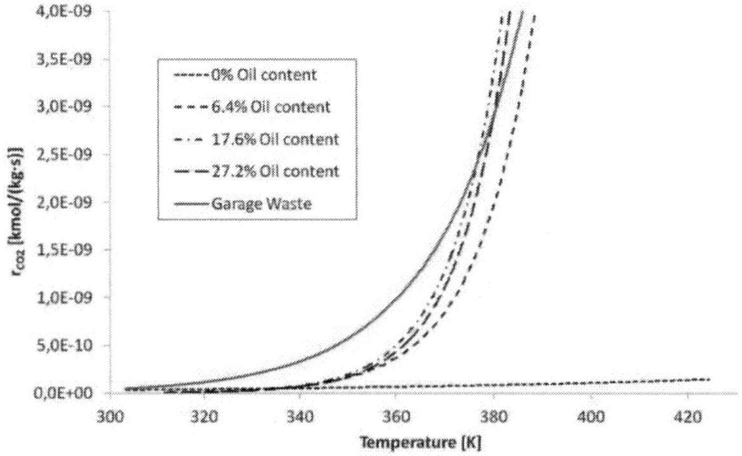

Figure 6: CO_2 production rates of cotton cloth with different oil contents

The kinetic data of the cotton cloth samples is presented in Table 3.

Table 3: Kinetic data of the cotton cloth sample material

	CO		CO_2	
	E_a [kJ/mol]	K_o Kmol/kg*s	E_a [kJ/mol]	K_o [Kmol/kg*s]
Cotton 0.0% Oil cont.	3.68E+04	1.66E-10	1.34E+04	1.99E-13
Cotton 6.4% Oil cont.	1.90E+05	2.90E+12	9.01E+04	1.52E-01
Cotton 17.6% Oil cont.	1.27E+05	3.14E+04	9.96E+04	4.82E+00
Cotton 27.2% Oil cont.	1.02E+05	1.80E+00	5.48E+04	9.78E-07
Garage waste	5.59E+04	3.35E-07	5.22E+04	1.14E-06

2.5 *Influence of iron dusts*

Due to normal workshop operations garage wastes always contain significant amounts of metal residues. To investigate the influence of metal residues on the oxidation kinetics small amounts of iron dust were added to a cotton cloth sample with an oil content of around 11%.

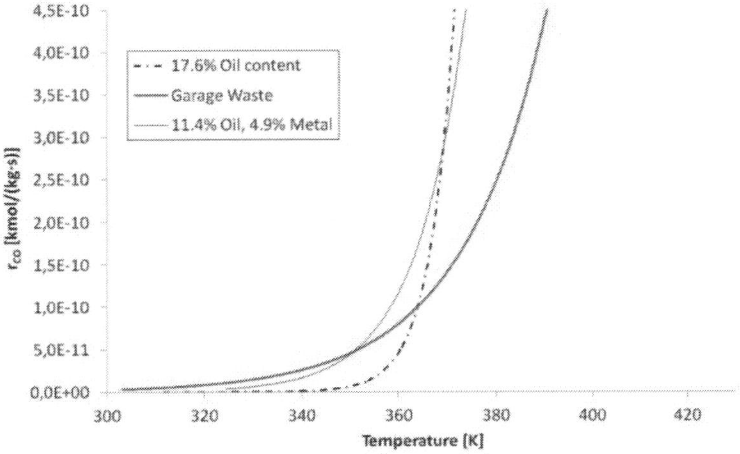

Figure 7: CO production rates of cotton cloth contaminated with oil and iron dust

Figure 7 shows the production rates for the cotton/oil/iron dust sample and also the results for pure cotton cloth with 17.6% oil content. The results show that the iron dust has an influence on the reactivity of the material. Both materials show comparable CO production rates even though the sample containing iron dust contains less oil. The same effect can be observed for the CO_2 production rate in Figure 8. The CO_2 production rate in general shows good accordance to the results of the garage waste sample which indicates the representativeness of the chosen sample materials.

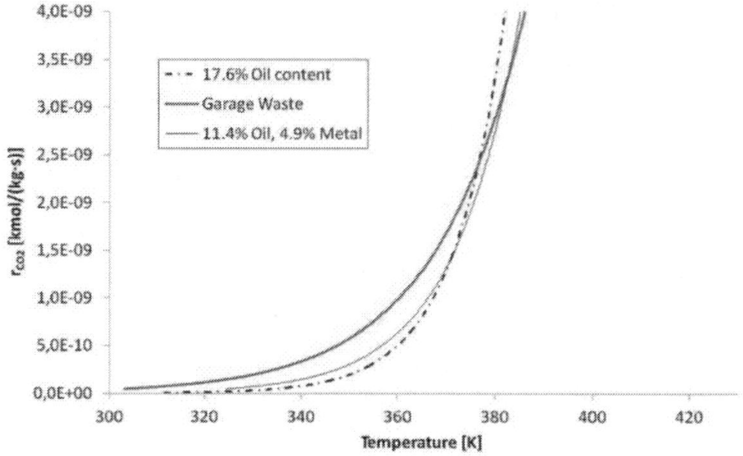

Figure 8: CO_2 production rates of cotton cloth contaminated with oil and iron dust

3. Conclusions

The results of the investigations of garage waste and different reference materials show a clear connection between the oil content of garage waste and its reactivity. Garage waste in general should be considered as highly reactive. Field investigations show, that a significant temperature rise can be observed within days from the beginning of the storage period. High contents of oil and metal residues lead to a significant fire risk. To determine the permissible storage conditions for garage waste CFD models could be applied. The results from measured kinetics can be used as the basis for CFD modeling in combination with the layout of the storage area and the further surrounding conditions [4].

4. References

[1] Wallner, S. (2004). Ein Modell zur Untersuchung des Selbsterwärmungsverhaltens von Schüttgütern. Dissertation, Technische Universität Graz

[2] Walkner, R. (2005) Untersuchung des Selbsterwärmungsverhaltens eines Lagers für aufbereiteten Substitut Brennstoff (ASB). Diplomarbeit, Technische Universität Graz

[3] Wallner, S., Walkner, R., Raupenstrauch, H. (2004). Untersuchung des Selbsterwärmungsverhaltens gelagerter Kunststoffmaterialien mittels numerischer Simulation. Endbericht für das FFF- Projekt „Entwicklung neuer Brennstoffqualitäten aus Abfall". Technische Universität Graz

[4] Raupenstrauch, H. (1997) Gasdurchströmte, chemisch reagierende Schüttschichten. Habilitationsschrift, Technische Universität Graz

Estimate Vibration Risk for Relief and Process Piping

Georges A. Melhem, Ph.D.
ioMosaic Corporation
93 Stiles Road | Salem, NH 03079
melhem@iomosaic.com

Prepared for Presentation at
American Institute of Chemical Engineers
2013 Spring Meeting
9th Global Congress on Process Safety
San Antonio, Texas
April 28 – May 1, 2013

UNPUBLISHED

Estimate Vibration Risk for Relief and Process Piping

Georges A. Melhem, Ph.D.
ioMosaic Corporation
93 Stiles Road | Salem, NH 03079
melhem@iomosaic.com

Keywords: Acoustic induced vibration (AIV), flow induced vibration (FIV), vibration risk, resonance, singing relief valve.

Abstract

Current API, AIChE/CCPS, and AIChE/DIERS pressure relief and flare systems guidelines and standards do not formally address vibration risk. They do not offer specific guidance on velocity limitations other than backpressure calculations and they do not offer guidance on acoustic induced or flow induced piping vibration fatigue failure. This paper provides a summary of experience based methods for the estimation of vibration risk in relief and process piping. In addition, this paper extends the applicability of the experience based methods to two-phase flow.

1. Introduction

Fatigue failure of relief and/or process piping caused by vibration can develop due to the conversion of flow mechanical energy to noise. Factors that have led to an increasing incidence of noise vibration related fatigue failures in piping systems include but are not limited to (a) increasing flow rates as a result of debottlenecking which contributes to higher flow velocities with a correspondingly greater level of turbulent energy, (b) frequent use of thin-walled piping which results in higher stress concentrations, particularly at small bore and branch connections, (c) design of process piping systems on the basis of a static analysis with little attention paid to vibration induced fatigue, (e) and lack of emphasis of the issue of vibration in piping design codes. Piping vibration is often considered on an ad-hoc or reactive basis. According to the UK Health and Safety Executive (HSE), 21 % of all piping failures offshore are caused by fatigue/vibration. Typical systems at risk include large compressor recycle systems and high capacity pressure relief depressuring systems. For relief and flare piping, flow induced turbulence and high frequency acoustic excitations are key concerns.

2. Flow Induced Vibration (FIV)

Fluid flow in pipes generates turbulent energy (pressure fluctuations). Dominant sources of turbulence are associated with flow discontinuities in the piping systems (e.g., partially closed valves, short radius, mitered bends, tees or expanders). The level of turbulence intensity is a function of pipe size, fluid density, viscosity, velocity, and structural support. High noise levels are generated by high velocity fluid impingement on the pipe wall, turbulent mixing, and if the flow is choked, shock waves discontinuities downstream of flow restriction. This can lead to high frequency excitation / vibration risk.

3. Acoustic Induced Vibration (AIV)

High frequency acoustic energy is often generated by a pressure reducing device such as a relief valve, control valve, or orifice plate. Acoustic induced piping failure is of a particular concern for safety related systems (e.g. relief and blowdown/depressuring). The severity of high frequency acoustic excitation is primarily a function of the pressure upstream and downstream of the pressure reducing device, pipe diameter, and the fluid volumetric flow. Acoustically induced piping failures are known to occur at non-asymmetric discontinuities in the downstream piping such as small bore and branch connections and welded supports.

4. Additional Causes of Vibration

Additional causes of piping vibration include mechanical excitation, pulsation, vortex shedding, surge or momentum changes associated with valves, cavitation, and flashing. Mechanical excitations are often associated with pipes connected with reciprocating compressors, pumps, or rotating machinery. Such connection machines cause vibration of the pipe and its support structure. Thermowells are intrusive fittings and are subject to static and dynamic fluid forces. Vortex shedding is the dominant concern as it is capable of forcing the thermowell into flow-induced resonance and fatigue failure. This is particularly important at high fluid velocities.

5. Relief and Depressuring Systems

Depressuring systems are often subjected to acoustic energy (rapidly fluctuating pressure forces) generated by flow turbulence which is accentuated by flow restricting devices within the flow path. The magnitude of pressure fluctuations depends on the mass flow rate, speed of sound, and density. For choked flow, intense noise due to large shock discontinuities and pressure fluctuations is generated. The generated noise is non-periodic due to the randomness of the pressure fluctuations. Choked flow typically leads to a wide frequency spectrum with peak values than can exceed 1000 Hz. Vibrations associated with small fittings and branch connections are of special concern because they introduce discontinuities and stress concentration points.

In many situations resonance can onset which can lead to magnification of static piping loads up to a factor of 50 times. The presence of discontinuities such as tees and welded pipe supports can further increase these loads.

6. Noise Generation

A pressure reducing device or relief device controls flow by converting internal energy into kinetic energy. Some energy is converted to heat through friction (viscous forces) by intense turbulence and shock formation. Some of the energy is also transferred to the pipe wall as vibration, and a portion of this is radiated as noise. The primary noise generating mechanism is the confined jet of fluid formed between the upstream and downstream locations. Flow noise can be modeled as noise of a confined jet. As a result, the noise-generation mechanisms are turbulent mixing, turbulence boundary interaction, shock, shock/turbulence interaction and flow separation.

Since the noise is generated downstream of a flow restriction, most of the acoustic energy is radiated to the downstream piping, which becomes the transmitting medium. As the noise travels downstream along the inside of the pipe in the fluid it radiates through the pipe wall along its entire length.

7. How to Calculate Sound Power Level

We can calculate the sound power level in a fundamental way by assigning an efficiency factor to provide the fraction of flow mechanical energy that is converted to noise:

$$W = \eta \frac{1}{2} \dot{m} u^2 \tag{1}$$

$$L_W = 10\log_{10}\left[\frac{W}{10^{-12}}\right] = 10\log_{10} W + 120 = 10\log_{10}\left[\eta \frac{1}{2} \dot{m} u^2\right] + 120 \tag{2}$$

where W is the flow mechanical power or energy, L_W is the sound power level in dB, and η is defined as the acoustical efficiency factor. If the flow is choked, then W becomes:

$$W = \eta \frac{1}{2} \dot{m} u_{sonic}^2 \tag{3}$$

If a value of η can be estimated for single and multi-phase flow, then the sound power level can be easily calculated not only for pressure reducing devices but also for pipe flow. Attenuation due to friction and temperature changes can be calculated from pipe flow equations in a more detailed manner. Computer codes such as SuperChems can then calculate the sound power level at every axial location for flow piping for single and multi-phase flow. For incompressible flow, the value of flow velocity for an ideal nozzle can be calculated from the mechanical energy balance:

$$u = \sqrt{\frac{2}{\rho_l} \Delta P} \tag{4}$$

Substituting the above equation for u in Equation 1 yields the following Equation for W for liquid flow through an ideal nozzle:

$$W = \eta \dot{m} \frac{\Delta P}{\rho_l} \tag{5}$$

For liquid flow, a typical value of η is approximately 10^{-8}.

All gas flow acoustic efficiencies have been reported to approach 1 % of the total flow mechanical energy for rockets[i]. This is illustrated in Figure 1. When the acoustic efficiency is plotted against the flow mechanical power Figure 1 is obtained. The measured curve indicates that the acoustic efficiency falls off as the flow mechanical energy gets larger while the calculated curve[ii] indicates increasing acoustic efficiencies.

Figure 1. Acoustic efficiency trends

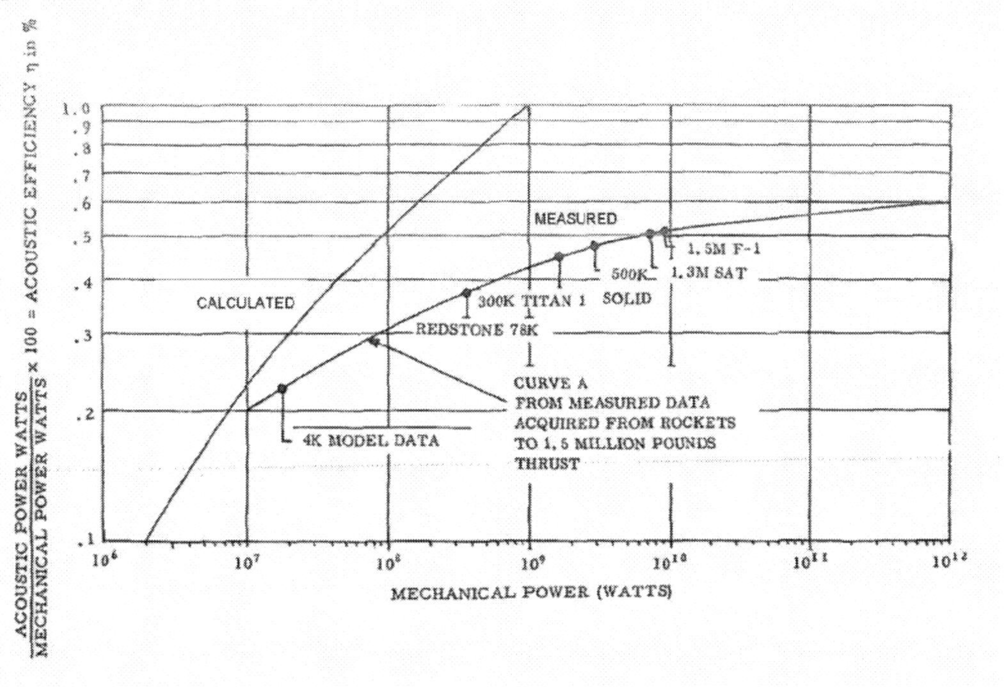

Two-phase flow sound power level can also be calculated using Equation 1. Recent work by Singh et al.[iii, iv] demonstrated that more attenuation of noise is exhibited by two-phase flow. Therefore, using the gas acoustic efficiency values will over predict sound power and noise levels for two-phase flow.

8. Existing Methods and Guidance

Current API, AIChE/CCPS, and AIChE/DIERS pressure relief and flare systems guidelines and standards do not formally address vibration risk. They do not offer specific guidance on velocity limitations other than backpressure calculations and they do not offer guidance on acoustic induced or turbulence induced piping vibration fatigue failure. The Marine Technology Directorate Limited (MTD) has published in 1999 "Guidelines for the Avoidance of Vibration Induced Fatigue in Process Pipework"[v]. A second edition of these guidelines were published in 2008 by the Energy Institute[vi]. The methods outlined in these guides have been incorporated in the SuperChems Expert ioVIPER modules. The MTD/Energy Institute guidelines provide qualitative and quantitative methods for the assessment of piping vibration failure risk and depending on the calculated risk level they provide generic guidance for the mitigation of vibration risk.

Many operating companies have established their own internal guidance for evaluating and minimizing piping vibration risk. Although these criteria vary from company to company, they all in general include a limit of flow velocity in some form.

A common criteria used in relief systems is to limit the value of flow Mach number to a value ranging from 0.3 to 0.9:

$$0.3 \leq M = \frac{u}{u_{sonic}} \leq 0.9 \tag{6}$$

Another common criteria often encountered is to limit the dynamic pressure component or kinetic energy of a flow stream to a value of 100,000 Pascals for gas flow:

$$\rho u^2 \leq 1 \times 10^5 \ \text{kg/m/s}^2 \tag{7}$$

and 50,000 for two-phase flow:

$$\rho u^2 \leq \frac{1}{2} \times 10^5 \ \text{kg/m/s}^2 \tag{8}$$

The 2006 fifth edition of the NORSOK Process Design Standard P-001 limits the flow velocities for all flare lines to $\rho u^2 \leq 200{,}000 \ \text{kg/m/s}^2$ for single and multiphase flow. The piping for flare headers and sub-headers are designed for a maximum Mach number of 0.6 and lines downstream of pressure relief valves to the first sub-header are designed for a maximum Mach number of 0.7. For process lines the maximum design velocity for gas pipes is limited to 60 m/s or $u \leq 175 \frac{1}{\rho^{0.43}}$ whichever is less. The maximum velocity for two-phase lines is limited to $u \leq 183 \frac{1}{\rho^{0.5}}$ or:

$$\rho u^2 \leq 183^2 \leq 33{,}489 \ \text{kg/m/s}^2 \tag{9}$$

9. The Method of Carucci and Mueller

Carucci and Mueller have published guidance for the estimation of sound power levels for control valves and pressure reducing devices[vii, viii].

$$L_W = 10 \times \log_{10}\left[\left(\frac{P_1 - P_2}{P_1}\right)^{3.6} \times \dot{m}^2 \times \left(\frac{T_1}{M_w}\right)^{1.2}\right] + 126.1 \tag{10}$$

$$=10\times\log_{10}\left[4\times\left(\frac{P_1-P_2}{P_1}\right)^{3.6}\times\dot{m}^2\times\left(\frac{T_1}{M_w}\right)^{1.2}\right]+120 \tag{11}$$

$$=36\log_{10}\left(\frac{P_1-P_2}{P_1}\right)+20\log_{10}(\dot{m})+12\log_{10}\left(\frac{T_1}{M_w}\right)+126.1 \tag{12}$$

where P_1 is the upstream pressure or source pressure, P_2 is the downstream pressure, T_1 is the source temperature, $\dot{m}$ is the mass flow rate, and M_w is the gas molecular weight.

Attenuation of noise due to friction and heat conduction losses is estimated from the following equation:

$$L_{W,At}=0.06\frac{L}{D_i} \tag{13}$$

where L is the pipe length and D_i is the internal pipe diameter. For an L/D of 50 the attenuation loss is 3 dB. Abrupt changes in flow area (expansion) in the piping can also lead to attenuation. The decrease in sound power level can be estimated from the following equation:

$$L_{W,Ex}=2\left(\frac{D_2}{D_1}-1\right) \tag{14}$$

where $D_2 > D_1$. A 3 dB reduction is typically applied to the flow leaving a tee in each direction or entering into a header or a large drum or vessel. The Carucci and Mueller Equation 12 is used by the Energy Institute Guidelines for the assessment of failure likelihood for high frequency acoustic excitation [vi]. Note that the Carucci and Mueller equation cannot be easily applied to complex piping systems, multi-phase systems, and relief piping with multiple chokes. Some companies also add 6 dB to the sound power level estimate when sonic flow exists at a branch connection to account for amplified dynamic strain response.

9.1 Analysis of the Carucci and Mueller Data Set

Careful analysis of the Carucci and Mueller data set (see Table 1) indicates that the source pressures were high enough to produce choked (sonic) flow through a flow limiting orifice or valve upstream of the failure point further downstream, often at a branch or line connection. This important point was missed by Eisinger who used the downstream (non-choked) flow velocity to establish his Mach number based failure criterion. We were also able to reproduce Eisinger's estimates of the original Carucci and Mueller data set based on his published paper.

Carucci and Mueller provided enough data in their original paper to allow the calculation of the upstream flow limiting flow area and choke point conditions (this is possible because they reported the actual flow rates) as well as the conditions downstream of the flow limiting device in the discharge piping. Note that the sound power level estimate upstream of the choke points should be based on the pressure difference (or pressure ratio) of the source pressure and the choke point. The sound pressure level downstream of the choke point (the primary cause of acoustic induced fatigue failure in downstream piping of the choked point) should be based on the difference (or pressure ratio) of the choke point and the shock discontinuity pressure downstream of the choke point. The original equation proposed by Carucci and Mueller used the pressure difference (or ratio) between the source pressure and the exit pressure downstream of the choke point. This yields an overestimate of the sound pressure. This overestimate is somewhat tempered by the fact that the pressure difference ratio to the source pressure is raised to the 3.6 power.

The following information was provided by Carucci and Mueller regarding the five failures and two high vibrations cases observed in Table 1.

A. Failure occurred during startup. Sonic velocity is achieved at the 6 inch branch connection of a 24 inch pipe run downstream of the recycle valve. The sound power level estimate considers the combined acoustic energy generated by the letdown valve and the sonic condition at the branch connection. The upstream pressure at the branch connection was estimated to be 98.8 psia.

B1/B2. This system consisted of a control valve letting down pressure to a safety valve / flare header system. The failure occurred after five to ten hours of operation as a crack at a 10 inch branch connection to 28 inch header. Sonic velocity was achieved at both the control valve and the 10 inch branch connection to the 28 inch flare header. As such, two pertinent data points are established:

1. Acoustic energy within the 10 inch line as generated by valve (not the failure point), and
2. Combined acoustic energy within the 28 in header as generated by valve and sonic condition at the 10 in branch connection (the failure point).

C. High vibrations, no failures.

D. Failure during startup. Cracks were observed in a 24 inch line downstream of the compressor recycle valve after twelve hours of operation. The cracks were near the branch connections of a 6 inch line and a 3/4 inch drain valve and at an I-beam support welded to the pipe at the elbow immediately downstream of the control valve.

E. High vibration, no failures.

F. Failure at severely undercut weld on a 300 mm (12 in.) line made to fasten conduit support clip. Points of high stress concentration were later eliminated.

G. This system consisted of six, parallel, high pressure steam letdown valves, each with a downstream, three pass contra-flow attemporator and an in-line silencer. Failure of the 18 inch attemporator shells occurred after four hundred hours of operation.

H. This system included four parallel desuperheaters located downstream of control valves. Several cracks developed in this system after two to three months of operation. Two to three of the four letdown valves were discharging steam through a 10 inch connection to a 20 inch header which swaged up to a 30 inch diameter. Longitudinal cracking occurred at the bottom of the 20 inch header at a transverse guide 0.5 meters downstream of the fourth 10 inch branch connection. In addition, a 1 inch bypass line for a block valve downstream of the third letdown valve cracked and the 20 inch header cracked around a pressure tap downstream of the transverse guide.

Table 1 compares the acoustic efficiencies implied in the Carucci and Mueller method (at the upstream choke point) vs. acoustic efficiencies established using IEC methods described later on in this paper. The acoustic efficiencies implied by the Carucci and Mueller method at the upstream choke point ranged from 0.1 to 5 percent while the IEC acoustic efficiency ranged from 0.67 to 1 percent. The Carucci and Mueller method is based on acoustic energy theories encompassing both jet and choked flow noise as well as test data from work performed by Exxon research and engineering. As discussed in later sections, the experience based failure criteria originally developed by Carucci and Mueller can only be used with sound power levels calculated by Equation 12.

The implied acoustic efficiency by Equation 12 is proportional to the mass flow rate and inversely proportional to u^2:

$$\eta = 8\left(1-\frac{P_2}{P_1}\right)^{3.6}\left(\frac{T_1}{M_w}\right)^{1.2}\left(\frac{\dot{m}}{u^2}\right) \tag{15}$$

If we consider the case of an ideal gas undergoing isentropic choked (sonic) flow through a restriction orifice, we can establish the following expressions for variables of interest at the choke point:

$$T_2 = 2\frac{T_1}{\gamma+1} \tag{16}$$

$$P_2 = P_1\left(\frac{2}{\gamma+1}\right)^{\frac{\gamma}{\gamma-1}} \tag{17}$$

$$\rho_2 = \frac{P_2 M_w}{R_g T_2} \tag{18}$$

$$u_2 = \sqrt{\frac{\gamma R_g T_2}{M_w}} \tag{19}$$

$$\dot{m} = \rho_2 u_2 A_o \tag{20}$$

Substituting the values of mass flow rate $\dot{m}$, choked flow velocity u_2, and choke pressure P_2 in Equation 15, we obtain the following simplified expression for the Carucci and Mueller acoustic efficiency after some algebraic manipulations:

Table 1. Vibration risk data set used by Carucci and Mueller and reproduced by Melhem

CASE	SERVICE	T_1, K	P_1, bara	T_2, K	P_2, bara	D_2, m	u2, m/s	Exit Mach Number, M_2	Mass Flow Rate, kg/s	Heat Capcity Ratio, C_p/C_v	Restriction Orifice / Valve Flow Speed, m/s	Carucci and Mueller Acoustic Efficiency, %	IEC Acoustic Efficiency, %	PWL_cm, dB	PWL_mel, dB	PWL_cm - PWL_mel, dB	PWL_mel (IEC), dB
A	Propylene Compressor Recycle	319.26	20.55	238.91	1.59	0.610	32.50	0.137	29.74	1.13	224.30	0.65	0.85	168.5	165.0	3.5	161.5
B1	Natural Gas Pressure	318.71	42.75	226.00	6.27	0.254	125.60	0.399	48.26	1.22	322.95	2.54	0.91	170.9	168.1	2.8	163.6
B2	Natural Gas Letdown to Flare	318.71	5.17	270.38	2.07	0.711	58.20	0.169	48.26	1.22	357.00	0.41	0.69	171.4	168.9	2.5	163.3
C	Natural Gas Letdown to Flare	308.15	47.57	195.68	4.14	0.406	64.40	0.218	47.63	1.23	311.29	4.17	0.95	171.8	169.8	2.0	163.4
D	Nitrogen Compressor Recycle	310.93	6.96	280.31	3.72	0.610	93.80	0.297	119.70	1.20	324.76	1.05	0.79	168.3	168.2	0.1	167.0
E	Nitrogen Compressor Recycle	310.93	16.82	267.59	7.03	0.457	60.10	0.194	84.80	1.21	328.35	0.79	0.66	168.8	165.6	3.3	164.8
F	Natural Gas Compressor Recycle	330.37	51.09	220.53	13.86	0.305	28.80	0.093	35.28	1.45	272.52	0.59	0.82	164.9	159.5	5.4	160.3
G	Desuperheater Steam	602.59	99.97	480.32	10.34	0.457	49.80	0.093	36.89	1.29	619.65	2.91	0.94	175.5	173.1	2.4	169.2
H	Desuperheater Steam	658.15	43.09	476.11	10.34	0.508	49.60	0.093	45.99	1.29	586.00	0.86	0.83	163.9	168.3	-4.4	168.9
1	Process Gas Compressor Recycle	308.15	13.17	204.98	1.03	0.406	122.80	0.461	27.09	1.19	256.81	2.22	0.95	165.9	164.2	1.7	160.6
2	Process Gas Compressor Recycle	311.48	38.75	265.88	10.00	0.305	14.20	0.048	13.86	1.13	258.54	0.32	0.83	156.6	151.8	4.8	155.9
3	Propylene Compressor Recycle	316.48	20.55	262.14	3.24	0.406	15.70	0.065	12.60	1.11	209.52	0.61	0.91	156.0	153.5	2.5	154.0
4	Process Gas Compressor Recycle	300.37	2.48	257.91	1.03	0.508	145.60	0.450	34.78	1.21	325.00	0.26	0.77	161.3	156.7	4.6	161.5
5	Process Gas Compressor Recycle	310.93	15.86	215.12	2.21	0.406	68.70	0.226	25.33	1.23	357.01	1.21	0.91	165.2	162.9	2.3	161.7
6	Process Gas Compressor Recycle	300.93	39.44	250.67	15.51	0.254	26.10	0.075	20.66	1.24	330.10	0.20	0.78	158.3	153.6	4.7	159.4
7	Propylene Compressor Recycle	310.93	16.69	233.88	1.38	0.356	131.30	0.575	38.05	1.13	216.35	3.52	0.95	166.6	165.4	1.2	159.3
8	Propylene Compressor Recycle	310.93	16.69	234.72	1.38	0.508	59.20	0.259	35.08	1.13	216.35	3.61	0.95	165.9	164.7	1.2	158.9
9	Waste Steam Dump	403.71	2.76	370.82	1.03	0.406	245.10	0.566	18.90	1.10	436.60	0.15	0.77	160.5	154.4	6.1	161.4
10	Process Gas Compressor Recycle	302.59	10.69	207.90	1.03	0.406	95.80	0.357	20.92	1.19	295.15	1.49	0.94	163.2	161.3	1.9	159.3
11	Process Gas Compressor Recycle	308.15	54.81	263.43	27.51	0.305	7.20	0.016	9.15	1.29	441.77	0.10	0.79	150.5	149.7	0.8	158.5
12	Natural Gas Compressor Recycle	308.15	3.79	282.54	2.48	0.914	28.40	0.075	39.69	1.26	315.99	0.19	0.73	155.4	155.7	-0.3	161.6
13	Natural Gas Compressor Recycle	308.15	6.48	279.42	4.00	0.914	17.30	0.046	39.44	1.25	334.77	0.24	0.75	157.2	157.2	0.0	162.2
14	Natural Gas Compressor Recycle	308.15	11.45	282.82	6.76	0.762	14.40	0.039	38.93	1.19	346.05	0.28	0.76	158.1	158.1	0.0	162.5
15	Natural Gas Compressor Recycle	308.15	21.79	266.72	11.65	0.660	10.10	0.027	36.04	1.30	359.11	0.33	0.78	159.4	158.8	0.6	162.6
16	Natural Gas Compressor Recycle	308.15	40.75	272.56	21.99	0.610	6.30	0.017	35.91	1.25	344.77	0.34	0.78	159.2	158.6	0.6	162.2
17	Natural Gas Compressor Recycle	308.15	80.12	271.44	40.89	0.508	4.80	0.013	35.41	1.23	324.89	0.40	0.78	160.1	158.8	1.3	161.6
18	Natural Gas Compressor Recycle	308.15	172.99	240.53	79.98	0.406	3.80	0.010	35.41	1.47	326.41	0.71	0.82	161.5	161.3	0.2	161.9
19	Natural Gas Compressor Recycle	308.15	3.86	280.10	2.34	0.914	27.60	0.074	37.17	1.24	340.74	0.24	0.75	157.3	157.1	0.2	162.1
20	Natural Gas Compressor Recycle	308.15	6.21	281.91	3.93	0.914	16.50	0.044	36.79	1.24	325.37	0.20	0.74	155.9	155.9	0.0	161.6
21	Natural Gas Compressor Recycle	308.15	10.62	274.09	6.27	0.762	14.30	0.038	36.04	1.29	345.37	0.26	0.76	157.4	157.4	0.0	162.1
22	Natural Gas Compressor Recycle	308.15	19.72	266.35	10.62	0.660	10.10	0.027	33.01	1.31	359.23	0.30	0.78	158.4	158.0	0.4	162.2
23	Natural Gas Compressor Recycle	308.15	36.82	263.14	19.79	0.610	6.40	0.017	32.89	1.34	346.54	0.31	0.78	158.4	157.9	0.5	161.9
24	Natural Gas Compressor Recycle	308.15	64.47	270.22	36.96	0.508	4.90	0.013	32.82	1.31	319.94	0.32	0.77	157.3	157.2	0.1	161.1
25	Natural Gas Compressor Recycle	308.15	139.48	320.30	64.54	0.406	4.20	0.012	32.76	0.95	311.67	0.56	0.80	160.7	159.5	1.2	161.1
26	Propane Compressor Recycle	308.15	12.07	234.24	0.97	0.457	143.50	0.644	50.40	1.12	217.76	4.62	0.96	168.9	167.6	1.3	160.6
27	Steam Desuperheater	558.15	43.09	522.78	10.34	0.762	22.00	0.041	45.99	1.19	554.55	0.51	0.83	161.2	168.3	-7.1	167.9

$$\eta = (4.7950\gamma - 2.5882)\left(\frac{M_w}{T_1}\right)^{0.3} P_1 A_o \tag{21}$$

where A_o is the effective restriction orifice flow area in m^2, P_1 is the upstream pressure in bara, T_1 is the upstream temperature in Kelvin, and η is the calculated Carucci and Mueller acoustic efficiency in percent. This expression shows that the Carucci and Mueller acoustic efficiency is directly proportional to upstream pressure, and the restriction orifice flow area (flow rate). It can be shown that the Carucci and Mueller acoustic efficiency produces unrealistic values for high pressure systems, large flow area, and/or mixtures with high molecular weights.

If we consider the case of methane discharging through a restriction with an effective flow area of 0.1 m^2, upstream temperature of 373.15 K, and an upstream pressure of 100 bara, the calculated acoustic efficiency is:

$$\eta = (4.7950 \times 1.282 - 2.5882)\left(\frac{16}{373.15}\right)^{0.3} \times 100 \times 0.1 \tag{22}$$

$$= 3.559 \times 0.388 \times 100 \times 0.1 = 13.8\% \tag{23}$$

The same value can also be calculated by calculating $\dot{m}$, u_2, P_2, and η directly from the above ideal gas flow equations and Equation 15:

$$T_2 = 2\frac{T_1}{\gamma+1} = 2\frac{373.15}{1.282+1} = 327.05 \text{ K} \tag{24}$$

$$P_2 = P_1\left(\frac{2}{\gamma+1}\right)^{\frac{\gamma}{\gamma-1}} = 100\left(\frac{2}{1.282+1}\right)^{\frac{1.282}{1.282-1}} = 54.9 \text{ bara} \tag{25}$$

$$\rho_2 = \frac{P_2 M_w}{R_g T_2} = \frac{54.9 \times 100000 \times 16}{8314 \times 327.03} = 32.306 \text{ kg/m}^3 \tag{26}$$

$$u_2 = \sqrt{\frac{\gamma R_g T_2}{M_w}} = \sqrt{\frac{1.282 \times 8314 \times 327.03}{16}} = 466.76 \text{ m/s} \tag{27}$$

$$\dot{m} = \rho_2 u_2 A_o = 32.306 \times 466.76 \times 0.1 = 1507.63 \text{ kg/s} \tag{28}$$

$$\eta = 8\left(1 - \frac{P_2}{P_1}\right)^{3.6}\left(\frac{T_1}{M_w}\right)^{1.2}\left(\frac{\dot{m}}{u^2}\right) \tag{29}$$

$$= 8 \times (1 - 0.549)^{3.6} \times \left(\frac{373.15}{16}\right)^{1.2} \times \left(\frac{1507.63}{466.76^2}\right) \tag{30}$$

$$= 8 \times 0.0568 \times 43.78 \times 69.2 \times 10^{-4} \tag{31}$$

$$= 0.1376 \text{ or } 13.76\% \tag{32}$$

A similar equation to 21 can be derived for the acoustic efficiency downstream of the choke point:

$$\eta = (68.63 - 4.836\gamma)\left[1 - (1.0397 + 0.6096\gamma)\frac{P_b}{P_1}\right]^{3.6}\left(\frac{M_w}{T_1}\right)^{0.3} P_1 A_o \tag{33}$$

where P_b is the superimposed backpressure downstream of the choke point. Applying Equation 33 to the same example above we calculate an acoustic efficiency of 227 percent at a backpressure of 1 bara:

$$\begin{aligned}\eta &= (68.63 - 4.836\times 1.282)\left[1 - (1.0397 + 0.6096\times 1.282)\frac{1}{100}\right]^{3.6}\left(\frac{16}{375}\right)^{0.3}\times 100\times 0.1 \\ &= 62.430\times 0.9359\times 0.388\times 100\times 0.1 \\ &= 226.70\%\end{aligned} \tag{34}$$

It is evident from the actual data reported by Carucci and Mueller and from the above theoretical proof that the acoustic efficiency used by Carucci and Mueller in their equation can produce unrealistic values, well in excess of 1 percent.

10. Failure Criteria

Several methods are now available for screening and analyzing the potential failure risk of piping caused by vibration. These methods include:

- Experience Based - D/t method or D Method
- Experience Based - Mach number method (Not recommended)
- MTD methods
- Detailed structural dynamics methods

The experience based methods center around correlating likelihood of failure based on actual experience and/or test data. The most widely cited reference is that of Carucci and Mueller (see vii and viii) for steel pipe.

Figure 2 illustrates the failure criteria developed originally by Carucci and Mueller and later modified by Eisinger. This failure criteria establishes a design sound power level vs. the ratio of pipe diameter to thickness. This criteria is fundamentally better than the other criteria based on pipe diameter only since thicker wall pipes are stronger than thinner wall pipes.

The original work by Carucci and Mueller suggests a limit provided by the following equation (Figure 2):

$$L_{W,limit} = 184.6 - 0.215\frac{2D_i}{D_o - D_i} = 184.6 - 0.215\frac{D_i}{\delta} \tag{35}$$

where δ is the pipe thickness.

The lower allowable limit developed by Eisinger also shown in Figure 2 is given by the following equation:

$$L_{W,limit} = 173.6 - 0.125\frac{2D_i}{D_o - D_i} = 173.6 - 0.125\frac{D_i}{\delta} \tag{36}$$

The NORSOK standard published guidance in 2006 using the same equation for $L_{W,limit}$ as proposed by Eisinger above.

Note that the above limits are based on sound power levels calculated using the source pressure and the downstream exit pressure. We have recalculated the sound power levels based on the correct pressure ratios downstream of the choke point. This data is shown in Figures 3, 4, and 5.

Another design limit criterion that was originally proposed by Carucci and Mueller is based on pipe diameter only for steel pipe. This criterion applies to pipe diameters ranging from 10 inch to 36 inch (approximately between 200 mm to 800 mm) and with wall thicknesses ranging from 0.219 in to 0.439 in (approximately 5.5 mm to 10 mm). The design limit is shown in Figure 3 for the corrected data and can be approximated by:

$$L_{W,limit} = 192.8 - 9.8\ln D \tag{37}$$

where D is the nominal pipe diameter in inches and L_W is the sound power limit in dB.

Recent theoretical and experimental work by Chiyoda Engineering[ix] indicates that AIV risk will increase for smaller pipe diameters at the same diameter to thickness ratio. As a result, the limiting criteria established above should be decreased by 6 dB for an 8 inch pipe, or by 20 $\log_{10}$ (16/D).

The design limit correlations developed in this paper can be used with computer programs such as SuperChems Expert which calculates the sound power level at every axial locations for single and multiphase flow to decide if the steel piping inside diameter to thickness ratio exceed the allowable limits as shown by Figure 6.

11. Other Methods for Calculating η

The simplest method for calculating η is to assume the same efficiency as an expanded jet (see Figure 7) and apply it equally to single and multi-phase flow. The efficiency shown in Figure 7 is the same as what is currently used by API for the estimation of flare noise.

The flow regime types are defined by varying jet shapes in the area upstream and downstream of the vena contracta. These jets change their shape when certain differential pressure ratios are exceeded. In regimes II to IV, higher Mach numbers arise downstream of the vena contracta. Yet, M at the vena contracta itself remains unchanged at 1. The IEC method produces equivalent values of efficiency to the data shown in Figure 7 (as $F_L \rightarrow 1.0$) as shown in Figure 9. Note that Figure 9 uses $x = 1 - P_2 / P_1$ as the X axis while Figure 7 uses $P_1 / P_2 = \frac{1}{1-x}$.

Figure 2. Piping failure limit as originally proposed by Carucci and Mueller and by Eisinger for steel pipe

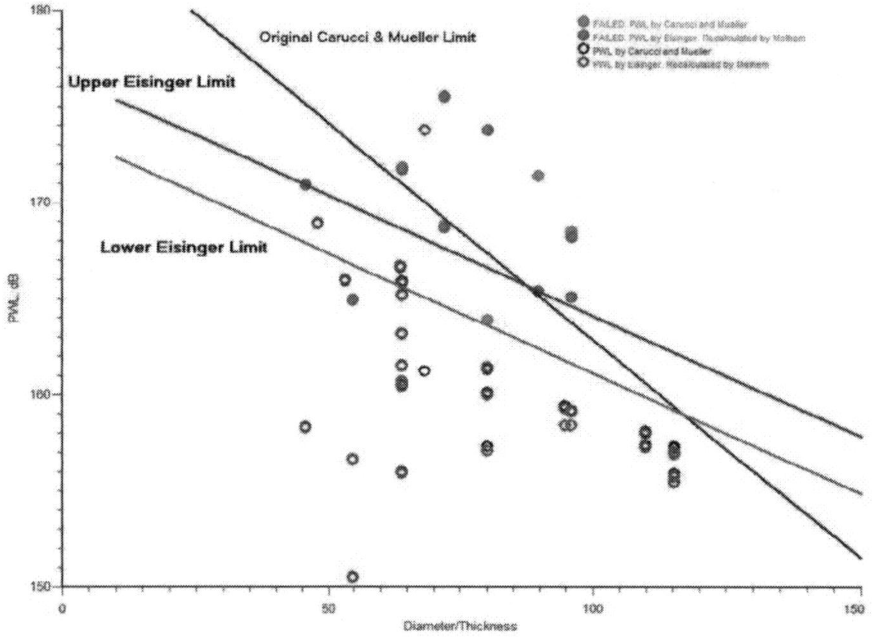

The efficiency is related to the flow pressure ratio and the flow regime as well. IEC calculates the efficiency based on five different flow regimes (see Figure 8) depending on the value of the downstream pressure, P_2:

Regime I The flow is sub-sonic. The sound generation has the character of a dipole jet. The highest Mach number is reached at the vena contracta, not exceeding Mach 1 at the maximum. Downstream of the vena contracta, the jet expands, leading to partial pressure recovery (F_L factor).

Regime II Sonic and supersonic flows exist together, which means that strongly turbulent flow and shock cell structure dominate. Pressure recovery drops until the top limit of regime II is reached.

Regime III The rise in pressure is non-isentropic. The flow is supersonic and shear turbulence predominates.

Regime IV The shock cells disappear and a Mach disk forms. The dominant mechanism is the interaction between shock cells and turbulence.

Regime V The acoustical efficiency is constant.

Figure 3. Piping failure correlation developed by Melhem using Carucci Mueller acoustic efficiency and corrected sound power level estimates for steel pipe. PWL vs. D

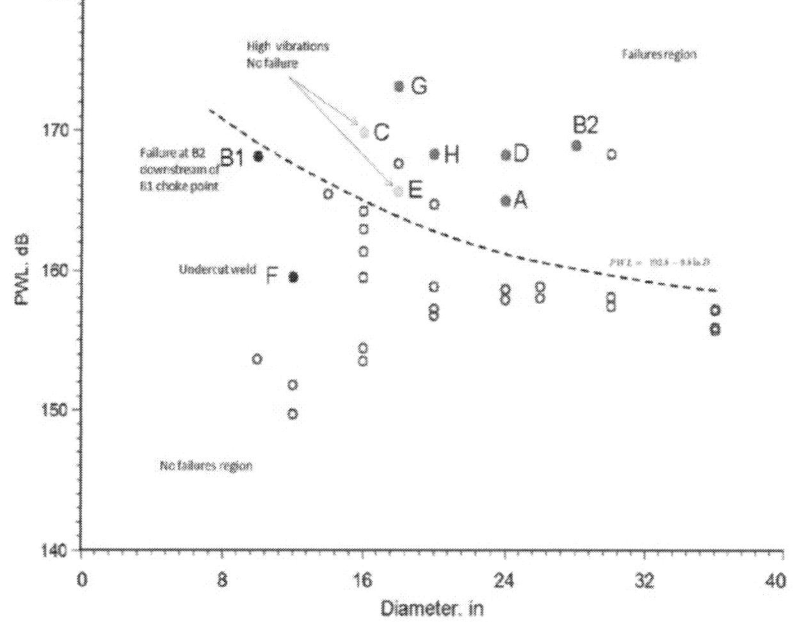

Figure 4. Piping failure correlation developed by Melhem using Carucci Mueller acoustic efficiency and corrected sound power level estimates for steel pipe. PWL vs. D/t

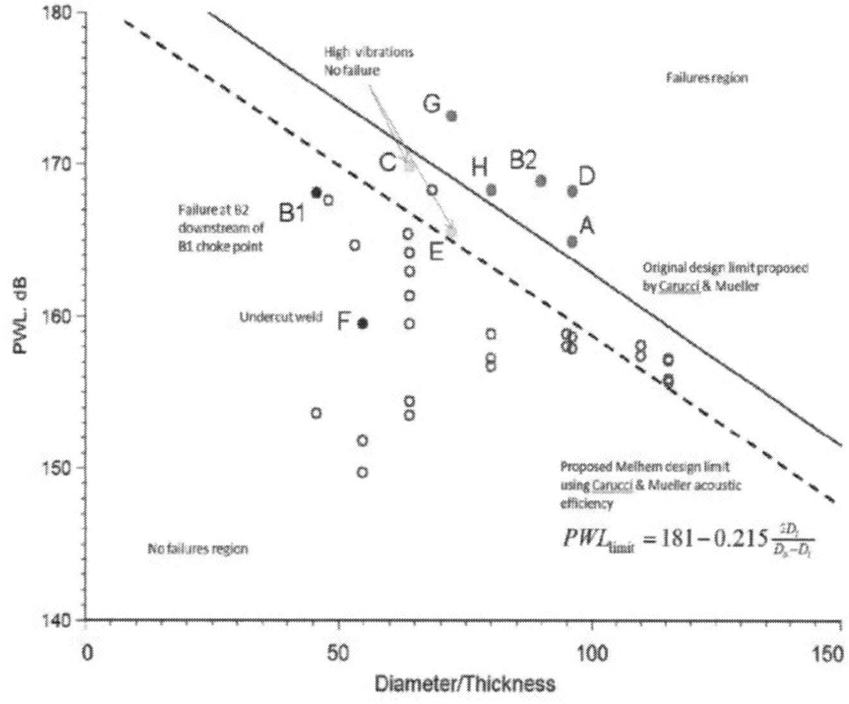

12. Vibration Frequencies

As a fluid moves through piping components there is typically a separation of the fluid from the constraining wall as the fluid changes flow direction. As a result a vortex is formed and then swept into the main stream. This vortex shedding occurs at fairly well defined dimensionless frequencies. The strength of the vortex varies but does not need to be very strong to cause damage especially if the shedding frequency couples with the natural frequency of the piping system. The shedding peak frequency for a vortex for subsonic and sonic flows (Regimes I and II up to a Mach number of 1.4) is given by:

$$f_p = \frac{N_{Str} u}{D} = 0.2\frac{u}{D} \tag{38}$$

Figure 5. Piping failure correlation developed by Melhem using IEC acoustic efficiency and corrected sound power level estimates for steel pipe. PWL vs. D/t

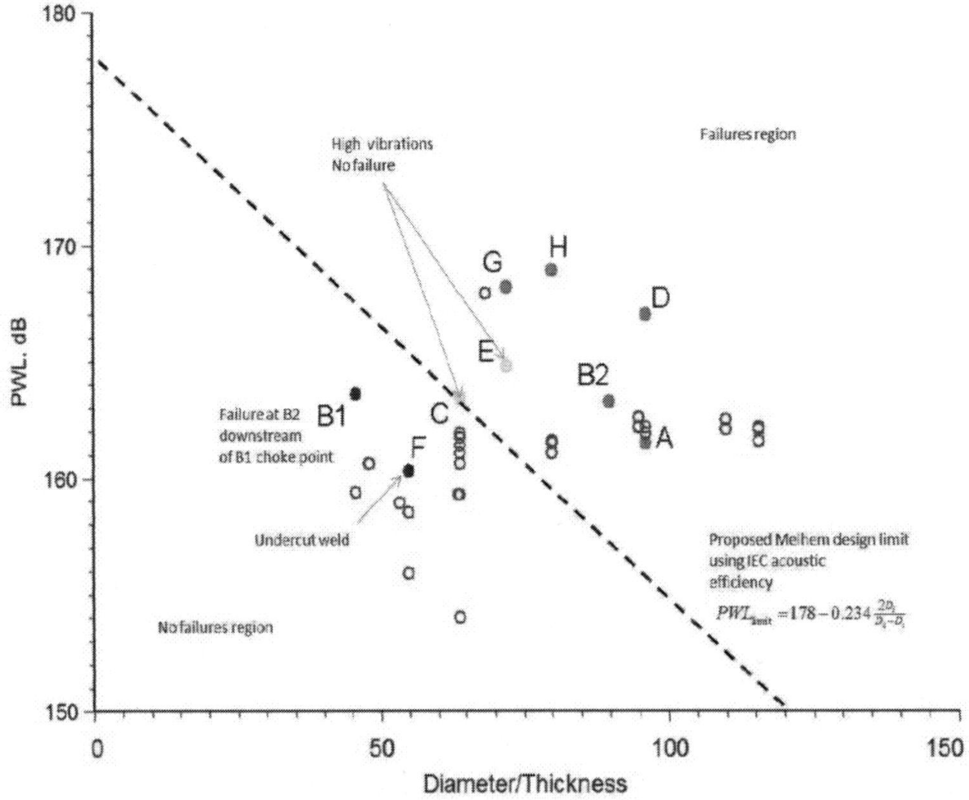

where u is the flow velocity in m/s, D is a characteristic flow dimension (perpendicular to flow), f_p is the peak frequency in Hz and N_{Str} is the Strouhal Number which varies depending upon the geometry causing the separation of the boundary layer. For a circular cylinder its value is 0.2 over a wide range of Reynolds numbers. It is usually between 0.1 and 0.3. Vortex shedding frequencies are generally more than 30 Hz [upper limit for most piping system natural frequencies].

For flows with Mach numbers larger than 1.4 ($M > 1.4$, Regimes III to V), the peak frequency f_p is given by:

$$f_p = \frac{0.4 u_{sonic}}{1.25 D \sqrt{M^2 - 1}} \tag{39}$$

For pipe flow, D is equal to the inside flow diameter. For a control valve, D is given by:

$$D = 0.0046 \sqrt{\frac{C_v F_l}{N_o}} \tag{40}$$

Figure 6. Sample piping sound power level calculated by SuperChems Expert v6.4mp vs. experience based allowable limit for steel pipe

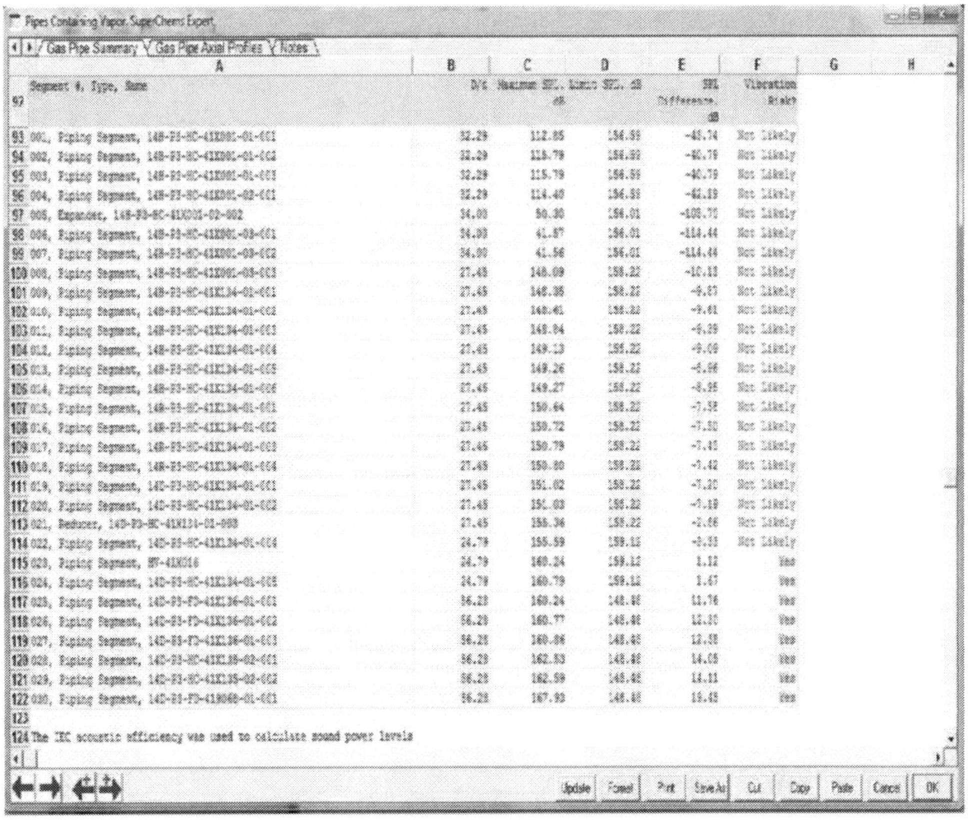

	A	B	C	D	E	F
92	Segment #, Type, Name	D/t	Maximum SPL, dB	Limit SPL, dB	SPL Difference, dB	Vibration Risk?
93	001, Piping Segment, 14B-P3-HC-41X001-01-001	32.29	112.85	156.59	-43.74	Not Likely
94	002, Piping Segment, 14B-P3-HC-41X001-01-002	32.29	115.79	156.59	-40.79	Not Likely
95	003, Piping Segment, 14B-P3-HC-41X001-01-003	32.29	115.79	156.59	-40.79	Not Likely
96	004, Piping Segment, 14B-P3-HC-41X001-02-001	32.29	114.40	156.59	-42.19	Not Likely
97	005, Expander, 14B-P3-HC-41X001-02-002	34.00	50.30	156.01	-105.71	Not Likely
98	006, Piping Segment, 14B-P3-HC-41X001-03-001	34.00	41.57	156.01	-114.44	Not Likely
99	007, Piping Segment, 14B-P3-HC-41X001-03-002	34.00	41.56	156.01	-114.44	Not Likely
100	008, Piping Segment, 14B-P3-HC-41X001-03-003	27.45	148.09	158.22	-10.13	Not Likely
101	009, Piping Segment, 14B-P3-HC-41X134-01-001	27.45	148.35	158.22	-9.87	Not Likely
102	010, Piping Segment, 14B-P3-HC-41X134-01-002	27.45	148.61	158.22	-9.61	Not Likely
103	011, Piping Segment, 14B-P3-HC-41X134-01-003	27.45	148.84	158.22	-9.39	Not Likely
104	012, Piping Segment, 14B-P3-HC-41X134-01-004	27.45	149.13	158.22	-9.09	Not Likely
105	013, Piping Segment, 14B-P3-HC-41X134-01-005	27.45	149.26	158.22	-8.96	Not Likely
106	014, Piping Segment, 14B-P3-HC-41X134-01-006	27.45	149.27	158.22	-8.95	Not Likely
107	015, Piping Segment, 14B-P3-HC-41X134-01-001	27.45	150.64	158.22	-7.58	Not Likely
108	016, Piping Segment, 14B-P3-HC-41X134-01-002	27.45	150.72	158.22	-7.50	Not Likely
109	017, Piping Segment, 14B-P3-HC-41X134-01-003	27.45	150.79	158.22	-7.43	Not Likely
110	018, Piping Segment, 14B-P3-HC-41X134-01-004	27.45	150.80	158.22	-7.42	Not Likely
111	019, Piping Segment, 14D-P3-HC-41X134-01-001	27.45	151.02	158.22	-7.20	Not Likely
112	020, Piping Segment, 14D-P3-HC-41X134-01-002	27.45	151.03	158.22	-7.19	Not Likely
113	021, Reducer, 14D-P3-HC-41X134-01-008	27.45	155.36	158.22	-2.86	Not Likely
114	022, Piping Segment, 14D-P3-HC-41X134-01-004	24.79	155.59	159.12	-3.53	Not Likely
115	023, Piping Segment, RF-41X016	24.79	160.24	159.12	1.12	Yes
116	024, Piping Segment, 14D-P3-HC-41X134-01-005	24.79	160.79	159.12	1.67	Yes
117	025, Piping Segment, 14D-P3-FD-41X136-01-001	56.23	160.24	148.48	11.76	Yes
118	026, Piping Segment, 14D-P3-FD-41X136-01-002	56.23	160.77	148.48	12.30	Yes
119	027, Piping Segment, 14D-P3-FD-41X136-01-003	56.23	160.86	148.48	12.38	Yes
120	028, Piping Segment, 14D-P3-HC-41X135-02-001	56.23	162.53	148.48	14.05	Yes
121	029, Piping Segment, 14D-P3-HC-41X135-02-002	56.23	162.59	148.48	14.11	Yes
122	030, Piping Segment, 14D-P3-FD-41X040-01-001	56.23	167.93	148.48	19.45	Yes
123						
124	The IEC acoustic efficiency was used to calculate sound power levels					

where D is in meters, N_o is the number of separate flow passages, C_v is the valve flow coefficient, $\frac{C_v}{N_o}$ is the channel flow coefficient, and F_l is the valve recovery factor. The pressure recovery factor is defined as:

$$F_l = \sqrt{\frac{P_1 - P_2}{P_1 - P_{vc}}} \quad (41)$$

where $P_1 - P_2$ is the pressure differential across the valve and P_{vc} is the pressure at the vena contracta (note that P_2 is larger than P_{vc} since the pressure at the exit of the valve recovers):

$$P_{vc} = P_1 - \frac{P_1 - P_2}{F_l^2} \quad (42)$$

For liquid flow, P_{vc} can reach the vapor pressure of the liquid and cause choked flow.

$$P_{vc} = F_f P_v \approx \left[0.96 - 0.25\sqrt{\frac{P_v}{P_c}}\right] P_v \quad (43)$$

Figure 7. Acoustic efficiency of shock noise generated by choked jets, η vs. jet pressure ratio P_1 / P_2x

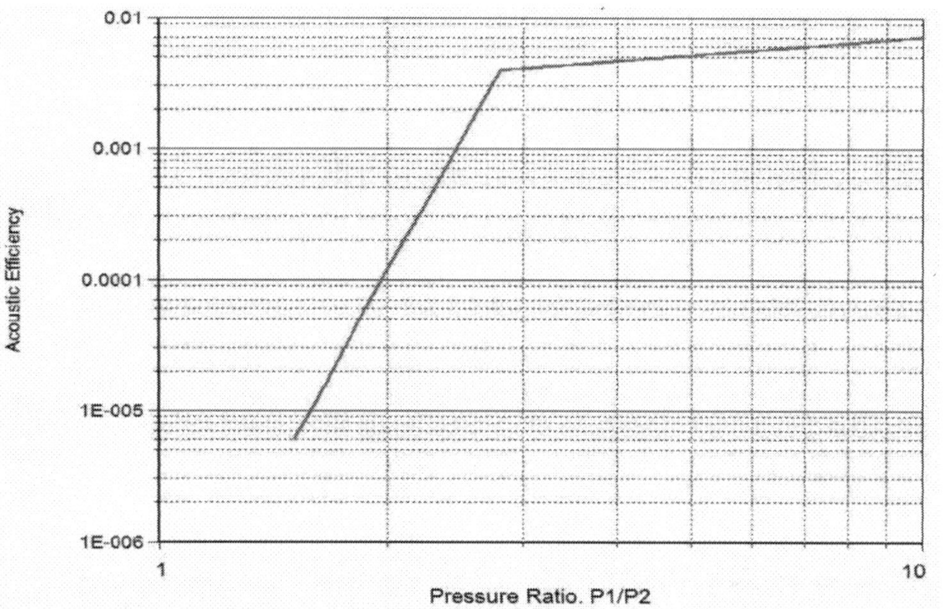

where F_f is the liquid critical pressure factor, P_v is the liquid vapor pressure, and P_c is the liquid critical pressure.

The lowest possible value of P_{vc} in a valve flowing liquid would be vacuum or 0. For this particular limiting case, Equation 41 can be solved for the pressure ratio P_1 / P_2:

$$P_d = \frac{P_1}{P_2} = \frac{1}{1 - F_l^2} \tag{44}$$

where P_d is the damaging pressure ratio. If the valve is operated at a pressure ratio exceeding P_d, the flow will be choked, noisy, and subject to excessive vibration. Table 2 shows typical valve values for F_l and P_d.

Within every flowing pipe there will also be a standing wave moving axially back and forth in the pipe. The frequency of this wave depends on the effective acoustic length of the pipe and the sonic velocity of the fluid in the pipe. The effective acoustic length of the pipe is the distance between obstructions or acoustic barriers. Examples of obstructions would be valves, pumps, and orifices. An acoustic barrier would be an opening into a larger pipe, a reservoir, the end of a pipe run such as a 'T' intersection where the branch of interest requires a right angle turn. Piping components such as expanders or reducers could be an obstruction. An assessment of vibration risk should consider both the frequencies with and without the presence of expanders as obstructions.

The frequency of the standing wave can be calculated as shown below and then compared to the natural frequencies of valve components and the piping system to determine if there is a potential for vibration and/or resonance.

Closed End Pipe $f = \frac{i * u_{ac}}{4L}$

Open End Pipe $f = \frac{i * u_{ac}}{2L}$

where $i = 1, 2, ...$ is the wave number, L is the length between acoustic barriers, and u_{ac} is a characteristic acoustic speed throughout the pipe contents defined as follows:

$$u_{ac} = \frac{u_{sonic}}{\sqrt{1 + \frac{dK}{\delta E}}} \tag{45}$$

where u_{sonic} is the speed of sound in the fluid, K is the bulk modulus of elasticity in the fluid, d is the pipe diameter, δ is the pipe thickness and E is the pipe material of construction modulus of elasticity.

The above equation is derived from a more general form that depends on the elastic properties of the pipe:

$$u_{ac} = \frac{u_{sonic}}{\sqrt{1+\frac{K}{E}\psi}} \tag{46}$$

where ψ is a function of the elastic properties of the pipe. Typical expressions of ψ are shown in Table 4. Typical material properties for ψ are shown in Table 3. Note that materials properties change with temperature as shown in Figure 10.

The speed of sound will change as a function of pressure, temperature, and composition. The presence of dissolved gas nuclei (such as air or other gases) in compressed liquids can significantly reduce the sonic velocity following a pressure drop which leads to the formation of gas bubbles. Up to 40 % reduction in sonic velocity has been observed[xi].

To control the vibration caused by a standing wave it is necessary to change the magnitude and/or the frequency of the standing wave or to change the natural frequency of the pipe or components being excited by the wave. The best approach is to address the magnitude of the standing wave. The magnitude is related to the fluid turbulent energy that is enforcing the wave. The most dominant source of this turbulence is the kinetic energy generated by the fluid jet exiting the valve trim. Thus a valve change with a trim that reduces this jet energy will eliminate this wave influence. Trying to change the frequencies is usually not beneficial. There is such a wide range of frequencies present in the turbulent flow that excitation can continue to establish a strong wave at the new frequency and continue the piping vibration.

Figure 8. Flow regimes considered for the estimation of acoustical efficiency

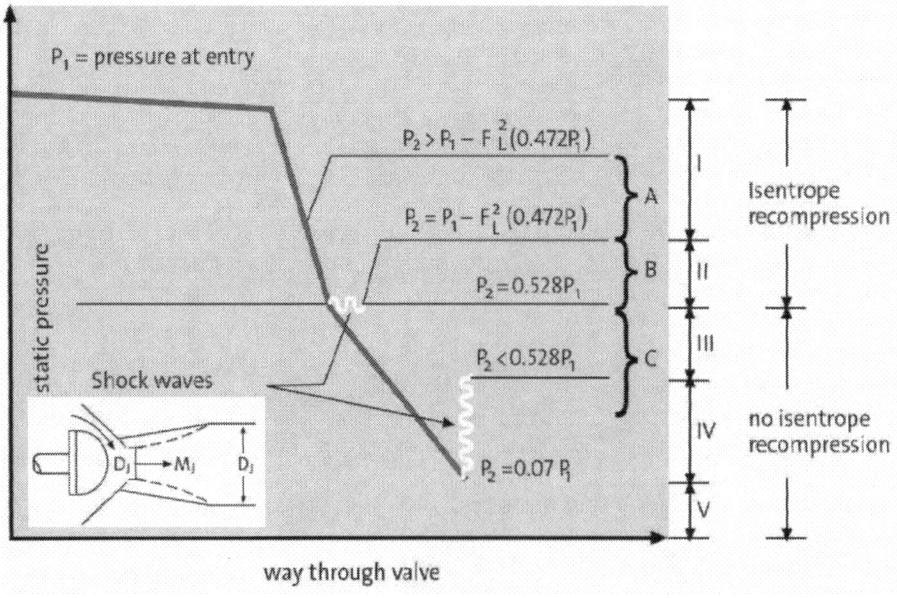

Figure 9. IEC flow acoustical efficiency as a function of F_L and differential pressure ratio

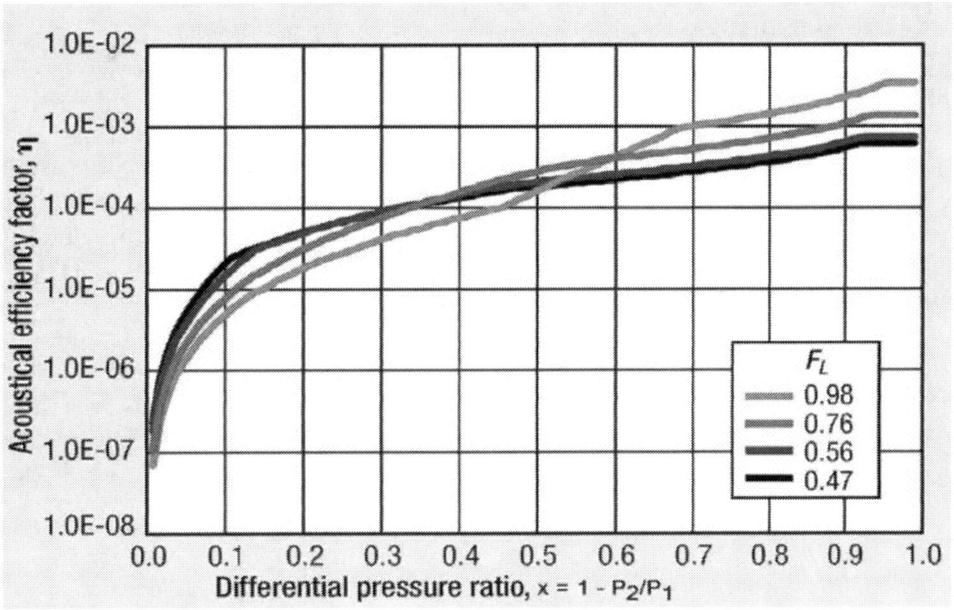

13. The Singing Safety Relief Valve Problem

As a result of increasing steam flow rates, several boiling water reactor (BWR) nuclear power plants have recently experienced the excitation of acoustic standing waves in closed side branches, e.g., safety relief valves (SRVs), due to vortex shedding generated by steam flow in the main stream lines. Flow past a valve entrance cavity excites a standing wave, resulting in noise and vibration[xii]. A similar tone is produced when air is blown across the mouth of a glass bottle.

Table 2. Typical valve values for F_l and P_d

Body	Trim	Flow Direction	F_l	$P_d = \frac{P_1}{P_2}$
Single Seat Globe	Cage	Open	0.90	5.3
	Cage	Closed	0.80	2.8
	V Plug	Open	0.90	5.3
	V Plug	Closed	0.90	5.3
	Contoured	Open	0.90	5.3
	Contoured	Closed	0.80	2.8
Double Seat Globe	V Plug		0.90	5.3
	Contoured		0.85	3.6
	Standard Bore		0.55	1.4
	Characterized		0.57	1.5
Angle	Cage	Open	0.85	3.6
	Cage	Closed	0.80	2.8

Ball	0.8 dia. Orifice		0.55	1.4
Butterfly	60°	Open	0.68	1.8
Butterfly	90°	Open	0.55	1.4

The amplitude of the acoustic pressure waves can be several times higher than the dynamic pressure present in the system (see Figure 12). The acoustic waves propagate in the steam lines, eventually reaching sensitive components such as steam dryers and turbine stop valves. In addition, the acoustic waves generated in the side branches may generate vibration problems locally and may lead to complications such as valve-seat wear. Therefore, the structural components are subjected to high-cycle fatigue loads, which over time may severely impact those components' functionality and safety.

Resonance occurs when the vortex shedding frequency coincides with the acoustic frequency of the standpipe or the valve components. The natural frequency of the standpipe/valve combination for a closed end pipe is given by the following equation:

$$f_a = \left(\frac{2n-1}{4}\right)\frac{u_{ac}}{L+L_e} = \left(\frac{2n-1}{4}\right)\frac{u_{ac}}{L+0.425d} \quad (47)$$

where n is the mode number (1 for 1st mode, 2 for 3rd mode, etc.), u_{ac} is the acoustic speed through the pipe contents as defined earlier, and L_e is an end correction corresponding to Rayleigh's upper limit.

The frequency of pressure oscillations (sound) created by vortex shedding, the energy source for the standing waves, is given by the following equation:

$$f_s = N_{str}\frac{u}{d+r} \approx 0.33(n-0.25)\frac{u}{d+r} \quad (48)$$

Typically peak oscillations occur at a Strouhal Number around 0.4 as shown in Figure 12. Note that the root mean square pressure amplitude shown in Figure 12 is the ratio of pressure oscillations divided by dynamic pressure ($\frac{1}{2}\rho u^2$). RMS begins increasing at a specific onset Strouhal Number and flow velocity depending on acoustic speed, pipe diameter, and pipe length, reaches a peak value and then decreases.

There are many similar installations of pressure relief valves in the process industries where the valves are mounted on large process lines such as overhead lines for distillation columns. In order to avoid fatigue failure from resonance caused by the coupling of normal flow vortex shedding frequency and the acoustic frequency of the standpipe ($f_s = f_a$), the normal flow velocity in the main line has to be limited to less than this critical value:

$$u_{max} < \frac{f_s}{N_{st}}(d+r) < \frac{1}{4}\left(\frac{u_{ac}}{L+L_e}\right)\left(\frac{d+r}{N_{str}}\right) \quad (49)$$

$$< \frac{1}{4}\left(\frac{u_{ac}}{L + 0.425d}\right)\left(\frac{d + r}{0.6}\right)$$

As shown in Figure 12 the pressure fluctuations start to increase at a Strouhal number of 0.6 and then decrease after they reach a peak value around a Strouhal number of 0.4. The same approach can be applied to the flow through the inlet and/or discharge line of a pressure relief valve:

$$u_{max} < \frac{1}{2}\frac{u_{ac}}{L}\frac{D}{0.6} \tag{50}$$

where L is the acoustic length of the inlet or discharge line. Resonance can also be checked by comparing an open pipe/contents frequency with the natural frequency of the pressure relief valve, f_n:

$$f_n = \frac{1}{\tau_n} = \frac{\omega_n}{2\pi} = \frac{1}{2\pi}\sqrt{\frac{K_s}{m_D}} \tag{51}$$

$$\tau_n = \frac{2\pi}{\omega_n} = 2\pi\sqrt{\frac{m_D}{K_s}} \tag{52}$$

where τ_n is the undamped natural period in s, and f_n is the undamped natural frequency in Hz where one Hz equals 1 cycle/second, K_s is the spring constant in $\mathrm{N/m}$, m_D is the mass of the valve disc and moving parts in kg, and ω_n is the undamped circular natural frequency in $\mathrm{radians/s}$.

14. Conclusions

A general method for the estimation of vibration risk for piping systems for single and multiphase flow is developed and implemented. An experience based vibration risk failure criteria is outlined using the work of Carucci and Mueller for gas flow. Because the Carucci and Mueller sound power level equation can produce unrealistic values of acoustic efficiency, well in excess of 1 percent for high pressure systems and/or systems with large mechanical flow energy, the revised experience based failure criteria developed in this paper (see Figure 5) based on the IEC acoustic efficiency is recommended for multiphase flow.

Figure 10. Temperature effects on material of construction

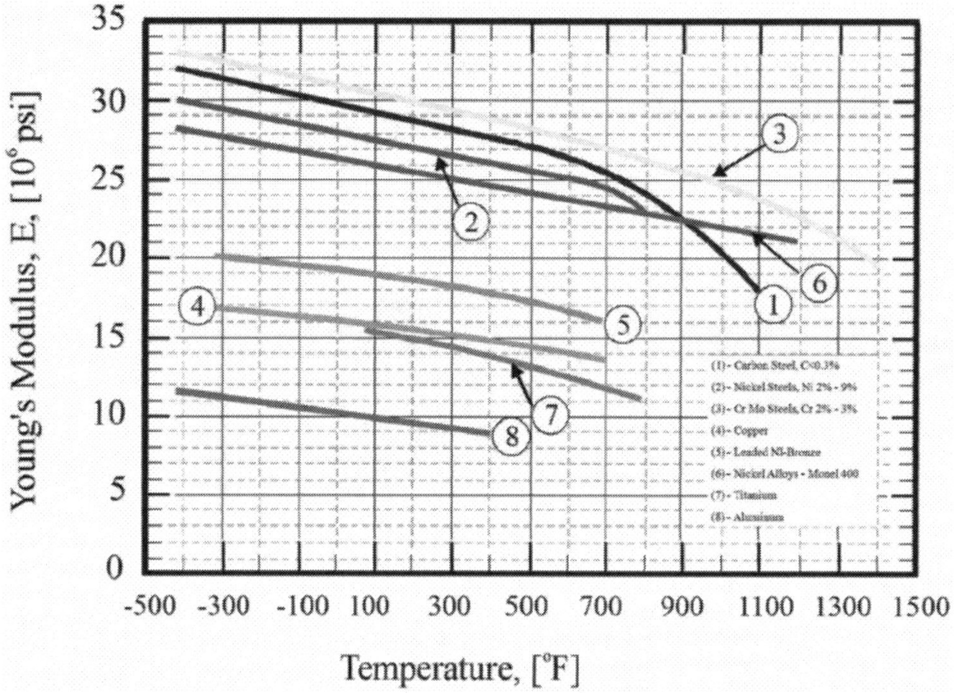

Figure 11. The singing safety relief valve[xiii]

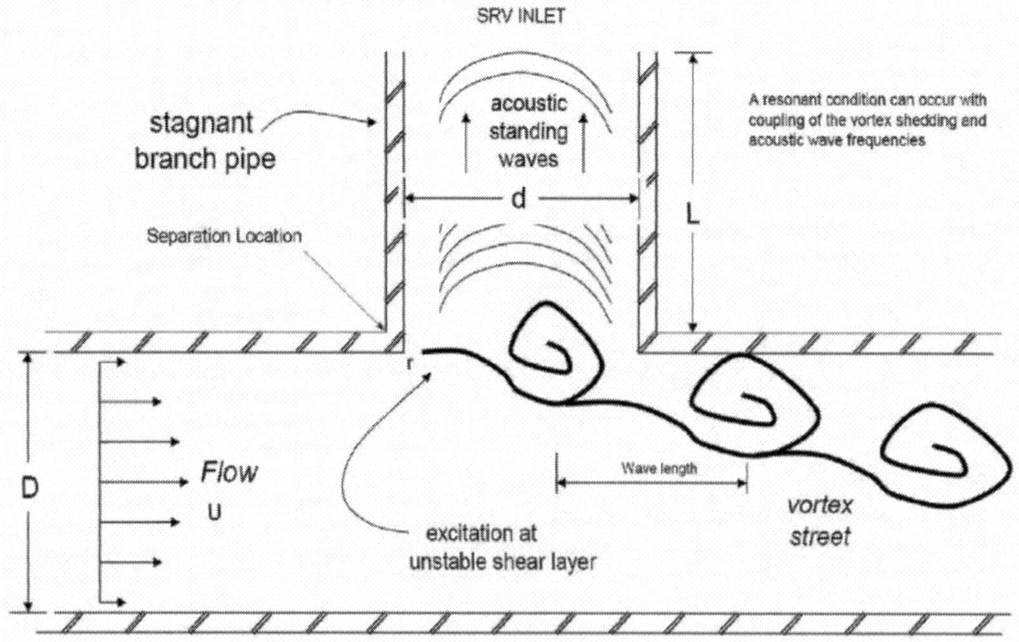

Figure 12. Comparison of calculated and measured pressure fluctuations as a function of Strouhal number[xiv]

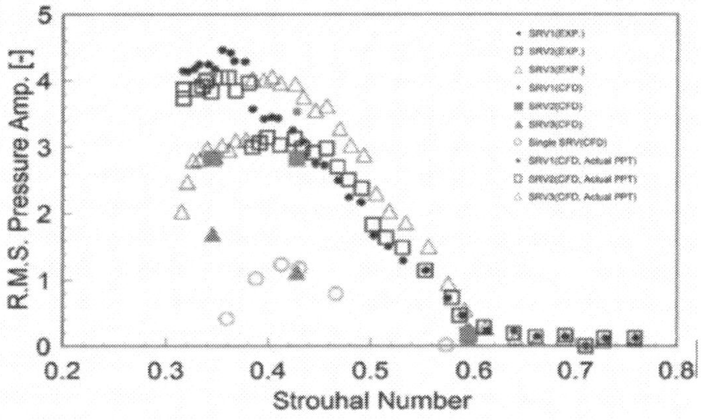

E is typically referred to as Young's modulus of elasticity

G is typically referred to as modulus of torsion, $G = \frac{1}{2}\frac{E}{1+\nu}$

Table 3. Typical data used in the estimation of sonic velocity in pipelines

Material	E (GPa)	Poisson's Ratio ν	$K = \frac{1}{\kappa}$ (GPa)	ρ in kg / m3
Aluminum	69	0.33		
Brass	78-110	0.36		
Carbon steel	202	0.303		
Cast iron	90-160	0.25		
Concrete	20-30	0.15		
Copper	117	0.36		
Ductile iron	172	0.30		
Fibre cement	24	0.17		
High carbon steel	210	0.295		
Inconel	214	0.29		
Mild steel	200-212	0.27		
Nickel steel	213	0.31		
Plastic / Perspex	6.0	0.33		
Plastic / Polyethylene	0.8	0.46		
Plastic / PVC rigid	2.4-2.75			
Stainless steel 18-8	201	0.30		
Water - fresh			2.19	999 at 20 C
Water - sea			2.27	1025 at 15 C

Table 4. Typical expressions for ψ

Pipe condition	ψ
Rigid	0
Anchored against longitudinal movement through its length	$\frac{d}{\delta}\left(1-\nu^2\right)$
Anchored against longitudinal movement at the upper end	$\frac{d}{\delta}(1.25-\nu)$
Frequent expansion joints present	$\frac{d}{\delta}$

References

[i] W.A. Skipwith. Acoustics technology. Survey SP-5093, NASA National Aeronautics and Space Administration, 1970.

[ii] C.M. Harris. *Handbook of Noise Control*. cGraw-Hill Book Co., Inc., 1957.

[iii] N.R. Miller, G.M. Singh, E. Rodarte and P.S. Hrnjak. Modification of standard aeroacoustic valve noise model to account for friction and two-phase flow. In *ACRC TR-162*, pages 1-11. Mechanical and Industrial Engineering Department, University of Illinois, 2000.

[iv] N.R. Miller, G.M, Singh, E. Rodarte and P.S. Hrnjak. Prediction of noise generated by expansion devices throttling refrigerant. In *ACRC TR-163*, pages 1-13. Mechanical and Industrial Engineering Department, University of Illinois, 2000.

[v] MTD. *Guidelines for the Avoidance of Vibration Induced Fatigue in Process Pipework*. Marine Technology Directorate Limited (MTD), 1999.

[vi] Energy Institute. *Guidelines for the Avoidance of Vibration Induced Fatigue in Process Pipework*. The Energy Institute, London, England, 2008.

[vii] V.A. Carucci and R.T. Mueller. Acoustically induced piping vibration in high capacity pressure reducing systems. In *82-/WAPVP-8*, pages 1-13. American Society of Mechanical Engineers, ASME, 1982.

[viii] F.L. Eisinger. Designing piping systems against acoustically-induced structural fatigue. In *PVP-VOL 328, Flow-Induced Vibration*, pages 397-404. American Society of Mechanical Engineers, ASME, 1982.

[ix] Hisao Izuchi, *Investigation of Pipe Size Effect Against Vibration Risk*, Presentation to ioMosaic Corporation, March 2013, Houston.

[x] I. Heitner. How to estimate plant noises. *Hydrocarbon Processing*, 47, 67-74, 1968.

[xi] V.L. Streeter and E.B. Wylie. Water hammer and surge control. *Annual Reviews in Fluid Mechanics*, pages 57-74, 1974.

[xii] T.M. Mulcahy, S.A. Hambric and V.N. Shah. Flow-induced vibration effects on nuclear power plant components due to main stream line valve singing. In *Ninth NRC/ASME Symposium on Valves, Pumps, and Inservice Testing, NUREG/CP-0152,* volume 6, pages 3B:49-3B:69. NCR/ASME, 2006.

[xiii] P. Sekerak. Potential adverse flow effects at nuclear power plants. In *16th International Conference on Nuclear Engineering*. ICONE16-48900, 2008.

[xiv] K. Okuyama, F. Inada, Y. Ogawa, R. Morita, S. Takahashi and K. Yoshikawa. Evaluation of acoustic and flow induced vibration of the BWR main stream lines and dryer. *Journal of Nuclear Science and Technology,* 48 (5): 759-776, 2011.

(44bt) Performance Metrics Development Management and Reporting Reaching Sustainability by Design

Robert C. Smith, ABB, Inc., Houston, TX

Overview

The saying is, "You can't manage, what you can't measure". This presentation will discuss various performance management measurement frame-works; How to use organizational goals and objectives to develop appropriate performance measures; Design effective performance measures to implement your organization's goals and strategies; Link performance measures to a variety of management processes; Develop a performance management framework that drives results. Presentation goals

Create a clear understanding of setting and managing key performance metrics surrounding highly hazardous environments.

- Create a practical base knowledge of leading and lagging metrics and how to practically apply them into your environment.
- Learn about the most common errors people make when selecting metrics and you'll gain new ideas around how metrics can be visualized better.
- This session will give you an in-depth look at the often confusing and misunderstood world of performance metrics. You'll learn the truth about metrics, and how an effective performance measurement system can help you make better decisions.
- Recognize management's role in leading and implementing performance metrics.
- Have a foundation to build metrics and scorecards by using the tips, techniques, and tools acquired in the presentation.

Who should attend?

This presentation is intended for any executive, manager or employee with responsibility for implementing or maintaining score carding performance management systems, and is interested in learning more about how to establish and measure metrics. What you'll learn:

- Detailed knowledge of leading and lagging metrics
- Measuring program success
- Key points continuous improvement and sustainability of measures

- Understand the purpose of process management
- How to analyze an existing metric system
- How to create well designed metrics that support strategy
- How to steer away from some of the classic pitfalls of metric development

(44bu) Alarm Management - the Human Factors Challenge

Robert C. Smith, ABB, Inc., Houston, TX

Refinery alarm systems are supposed to support the operator, yet in several notable accidents they have proven to be ineffective. Part of the problem is that these systems are not being designed or managed with human needs in mind. Do alarm systems help the operator or are they a hindrance? This paper is a short review of how human factors contributed to several incidents including:

- July 1994 explosion at Texaco, Milford Haven, UK
- September 1998 explosion at Esso, Longford, Australia
- March 2005 explosion at BP, Texas City, US.

Human operators are needed to ensure plants and processes continue to operate safely. The control room and control system interfaces must therefore be built with human needs in mind. Includes a review of how implementation of ISA 18.2 and/or EEMUA 191 can improve alarm management practice.

Dust Flame Propagation in Industrial Scale Piping
Part 2: CFD Study of a Conveying Vessel-Pipeline System

Jim Vingerhoets
Fike Europe
Toekomstlaan 52, 2200 Herentals, Belgium
jim.vingerhoets@fike.com

Thomas M. Farrell
Fike Corporation, Industrial Protection Group
704 SW 10th Street, Blue Springs, MO 64015, USA
tom.farrell@fike.com

Jef Snoeys
Fike Europe
Toekomstlaan 52, 2200 Herentals, Belgium
jef.snoeys@fike.com

Prepared for Presentation at
American Institute of Chemical Engineers
2013 Spring Meeting
9th Global Congress on Process Safety
San Antonio, Texas
April 28 – May 1, 2013

UNPUBLISHED

Dust Flame Propagation in Industrial Scale Piping
Part 2: CFD Study of a Conveying Vessel-Pipeline System

Jim Vingerhoets
Fike Europe
Toekomstlaan 52, 2200 Herentals, Belgium
jim.vingerhoets@fike.com

Thomas M. Farrell
Fike Corporation Industrial Protection Group
704 SW 10th Street, Blue Springs, MO 64015, USA
tom.farrell@fike.com

Jef Snoeys
Fike Europe
Toekomstlaan 52, 2200 Herentals, Belgium
jef.snoeys@fike.com

Keywords: Dust explosion, flame propagation, vented vessel, pipelines, CFD

Abstract

The use of pipes to connect vessels and to transport hazardous materials, such as flammable gases or dusts, is a common practice in the process industry. The potential of fires and explosions has been recognized for decades and is addressed in standards such as NFPA 68 [1] and NFPA 69 [2]. Unlike for gases, the experimental evidence on dust flame propagation is scarce, and this study is particularly aimed at expanding the knowledge in this field. The purposes of this paper are to present the results of a parametric computational fluid dynamics (CFD) study validated by experimental evidence and to further develop a detailed understanding of combustion propagation in industrial scale piping.

The results of an ongoing research program which involves a piping system connected to a 5 m3 vented vessel are presented in the accompanying paper [3]. CFD was used to broaden the field of application beyond the tested configurations, and to study the influence of vessel volume and pressure, fuel characteristics and flow conditions.

Upstream and downstream explosion propagation behavior is studied as a function of fuel reactivity and type, vessel fuel concentration, pipeline fuel concentration, flow velocity, vessel volume, vent area, vent location, ignition location, pipe diameter and length.

1. Introduction

The dust explosion hazard continues to represent a constant threat to process industries that manufacture, use and/or handle powders and dust of combustible materials. In the process industry, the majority of process equipment is connected via pipes or ducts, and accidents have shown that the propagation of the explosion through the connection pipe work worsens the consequences of the explosion. Understanding the mechanisms of explosion flame propagation is a prerequisite for designing effective explosion protection solutions.

The problem of dust explosions in large-scale connected equipment has been much less investigated than isolated equipment, mainly due to cost and complexity to conduct experiments. The information in the public domain on dust explosion propagation is scarce; refer to the bibliography section at the end of the paper for relevant contributions. An alternative to large scale testing is the use of mathematical modeling and simulating the problem using CFD - computational fluid dynamics. CFD uses numerical methods and algorithms to simulate the combustion phenomena defined by boundary conditions. Although the use of CFD in gas and dust explosion applications is relatively new, this technique is irrefutably gaining interest.

The present research investigates, by experiment and CFD, flame propagation from a vented vessel through connected pipes. The aim is to examine the mechanisms that govern the velocity and acceleration of flame through pipelines, thus providing valuable information for the design of explosion protection solutions and systems. Once sufficient correlation between experimental and numerical results was demonstrated, the scope was extended with a numerical sensitivity study to quantify the influence of several application parameters on the flame propagation behavior. These parameters include:

- fuel type and reactivity,
- fuel concentration in vessel and pipelines,
- flow velocity and direction,
- vessel volume, pipe diameter and length,
- vent area, location,
- ignition location.

2. Simulation code

The numerical explosion simulations have been performed using DESC - Dust Explosion Simulation Code. DESC resulted from a research project supported by the European Commission under the Fifth Framework Program and was reviewed by Skjold in 2007 [4]. The main purpose of the project was to develop a simulation tool based on computational fluid dynamics (CFD) that could predict the potential consequences of industrial dust explosions in complex geometries. DESC is presently based on the existing Flame Acceleration Simulator (FLACS) code for gas explosions. The dust particles are modeled as gaseous components that follow the flow of the other gaseous components in the system. The combustion model used in FLACS and adopted by DESC is the so-called β flame model developed by Arntzen [5]. The turbulent burning velocity used by the β model is a function of the root mean square of the turbulent velocity fluctuations and the Karlovitz stretch factor. More specifically, the turbulent burning velocity is formulated as:

$$S_{\mathrm{T}} = 15.1 S_{\mathrm{L}}^{0.784} u'^{0.412}_{\mathrm{rms}} l_{\mathrm{I}}^{0.196}$$

with S_{T} the turbulent burning velocity, S_{L} the laminar burning velocity, u'_{rms} the root mean square of the turbulent velocity fluctuation, also called the turbulence intensity and l_{I} the turbulence integral length scale. The inputs for the combustion model are laminar burning velocities extracted from the measured K_{st} and P_{max} values in standardized 20 L sphere tests. The code has been validated against large scale tests reported in the public domain. It was found that DESC is sensitive to the numerical grid resolution, the presence of bends and interconnected vessels in general. The grid resolution in the current research has been chosen based on recommendations from this validation, and bends were excluded from the study.

3. Experimental setup

A test rig was built at the test site of Fike Corporation consisting of an inlet pipeline, a vessel, and an outlet pipeline, arranged as shown in the following image. The inlet and outlet pipelines were oriented 90° to each other, with the outlet pipeline aligned with the major axis of the vessel. All explosion venting was constrained to a single panel on the vessel, on the wall opposite the outlet pipeline interconnection. The system was equipped with a fan, enabling adjustable pipe flow velocities. By using separate constant mass flow rate dust injectors on the pipelines in addition to traditional dust injectors on the vessel, the vessel dust concentration and pipeline dust concentration could be varied independently of each other.

Figure 1: Photo of experimental setup

In order to cover several industry branches, different fuel types were used to conduct the research. The properties of the fuels used in this program are listed in Table 1.

Table 1: Closed vessel explosion characteristics of the dust fuels used

Dust type	Dust concentration (g/m^3)	20 L sphere	
		K_{st} (bar.m/s)	P_{max} (barg)
Wood flour	250	50	5.7
	500	108	7.0
	750	109	7.2
Phenolic dust	250	203	7.3
	500	217	6.9
	750	200	6.6
Corn Starch	250	93	6.6
	500	199	8.7
	750	205	8.6

Many tests were conducted to determine explosion pressures and flame arrival times as a function of venting area, vessel fuel concentration, pipeline fuel concentration and conveyance velocity. More information can be found in the accompanying paper [3].

4. Simulation setup

Figure 2 shows the basic setup of the explosion simulations. It represents as much as possible the experiment setup. The vessel itself consists of two identical cubes of 2.55 m^3 each, attached to each other without a partitioning wall in between. The cube assembly is equipped with dust jet injectors, 4 vent panels, a 10 kJ igniter and flanges to connect two pipelines. Each of the 4 vent panels can be closed off (to vary the total vent area) or opened at a controlled burst pressure. In some simulation scenarios, the location of the vent panels was moved to another wall of the vessel.

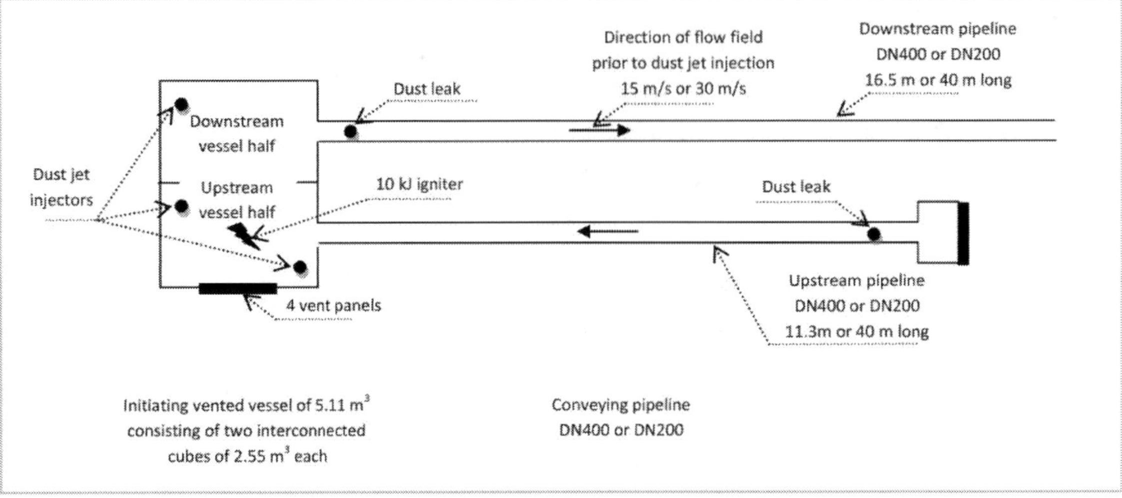

Figure 2: Sketch of basic setup of the explosion simulations.

Figure 3: Three dimensional CAD drawing picture of the simulation setup

The default location of the igniter is in the centre of the upstream vessel half, but in several simulation scenarios the igniter was moved to the centre of the downstream vessel half or to the interconnection with one of the pipelines. The delay time between start of dust jet injection and ignition was identical to the experimental program. The pipelines were either 400 mm or 200 mm in diameter and 11.3 m, 16.5 m or 40 m long. In order to restrict the simulation domain and associated computer processing time, the downstream pipeline was rotated 90° with respect to the experimental setup. Flow of 15 m/s or 30 m/s was established in the pipelines, depending on the simulation scenario. Dust leaks were programmed to fill the pipelines with a dust/air mixture of known concentration. Other simulations were done in a 62 m^3 vessel with proportionally scaled vent panels.

All simulation scenarios have been numerically solved on cubical grid cells with optimized size conforming to DESC guidelines.

5. Comparison between experiments and numerical simulations

Figure 4 to Figure 6 display comparative plots of vessel pressure and flame position as a function of time for three example scenarios including each of the three dusts of the program. Flames propagating towards the blower are represented by negative pipeline distances.

Table 2: Comparison between experiments and simulations giving the main characteristic values of the explosion propagation

Comparison of main characteristic values		Example 1		Example 2		Example 3	
		Exp	Sim	Exp	Sim	Exp	Sim
Vessel peak pressure		3.2 psig	2.0 psig	4.2 psig	6.7 psig	8.6 psig	2.3 psig
Time to reach 1,5 psig in vessel		64 ms	161 ms	46 ms	56 ms	54 ms	110 ms
Average flame speed in first 10 meters of pipe	Downstream	90 m/s	36 m/s	90 m/s	116 m/s	154 m/s	116 m/s
	Upstream	79 m/s	55 m/s	116 m/s	143 m/s	221 m/s	112 m/s
Time to reach 2,7 meters in the pipe	Downstream	110 ms	382 ms	114 ms	117 ms	88 ms	214 ms
	Upstream	127 ms	233 ms	106 ms	95 ms	85 ms	135 ms

The simulations under-predict the explosion peak pressure and pressure rise time in the vessel for two out of the three scenarios. The modeled explosion peak pressure for example 2 is higher (6.7 psig) than its experimental equivalent (4.2 psig), with similar pressure rise time (1.5 psig @ 56 ms and 1.5 psig @ 46 ms). Similar observations are made for the flame front pipeline propagation: the modeled explosions for example 1 and 3 propagate at lower speed (36 m/s downstream, 55 m/s upstream and 116 m/s downstream, 112 m/s upstream respectively), whereas the modeled dust explosion of example 2 propagates faster. Finally, the modeled explosions of example 1 and 3 enter the pipes later whereas the modeled dust explosion pipe entrance time of example 2 is similar to the experiment.

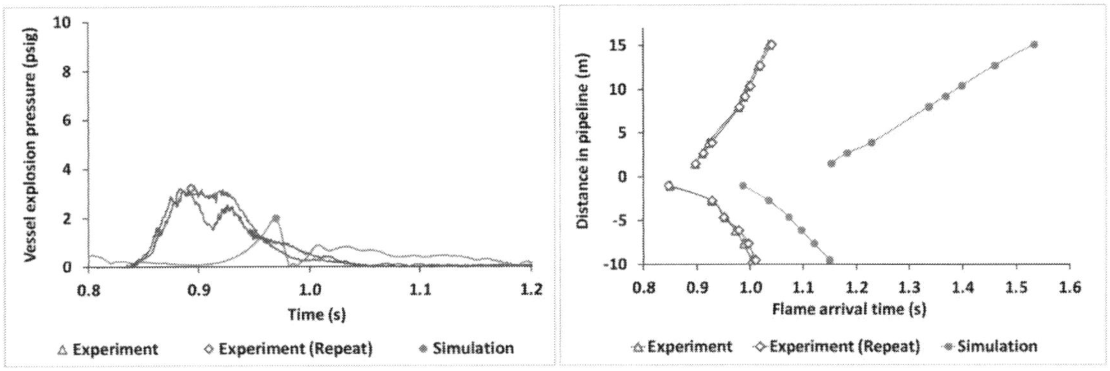

Figure 4: Comparison experiment and simulation: example 1

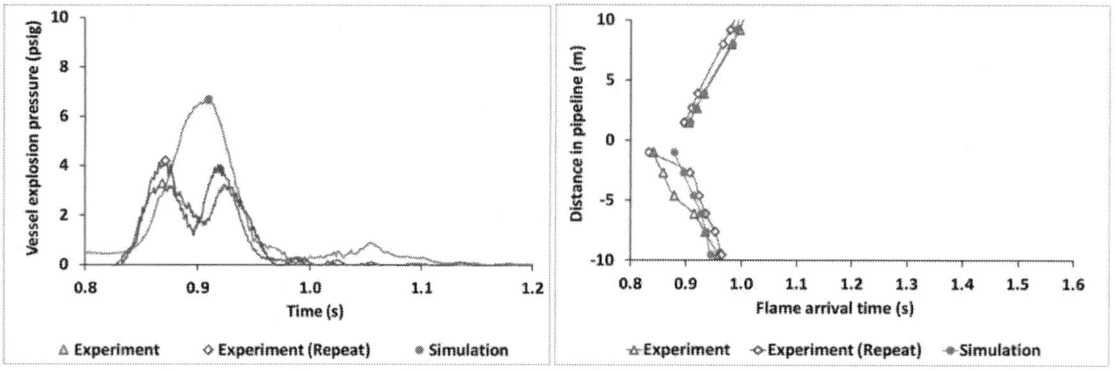

Figure 5: Comparison experimental and simulation: example 2

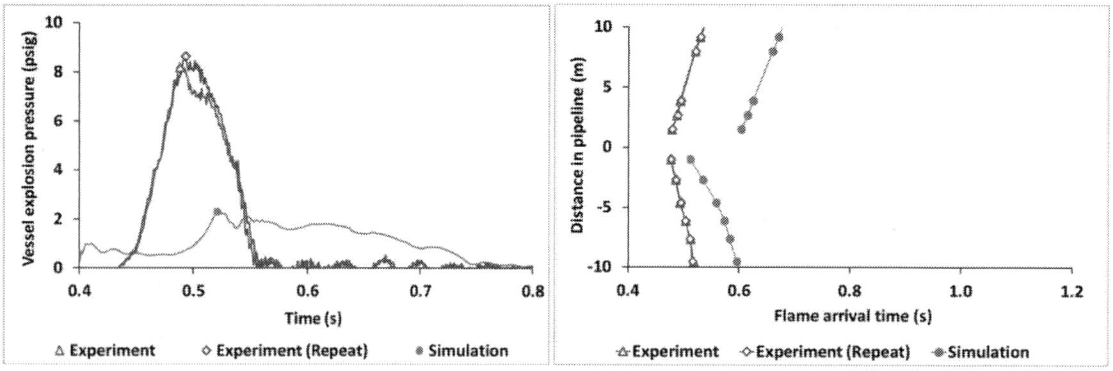

Figure 6: Comparison experimental and simulation: example 3

A likely explanation for the divergence between experimental and simulated results is that the modeled dust jet injectors generate different turbulence conditions than the injectors used in the experiments. This is a plausible explanation since the dust particles are modeled as gaseous phase fuel. Figure 7 shows the turbulence intensity in the centre of the upstream vessel half during simulated dust injection for different vessel dust concentrations and for a pipe flow of 15 m/s. The simulated dust jet injectors create an average turbulence intensity level between 2.0 and 3.4 m/s in the first 200 ms after ignition. 200 ms is the typical duration of an explosion in a 5 m^3 vessel. Although slightly higher than the reported turbulence in 20 L and 1 m^3 tests, the numbers are quite reasonable. Unfortunately, the turbulence could not be measured during the experiments.

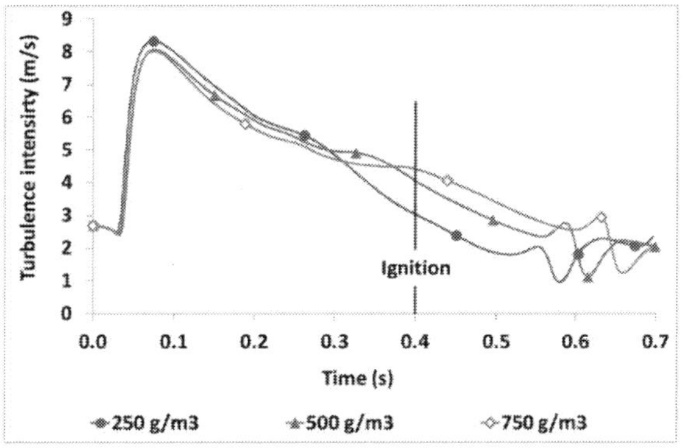

Figure 7: Turbulence intensity in the centre of the upstream vessel half during simulated dust injection for different vessel dust concentrations; initial pipe flow = 15 m/s

Another possible reason is that the 20 L flame data used by DESC do not scale up well to larger volumes. Several researchers postulated scaling problems related to turbulence and the powerful igniter used in the closed vessel 20 L and 1 m^3 standardized tests. This problem was investigated by additional explosion experiments in a closed 1m^3 vessel with varying ignition delay [3]: the K_{st} of two of the three dusts were more or less independent of volume, although strongly affected by ignition delay. The other dust showed an opposite behavior and is relatively insensitive to decreased ignition delay.

Figure 8 shows the effect of increasing the laminar burning velocities which was used as a surrogate for boosting the turbulence. The agreement becomes much better which indicates that the prediction potential of DESC is improved, if both the fundamental combustion parameters (the laminar burning velocity derived from K_{st} and P_{max}) and the initial process turbulence are estimated correctly.

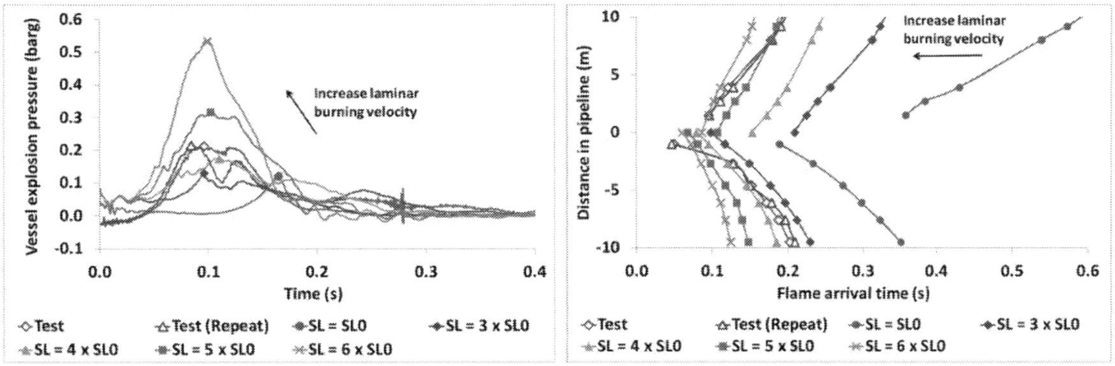

Figure 8: Sensitivity of vessel explosion pressure and flame propagation to the modeled laminar burning velocity and comparison with experiment (example 1)

In order to deal with the uncertainties in turbulence and combustion parameters for different vessel volumes and injector assemblies, simulations were performed in the reference test volume with all vent panels and connections to the pipelines closed off, and the K_{st} and P_{max} values were recorded. The results obtained by DESC should be referenced to K_{st} and P_{max} as obtained by DESC for the closed vessel. By following this approach, together with additional specific considerations as pointed out further, it is believed that simulations can represent industry reality with similar level of confidence as experiments.

6. Mechanisms of flame propagation

A thorough analysis of 94 explosion simulations for different fuel types, vessel and pipeline fuel concentrations, vent areas, flow velocities, significantly contributes to our understanding of the fundamental mechanisms which govern the propagation of dust flames through pipelines which are connected to a vented vessel. For all these scenarios the vessel and pipeline explosion pressures have been recorded as well as the flame arrival times at several distances along the pipelines. Besides pressure and flame, other physical variables, such as flow velocity and dust concentration have also been recorded. In addition to dust fuels listed in Table 1, also other dust fuels were used in the simulation study, so covering a wider range of K_{st} values, from 28 bar.m/s up to 281 bar.m/s.

In all scenarios the flame propagated with high speed up until the end of both pipelines, including the ejection of a large fireball, even against strong initial flows of 30 m/s. The highest recorded flame speed was 434 m/s. This was at the end of the downstream pipeline for the dust and concentration with highest reported K_{st} value and with the smallest venting area programmed. The lowest recorded flame speed was 34 m/s, for the dust and concentration with lowest K_{st} value and the largest vent area programmed.

It was confirmed that flame enters both pipes and propagates through a dust cloud that is pushed from the vessel into the pipes, as shown in Figure 9 below. In this scenario, the pipelines were not filled with dust fuel prior to ignition.

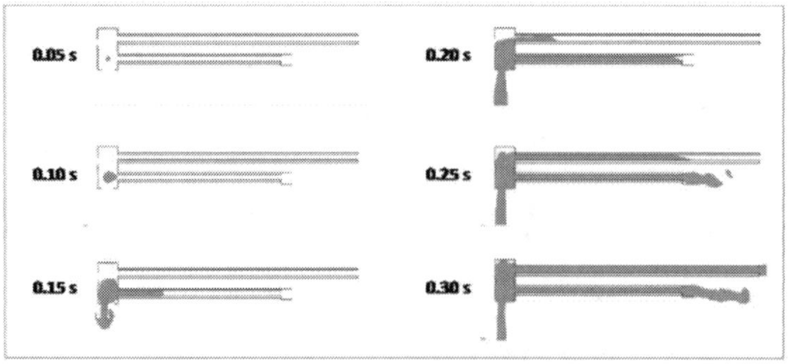

Figure 9: Flame propagation shown at subsequent time steps during an example explosion. No flow through pipe, no dust in pipe. 2 vent panels open. Vessel dust concentration = 500 g/m^3.

Figure 9 also illustrates that flame accelerates as it enters the pipe: It takes 0.1 seconds for the flame to reach the pipe and in only another 0.1 seconds it manages to travel the full pipeline distance.

In the vessel the freely expanding flame front propagates at a moderate speed of about 5 to 10 m/s. When the explosion flame enters the connected pipes, it will accelerate because the flame front will travel along with the combustion gases flowing at high speed out of the vessel:

$$S_{\mathrm{f,\,pipe}} = S_{\mathrm{g}} + S_{\mathrm{T}}.E$$

The pipe gas flow velocity induced by the explosion pressure in the vessel (S_g) is often a lot larger than the speed of turbulent burning and associated local expansion ($S_T.E$) and can reach up to 300 m/s. For low to moderate explosive reactive materials the vessel pressure induced gas flow will dominate the pipe flame speed.

Figure 10 shows the correlation between gas flow velocity in the pipelines and flame front velocity, for three different fuels and for three different dust concentrations. The more reactive the fuel and the larger the gas flow velocity, the more the flame velocity exceeds the gas velocity. This is because the turbulent burning velocity increases with fuel reactivity and with increased flow turbulence. The flow turbulence is higher at higher gas velocities.

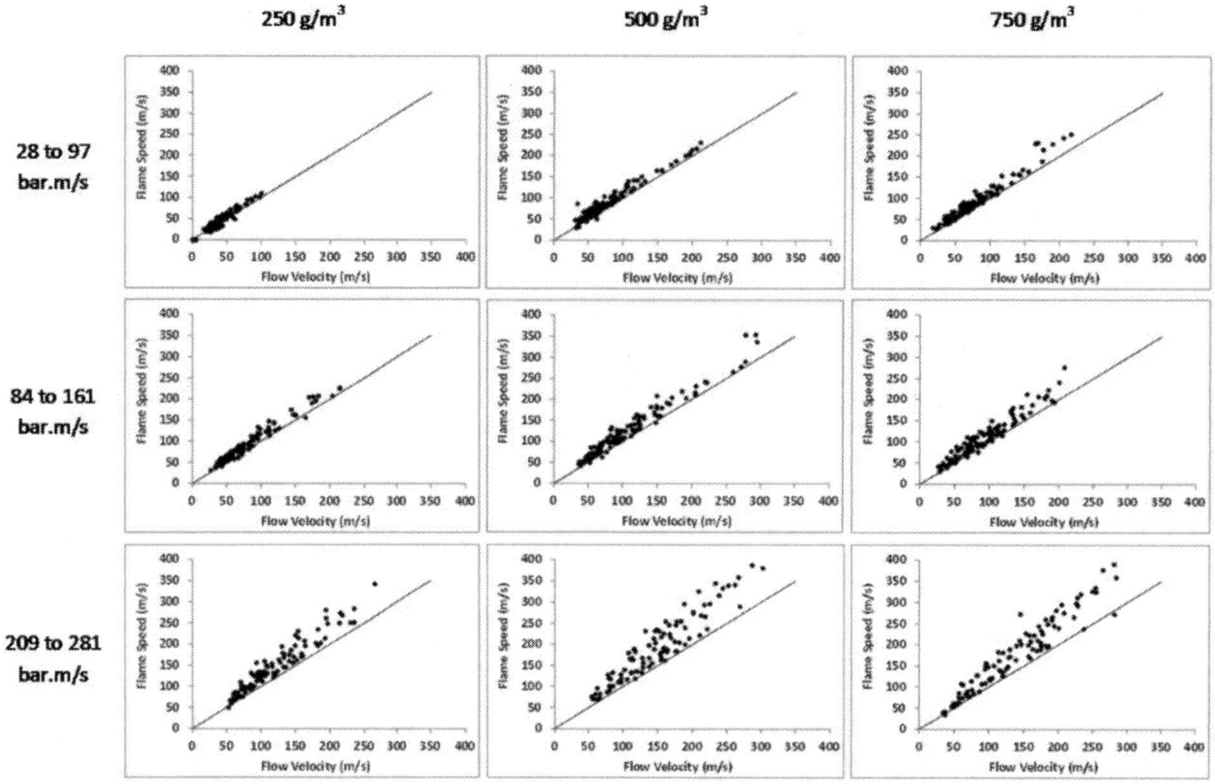

Figure 10: Correlation between flame speed and explosion induced gas velocity in the pipelines for different fuel types and dust concentrations

7. Sensitivity study

To enable extrapolation of the findings to real industry applications, 172 additional scenarios have been simulated in which the flow direction, vessel volume, vent area, vent location, ignition location, pipe diameter and length were varied.

7.1. Effect of vent area

In all cases, the flame speed decreases with larger vent areas. Figure 11 shows example simulation results for two dust types. This finding was consistently confirmed by experiments [3] and is attributed to lower vessel reduced pressures associated with larger venting areas. The decrease in flame speed is less pronounced in the upstream pipe, because the ignition was located closer to the upstream pipe, so that the flame could enter the upstream pipe before the vent panel opened.

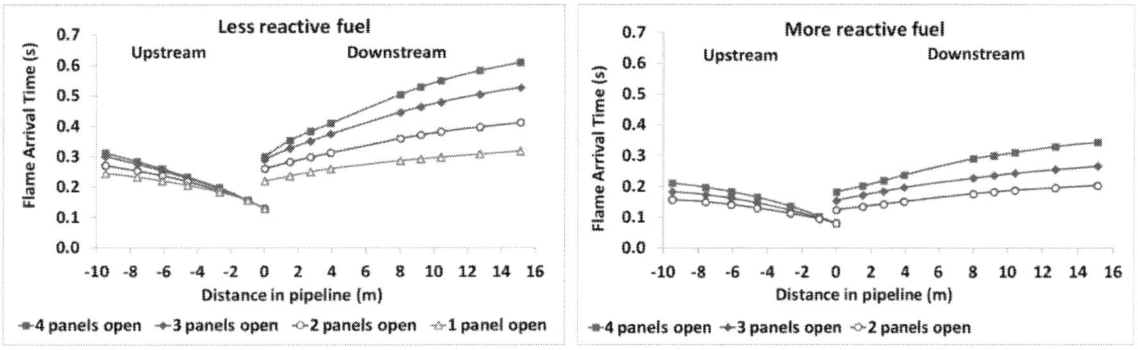

Figure 11: Effect of vent area on flame propagation. Flow = 15 m/s; Vessel and pipeline dust concentration = 750 g/m^3; Ignition location = centre of upstream vessel half.

7.2. Effect of pipeline dust concentration

Many different types of dust transport exist in the industry, including dust aspiration and dust extraction and lean to dense phase transport. Some dusts will stick to the pipe wall and build layers that may become airborne by the explosion propagating through the pipeline, resulting in combustible dust-air cloud generation in transport systems that typically operate at or below the minimum explosible concentration. It is therefore interesting to study the effect of pipeline dust concentration on the explosion propagation. Figure 12 shows examples for two dust types with the pipeline dust concentration changed from 500 g/m^3 to 50 g/m^3. The vessel dust concentration was kept constant at 500 g/m^3. Although the fuels are more reactive at the higher dust concentration, flames are slower at high pipeline dust concentrations. This effect is attributed to the change of the inertia of the fluids in the pipelines. Also this finding was confirmed by the experiments [3].

Figure 13 demonstrates that the combustion in the first meters of the pipeline is fueled by dust ejected from the exploding vessel and not so much by the pipeline dust.

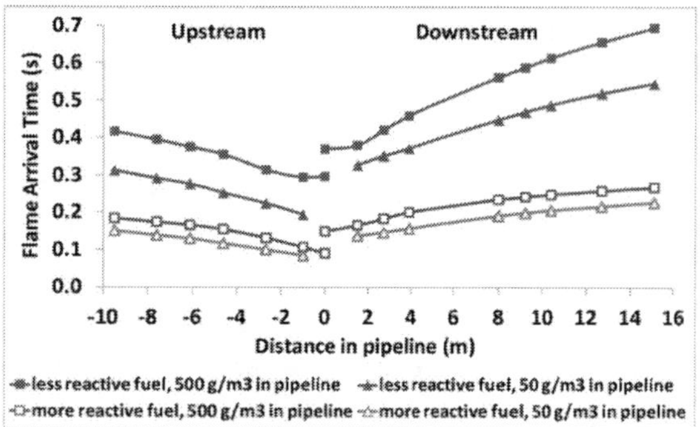

Figure 12: Effect of pipeline dust concentration on flame propagation for two dust fuels; Vessel dust concentration = 500 g/m^3.

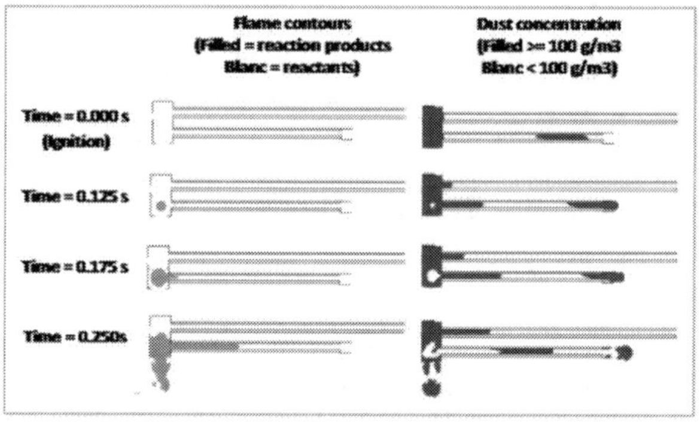

Figure 13: Illustration of how dust is ejected out of the initiating vessel before the flame front arrives for an example explosion; Vessel dust concentration = 500 g/m3; Pipeline dust concentration = 50 g/m^3.

7.3. Effect of ignition location

Scenarios testing the effect of ignition location were run with the igniters located in the three positions indicated in Figure 14 below, including ignition near the vent (1), at the interconnection with the upstream pipeline (2), and in the center of the downstream vessel half, far from the vent (3)

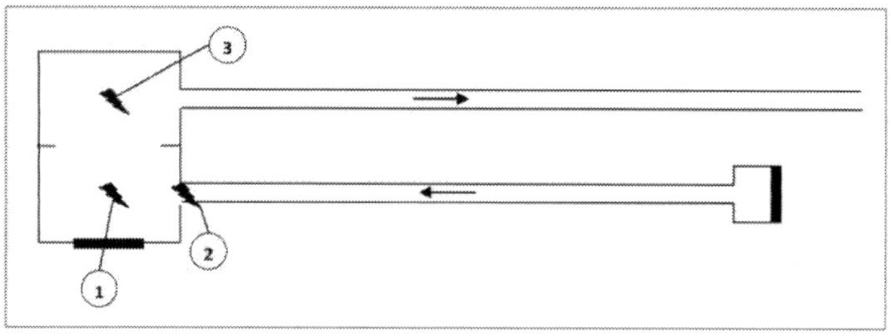

Figure 14: Ignition locations

The effect of the ignition location is twofold: the closer the ignition point is located towards the pipe entry, the sooner the explosion flame will enter the pipeline; and the further the ignition point is located from the pressure venting device, the higher will the explosion pressure in the vessel be, with faster flames as a result. These phenomena are illustrated in Figure 15, where for examples with two dust types, the flame arrival times are compared for three explosion scenarios with varying ignition location. It is clearly shown that flames enter the pipeline earlier for pipeline ignition (2), and that ignition far away from the vent panel (3) generates significantly faster flames

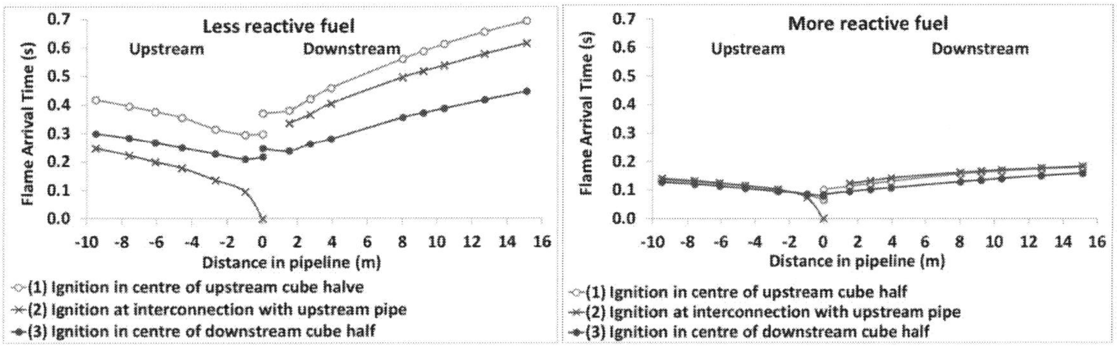

Figure 15: Effect of ignition location on flame propagation. Examples for two dust fuels.

7.4. Effect of vent location

Ideally, the explosion vent panel should be positioned as close as practically possible to the expected ignition location. This will provide an early escape of the combustion gases and will limit the resulting explosion overpressure in the vessel and consequently the flame speed in the pipeline. This is illustrated in Figure 16 and Figure 17: the further the vent panel is moved away from the ignition position (– with vent location alternative 3 further than alternative 2, and alternative 2 further than alternative 1), the faster the explosion flame propagates in both pipelines. The driving factor is the increased explosion intensity in the initiating vessel.

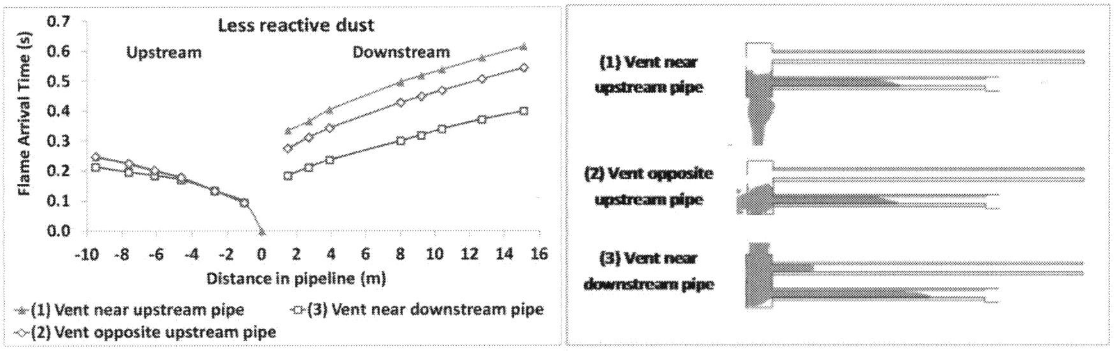

Figure 16: Effect of vent location on flame propagation for an example explosion scenario. Ignition location = interconnection of vessel with upstream pipeline. Left: flame arrival times. Right: flame contours at 0.2 s after ignition.

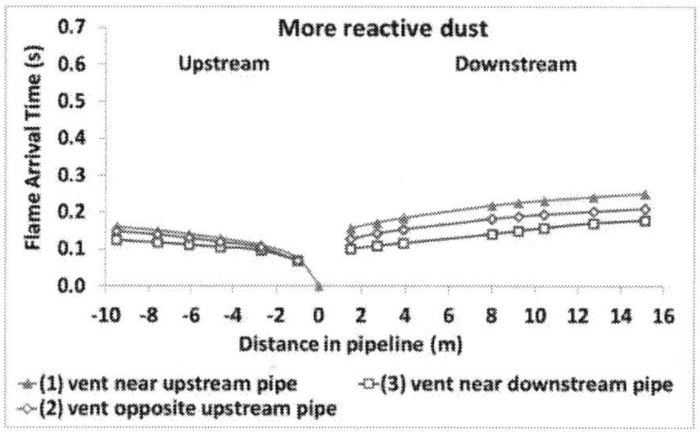

Figure 17: Effect of vent location on flame propagation for another example explosion scenario. Ignition location = interconnection of vessel with upstream pipeline.

7.5. *Effect of pipe length*

Longer pipes will increase the flow resistance and gas column inertia, resulting in a less aggressively accelerating flame along the pipeline. The effect is most pronounced for cases with low vessel explosion pressure, where flow resistance and gas column inertia play a relatively stronger role, and for cases with low K_{st} value, where the flame velocity is stronger linked with the gas flow velocity. Figure 18 shows the simulated flame arrival times for three example explosions, where the length of the piping was varied from the 11.3/16.5 m of the basic configuration to 40 meters.

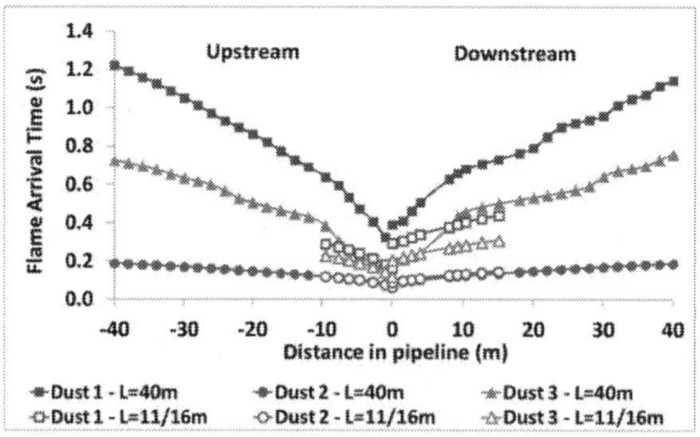

Figure 18: Effect of pipe length on flame propagation for three example explosions.

7.6. *Effect of pipe diameter*

As with the pipe length, changing the pipe diameter affects the flow resistance. But, in addition to this, the pipe diameter also changes the Reynolds number and thus the gas turbulence in the pipeline. Flames will accelerate faster in smaller pipes as a result of higher turbulence. Note that heat losses of the hot burning gases to the cold pipe wall material have not been modeled. Figure 19 shows the effect on flame propagation when the pipe diameter is decreased from 0.4 m to 0.2

m. For the example with lowest fuel reactivity and generating relatively low vessel explosion pressures, the flame propagates slower in the smaller diameter pipes than in the larger diameter pipes. For the example with highest fuel reactivity and generating relatively high vessel explosion pressures, the flame propagates faster in the smaller diameter pipes than in the larger diameter pipes. For the example with medium fuel reactivity, it is seen that in the initial phase of the explosion, the flame propagates slower in the smaller diameter pipe geometry, but later on, when the explosion pressure develops and the fluid turbulence enhances, the flame will propagate faster.

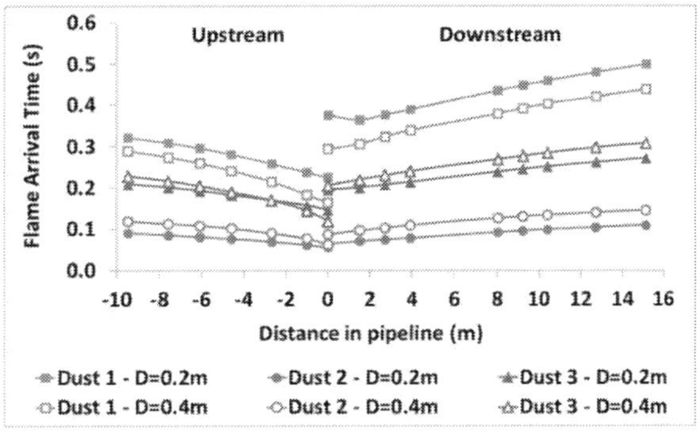

Figure 19: Effect of pipe diameter on flame propagation for three example explosions.

7.7. Effect of vessel volume

When the volume of the initiating vessel is increased, delayed pipeline flame propagation can be expected, because the explosion pressure rise in the vessel will be slower. This is clearly illustrated in Figure 20, where for three example explosions the flame propagation is compared for the cases when initiated in a 5 m^3 vessel and in a 62 m^3 vessel, all other parameters either kept identical or scaled with the volume. The ignition was located in the centre of the 5 m^3 upstream vessel half, which is 0.68 m away from the interconnection with the upstream pipeline, which is then of course out of the centre for the 62 m^3 vessel half. To make the comparison explosion pressure independent, the vent area on the 62 m^3 vessel was increased by factor of 4, which approaches the factor $\left(\frac{62}{5}\right)^{2/3} = 5.36$ which would be required theoretically. Initially the pressure rise is slower which leads to a delayed flame entry into the pipes, but as soon as the vessel pressures in the 62 m^3 exceed the ones recorded in the 5 m3, the flame acceleration rate for the 62 m^3 vessel cases becomes higher.

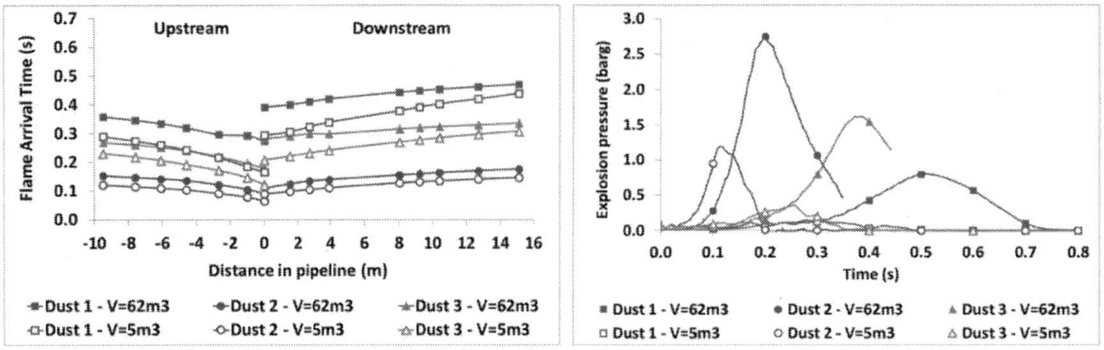

Figure 20: Effect of vessel volume on flame propagation for three example explosions. Left: flame arrival times in pipelines. Right: vessel explosion pressures.

8. Summary

DESC, a CFD code with an integrated combustion model, has been used to study the behavior of propagation of dust explosions through pipelines connected to vented vessels. A 5 m^3 vented vessel in which explosions were initiated, was connected to two DN400 pipelines, respectively 11.3 m and 16.5 meter long. The system was equipped with a fan to establish a flow field prior to dust injection and explosion ignition. Ignition always occurred in the vessel; neither the present study nor the accompanying empirical program [3] considered ignition in the pipeline. Input for the combustion model are laminar burning velocities extracted from K_{st} and P_{max} values determined by standardized 20 L tests.

Comparison between experimental and simulation results revealed that for two out of the three dust fuels, the experimental dust explosions were more intense than the simulated equivalents. A method was devised to improve the confidence level of DESC simulations. By following this approach, and by understanding the underlying physics, it is believed that simulations can represent industry reality with similar level of confidence as experiments.

A total of 94 explosion simulations have been performed in the basic configuration, where the fuel type, vessel and pipeline fuel concentration, vent area and flow velocity were varied. The effects of the parameter variations on flame arrival times in the pipelines were fully in line with the experimental results [3]. In all cases the flame propagated along the full length of the pipes, both upstream and downstream. A high degree of correlation was revealed between explosion induced expansion flow and flame speed.

172 additional simulations were performed in which flow direction, vessel volume, vent location, ignition location, pipe diameter and length were varied. This extended scope enabled to confirm the pronounced effect of the vessel reduced explosion pressure and pipeline geometry.

9. Conclusion

Understanding the mechanisms of explosion flame propagation is paramount for designing effective explosion protection solutions for interconnected process equipment that handle flammable substances. The present research has demonstrated that dust explosions do propagate through pipelines, even against strong flows and that many different application parameters besides K_{st} and P_{max} considerably influence the speed of dust flame propagation. When these parameters are not taken into account, the protection system may fail to protect against the explosion hazard. It has been irrefutably demonstrated that controlling the explosion pressure in the initiating vessel is very effective in controlling the speed of flame propagation. The explosion protection design should therefore always consider the explosion protection problem in its totality: a combination of vessel explosion pressure protection and pipeline explosion propagation protection.

10. References

[1] NFPA 68, Standard on Explosion Protection by Deflagration Venting, 2007

[2] NFPA 69, Standard on Explosion Prevention Systems, 2008

[3] T.M. Farrell, J. Vingerhoets, J. Snoeys, J.E. Going, "Dust Flame Propagation in Industrial Scale Piping - Part 1: Experimental Study in a Conveying Vessel-Pipeline System", Submitted for 9th Global Congress on Process Safety, San Antonio, April 28 - May 1, 2013

[4] T. Skjold, "Review of the DESC project", *Journal of Loss Prevention in the Process Industries*, Volume 20, p291-302, 2007

[5] Arntzen, B.J., "Modeling of turbulence and combustion for simulation of gas explosions in complex geometries", Dr. Ing Thesis, NTNU, Trondheim, Norway, 1998

Additional references not cited:

[6] A.E. Dahoe, J.F. Zevenbergen, S.M. Lemkowitz, B. Scarlett, "Dust explosions in spherical vessels: the role of flame thickness in the validity of the cube-root law", *Journal of Loss Prevention in the Process Industries*, Volume 9, p33-44, 1996

[7] A.E. Dahoe, R.S. Cant. and B. Scarlett, "On the decay of turbulence in the 20-litre explosion sphere", *Turbulence and combustion*, Volume 67, p159-184, 2001

[8] A.E. Dahoe, K. van der Nat, M. Braithwaite and B. Scarlett, "On the sensitivity of the maximum explosion pressure of a dust deflagration to turbulence", *KONA*, Volume 19, p178-195, 2001

[9] G.A. Lunn, "Dust explosions in totally enclosed interconnected vessel systems", *Journal of Loss Prevention in the Process Industries*, Volume 9, p45-58, 1996

[10] P. Holbrow, S. Andrews, G.A. Lunn, "Dust explosions in interconnected vented vessels", *Journal of Loss Prevention in the Process Industries*, Volume 9, p91-103, 1996

[11] M. Roser, A. Vogl, S. Radandt, W. Malalasekara, R. Parkin, "Investigations of flame front propagation between interconnected process vessels. Development of a new flame front propagation time prediction model", *Journal of Loss Prevention in the Process Industries*, Volume 12, p421-436, 1999

[12] A. Vogl, "Explosionsübertragung aus Behältern in Rohrleitungen", *VDI Berichte*, Volume 1272, p215-236, 1996

[13] J. P. Pineau J.P., G. Ronchail, "Propagation of dust explosions in ducts", Symposium on the control and prevention of dust explosions, Bâle, 1982

[14] A. Vogl, S. Radandt, "Explosionsübertragung durch dünne Rohrleitungen", FSA Report F-05-9903, 2005

[15] U. Bielert, M.Sichel, "Numerische Simulation von Staubexplosionen in pneumatischen Saug-Flug-Förderanlagen", *VDI Berichte*, Volume 1601, p449-472, 2001

[16] W. Jian, L. Xinguang, Z. Shengjun, S. Randandt, W. Fuli, R. Chunli, "Flame propagation through potato starch/air mixtures in pipe of interconnected vessels", *Measurement and Control of Granular Materials*, Volume 8, p376-379, 2009

(44c) The Use of CFD to Evaluate the Interactions Between Multiple-Leak Sources and to Assess Effectiveness of Integral Modeling Techniques in a Multiple-Leak Scenario

Joshua Richardson, Ha Nguyen and M. Sam Mannan, Mary Kay O'Connor Process Safety Center - Texas A&M University, College Station, TX

In many practical applications within the chemical industry, it is of vital importance to know the concentration profile that may occur from the dispersion of a material. For example, effect footprints for the release of a toxic material are needed to properly assess risk to both a facility and its surrounding communities and are useful for emergency response planning. Likewise, knowing the concentration gradient of a flammable gas can affect, among many other things, the layout of a facility to prevent or lessen the consequence of explosion and fire scenarios or for the placement of detection systems.

There are two main tools for the modeling of dispersion scenarios. Computational fluid dynamics (CFD) is a useful tool for accurately assessing the dispersion of materials, whether toxic, flammable, pollutant, or otherwise. It has found extensive use in academic as well as industrial settings when scenarios and geometries are well-defined because it takes into account the surroundings in the release environment. However, it is relatively slow and computationally expensive to use.

Also used extensively in the chemical industry are integral consequence modeling tools (such as Phast) which are generally more user-friendly and have the advantage of delivering results in a more time-efficient manner, but cannot take into account complex geometry considerations. A particularly interesting problem to consider in practice is that of multiple simultaneous release sources. This could be caused by a common failure of two units causing concurrent release, a domino effect where an event occurs that causes another event leading to multiple releases, or even an intentional event such as a terrorist attack. While CFD has the capability to resolve the possible interactions between the multiple release sources, integral modeling does not.

A simple current approach to the problem of multiple-source scenarios is to model them independently and superimpose the results upon one another. This may be a good approximation in some cases, but it does not take into account the complex interactions that may take place between the sources. The practical question then becomes whether it is acceptable in certain cases to model a certain set of simultaneous releases as an independent set of sources or whether it could be possible to model them as a single combined source.

This study aims to utilize CFD modeling of a set of varying scenarios to classify multiple-release dispersion scenarios into one of three different categories:

- Releases that are independent and may be modeled as such
- Releases that may be combined and modeled as one source

Releases that must be considered together due to inherent interactions between them

Based upon criteria such as material, distance between sources, dispersion rate and duration, weather conditions, and geometry considerations, a general methodology for classifying multiple-release scenarios into these three different categories will be proposed so that these releases can ultimately be modeled in the appropriate manner.

(44ca) Understanding, Managing, and Communicating High Consequence Risk

Charles A. Soczek, DuPont Sustainable Solutions, Wilmington, DE

This presentation discusses the relationship between the tools used to identify, understand, mitigate and communicate high consequence risks. The understanding and reporting of high consequence risks, along with providing management, senior leadership, and the Board of Directors with the information necessary to ensure successful risk management, are fundamental outputs to a risk management process. Tools such as the Risk Matrix, Bow Tie Analysis, and Risk Register will be discussed along with key issues such as risk mitigation and risk information flow from the operating level to senior leadership to the Board of Directors. Fundamental to this discussion is the concept that all hazardous events are managed to an acceptable risk through engineering and administrative controls, and a robust Process Safety Management system.

Frequencies of Accidents as Functions of Design Variables

Manuscript ID: 295029

Esteban Bernechea
Centre for Technological Risk Studies (CERTEC). Department of Chemical Engineering. ETSEIB. Universitat Politècnica de Catalunya.
Av. Diagonal, 647. 08028-Barcelona (Spain).
esteban.bernechea@estudiant.upc.edu

Cristian Massaguer
Department of Chemical Engineering. ETSEIB. Universitat Politècnica de Catalunya.
Av. Diagonal, 647. 08028-Barcelona (Spain).
cristian.massaguer@gmail.com

Josep Arnaldos
Centre for Technological Risk Studies (CERTEC). Department of Chemical Engineering. ETSEIB. Universitat Politècnica de Catalunya.
Av. Diagonal, 647. 08028-Barcelona (Spain).
josep.arnaldos@upc.edu

1. Introduction

The mainstream definition of Risk is that it is a measure of the damage caused on humans, the environment or material property, in terms of the probability of occurrence of an accident and the magnitude of the damage it causes. This can be expressed mathematically as follows:

$$Risk = Frequency \cdot Consequences \quad \text{[Eq. 1]}$$

Risk is then, a function of the frequency of occurrence and consequences of a possible accident.

Often, a Quantitative Risk Assessment (QRA) of a specific equipment or installation has to be performed; in Europe the Seveso Directive makes it compulsory for installations that work with a certain amount of dangerous materials. In order to perform a QRA it is necessary to define initial Loss of Containment Events (LOCs) in the studied equipment, to later estimate the frequencies of the accidents that may result from these releases, and the consequences of said events. These values can finally be combined in order to obtain a measure of the risk that the installation poses over individuals or communities.

There is a wide variety of techniques that can be applied to estimate the frequency of an accident, like historical analysis, failure trees or event trees, subjective estimation, Bayesian analysis or Monte Carlo simulations. In Catalonia and Spain it is common to use the values of frequency provided by "The Purple Book" (CPR18E) [1] for the possible LOCs that can occur in different equipment. Later, event trees which depend on the probabilities of occurrence of certain events (like the immediate or delayed ignition of a flammable release) are used to obtain the final frequencies of the set of accidents that will be studied in the QRA.

However, the base frequencies presented in The Purple Book do not explicitly include the possibility that the LOCs will occur due to a domino effect; this phenomenon is defined as a cascade of events in which the consequences of previous accidents are increased both spatially and temporally by following ones, thus leading to a major accident [2]. The complexity of domino effect makes it difficult to model, and more so to include in the calculations performed in a QRA. This leads to the problem of how to take into account a phenomenon that can have a significant impact on the risk associated to the installation, but that is virtually impossible to include (without significantly increasing the number of scenarios to be considered). In some cases in which there are various similar equipment, the common criteria is to select a representative scenario, and multiply its frequency by the number of units and a factor of 2. In this way, domino effect may be included in the calculations, but not in a rigorous way, which could lead to the under or over-estimation of risk.

In this work, a methodology proposed by the authors to produce accident sequences in a domino effect has been exploited to obtain equations that describe frequencies of accidents in a specific installation. The studied plant stores pressurized liquefied propane in horizontal cylinders. The equations will depend on the number of units and the mass that is stored in the installation; they will also depend on the layout, which in turn depends on the size of the tanks, and therefore on the mass and number of equipment. Distances between tanks were defined by using the NFPA58 standard [3].

By finding equations that can be used to estimate the frequency with which a certain accident will occur in the installation, the time required to include domino effect into a QRA could be decreased importantly; only the frequency obtained through the equations, applied to a representative scenario would be necessary to take the phenomenon into account.

The main objective of this work is to use a mathematical model to find equations that relate the frequency of accidents in an installation to some of its design variables. This could be useful not only at the moment of performing QRA, but also at the moment of designing the installation; also, it can help understanding how the frequency of the diverse accidents is affected by the design of the plant.

In the next sections, the model that allows developing the accident sequences is briefly explained, after this, the design of the experiment, which is how the mentioned model was exploited, is presented; later, the results obtained are shown, and finally, some conclusions are reached.

2. Model for the development of accident sequences

The model used to develop the domino effect sequences [4] is based on the use of event trees in a sequential way; it evaluates the damage suffered by one tank in the case that an accident occurs in another, and establishes the possible accidents that may occur afterwards. The LOCs and event trees used to develop the sequences are those presented in the Purple Book [1], and are shown in Table 1 and Figures 1 and 2.

Table 1: LOCs used to define initial accidents

Code	Definition
G.1	Instantaneous release of the complete inventory.
G.2	Continuous release of the complete inventory in 10 min at a constant rate of release.
G.3	Continuous release from a hole with an effective diameter of 10 mm.

The first step to produce the accident sequences is to select one of the units, choose an initial LOC and define the possible accidents that may occur. After this, one of the accidents must be selected, and its effects must be calculated on all surrounding units and compared against the threshold values presented in Table 2.

Depending on the effects of the previous accident each of the affected tanks will suffer a specific LOC, which will in turn, generate new possible accidents; the sequence will continue to unfold until all the tanks have suffered catastrophic events, or the frequencies of the accidents become lower than 10^{-9} y^{-1}. This procedure has to be repeated for each of the possible accidents and for every one of the initial LOCs, starting the sequence in each of the tanks. Finally, the frequencies of all the occurrences of a specific accident in all the sequences, (for example, a flash fire derived from an instantaneous release), has to be summed, in order to know the number of occurrences of the incident.

The frequencies of the accidents down the sequence are calculated using the event trees and probabilities of accidents; each time an accident occurs caused by a previous incident, its initial frequency will be that of the event that caused its failure, multiplied by its probability of occurrence.

Figure 1. Event Tree for the Instantaneous Release of a Pressurized Flammable Gas

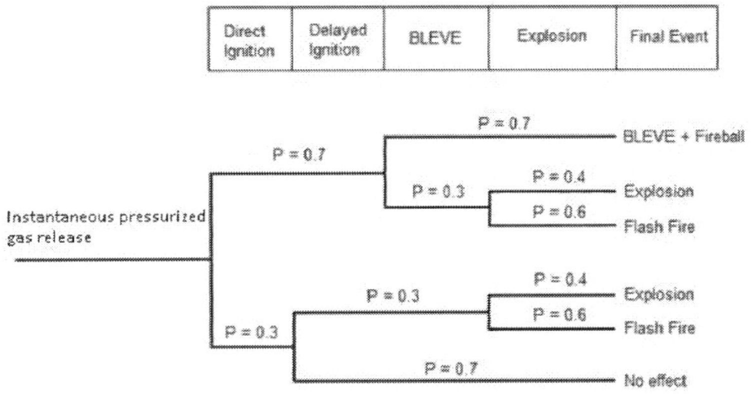

Figure 2. Event Tree for the Continuous Release of a Pressurized Flammable Gas

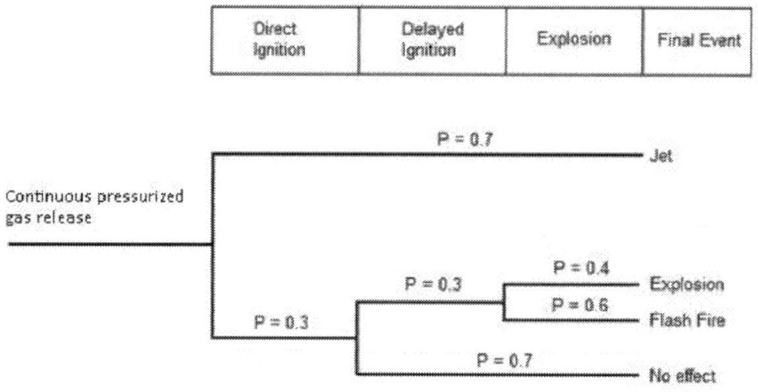

3. Experimental design

The equations that describe the frequencies of the different accidents will be developed for storage terminals of pressurized liquefied propane that use horizontal cylinders. The equations will be functions of the number of tanks used and the total mass of propane stored in the terminal. The number of tanks will be varied from 1 to 4, while 5 values of stored mass will be used: 5,000 kg, 10,000 kg, 20,000 kg, 50,000 kg and 100,000 kg; these possibilities for number of equipment and stored mass will be used to develop different layouts (for example 3 tanks which store 50,000 kg of propane) in which the total mass stored will be equally divided amongst all the tanks (all of which will have the same dimensions). The propane will be said to be stored at the conditions presented in Table 3.

Table 2: Threshold values for tank damage and associated LOCs

Accident	Effect	Threshold Values		Consequence
		Pressurized Tanks	Atmospheric Tanks	
BLEVE	Thermal radiation (kW/m2)	≥ 100	≥ 100 [5]	G.3 type release
		< 100	< 100 [5]	No consequence
Explosion	Overpressure wave (kPa)	≥ 100	≥ 25 [6]	G.1 type release
		$53 < \Delta P < 100$ [7]	$20 < \Delta P < 25$ [7]	G.2 type release
		$30 \leq \Delta P < 53$ [7]	$14 \leq \Delta P < 20$ [7]	G.3 type release
		$\Delta P < 30$ [5]	$\Delta P < 14$ [8]	No consequence
Jet fire	Jet length (m) Heat Radiation (kW/m2)	$dj \geq d$ [5]	$dj \geq d$ [5]	G.1 type release
		$dj < d$ & $I \geq 40$ [5]	$dj < d$ & $I \geq 10$ [5]	G.3 type release
		$dj < d$ & $I < 40$ [5]	$dj < d$ & $I < 10$ [5]	No consequence
Flash fire	-	-	-	No consequence
Pool fire	Thermal radiation (kW/m2) Pool duration (min)	$dp \geq d$ [5]	$dp \geq d$ [5]	G.1 type release
		$60 \leq I$ & $tp \geq 10$ & $dj < d$ [9]	$15 \leq I$ & $tp \geq 10$ & $dj < d$ [9]	G.2 type release
		$I \geq 8$ & $dj < d$	$I \geq 8$ & $dj < d$	G.3 type release
		$I < 8$ & $dj < d$	$I < 8$ & $dj < d$	No consequence

dj = jet length; dp = pool diameter; d = distance between tanks; tj = duration of jet fire; tp = duration of pool fire

As the frequency of the accidents will depend, not only on the number of tanks and the mass stored, but also on the distances between the units, this parameter has also been taken into account; this was done by using the distances proposed in the NFPA58 [3], which depend on the dimensions of the tank.

Figure 3 shows an example of the used layouts. In it, L is the length of the cylinder, while ϕ stands for the diameter; B represents the sum of the diameters of the tanks and the distances between them (d). The complete width of the facility will come from multiplying B by a factor of 1.1, and its length, from multiplying L by an equal factor.

Table 3: Atmospheric and Storage Conditions Used for the Application of the Model

Parameter	Value
Ambient Temperature (K)	298.15
Relative Humidity (%)	70
Atmospheric Stability Class	D
Wind velocity (m/s)	5
Ground Roughness Coefficient (cm)	10
Tank Pressure (bar)	9.51
Tank Temperature (K)	298.15

Since the substance stored in the facility is a liquefied pressurized flammable gas, the LOCs presented in Table 1, as defined in The Purple Book are used. The base frequencies of the different accidents for each tank are obtained by using the event trees presented in Figures 1 and 2 for the instantaneous and continuous releases.

Figure 3. Layout example

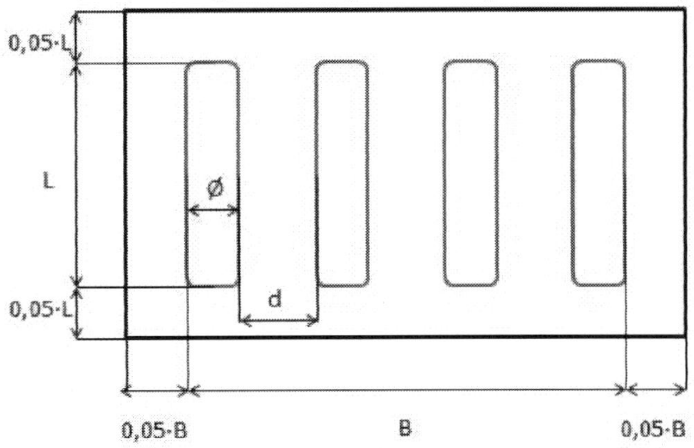

4. Results

It was necessary to use the model a total of 180 times to develop accident sequences for every possible initial accident, combining the number of tanks and values of stored mass. However, since all the proposed layouts are symmetrical, it was only necessary to perform 90 simulations.

Figure 4. G.1 accident sequences for two tanks storing 50,000 kg each

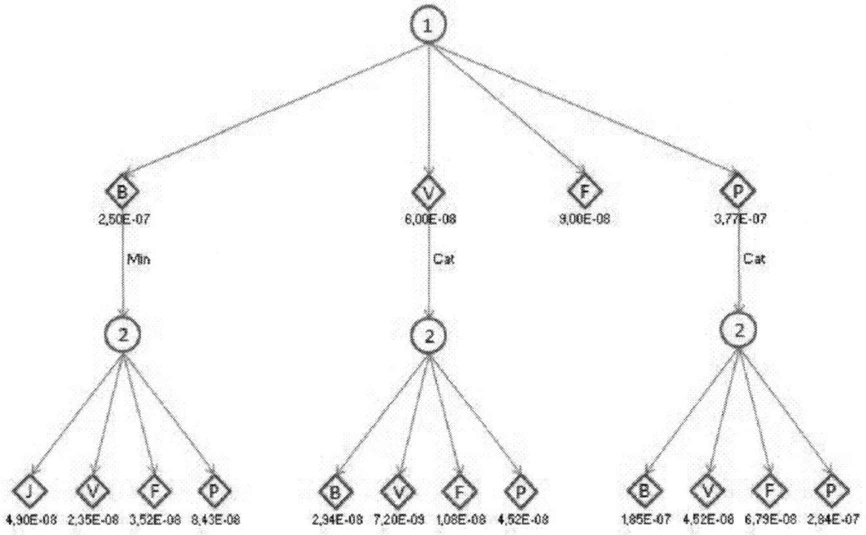

Figure 4 presents the possible accident sequences obtained by applying the model for the case of an instantaneous release from one of two tanks, each storing 50,000 kg of propane. In it, it can be seen that four possible accidents can occur after the G.1 release, a BLEVE (B), a VCE (V), a flash fire (F) or a pool fire (P); the frequencies with which these accidents occur is presented, as well as the level of damage they cause on the second tank, and the possible accidents that may occur in it, due to the initial ones.

The final values of frequency obtained for the case of BLEVE, depending on the mass and number of tanks are presented in Table 4. They serve as examples of the complete results obtained when the model is applied.

After the final frequencies for all possible combinations of design variables have been obtained, they can be plotted in 3D graphs, (in which one of the axis represents the number of tanks, another, the stored mass, and the last one, the frequency of occurrence). The plots obtained for all the studied accidents are presented in Figures 5 through 7.

Table 4: Final frequencies for the BLEVE case

Mass (kg)	Number of Tanks			
	1	2	3	4
5,000	1.75×10^{-7}	1.90×10^{-6}	5.77×10^{-6}	1.04×10^{-5}
10,000	1.75×10^{-7}	1.90×10^{-6}	4.12×10^{-6}	6.89×10^{-6}
20,000	2.45×10^{-7}	3.44×10^{-6}	5.13×10^{-6}	6.87×10^{-6}
50,000	2.45×10^{-7}	3.44×10^{-6}	9.95×10^{-6}	2.21×10^{-5}
100,000	2.45×10^{-7}	3.44×10^{-6}	1.18×10^{-5}	2.10×10^{-5}

Figure 5. Frequency vs. Number of tanks and Stored Mass for G.1 accidents

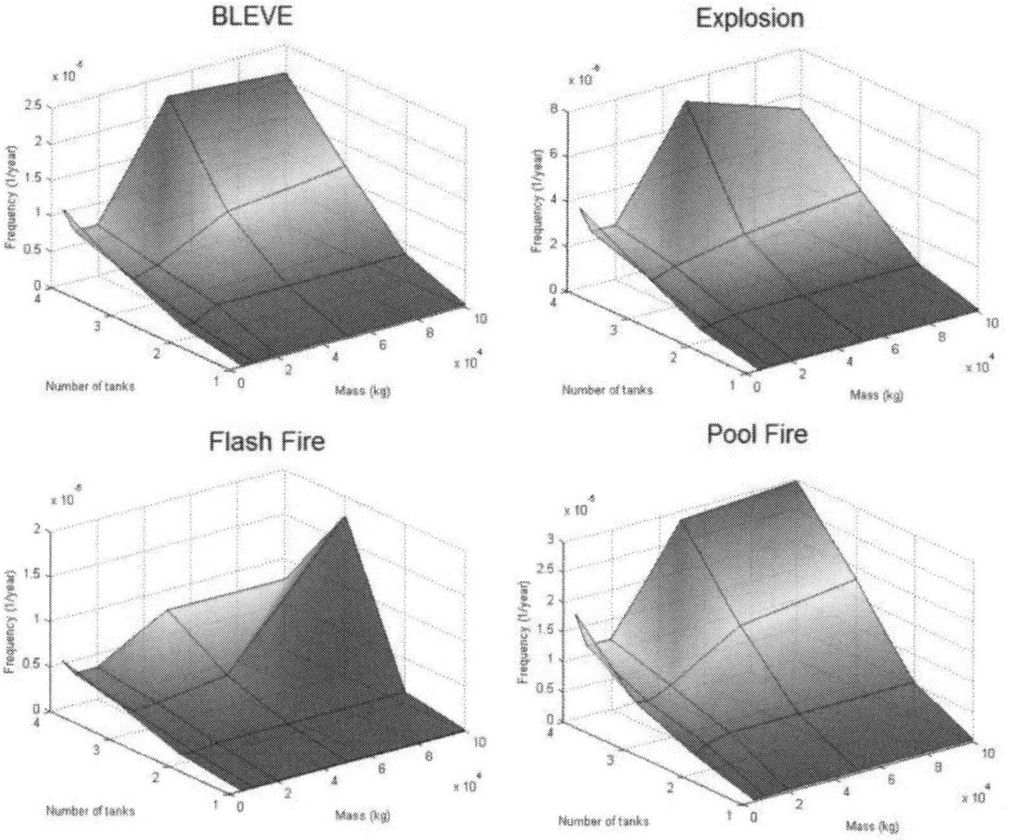

After the plots were obtained, they were fitted using different types of equations, like power, exponential or logarithmic among others. Finally, a Gaussian equation of the form presented in Eq. 2 proved to be the best fitting.

$$f = A \cdot exp\left(\frac{(n-B)^2}{C} + \frac{(m-D)^2}{E}\right) \quad \text{[Eq. 2]}$$

Where n is the number of tanks, m the stored mass in kg and A, B, C, D and E are dimensionless coefficients; these coefficients for all the studied accidents, as well as the values of R^2 obtained for the fitting are presented in Table 5, and the plots obtained when using the Gaussian equation with the presented coefficients are shown in Figures 8 through 10.

Figure 6. Frequency vs. Number of tanks and Stored Mass for G.2 accidents

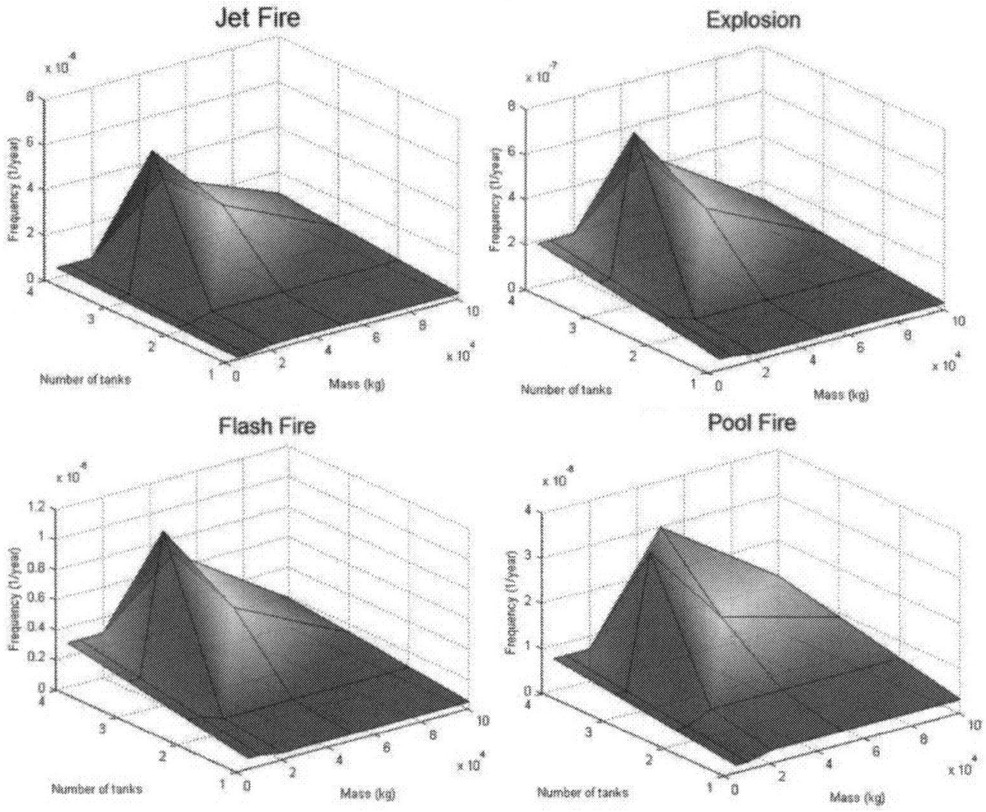

Table 5: Coefficients for the Gaussian equation for all accidents

Accident	Coefficients					R2
	A	B	C	D	E	
BLEVE	2.8x10-5	4.79	-4.15	7.7x104	-4x109	**0.94**
G.1 Explosion	9.9x10-6	5.37	-5.10	7.0x104	-4x109	**0.91**
G.1 Flash Fire	2.8x10-2	3.33	-0.8	1.0x102	-1.2x1011	**0.85**
G.1 Pool Fire	3.4x10-5	4.57	-3.96	8.3x104	-6.3x109	**0.95**
G.2 Jet Fire	2.2x10-5	3.14	-0.56	3.4x104	-1.5x108	**0.84**
G.2 Explosion	1.1x10-6	3.28	-0.82	3.3x104	-3.9x108	**0.77**
G.2 Flash Fire	1.6x10-6	3.27	-0.82	3.3x104	-3.9x108	**0.77**
G.2 Pool Fire	2.7x10-6	3.53	-2.17	5.5x104	-2.3x109	**0.57**
G.3 Jet Fire	1.5x10-4	6.13	-0.15	2.2x106	-1.8x1012	**0.98**
G.3 Explosion	7.8x10-5	6.04	-0.15	2.3x106	-1.9x1012	**0.99**
G.3 Flash Fire	9.2x10-5	6.10	-0.15	2.1x106	-1.7x1012	**0.98**
G.3 Pool Fire	2.2x10-4	6.10	-0.15	2.1x106	-1.7x1012	**0.99**

Figure 7. Frequency vs. Number of tanks and Stored Mass for G.3 accidents

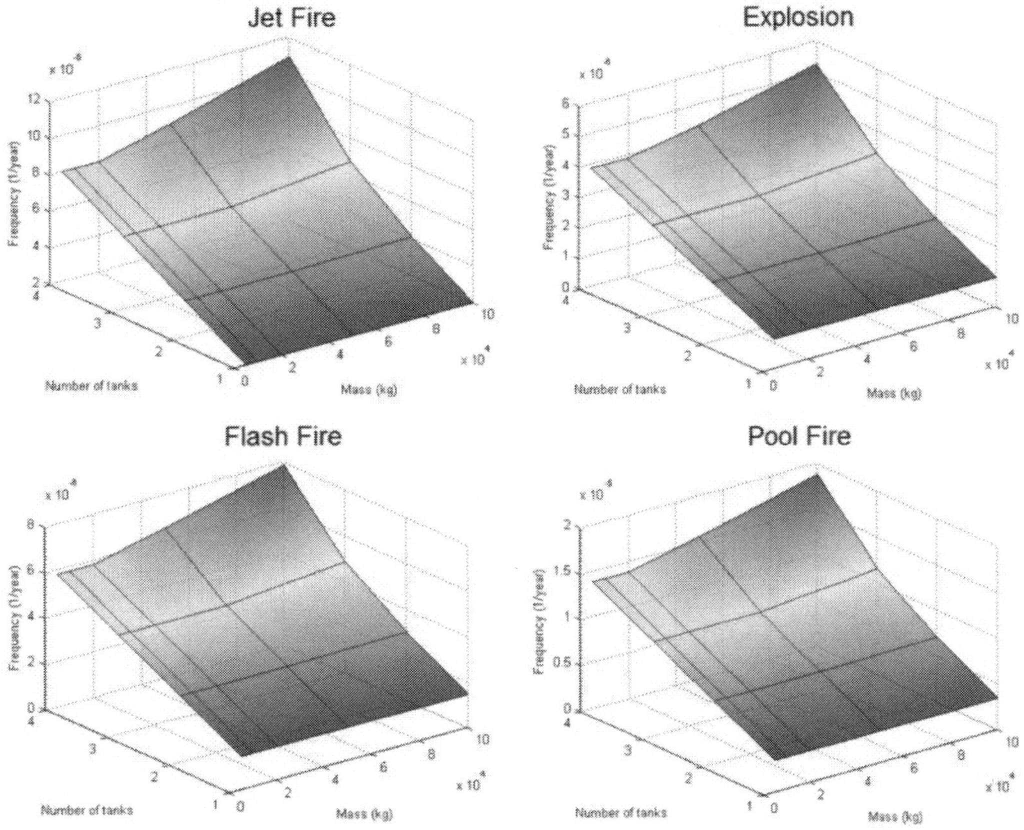

5. Conclusions

Equations that depend on the number of tanks and mass stored in the installation have been found, which, in many of the cases, present good tendencies in replicating the data obtained in the experiment.

The equations obtained could be used to take domino effect into account when performing a QRA on a storage installation of characteristics similar to those presented in the experiment, without increasing significantly the time required to complete the study.

Also, equations of this kind could be used during the design stages of a storage installation to help decide the number of equipment and quantities of substance stored; they could be used to know how risk behaves depending on different designs, and which design is inherently safer.

Figure 8. Frequency vs. Number of tanks and Stored Mass for G.1 accidents using the Gaussian Equation

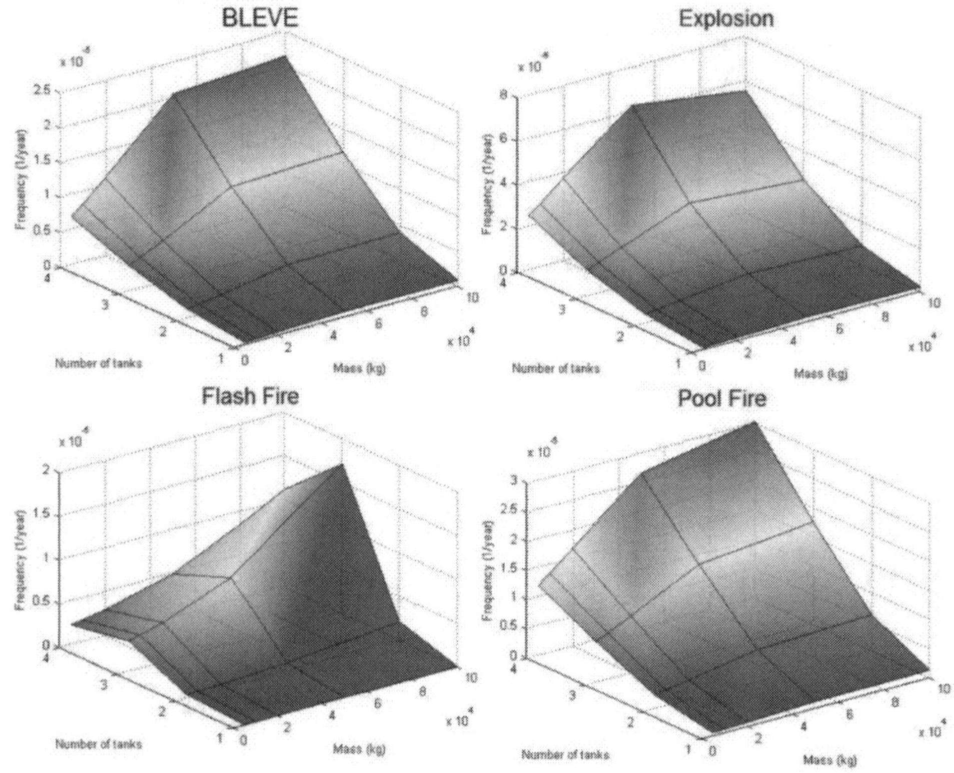

Figure 9. Frequency vs. Number of tanks and Stored Mass for G.2 accidents using the Gaussian Equation

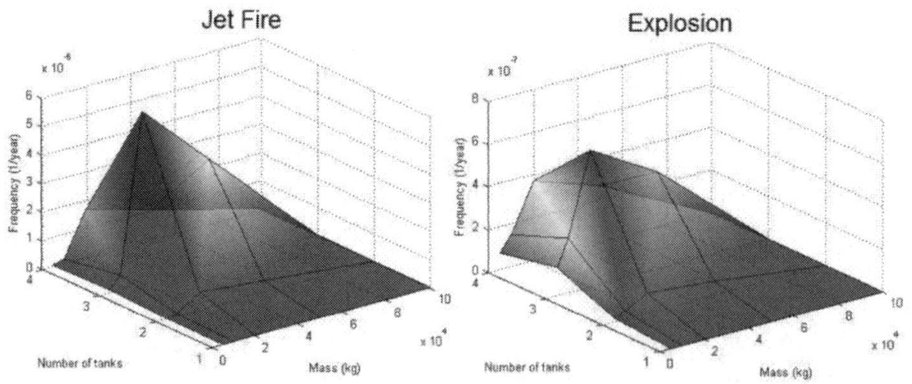

Figure 9. Frequency vs. Number of tanks and Stored Mass for G.2 accidents using the Gaussian Equation (Cont.)

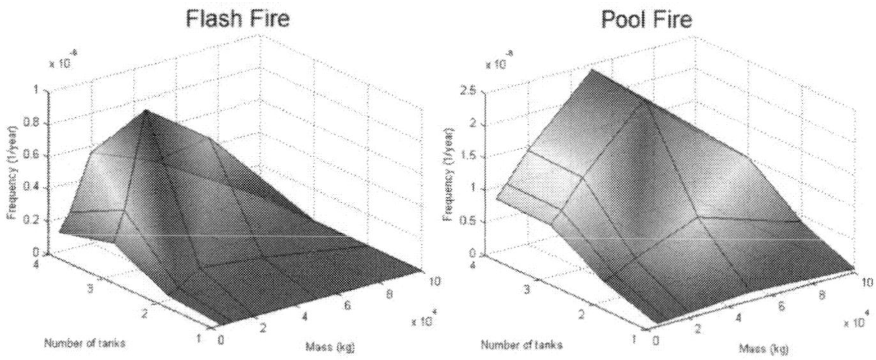

Figure 10. Frequency vs. Number of tanks and Stored Mass for G.3 accidents using the Gaussian Equation

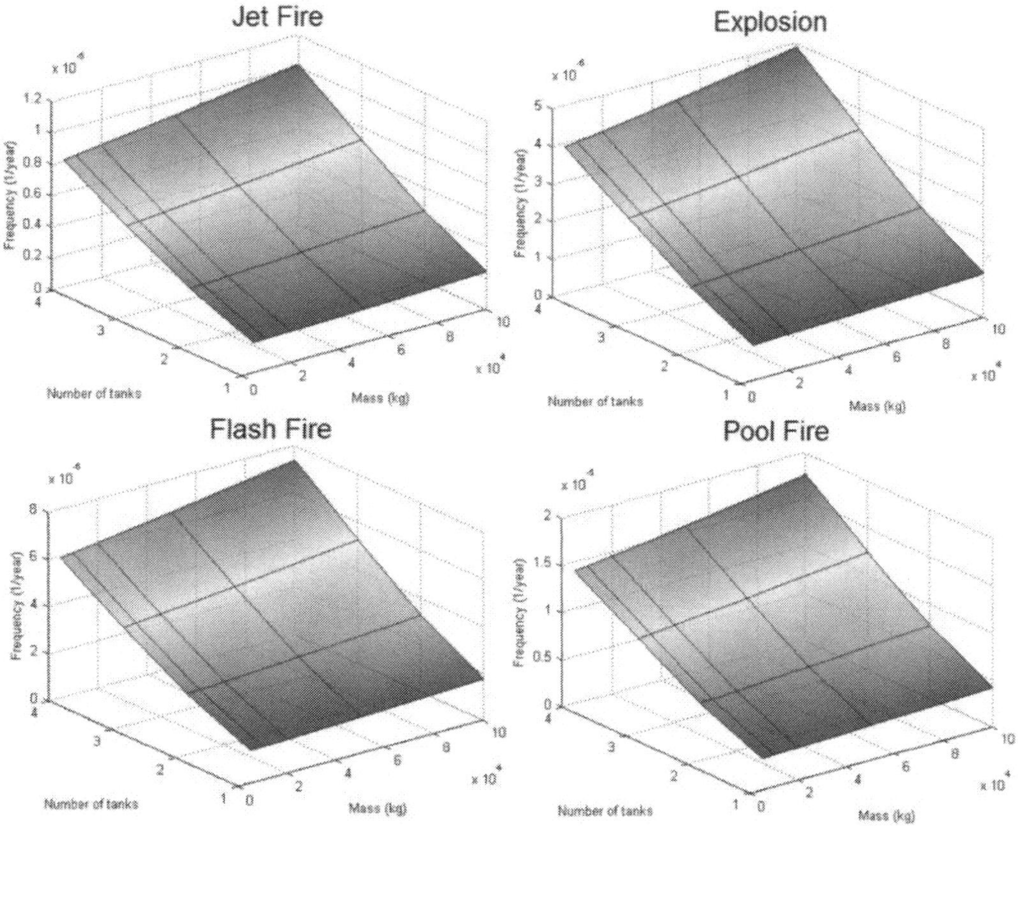

References

[1] CPR 18E, 2005. Guidelines for Quantitative Risk Assessment (Purple Book). Directorate General for Social Affairs and Employment, The Hague, Netherlands.

[2] Delvosalle, C, "Domino Effects Phenomena: Definition, Overview and Classification", First European Seminar on Domino Effects, Leuven, 1996.

[3] "Liquified Petroleum Gas Code". National Fire Protection Association (EEUU), 2001.

[4] Bernechea, E.J., Vílchez, J.A., Arnaldos, J., 2012. A model for estimating the impact of the domino effect on accident frequencies in quantitative risk assessments of storage facilities. Process Safety and Environmental Protection, http://dx.doi.org/10.1016/j.psep.2012.09.004

[5] G. Antonioni, G. Spadoni, V. Cozzani, Application of domino effect quantitative risk assessment to an extended industrial area, Journal of Loss Prevention in the Process Industries, 22, 614-624, 2009.

[6] F. I. Khan, S. A. Abassi, Estimation of probabilities and likely consequences of a chain of accidents (domino effect) in Manali Industrial Complex, Journal of Cleaner Production, 9, 493-508, 2001.

[7] V. Cozzani, E. Salzano, The quantitative assessment of domino effects caused by overpressure: Part I. Probit models, Journal of Hazardous Materials, 107, 67-80, 2004.

[8] V. Cozzani, E. Salzano, Threshold Values for domino effects caused by blast wave interaction with process equipment, Journal of Loss Prevention in the Process Industries, 17, 437-447, 2004.

[9] G. Landucci, G. Gubinelli, G. Antonioni, V. Cozzani, The assessment of the damage probability of storage tanks in domino events triggered by fire, Accident Analysis & Prevention, 41, 1206-1215, 2009.

(44ce) Routes Optimization for Hazardous Materials Transportation in Bogotá, Colombia

Jorge Victoria, Universidad de los Andes, Bogota, Colombia, Nubia Velasco, Universidad de los Andes, Eliecer Gutierrez, Industrial Engineer, Universidad de los Andes, Bogota, Colombia and Felipe Munoz, Universidad de los Andes, Bogota D. C., Colombia

The present work presents an optimization approach to design the set of routes for the transport of hazardous materials (HazMats) at Bogotá. HazMats transportation importance lies in the social, environmental and economic impact that might happen when an accident, during the transportation occurs. This paper is focused in developing a decision support system supported on optimization tools with the purpose of find the routes, -from gas distributor to gas stations-, that minimize: the total travel distance of heterogeneous tank trucks, the social risk and the infrastructure risk around the routes created.

The problem is modeled as a vehicle routing problem and it is solved using the principle of Route first, Cluster second. In a first stage the algorithm generates a big tour where all clients are visited, ignoring the vehicle capacities. Secondly, thanks to a Split procedure the tour is segmented considering the capacities. Then routes are improved with a MultiObjective Metaheuristic called MultiObjective Variable Neighborhood Search (MO-VNS). The algorithm is tested on two set of instances, the first one randomly generated and the second from Bogotá Geographic Information System (GIS). Several approaches to generated solution are proposed.

(44cf) Mejores Prácticas En Los Procesos De Gestión Del Cambio Para Seguridad De Procesos y La Evaluación De Seguridad Previa Al Inicio (PSSR)

J. Luis Aguirre, KMS, Optimized Systems and Solutions, Houston, TX

La Gestión del Cambio continúa siendo el elemento más retador/complicado de la Gestión de Seguridad de Procesos (PSM), del Programa de Gestión de Riesgos (RMP) y de los requerimientos regulatorios de los Sistemas de Gestión de Medio Ambiente y Seguridad (SEMS). A pesar de que el estándar PSM ha existido desde 1992 y que la industria ha gestionando cambios desde hace mucho tiempo, aún es común no obtener el resultado deseado. Los esfuerzos conscientes de muchas personas trabajando en conjunto son necesarios para garantizar que el cambio sea identificado, analizado y ejecutado con calidad. Justo cuando logramos capturarlo en papel y mantener informado al equipo de trabajo involucrado, se presentan cambios, negligencias o las regulaciones cambian. Nuevos involucrado y responsabilidades dificultan la gestión del cambio como sabemos que debe hacerse.

Sin embargo, la tecnología ayuda en todas las etapas del flujo de la Gestión del Cambio (MOC) al permitir la recopilación de ideas, formalizando el análisis de las mismas, su aprobación y ejecución del cambio. También puede ayudar a proporcionar visibilidad del estado actual y análisis de tendencias, así como la identificación de peligros a los que están expuestos los diferentes equipos, sitios, y bienes. Correctamente aplicada, la tecnología podría ayudar a predecir el futuro. Además, la tecnología facilita y ayuda a implementar sus estándares resultando en Mejores Prácticas de Gestión del Cambio, Evaluación de Seguridad Previa al Inicio (PSSR) y el seguimiento al cumplimiento de acciones, notificaciones de alerta y el cierre.

(44cg) Lesha – Multi-Layer Progressive Stress & Impact Assessment On Health & Behavior

Salvador Ávila Filho, PEI TECLIM, UFBA, Salvador, Brazil

The chemical process industry seeks, constantly, to achieve the demands social and environmental (mandatory and voluntary) within the international economic chain. The Integrated Management Systems meet issues in the areas of quality, environment, safety, health, social responsibility, now including risk management (ISO31000). This excess of valuation in integrated management system just prioritizes management functions in place of the functions related to core activity. By integrating management activities into the routine of work, the task will be performed in a more committed form avoiding the domain of management system in place of production skills.

Mandatory Standards for work in the areas of health, hygiene and safety are concerned in dealing the most damage, searching through punishment, correct managerial failure, or correct the ignorance. The preventive tools are not fully developed leading to leaders to be worried about maximum allowable values in the areas of safety, health, hygiene, and also the environment.

The Voluntary activities are not developed due to difficulties initiating deployment of management systems that were born in a TOP-DOWN philosophy. The human and operational reliability was borned in a shape BOTTOM-UP avoiding the occupation of operator, most part of time, doing management systems reports. Operational risks are analyzed including the role of man, the role of group in the planning and execution of tasks. The techniques involve subjectivities as effect of cultures fight in the body of work(social - mass culture, regional culture, organizational, and security). The analysis of dominant cultures involves identifying the various vectors impacting cultural influences in the workplace. Psychological stress causes physical and behavioral changes (affecting the decision) and in the body, both, can cause the loss of quality of life.

Various techniques were developed and presented in GCPS to assess the hazard and the risk of social conflicts and that can cause human error. This error initiates or increases the process of failure in equipment or processes through the intervention performed on the steps of the task. Thus, social conflict in new business (APRSH), cognitive traps and human error in critical task (Social HAZOP), causality analysis of cognition / decision, task and process losses (COGNOSP), analysis of organizational and cultural factors as root cause for accidents, decision analysis operation in progressive stress (LODA), and analysis of cognitive failure of operator in the control of equipment and components (FMEAH).

At the present moment we present a technique that try to study the effect of progressive psychological and physical stress on the body and quality of life, here dubbed as LESHA (Stress Analysis & Progressive Effect on Human Health and Behavior). This study highlights the importance of analyzing the life cycle of faults that impact on health by studying the trend of biological indicators resulting from psychological and physical stress. We also emphasize the search for signs of failure before occupational diseases being installed, showing the most suitable installation of barriers against failure, bringing the barrier closer to the root cause.

The LESHA technique therefore intends to monitor the level of physical and psychological stress of environmental factors, and stress from events related to task, which can turn a pathogen resident in symptom (Holnagel and Reason) and hence in occupational disease.

The events that initiate failures, and the stressed environments can cause stress progressive state on individuals, activating, or causing the trigger of failure through the stages: pathogen resident; uncontrolled organic process; symptom; disease; replication of the disease; and fatality or permanent labor shortage. Each level of stress is analyzed for signs and / or indicators of failure that impact over worker health permitting taking preventive or corrective action (biological safety barriers).

The following levels of analysis are proposed: Level 1 - Identification of Human Type, and the respective pathogen resident or signs that may cause the future disease in the body and in the behavior. Level 2 - initiators of failure in the task, or, bad social relationship that can cause the lack/ fault of organic metabolism, generating growth signals, variables still toward the disease. Signs of lost productivity interlinked with organic symptoms ... Level 3 - Shares of the social environment (society) or technical (technical knowledge), and secondary causal events and their impact on the body by increasing the metabolism of organic lack/ fault of symptom onset with fixed initial stage of the disease. Signs of lost productivity interlinked with organic symptoms...

Level 4 - Emergence of social ills (irritation, and communication difficulties), emergence of occupational diseases (slips related to the task and can cause accidents), and biological diseases like ulcer, high blood pressure and other diseases related to workplace using indirect biological ways (early onset of diabetes or other). Level 5 - Replication of the disease by adopting bad habits in the workplace where colleagues, "introject" behavior and even biological processes of "friends." Level 6 - fatality or permanent labor shortage causing low morale in the team, and developing or activating resident pathogen not active or nonexistent.

Understand the signs of this biological process is challenge of safety, hygiene and occupational medicine in conjunction with studies of human reliability. Different disciplines come together to prevent the disease from the body (regardless of the root cause) had been evident. The diseases are originate from labor, directly, or resulting from the influence of the job, and, from society influence. One should avoid firing the trigger of failure and the development of the pathogen resident.

The result of this analysis allows build the programs to avoid diseases. Prevention and correction programs depend on the intensity of the interference signals and the stress level reached in this process. It is proposed for each level and intensity of the biological and behavior signals following through risk mapping:
* Intensity Level X frequency.

Building Design for Resilience Beyond Design Basis Blast Loads

Michael A. Polcyn
Baker Engineering and Risk Consultants, Inc.
3330 Oakwell Court, Suite 100
San Antonio, TX 78218
MPolcyn@BakerRisk.com

Raymond H. Bennett, Ph.D., P.E.
Baker Engineering and Risk Consultants, Inc.
3330 Oakwell Court, Suite 100
San Antonio, TX 78218
RBennett@BakerRisk.com

Prepared for Presentation at
American Institute of Chemical Engineers
2013 Spring Meeting
9th Global Congress on Process Safety
San Antonio, Texas
April 28 – May 1, 2013

UNPUBLISHED

Building Design for Resilience Beyond Design Basis Blast Loads

Michael A. Polcyn
Baker Engineering and Risk Consultants, Inc.
3330 Oakwell Court, Suite 100
San Antonio, TX 78218
MPolcyn@BakerRisk.com

Raymond H. Bennett, Ph.D., P.E.
Baker Engineering and Risk Consultants, Inc.

Keywords: blast resistant design, consequence-based design, risk-based design

Abstract

One of the guiding principles of API RP-752 is that buildings intended for occupancy are to be designed, constructed, installed, modified, and maintained to protect occupants against explosion, fire, and toxic material releases. Owners/operators may use either a consequence-based approach or a risk-based approach to evaluate the siting of both new and existing buildings for blast loads. While a consequence-based approach should be based on a maximum credible event (MCE) for each building and type of hazard considered, a risk-based approach may consider numerous scenarios, accounting for both the consequences and the frequencies of the scenarios. However, when designing either new buildings or upgrades to existing buildings, designers are typically provided only a single set of blast loads and a single response criteria (for example, medium response per ASCE "Design of Blast-Resistant Buildings in Petrochemical Facilities"). A lower response level implies lower occupant vulnerability, while higher response levels imply higher occupant vulnerability. A design for higher magnitude loads to low response criteria would result in very conservative design. On the other hand, a design for lower magnitude loads to higher response limits would result in a building without capacity to protect occupants in the event of a more severe explosion.

A better approach would be to consider multiple levels of blast loads when developing a design. Lower magnitude loads (usually from higher probability events) should require designs to meet the lower response limits, while higher magnitude loads (usually from lower probability events) should require designs to meet the higher response limits. This paper demonstrates this approach by evaluating typically designed building systems for a range of blast loads commonly encountered at refining and chemical plants. Recommendations are provided for the design of new buildings or upgrades to existing buildings. By including the concept of resilience in the design process, the designer may provide protection from future yet-to-be-determined hazards or scenarios.

1 Introduction

The structural design of buildings for conventional loads, including wind, snow, and seismic loads, is a well-defined process that is followed by engineers on a routine basis. Loads are specified by various codes or design manuals (e.g., International Building Code [1] or ASCE 7-05 [2]). Similarly, the design approach for determining the capacity of various types of structures is defined in corresponding design manuals. For example, the American Concrete Institute publishes "Building Code Requirements for Structural Concrete" [3], and the American Institute of Steel Construction publishes the "Manual for Steel Construction" [4].

In contrast, the design of buildings subjected to blast loads is not as formalized. Design case blast loads are not specified in any guidance documents and are left for the owner to define appropriate loads based on the processes and materials in place at their facilities. There are several guidance documents that describe approaches for designing and analyzing structures subjected to blast loads, which include dynamic analysis methodology, material properties, and structural response limits. Response or deformation limits, such as those defined by the American Society of Civil Engineers [6], are given ranges of response, where Low Response refers to a component with no more than slight visible permanent damage, Medium Response corresponds to a component with some permanent deflection, but is generally repairable, and High Response indicates that the component has not failed, but has undergone significant permanent deflections and is likely unrepairable following the design case loads. It is up to the owner to select the response level, and thus the level of protection to building occupants, for their building design.

This paper provides an overview of the design process, highlighting the differences between a consequence-based design approach and a risk-based design approach. The discussion continues by presenting features to the design that will enhance the structure's ability to be usable following an explosion event and to provide some degree of protection to occupants for events exceeding the selected design load cases. It should be noted that only explosion hazards are addressed in this paper; building and upgrade designs should also consider hazards from fire and toxics.

2 Dynamic Analysis Method

When designing buildings or building upgrades for blast loads, structural components are usually assessed using nonlinear dynamic Single-Degree-of-Freedom (SDOF) analytical methodology. The SDOF methodology summarized in Biggs [6] is a widely accepted approach for estimating damage to structural components subjected to time-varying blast loading. The SDOF approach is a government and industry accepted methodology that has been described in many guidelines such as UFC 3-340-01 [7], UFC 3-340-02 [8], and ASCE [5].

The dynamic response of structural components from prescribed blast loads is determined by modeling the components as an SDOF system, such as the equivalent spring-mass system shown in Figure 1. Structural components such as walls, windows, beams, doors, and panels will deform and vibrate when loaded with a shock wave. The SDOF model for each component is constructed using the component's physical structural properties so that the model will exhibit the same displacement history as the point of maximum deflection in the component. The displacement history of the SDOF model is obtained with numerical integration techniques using a time-stepping computer program to solve the equation of motion of the equivalent system at discrete time steps.

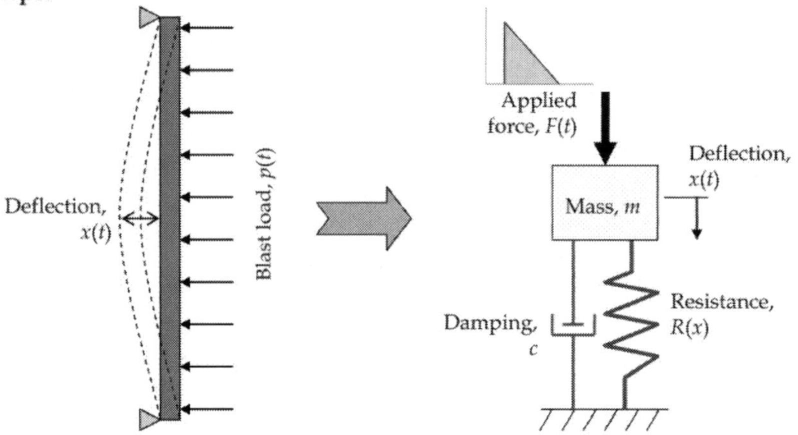

Figure 1. Equivalent Spring-Mass SDOF System

The calculated peak deflection (see Figure 2 for example) is used to obtain the support rotation and ductility ratio, which represent the deformation limit criteria most commonly used in blast design. As shown in Figure 2, the support rotation is the angle between the original un-deflected shape of a component and a straight-line segment between the point of maximum deflection and the support. The ductility ratio expresses the maximum deflection in terms of the minimum deflection of the component when it reaches its ultimate capacity. Therefore, ductility ratios that are greater than unity (1.0) indicate that permanent deformations have been sustained.

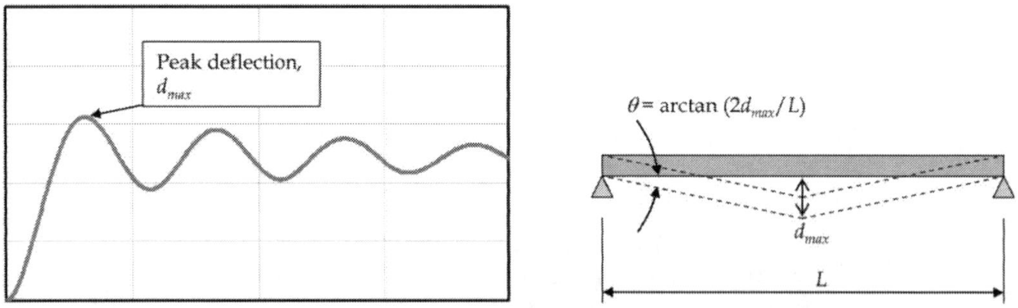

Figure 2. Calculation of Support Rotation from Peak Deflection

The design process for blast loads is an iterative process, where the pertinent design blast loads and response criteria are defined, "trial" structural members are selected based on the building geometry, and the structural components are analyzed using the dynamic analysis methods discussed above. If the response of the component does not meet the required response criteria, a heavier section is evaluated or the building geometry is modified until the analysis shows that the response criteria are met. Connections are typically designed to resist the maximum end reactions predicted from the dynamic analysis.

Pressure-impulse (P-i) diagrams are commonly used to assess building or component response to multiple sets of blast loads. P-i diagrams are plots of constant damage levels for various combinations of blast overpressure and impulse. Each individual curve represents the suite of all pressure and impulse pairs that will result in the same structural damage to a given component. The combination of the three individual curves forms a complete component P-i diagram to define Low, Medium, and High Damage Levels. An example of a P-i diagram is provided in Figure 3. As shown, the region above the upper curve would imply component failure.

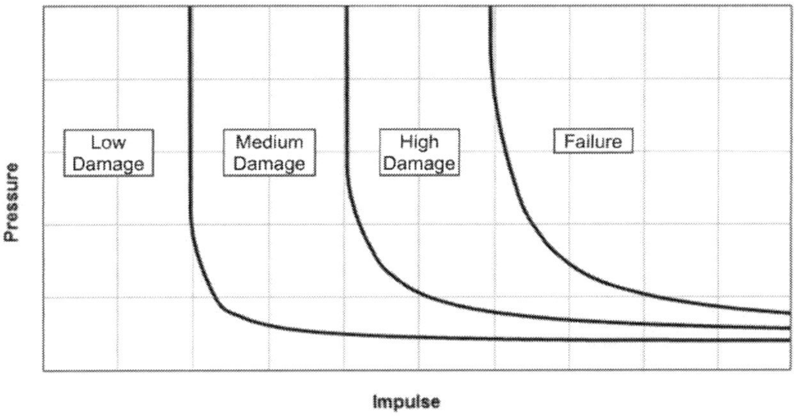

Figure 3. Example of a Pressure-Impulse Diagram

The use of P-i diagrams allows for quick consideration of a variety of blast loading situations. If the blast load of interest falls below or to the left of a particular damage curve, then the given component is predicted to receive damage less than or equal to the level represented by that P-i curve. In addition, component curves can be combined to develop a set of building P-i damage curves, or BDL curves.

3 Consequence-Based vs. Risk-Based Design

Traditionally, design of buildings or upgrade to buildings has been performed using a consequence-based approach where the frequency of events is not considered. According to API RP-752, a consequence-based approach is to be based on maximum credible events (MCEs) for each building. For example, a precast concrete wall for a control room could be designed to meet the ASCE Medium Response criteria when subjected to the applied blast loads from the MCE determined for that building. The resulting P-i curves for the wall would be similar to those shown in Figure 4.

The figure also demonstrates the concept of resiliency in the design. Four pressure-impulse pairs are shown, corresponding to two Design Load Cases and two Potential Overload Cases. The potential overload is 1 psi greater than the design pressure for both cases, while the impulse is held constant. In both Design Cases, the component just meets the Medium Response criteria. However, the curves show that for Design Load 1, the component has the capacity to resist the overload, albeit with larger deformations, but without failing (the component undergoes a high level of response). For low probability events, such behavior might be acceptable. However, the implied resiliency of the component cannot be determined solely by setting the response limits. For example, in Design Case 2, the corresponding overload will result in the failure of the component, even though the Design Case load is farther below the boundary between medium and high response. Thus the true resiliency of a design has to be evaluated in light of the potential for an overload.

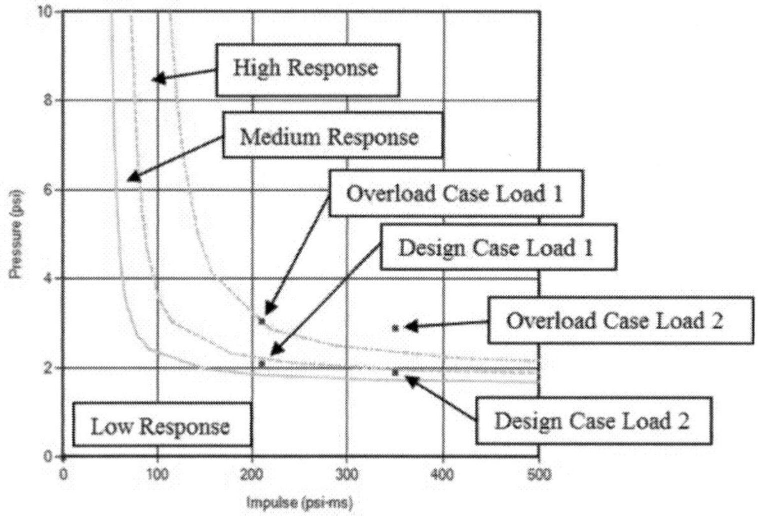

Figure 4. Component P-i Curves

In more recent years, companies have conducted quantitative risk assessments (QRAs) in support of facility siting study efforts. Such studies include estimates of the expected frequency and consequences of potential accidents, typically resulting in an estimate of risk, either as an aggregate risk, an individual risk, or a relationship of the expected frequency to the consequence (e.g., an F/N curve). This approach, which considers a large number of possible explosion scenarios, can be extended to a risk-based design approach for buildings or building upgrades.

When using a risk-based approach, it is necessary to consider the overall building damage rather than the response of the individual components independently. By doing so, the predicted building damage for each scenario can be related to the expected occupant vulnerability [9], which is used to perform the risk calculations. For example, the Building Damage Level (BDL) curves in Figure 5 show that the building is expected to undergo moderate damage, or BDL 2a, when the components are designed to meet the ASCE Medium Response criteria. Note that the inflection points in the BDL curves are caused by a change in the governing building component. For example, the upper left portion of the BDL 3-4 curve may be set by roof purlin response and the lower left portion may be set by a roof girder response. If the design case loads are defined by the MCE for all of the postulated explosion scenarios, the BDL plot in Figure 6 can be generated showing the loads from “all” postulated scenarios with the BDL curves. As shown, none of the scenarios will result in greater than moderate damage (BDL of 2a), and the resulting risk will be very low. It can be concluded that such a design has significant resiliency to withstand loads greater than those postulated.

Alternatively, the building could be designed for higher response limits under the design load cases, possibly allowing partial building failure under the most severe load cases. The BDL curves in Figure 7 demonstrate a case where the components were designed to meet the ASCE High Response limits for the MCE and ASCE Medium Response Limits for the majority of the load cases. As shown, the expected damage for the MCE approaches higher BDLs, but building collapse (BDL of 4) is not expected. In this case, the expected risk to building occupants would still be relatively low, although somewhat higher than the risk demonstrated in Figure 6. However, there is less resiliency in the design should there be an overload case. Note particularly that if the scenario shown in Figure 5 were 0.5 psi greater, the BDL for the building portrayed in Figure 6 would still be 2A, while the BDL for the design represented in Figure 7 would be increased to BDL of 3.

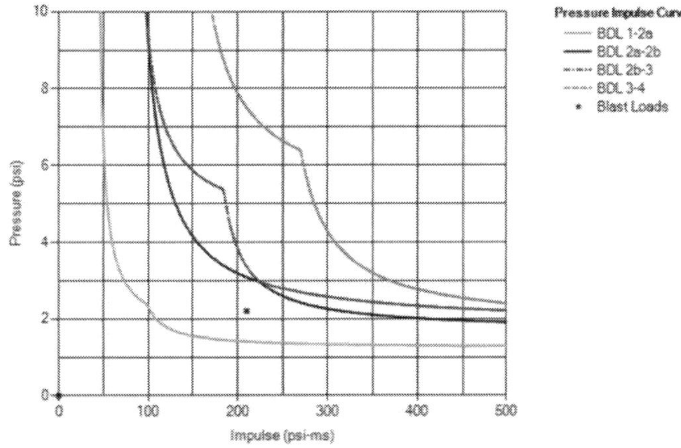

Figure 5. Building Damage Level (BDL) Curves

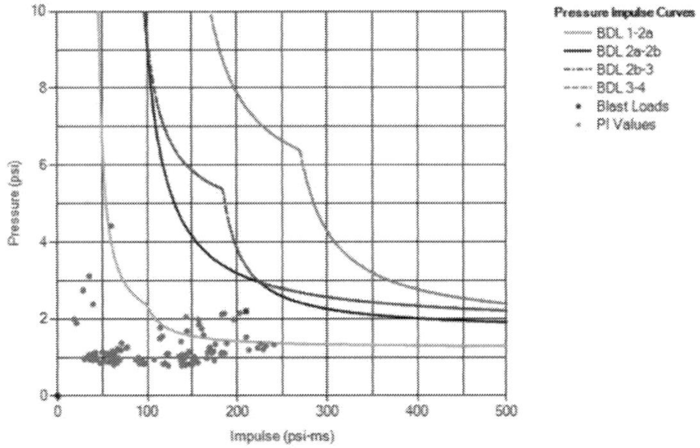

Figure 6. BDL Curves with Loads from "All" Postulated Scenarios

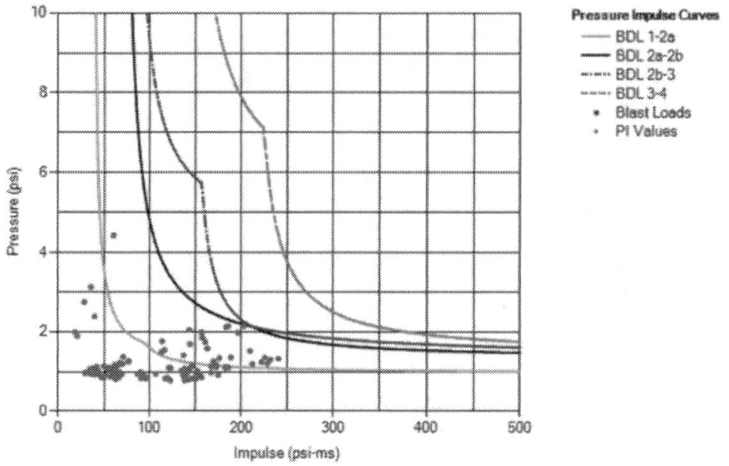

Figure 7. BDL Curves for "Weaker" Building

4 Resiliency in Design of Occupied Buildings

The examples discussed in the previous section demonstrate three approaches to providing a resilient structural design:

1. Structural systems that can develop a high degree of ductility are greatly preferred over structural systems that undergo a more brittle response. Most materials commonly used for blast resistant structures (steel and concrete) can undergo significant plastic deformations, typically well beyond those associated with published response limits, before failing. This plastic deformation allows the building to absorb the energy applied by the blast load. In addition, structural components designed to undergo large plastic deformations will have lower support reactions than those designed to behave elastically or with small plastic deformations, thus resulting in lower demands for connections and the building lateral load-carrying systems.

2. When selecting response limits, the likelihood of the selected design loads should be considered. Typically, high probability scenarios result in lower magnitude loads, while more severe load cases are the result of lower probability scenarios. This would imply that higher response limits could reasonably be used when designing for low probability, high magnitude load cases. Similarly, when designing for higher probability / lower magnitude loads, or loads with considerable uncertainty associated with their prediction, lower response limits should be considered.

3. Since one of the objectives of a blast resistant building is to protect building occupants, the layout of the building and the location of occupants relative to the sources of the potential explosions should be considered. By placing unoccupied areas of the building on sides facing the explosion source, localized failure of non-loadbearing components may be acceptable.

5 Conclusion

The design of blast resistant buildings does not follow an approach that is as prescriptive as the approach used for conventional design. Blast loads must be defined based on the conditions at the facility. Appropriate response limits must be selected for the components comprising the building. In addition, process conditions and building use may change to meet future requirements. Due to the uncertainty that is inherent in this design process, it is important to provide a building design that protects building occupants with a degree of resiliency. This can be developed by selecting the building structural system, building layout, and required response limits in a manner that is consistent with the postulated loads and the expected potential for these loads. In addition, the potential for more severe loads should also be considered in the design process.

6 References

[1] 2012 International Building Code, published by the International Code Council, Inc., May 2011.

[2] ASCE/SEI 7-05, "Minimum Design Loads for Buildings and Other Structures," published by the American Society of Civil Engineers, 2006.

[3] ACI 318-11, "Building Code Requirements for Structural Concrete and Commentary," reported by ACI Committee 318, published by the American Concrete Institute, 2011.

[4] "Steel Construction Manual," 14th Edition, Second Printing, published by the American Institute of Steel Construction.

[5] "Design of Blast Resistant Buildings in Petrochemical Facilities, Second Edition," Task Committee on Blast Resistant Design, American Society of Civil Engineers, 2010.

[6] Biggs, J., Introduction to Structural Dynamics, McGraw-Hill, Inc. New York, NY, 1964.

[7] UFC 3-340-01, "Design and Analysis of Hardened Structures to Conventional Weapon Effects," June 2002. (Formerly Technical Manual for Army TM 5-855-1, Air Force AFPAM 32-1147(I), Navy NAVFAC P-1080, and Defense Special Weapons Agency DAHSCWEMAN-97, August 1998.)

[8] "Structures to Resist the Effects of Accidental Explosions," UFC 3-340-02, 5 December 2008.

[9] Oswald, Charles J. and Baker, Quentin A., "Vulnerability Model for Occupants of Blast Damaged Buildings," presented at the 34th Annual Loss Prevention Symposium, March 6-8, 2000, Session 3.

(44g) Reliability Analysis of Man-Machine Response to an Emergency Situation Based On a Detailed-Action Plan

Najmeh Vaez, North America Risk, Environmental Resources Management (ERM), Houston, TX; Environmental Science and Safety Department, University of Houston Clear Lake, Houston, TX, Farshad Nourai, Aftab Imen Parto Consulting Engineers Co., Tehran, Iran, Yaneira Saud, North America Risk, ERM Southwest, Houston, TX and Suresh Yelisetty, Americas Risk Practice, Environmental Resource Management (ERM), Houston, TX

The primary emergency response actions in oil and gas industries are the functions of the process safety control system and the actions completed by the control room operators. The performance of the safety control system and the control room operating crew has a significant role in proper emergency response. To overcome each critical scenario, an emergency response operating procedure is provided with specific instructions, that in this research paper it is named a detailed action plan (DAP). This term is recommended for this purpose by the Center for Chemical Process Safety (CCPS). The reason of using a DAP is to reduce the probability of error in the emergency response of control room operators by providing scenario-specific written instructions. However, most of the DAP's are developed during design phase when there is no opportunity to test them through emergency drills. Therefore, a framework is proposed to analyze the reliability of the DAP's. This framework should be able to determine reliability of every single human and equipment action as well as the combined human-machine reliability, modeled by means of a Reliability Block Diagram to calculate the reliability of integrated automatic-operator emergency actions. The framework focuses on incorporating the operational and cognitive operator's errors in the process/equipment reliability analysis. For this purpose, a human reliability analysis method named SPAR-H was utilized. The proposed framework was applied for analyzing DAPs of an emergency well control operation in order to check and improve the applicability of this framework. The presented framework facilitates the DAP assessment process by performing task analysis modeling the detailed tasks, and completing a human-machine reliability analysis. Besides, the application of the presented framework, its findings and recommended actions enables a considerable improvement in the reliability of DAPs and therefore proves its efficiency and effectiveness.

(44i) Hazards of Gasoline Contamination – Case Study of a Tanker Truck Explosion Resulting From Switch Loading

Scott G. Davis, GexCon US, Bethesda, MD and Kees Van Wingerden, GexCon AS, Bergen, Norway

When a tanker truck is drained of a Class I liquid (e.g., gasoline) and refilled with a similar liquid, there is no flammable mixture on the surface of the rising fluid because Class I fluids create mixtures at their surface that are too fuel rich to burn. Similarly, if a tanker truck is drained of Class II liquid (e.g., diesel) and refilled with a Class II liquid, there is no flammable mixture created because Class II liquids do not produce ignitable mixtures at ordinary temperatures. An extremely hazardous condition occurs, however, if the liquids are *switch loaded*, which is when a Class II or III liquid is loaded into a tank vehicle that previously contained a Class I liquid. Under these conditions, the atmosphere above the fluid surface can become ignitable. An explosion can occur if a spark occurs from the fluid surface to another object, igniting the flammable mixture in the headspace of the tank.

This paper describes a recent incident involving an explosion during the filling of a tank of a transport truck with diesel fuel. This particular tank was previously filled with gasoline and was not completely emptied of the gasoline prior to the switch loading. After only filling one-quarter of the programmed load, a spark ignited the flammable mixture within the headspace. The resulting explosion and overpressure caused significant damage to the truck and neighboring tanks. The study will address issues and lessons learned regarding spark promotion with the tank, reduction of static generation by the fluid being filled into the tank, and elimination of the flammable mixture within the tank prior to switch loading.

(44j) Is Your MOC Process Doing All That It Should Be?

William Fink and Carolina Herrera, Sage Environmental Consulting, L.P., Austin, TX

OSHA 1910.119 Process Safety Management of Highly Hazardous Chemicals was the driver for refining and chemical industry to implement Management of Change (MOC) programs some 20+ years ago. However, companies have now recognized the benefits of implementing and maintaining an effective MOC program beyond regulatory compliance. Effective MOC programs enhance a company's ability to respond to changing "market" or processing needs with up to date process safety information. Management of Change has even been incorporated into staffing levels and personnel expertise by some Companies.

Management of Change programs typically focus on changes that are "not in kind" changes. Significant effort is placed on proper evaluation of these changes and the potential impacts on the operation. This presentation emphasizes both the typical and non-typical areas of MOC that have been found to be overlooked. MOC programs rigorously evaluate the potential to introduce new hazards of the process such as fire, explosion, toxicity, and environmental impacts. Is that far enough to ensure that you are meeting all of your performance needs?

Focus on environmental impacts such as water contamination, air pollutants, and hazardous waste generation may not be going far enough. Gradual increases in highly hazardous material inventories may cross the lines of "threshold quantities" and introduce new OSHA or EPA requirements. Does your MOC program look at this potential? In addition, management of "in kind" changes may also not be keeping pace with the changes required by the EPA Leak Detection and Repair (LDAR) requirements or the Benzene Waste Operations NESHAP (BWON) requirements. These requirements as applied to MOC are further explained in this presentation. Several "gaps" in MOC programs will also be examined and suggestions for improvements provided.

*** Formatting Issue - Best Available Paper/Graphic ***

(44l) Beyond DSC, Improved Thermal Hazards Screening Methodology

Peter J. Ralbovsky IV, Netzsch Instruments North America, LLC, Burlington, MA

Thermal Hazards Screening is a primary part of any chemical process safety program. Differential Scanning Calorimetry (DSC) is used extensively in studying the thermal properties of materials and is often used to screen for thermal hazards. This method is quick and inherently safe as it can be done with a small amount of material (typically milligrams). These two benefits cannot be understated as there may be hundreds of screening test which a lab may have to do in a given year and sometimes little or nothing is known about its thermal behavior, its toxicity, or the toxicity of the decomposition products. DSC also has some short comings as it is unable to measure pressure, which can be a significant hazard, even when the heat release is relatively low. Also it can be difficult to handle multiphase samples or samples which react together at room temperature.

The Multiple Mode Calorimeter (MMC) is an excellent tool to be used in conjunction with the DSC to provide a more comprehensive look at thermal hazards. The MMC uses samples sizes up to about 3 grams so it is possible to do injection and stirring. In the MMC temperature and pressure are recorded as a function of time. Sample energies measurement of both endotherms and exotherms is possible. Heat capacity can be calculated as a function of temperature. This method offers a quick and inexpensive way to perform isothermal tests which can be used to help design larger scale tests done in 2L+ reaction calorimeters or pilot plant operations.

The Multiple Mode Calorimeter consists of an electronic base and a detachable calorimeter assembly. There are multiple types of calorimeters available which can be operated by the electronic base. This allows the user to select the right calorimeter for the type of experiment that is needed. The calorimeters can also be tethered to the electronic base with an electronic cable. This allow for the calorimeter to be placed remotely from the electronics. The calorimeter can then be easily placed in the hood and/or behind explosion barriers or even placed inside a freezer for sub-ambient operations. This instrument has been described in detail elsewhere.

There are a number of calorimeters available; the best two for Thermal Hazards Screening is shown in Figure 1 & 2. The first calorimeter (Figure 3) is capable of operating as a low phi adiabatic calorimeter but more importantly, for thermal hazards screening, can run in constant power or constant temperature rate mode. The MMC external heater module does not run in adiabatic mode but can be operated in scanning mode. Both MMC modules incorporate the VariPhi technology and can run true isothermal calorimetry tests. These MMC modules can measure endotherms and exotherms as well as calculating heat capacity during those periods where no transitions or reactions are occurring. The MMC measures sample pressure as a function of time/temperature which is critical in screening for thermal hazards.

Figure 1 Figure 2 Figure 3

***** Formatting Issue - Best Available Paper/Graphic *****

MMC VariPhi Calorimeter MMC External Heater Netzsch DSC 200 F3 *Maia®*

DSCs, like shown in Figures 3, have been used routinely as an initial step in the thermal hazards screening process. They have some unique advantages which make them suitable for this task. Scanning tests can be done very rapidly and with an auto-sampler they can be used to screen many samples efficiently. They sample size is small which makes them intrinsically safe in case of explosion or toxic gas production. These systems are small, inexpensive and have low operating costs. There are numerous DSCs available on the marketplace and they are used in many different applications. Two features are important from the thermal screening application. The first is an auto-sampler as discussed earlier. The second is the material of construction used in the DSC. It is likely that a DSC system used for thermal hazard screening will see explosions and the release of corrosive materials. DSC systems that can run hundreds of polymer melt tests without fail may have a limited life in the thermal hazards screening field. This is often unlooked. Figure 5 shows the cut-away of a robust DSC system.

Their continues to be a need to be able to safely and effectively screen thermal hazards for organizations involved in the manufacture, processing, storage or transportation of chemicals. Numerous screening tools are available in the marketplace, perhaps the most widely used is the Differential Scanning Calorimeter (DSC). This method has many advantages but the largest disadvantage is that the sample pressure is not measured. Thermal hazards are normally associated with pressure increases and these pressure increases are what ultimately provide the hazard to equipment and personnel. A storage drum does not melt due to a thermal runaway it ruptures due to the pressure increase because of the thermal runaway. Screening tools should be sensitive enough to see small exotherms that in near adiabatic conditions can fuel a runaway, they should be easy to use, provide quick results, and be as flexible as possible to handle different scenarios.

[1] Chippett, A New, Low Φ Factor, Scanning, Adiabatic Calorimeter, 18th Annual CCPS International Conference, 2003

[2]Chippett, US Patent Number 7,021,820 B2, April 4, 2006

Using Explicit Finite Element Analysis to Simulate the Effects of External Chemical Explosions on Single and Double-Walled Storage Tanks

Phillip E. Prueter
The Equity Engineering Group Inc., Shaker Heights, Ohio
pepprueter@equityeng.com

David J. Dewees, P.E.
The Equity Engineering Group Inc., Shaker Heights, Ohio
djdewees@equityeng.com

Prepared for Presentation at
American Institute of Chemical Engineers
2013 Spring Meeting
9th Global Congress on Process Safety
San Antonio, Texas
April 28 – May 1, 2013

UNPUBLISHED

Using Explicit Finite Element Analysis to Simulate the Effects of External Chemical Explosions on Single and Double-Walled Storage Tanks

Phillip E. Prueter
David J. Dewees, P.E.
The Equity Engineering Group Inc., Shaker Heights, Ohio
pepruteter@equityeng.com

Keywords: Explicit Finite Element Analysis, Blast Loading, CONWEP, Accidental Explosions, Storage Tanks

Abstract

Accurately simulating the overpressure or shock wave associated with a given far-field chemical explosion is extremely valuable in assessing the structural response and possible failure modes of critical process equipment such as storage tanks, piping, or pressure vessels. Furthermore, assessing the potential damage from an overpressure wave can provide valuable information about protecting structures from external explosions and improving designs that advocate blast damage mitigation. Such analyses become all the more important should the contents of the tank or vessel exposed to such a potentially catastrophic event pose a risk to humans or the surrounding environment. This paper discusses the underlying theory and examines the practical application of multiple finite element based explicit computational techniques for simulating the load acting on a structure due to an external blast. Explicit three-dimensional blast analysis of single and double-walled storage tanks that carry an extremely high consequence of failure is performed. The structural response of the tanks due to postulated accidental explosions is investigated and likely failure modes are discussed.

The two explicit computational blast loading methods discussed in this paper are the incident wave loading model and the Conventional Weapons Effects Blast Loading Model or CONWEP. The CONWEP model is appropriate for detonations of conventional explosions, and in this case automates many crucial features of analysis, such as developing the overpressure time history (positive and negative phases), defining overpressure spatial decay, and accounting for spatially varying reflection effects. Large vapor cloud explosions are known to behave very differently from conventional explosives, and for these cases, the more manual, but also general, incident wave loading approach permits the user to define their own time-varying overpressure amplitude. Additionally, this method also allows for coupled structural-acoustic analysis (should reflection effects need to be considered in detail). While it is not necessary to model the fluid medium using acoustic elements when using the incident wave loading model, it does become essential if reflection effects dominate the response of the given problem. If the fluid medium is modeled (air in this case), the acoustic elements can be coupled to the structural elements and the loading surface becomes the boundary of the acoustic mesh closest to the blast source. While this approach can be accomplished, it is extremely resource and time intensive, and is generally not necessary (which will be discussed later in this paper). Three-dimensional examples of the above blast loading approaches are presented in this paper; incident wave, and CONWEP. Comparisons are made between these methods and the corresponding computational results.

Furthermore, commentary on the structural response and likely failure modes of the storage tanks is provided and a discussion regarding design features that supplement blast damage mitigation and process safety is rendered. The advanced computational methods discussed herein permit realistic evaluation of complex, three-dimensional process equipment subjected to accidental explosions. Simulating and understanding the possible failure modes due to blast loading of critical process equipment in the petrochemical and related industries is very beneficial in not only improving process safety, but also providing plant operators and safety personnel with crucial information regarding the possibility of equipment failure should a catastrophic explosion occur.

1. Introduction

An explosion in air creates a blast wave as a result of the atmosphere surrounding the explosion being pushed back. In general, the pressure of the compressed air at the blast wavefront decays as it moves away from the blast source. As discussed in references [1-6], a typical blast wave, as observed at a location removed from the center of a chemical explosion, reaches a peak value of overpressure and begins to decay exponentially (as shown in Figure 1), eventually decreasing below atmospheric pressure. This peak overpressure decreases as the distance from the explosion source increases. The amount of time it takes the wavefront to reach a given point is known as the arrival time. The overpressure profile can be divided into the positive pressure portion (positive phase) and negative pressure portion (negative phase). Furthermore, the amount of time it takes the peak overpressure to decay below atmospheric pressure is known as the positive phase duration and is closely related to the damage capabilities of a given blast wave. Similarly, the amount of time it takes for the negative pressure to return to atmospheric pressure is the negative phase duration [1]. The area under the curve during the positive phase duration is the positive blast wave impulse.

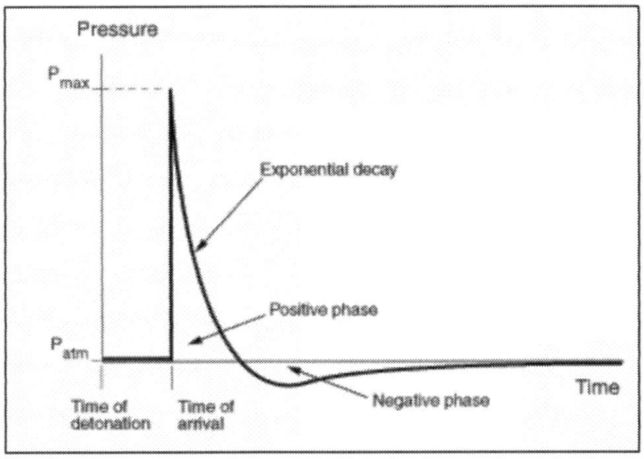

Figure 1: Typical pressure-time amplitude for a chemical explosion [7].

In some cases, the positive phase impulse is calculated and an equivalent triangular impulse with the same peak overpressure is used to approximate the realistic overpressure amplitude shown in Figure 1. In many cases, this approximation is appropriate because the positive phase portion of the incident pressure wave is typically the most damaging. Generally, structural response is more sensitive to capturing the correct incident wave reflection. Accurately simulating a realistic blast wavefront on a structure for a given explosive scenario is challenging because the peak overpressure, positive phase duration, amount of explosives, and distance from the explosion source all affect the overpressure amplitude. Furthermore, accounting for reflection effects on a loading surface adds further complication to the blast loading model.

Two different explicit finite element techniques used to simulate blast loading due to external chemical explosions in air will be discussed herein; incident wave loading and the Conventional Weapons loading model (CONWEP) [8]. CONWEP is well suited for simulating the overpressure wave associated with the detonation of conventional explosives while incident wave loading permits reasonable analysis of vapor cloud explosions. The underlying methodology of each approach will be discussed as well as the advantages and applications of each. Furthermore, both computational methods will be compared for the explicit dynamic simulations of realistic explosive overpressure waves (from a postulated concentrated explosion source and a vapor cloud) impacting single and double-walled storage tanks that carry an extremely high consequence of failure.

2. Blast Curve Generation

In order to develop the blast parameters required to generate realistic overpressure curves for conventional explosives, the equivalent TNT method is most commonly used. TNT equivalency of an explosive is simply the ratio of the mass of TNT to the mass of explosive such that both yield equal pressure and impulse [9]. In order to compare shock wave parameters to available published data, a scaled distance from an explosion source to the loaded structure may be used. As discussed in [10], the most common form of blast scaling is the Hopkinson-Cranz or cube-root scaling method following

$$Z = R/W^{(1/3)}$$

where Z is the scaled distance, R is distance from the explosion source to the loading point, and W is the weight of TNT. This formulation implies that self-similar blast waves are produced at identical scaled distances when two different sized explosive charges (of similar geometry and chemical composition) are detonated in the same atmosphere. Using published empirical data, this scaled distance from the explosion source to a loading surface can be used to determine specific blast parameters. The CONWEP blast loading model discussed herein is based on a TNT explosion (assumed surface blast). Thus, when using CONWEP, it is necessary to convert the mass of explosives in question to an equivalent amount of TNT.

As outlined in [11], a major limitation associated with the equivalent TNT method is that the blast yield is a function of the amount of explosives and not the combustion mode; making this method not preferred when analyzing vapor cloud explosions. As discussed in References [12-

16], vapor cloud explosions differ from conventional explosive detonations in that flame accelerations, which can be exacerbated by confinement due to intervening structures, define the overpressure associated with a given explosion more so than simply the amount of fuel ignited. In order to quantify blast curves associated with vapor cloud explosions several widely accepted methods have been developed; TNO [15] blast curves and Baker-Strehlow [17] blast curves among others. In this study, triangular impulses of varying peak overpressures are used for simplicity when analyzing the structural response of a double-walled storage tank subjected to a postulated vapor cloud explosion (discussed below). This also permits investigation into the sensitivity of structural response to varying blast wave parameters.

3. CONWEP and Incident Wave Loading

A significant challenge associated with computational blast modeling is accounting for reflection effects. The peak blast pressure that a structure is subjected to is greater than the peak overpressure and is known as the reflected pressure. This pressure is a function of the peak overpressure and the angle of incidence (the angle between the loading direction and the surface normal of a point on the loading surface). Neglecting incident wave reflection can yield non-conservative results, particularly for high peak overpressures, while assuming the maximum reflection effect at all points on a (cylindrical) structure can potentially be overly conservative.

The CONWEP blast loading model is based on a collection of conventional weapons effects calculations from the equations and curves of TM 5-855-1 [8] and is implemented into the commercial finite element program, Abaqus. The main advantage of using CONWEP over the incident wave loading approach is that realistic overpressure amplitudes (including positive and negative phase) and other blast wave parameters are inherently calculated based on a defined amount of TNT at a given distance from the explosion source. Furthermore, the CONWEP blast loading model does not require modeling the fluid medium (air) to account for reflection effects.

Incident wave loading is an explicit finite element technique in which an incident wave overpressure at a point can be defined as a function of space and time. This pressure amplitude is defined at a standoff point, located between the detonation point and the loaded structure. When simulating blast loading, realistic overpressure amplitudes need to be defined by the user. An incident wave can be defined as planar or spherical. A schematic showing the incident wave loading model in Abaqus is given in Figure 2 below. For a more rigorous overview of the underlying theory behind both the CONWEP blast loading model and the incident wave loading as well as a benchmark of the overpressure reflection calculations of CONWEP, refer to Reference [18].

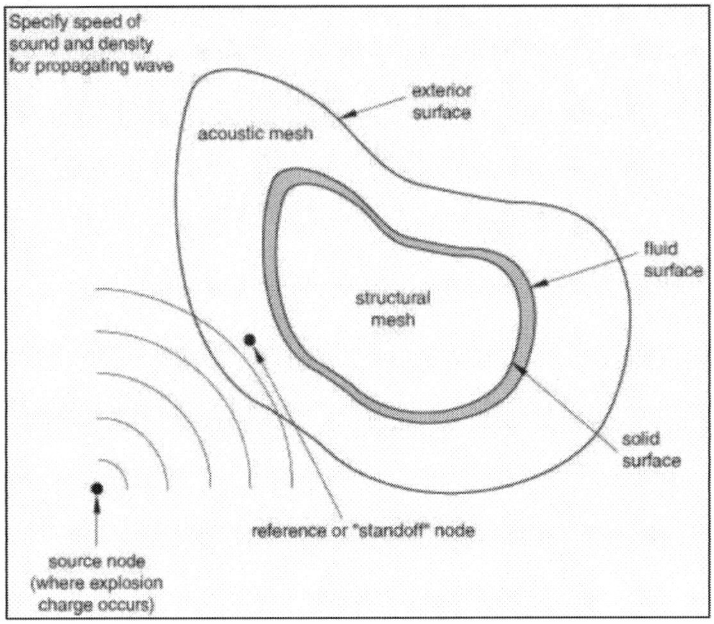

Figure 2: Incident wave loading model [7].

4. Explicit Analysis of a Storage Tank Subjected to a Vapor Cloud Explosion

As discussed in [19, 20] the consequences of storage tank failure in chemical or petrochemical facilities can be dire. Furthermore, it is estimated that 85 percent of storage tank accidents are due to fire and explosion [21]. Analyzing the structural response of storage tanks subjected to potential blast loading from conventional detonations or vapor cloud explosions is of great interest to the process safety community. The explicit finite element analyses discussed below provide real-world examples of the usefulness that computational blast analysis provides in the petrochemical and related industries.

The first computational blast analysis examined herein relates to a double-walled storage tank subjected to a postulated external vapor cloud explosion. The rough size of the outer tank is a 140 foot diameter with a shell height of approximately 125 feet. A complete, elastic-plastic, three dimensional finite element model was constructed using linear shell elements. Figure 3 shows the half-symmetry model with detailed geometric features. The baseline finite element model includes all major structural features, including the suspended deck, bracket supports, and the lower support straps. Quadratic beam elements are used to model the suspended deck hanger rods and the strap supports, and both the inner and outer tanks are modeled for completeness. Contact is defined between the suspended deck and the shell wall.

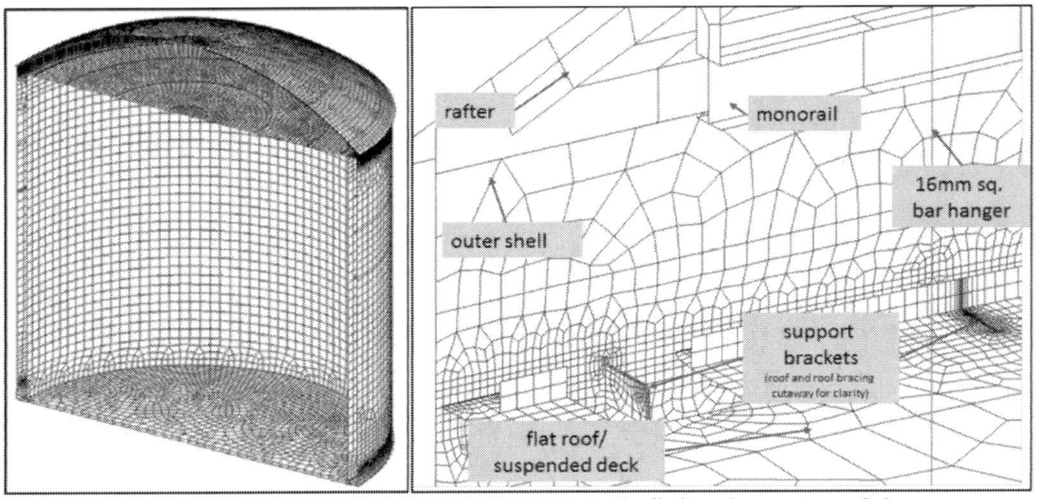

Figure 3: Vapor cloud explosion analysis finite element model.

A simplified, lumped-mass model is also analyzed. In this model, the mass associated with the suspended deck is lumped into the domed roof and the mass of the inner tank shell is lumped into the outer tank shell. The mesh is coarsened in the simplified model to aid in computational efficiency. Incident wave loading is used to analyze a range of peak overpressures (assumed to be peak reflected pressure because acoustic elements are not used) applied as planar triangular impulses. The effects of gravity (self-weight) are included.

In this case, because of the range of potential vapor cloud explosive scenarios, iterations are performed to determine the maximum permissible overpressures and loading durations at the tank to ensure containment and safe operability. That is, the focus is on establishing a sort of blast rating for the structure independent of a specific explosion scenario. This provides an essential characteristic of the structure that can be referred to for any future scenarios or concerns that might arise, and is similar in some sense to the maximum allowed working pressure (MAWP) of a pressure vessel. Peak reflected pressures of 5, 8, and 11 psi are applied for an impulse duration of 50 milliseconds to the lumped model. Figure 4 below compares maximum deflections and stresses in the lumped model from the domed roof (opposite impact side) for peak applied overpressures of 5, 8 and 11 psi. This figure demonstrates the relatively constant nature of the largest displacements. In the 5 psi case, maximum deflections are generally less than 12 inches with relatively little accumulation in plastic strain.

These results indicate that a peak reflected pressure of 5 psi is near the limit to achieve tank containment following blast loading. Furthermore, for the 8 and 11 psi overpressure cases, results indicate that plastic collapse of the domed roof and tank shell would be likely and containment of the vessel contents would be jeopardized. Figure 4 also shows an example transient deformation corresponding to the full model results for 5 psi. In general, there is reasonable agreement between the simplified, lumped parameter model and the complete model with the suspended deck across cases. However, the suspended deck is a critical feature of the tank, and is also an indicator of safe operability. As Figure 4 illustrates, the compete model indicates that the suspended deck is one of the weakest points of the tank internals; results from the complete model series of analyses show that a maximum peak reflected pressure of 2 psi is advisable to ensure safe tank operability (again, impulse duration assumed to be 50 milliseconds). This corresponds to a maximum side-on overpressure of 1 psi (uniform/constant reflection coefficient of 2).

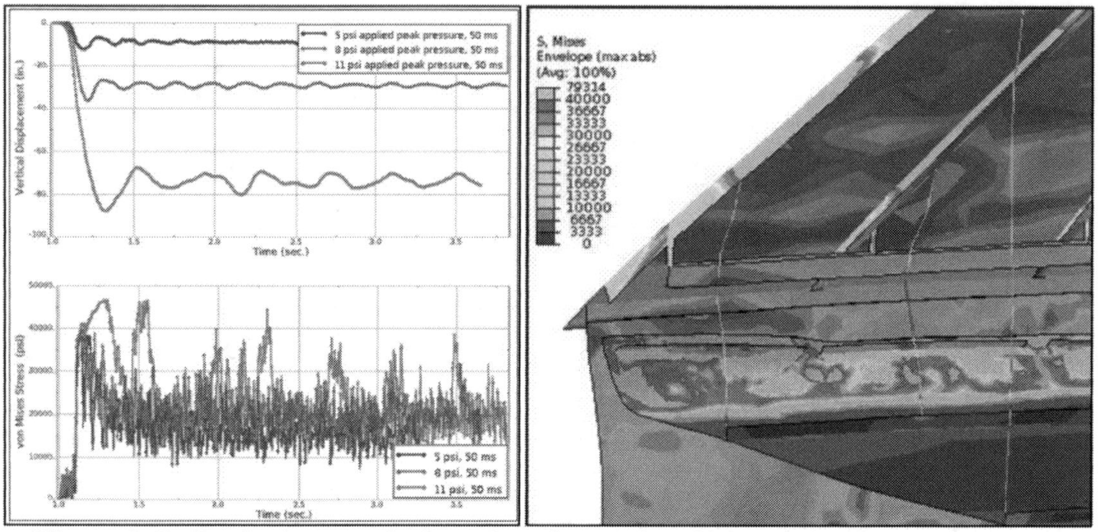

Figure 4: Deflection/stresses in domed roof opposite of impact side for varying peak overpressures and suspended deck deformation.

Additionally, in order to compare the effect of varying blast wave impulse durations, a simulation at 11 psi peak applied overpressure with an impulse duration of 10 milliseconds is carried out. Figure 5 compares maximum domed roof deflections and stresses for the 50 and 10 millisecond loading durations. As expected the duration of the impulse has a significant effect on the overall structural damage. It is noted that much lower peak overpressures acting at longer durations could also be damaging.

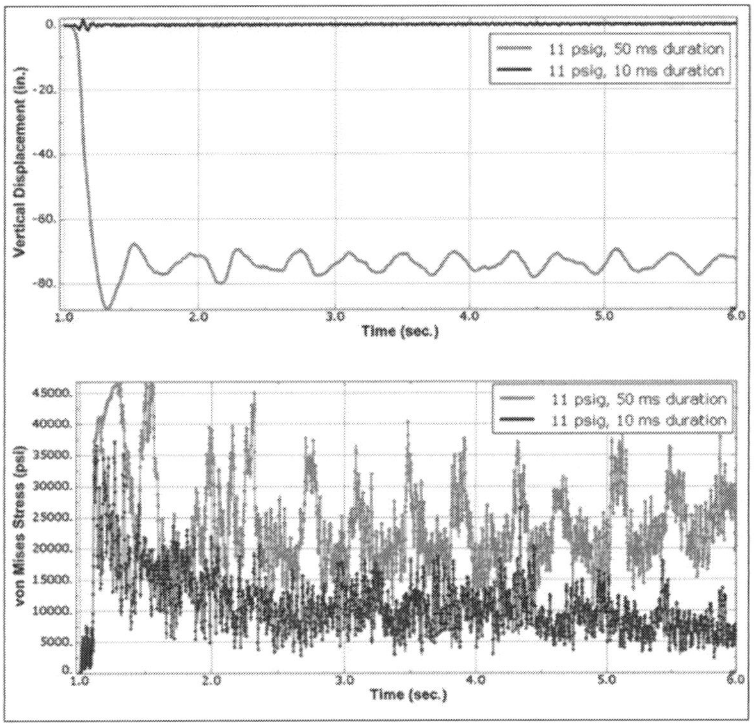

Figure 5: Deflection/stresses in domed roof opposite of impact side for varying impulse duration.

5. Explicit Analysis of a Storage Tank Subjected to Conventional Explosion

A different double-walled storage tank is analyzed; this time the postulated blast loading is due to accidental detonation of a known amount of concentrated, combustible fuel. The approximate outer tank diameter is 100 feet, and the outer tank height is roughly 75 feet. Again a complete, elastic-plastic, three dimensional finite element model is constructed using linear shell elements. Figure 6 shows a cutaway view of the model with contours of shell element thickness. The inner tank has four stiffening rings that protrude into the annular space between the inner and outer tank. The outer tank roof support structure consisting of channels and angles is also modeled in detail and included in the finite element model. Contact is modeled in between the tank structures and the effects of self-weight are included.

An initial analysis is performed on just the outer tank to investigate the plastic strains and wall deflections due to the applied blast wave. Both the CONWEP and incident wave loading models are compared (the equivalent TNT method is used to determine the overpressure profile in accordance with CONWEP). Due to the size of the finite element model and the potentially high computational cost, the fluid medium is not explicitly modeled for the incident wave simulation. An overpressure amplitude that accounts for reflection effects is used (peak reflected pressure applied in the incident wave model). Furthermore, in order to mimic the behavior of the CONWEP simulation as much as possible, the overpressure profile defined for the incident wave simulation as a function of time does, in fact, consist of positive and negative phases and decays exponentially once it reaches its peak value. The side-on reflection coefficient for the overpressure magnitude in this case is 2.

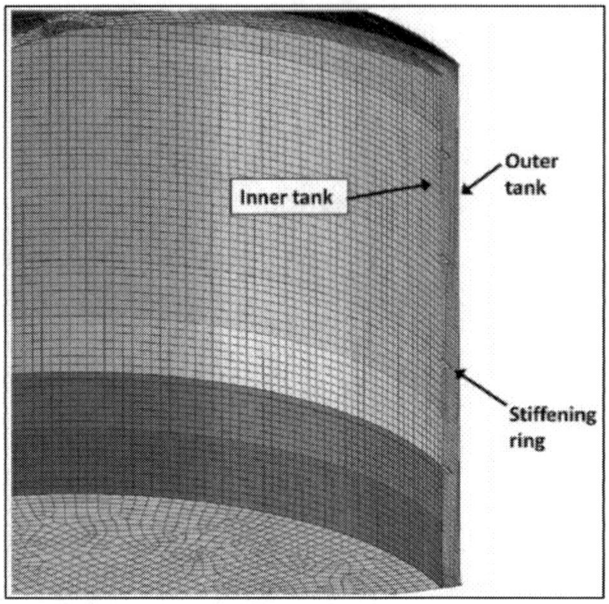

Figure 6: Cutaway view of the complete finite element model with shell thickness contours shown.

Even for the relatively small peak reflected overpressure of roughly 1.9 psi, the outer tank wall deflections prove to be significant, to the point of local collapse. A static post-buckling analysis (using the Riks solver), with surface pressure applied using an analytical field definition to match surface pressure in the dynamic simulations confirms that the outer tank wall buckles before the peak reflected pressure is reached. Riks post-buckling analyses are discussed in more detail below.

For comparison, Figure 7 shows contours of outer tank plastic deformations (identical scale) due to the overpressure wave for both methods. As expected, the incident wave approach produces higher deflections than CONWEP because the applied peak reflected overpressure does not decrease as angle of incidence increases. However, the general agreement between methods is actually quite good. This suggests that the simplification of a single maximum reflection coefficient in this case for the incident wave approach is not excessively conservative, and that reasonable results were obtained without modeling the entire acoustic domain (in which case reflection effects would be rigorously accounted for).

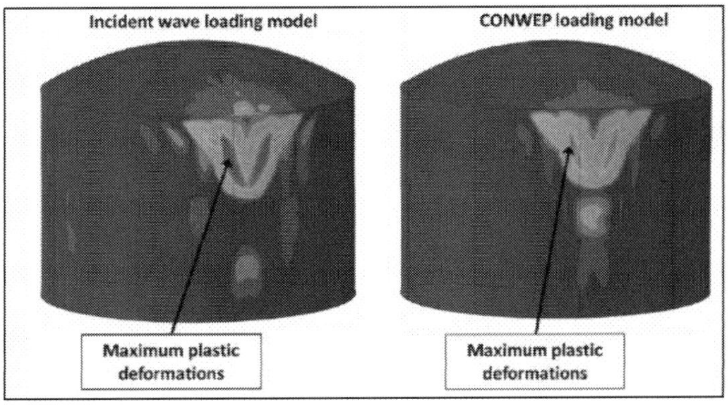

Figure 7: Outer tank plastic deformations due to blast loading using the CONWEP and incident wave loading models (contour scales identical).

For the blast analysis of the complete finite element model that includes both the inner and outer tanks with contact defined between shells, the outer wall deflections are much less significant because the outer tank impacts the inner tank stiffeners. Essentially, the inner tank stiffening rings absorb the impact from the outer tank wall and plastically deform. Furthermore, the deflections in the inner tank wall are minimal. This is demonstrated in Figure 8, which shows the maximum deflection during blast loading from the CONWEP simulation of the inner tank and the outer tank. The contour scale for the inner tank corresponds to 10 percent of the outer tank contour scale. Additionally, the maximum inner tank wall deflection is less than 10 percent of the maximum outer tank wall deflection, and the maximum outer tank deflections occur in the regions in between the inner tank stiffening rings. The outer tank wall never directly impacts the inner tank wall. The minimal inner tank deflections suggest that the sloshing effects of the liquid contained by the inner tank should also be minimal.

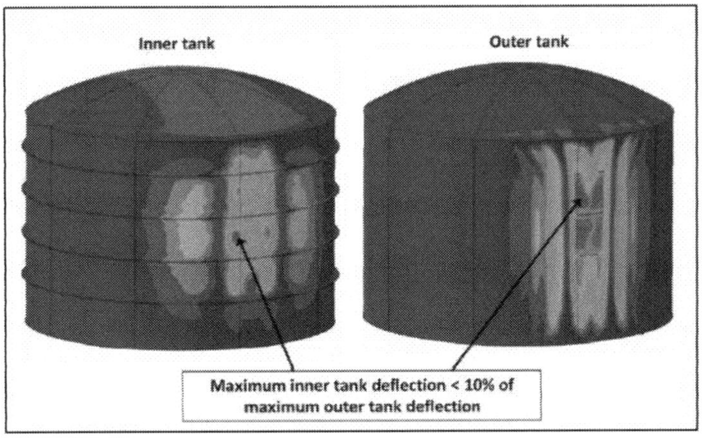

Figure 8: Maximum deflection of the inner and outer tanks due to blast loading from the CONWEP model (inner tank contour scale corresponds to 10 percent of outer tank scale).

These results indicate that from a blast damage mitigation perspective, the four inner tank stiffening rings prove to be very beneficial in reducing the severity of the impact between the inner and outer tanks. While both explicit computational techniques used to analyze this complete, three dimensional model give reasonably similar results, from a convenience standpoint, CONWEP proves to be advantageous for conventional detonations because the appropriate overpressure and spatial decay are automatically calculated for the given equivalent amount of TNT. Additionally, due to more rigorous accounting for reflection effects, some conservatism is removed in the CONWEP case, making it the natural choice for well-defined, conventional explosive detonation scenarios.

6. Explicit Analysis of a Cluster of Single-Walled Storage Tanks

The postulated blast source in this case is the same source as discussed in the previous example (conventional detonation). However, this cluster of single-walled tanks is located much closer to the blast source, meaning peak overpressures are much higher than the above example. All three tanks are roughly 30 feet in diameter and 35 feet high. The tanks also have self-supporting roofs and are anchored to their respective foundations via hold down clips and hold down bolts. As is the case for the aforementioned storage tanks in this study, these tanks carry an extremely high consequence of failure due to their process contents.

Three dimensional, elastic-plastic finite element models of each tank are constructed using linear shell elements and explicitly analyzed using CONWEP. Peak reflected overpressures at each tank as a function of time are provided in Figure 9 below. Again, the effects of gravity are included. A node at the center of each hold-down clip is fixed in all degrees of freedom and then coupled to the surrounding bolt hole to simulate the restraint of the bolts as shown in Figure 10. A rigid surface representing the tank foundation is constructed and included in the model, and contact between this rigid surface and the tank bottom is defined. This ensures that the tank will not unrealistically sag due to gravity alone (essentially penetrate the foundation due to self-weight and induce non-realistic stresses in the tank floor/walls) while allowing the tank bottom to physically lift off its foundation when the blast load is applied. In order to account for the weight of the process fluid, an equivalent pressure is applied to the tank bottom that corresponds to the weight of fluid up to the maximum fill height. This pressure is applied throughout the dynamic simulation (sloshing effects of the fluid are not considered; this assumption is discussed below).

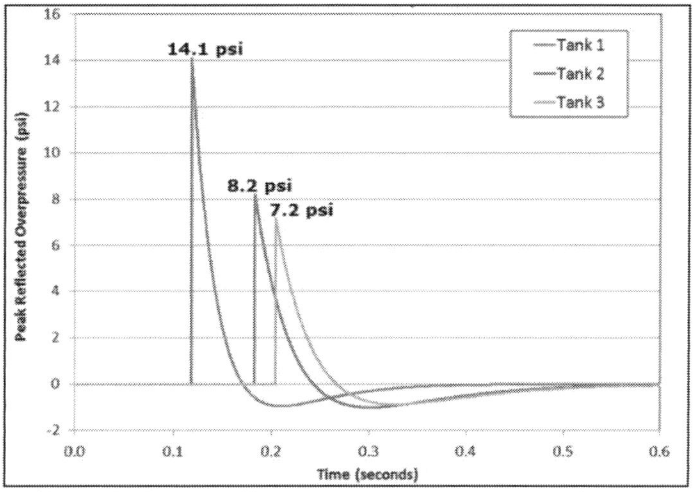

Figure 9: Peak reflected overpressures at each tank.

After the onset of the blast wave, the wall and the roof of each tank significantly deform. In order to better visualize the tank behavior following the onset of the blast wave impact, Figure 11 shows contours of Tank 1 deflection as time elapses (time scale given is seconds after detonation). These large deflections are more indicative of plastic collapse like behavior than a dynamic or vibratory failure. Furthermore, the stresses observed in the tank shortly after blast wave impact are significant.

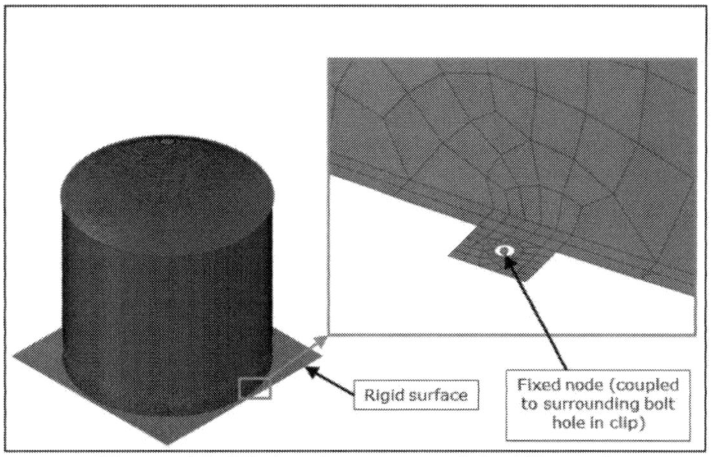

Figure 10: Finite element model of Tank 1.

Additionally, the stresses in the tank at the hold-down clips and reaction forces at the assumed bolt locations predict likely failure of either the bolts themselves or the hold-down clips. Figure 12 shows equivalent plastic strains in Tank 1 after the impact of the blast wave (contour scale capped at 5%). The hold-down clips and portions of the tank wall near the hold-down clips exceed 100% equivalent plastic strain. Results for Tanks 2 and 3 are similar.

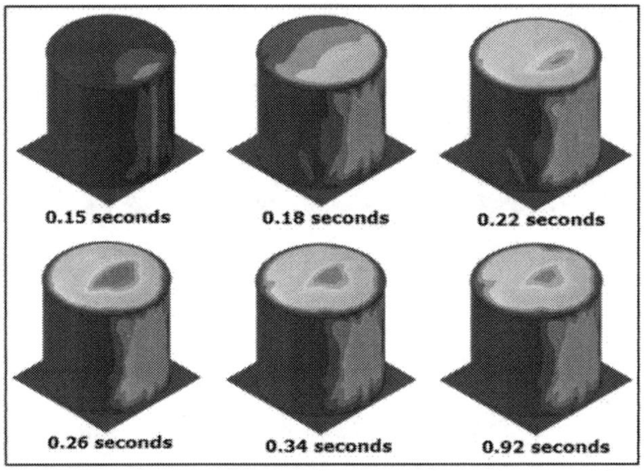

Figure 11: Tank 1 deflections during blast loading.

In order to further investigate the possibility of plastic collapse or buckling, static Riks post-buckling analyses are performed on each tank (a static pressure distribution is applied to mimic peak blast loading pressures). A Riks analysis is useful for simulating non-linear instability problems, and includes the effects of large deformations and elastic-plastic material behavior. It can also be used to look at post-buckling or collapse shapes. The applied load in a Riks analysis is actually a variable and part of the solution for each increment, and is always relative to the current deformed configuration of the structure. If no converged solution is possible for an incremental increase in load, then the equilibrium solutions can be searched for with a decreased load. This is the mechanism that allows post-buckled shapes to be predicted and visualized. In extreme cases, no equilibrium solution can be reached, even with decreased load, and only load reversal is predicted to lead to stability. This is actually the case for the tanks analyzed here, where once collapse has initiated, displacements are predicted to be essentially unbounded for any amount of positive load and thus, surface pressure becomes negative before peak overpressure values are achieved.

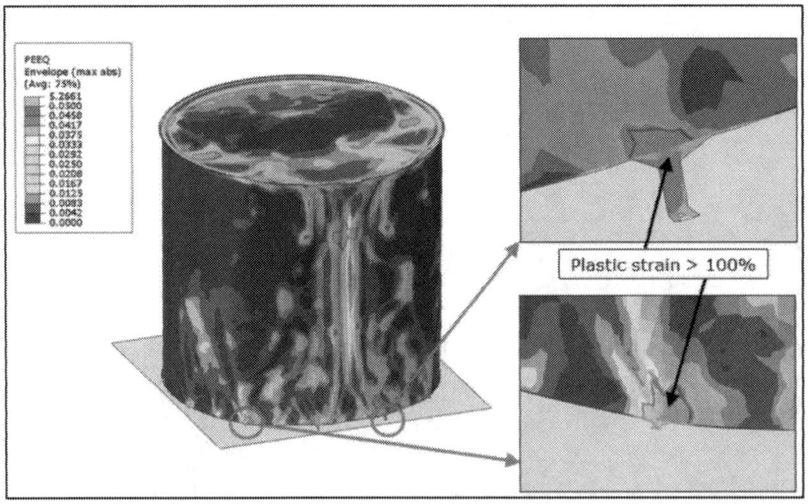

Figure 12: Equivalent plastic strain in Tank 1 due to blast loading (scale capped at 5%).

Based on the results discussed above, the three storage tanks subjected to the assumed postulated blast load are likely to fail via several possible failure modes. The first likely failure mode is plastic collapse of the tank walls and/or roofs. All three tanks see large deflections (the most severe occurring in Tank 1). Riks post-buckling analyses confirm that the magnitude of deflections seen in the tank walls would likely result in unstable collapse behavior before the peak blast overpressure is achieved. A second likely failure mode is extreme damage in the tanks at the hold-down bolts and/or regions of the tanks that are connected to the bolts. In all three tanks, equivalent plastic strains exceed 100% in these regions. Failure of multiple hold-down bolts or hold-down clips could cause a given tank to slide off its foundation and/or overturn. Furthermore, sloshing effects of the fluid inside the tanks are neglected in the simulations discussed herein, but would increase the likelihood of failure.

In short, the close proximity of all three tanks to the postulated blast source causes the corresponding overpressures acting on each tank to be relatively severe. Design changes to the tanks to supplement blast damage mitigation would likely have to be drastic in order to achieve a significant reduction in the probability of failure due to the postulated explosion considered in this study. Finally, in order to achieve practical simulation times, all of the explicit analyses discussed herein are conducted across multiple CPUs (often 16 or more) simultaneously. Utilizing parallel computing permits evaluation of complex three dimensional structures, such as the tanks discussed above, in reasonable amounts of time.

7. Conclusions

This study compares two different explicit computational methods for modeling blast loading on structures; incident wave loading and the CONWEP charge property. The underlying methodology for each method is discussed. CONWEP is well suited for conventional detonations and uses the equivalent TNT method, and it inherently calculates the appropriate blast wave parameters for a given explosive event. Additionally, CONWEP achieves realistic overpressure amplitudes, spatial decay, and reflection effects without having to model the fluid medium, as is the case for incident wave loading. Incident wave loading provides the user with the flexibility to define a time-varying overpressure amplitude. This is useful when performing sensitivity studies to blast loading parameters or when attempting to characterize the blast wave associated with a vapor cloud explosion that does not behave like a conventional detonation. The development and implementation of the CONWEP and incident wave blast loading models into commercial finite element programs makes accurately analyzing the response of complex three dimensional structures subjected to blast loading feasible.

Additionally, explicit, three dimensional finite element analyses of both single and double-walled storage tanks that carry an extremely high consequence of failure are discussed herein. Incident wave loading is used iteratively to determine the maximum permissible overpressure for a double-walled storage tank subject to a postulated vapor cloud explosion. The CONWEP blast loading model is used to analyze single and double-walled storage tanks subjected to a postulated chemical explosion (conventional detonation). The explicit computational methods, such as the ones examined herein, can provide valuable insight into the blast resistance of existing structures and can guide design decisions regarding how to mitigate blast damage for new structures in the petrochemical and related industries. Gaining a better understanding of the possible failure modes due to blast loading of critical process equipment is very beneficial in not only improving process safety, but also in enhancing safety protocol for catastrophic accidental explosions in industrial environments.

8. References

1. DOD, "Unified Facilities Criteria (UFC) 3-340-02: Structures to Resist the Effects of Accidental Explosions," *Department of Defense, United States of America*, 2008
2. Smith, P., and Hetherington, J., "Blast and Ballistic Loading of Structures, 1st ed," *Butterworth-Heinemann*, Oxford, 1994.
3. Kinney, G., and Graham, K., "Explosive Shocks in Air, 2nd ed," *Springer-Verlag*, Berlin, 1985.
4. Dusenberry, D., "Handbook for Blast-Resistant Design of Buildings, 1st ed," *John Wiley and Sons, Inc.*, Hoboken, New Jersey, 2010.
5. ASCE, "Design of Blast-Resistant Buildings in Petrochemical Facilities, 1st ed," Task Committee on Blast-Resistant Design, *American Society of Civil Engineers.*, Reston, Virginia, 1997.
6. Bulson, P., "Explosive Loading of Engineering Structures, 1st ed," *Taylor and Francis Group*, London, 1997.
7. Abaqus 6.10-1, DSS Simulia, "Analysis Users' Manual," 2010.
8. Hyde, D., "Users' Guide for Microcomputer Programs CONWEP and FUNPRO – Applications of TM 5-885-1," *U.S. Army Engineer Waterways Experimental Station*, Vicksburg, VA, 1988.
9. Huntington-Thresher, W., and Cullis, I., "TNT Blast Scaling for Small Charges," *19th International Symposium of Ballistics*, 2001.
10. Baker, W., "Design Considerations for Toxic Chemical and Explosives Facilities – Chapter 1, Blast Pressure Effects: An Overview," *American Chemical Society*, 1987.
11. Tang, M.J., and Baker, Q.A., "A New Set of Blast Curves from Vapor Cloud Explosion," *Process Safety Progress*. Vol. 18, No. 3, 1999.
12. American Institute of Chemical Engineers (AiChE), "Guidelines for Evaluating the Characteristics of Vapor Cloud Explosions, Flash Fires, and BLEVEs," *Center for Chemical Process Safety of the American Institute of Chemical Engineers.*, New York, 1994.
13. Lees, F.P., "Loss Prevention in the Process Industries," *Butterworth-Heinemann*, Oxford, 1996.
14. Mercx, W.P.M., van den Berg, A.C. and van Leeuwen, D., TNO report PML 1998-C53, "Application of correlations to quantify the source strength of vapor cloud explosions in realistic situations, Final report for the project: GAMES," *Netherlands Organization for Applied Scientific Research (TNO)*, 1998.
15. Van den Bosch, C.J.H. and R.A.P.M. Weterings, eds., "Methods for the calculation of physical effects: due to releases of hazardous materials (liquids and gases) ("TNO Yellow Book"), CPR 14E, Parts 1 and 2, 3rd Edition," Director-General for Social Affairs and Employment, *Committee for the Prevention of Disasters*. The Hague, 1997.

16. Woodward, J. L., "Estimating the Flammable Mass of a Vapor Cloud," *Center for Chemical Process Safety of the American Institute of Chemical Engineers*, New York, 1998.
17. Strehlow, R.A., Luckritz, R.T., Adamczyk, A.A., and Shimpi, S.A. "The Blast Wave Generated by Spherical Flames," *Combustion and Flame*, Vol. 35, 297-310, 1979.
18. Prueter, P.E., Dewees, D.J., and Brown, R.G. "Examining Computational Techniques to Simulate the Effects of External Chemical Explosions Using Explicit Finite Element Analysis," *In ASME Pressure Vessels and Piping Conference*. PVP2012-78566, 2012.
19. Maremonti, M., Russo, G., Salzano, E., and Tufano, V., "Post-Accident Analysis of Vapor Cloud Explosions in Fuel Storage Areas," *Trans IChemE*, 77(B), 1999.
20. Buncefield Investigation Board, "The Buncefield Incident 11 December 2005: The Final Report of the Major Incident Investigation Board," Volume 1, 2008.
21. Chang, J., and Lin, C., "A Study of Storage Tank Accidents," *Journal of Loss Prevention in the Process Industries*, 19(1) 51-59, 2006.

(44o) The Hidden Risks of Deposits, by-Products, and Residues On Exhaust and Waste Streams

Carlos A. Barrera and Abid Kemal, Thermal Sciences, Exponent, Menlo Park, CA

The safety risks associated with reactive compounds are usually captured and addressed during the conceptual design and pre-construction process safety reviews as, and for well-established chemistries these risks are well understood for the main processes and reactions. Lest time, if any, is devoted to the analysis of the reactive nature of the compounds present waste streams or exhaust lines. These lines are often overlooked during HAZOPs because of their secondary nature, or because they go to a scrubber or flare, which erroneously is considered a "catch-all" device that will "take care of everything." Waste streams and exhaust lines from certain physicochemical processes contain highly reactive compounds in the form of residues or by-products that in the right conditions can accumulate and form dangerous deposits. Some examples of such deposits are pyrophoric, shock-sensitive, and corrosive compounds. The purpose of this paper is to create awareness in the process safety community about the importance of a complete understanding of the chemistry of the reactive species present in these streams and the possible physical mechanisms than can lead to the formation of dangerous deposits, as well as the importance of including these streams in the process hazard analysis. Examples drawn from multiple processes will be presented.

(44q) Explosion Characteristics of Carbon Nano-Fibers

Jiaqi Zhang, Artie McFerrin Department of Chemical Engineering, Texas A&M Univerisity, Mary Kay O'Connor Process Safety Center, College Station, TX, Yi Liu, Mary Kay O'Connor Process Safety Center, Department of Chemical Engineering,Texas A&M University, College Station, TX and Dr. M Sam Mannan, Artie McFerrin Department of Chemical Engineering, Mary Kay O'Connor Process Safety Center, College Station, TX

Dust explosion is a hazard with serious consequences. While the main principle of dust explosion is still being exploring, the gap of dust explosion between particles of micro-size and those of nano-size also needs tremendous efforts. The explosion characteristics of CNFs are studied in Mary Kay O'Connor Process Safety Center. These experiments are conducted in a 36-L explosion vessel, which has been calibrated with a standard 20-L explosion vessel. Among the experiments, different kinds of CNFs of different grades and different conditions (milled and unmilled) are used. Compared with collected published data of graphic particles of micro-scale, the experimental data shows that the carbon nano-fibers create a higher Kst than the micro-scale ones. The results also show that explosion of the XT-PS grade (Pyrolytically Stripped) is much stronger than that of the XT-LHT grade (Low Temperature Heat-treated) and XT-HHT grade (High Temperature Heat-treated). Also, the explosion of milled CNFs is stronger than the unmilled ones. Further study to seek the causes is undergoing.

(44r) High-Fidelity Combined Dynamic Modeling of Depressurizing Vessels and Flare Networks: Accurately Assess the Effects of Process Changes On Mechanical Integrity and System Capacity

James Marriott, VP PSE Oil & Gas, Process Systems Enterprise Ltd, London, United Kingdom

The use of high-fidelity dynamic modeling for combined depressurization and flare network analysis can result in identification of additional capacity in flare networks without compromising process safety, presenting potentially significant opportunities for reducing or saving capital expenditure (CAPEX). The same analysis can simultaneously provide accurate assessment of temperatures in vessels and pipework, to identify areas where close approach to or violation of minimum metal temperatures may occur, thus compromising vessel and pipework mechanical integrity.

The flare networks for major plants represent a non-negligible part of the overall capital investment. Current industrial practice for their design is primarily based on steady-state analysis, as described in standards such as API 521 and supported by widely-used software tools. However, it is a widely recognized fact that the application of steady-steady considerations to what is fundamentally a dynamic system inevitably requires the use of conservative assumptions which often result in significant oversizing of flare headers and other components of the network. Another major contributor to capital cost is the use of special materials for the parts of the system that may be exposed to low-temperature fluids – typically the tail pipes attached to high-pressure process vessels – and which are therefore at risk of embrittlement. The key to being able to limit the additional capital expenditure is the accurate estimation of the length of piping that is subject to "abnormal" temperatures.

In addition to dynamic flare network analysis, the detailed dynamic modeling and simulation of the rapid depressurization ("blowdown") of high-pressure vessels is a key factor in determining both the load imposed on the pressure relief system (e.g. flare network) and also the temperatures of the vessel and downstream pipework walls. Current methods typically use 'lumped' equilibrium approaches that do not adequately predict these quantities; indeed, in contrast with the current over-conservative practice in flare network design, they may provide non-conservative solutions. By contrast, the high-fidelity approach incorporates a multi-phase non-equilibrium representation of the vessel fluid contents, as well as a 3-dimensional model of the metal walls that takes into account the transfer of heat between regions of the wall in contact with different phases. This allows a much more accurate estimation of the wall temperatures, and the direct computation of thermal stresses.

In the past it has been difficult to assess these important design criteria effectively because of the challenges in modeling such complex systems. This paper describes how the combination of high-fidelity depressurization capabilities and an advanced model-based system for flare system network design, coupled with evolving best practice in using these tools, represents a transformation for process safety management. The new techniques allow accurate assessment of the effects of process changes on mechanical integrity, thus enabling informed safety and CAPEX decision support.

An Engineering Method to Mitigate the Impact of Regulatory Focus on Relief System Installations by Prioritizing Risk

Dustin Smith
Smith & Burgess
5700 NW Central Dr. Ste 301 Houston, TX 77092
dustin.smith@smithburgess.com

Prepared for Presentation at
American Institute of Chemical Engineers
2013 Spring Meeting
9th Global Congress on Process Safety
San Antonio, Texas
April 28 – May 1, 2013

UNPUBLISHED

An Engineering Method to Mitigate the Impact of Regulatory Focus on Relief System Installations by Prioritizing Risk

Dustin Smith
Smith & Burgess
5700 NW Central Dr. Ste 301 Houston, TX 77092
dustin.smith@smithburgess.com

Keywords: Pressure Relief Valve Installation, API 3% Rule, Chatter, Relief Device Stability

Abstract

Since around 2005, regulators in the United States have put greater emphasis on relief device installations meeting the 3% rule. Spending large amounts of money to "fix" relief device installations that pose no safety risk decreases the plant's overall safety. To date, more incidents have occurred during facility construction / modifications than due to relief device chatter. This paper presents a method to assist engineers in determining if relief devices are susceptible to chatter. The methodology in this paper provides an engineering study to determine if existing installations are safe, which are allowed in the relevant engineering standards (API STD 520), and shows the research data by which it was validated. The model is used as a screening method that places the relief devices into two categories: (1) those installations that may chatter and (2) those installations that need no further review. The goal of any experimental comparison is that the model will error on the side of predicting chatter, but will be reliable enough to screen valves.

In addition to presenting the model, this paper will compare instances of known chatter to research conducted by API and work done by the Electric Power Research Institute in the 1980s. Thus far, based on research and acquired information, the method predicted all instances of chatter known to the authors. By providing a screening methodology that is supported by experimental data, plants can focus their spending on fixing real safety issues by identifying which relief installations are not expected to chatter. The paper will close by giving a brief explanation of the on-going research in relief valve stability.

1. Introduction

In November 2011, *Hydrocarbon Processing* published a paper that documented a method to determine if relief devices were susceptible to chatter. Other methods are being developed to determine the chances of chatter for a specific installation; however, the model discussed in the published paper is the only screening method that places the relief devices into two categories: (1) those installations that may chatter and (2) those installations that need no further review. The goal of any experimental comparison is that it will error on the side of predicting chatter, but will be reliable enough to screen valves. Since the publication of that article, the Oil & Gas industry has continued to struggle with the issue of relief device stability, so much so that API delayed issuance of *API STD 520 Part II Sizing, Selection, and Installation of Pressure-Relieving Devices in Refineries-Part II, Installation.* This paper compares instances of known chatter to research conducted by API, and uses this comparison to evaluate the model. Thus far, based on research and all acquired information, the method predicted all instances of chatter known to the authors.

This paper summarizes the analysis methods and validation for vapor service valves. The November 2011, *Hydrocarbon Processing* paper documents analysis methods for two-phase and liquid services. The focus on vapor / gas service valves is based on the validation data for these services analyzed herein.

2. Summary of the Stability Screening Method

The following is a summary of items used to confirm if chattering is likely to occur. Refer to the published paper for an extensive discussion of the methodology:

1. Confirm that the inlet line length is within the criteria to ensure that the pressure wave decay is not high enough to cause valve instability. The following equation (from Frommann & Friedel, Source 4) is used for this criteria:

$$L_i < 45{,}390 \frac{d_i^2}{w_{\%O}} \left(\frac{P_S - P_{RC}}{P_S} \right) (P_S - P_B)\, t_o \qquad \textbf{Equation 1}$$

2. Confirm that the system's frictional and acoustic losses are less than the difference between the operating pressure and the valve's closing pressure. The formula for estimating the acoustical losses (from Singh, Source 5) is presented below:

$$\Delta P_{Acoustic} = \frac{L w_{PSV}}{12.6 d_i^2 t_O} + \frac{1}{10.5\rho} \left(\frac{w_{PSV} L}{c d_i t_O} \right)^2 \qquad \textbf{Equation 2}$$

3. Ensure that stability will not be compromised due to vortex shedding in the inlet line (this phenomenon occurs during normal operation and is generally not associated with relief conditions).

$$L_i < \frac{d_i c}{2.4U}$$ **Equation 3**

4. Ensure that the relief rate or system size is at sufficient capacity to confirm that chattering will not occur from being oversized. For vapor service relief devices, this is assumed to not occur if the open-close cycle time for the system is greater than 1 second. This cycle is broken into two components: the time it takes for the system pressure to increase from the relief devices' blowdown/closing pressure to the set pressure and the time it takes for the system to depressure through the relief valve from the set pressure to the blowdown/closing pressure. This is represented by the following equation.

$$1 > t_{cycle} = t_{P_{RC} \to P_S} + t_{P_S \to P_{RC}}$$ **Equation 4**

5. Ensure that the relief device is installed per the manufacturers' and/or general installation guidelines. These guidelines are shown in the catalogs, ASME B&PVC Section VIII, API STD 520, etc.

6. Review the valve's operational history to ensure that there is no known history of chatter.

For a complete discussion on the methodology of this analysis, the reader is referred to the work published by Smith Et. al, as the discussion is not further expanded in this paper.[1] Note that equation 17 does not follow the verbiage, and the *less than* sign should be replaced with a *greater than* sign.

3. Acceptance Criteria

This model is the only screening method to place the relief devices into two categories: (1) those installations that may chatter and (2) those installations that need no further review. The goal of any experimental comparison is that it will error on the side of predicting chatter, but will be reliable enough to screen valves. The author considers this model to be reliable as an engineering screening tool if the following criteria are met:

- *High Correlation to Experimental Data* – If the method predicts that valves will chatter and valves won't chatter most of the time, then it is a viable screening tool given the low cost of performing the analysis compared to making piping modifications.

- *No False Negatives* – The screening tool will be considered valid so long as it does not predict any installations as safe (no chatter) when in fact chatter was experimentally found.
- *Limited False Positives* – Given that the method is a conservative screening tool, some level of false positives are to be expected. This includes instances when the model would indicate the possibility of chatter when experimental results show valve stability.
- *Screens Valves as Acceptable* – The model will only be of value if the results of the model result in existing installations not requiring further modification. Based on the results presented in the referenced paper, 50% of the installations with inlet pressure losses greater than 3% were deemed acceptable as is.

The model, therefore, will be deemed conservative and reliable if the previous four criteria are met.

4. Comparisons / Validations

To confirm the validity of the procedure, the procedure has been reviewed against all cases of chatter known to the authors. The cases are divided into three different categories:

1. ***Known Installations*** – These installations are those that are known by the authors to have failed with a loss of containment and those installations that have sufficient information to perform the analysis. The information for some of the installations is not available in the public domain, as the confidentiality of these installations has been preserved.
2. ***ASME Studies*** – In the 80's, Zahorsky performed a set of research for the nuclear industry in which they experimentally set the blowdown as the minimum percentage needed for stable operation[2]. In order to meet the acceptance criteria, therefore, this method should determine that these installations are unstable. The results were re-checked with an increased blowdown of 2% above what the author listed to see what would happen.
3. ***API PERF Studies*** – In 2011 the results of the API PERF study were presented at the API 520 Committee meeting. In this work, the API subgroup tested 18 different valves with three different inlet line lengths to determine when chatter would occur. The results of the 54 trials were introduced in their presentation (some combinations were not tested).

5. Results

The following are the results of the comparisons:

Table 1 lists the results of the model comparison with the valves that have had a loss of containment where inlet line chatter was the cause or a contributor to the incident. The model predicted the potential for chatter for each of these installations.

Table 1: Summary of known installations that have chattered

Installation	**Service**	**Chatter Predicted**	**Data Source**
Refinery, North America	Liquid	Yes	Internal company incident investigation
Gas Plant, Middle East	Vapor	Yes	Internal company incident investigation
Refinery, North America	Liquid	Yes	Internal company incident investigation
Refinery, North America	Liquid	Yes	API Published Document [3]

Table 2 lists the blowdown at which the relief devices do not chatter as experimentally determined by Zahorsky. The table also lists the blowdown at which the model would predict stability. The difference between the two values can be taken as an estimate to which the model is conservative. Since the blowdown was experimentally set as the minimum percentage, the model is accurate, as it predicts false positives for these devices.

Table 2: Comparison of minimum predicted blowdown to the results experimentally derived by Zahorsky

Run Case	**Experimentally Determined Blowdown**	**Predicted Acceptable Blowdown**	**(Δ Blowdown) Predicted – Expermintal)**
1	3.9%	4%	0.1%
2	3.9%	5.6%	1.7%
3	5.6%	9.7%	4.1%
4	8.4%	16.7%	8.3%
5	8.3%	12.6%	4.3%
6	4.3	5.3	1.0%

Table 3 lists the comparison of the model results to the experimental PERF results. The model had agreement ~72% of the time if the cases that were not tested were excluded. If one assumes that the results for the untested cases would have chattered, the agreement is ~76%. Of the cases that the model did not agree with the experimental results, the model always predicted chatter on stable valves; thus, the model is accurately screening and is slightly conservative.

Table 3: Comparison of the model to the API PERF Study Results

Model Correlation	PERF Results	Model Prediction	No. Of Cases
Agreement	Chatter	Chatter	9
Agreement	Stable	Stable	25
False Negative	Chatter	Stable	0
False Positive	Stable	Chatter	15
Agreement[1]	Not Tested	Chatter	7

Note 1: There are a number of cases that were not tested, but were assumed to chatter as the reason for not being tested was not included.

6. Conclusion

The model results always predicted when valves may chatter (need modifications). There are no instances of valves that have chattered where the model predicted stability.

- *High Correlation to Experimental Data* – There is a 70% to 75% agreement with the API PERF Studies and all industry installations with known chatter where identified with the model. This model predicted if a valve would chatter accurately 75% of the time (see the next item on false negatives).
- *No False Negatives* – No cases of chatter were found when the model predicted stable operation.
- *Limited False Positives* – 25% of the PERF study valves that were stable indicated that chatter was possible. Also, the work by Zahorsky indicated that the model predicts stable operation with an inlet blowdown generally 2% to 4% greater than needed.

The model, therefore, is a reliable method to screen relief devices based on the comparison to 60+ experimental and industrial data points.

7. Notation

The equations presented in this paper are dimensional and the following listing explains the variables and unit set.
c = speed of sound (ft/s)
d = diameter (in)
L = length (ft)
P = pressure (psig)
t = time (s)
U = process fluid velocity as it passes the PSV nozzle (ft/s)
w = mass flow rate (lb/s)

Greek Letters
ρ = fluid density (lb/ft^3)

Subscripts
B = backpressure on relief device
i = inlet
o = opening
PSV = process safety valve
RC = valve reclosing pressure
S = relief device set pressure
%O = flow rate at the valves percent open

8. References

[1] Smith, Dustin P.E., John Burgess P.E., and Craig Powers PhD. "Relief Device Inlet Piping: Beyond the 3% Rule." *Hydrocarbon Processing.* 2011.

[2] Zahorsky, J.R. "Testing and Analysis of Safety/Relief Valve Performance." The ASME Winter Annual Meeting. November 15-19, 1982. Atlanta, GA

[3] Myers, DF, Conoco. "Denver Refinery Fire." *API.* 1986.

[4] Frommann, O. and Friedel, L. "Analysis of safety relief valve chatter induced by pressure waves in gas flow." *Journal of Loss Prevention in the Process Industries.* 1998.

[5] Singh, A. "On the Stability of a Coupled Safety Valve-Piping System." The ASME Winter Annual Meeting. November 15-19, 1982. Atlanta, GA.

9TH GLOBAL CONGRESS ON PROCESS SAFETY

Making Relief Load Estimates Match Reality

Nicholas N. Cristea
Smith & Burgess, LLC
5700 Northwest Central Dr. Ste 301 Houston, Texas 77092
Nicholas.Cristea@smithburgess.com

Dustin Smith
Smith & Burgess, LLC
5700 Northwest Central Dr. Ste 301 Houston, Texas 77092
Dustin.Smith@smithburgess.com

Prepared for Presentation at
American Institute of Chemical Engineers
2013 Spring Meeting
9th Global Congress on Process Safety
San Antonio, Texas
April 28 – May 1, 2013

UNPUBLISHED

Making Relief Load Estimates Match Reality

Nicholas N. Cristea
Smith & Burgess, LLC
5700 Northwest Central Dr. Ste 301
Houston, Texas 77092
Nicholas.Cristea@smithburgess.com

Dustin J. Smith, P.E.
Smith & Burgess, LLC

Keywords:
Relief load estimates, relief system analysis, process simulations, reduced duty, inlet control valve failure, column systems

Abstract

A pressure relief system designer must make some assumptions when looking at a relief systems analysis. Some assumptions may result in costly modifications of the system, so knowing when to apply certain assumptions is paramount with the ability to accurately size a relief valve. When looking at systems, industry standards should be consulted before taking certain credits based on the given scenario. For example, the designer may be able to reduce the relief load by looking at a reduced reboiler duty and a temperature pinch during an overpressure event. When evaluating an inlet control valve failure, some credits may be taken for the minimum turndown for the outflow. However, when looking at how a downstream system may respond to the overpressure event, credit for flow continuing may not be acceptable and flow may be assumed to be lost. The application of assumptions is common practice in relief systems analysis; however, an assumption that does not match the true operational limits may lead to problems with the relief system that do not actually exist. These invalid assumptions could cost millions of dollars to fix the problem.

1. Introduction

When doing a Relief Systems Documentation Analysis, the ability to accurately model a "worst case" relief scenario is paramount to ensuring the system is protected from the overpressure event. The engineer is undoubtedly limited to the availability of equipment data and accuracy of the process simulation developed to predict upset conditions. However, if some conservative yet unrealistic assumptions are used, the relief device can be labeled as providing inadequate protection in the event of the initiating event. [1] Physically modifying the system to fix the problem, as well as the loss of profit from having to shut down the unit to perform the update, can cost the facility large sums of money. However, performing a detailed analysis on the system may better predict the "worst case" scenario, avoiding unnecessary modifications. The ability to make relief load estimates match what may happen out in the field can still be accomplished, but knowing the difference between a realistic conservative assumption and an unrealistic conservative assumption is paramount. The pressure relief system designer should consult industry standards (e.g. API 521or NFPA) to see how to estimate relief loads for given overpressure scenarios and what assumptions for mitigating items/actions are acceptable. This paper will look at two cases where relief load estimates were reduced and one where by matching the physical constraints of the system to reality resulted in a larger relief load estimate.

2. When Realistic Assumptions Lower Relief Estimates

Oftentimes relief estimates are based on normal operating flow rates, heat exchange, and/or flow paths continuing during the upsets. With the amount of optimization and debottlenecking, these assumptions may not be realistic. Equipment capacities that are right on the design limit may not be able to produce the same flow rate or heat transfer under upset conditions. Flow paths that are available under normal operation may be isolated or otherwise not be available during upsets. The following are three examples that show these effects in a typical plant situation. The first two are examples of assumptions that tended to increase the relief rate while the last example illustrates an optimistic assumption that under-predicted the relief requirements.

2.1 Reduced Reboiler Duty

When looking at a column system, the primary relief loads tend to consist of a combination of continued heat input with the loss of cooling. The relief load estimates can be complex due to the very nature of the system. The ability to do a detailed analysis of the column may be able to better predict upset conditions. The authors have seen many cases where a detailed column analysis prevented the relief device modifications from being made based on inaccurate sizing basis. The ability to accurately model the relief scenario is paramount because an error could lead to events such as an unnecessary facility modification or a loss of containment from an undersized relief device. When a pressure relief system designer looks at the loss of overhead cooling on a column system, the scenario is a transient event. The case can start with the loss of overhead condensing with the reflux continuing for a short duration. After some time, the loss of condensing and continued overhead liquid product output results in the loss of level and reflux flow. To estimate this relief condition, the column system is modeled as two steady state cases to bracket the relief load. Under either of these conditions, the bottoms temperature may increase due to the increase in system pressure, but this phenomenon may be offset by the reduction in separation; therefore, a reduced duty calculation should be performed for both calculation sets to ensure that the reboiler is capable of transferring the same heat into the column system as during normal operating conditions. When looking at the loss of overhead cooling with the reflux continuing, the reboiler duty may not be able to be reduced due to increased light component traffic in the tower (as compared to the top tower reflux failure) when compared to the high temperatures in the reboiler and lower duties expected during the top-tower reflux failure scenario due to the loss of the light ends liquid traffic in the column. Taking credit for the reduced duty is still a reasonable assumption as an upper conservative limit when the assumption is based on the maximum duty possible for the installed reboiler and is limited by equipment capacity. The design equation for heat exchangers is given below [2]:

$$Q = UA\Delta T_{LM}$$

Reduced duty calculations, limited to process/process or process/utility exchangers (e.g. shell and tube exchangers), are not typically performed on fired heater reboilers. The exchanger data sheet, if up-to-date, is the best resource for determining the heat transfer characteristics of an exchanger. Most process simulators can simulate the duty of shell and tube exchangers by specifying the clean UA and, if applicable, the utility stream (e.g. 50# steam). The duty for the exchanger may change as reflux is lost to the system, which the upset may result in hotter bottoms temperature to the reboiler. The reduced duty may reduce the relief load estimate, as there is less vapor generation as compared to using the process duty under normal conditions. Figure 1 shows a screenshot from a process simulator that estimates the rate for a column system with continued feed and heat input, but a loss of cooling and reflux.

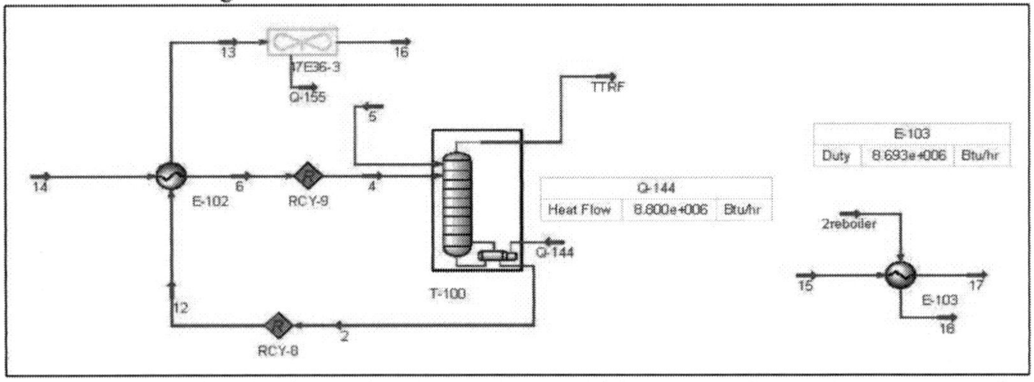

Figure 1: Example of a simulation for a column with loss of cooling and top-tower reflux failure with a reduced duty calculation performed.

2.2 *Using the correct composition and physical properties*

Generally the capacity of the reboiler (e.g. the reduced or LMTD pinched duty) will vary greatly if the calculation is based on the normal bottoms product as compared to the overhead reflux/liquid product. In the case of system upsets, the actual liquid inventory in the bottom of the tower may be lighter than normal. The pressure relief system designer is cautioned to use a composition in the reboiler to estimate the reduced duty that is consistent with the overpressure scenario being reviewed.

2.3 *Inlet Control Valve Failure Credits*

Another common overpressure scenario is the failure of automatic controls. Per API 521, this scenario is the complete failing open of a control valve (irrespective of its fail safe position).[3] If this upset condition results in overpressure, a relief device may be required to protect downstream equipment. When looking at an inlet control valve failure, the first common assumption pressure relief system designers make is to estimate the full capacity of the control valve and then verify if the relief valve can pass that fluid at the appropriate overpressure. The designers verify if the capacity of the relief device is greater than the control valve. If this estimation results in insufficient relief valve capacity, then some credits may need to be taken to make the relief load estimate more realistic.

> Per API 521§ 5.10.2,
> "in evaluating relieving requirements due to any cause, any automatic control valves that are not under consideration as causing a relieving requirement and that would tend to relieve the system should be assumed to remain in the position required for minimum normal processing flow. In other words, no credit should be taken for any favourable instrument response." [3]

For example, if the outlets from the system are expected to remain open, the calculation may assume that some of the fluid will flow through the normal path, see Figure 2. This credit is usually at the minimum turndown rate, as common industry practice is to assume that other control system remains in place and does not respond (unless it exasperates the single failure).[3] If the relief scenario is liquid, credit can be taken for the liquid outlet to reduce the fluid from the control valve that needs to be relieved by the pressure relief device by the minimum turndown rate. If the liquid outlet does not provide enough credit, the vapor outlet can be sized to take some relief as long as the vapor line is not lost by passing the liquid. A common example of this would be that a high-level alarm would trip a compressor; in this case, the relief device would have to pass both the liquid and the vapor feed to the vessel as the increased liquid level caused multiple failures. As long as the pressure relief system designer verified the liquid could flow through the vapor line and accumulate in the downstream system, credit can be taken through the vapor line. If the relief scenario is vapor, credit may be taken for flow through the vapor line at the minimum turndown rate (though the potential for credit for vapor flow through the normal liquid outlet is normally not considered). Consideration should be given for the difference in the relief fluid and the normal fluid and to adjust the credit based on the actual volume rate.

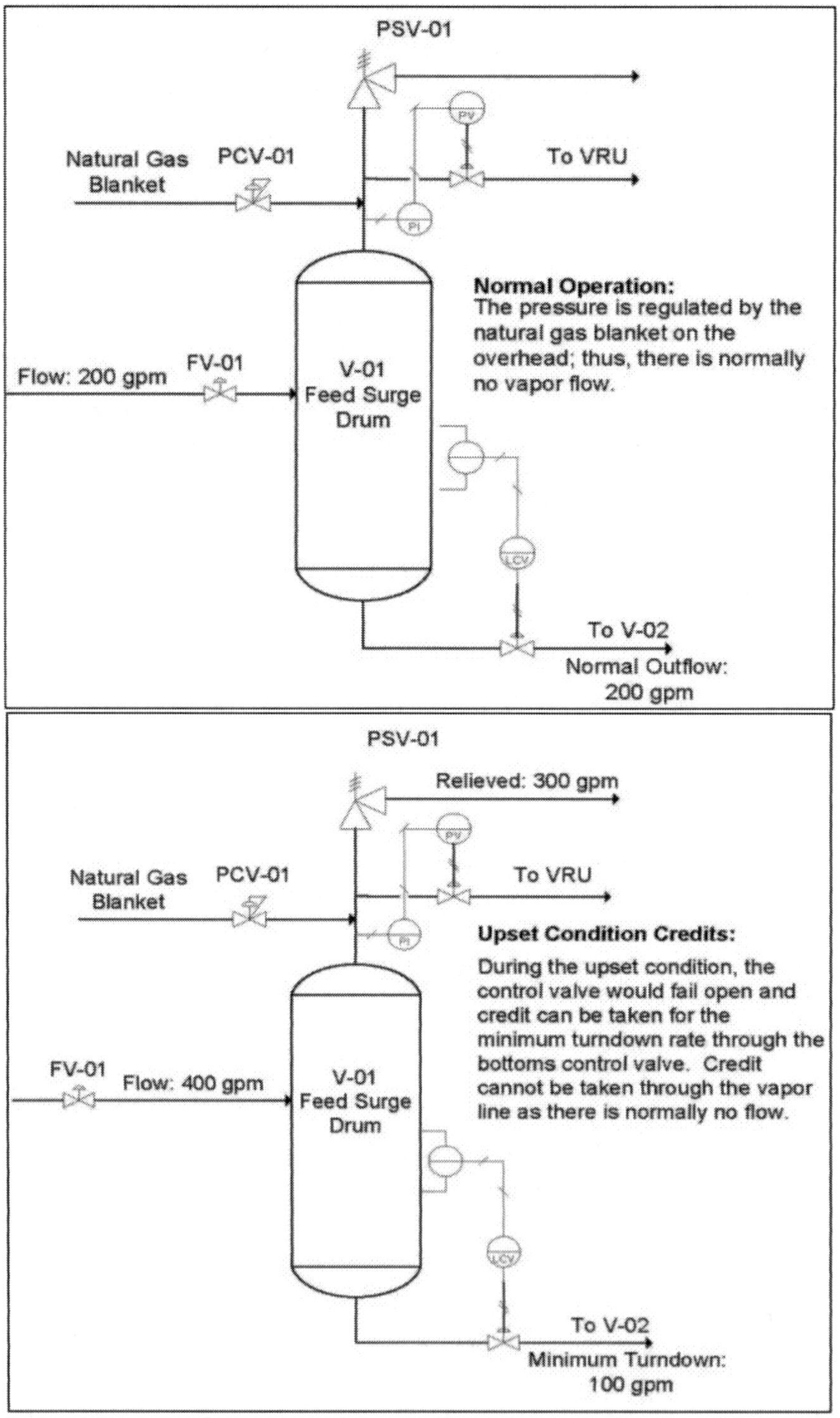

Figure 2: Looking at a feed surge drum during inlet control valve failure can lead to credits to be taken as long as they are reasonable for the system. Credit cannot be taken through the vapor line as there is normally no flow.

There is a special control valve failure case that pressure relief system designers need to be aware of. When a liquid control valve fails open, the vessel may de-inventory and gas may flow through the liquid control valve into the downstream vessel. This scenario, typically called a gas blow-by, should be reviewed for all control valve failures where there is liquid and gas in the upstream vessel. Figure 3 shows a typical system where gas blow-by may occur.

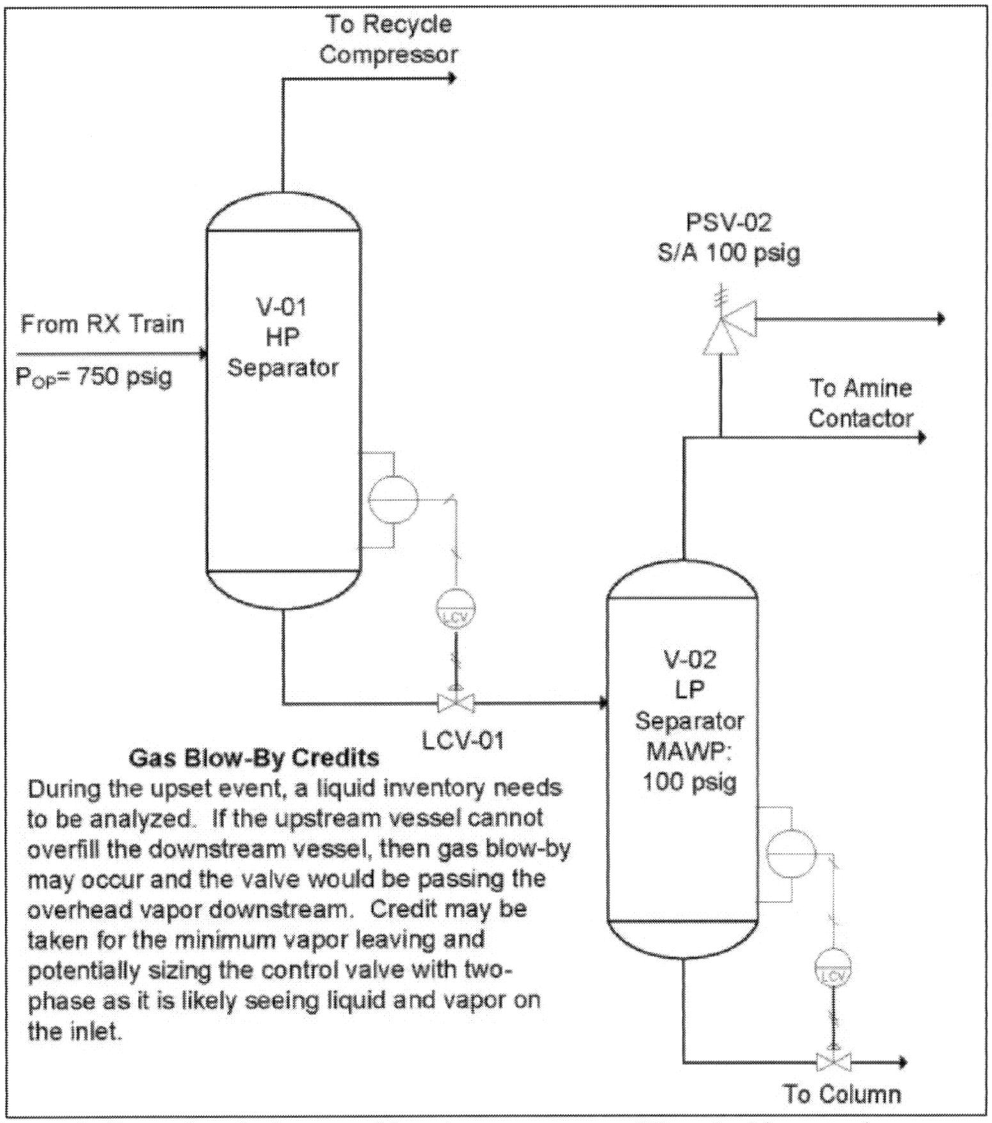

Figure 3: Example of when gas blow-by may occur. When looking at a low pressure separator, the sizing event should be vapor breakthrough the control valve from the upstream high pressure separator.

3. When Realistic Assumptions Make the Estimates Worse

3.1 Compressors Downstream of the System

In some instances, making relief load estimates match closer to reality may lead to a larger relief load estimate. In these cases, the pressure relief system designer must ensure that the relief system is accurately modeled and that the designer uses the larger rates and does not allow designs based on nonsensical assumptions to be used as the basis. There are many different assumptions that could cause this case to be true; the example that we are going to highlight is when the cascading effect of one failure causes the normal plant control system to take an action and increase the relief requirements for a pressure relief device. Oftentimes, a pressure relief system designer may want to take credit for some of the relief effluent to outflow from the system through a compressor. As stated above, this would not be a viable flow path for liquids, as almost all compressors have a high level trip on the inlet drum, but what about vapors? Depending on the scenario, the compressor may pass vapors at the upset conditions or the compressor could trip and result in a blocked outlet. Examples that may cause a compressor to trip under upset conditions include a higher inlet gas temperature trip, a high amp / power requirement trip (due to the increased suction pressure), high temperate trip (due to the potential loss of cooling to the compressor), vibration trips, and etc.[4,5] It is useful for the pressure relief system designer to discuss the scenario with personnel who are familiar with the operation of the compressor (usually either the operators or unit process engineer) to ensure that any assumptions made are consistent with operational history. Figure 4 shows a case where a compressor trip would not result in a larger relief load.

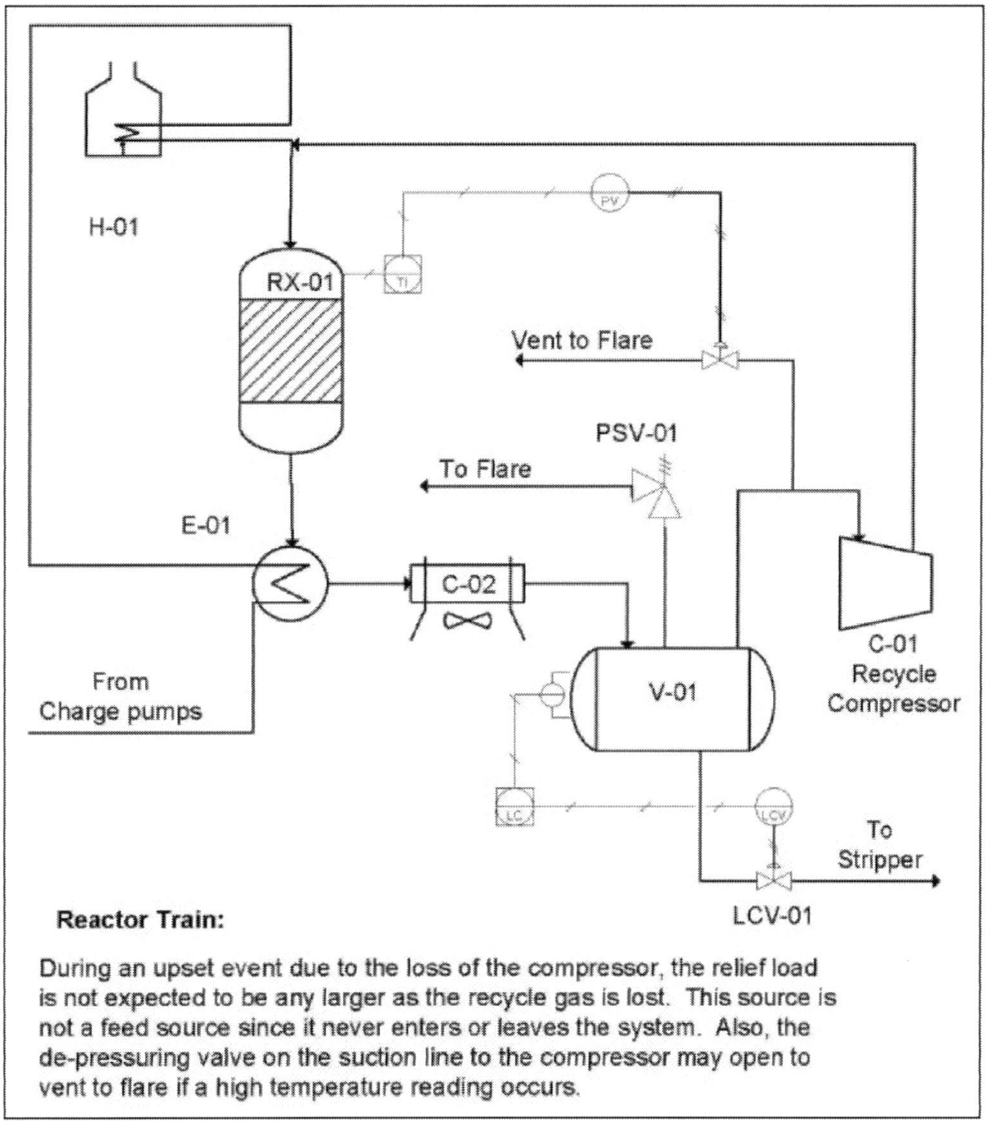

Figure 4: An example of how compressors can affect the relief load. The loss of the compressor would not result in any excess relief as the recycle gas is lost.

3.2 Example of Compressors Shutdown Analysis

A larger estimated rate may occur in a scenario where there is a partial loss of cooling in a column system that uses a compressor to move the off gas from the overhead accumulator to the downstream system (e.g. a FCC Main Fractionator, Coker Bubble Tower, Atmospheric Tower, and etc.). After the simulation is developed to model the column operation during the partial loss of cooling, the modified characteristics of the off gas (e.g. temperature, density) are consistent with conditions that trip the compressor.

Additionally, the pressure relief system designer must ensure that the scenario itself (e.g. a unit wide power or cooling tower failure) could result in a compressor trip and result in this outlet being blocked. The relief load estimate would then need to include the cascading effects of the compressor shutting down and may be much larger than would be otherwise predicted (see Figure 5).

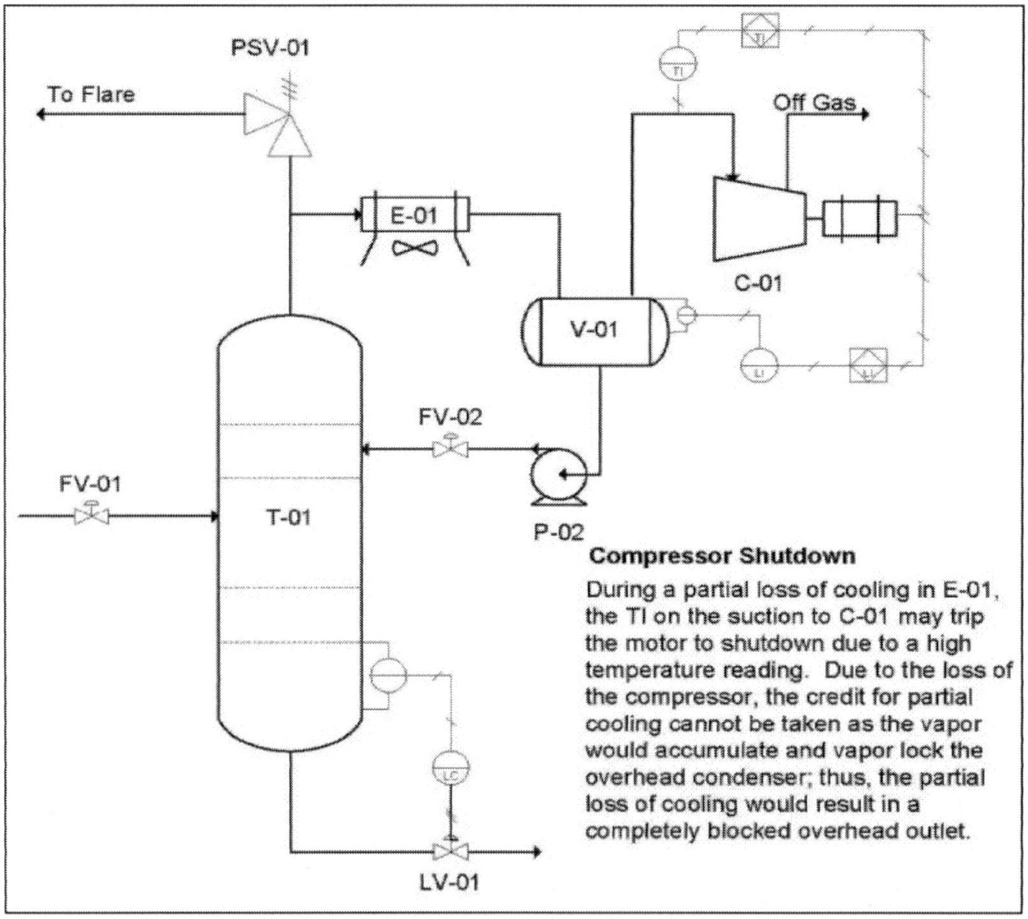

Figure 5: Example of how compressors can affect the relief load. On the right, the loss of the compressor would result in a larger relief as the vapor outlet is completely blocked.

4. Conclusion

To realistically represent the relief load estimate, the pressure relief system designer must understand the system, how it is operated, the limits and shutdowns of the associated equipment, and how these parameters interact during the specific overpressure event being reviewed. The pressure relief system designer must also ensure that all assumptions are consistent with industry, company, and regulatory requirements for the facility. There are many ways to check to see if the relief load estimate is based on realistic assumptions or on conservative/simplifying assumptions. In many cases, conservative/simplifying assumptions are acceptable as long as the assumptions don't result in deviations from standards or practices that result in spending more money. Oftentimes, conservative/simplifying assumptions do result in deviations, and the pressure relief system designer must then use more realistic assumptions to generate the relief load.

The pressure relief system designer must also be vigilant to not use simplifying assumptions that are not conservative. These assumptions tend to result in relief rates that may under-predict the true requirements in an overpressure event. Oftentimes industry, company, and regulatory requirements for the facility are set up to eliminate these scenarios, but the example in Section 3.2 illustrates the pressure relief system designer must check these assumptions (many of which can be implicit).

5. References

[1]. Smith, Dustin and John Burgess, "Relief Valve and flare action items: What plant engineers should know," *Hydrocarbon Processing*, Volume 91, NO.11, 2012

[2]. Middleman, Stanley, *An Introduction to Mass and Heat Transfer: Principles of Analysis and Design*, John Wiley and Sons, New York, NY. 1998

[3] ANSI/API Standard 521, "Pressure-relieving and Depressuring Systems," American Petroleum Institute, 5th ed., 2008.

[4]. GPSA, *Engineering Data Book: Volume I 10th Edition*, Gas Processors Suppliers Association, Tulsa, OK. 1987

[5]. Lieberman, Norman and Elizabeth Lieberman, *A Working Guide to Process Equipment 3rd Edition*, McGraw-Hill, New York, NY. 2008

Explosion incident simulation for Petrochemical plant

Zhao Xiangdi
State Key Laboratory of Safety and Control for Chemicals, SINOPEC Safety Engineering Institute
No.218, Yan'an San Road, Qingdao, P.R.China 266071
cheeryde@gmail.com

Yuan Jiwu
State Key Laboratory of Safety and Control for Chemicals, SINOPEC Safety Engineering Institute
No.218, Yan'an San Road, Qingdao, P.R.China 266071
Yuanjw.qday@sinopec.com

Jiang Chunming
State Key Laboratory of Safety and Control for Chemicals, SINOPEC Safety Engineering Institute
No.218, Yan'an San Road, Qingdao, P.R.China 266071
Jiangcm.qday@sinopec.com

Wang Zheng
State Key Laboratory of Safety and Control for Chemicals, SINOPEC Safety Engineering Institute
No.218, Yan'an San Road, Qingdao, P.R.China 266071
wz.qday@sinopec.com

Prepared for Presentation at
American Institute of Chemical Engineers
2013 Spring Meeting
9th Global Congress on Process Safety
San Antonio, Texas
April 28 – May 1, 2013

UNPUBLISHED

Explosion incident simulation for Petrochemical plant

Keywords: numerical simulation, explosion incident, TNT equivalent, Multi-Energy

Abstract

A new method of explosion simulation which is based on FLACS software was used for simulating and rebuilding an explosion incident which occurred in a coal gasification company. The other two commonly used methods which are based on TNT equivalent method and Multi-Energy method were considered to calculate the explosion overpressure also. Simulation results indicated that this new method can calculate explosion overpressure in three dimensional. It can make up for the shortfall of traditional assessment techniques. The result is more stable and reasonable when compared with the scene of the incident.

1. Introduction

Petrochemical plants may happen explosion incident easily than other facilities since they have many equipments and devices with lots of hazard material. Once the device exploded, it will cause large casualty and property loss. For example, Alon big spring refinery(2007) and Guangwei(2008) occurred fire and explosion incidents and caused many people dead. So it is important for our to calculate plant's capability against explosion. We could use the results to determine whether those plants should be firmed and changed or not. Researchers around the world found a series special assessment methods based on incident consequences or risk. Someone used those methods to design and renovate process safety of petrochemical plant[1][2]. However, researches on those technology were lags behind in China. We used traditional assessment technology which based on empirical formula. Those traditional methods did not consider the actual situation of factory. It couldn't show the those factor's effect on explosion incident such as the layer equipment, wind, fire protection system and other condition [3][4][5].

Nowadays the Computational Fluid Dynamics analysis technique is becoming more sophisticated. Someone in industry and academia studies explosion simulation used CFD. Herrmann, Hansen studied the consequences of gas cloud explosion using FLACS computational fluid dynamics software, then compared results with experiment, and found the results matched each other well[6][7]. Windhors, used AutoReaGas software to simulate the gas explosion and determine safe distance between petrochemical cracker and control room [8]. P.Hoorelbeke studied explosion risk assessment techniques commonly used [9]. However, there were only a little study on numerical simulation model for explosion incidents. Therefore, assessment on explosion incident by using FLACS were performed in the work.

2. Explosion incident and Simulate model

2.1 Explosion incident

23 December 2008 an explosion incident occurred in Hunan province in China. Gas released from one pipe in the carbon monoxide shift.

Then the gas cloud was ignited and caused explosion. The scenes after this incident with the insulation layer damaged was shown in figure 1.

(a)F2012， F2013, F2014 (b) T2101， R2103

(c) Office building with windows and glass damaged

Figure 1: The scene of the explosion incident with the insulation layer damaged

The gas which flowed in the pipeline was hydrogen. The loss rate of contain was 562kNm3/h in the release phase. The wind speed in-situ was about 3m/s with the direction was NE. The key equipment in this unit for simulation were shown in table 1.

Table 1. Information for the key equipment

Number	Name	Size(m)	Description
F2102	Gas-liquid separator	Φ4.2×13.4	40% of insulation damaged
F2103	Gas-liquid separator	Φ3.7×11.6	25% of insulation damaged
F2104	Gas-liquid separator	Φ3.7×15.6	Little effect
T2101	Regulate tower	Φ3.2×26.0	70% of insulation damaged

R2103	Converter	Φ4.9×13.8	50% of insulation damaged
R2104	Converter	Φ4.9×11.7	30% of insulation damaged
Control room		15.0×46.8×4	Some of the windows damaged; Lots of the windows broken
Office building		15×46.8×19.5	Some of the windows damaged; Lots of the windows broken

2.2 Model for simulation

3-D model which was based on the factory's actual layout was shown in figure 1. This model was built using the actual sizes of the carbon monoxide shift unit. The total size for simulation was 100m×80m×40m. Ignition position was in the center of carbon monoxide shift unit nearby leak point. The distance from ignition position to control room and office building was 65m. Shapes of equipment were modified for simulation. Monitor points were set for key equipments, as shown in figure 2.

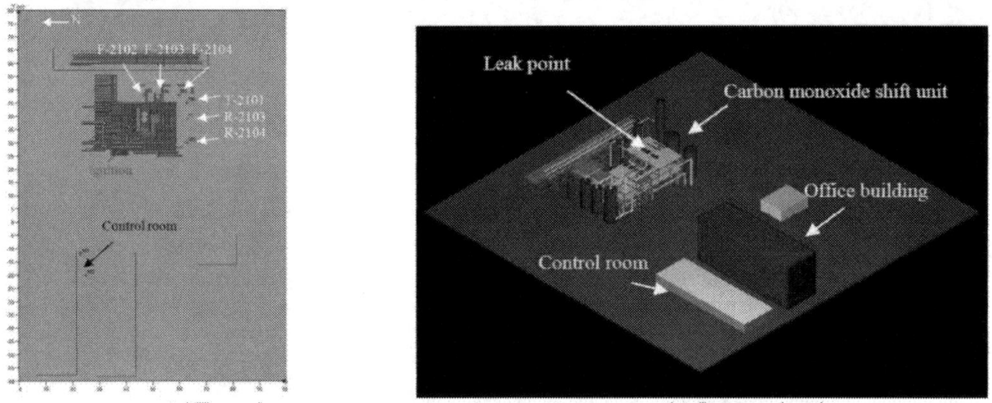

(a)Top view (b) Isometric view

Figure 2: Three dimensional model for simulation

2.3 Results and discussion

2.3.1 Leak and dispersion

The material in pipeline was hydrogen. Loss rate for contain was 562kNm^3/h. The time was 24 second between leak started and gas ignited. Wind speed was about 3m/s with direction was NE when the incident occurred. The new method which was based on FLACS was used to rebuilt dispersion process for hydrogen. Simulation results of fuel size after dispersion were shown in figure 3. The whole volume of the hydrogen was about 10×13×7 m^3 in a mixture concentration with equivalence ratio(ER).

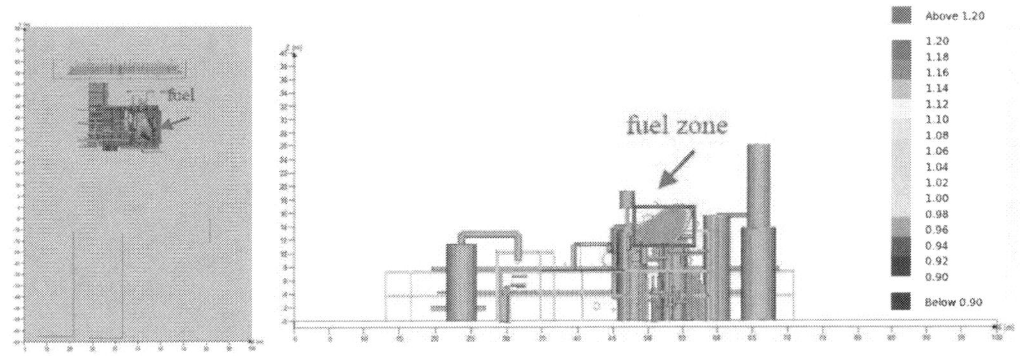

(a) top view (b)side view

Figure 3: The fuel size in this unit after dispersion (0.9~1.2 ER)

2.3.2 Explosion

Simulation assumes that hydrogen and air were in a mixture concentration with equivalence ratio. TNT equivalent method, Multi-Energy method and FLACS method were used to assess the process for gas explosion impact on the equipments and facilities. However, TNT equivalent method was simple and did not consider the effect of actual situation. Although the Multi-Energy method considered the explosion strength as the effect of actual situation, but it also did not consider reflection of complex equipment and did not predict the effect of risk mitigation system such as water spray.

Results of explosion overpressure using those three method were shown in table 2. It can be seen that TNT equivalent method and Multi-Energy method could only take into account explosion overpressure in two-dimensional. For example, overpressure calculated in the position 48m from center of explosion was 5.2 kPa using TNT equivalent method, however, the results increased to 7.9 kPa using Multi-Energy method. But, simulation results using FLACS were different. It could give overpressure in three-dimensional. The overpressure in position 48m from center of explosion was 6.7 kPa with 10m above ground, and 4.7 kPa with 19m above ground , as shows in table 2.

Table 2. Results of explosion overpressure using different methods.

		Overpressure in the different position from the ignition point (kPa)		
		48m (office building)	65m(control room)	74m
TNT equivalent method		5.2	3.48	2.95
Multi-Energy (blast strength 5~10)		7.9 ~ 24.7	5 ~ 16	5.1 ~ 13.4
FLACS method the height above ground(m)	1m	7.8	6.5	3.9
	10m	6.7	-	3.6
	19m	4.7	-	1.9

The overpressure in different monitor points(as shown in figure1) were shown in figure 4. The maximum explosion overpressure in carbon monoxide shift unit was about 62kPa, as well as, the overpressure was about 6.5kPa near the control room and office building.

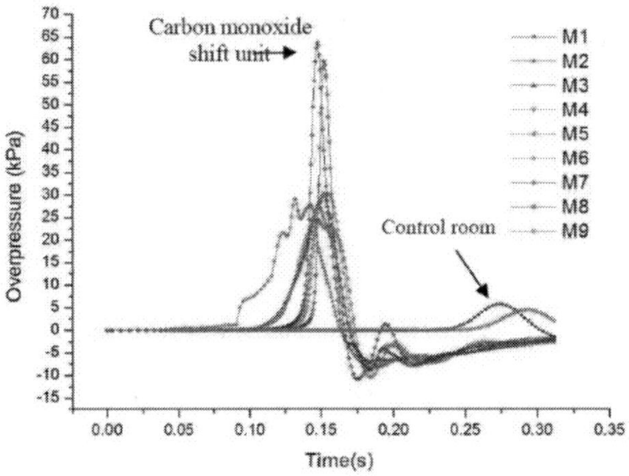

Figure 4: Overpressure of explosion in the different monitor points

The process for explosion overpressure wave affecting on equipments and facilities were shown in figure 5. As soon as explosion starting, a strong shock wave (a), causing major damage to the surrounding equipments, then it would gradually expand and weak (b-c), and subsequently affected office building and control room (d).

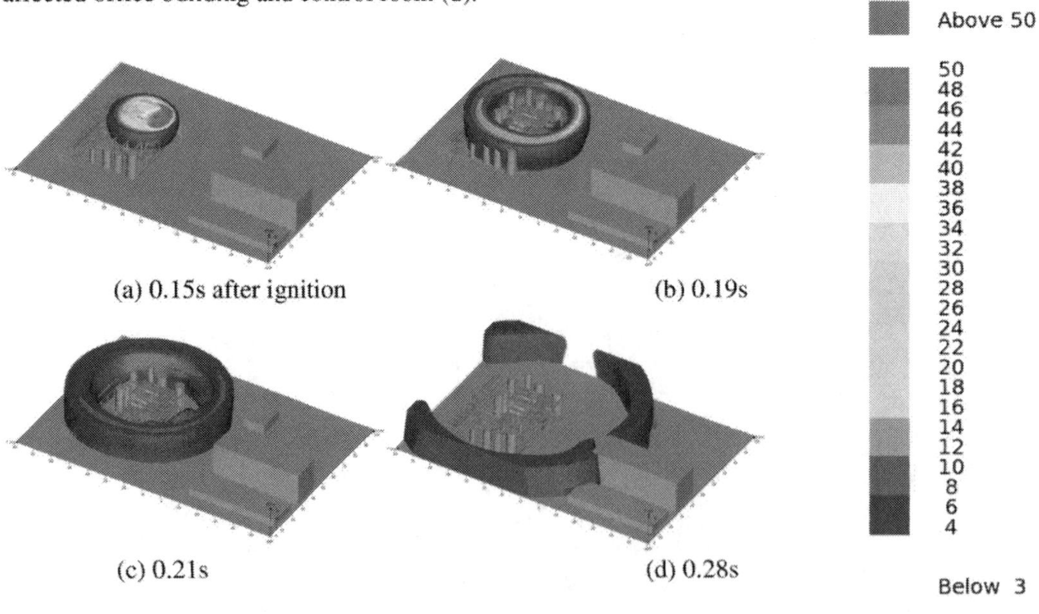

Figure 5: Explosion overpressure wave changed and expanded process (3-50 kPa)

3. Conclusion

The shortfall of traditional assessment techniques is that they can't predict the explosion effects with the layer equipment, wind, fire protection system and other condition's effect on the flame acceleration. A method of explosion simulation which is based on FLACS software for petrochemical plants was introduced in this paper. The results were as follows:

This method can calculate all parts of the space after explosion, and can make up for the shortfall of traditional assessment techniques. The overpressure simulation results using this method are between the results calculated by TNT equivalent method and Multi-Energy method. The result is more stable and reasonable when compared with the scene of the incident.

4. References

[1] D.Bjerketvedt, J.R.Bakke and K. Wingerden(1997), "Gas explosion handbook ,"*Journal of Hazardous Materials*, Volume 52, pp.1-150.

[2] S.B. Dorofeev(2007), "Evaluation of safety distances related to unconfined hydrogen explosions ," *International Journal of Hydrogen Energy*, Volume 32,p.2118-2124.

[3] B.H.Hjertager, I.Moen, J.H.S.Lee, R.k.Eckhoff, k.Fuhre, O. krest,(1981), "The influence of obstacles on Flame propagation and pressure development in a large vented tube ," *CMI Report No. 803403-2*, Chr. Michelsen Institute, Bergen, Norway.

[4] "Guidelines for chemical process quantitative risk analysis,"*American institute of chemical engineers*, NY,2000.

[5] A.C. van den Berg (1985), "The multi-energy method. A framework for vapour cloud explosion blast prediction ,"*Journal of Hazardous Materials*, Vol.12, pp.1-10.

[6] Herrmann, D.D. (1999), " Improved Vapor Cloud Explosion Predictions by Combining CFD Modeling with the Baker-Strehlow Method ," *International Conference and Workshop on Modeling the Consequence of Accidental Releases of Hazardous Materials*. CCPS/AIChE, Sep. 28 ,pp.479-494.

[7] Hansen, O.R., Talberg, O. and Bakke, J.R (1999), "CFD-Based Methodology for Quantitative Gas Explosion Risk Assessment in Congested Process Areas: Examples and Validation Status ," *International Conference and Workshop on Modeling the Consequence of Accidental Releases of Hazardous Materials*. CCPS/AIChE, Sep. 28 ,pp.457-477.

[8] Windhorst, J.C.A (1999), " Deriving Engineering Specifications from Release Dispersion and CFD Explosion," *International Conference and Workshop on Modeling the Consequence of Accidental Releases of Hazardous Materials*. CCPS/AIChE, Sep. 28 ,pp.495-514.

[9] P.Hoorelbeke, C.Izatt, J.R.Bakke, J.Renoult, R.W.Brewerton(2006), "Vapor cloud explosion analysis of onshore petrochemical facilities," *American Society of Safety Engineers Middle East Chapter 7th Professional Development Conference& Exhibition kingdom of Bahrain*. March 18-22,pp.1-15.

9TH GLOBAL CONGRESS ON PROCESS SAFETY

Using Reliability Centered Maintenance (RCM) to Reduce Fire Risk in High Temperature Oil Handling

Marc W. Yarlott, P.E.
Veolia Water North America
200 E Randolph St,
Chicago, IL 60601
marc.yarlott@veoliawaterna.com

Ryan Johnson
Veolia Water North America
200 E Randolph St,
Chicago, IL 60601
ryan.johnson@veoliawaterna.com

Steven Waters
Veolia Water North America
200 E Randolph St,
Chicago, IL 60601
steven.waters@veoliawaterna.com

Prepared for Presentation at
American Institute of Chemical Engineers
2013 Spring Meeting
9th Global Congress on Process Safety
San Antonio, Texas
April 28 – May 1, 2013

UNPUBLISHED

Using Reliability Centered Maintenance (RCM) to Reduce Fire Risk in High Temperature Oil Handling

Marc W. Yarlott, P.E.
Veolia Water North America
200 E Randolph St,
Chicago, IL 60601
marc.yarlott@veoliawaterna.com

Ryan Johnson
Veolia Water North America
Steven Waters
Veolia Water North America

Keywords: reliability, maintenance, RCM, critical spares, uptime

Abstract

Veolia Water North America contracted for the technically difficult Biosolids handling process Design, Build, Operate project with the Metropolitan Water Reclamation District of Greater Chicago at the Stickney Water Reclamation Plant. The plant design uses a high temperature oil to operate the Biosolids drying equipment. Several other plants using a similar high temperature oil system have had fires related to leakage of the oil. Having successfully employed the RCM analysis at many of Veolia's projects world wide, Veolia applied the RCM process to gain improvements in the overall reliability of the plant, and particularly the high temperature oil system, by optimizing maintenance strategies and identifying spare parts, procedures, and plans to maximize the uptime of the system. The RCM process resulted in identification of 425 spare parts at a cost of $525,000 which mitigated over $33 million dollars worth of potential risk. For a very technical process with significant risk of failures, the value of a complete RCM study will have a very high return on investment..

1. Introduction

Veolia Water North America (Veolia) operates a 150 dry ton per day biosolids pelletizer operation within the Stickney Water Reclamation plant in Chicago, Illinois. The facility is comprised of 4 independent pelletizer trains. The facility receives thickened solids (22-28%) and the solids are pumped to one of 4 large Seghers pelletizer units. The facility utilizes a gas fired boiler to heat the thermal oil that is used to transfer the heat needed to evaporate the moisture from the biosolids. The systems are very complex in nature and as with most solids processing facilities these systems are a challenge to operate and maintain.

As with any major start-up and commissioning, the operations staff gained a great deal of knowledge during the commissioning and start-up of the system. The process of converting the solids to a pellet is not just about heat transfer. Veolia's contract is a performance based one with plant uptime being the most significant driver for success. Veolia successfully completed its performance testing in late 2009. The management team at Veolia carefully evaluated each part of the performance test data and determined that it would be prudent to focus technical resources to build an uptime focused maintenance strategy.

In the Spring of 2010 Veolia initiated a comprehensive Reliability Centered Maintenance (RCM) review at the facility. This review was facilitated by internal Veolia employees and included the valuable input of the operations and maintenance staff at the facility. Veolia also brought in staff from other projects to help in analyzing specific functional areas. This RCM review incorporated a critical spares analysis that would be used to make cost/risk based storage decisions.

2. METHODOLOGY

2.1 Reliability Centered Maintenance

RCM is "a process used to determine the maintenance requirements of any physical asset in its operating context" [1]

Reliability Centered Maintenance has its roots in the civil aviation industry. In the late 1960s United Airlines invested a great deal of time and money in studying how and why equipment failed. The United Airlines management team understood that safety and reliability were of paramount importance to the operational mission. United Airlines mapped out a plan to better understand equipment failures, and focused the project on improving overall reliability of its aircraft. This analysis resulted in a startling conclusion that "there was no correlation of equipment age and failure" and that "equipment failures could not be prevented." In 1978, United Airlines published its findings in a document titled Reliability Centered Maintenance [2]. This publication was a significant document that challenged maintenance professionals to take a hard look at how and why maintenance tasks were performed.

The study concluded that equipment failure conformed to one of six failure patterns, see Figure 1 [1]. These curves plot the probability of failure on the y-axis and time on the x-axis. The failure curve analysis yielded very surprising results in that the statistics proved conclusively that equipment failures were not primarily dependent upon equipment age. The equipment failure distribution of these patterns is as follows:

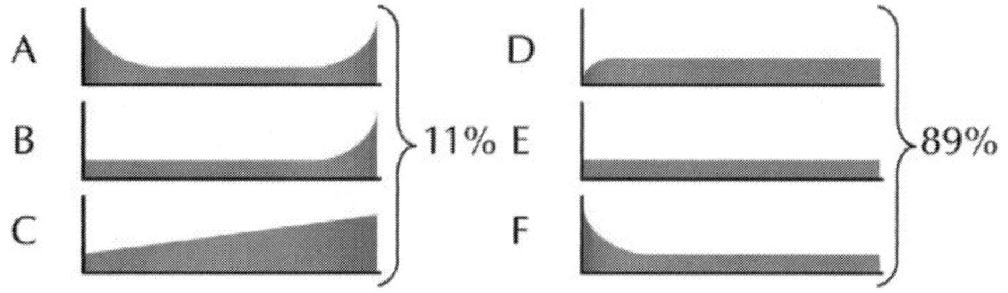

Pattern A	**4%**
Pattern B	**2%**
Pattern C	**5%**
Pattern D	**7%**
Pattern E	**14%**
Pattern F	**68%**

Figure 1, Six Patterns of Failure [1]

Based on this data, government regulators and the airlines decided to revamp the maintenance strategy of the civil aviation industry. In reviewing the failure patterns, one can quickly deduce that patterns A, B, and C have some dependency on equipment age, while patterns D, E, and F are definitely not age related.

It became quite obvious that performing calendar time based overhaul/intrusive maintenance was actually contributing to equipment failures rather than preventing them as planned. This realization resulted in drastic change to the airline maintenance programs and aircraft designs. This was also the beginning of the reliability-centered maintenance movement.

RCM will result in the correct maintenance mix of preventive maintenance (PM), condition assessment, repair (also called reactive maintenance) and proactive maintenance techniques to ensure an increase in the probability that a machine or component will function in the required manner over its design life cycle with a minimum of maintenance. RCM requires that maintenance decisions be based on maintenance requirements supported by sound technical and economic justification. As with any strategies, there are many ways to achieve the end goals.

It is important to note that RCM is not a "silver bullet" that can be applied in a lackadaisical manner. RCM requires discipline and attention to detail in the evaluation process. An RCM analysis is based on a detailed Failure Modes and Effects Analysis (FMEA) and includes probabilities of failure and system reliability calculations. The RCM process ensures that all the critical failure modes are being managed. The analysis is used to determine appropriate maintenance tasks to address each of the identified failure modes and their consequences. RCM is a proven approach that reduces risk and focuses limited maintenance resources on important assets.

In its simplest form the RCM process requires that a group knowledgeable in the operation and maintenance of the facility review systems in a methodical and organized manner. The goal is for the Operations and Maintenance (O&M) staff to learn from one another and freely exchange information related to the process, equipment, reliability, and operation. The RCM process requires that the O&M staff evaluate the following seven questions:

- **Functions:** What does this asset /system do (what do the users want this asset to do)?
- **Functional Failures:** In what ways can the asset/system fail to deliver the functions described above?
- **Failure Modes:** What causes the functional failures to occur?
- **Failure Effects:** What happens when the failure occurs?
- **Consequences:** Does this failure even matter?
- **Proactive Tasks:** Is there anything that can be done to predict or reduce the likelihood of the failure?
- **Default actions:** Otherwise, what other action is available to improve reliability?

Each failure mode is put through a decision tree (Figure 2) to determine its impact on environmental, safety, and financial risk. All of the critical failure modes are subsequently examined and assigned a proactive task (PM, PdM, or failure finding) to manage the risk.

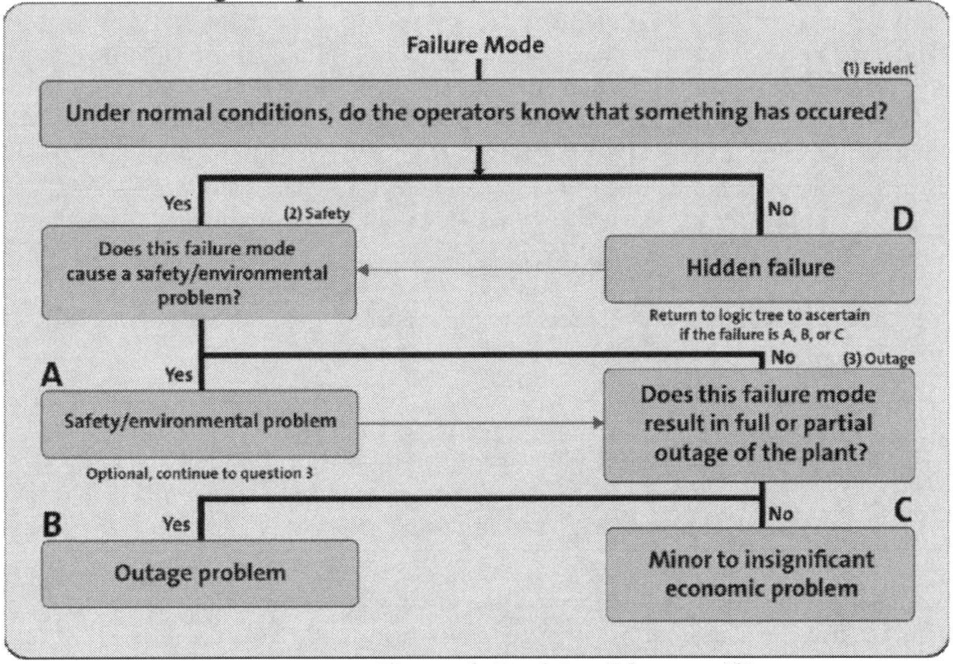

Figure 2, Decision Diagram [3]

3. RESULTS

3.1 Implementation of RCM

Prior to performing the RCM review, the project team was experiencing mechanical failures of system components that were resulting in some periods of downtime. The facility uptime was being negatively impacted by asset failures, and the need for rapid repairs resulted in little to no root cause analysis. The facility uptime had dropped off after performance testing and uptime was running on average of 50% June 2010. Staff worked out a plan that prioritized the RCM reviews based on the principle of analyzing problem systems first.

Initially the RCM project focused on one train of the processing facility. The train was further separated into 13 systems. Over a 16 week period, staff completed 13 classical RCM Reviews. In total over 1784 failure modes were analyzed in total.

One of the benefits of conducting the RCM review was that it also gave the project staff the opportunity to understand in more detail where the system choke points were, and why the failures were occurring. Many useful suggestions on improving uptime and reliability were recorded and noted in the RCM review as “items of interest”. In total over 104 items were identified as recommended improvements to the existing installation. For example, one item of interest was related to the biosolids cake pumps. These pumps had some highly unusual failure modes and had become quite problematic from an O&M perspective. During the RCM review there were several failure modes that pointed the review team to questions if the hydraulic unit that was provided with these pumps was adequately sized. The hydraulic unit reservoir was running hot, and there were wide fluctuations and pressure swings in the system. After the RCM review the facility staff worked with a local engineering company to resize the drive unit. This change has significantly improved the uptime of the pumps and eliminated many of the “unusual” failure modes. Staff at the project are now working their way through the “items of interest” and implementing many of the group’s recommendations.

A second benefit of the RCM review was a success directly related to the high temperature oil system. During the RCM Review the staff identified Failure Mode #61:

61	***Primary Oil***	Oil heater High Pressure alarm: Staff needs to confirm that there is indeed a High Pressure alarm on the oil heater when the PRV fails closed. This is a safety issue that can be mitigated with existing instrumentation

In following up on this item the plant staff confirmed that there was indeed a high pressure alarm; however, in checking the alarm, they discovered that the Pressure Relief Valves (PRV)’s were not the correct for this service. The Original Equipment Manufacturer (OEM) had supplied air service PRV’s instead of liquid service. The process was shut down and immediately new PRVs were ordered and installed on the Heaters. The RCM review, allowed the staff to step through the features and failure modes of the equipment and identify and prevent a failure before it ever happened.

3.2 Critical Spare parts

There are many maintenance management practices that can help to properly manage risk. One area of maintenance management that must be included in a sound asset management program is the selection/identification of "critical spares." To achieve the desired availability of equipment, it is important to give proper consideration to having some spares parts/components available on-site. This is often implicitly considered by maintenance teams as an easy way to mitigate down time risks, thus leading to unnecessary amount of spares. One must be careful when evaluating critical spare parts. It is important that each spare part be properly vetted and justified through a proper risk based method.

Over the past 10 years equipment manufacturers have changed the way that they support their products. The days of having spares readily available for purchase have ended, in favor of the "just in time" manufacturing philosophy. In addition to less available stock, manufacturers have set some limitations on the timeframe that equipment is supported (planned obsolescence). These changes have created a need for operations and maintenance staff to better understand the challenges involved in obtaining needed parts.

Indeed, to be truly cost-effective the proper selection of critical spares has to be developed based on a methodical process. This process must take into consideration a cost-benefit/risk analysis that is aligned with the facility's goals and objectives. This analysis needs to include the following:

- Failure mode identification
- Consequences (safety, environmental, and economic) of not having the spare
- Lead time to procure
- Availability of spare (i.e. obsolete)
- Probability of failure
- Predictability of the failure
- Shelf life
- Storage considerations (space, environmental conditions, etc.)
- Cash flow

Each of these factors will now be considered in more detail.

3.3 Failure Mode Identification

A "failure mode" is an event that is reasonably likely to occur and that would cause the loss of an asset's functionality [1]. When developing a potential list of spare parts for an asset, an understanding of the equipment at the "failure mode" level is necessary.

The Chicago project identified 1784 failure modes in total. During the RCM review, staff decided which parts would be considered for spares. It is important that experienced operations and maintenance staff participate in the development of the failure modes. Special attention should be given to failure modes that are wear-related.

3.4 Consequences

Not all failure modes have the same impacts to the operations at a facility. Great care should be taken in evaluating the consequence of failure, utilizing a criticality review process (Figure 2). One should make an effort to monetize the risk of lacking a spare part when needed. For Chicago a project-specific matrix was developed and used to quantify lack of equipment availability into money. This allowed the staff to factor in items like cost of downtime, contract issues and safety into economic terms. This provided the framework upon which a proper spare part evaluation can be performed.

3.5 Lead Time

It is critically important that we understand the supply chain when it comes to spare parts. The availability and lead time for obtaining spare parts varies significantly based on the manufacturer/supplier and geographical location of the facility. When conducting a critical spares evaluation, one must always fully understand how long it will take to procure an item. Oftentimes excessive lead times may be reason enough to have a spare available at the facility. Initially a decision may be made not to stock a spare because a redundant asset is available. This decision may have to be reconsidered if one has to rely on solely the redundant system for a long period (while one waits for the spare to arrive). Many of the assets at the Chicago plant were custom built and manufactured overseas. Staff made contact with the vendors and determined the lead times on each of the items identified during the review.

3.6 Availability

Oftentimes equipment vendors discontinue product lines and support for products. Some of these changes are the result of market-driven forces, such as mergers and acquisitions within the manufacturing industry itself, while others are purely technologically-driven. Special consideration should be given to electronic components such as Variable Frequency Driver's (VFD) and I/O cards. In the Chicago review we discovered that some of the electronic components were no longer supported by manufacturers.

3.7 Probability of Failure

A proper critical spares evaluation needs to factor the probability of occurrence into the equation. This is usually done by experienced staff familiar with the common equipment in a similar operating context. There are published failure statistics that can also be used to benchmark this type of probability. Veolia has an enterprise Computerized Maintenance Management System (CMMS) that it uses to quantify the Mean Time Between Failures (MTBF). Understanding the failure rates will help in making a sound business decision as it relates to procuring spares.

3.8 Predictability of Failure

Some equipment/components will fail over time due to wear or age/fatigue. It may be possible to properly manage this type of failure by conducting periodic inspections to measure the overall condition of the asset. We identified components like chains, belts, sprockets, liners, and screws can be evaluated for wear and new parts can be ordered prior to loss of functionality.

Properly set up and monitored, condition assessment programs can minimize the amount of spares needed at a facility. Items like fuses, shear pins and electronics fail with little warning and should always be stocked if the asset is critical.

3.9 Shelf Life

Some components deteriorate with time (like probes, sensors, ball bearings, batteries and gaskets). These timing considerations should be factored into the decision-making process.

3.10 Storage Considerations

Prior to procuring any spare parts one must evaluate his/her location to determine if the facility has adequate storage facilities on site.

Consideration should be given to environmental conditions (humidity, temperature, sulfides, etc.) and the available space. Staff also needs to ensure that it has the proper systems in place to track the spares and to perform PM maintenance tasks on the inventory if needed (i.e., a spare pump may require that the shaft be rotated periodically to distribute lubrication and prevent shaft deflection). In Chicago we decided to store some of the spares on site and some in off site storage facilities. We had very good success in working with vendors in some cases to store items for us.

3.11 Critical Spares Methodology

Veolia Water utilized its internally developed critical spares methodology during the RCM review. The following is a small section of the spares analysis that was based on some contract terms and downtime estimates.

The contract is production-based and includes liquidated damages for downtime and loss of availability. What follows is a general description as to how the process was used to determine the need to stock a spare. For presentation purposes, only three scenarios are included.

3.11.1 Step One: "Monetize the Risk"

There are 4 process trains at this facility and three are needed to meet full capacity. The group determined that if all 4 trains were unavailable there would be a $1,250/hour cost of downtime. The group then evaluated loss of capacity and assigned a lower dollar number for some unit operations ($450/hr). After debating the liquidated damages clause of the contract and its impact on the business, the group settled on $225/hr as being the cost of loss of redundancy. This process was extremely rigorous and debated by the review team. The assumptions were sent in a memorandum to the senior management team.

3.11.2 Step Two: Failure Mode Evaluation

	Failure Mode	Local	System	Plant
1	Bearing seizes due to lack of lubrication	Bearing wears, gets hot and eventually fails	Bearing wears, gets hot and eventually fails. Standby pump starts and carries load	Loss of redundancy
2	Temperature transmitter fails	Transmitter fails and thermal oil pumps stop pumping	Oil system shutdown and OLC initiates a system shutdown	Loss of one train
3	Heat trace thermostat fails	Heat trace fails, water line freezes, resulting in loss of flow	Biosolids pumps trip on loss of seal water.	Complete plant shutdown

3.11.3 Step Three: Match Critical Failure Mode to Spare

Failure Mode	Spare
Bearing seizes due to lack of lubrication	Bearing
Temperature transmitter fails	Transmitter
Heat trace thermostat fails	Thermostat

3.11.4 Step Four: Determine Spare Cost and Lead Time and Calculated Downtime Cost

Spare	Lead time (in operating hours)	Downtime cost/hour	Cost of downtime due to lack of spare
Bearing	48	$225	$10,800/event
Transmitter	24	$225	$5,400/event
Thermostat	24	$1250	$30,000/event

3.11.5 Step Five: Adjust Downtime Cost for Failure Probability

Spare	Downtime Cost (DC) due to lack of spare	Failure Probability (FP)**	Downtime Cost adjusted (DC * FP)	Cost of spare and storage
Bearing	$10,800/event	.2	$2,160/year	$1,200
Transmitter	$5,400/event	.1	$540/year	$250
Thermostat	$30,000	.1	$3,000/year	$600

** The Failure Probability (FP) is based on the number of failures anticipated in a ten-year period (e.g. a .1 signifies a 10% chance of at least one failure in a 10-year period). This can be based on past failure data or available MTBF data from other sources. One must consider contractual terms and residual contract years when making spare part decisions. For example, if the contract end date is within a two-year period and the risk of not carrying a spare is tolerable, one should not stock the item.

3.11.6 Step Six: Step Six: Evaluate Cost of Spare versus the Downtime Costs

Spare	Cost of spare and storage	Downtime cost adjusted	Spare required (yes or no)
Bearing	$1,200	$2,100	Yes
Transmitter	$250	$540	Yes
Thermostat	$600	$1250	Yes

As with any type of analysis it is important that the review group uses a bit of common sense when making spare part decisions. One must factor into the decision-making process contractual obligations, contract term, and the practicality of storing the item. Oftentimes the decision may come down to a very small risk (<$1,500 can be mitigated by spending about the same amount of money). One needs to use sound judgment before money is spent on spares.

In total, we analyzed 1784 failure modes for critical spare purposes. In total we decided to stock 425 spares parts at a cost of $525,000. These spares mitigated over $33 million dollars worth of potential risk.

4. CONCLUSIONS

Veolia Water has introduced a number of innovations to the standard framework for delivering RCM that have improved the business case for delivery. These innovations include techniques for prioritizing and simplifying the analysis, as well as more reliable means for estimating maintenance frequencies and estimating the true cost-benefit of critical spares. The use of RCM in the water/wastewater industry has been adapted from the original methodology used by United Airlines. The changes to the process allow for speedier delivery and execution of RCM. Implementation of RCM on 100% of the assets is not practical, nor cost-effective in our industry.

Veolia Water has been able to demonstrate a link between doing RCM well and delivering excellence in asset management. Overall, the benefit to the business and our customers that result from implementing the Veolia Water RCM process is as follows:

- Better identification of risks. Many new, and often significant, operational risks are identified with recommendations to mitigate them.

- Improved maintenance plan. New maintenance plans are developed to specifically manage the risks of asset failure. Frequencies and resource levels are optimized.

- Gains in operational efficiency. Opportunities for improvement are proposed for operating scenarios that offer risk and efficiency benefits.

- List of critical spares. Spares needs are determined, based on cost-benefit assessments, including financial estimates of risk.

- Prioritized capital improvements. Investigations are recommended for a number of potential capital works items designed to enhance contract performance and minimize risk.

- Effective knowledge capture. The RCM captures both documented information as well as the tacit knowledge held by staff about how the systems work.

- Tool for operations training. Participating staff are trained in the methodology and learn in detail how their systems work.

The project staff has been busy implementing the recommendations that came about from the RCM review. The identification of the wrong PRV's on the high temperature oil system was a very important safety related finding, but the benefits of the program continue to add up. The project has already seen drastic improvements in equipment availability and uptime. Each maintenance task, spare part, and equipment check has been developed to manage risk and improve uptime. The staff at the plant continues to learn from equipment failure and has validated that the time spent on the analysis and spares has tremendously improved the overall operation. The following, Figure 3, is a trend of uptime of one train since the RCM reviews were completed.

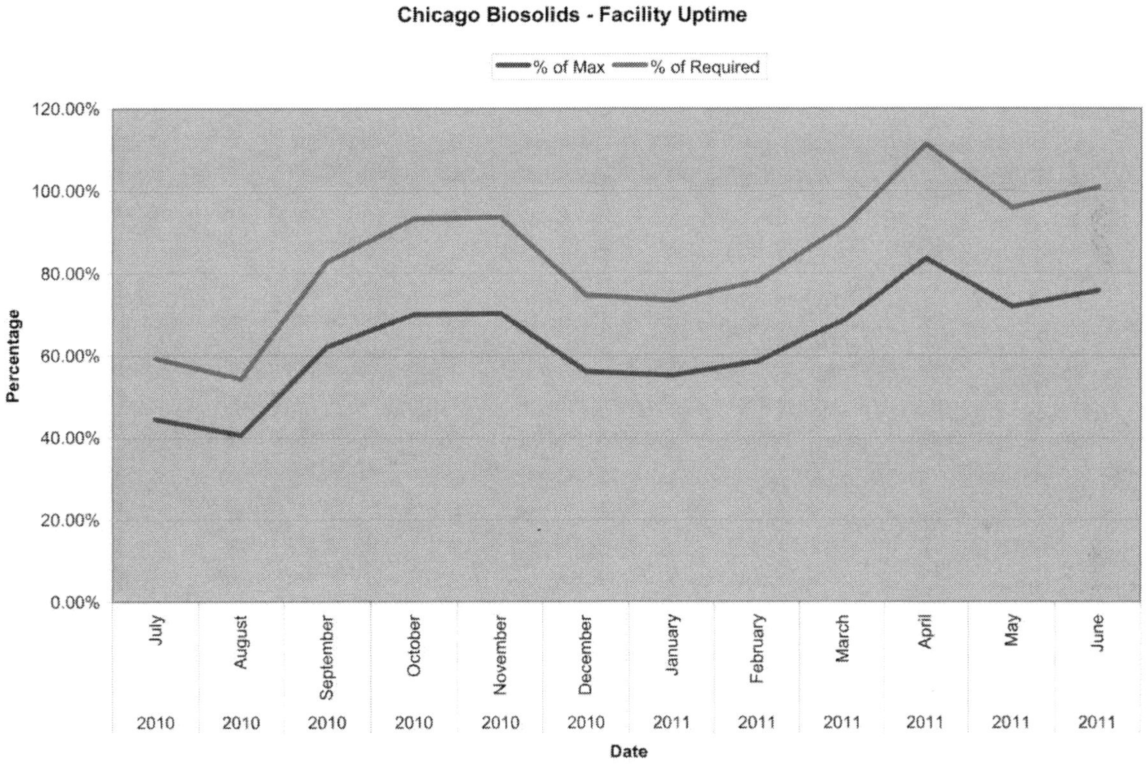

Figure 3, Uptime Performance Through June 2011

5. REFERENCES

[1] Moubray, John, *Reliability-centered Maintenance*, 2nd Edition, Industrial Press Inc., New York, New York, 1997.

[2] Nowlan, F. S.; Heap, H. F., *Reliability-centered Maintenance*, United Airlines and Dolby Access Press, sponsored and published by the office of Assistant Secretary of Defense, San Francisco, California, 1978.

[3] Smith, A.M.; Hinchcliffe, G.R., *RCM Gateway to World Class Maintenance*, Elsevier Butterworth Heinemann., Burlington, Massachusetts, 2004.

(44x) Characterization of the Emissive Power of Hybrid Methane-Air-Dust Flames

Brian Elias and Ali S. Rangwala, Fire Protection Engineering, Worcester Polytechnic Institute, Worcester, MA

Combustible dusts are commonplace in a wide variety of industrial settings ranging from coal mines to manufacturing facilities. The ignition of a dust cloud formed by the mixing of these combustible particles with the surrounding air may result in a deflagration or flash fire. The purpose of this study is to quantify the emissive power of the flame in order to better understand the combustion dynamics and the risk of burn injury associated with a dust flash fire.

Experimental tests are conducted by feeding combustible dust samples into a premixed dust-air burner apparatus to create a stabilized flame for analysis. The two dust types investigated in this study are Pittsburgh seam coal, and iron dust – both with sizes ranging from 70 to 90 um. In addition to the composition of the dust, other controlling parameters including dust concentration and turbulent intensity are also varied and their respective effect on the heat flux is reported. The total emissive power of the flame is calculated from the heat flux measurements in combination with flame surface area and an appropriate view factor.

Are You in Control of Process Safety?

Basis of Safety Assurance Can Provide the Answer

Graeme Ellis, Principal Lead Consultant
ABB Consulting
Daresbury Park, Warrington, WA4 4BT, UK
graeme.ellis@gb.abb.com

(Acknowledgement: Based on paper written by Phil Eames, ABB Consulting)

Prepared for Presentation at
American Institute of Chemical Engineers
2013 Spring Meeting
9th Global Congress on Process Safety
San Antonio, Texas
April 28 – May 1, 2013

UNPUBLISHED

Are You in Control of Process Safety?

Basis of Safety Assurance Can Provide the Answer

Graeme Ellis, Principal Lead Consultant
ABB Consulting
Daresbury Park, Warrington, WA4 4BT, UK
graeme.ellis@gb.abb.com

(Acknowledgement: Based on paper written by Phil Eames, ABB Consulting)

Keywords: Basis of Safety, Assurance, layers of protection, drill-down, vital signs, audit, safety critical elements

Abstract

Serious process safety accidents continue at an unacceptable frequency in the process industry, and senior managers are right to be concerned about whether all reasonable steps are being taken to control risks. Traditional methods for revalidation of Process Hazards Analysis and auditing of Process Safety Management systems can involve considerable time and resources. This paper will describe a 'Basis of Safety Assurance' approach that has been developed to provide a rapid view on the 'vital signs' related to the state of health of measures to prevent and mitigate major accident hazards. A representative set of scenarios for the site are selected and evidence gathered from documentation and site visits to confirm the robustness of the key layers of protection. Deficiencies related to plant, processes or people are identified resulting in a risk targeted set of improvement recommendations.

1. Introduction

Effective process safety management requires a thorough understanding of process risks so that measures to prevent and control them can be implemented and maintained. Structured techniques such as hazard identification (HAZID) and hazard and operability study (HAZOP) are used to identify major accident scenarios and decide how they will be prevented, controlled and mitigated. The description of major accident scenarios and the agreed layers of protection (or barriers) to reduce the risks to an acceptable level, is often termed the basis of safety.

The purpose of a process safety management system (PSMS) is to provide a framework of high level procedures (or risk control systems), in order to maintain protective equipment and procedures in a healthy state. Therefore many organizations, when they seek to assure themselves that process safety is being managed appropriately, base this assurance on detailed auditing of the PSMS.

However, process safety incidents are usually caused by failures of multiple risk control systems at a detailed level. The potential interactions between risk control systems are often not visible at the level of the PSMS, which is constructed of distinct generic elements (mechanical integrity, incident investigation, management of change etc.). Auditing of distinct elements may not identify the potential for an incident due to such interactions. Moreover, modern PSMS are tending to become more integrated with occupational safety management systems, environmental management systems and quality systems, making their role in the prevention of process safety incidents potentially less visible at system level.

Given that the prime concern in process safety management is to prevent process incidents from occurring, it makes sense to devote some assurance effort directly to the scenarios themselves and to their specific layers of protection. This is the essence of Basis of Safety Assurance. Whereas auditing addresses the component elements of the management system, Basis of Safety Assurance seeks to challenge the effectiveness of the process hazards analysis and robustness of the layers of protection for specific scenarios, thereby providing direct assurance that the risk reduction measures are in place and working.

This is not to suggest that auditing of the PSMS is not a necessary or effective tool. But focusing on an organization's key process incident scenarios can be a very direct and cost-effective way to provide assurance that process safety is being managed appropriately, complementing traditional audit-based assurance processes.

This paper describes the Basis of Safety Assurance technique and the experience of its use across a number of sectors in the process industry. It explains how the technique can be used to enhance an organization's process safety management audit and review program.

2. Limitations of Audit Based Assurance

Traditional process safety assurance or governance approaches are based on audits of the process safety management system (PSMS). However, in many organizations the components of the PSMS form part of a larger safety, health, environmental and possible quality management system, and may not be distinguished from components designed to control personal safety and health risks. This means that focus on process safety may diluted by or obscured by the larger management system.

Even if the PSMS is a stand-alone system, or is visible within a larger system, auditing focuses on the system as a whole and not on the control of specific risks. Auditing is often undertaken on an element-by-element basis, which is likely to miss the potential interactions between elements that characterize complex process safety events, typically represented by the "Swiss cheese" model. Moreover, even within a single element significant audit time is normally spent on understanding how the element is managed rather than on detailed field verification. Field verification on audits typically aims to confirm that the system is in compliance with requirements, for example that trip testing is being carried out to plan. It generally does not test that the system is effective in protecting against the risk of a specific scenario, for example the trip is identified as a layer of protection in HAZOP, there is a written test method, the trip test method is an "end to end", faults identified in previous tests have been corrected, etc.

3. Layers of protection and process safety "vital signs"

In his book on the management of complex process risks, James Reason [1] wrote of the need for proactive checking of the process or plant, or the 'system' as he describes it:

"*In the absence of bad outcomes, the best way - perhaps the only way - to sustain a state of intelligent and respectful wariness is to gather the right kinds of data. This means creating a safety information system that collects, analyses and disseminates information from incidents and near misses as well as from regular proactive checks on the system's vital signs. All of these activities can be said to make up an informed culture - one in which those who manage and operate the system have current knowledge about the human, technical, organisational and environmental factors that determine the safety of the system as a whole. In most important respects, an informed culture is a safety culture*".

The state of health of safety critical layers of protection is very much vital signs for the process in relation to major accident scenarios. If they are specified and designed correctly, maintained and tested, then the process is safe. The proactive checking that layers of protection are in place and healthy provides exactly the current knowledge of the state of the plant and the organization that Reason holds to be so important.

4. The 'Basis of Safety' Event

The testing of the state of health of layers of protection for specific major accident scenarios is the focus for a Basis of Safety Assurance event, a generic agenda for which is shown in Table 1.

Table 1. Basis of Safety Assurance Agenda

Phase	Day	Activity
Preparation	**0**	**Arrange for immediate access to key site personnel and information during time on site**
Introduction	**1**	**Introduction to Process Safety on site.** **Review of major accident hazards** **Site tour** **Examination of Process Hazard Analyses** **Selection of 'Drill Down' scenarios**
Drill Downs	**2-N**	**Detailed drill downs** **Approx 4 scenarios per day including field visits**
Feedback	**Final**	**Preparation and delivery of feedback**

The initial part of the event aims to understand the major accident hazards associated with the process by discussion, field visits and review of the process hazards analysis (HAZID and HAZOP documentation). This would typically take one day, during which time an assessment of the quality of the site's process hazards analysis and documentation of the basis of safety is also made. It culminates in the selection of specific major accident scenarios to be subjected to detailed assessment, sometimes termed a "drill-down" or "deep dive".

It is useful to select a range of different types of scenario where possible (fire, explosion, toxic release, environmental, etc.), so that as many different types of safeguard are subject to assessment. These then include for example; pressure relief systems, safety instrumented systems, passive and active fire protection systems, secondary and tertiary containment systems, etc.

The main part of the Basis of Safety event comprises the scenario assessments. These typically take around 2 hours per scenario, depending on the quality of the basis of safety documentation, the knowledge of the team and the speed of data retrieval, for example how fast plant records can be accessed to verify that maintenance and testing is taking place.

Findings from the scenario assessments provide the required assurance that layers of protection are healthy or lead to the generation of recommendations for improvement. Common findings across a number of scenario assessments can provide a useful indication of systemic issues, for example weaknesses in proof test procedures or compliance with critical equipment testing regimes. This part of the event complements conventional PSMS auditing.

5. The Scenario Assessment Technique

Most value can be gained from the scenario assessment technique if it is conducted as a team exercise. Experience has shown that the best results can be obtained by deploying two process safety specialists as assessors; one with a PSMS and operations background and the other with a plant engineering and asset integrity management background. This approach also helps from a leading and scribing standpoint since the event is recorded in a similar way to a process hazards analysis. In addition to the assessors, the involvement of process engineers, operating managers and maintenance engineers from the site is very important. This provides an understanding of the scenario and the location of information to verify that the identified layers of protection are in place and being maintained.

Since scenarios are normally selected from the site's process hazards analysis or basis of safety documentation, a scenario assessment normally starts with this documentation. The scenario is described and the relevant records are referenced, shown in the upper part of the recording form in Table 2.

It is important to note that the existing documentation may not fully describe the scenario, since the quality of HAZID and HAZOP documentation is heavily influenced by team composition, knowledge available, leader competence, availability of time and recording policy.

Table 2. Scenario Assessment Template (example)

Major Accident Hazard Drill Down Record
Scenario: Storage Tank V7706 (150m3) overfilled during (continuous) run-down and a large volume of solvent released into bund from the breather/emergency vent, subsequent ignition resulting in a flash fire/pool fire and potential for several fatalities.

PHA Reference: No specific scenario was identified for this material as incidents involving acetone were taken as the representative set. The acetone scenario for overfill was estimated to result in two fatalities.			
Safeguard	**Verification**	**Findings**	**Conclusions**
Monitoring of level transmitter by operators and dependent alarm function.	1. Calibration and maintenance of level transmitter and alarm LIA-7706. 2. SOP requiring level monitoring.	1. No regular maintenance is now performed on the level transmitter following the installation of the independent high level switches. Corrective maintenance is performed based on notifications raised by the operators. 2. There is no specific SOP relating to the operation of this solvent storage tank.	Consideration implementing routine level transmitter calibration on solvent storage tanks.
Independent high level switch with trip of run-down line.	1. Proof testing of high level trip LZA-7706H.	1. The trip is proof tested annually 2. The last proof test of this instrument was carried out on the 27/4/10. The proof test is an end- to-end test to a pre-defined test method including checklist. 3. No SIL assessment has been conducted.	Controls appear to be effective. However, conduct a SIL assessment for this scenario to verify that the proof test interval is adequate to provide the required level of protection.

It is therefore important that the assessors work with the team to identify all the relevant safeguards, which are recorded in the left hand column of the template shown in Table 2. Note that the term "safeguard" is used rather than "layer of protection" because the aim is to capture safeguards such as procedures and emergency plans as well as physical independent protection systems that are the focus of formalized Layers of Protection Analysis (LOPA) techniques.

Once safeguards have been listed the assessors ask - for each safeguard in turn - what can we check is in place? This is recorded in the "verification" column. Examples include procedure number and latest version, alarm test methods, maintenance routines and pressure safety valve test records.

Evidence for each verification is then sought and the findings recorded in the "findings" column of the template. It is important that documentary evidence is collected to ensure that the assessment is suitably probing and detailed.

Conclusions are then drawn for each safeguard. This may be a simple "Okay" if everything expected is in place, but may be a recommendation for improvement, a corrective action or a recommendation to consider whether a finding may be evidence of a systemic weakness. In viewing the robustness of the safeguards as a whole the team may determine that additional risk reduction is appropriate and make a recommendation to identify additional risk reduction measures. This would be required in a UK legislative context to demonstrate that risks have been reduced to "as low as reasonably practicable" or ALARP.

It is important to include a field visit for each scenario. This has two purposes, firstly to physically check that safeguards are installed in line with the design intent, for example that relief devices are in place and that passive fire protection and secondary containment are in good condition. Photographs of deficiencies provide high-impact evidence to site management. Secondly to talk to operators and maintenance technicians about their understanding of the process and of the major accident hazard that is being assessed. This gives an insight into the quality and experience of staff and the quality of process-specific training that they have received.

6. Experience with the Technique

There is now a significant body of experience of Basis of Safety Assurance across a number of process industry sectors. Companies have found it effective regardless of sector or scale of process hazard. Major hazards facilities are more familiar with the concepts but it is readily applicable to lower level process safety hazards, including those in the pharmaceutical and power generation sectors.

Some important advantages of using the scenario assessment "drill-down" technique are as follows:

- Drilling down into specific scenarios can identify significant risk exposures and weaknesses at a very detailed level, for example:
    - Trips tested only at commissioning;
    - Absence of detailed test methods and reliance on technicians' experience;
    - Trip testing not from sensor to final element (end-to-end);
    - Reliance on inexperienced 3rd parties for proof testing;
    - Demands on trips more frequent than the design basis.
- Examining multiple scenarios can provide evidence of whether there are systemic weaknesses.
- The technique is useful to identify weaknesses in process hazards analysis and in keeping records and basis of safety documentation up-to-date.
- The technique is a useful vehicle for providing training materials for specific process hazards.

7. Conclusion

Because it focuses on specific risks, Basis of Safety Assurance complements conventional PSM system audit activities. It helps to maintain a focus on major accident hazards and improve the understanding of site personnel and standard of documentation. It aligns with line management responsibilities for specific processes and assets, and it directly exposes risks of major accident hazards, thus providing real time assurance to senior management.

8. References

[1] Reason, J.T (1997), Managing the Risks of Organizational Accidents, Ashgate Publishing, Aldershot, UK.

9TH GLOBAL CONGRESS ON PROCESS SAFETY

Considerations for Effectively Managing Emergency Relief Systems Design

Thomas D. Watson
KBC Advanced Technologies, Inc.
15021 Katy Freeway, Suite 600
Houston, TX 77094

TWatson@kbcat.com
thosdwatson@yahoo.com

Prepared for Presentation at
American Institute of Chemical Engineers
2013 Spring Meeting
9th Global Congress on Process Safety
San Antonio, Texas
April 28 – May 1, 2013

UNPUBLISHED

Considerations for Effectively Managing Emergency Relief Systems Design

Thomas D. Watson
KBC Advanced Technologies, Inc.
15021 Katy Freeway, Suite 600
Houston, TX 77094

TWatson@kbcat.com

Keywords: Process Safety Management, Relief System, Relief Valve, Process Safety

Abstract

Emergency relief systems (ERSs) that mitigate all credible scenarios are a fundamental objective of ERS design, and ensure a safe and profitable facility. ERSs are a key layer of protection against loss of containment at a chemical process facility. Consequently, organizations are placing renewed emphasis on their ERSs design and documentation to make sure their facility and community are safe, and their organization remains profitable. The integrity of the ERS as a layer of protection, however, must continue after "completing" a project to update and document the facility wide ERS design basis. In order for the ERS to continue to serve as a key layer of protection, an operating facility must manage the facility wide ERS design basis documentation, as well as integrate the ERS design basis into the overall process safety management (PSM) system. An organization's management must realize the full "lifecycle" cost of integrating and maintaining the facility wide ERS design basis (a PSM requirement) to address the challenge adequately.

This paper discusses key considerations for an organization to ensure their ERS not only protects their facility, but also serves as a critical input into their process hazard analysis (PHA) process and other provisions of PSM (management of change, and pre-startup safety reviews, for example). Considerations include implementing and managing a facility wide ERS design basis documentation project (a "flare study"), maintaining the ERS design documentation after the project, and effectively integrating the ERS design documentation into other elements of PSM. Abstract here. Say what the paper says, but say it very briefly.

1.0 Introduction

Emergency relief systems (ERSs) that mitigate all credible scenarios are a fundamental objective of ERS design, ensuring a safe and profitable facility. ERSs are a key layer of protection against loss of containment at a chemical process facility. The integrity of the ERS as a layer of protection, however, must continue after "completing" a project to update and document the facility wide ERS design basis. In order for the ERS to continue to serve as a key layer of protection, an operating facility must maintain, update, and integrate the ERS design basis into the overall process safety management (PSM) system. This continued effort is a challenge for any organization because the effort really begins after the facility wide ERS update project is complete. An organization's management must realize the full "lifecycle" cost of integrating and maintaining the facility wide ERS design basis (a PSM requirement) to address the challenge adequately. Two key management objectives are:

- Managing and implementing a facility wide ERS design documentation project (a "flare study")
- Effectively integrating the ERS design documentation into other elements of PSM in order to support PSM and maintain the ERS design documentation.

This paper discusses the above two objectives in order to help an organization's management ensure their facility wide ERS system not only protects their facility, but also serves as a critical input into other provisions of PSM (process hazard analysis, management of change, and pre-startup safety reviews, for example).

2.0 Manage and Implement a Facility Wide ERS Design Basis Documentation Project

Management planning a facility wide ERS design project (a "flare study"), must ensure the project effectively utilizes people, tools, and work processes to achieve a technically correct ERS design that meets an organization's acceptable risk criteria in the most cost effective manner. Consequently, management must:

- Incorporate the facility's process technology, equipment, and operations information into the ERS design
- Utilize recognized and generally accepted good engineering practices ("RAGAGEP") with regard to ERS design and facility operation
- Properly consider personnel resourcing requirements
- Properly schedule the ERS design activity considering other facility work activities, processes, and timelines

The following subsections discuss management considerations for the above aspects of a successful ERS design project.

2.1 Considerations for Incorporating Facility Process Technology, Equipment, and Operations Information

OSHA 29CFR1910.119 "Process Safety Management of Highly Hazards Chemicals" defines process safety information (PSI) as information regarding a process' hazards, technology, and equipment.[1] Among other items, PSI includes:

- Process chemistry

---

[1] Occupational Health and Safety (OSHA) "Process Safety management of Highly Hazards Chemicals" 29CFR1910.119(d).

- Material and energy balances
- Safe operating limits
- Piping and instrumentation diagrams (P&IDs)
- Relief systems design and design basis
- Design codes and standards employed
- Safety systems

An ERS design basis documentation project will utilize all of the above PSI (see OSHA PSM for a fuller list of PSI). Facility management must consider the current state of the PSI at the facility and address or mitigate any concerns that may lessen the ERS project's effectiveness. The following subsections discuss considerations regarding some of the above noted PSI items in more detail.

2.1.1 Process Chemistry

Considerations for how the process chemistry will affect the ERS design basis documentation are many. For example, understanding the process chemistry provides insight into the process conditions and hazards associated with the process. Understanding the process conditions and hazards, along with materials of construction data, in turn helps establish safe operating limits. Additionally, a clear understanding of the process chemistry requires fully understanding the reactive hazards associated with the process chemistry. This in turn requires a reactive model that accurately represents all relevant physical and chemical phenomena associated with the process. Management should assess and take steps to address any deficiencies in their understanding of the process chemistry before embarking on an ERS design basis documentation project. For additional considerations regarding reactive systems modeling and its relationship to ERS design, The Design Institute for Emergency Relief Systems (DIERS) Users Group is a valuable resource.

2.1.2 Material and Energy Balances and Safe Operating Limits

Process units that have multiple operating conditions (for example, start of run, end of run, and/or multiple feed conditions) will require due consideration of the process' operating conditions in order to assess the potential impact on the basis for the material and energy balance. The material and energy balance is a key input to the ERS design. Furthermore, safe operating limits of the process unit frame the process operating conditions. A failure to understand and establish the various operating conditions and limits can result in material and energy balances that inaccurately reflect the facility operating modes and limits, as well as incorrect expectations regarding the consequences of deviation from the operating limits. Consequently, the ERS design basis development does not correctly reflect the process unit either. Facilities managers must ensure the project considers not only how a process unit may run, but also what constraints the overall facility places on the individual process unit. Then the project must incorporate these considerations into the overall project basis. One such example is setting a facility wide basis for the flare system. Analyzing individual units at maximum processing conditions is necessary, but for an entire facility, it may not be necessary to analyze the flare system for the simultaneous maximum processing conditions of all process units.

2.1.3 Piping and Instrumentation Diagrams (P&IDs)

If P&IDs are not current in a process (i.e., a P&ID update is underway) then utilizing outdated PSI will result in project rework, incorrect ERS design basis documentation, and an unsafe facility. Management should consider the P&ID maintenance schedule for process units, as well as units that may undergo significant changes when developing a facilities wide ERS design basis update project.

2.1.4 Other PSI Considerations

The above examples are just a few of the considerations necessary. Facility managers must review and understand their current PSI status and then develop a plan to incorporate it into the project. While not every example can be discussed in this paper, other considerations include, how will the current ERS design basis will be utilized, and what design codes and standards will be employed, especially if the facility is older and various codes and standards have been previously employed.

2.2 Utilize Recognized and Generally Accepted Good Engineering Practices ("RAGAGEP")

Again, the overall objective of a facility wide ERS design basis documentation project is to help protect the facility and serve as a critical input into the PSM system. Another aspect of the facility management's objective to manage and implement the project is to ensure the project utilizes RAGAGEP. While there are many aspects of this, for the benefit of discussion, this paper explores three key considerations:

- Industry and facility standards
- Methodologies
- Tools and software

The following subsections discuss aspects of each.

2.2.1 Industry and Facility Standards

A facility wide ERS design basis project will utilize facility and industry standards to comply with RAGAGEP. A few points to consider follow.

First, facility management must recognize and consider how the current facilities reflect past and current facility standards. Management must be able to assess their standards relative to current industry standards recognized as RAGAGEP. Facilities, especially smaller ones, that don't keep an accurate accounting of facility standards employed wind up having to default to current industry standards considered RAGAGEP regarding ERS design.

Second, RAGAGEP in some instances provides latitude in ERS design subject to facility practice and facility judgment. Facility management must recognize this and be prepared to address the latitude through facility standards or development of project specific guidelines that may evolve into facility standards (see "Methodology Considerations").

Third, the ERS design will require general design guidance regarding the facility (for example, utilities and infrastructure) that a facility standard should provide. If this type of guidance is not available, the facility should consider developing such a standard.

Some typical industry standards considered RAGAGEP include, but are not limited to those listed below. Facility management should ensure their staff has access to the latest versions of industry standards.

- OSHA 29CFR1910.119, Process Safety Management of Highly Hazardous Chemicals
- OSHA Title 29, Chapter XVII, Part 1910, Subpart H – Hazardous Materials
- ASME Boiler and Pressure Vessel Code Section VIII – Pressure Vessels
- ASME Boiler and Pressure Vessel Code Section I – Power Boilers
- ASME B31.1 and B31.3 – Power Piping and Process Piping, respectively
- API STD 520 – Sizing, Selection, and Installation of Pressure Relief Devices in Refineries (Part I – Sizing and Selection)
- API-RP 520 – Sizing, Selection, and Installation of Pressure Relief Devices in Refineries (Part II – Installation)
- API-STD 521/ISO 23251:2006 – Guide for Pressure-relieving and Depressuring Systems
- API-STD 526 – Flanged Steel Pressure-relief Valves
- API-STD 2000 – Venting Atmospheric and Low-pressure Storage Tanks
- NFPA-30 – Flammable and Combustible Liquids Code
- CCPS, Guidelines for Pressure Relief and Effluent Handling, AIChE/CCPS, New York, 1998.
- Fisher, et.al., Emergency Relief System Design Using DIERS Technology: The DIERS Project Manual, AIChE, New York, 1992.

2.2.2 Methodology Considerations

A successful facilities wide ERS design basis project requires all organizations participating in the project (the facility and any contractors involved) follow a rigorous methodology. Below is a summary of key methodology considerations:

- Review and align on the relief study guidelines to be used for the study
- Clarify and define the basis
- Develop energy and material balances
- Collect and verify data
- Develop credible relief scenarios
- Calculate relief loads and relief stream properties
- Identify controlling scenarios for the flare system analysis
- Model the flare system
- Review the adequacy of the flare system
- Document the analysis and findings

- Address deficiencies

The following paragraphs discuss some key management considerations of each topic.

Review and Align on Relief Study Guidelines

A successful project requires establishing and maintaining client and project specific guidelines. These guidelines are in addition to RAGAGEP such as API 520/521. The guidelines document will cover various aspects, for example:

- Alignment between industry standards and client specific standards
- Clarification of industry standards where needed
- Methodology and guidelines for evaluating relief loads
- Scenario and equipment specific considerations

Clarify and Define the Basis

The project must ensure full discussion and consideration of the facilities' operating basis prior to proceeding with the analysis. For example:

- Are all relevant modes of unit operation (feed or product line-ups, feed conditions, alternate operations) considered?
- Are pending facilities changes via projects or procedures identified?
- Are energy and material balances consistent with the identified operating modes?
- Are special and/or emergency operating procedures identified?
- Are relationships between units and the overall facility's capabilities identified and their impact on the analysis understood?

Develop Energy and Material Balances

Energy and material balances for the process units must be available to the project, typically using process simulation software. Sometimes, energy and material balances are already available. Their basis, however, should be consistent with the intent of the study. Furthermore, the balances and simulation models must represent the unit sufficiently to allow for development of all potential relieving streams. Ensure the project considers alternate operations in the model development and develops energy and material balances accordingly.

Data Collection and Verification

ERS design basis analysis requires extensive process safety information (PSI). Summarized below are general data requirements. Facilities management should consider how the project would process data requirements. Field verification may also be required and clear alignment on resourcing and scheduling to complete collection and verification is necessary. Ultimately, the facility must accept responsibility for the validity of all data. Consequently, the facility management must ensure the process by which data is collected, stored, and maintained is acceptable. Data includes:

- Energy and material balance(s)/model(s)
- Process flow diagrams and piping and instrumentation diagrams (P&IDs)
- Drawing standards for symbols, equipment and instrument designations, line designations and ratings
- Isometric drawings for the flare system piping from the unit headers to the flare tip
- Isometric drawings for the inlet and outlet piping of relief valves
- Utility system operating and design data
- Plot plans and drainage information for units and flare/vent stacks
- Fire water system information, if available
- Electrical one-line diagrams
- Equipment specifications sheets for vessels, heat exchangers, relief devices, pumps, control valves, and any other necessary equipment that may be involved in the identified scenarios
- Vessel drawings showing orientation and elevation
- Equipment and piping insulation information
- Operating curves for equipment (pumps and compressors)
- Minimum, normal, and maximum operating process conditions, including but not limited to, pressures, temperatures, flows, and levels
- Emergency shutdown system and safety instrumented system information, including, but not limited to cause and effect diagrams
- Procedures, as needed, for relevant equipment and processes

Develop Credible Relief Scenarios

Scenarios considered should include those published in ANSI/API Standards 520 and 521, as well as scenarios developed through the ERS project team's collective experience in consultation with the facility. Utility failure scenarios developed should include both global and partial credible utility failures. Facility management must fully consider their organization's level of involvement in developing and reviewing scenarios and calculations. Leveraging these activities with the overall project team is one of the most important considerations for the project.

Calculation Development, Review, and Approval

Upon endorsement of the scenarios, the project will complete ERS evaluations and calculations. In some instances, scenario development and calculations may occur simultaneously, but the facility should duly consider the scenario basis before calculations occur. Base calculations on the *endorsed* scenarios and the PSI provided. Documentation should consider and follow RAGAGEP, such as ANSI/API Standard 521. Calculations should at least include:

- Relief device, including conservation vents, adequacy (required relief/vent rate and required relief device orifice area), actual relief device capacity (both API and vendor actual) for all credible scenarios

- Inlet line pressure drops (between protected equipment and the relief device) and outlet pressure drops (between the relief device and the main unit header) based on actual (vendor specific) expected relief device capacities
- Hydraulic and other calculations as needed for scenario development (control valve and piping flow cases, for example)
- Unbalanced heat load calculations, or alternate calculation approaches
- Any relevant flashes or other process simulations, as needed
- Atmospheric vents, noise levels and dispersion modeling as necessary.

Again, facilities must fully consider their organization's level of involvement in developing and reviewing scenarios and calculations. Leveraging these activities with the overall project team is one of the most important considerations for the project.

Identify Controlling Scenarios for the Flare System

Upon endorsement of the calculation results, the project team should identify and agree upon the controlling scenario(s). The intent is to define the limiting scenarios necessary to evaluate the flare system. Not all scenarios need evaluation in the flare model, as some will have negligible impact on the flare system. Management must understand the controlling scenario identification process as clearly as possible as potentially significant facilities changes may be involved.

Model the Flare System and Review its Adequacy

Most facilities use commercially available software for relief-system network-flow simulation, radiation, and dispersion calculations (as well as individual relief device design). The project should build the models / tools in such a way as to facilitate needed changes and additions in the future. Build the flare model from the interested relief valves involved in the controlling scenario(s) to the flare knock out drum(s) and the flare tip. With the model, evaluations include the following:

- Backpressures in the flare system and their impact on equipment required to operate in the event of a defined scenario
- Excessive velocities in the flare system
- Knock-out drum separation efficiencies and liquid handling requirements
- Flare tip pressure drop
- Maximum flare tip radiation levels at grade or at elevated structures in the area

Document the Analysis and Findings

Formal documentation of the ERS design basis and flare evaluation should include:

- A summary table highlighting aspects of the relief valve or flare system that do not meet RAGAGEP
- Unit pressure relief valve (PRV) summary table (listing all PRVs evaluated)
- Scenario descriptions for each identified scenario and relevant individual PRVs

- Flare system documentation that summarizes the evaluation of the agreed upon controlling scenarios
- Recommended upgrades to the flare header/burner/KO drum
- Individual PRV/conservation vent documentation that includes the following for all credible scenarios:
    - PRV summary sheet and concerns
    - PRV data sheet
    - PRV calculation details
    - Unbalanced heat load calculations
    - Relief stream properties
    - Inlet line pressure drops
    - Outlet line pressure drops
    - Nomenclature
    - Recommended upgrades for inadequately sized valves
- Unit/refinery flare routing plot plan
- Process flow diagrams (PFD) reflecting summary scenario information, similar to those developed for the global utility failure work
- Electrical one-line summary diagrams for the unit (and utility equipment impacting the unit)

Various industry guidance documents elaborate on documentation considerations. Facility management must consider the broader implication of documentation for the entire PSM systems. Subject matter expertise is critical in these considerations as rework can be extensive, otherwise.

Address Deficiencies

After identifying deficiencies, the project team should collectively work to address them. In some instances (in particular deficiencies of a more significant nature such as flare system deficiencies), addressing the issues may have to wait until the refinery wide analysis is complete to ensure the controlling scenarios are addressed. Several considerations and activities may affect a final solution, depending on the deficiency and scope of the project. Activities may range from a further review of the controlling scenario to identifying conservatism that may not be warranted, dynamic simulation of equipment or the overall flare loads to the system, risk analysis, and evaluation of safety instrumented systems to name a few. Facilities management and the project team must collectively work to address the deficiencies. This invariably requires the facilities management to have a general understanding of the methodology employed and ensuring the methodology is completed. Furthermore, ensuring the original ERS project resources address deficiencies is the most efficient next step, if not critical to the overall success of mitigating deficiencies (see "Personnel Resourcing Considerations"). Finally, ERS design basis documentation is a key component of effective Process Safety Management. When addressing the deficiencies, facilities management must ensure processes are in place to maintain the ERS design basis documentation.

2.2.3 Tools and Software Considerations

Facilities management needs to consider and define how the project will manage all of the information, analysis, and findings involved. As just discussed, these considerations arise not only during the project, but also after the project is "complete" when changes require updates to the ERS design basis. There are extensive calculations and analysis required to complete a site wide ERS design basis documentation project. The project must manage data in an orderly fashion. Analysis must be readily available and accessible for updating on a regular basis.

Consequently, most facilities use commercially available software for process simulation, individual relief device analysis, relief-system network-flow simulation, radiation, and dispersion calculations (as well as individual relief device design). Even relatively small facilities find substantial benefit using software applications rather than spreadsheet or hand calculation resources due to the repetitive nature of the analysis and the requirement to maintain the ERS design basis information (keep it "evergreen"). Additionally, the processes that analyze and track concerns and their resolutions lend themselves to software applications similar to what process hazard analyses utilize.

Individual software capabilities do not encompass all of the analysis and documentation requirements necessary to complete a facility wide ERS design basis documentation project. Consequently, facility management must understand the limitations and interactions of the different software they will utilize. Management's considerations regarding personnel resourcing are extremely important and are integral when considering the tools and software applications. Tools and software require training, familiarity, and expertise. All three factor into resource requirements. Facilities have made decisions regarding tools, software, and analysis that have later added unanticipated constraints to resourcing requirements.

2.3 Personnel Resourcing Considerations

Because facility wide ERS projects incorporate so much PSI and require such an extensive use of RAGAGEP, management must identify and ensure the appropriate personnel support the project. Fundamental to a successful project is the active participation of the facility's organization. Specifically, management must consider what personnel should be involved in the ERS design process and when to bring those resources into the process. Some of the key responsibilities will include:

- Collecting, processing and maintaining PSI in the ERS design process
- Developing ERS design basis documentation
- Assuring quality ERS design basis documentation
- Follow up for facilities changes required to address ERS design deficiencies

The following subsections discuss considerations regarding each of the above points.

2.3.1 Collecting, Processing and Maintaining PSI

As noted previously, PSI includes a large amount of information. Management must consider what personnel resources should be involved in the collection, processing, and maintenance of PSI. Should facility personnel or contract staff be responsible for all PSI or just specific components? Who will communicate changes in PSI and maintain the information in the ERS design basis documentation after the project to ensure the ERS design documentation is up to date? Significant effort is required to collect, process, and incorporate the basic equipment information into the ERS project. Examples include equipment datasheets and P&IDs. Another example is a material and energy balance. Personnel must define, possibly develop, and validate material and energy balances. Management must review the PSI collection, processing, and maintenance responsibilities, and then identify and plan for the appropriate personnel to meet those responsibilities. Typical personnel resources include:

- Data collectors and processers
- Draftsmen
- Operations personnel
- Technical personnel

2.3.2 Developing ERS Design Basis Documentation

Management must consider whom, how, and when, the project will utilize site engineering resources and/or contract resources to develop ERS design basis documentation. There are beneficial considerations and challenges to all options. Relieving scenarios developed as part of the ERS design start with ***individual*** equipment, are fairly standard, and are outlined in RAGAGEP such as API Standard 521 / ISO 2325 "Pressure-Relieving and Depressuring Systems". The necessary process technology, equipment, and operations understanding can be rudimentary at this point. A contract organization is often the most effective resource in executing this phase of the project. However, scenarios involving multiple equipment items in an operating unit, or multiple operating units in a facility will present themselves and require extensive and firsthand knowledge of the process, equipment, and operations to avoid incorrect overall scenario development. Incorrect overall scenario development can lead to oversized ***or*** undersized ERS design, and facilities modifications that may not be required. Consequently, management must identify knowledgeable engineering and operations staff to participate to some extent in the ERS project to ensure overall project integrity and quality. When facility management chooses limited site involvement in the scenario development, facility management should ensure site resources have at least some involvement in critical activities. Critical activities include identifying and endorsing the facility process basis (for the entire facility as well as individual units and equipment), and reviewing relevant key relieving scenarios and their subsequent concerns (in particular "global" scenarios involving multiple relief devices and process areas). Smith and Burgess offer additional insight into why site personnel with appropriate site knowledge are important in an ERS design project[2].

Management must consider if the facility has available technical resources to develop the ERS design basis AND are those staff knowledgeable in ERS design RAGAGEP (see section 2.2 for further discussion regarding RAGAGEP). If management includes contract personnel, they must evaluate the contractor's experience, capability, oversight, and toolset. Additionally, at what point should operational staff become involved in the scenario development and/or review

[2] D. Smith and J. Burgess, "Relief Valve and Flare Action Items: What Plant Engineers Should Know." *Hydrocarbon Processing*, November 2012, 42.

process? These are just some of the many points management must consider for the ERS design basis activity.

2.3.3 Assuring a Quality ERS Design Basis Documentation

As a continuation of the above discussion on ERS design, a quality ERS design requires correct application of RAGAGEP with regard to ERS design, appropriate incorporation of process technology, and inclusion of operational knowledge. Management must provide resources to ensure all three, as well as set expectations of time involved. Typically, the three resources involved for the respective responsibilities are:

- RAGAGEP ERS design - a subject matter expert in ERS design
- Incorporation of process technology – senior plant engineering staff
- Inclusion of operational knowledge – senior operations staff

With regard to an ERS subject matter expert, some facilities may not have internal staff knowledgeable enough to serve in this capacity so management may consider using contract staff, from either the contractor performing the ERS design, or an independent contractor. With regard to senior plant engineering and operations staff, their time allocation and points of interaction with the ERS design basis project are usually the key considerations. How the project utilizes RAGAGEP usually defines how all three ERS quality assurance positions will be involved. Finally, facility management must define the ERS design basis review and approval process to ensure a quality result. If management does not address these considerations, the project risks outcomes that the facility neither understands nor agrees with. A significant amount of recycle will result.

2.3.4 Follow up for facilities changes required to address ERS design deficiencies

Once the ERS design basis is developed and reviewed, it identifies deficiencies. The deficiencies may involve individual pieces of equipment (such as vessel design temperatures and pressures), individual relief valves, piping systems, instrumentation, and/or major flare system components. Personnel knowledgeable with the ERS design basis will be a critical component to resolving the identified deficiencies appropriately, as they are the most familiar with the scenario and the deficiencies. Management must anticipate and plan for an appropriate level of continued involvement by site and contract personnel resources to resolve the concerns and maintain the ERS design basis documentation. Otherwise, the project does not address deficiencies in a timely fashion, jeopardizes facility safety, and the ERS design basis quickly becomes outdated. Consequently, personnel resourcing decisions made during the early phases of the project now impact decisions regarding deficiencies resolution and ERS design basis maintenance. Management who fully realize and consider this relationship can make better resourcing decisions at the outset of the project. Management who do not consider this relationship may arrive at situations where unanticipated resourcing requirements (personnel as well as tools and software) surface.

2.4 Considerations for Scheduling the ERS Design Basis Documentation Project

Once facility management has considered how to incorporate PSI into the project, how the project will utilize RAGAGEP, and what personnel resources will be required, management can begin to plan and schedule the facility wide ERS design basis project. The ERS project schedule should give due consideration to other facility activities, processes, and timelines that may influence the project. Forcing an expedited project to meet an ill planned deadline will jeopardize the quality of the facility's ERS design basis and a critical layer of protection that supports a safe facility.

Below is a suggested sequence and outline of points for management to consider when developing a schedule for a facility wide ERS design basis documentation project:

- Having understood the RAGAGEP methodology outlined previously, define the level of internal resourcing the facility can provide. This will help establish an acceptable "rate" of process unit ERS design basis activity the facility can handle.
- Identify process units with the most up to date PSI that can serve as the first process units the ERS design basis project will address. Address other process units' PSI deficiencies before starting their ERS design basis development.
- From the above units, identify those units having enough time before the next turnaround to complete the unit ERS design basis and incorporate required facilities changes into the turnaround planning activity.
- From the above units, identify those units with staff available to support the ERS design basis activity.

Utilize the above evaluation outline to establish a first pass priority and schedule for respective units. Work process units first that have the highest potential for maintaining a schedule and effective outcome. The priority and schedule will then form the basis for subsequent analysis and iterations on the schedule. This evaluation outline also allows facility management to address individual process unit concerns to the extent practical and evaluate the global relieving scenarios when they are fully developed. It also allows the facility to integrate individual process units' ERS design basis documentation into the other PSM elements on a ratable basis and more effectively maintain the ERS design basis documentation going forward.

Other considerations will undoubtedly be required. For example, a process unit that has high risk hazards already identified (say from a PHA) may require a higher priority to address. The evaluation outline, however, will still benefit management in prioritizing and considering what additional action is required to work the identified process unit.

3.0 Effectively Integrate the ERS Design Documentation into other PSM Elements and Maintain the ERS Design Basis Documentation.

A fundamental objective of ERSs is to mitigate all credible scenarios. ERSs serve as a key layer of protection against loss of containment at a chemical process facility. The integrity of the ERS as a layer of protection, however, must continue after "completing" a project to update and document the ERS design basis. In order for the ERS to continue to serve as a key layer of protection, an operating facility must effectively integrate the ERS design basis documentation into the process safety management (PSM) system and maintain the ERS design basis documentation going forward. Facilities are dynamic, ever changing places that continue to evolve. Consequently, the ERS design basis must evolve as well. Facilities change for various reasons, including:

- Maintenance activity on units
- Operating changes
- Process hazard analyses
- Safety initiatives
- Project activities
- Information systems
- Management of change process

Facilities management that effectively integrate the ERS design basis documentation into other PSM elements provide a foundation for effectively maintaining the ERS design basis documentation and improving process safety and profitability. The following subsections discuss just some of the considerations for ERS design basis documentation as it relates to the following PSM elements:

- Management of Change
- Process Safety Information
- Process Hazard Analysis and Incident Investigation
- Operating Procedures
- Employee Participation

There are additional considerations for each of the above, as well as considerations for the other PSM elements not listed. The subsections that follow should promote management thinking as to how the facility can improve safety and profitability with a more complete integration of ERS design basis documentation.

3.1 Management of Change

Facilities should already have a management of change (MOC) process that ensures any change or modification to the facility's design, operation, equipment, etc…, requires a thorough change management review. The management of change process often involves an "MOC Checklist" designed to highlight and identify relationships between facility processes, systems, training, etc. The MOC checklist should also identify the potential relationship of the change with the ERS design basis documentation. Management should review their MOC process at the outset of the ERS design basis documentation process and ensure it properly captures changes to the ERS design basis. This may include an additional checklist. A word of caution, the checklist requires a knowledgeable person reviewing it, and enough of a description to highlight the relevance adequately. Some organizations develop an extensive checklist of issues, but the checklist often does not cover all aspects. Management should consider assigning some MOC review responsibilities to a qualified ERS subject matter expert (perhaps part of the ERS design basis project team) as the project progresses, especially when the change involves process units where the ERS design basis update is underway. Organizations who do not consider this, risk turning the ERS design basis for the process unit into an outdated document almost immediately.

3.2 Process Safety Information

OSHA classifies ERS design as part of PSI, but it is entirely interrelated to all other components in PSI. Consequently, most changes in PSI could cause a change in the ERS design basis. Management must recognize this relationship prior to the outset of the ERS design basis project and take steps to ensure the facility integrates PSI changes into the ERS design basis via the MOC process. Furthermore, with the ERS design basis established, management should communicate with the rest of the organization the interrelationship between the ERS design basis and PSI, as well as the benefit for utilizing the ERS design basis when considering changes to PSI. For example, if the facility is considering a project to modify the facility, utilizing the ERS design basis documentation during facility screening will help identify potential ERS deficiencies the modification project may need to consider and include.

3.3 Process Hazard Analysis and Incident Investigation

Some organizations integrate their ERS design basis documentation into the process hazard analysis (PHA) process by utilizing the ERS scenario summary to confirm the ERS addresses potential hazards. If the ERS design basis documentation does not indicate it addresses a potential hazard, the PHA develops an action item to update the ERS design basis to ensure the ERS design is adequate. Alternatively, if the ERS design basis documentation addresses the potential hazard, the PHA does not explore further.

The relationship between incident investigations and PHAs is common and utilized in many organizations to improve safety. PHA teams often review incident investigations for a process area as part of the area's PHA. Similarly, there is a relationship between incidents and the ERS design basis as both identify credible scenarios, which when addressed, improve safety. Facility management can strengthen both PSM elements (PHA and incident investigation) by promoting the communication and process interactions between their incident investigation teams, PHA teams, and ERS design basis maintenance teams.

3.4 Operating Procedures

Operating procedures involve startup, normal operation, alternative operation, normal shutdown, and emergency shutdown. The ERS design basis should cover all of these operations. As part of an operating procedure review, consider identifying and reviewing the relevant ERS design basis scenarios with their corresponding operating procedures. This helps identify potential process safety impacts of the operating procedures and their changes. For example, changing pump line-ups, or modifying equipment "shed lists" could both influence the ERS design basis. Furthermore, incorporating relevant safe operating limits, as defined by the ERS design basis into the operating procedures and alarm systems promotes operational awareness of the process safety implication in the procedure.

3.5 Employee Participation

Just as employee participation in PHAs is critical to identifying, mitigating and communicating potential process hazards, employee participation in ERS design basis development is also critical for the same reasons. Employees gain an increased awareness of their role and the safeguards that facilities design incorporates as additional layers of protection for the process area personnel work in. Personnel also increase their awareness of where deviations in policies, procedures, and operating conditions can affect their personal safety. Examples are numerous but include alternate equipment line-up and equipment isolation, double pumping practices, use of control valve bypasses, and fluid property changes. One way to improve process safety awareness at a facility is to promote interactions between employees (operations and technical) and the ERS design basis development process. Ensuring both operations and technical participate in the reviews and updates improves operational knowledge, procedure development, and design.

4.0 Conclusions

Emergency relief systems (ERSs) that mitigate all credible scenarios are a fundamental objective of ERS design, and ensure a safe and profitable facility. ERSs are a key layer of protection against loss of containment at a chemical process facility. In order for the ERS to continue to serve as a key layer of protection, an operating facility must manage the facility wide ERS design basis documentation. Such an effort requires a tremendous amount of information, extensive work processes that comply with RAGAGEP, and a significant amount of personnel resources. To effectively maintain a completed ERS design update, a facility must integrate the ERS design basis into the overall process safety management (PSM) system. This paper has presented considerations for all of the above aspects to assist an organization's management in understanding and considering the full "lifecycle" cost of integrating and maintaining the facility wide ERS design basis documentation.

Balancing Mechanical and Instrumented Protections in Brownfield Projects

Irfan Shaikh
MMI Engineering
11490 Westheimer Road, Suite 150
Houston, Texas 77077
ishaikh@mmiengineeering.com

Prepared for Presentation at
American Institute of Chemical Engineers
2013 Spring Meeting
9th Global Congress on Process Safety
San Antonio, Texas
April 28 – May 1, 2013

UNPUBLISHED

Balancing Mechanical and Instrumented Protections in Brownfield Projects

Irfan Shaikh
MMI Engineering
11490 Westheimer Road, Suite 150
Houston, Texas 77077
ishaikh@mmiengineeering.com

Keywords: safety instrumented, protection layer, pressure relief, risk reduction, IPL, SIS,

Abstract

There are many oil and gas facilities around the world that were built at a time when current acceptable engineering requirements were different from today's acceptable engineering best practices and requirements. As a result, many of the existing facilities are grandfathered from new regulations unless major changes in design or services are initiated. From the operating company's perspective, this also implies that older facilities may not be at same acceptable risk levels as newly built facilities. This paper will describe utilizing overpressure and overfilling protection methods commonly employed by industry (such as ISO/EN 4126, API 521, IEC 61508 / 61511, ISA 84.01 etc.) on brownfield projects for mechanical and instrumented protection systems to achieve an acceptable operating company risk levels on brownfield projects. One way to achieve this is by balancing both prescriptive and risks based approaches to establishing acceptance criteria that takes into account operability and reliability framework necessary for the operating company to justify the engineering solutions. Some of the issues this paper intends to deal with include pressure relief with header tie-ins and modifications to flare or vent recovery systems, safety instrumented systems independence and physical separation from process control systems, and issues related to location and infrastructure challenges.

1. Introduction

Many of the oil and gas operating facilities in the world are not new. Since acceptable engineering practices have evolved substantially over the last decades, major changes at the plant level require upgrading existing systems to meet the new standards. In addition, in some countries there is a minimum design basis requirement for hazardous operations. This may lead to a requirement to evaluate and modify existing safety equipment. The paper discusses options available to facilities to update their systems and how to develop a consistent gap closure plan based on the plant requirements and constraints.

2.1 Brownfield Projects

Brownfields are real property, the expansion, redevelopment, or reuse of which may be complicated by the presence or potential presence of a hazardous substance, pollutant, or contaminant. Cleaning up and reinvesting in these properties protects the environment, reduces blight, and takes development pressures off green spaces and working lands. There are usually higher number of permitting hurdles to climb related to greenfield developments versus brownfield re-developments in most every regulatory environmental around the world.

Companies looking to increase their business offerings may look toward increasing capacity at their facilities or adding on new product lines. During feasibility studies for the project, existing equipment is reviewed for its potential to be used in these projects. Since at the early stages of a project the process details may not be available, this feasibility review is usually cursory and for estimation purposes only. As the project moves toward detailed design, a more detailed review of the equipment will be necessary to confirm that it is fit for purpose for the project and that it meets current regulations and Recognized and Generally Accepted Good Engineer Practices (RAGAGEP). Typically there are several paths to closing the gaps discovered during this review process. They may include relief valve changes, safety instrumented system changes and/or operating changes. This is when a clearly defined set of options is useful to the project team.

In order to implement a major change at a brownfield plant successfully from both a safety and business perspective, the following steps should be followed. This process should be started in early project development. In our experience the most effective results at each step are produced by cross functional teams:

- Develop initial design parameters for the change
  - Chemical composition
  - Process Conditions – Temperature, Pressure, Flow, etc.
- Complete a Hazard review around the proposed changes and the effects on equipment using acceptable risk criteria/tools. Findings may include
  - Pressure may increase - Evaluate pressure vessel relief valve sizing
  - An interlock will be required to mitigate an event – use an existing safety shutdown or add a shutdown
  - Operators can mitigate a hazardous event through action - add an additional alarm
- Review the existing equipment specifications against the required capabilities and the company engineering standards for the equipment to identify potential issues
  - Pressure vessel design requirements – analyze whether the vessel meets current code requirements
  - Piping standards vs. new service
  - Electrical Classification standards
  - Facility Siting standards
  - Interlock requirements
  - Etc.
- Develop options for closing gap for both identified hazard mitigation and to meet the current company standards.

1.2 Regulatory

Regulatory process safety requirements vary by region, country, and locally. There are broad regulations that cover maintaining a safe operation that are performance based. This includes in the US, OSHA 29CFR 1910.119 and EPA 40 CFR68. Both of these regulations require a safety management system to address the many facets of process safety. However, the implementation is at the discretion of the operating organization.

On the other side of the spectrum are the less prescriptive regulations with risk based decisions as in regulatory regime for safety case in many international countries like UK, Norway, Denmark, Netherlands, Australia etc., and sometimes with performance based demonstration included as in Australian regulations.

In either case, balancing both prescriptive and risks based approaches to establishing acceptance criteria that takes into account operability and reliability framework necessary for the operating company to justify the engineering solutions and satisfy appropriate regulatory regime.

1.3 Industry Practices

The majority of codes that are used in the United States are not laws. However, meeting them may be required by local laws. Even if they are not specifically required, they are considered the minimum standard that must be met from an engineering perspective. These include the NFPA codes, the National Electrical Code, and the ASME Boiler and Pressure Vessel code. These codes have been developed by teams of industry knowledgeable personnel.

In addition to codes, there are recommended industry practices, which are also developed by teams of industry knowledgeable personnel to support companies drive toward best practices. These are becoming more important to understand since they are considered Recognized and Generally Accepted Good Engineer Practices (RAGAGEP) by OSHA. This means that the organizations must address the issues outlined in the practice. The practices are typically not prescriptive, but where the recommendations are prescriptive, they must be addressed.

2. Risk Management

To determine the risk, the failure probability and severity shall be determined for each possible event (e.g. local fire, vapor cloud explosion, pool fire). A systematic approach is utilized here to determine the development of the scenario on the system, including Consequence and Likelihood analysis in relation to a corporate risk matrix.

2.1 EHS Risks

Environmental Health and Safety Risks are based on the potential consequences that could occur if there is a problem during operation of the facility. EHS Risks are defined during the hazard reviews of the process and process area.

2.2 Operability Risks

Operability risks are based on the potential for reduced reliability or availability of the production facility. These may be directly affected by the solutions to mitigate EHS Risks, usually in the opposite direction. Operability risks directly affect the monetary bottom line of the company.

2.3 Compliance Risks

Compliance risks are based on the potential that the facility is not meeting current minimum standards required by regulations. This can lead to citations, fines, and increased scrutiny from the public and other governmental agencies and potentially a loss of the right to operate.

2.4 Risk Assessment Process

The risk assessment processes for a brownfield projects must take into account all of the potential risks. Determining the EHS risks and compliance risks should be a concurrent exercise. It is important that Subject Matter experts are available throughout the process to confirm the correct path forward. The business or operability risks are analyzed in more detail as the potential risk mitigation for compliance and EHS issues change the capability of the process. During the risk assessment step, the recommendations for solutions should be based on company standardized solutions. For example for a typical overpressure scenario of a pressure vessel, there will be a requirement for a pressure relief valve, and/or a safety instrumented system, and/or an alarm that can be responded to by an operator action.

3. Independent Protection Layers Solutions

The classic description of an independent protection layer (IPL) is systematic device or human administrative function, in which the performance of a protection of a perceived risk reducing layer is not affected by the identified initiating cause of a hazardous concern leading to an undesirable event or by the failure of other perceived risk reducing protection layers. Additionally is also capable of mitigating an initiating cause from escalating to an undesired event. The following are some of the common examples of types of protection layers:

- Mechanical Protection Layer: The two most common types of mechanical protection layers are pressure relief devices (PRD) and mechanical interlocks. The Pressure Relief Devices (PRD) is address in the next section as the primary mechanical safeguard. The PRD is designed to mitigate scenario for all potential identified hazardous scenarios. The design can be based on prior history in similar services and includes specially designed devices, inlet and outlet header purges, and rupture disks.

- Instrumented Protection Layer: The two most common types of instrumented protection layers are process controls and safety systems. The Safety Instrumented System (SIS) is a subset of safety systems and is address in the next section as the primary instrumented safeguard. SIS is defined as a designed and managed protection to achieve a perceived risk reduction factor. The SIS IPL is implemented using relays, trip amplifiers, solid state systems, or safety-configured programmable logic solvers. Moreover, the SIS loop is typically a dormant function that operates only when there is a process excursion outside the normal operating envelope and is comprised of includes a sensors, logic solvers, final elements, and support systems.

- Other Protection layers: Although not desirable but administrative controls are used as an IPL when documented and implement and managed over the lifecycle with training to restrict access to undesirable hazards.

4. Developing a Path to Close Risk Gaps

Before identified gaps can be closed and implemented with lifecycle management and facility operations, it is important to address a path to closure.

4.1 Gap Assessment

Review the existing equipment specifications against the required capabilities and the company engineering standards for the equipment to identify potential issues. Some of the issues for example can be:

- Pressure vessel design requirements – analyze whether the vessel meets current code requirements
- Relief valve inlet and outlet header sizing requirements
- Overpressure relief system capability to upgraded design
- Instrumentation capability with Input/output (I/O) counts

4.2 Gap Closure Philosophy

Developing a gap closure philosophy is the next step:

- Document the basis or rationale for taking the approach to gap closure for both mechanical independent protection layers (IPL) and adding IPL's to instrumented or safety instrumented systems (SIS) solutions.
- This will provide a consistent approach for closing gaps, as well as providing a 'touchstone' document for future decisions.
- The development of this closure philosophy is managed together for both mechanical and instrumented solutions even if it is engineered separately. See Figure 1 for a typical philosophical approach.

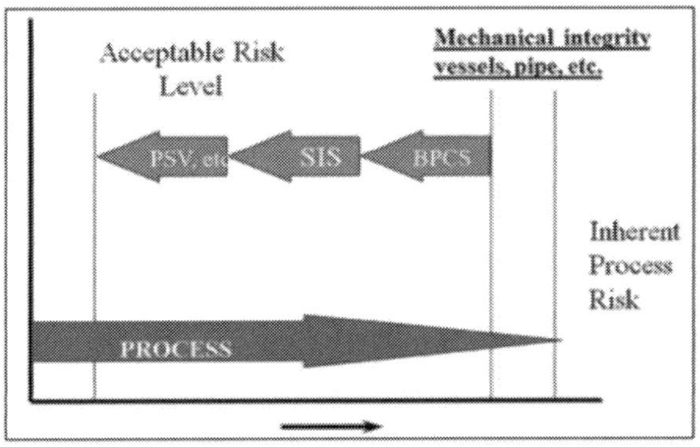

Figure 1 - Typical Risk Reduction Philosophy

As an example, the traditional approach to assure adequate overpressure protection is the proper application of pressure relief devices. In certain situations the use of safety instrumented systems is an alternate means to provide the overpressure protection.

Sometimes, the results of an overpressure protection system such as cold vent or flare system can influence the philosophy the project would choose to follow in closing gaps found. So, first is to close gaps by increasing pressure relief system capacity as needed. The second potential approach would be to reduce pressure relief capacity demand by eliminating new specified load with Safety Instrumented Systems solutions to address overpressure scenarios.

4.3 Risk Reduction Measures

Risk reduction measures for closing gap assessment should follow inherently safer design (ISD) principles and risk reduction traditionally categorized as below in Table 1.

Table 1. Risk Reduction Effectiveness

Categories for Risk Reduction in Order of decreasing Effectiveness	
1	Eliminated by use of substitution to reduce the exposure to hazards
2	Prevented at source
3	Controlled through design features or administrative procedures
4	Mitigated by consequence reduction
5	Mitigated through effective Emergency Response

Using the above example, developing options for closing gaps requires a risk assessment and a prioritization approach. Issues with increasing overpressure protection system capacity may require addressing the below:

- Potential community perceptions (e.g. location with respect to off-site neighbors)
- High capital costs (e.g. depending on distance or type of vent or flare system)
- Environmental delays (e.g. regulatory delays with permitting)

And similarly, issues with addressing overpressure scenarios with SIS may require addressing:

- Non-traditional approach (e.g. perspective with culture change for operations)
- Complexity of the systems (e.g. increased monitoring, troubleshooting, spurious tripping concerns)
- On-going costs for system testing / verification

Other concerns that may result from the above chosen engineering path may result in the following:

- Changing relief valve settings
- Routing relief valve to different path
- Piping changes
- Administrative controls
- Using instrumented systems to prevent relief, or reduce the likelihood of relief
- Analysis of safety and business impact for each option
  - Cost to implement
  - Cost to maintain
  - Potential impacts of failure of equipment
  - Permit e.g. environmental permits
  - Challenges due to plant constraints – physical space limitations, infrastructure
  - Current plant design for systems such as existing relief paths (vent, flare, vapor recovery)
  - Existing electrical, instrumentation and control and safety systems.

4.4 Temporary Solutions

In the meantime, it is important to address temporary solutions or interim measures to address the identified risks as the facility had identified hazards. These can be engineered or administrative controlled solutions similar to protective layers except serve a temporary purpose with a limited acceptable span. Interim solutions can be made into permanent solutions but this should be carefully assessed and designed as they are not intended to be for a limited lifespan.

Examples of interim controls include changes in operating limits (e.g. reducing equipment demand), critical alarms, process design and process control changes to reduce demand, barriers, physical protection devices (e.g., relief valves, vents, blast/ fire walls, dikes), and additional testing and inspection.

Examples of interim administrative controlled solution include training, certification, revised operating procedures, plant communication, fire protection, and plant emergency response.

4.5 Applying Temporary Solution as Risk Reduction

Using the above example, where a vessel overpressure gap is identified as a "high risk", and is to be closed by two protection layers credits using mechanical relief devices and two protection layer credits using instrumented solution. It is advisable to develop interim measures to meet independent protection layer (IPL) gap for risk reduction such as:

- Engineering calculation and documentation is required to establish a failure rate to use the mechanical device(s) as an interim IPL for some overpressure scenarios as it is likely that partial credit could be taken for existing mechanical relief devices on brownfield facilities.
- Another temporary solution would be to develop procedures to limit buildup of pressure in the vessel to reduce demand on the overpressure scenario e.g. limit usage of this vessel.
- A third option may be to establish critical alarm with defined operator action assigned at a critical pressure setpoint. This action supersedes standard operating procedures and operator does not get into "trouble" for "pushing an emergency dump valve or block valve button".

Key criteria for interim measure or temporary solution to reduce risk for the above example are based on the following principles:

- If identified scenario risk is "very high" to "high" risk, and no IPL can be adequately established and no interim measures are acceptable, then the operator must consider safe shutdown of the vessel as a primary option.
- If identified scenario risk is "very high" to "high" risk, and no IPL presently exists but the identified measure(s) can achieve a combined probability of failure on demand (PFD) of at least 0.01 or whatever it takes to bring the risk down into "tolerable" region.
- If identified scenario risk is "very high" to "high" risk, and one IPL exist but the identified measure(s) can achieve a combined probability of failure on demand (PFD) of at least 0. 1 or whatever it takes to bring the risk down into "tolerable" region.
- If identified scenario risk is "very high" to "high" risk, and multiple IPL exist then the identified measure(s) should bring the risk down into "tolerable" region without establishing PFD criteria.

Note: This above example using an order of magnitude approach for risk reduction and establishing IPLs. This process is very company and approach specific.

Another criterion for interim measure to consider is to establish time specific actions such as:

- Short term interim measures (ranging from immediate, 2 weeks, 3 months, and up to 6 months)
- Long term interim measures that do not require turnaround (usually greater than 6 months up to a year)
- Long term interim measures that require turnaround activities (usually greater than 6 months up to 2 years)

Note: Risk prioritization should define length of time to implement interim measures or . This process is again very company and approach specific.

5. Conclusion

It is important to note that this process of gap identification, philosophy, gap risk assessment, prioritization and temporary measures is an engineering project and may times often than not, this can result increased costs to brownfield upgrades and run into resistance from projects, management, operations and overall corporate culture.

Also, facility needs to incorporate and address interim measures on identified hazards whether the project goes forward or not.

The gap closure process will be an "ongoing" process which occurs after each field correction is completed. The elements of the closure process for IPL gaps should be documented, interim measures continuously monitored, and actions tracked as the project moves forward.

6. Acknowledgement

This author is greatly appreciative of the advice and review provided by Sheri Sammons.

7. References

[1] U.S. Environmental Protection Agency (EPA), 40 CFR Part 68: Accidental Release Prevention Requirements: Risk Management Programs Under the Clean Air Act Section 112(r)(7)

[2] Occupational Safety and Hazards Administration (OSHA), CFR 1910.119: Process Safety Management of Highly Hazardous Chemicals.

[3] American Petroleum Institute Recommended Practice, API RP 521, "Pressure-relieving and Depressuring Systems", 2008

[4] BS EN ISO 4126-1:2004, Safety devices for protection against excessive pressure. Safety valves

[5] ISO 23251, Petroleum and natural gas industries—Pressure-Relieving and Depressuring Systems", 2006

[6] IEC 61511 Functional Safety – Safety Instrumented Systems for Process Industry Sector, Parts 1 – 3, 2004

[7] ANSI/ISA S84.00.01-2004 - Application of Safety Instrumented Systems for the Process Industries

[8] ASME Code Case 2211 "Pressure Vessels with Overpressure Protection by System Design", 1999

[9] Tools for Making Acute Risk Decisions, with Chemical Process Safety Applications, CCPS, American Institute of Chemical Engineers, New York, 1995

[10] Guidelines for Risk Based Process Safety, Center for Chemical Process Safety, CCPS, American Institute of Chemical Engineers, New York, 2007

[11] Guidelines for Developing Quantitative Safety Risk Criteria, CCPS, American Institute of Chemical Engineers, New York, 2008

[12] IMO, 2000: Formal Safety Assessment – decision parameters including risk acceptance criteria, MSC 72/16, Norway, International Maritime Organization, 2000

[13] UK Offshore Operators Association, UKOOA, "Industry Guidelines on a Framework for Risk Related Decision Support", Issue 1, April 1999

[14] OSHA, US Department of Labor v BP Products North America, Inc., OSHRC Docket Nos. 09-1788 , Partial Stipulation and Agreement

ISSUES IN DEVELOPING AND USING RISK TOLERANCE CRITERIA

Dr. Paul Baybutt
Primatech Inc.
50 Northwoods Blvd., Columbus, Ohio, USA
paulb@primatech.com

Prepared for Presentation at
American Institute of Chemical Engineers
2013 Spring Meeting
9th Global Congress on Process Safety
San Antonio, Texas
April 28 – May 1, 2013

UNPUBLISHED

ISSUES IN DEVELOPING AND USING RISK TOLERANCE CRITERIA

Dr. Paul Baybutt
Primatech Inc.
50 Northwoods Blvd., Columbus, Ohio, USA
paulb@primatech.com

Keywords: Risk tolerance criteria, risk analysis, process safety.

Abstract

Ultimately, decisions on process safety must be made with reference to risk tolerance criteria. Increasingly, risk analysis methods and codes, standards, and regulations around the world are moving towards the use of numerical criteria, for example, with Layers of Protection Analysis (LOPA) and standards for safety instrumented systems such as IEC 61511 / ISA 84. Various issues must be addressed if such criteria are to be developed correctly and used appropriately. These issues include the entity to which criteria apply, e.g. scenario, hazardous event or process risk; individual versus group risk and their relationship; suitable allocation and scaling of tolerable risk; uncertainties in risk estimates; aggregation of risk from different consequence levels; and risk qualities.

This paper discusses these issues and others and provides guidance to avoid the pitfalls that are often encountered in using numerical criteria.

1. Introduction

Risk has always been of concern in the process industries where hazardous materials are handled, and fires, explosions and toxic releases may occur. Various definitions of risk are used by risk analysts [1]. In process safety, risk is most often defined as the combination of the frequency of an event with the severity of its consequence, usually the simple product [2]. Consequence severity is the impact on receptors such as people, the environment, and property. Risk tolerance criteria express the level of risk that is tolerable, given the benefits provided, according to the collective standards of stakeholders in a facility (e.g. owners, employers, employees, the public, regulators, and insurers) [3]. The criteria express the maximum tolerable frequency of a specified level of harm to a receptor. Risk tolerance criteria are used to determine if existing risks are tolerable by comparison of estimates of the risks with the criteria to determine if there is a risk gap. If a gap exists, recommendations are developed for risk reduction measures to close the gap. For a valid comparison to be made, the type and form of risk estimates and risk tolerance criteria must match.

Regulators have established risk tolerance criteria motivated, for example, by adoption of the As Low As Reasonably Achievable (ALARP) principle within the Tolerability of Risk (TOR) framework of the United Kingdom Health and Safety Executive (UK HSE) [4]. Industry has moved towards the use of risk tolerance criteria motivated, for example, by standards such as IEC 61511 / ISA 84 for Safety Instrumented Systems [5, 6, 7].

Unfortunately, considerable confusion exists with the development and use of risk tolerance criteria. This paper identifies key issues and describes how they should be addressed.

2. Forms and Types of Risk Tolerance Criteria

Various forms and types of risk measures can be used [8]. Some forms distill risk estimates into simple measures such as risk indices which are single numbers. Other forms provide more detailed information presented as tables, graphs and plots. An understanding of the different forms and types of risk measures is needed to avoid errors in developing and using risk tolerance criteria.

For people, both risk to individuals and risk to groups of people are important risk measures and should be addressed in risk analysis. *Individual risk* is the frequency at which an individual may experience a given level of harm as a result of exposure to one or more hazards and *group risk* is the relationship between frequency and the number of people in a given population experiencing a specified level of harm from exposure to one or more hazards [2]. Individual risk does not describe completely situations where one or more events could cause multiple casualties, hence the need for group risk. Consequently, group risk measures are used to represent this aspect of risk to people. Individual risk depends on the hazards to which an individual is exposed and the duration of the exposure. In addition to these factors, group risk also depends on the number of people exposed which is a function of population density. Individual and group risk tolerance criteria specify limits to these risks for individuals and groups of people. Individual risk criteria protect any single individual from bearing too large a share of the risk. Group risk criteria protect populations of people from bearing undue risk. Both individual and group risk measures are calculated from the same set of frequency and consequence data for the set of hazard scenarios used in a risk analysis [2]. They are related through the individual risk to which receptors are exposed. For group risk, the population at risk must be specified.

Individual risk depends on where people are located but not on the number of people present. For example, if two operators work in the same location at the same time for the same duration, they are exposed to the same risk. The individual risk does not increase because two people are present. However, the group risk does increase. In simple terms, for illustrative purposes, if the fatality risk to an individual from a facility is 1×10^{-1} per year, an individual would expect to be killed within a 10 year period. However, if there are 10 individuals working at the facility, the company could expect to see an average of one fatality per year. Furthermore, some hazards may impact more than one person at a time. Clearly, individual and group risk are quite different. The scale of an incident, i.e. the number of people impacted, affects group risk but not individual risk. Group risk provides a measure of the overall risk to a population but only individual risk provides information about the distribution of risk within the population. Individual risk criteria help to prevent any one individual being exposed to an inequitable share of risk. Group risk criteria help to ensure the risk to groups of people is tolerable.

A facility may meet individual risk tolerance criteria but not group risk criteria, or vice versa. For example, individual risk may be low but group risk high if there is a large population of people exposed to the risk. Conversely, group risk for facility personnel may be low but individual risk may be high for personnel working in a process unit at the facility that contains a large amount of hazardous material. Such situations would not be tolerable.

2.1 Individual Risk Measures for People

Various types of individual risk are used. It may be the risk to actual individuals or hypothetical people. A hypothetical person is a theoretical construction used to simplify the calculation of risk. Various types of hypothetical individuals are used [8]. For example, a person who is present outdoors 24/7 with no protection. This is called *location* or *geographic* risk because it is applied to the risk at a location within a facility regardless of whether an individual is actually present there.

Arguably, the preferred measure of individual risk is the actual risk to each individual determined by calculating individual risk at every geographical location where people are present and accounting for time spent at the location. Such a measure allows risk estimates to reflect variations in exposure durations, protections available, individual susceptibility to harm, etc. However, when performed as part of Quantitative Risk Analysis (QRA), many calculations are needed to address the risk to each exposed individual and the resources required quickly become prohibitive and the analysis of the results becomes complex. Consequently, practitioners use the concept of a hypothetical person to provide a reasonable representation of people actually exposed to the risk, for example, a plant operator with a defined occupancy profile for plant areas during a typical working period. A number of hypothetical persons may need to be constructed to represent different categories of people with different exposures, protections, susceptibilities, etc. Note that this is not the same as the hypothetical person used to give meaning to location individual risk. Location risk is a theoretical construction that does not account for the presence of people and is not a measure of the risk actually experienced. However, it does represent the maximum possible risk that an individual could experience at a particular location.

Location risk criteria are used in the Netherlands for public risk [9]. This usage is understandable from the perspective that no one controls the movements of the public, some members of the public may indeed spend the majority of their time in one location, and a measure of conservatism in public risk criteria is appropriate given that the risk is involuntary. In contrast, the movements and locations of facility workers can be controlled and, usually, no individual is located at a facility 24/7. Consequently, location risk is less applicable to worker risk. The Netherlands does not use risk criteria for workers.

Individual risk calculated for compliance with regulations in the United Kingdom (UK) does account for the presence of people, albeit hypothetical individuals, and risk tolerance criteria are used for both the public and workers [10]. The UK Health and Safety Executive (HSE) views the concept of a hypothetical person as advantageous because it avoids the assessment of risk to people actually exposed to the risks [10]. This approach prevents the adoption of inappropriate strategies for regulatory compliance, such as limiting the time exposure to a high hazard, for example, by rotating personnel. However, these factors are part of the risk analysis model regardless of whether actual or hypothetical people are used, and they can be controlled through the assumptions made in the analyses.

Different types of individual risk measures can be used, for example, the maximum geographic individual risk and the maximum individual risk. The maximum geographic individual risk is the maximum value of the geographic individual risk. The maximum individual risk is the individual risk to the person exposed to the highest risk in an exposed population. Maximum individual risk can be determined by calculating the individual risk at every location where people are actually present and selecting the maximum value. It is equivalent to the sum of the geographic individual risks at the locations where people are present for the maximum exposed person. Individual risk is usually expressed as a single number, risk contour plot, or risk profile / transect [2] according to the type of individual risk measure used. It can be determined for different levels of harm, e.g. first-aid case, severe injury, and fatality. Often, the latter is used.

The risks of all scenarios that could impact an individual, regardless of the number of people impacted by the scenario, contribute to individual risk. For example, if a hazard scenario can result in the single fatality of Operator A and another hazard scenario can result in 5 fatalities, one of which is Operator A, both scenarios would contribute to the individual risk to Operator A.

Similar types of group risk are used because they are summations of individual risks.

2.2 Group Risk Measures for People

Historically, group risk has been called *societal risk*, although that is a misnomer as it does not relate to society at large. It was first used as a risk measure for the public, hence its name. It can also be used for personnel at a facility. Group risk is defined for the population exposed to hazards from facilities. It can be calculated for the public and facility personnel separately, or in combination. Group risk for facility personnel has been addressed infrequently, although logically it should be included in risk-informed decision making.

Group risk is often expressed as f-N or F-N curves [2], although F-N curves are more commonly used. f-N curves display the frequencies of all events that result in N casualties. F-N curves display the cumulative frequencies of N or more casualties from a single event summed for all events that lead to N or more casualties (Figure 1). The measure used for N must be the same, e.g. fatalities. Such curves cannot be used to display different types of casualties such as fatalities and injuries within the same curve. Both curves are functions of N, i.e. f-N = f(N) and F-N = F(N). Usually, both types of curve are plotted on a log-log scale since f, F, and N cover orders of magnitude. f-N and F-N curves are usually displayed for intervals of casualty numbers because risk analysis methods usually organize the consequence severities into bins for computational convenience and uncertainties in the analysis obviate the need for more precision. When f-N or F-N curves are used for group risk tolerance criteria they are called limit lines [11].

In a f-N curve, the frequencies of events with particular numbers of casualties are plotted separately. This can be misleading. For example, if fatality events are being plotted, the frequency of all single fatality events is plotted, followed by the frequencies of multiple fatality events. However, the multiple fatality events are composed of individual fatalities that occur together. F-N curves account for this relationship by showing the frequencies of all events that result in one fatality or more, two fatalities or more, etc. Thus, the data point for a single fatality is a summation of the frequencies of all scenarios in which an individual is killed, regardless of how many more people are killed at the same time. Similarly, the data point for two fatalities is a summation of the frequencies of all scenarios in which two or more individuals are killed, regardless of how many more people are killed at the same time. This display provides a clearer depiction of the risks and helps to account for the preference of F-N over f-N curves.

F-N limit lines are defined by these parameters:

- Base point (the cumulative frequency of 1 or more casualties, also called the exceedance frequency for a single casualty), or another anchor point on the line
- Slope
- Cut-offs for cumulative frequency and/or consequences.

A shorthand notation is used to describe F-N limit lines using an anchor point and slope. For example, the triplet (10, 1×10^{-4}, -1) represents a limit line for which N = 10 at F = 1×10^{-4} with a slope -1, i.e. a straight line passing through the anchor point (10, 1×10^{-4}). The slope determines how quickly cumulative frequency falls off with increasing consequence. For slopes of -1, cumulative frequency decreases by an order of magnitude as N increases by an order of magnitude. F-N curves are defined by the equation:

$$F(N) = k / N$$

where k, a constant, is the intercept with the cumulative frequency axis, i.e. F(1).

A consequence cut-off is used when the entire population at risk has been addressed and there is no one else who could be affected. For example, the group risk for personnel in a facility is limited to the number of people present in the facility, although it is possible that no hazard scenario exists that would impact all personnel. Group risk for members of the public is limited to the population of people who may be exposed to hazards from a facility at a level that may cause casualties of concern. Frequency cut-offs are used less often. They allow a higher frequency for low casualty nuisance events than would otherwise be permissible. Clearly, such a cut-off would not be used when considering fatalities. Zero casualties are not displayed in F-N curves as they are not of concern and have a far higher cumulative frequency than for fatalities. The higher cumulative frequency would require that the curve display an extremely wide range of frequencies.

Group risk measures are a function of the number of casualties and this raises the issue of aversion to higher consequence casualties. Usually, the risk of a scenario that results in say 10 fatalities is viewed as less tolerable than the risk from 10 scenarios that individually have single fatalities. For example, far more people are killed in automobile accidents than in airplane crashes every year but the latter receive much more publicity owing to the larger number of people impacted. Early F-N limit lines used slopes of -1 in the belief that this was risk neutral (the higher the consequence, proportionately, the lower the frequency) and that consequence casualties should drop off at least as fast as the frequency. When risk is defined as the product of frequency and magnitude of harm, for slopes of -1, the same risk value will apply to high frequency / low harm events as for low frequency / high harm events. However, usually, people are more averse to the latter and do not perceive the risk to be the same. Therefore, scale aversion, sometimes called risk aversion, was introduced using slopes between - 1 and - 2 (the higher the consequence, the greater the reduction in frequency). However, it is now recognized that F-N limit lines with a slope of -1 actually incorporate a degree of scale aversion owing to the relationship between f-N and F-N limit lines [12]. Factors that influence scale aversion change over time, sometimes very quickly such as when a catastrophic accident influences public opinion. Thus, risk that may be judged tolerable at one point in time may become intolerable at another point in time, even though the actual risk does not change.

Scale aversion can be shown in either F-N space or f-N space but it is defined in terms of f-N space. Consequently, a f-N limit line with slope -1 is risk neutral but when this is transformed into a F-N limit line, lower scale aversion is displayed for most N values owing to the mathematical properties of F-N curves which are not apparent to the casual observer. F-N limit lines with inflections can be used to incorporate more scale aversion for higher consequence events by changing to a higher gradient line after some large number of casualties has been reached. The equation for a limit line including scale aversion is:

$$F(N) = k / N^{\alpha}$$

where α is a constant ≥ 1. It is the negative slope of the F(N) curve in log space.

Manipulation of F-N curves can be tricky owing to their mathematical properties [11, 13]. Caution must be exercised for several reasons:

- Cumulative frequency plots are more difficult to manipulate than frequency plots.
- F-N curves are usually shown in log - log form which complicates their mathematics.
- Often a consequence cut-off is used and sometimes also a frequency cut-off. These end effects must be addressed in mathematical representations of the curves.
- F-N curves are usually shown as continuous lines, even though casualty numbers are discrete. Thus, the mathematics of the curves can be addressed in continuous or discrete space [11].

Group risk can be expressed in simpler forms similar to individual risk, for example, the frequency of 10 fatalities occurring from a particular facility. Also, group risk can be reduced to a single number, or index. For example, the expectation value for casualties can be calculated by summing f(N) x N for all events at a facility. For fatalities, the expectation value is known as the *Probable Loss of Life (PLL)* or *average rate of death (ROD),* or *average societal (group) risk.* Of course, as simpler group risk measures are used, information is lost about the distribution of group risk. For example, PLL values do not differentiate between low likelihood / high consequence events and high likelihood / low consequence events. However, some of the simpler risk measures are easier to understand and use.

Integrals of the areas under F-N curves can also be used as measures of group risk. For example, the UK HSE has developed a Scaled Risk Integral (SRI) as a measure of group risk [14]. It is used as a screening tool for land use planning in the vicinity of hazardous facilities.

A group risk measure can be calculated for individual consequence severities, e.g. single fatality events. However, this is not the same as individual risk for a single fatality. In other words, f(1) and F(1) are not the same. The difference is that the group risk measure represents the total frequency of all single fatalities in an exposed population, regardless of the individuals involved, while the individual risk of a fatality is for specific individuals. For example, if there are 100 people who work in a plant, the group risk of a single fatality is the sum of the frequencies of all hazard scenarios that can result in a single fatality among the 100 people, regardless of the individual. However, the individual risk of a fatality is the sum of the frequencies of only those hazard scenarios that can result in a fatality for each individual among the 100 people. Individual risk will vary according to the individual and their job, location, etc.

F-N curves can be calculated using techniques such as QRA and LOPA. They can also be constructed from historical data where they are available. Not only can F-N curves be used for comparison with limit lines but they can also be used to compare the risks of facilities or to benchmark a new facility with an existing one.

2.3 Risk Measures for Other Receptors

For other types of receptors, such as the environment and property, distinctions usually are not made between impacts on individual and multiple receptors. Thus, if a process contains multiple pumps, no distinction would be made between the tolerable risk of ten pumps being destroyed in individual scenarios versus ten pumps in a single scenario, so long as the risk was the same, that is the total loss was equally likely for both cases. Naturally, in the case of ten pumps being destroyed at the same time, the actual loss may be greater than the replacement cost of the pumps owing to the impacts of such a loss on the process. Such effects can be captured in the estimation of the overall damage costs and the tolerability of such losses is expressed using criteria that define tolerable frequencies as a function of the monetary loss.

Similarly, the concept of individual risk for people is usually not extended to other receptors. For example, normally, the risk to individual pieces of equipment or property is not of concern; rather, it is total risk of harm that matters.

Risk measures for hazard scenarios that impact receptors such as the environment and property are usually expressed by converting impacts such as property losses and environmental remediation into monetary terms and specifying risk as the frequency of the monetary loss. Difficulties may be encountered if the impacts cannot be converted into monetary terms, for example, in the case of permanent environmental impacts and the loss of irreplaceable property. There are no commonly-used risk tolerance criteria for these receptors.

2.4 Expression of Risk Tolerance Criteria

Hazard scenarios may result in multiple consequence types with different consequence severities. Therefore, risk tolerance criteria are usually established for each possible consequence type and severity. Individual risk tolerance criteria take the form of single numbers for different receptor classes such as people onsite and people onsite, and different entities such as a hazard scenario. For example, the tolerable frequency for a single fatality to facility personnel from an individual hazard scenario may be set at, say, 1×10^{-5} per year. Obviously, the greater the magnitude of the consequence severity the lower should be its tolerable frequency.

Consequences are expressed using different consequence endpoints representing events that occur in the scenario event sequence after loss of containment of hazardous materials or energy (Table 1). Consequence severities are set using the measures shown in the table. For example, if releases are used as the endpoint, release quantities can be used to define consequence severities that could occur. When impacts are used as the scenario endpoint, consequence types must also be specified to identify the type of receptors impacted, e.g. people (workers, the public), the environment, property / equipment, the process, the company, etc., and consequence severities denote the magnitude of the impact for the consequence types addressed, e.g. multiple fatalities, single fatality, injuries. Only risk tolerance criteria for impacts have real meaning. Some practitioners use criteria for other endpoints but ultimately they must be related to actual harm, i.e. impacts.

Group risk tolerance criteria can be presented as an F-N limit line or as a tabulation of the numbers of casualties and their corresponding maximum tolerable frequencies. Data are usually displayed for categories representing intervals of casualty numbers, for example numbers of people impacted, e.g. 1, 2 - 9, 10 - 99, etc. because risk analysis methods usually organize consequence severities into bins for computational convenience and uncertainties in the analysis obviate the need for more precision. They can be displayed in the form of a histogram. Commonly, fatalities are used as impacts although other impacts such as injuries can be used. Such a tabulation for group risk tolerance criteria can include any extent of scale aversion.

2.5 The ALARP Principle and Risk Tolerance Criteria

The principle that the risks for a facility should be reduced to as low as reasonably practicable (ALARP) is increasingly used around the world (Figure 2) [4]. The ALARP principle can be used to define two sets of risk tolerance criteria, a minimum requirement (the de manifestus level) that must be achieved and a target value (the de minimis level) that is the goal but may not be reached. The divisions between risk regions are used to set risk tolerance criteria. For example, the UK HSE has suggested the following values for individual annual fatality risk from all hazards present at a single major industrial activity (i.e. a facility) [10]:

de manifestus		de minimis	
Workers	Public	Workers	Public
1×10^{-3}	1×10^{-4}	1×10^{-6}	1×10^{-6}

Note that the word "public' has been used and not "societal" or "social". The latter terms are associated with risk to groups of people. The "public" numbers in the table are for individual members of the public located outside a facility. The UK HSE defines a single major industrial activity as:

> "... an industrial activity from which risk is assessed as a whole, such as all chemical manufacturing and storage units within the control of one company in one location or within a site boundary ..."

The ALARP principle also can be used to establish group risk tolerance criteria. In using the ALARP principle with F-N curves to depict group risk, upper and lower limit lines are defined (Figure 3). The upper line represents the de manifestus risk level that cannot be exceeded and the lower line represents the de minimis risk level that is the goal.

The UK HSE is considering the development of group risk tolerance criteria [15]. However, few F-N curves are available showing actual group risks that are tolerated. Also, there are few precedents owing to the need to scale group risk. However, the UK HSE has suggested an anchor point for an F-N curve of a cumulative frequency of 1 in 5,000 per year at a value of 50 fatalities for N in a single event as intolerable risk for the public [10]. Independently, the UK HSE has endorsed a slope of -1 for a group risk tolerance limit line [16]. The UK Hazardous Installations Directorate (HID), part of the UK HSE, has adopted the limit line (50, 2×10^{-4}, -1) as a de manifestus risk level and a limit line two orders of magnitude below and parallel to this line, i.e. (50, 2×10^{-6}, -1), as a de minimis risk level [17] (Figure 3).

3. Issues in Developing Risk Tolerance Criteria

3.1 Types of Receptors

Receptors are the entities affected by a hazard scenario, i.e. people, the environment, property, etc. People include those who are onsite at a facility (e.g. workers and contractors) and those who are offsite (i.e. members of the public). Property includes equipment and buildings onsite as well as property offsite that is owned by others such as members of the public. The receptors for which risk criteria are needed must be defined. They are determined by the scope and objectives of a risk analysis study, but, at a minimum, studies usually address risks to people onsite and offsite, and the environment.

3.2 Nature of Risks

Risk tolerance criteria must specify the nature of the risks. By definition, risk involves harm to receptors. Therefore, at issue, is the nature of the harm for process safety events. For people, it is some form of casualty, ranging from injury to fatality. For the environment, it is contamination, pollution, and / or damage to flora and fauna. For property, it is damage or destruction. The nature of the risks to be addressed must be specified for the receptors to be included in a risk analysis.

3.3 Sources of Risk

In process safety, the concern is with risk from hazardous processes in facilities where catastrophic accidents can result from the realization of flammable, explosive, reactive and/or toxic hazards (so-called *major hazards*). Thus, sources of hazard scenarios for risk analysis studies, such as process hazards analysis (PHA), often address only scenarios that realize major hazards. Other hazards are not addressed. However, facilities may also pose risks to people from such other hazards as working at height, confined space entry, asphyxiants, corrosives, hot gases, hot liquids, cryogenics, electricity, pinch points, vehicle accidents, etc. Overall facility risk tolerance criteria are usually intended to address risks from all hazards at a facility. Consequently, the sources of risk for risk analysis studies should be specified and appropriate risk tolerance criteria used.

Overall facility risk tolerance criteria should be offset to account for casualties from any sources excluded from a risk analysis. For example, in risk analysis studies for major hazards, data should be available for occupational risks of the types listed above that can be used to reduce the overall facility tolerable risk appropriately so that the remaining tolerable risk is allocated to harm from hazards addressed in the risk analysis. For example, if an individual fatality risk tolerance criterion of 1×10^{-3} is set for workers in a facility, and it is known that the existing fatality rate from occupational accidents, i.e. non-process-safety accidents, in the facility is 0.9×10^{-3}, the tolerable risk from process safety accidents is 1×10^{-4}, i.e. an order of magnitude lower than the overall facility individual fatality tolerable risk criterion which will have a major impact on risk reduction measures needed.

Usually, the risk posed by a facility to people offsite derives principally from major hazards. However, to the extent that other sources of risk impact people off-site, they should be addressed by risk analysis. For example, the Occupational Safety and Health Administration's (OSHA's) Process Safety Management (PSM) standard excludes hydrocarbon fuels used solely for workplace consumption as a fuel that are not a part of a process containing a highly hazardous chemical covered by the standard [18]. However, it is conceivable that an accident involving such a fuel could have both onsite and offsite impacts on people. Furthermore, there may be risks to receptors other than people from sources other than major hazards. Those risks can be handled in a similar way to handling risks to people.

3.4 Entity to Which Risk Tolerance Criteria Apply

Facilities may contain multiple processes and units. Risks can be evaluated for entities such as individual processes, units, hazardous events or hazard scenarios. However, individuals who work in a facility or members of the public living in the vicinity of a facility each will be concerned about the total risk to which they are exposed from all hazards within the facility, not just the risk from one hazard scenario, hazardous event, unit or process. Companies will be concerned about the risk to all employees and members of the public from all hazards within the facility, as well as the risk to individuals to ensure no one person is exposed to undue risk. The determination of these risks requires that the risks of hazard scenarios be aggregated over all hazard types, processes, process units and process modes for the facility. In other words, cumulative risk estimates are needed for comparison with cumulative risk criteria, i.e. overall facility risk tolerance criteria [19]. Indeed, these are the type of criteria used by regulators, such as the UK Health and Safety Executive (HSE) [10]. Only total facility risk has real meaning. Such overall facility risk tolerance criteria should not be applied to individual processes, hazardous events or scenarios without suitable allocation. Clearly, less risk is tolerable from individual processes, units, hazardous events, and hazard scenarios than from all processes, units, hazardous events, and hazard scenarios at a facility. Unfortunately, risk tolerance criteria documented in the literature often do not clarify the entity to which they apply.

3.5 Basis for Values

Values of risk tolerance criteria can be based on regulatory and industry precedents, and public decisions on the tolerability of major hazards. Usually, decisions on individual risk tolerance criteria for facility personnel are made with reference to the risk levels from accidents that actually have been tolerated in the workplace and decisions on individual risk tolerance criteria for the public are made with reference to the risk levels from non-work-related accidents that actually have been tolerated, although with a reduction factor, typically 1 percent, to account for the involuntary nature of the risk.

For example, values of individual risk used by UK regulators derive from statistics on occupational fatalities. The value of 1×10^{-3} for maximum tolerable individual fatality risk for workers was established as an approximation of the risk that many workers experience in the riskier groups in risky industries [20]. Total fatalities in those industries were averaged over the population exposed to the risks to establish statistics on the fatality risk. Presumably, the actual risk to any individual in those industries varies according to the nature of their work and the circumstances they experience but these factors are averaged out. Thus, there may be people actually exposed to higher or lower fatality risk than the average fatality rate. However, such reference points are used to establish a tolerable risk criterion, as the *maximum* tolerable risk to an individual rather than an average over a population as it would be unreasonable to take the position that tolerable risk can be achieved by compensating for higher-than-average risk to some individuals by lower-than-average risk to other individuals.

For group risk, historical data on the frequency of major accidents show an inverse relationship between the severity of accident consequences and their frequencies [21]. Of course, anything else would not be tolerated and processes have been engineered over the years to ensure this relationship holds true.

The precise relationship between individual and group risk varies with the specific circumstances, but a commonly-used guide based on experience with numerous risk assessments is that if the group risk of 10 or more fatalities has the value R, the maximum individual fatality risk will be of the order of 10 x R [12]. This can be used to set the anchor point for a F-N curve. For example, if the maximum individual risk for the public is set at 1×10^{-4} fatalities per year, the F-N curve anchor point can be set at $(10, 1 \times 10^{-5})$ for the maximum public group risk limit line. F-N limit lines also can be derived from the individual risk to which a population is exposed [22].

3.6 People at Risk

An egalitarian view of risk would suggest that the individual and group risks of comparable process facilities should be the same, and some people may advocate for the same risk tolerance criteria regardless of the people exposed to the risk. However, ultimately, the decision on the tolerability of risk must rest with those exposed to the risk, i.e. facility personnel and members of the public. Groups of people may be distinguished in various ways, e.g. vulnerability, generation, income, culture, nationality, etc. [23], and it is possible that different risk tolerance criteria may be applicable to different groups. Indeed, regulators in the Netherlands have addressed vulnerable populations in their criteria [9].

As laudable as may be the use of global criteria, the reality is that different communities and countries accept different levels of risk and the decision is theirs to make. Many companies operate in numerous countries and communities and, while the same risk tolerance criteria could be used throughout such operations, it is likely that local adjustments will be desirable or necessary, although there are ethical, legal, social and other factors that must be addressed in doing so. If local conditions are not addressed, and the same risk tolerance criteria are used, all receptors are treated equally. If local conditions are addressed, risk tolerance criteria are set for different classes of receptors. Similarly, the same or different risk tolerance criteria can be used for other types of receptor, e.g. property, the environment, etc.

Consistent and justifiable risk tolerance criteria can be set for different countries and communities by setting individual risk tolerance criteria for facility personnel and members of the public with reference to the risk levels from, respectively, workplace and non-work-related accidents that actually have been tolerated, although in the latter case with a reduction factor of as much as 1 percent to account for the involuntary nature of the risk to the public from hazardous facilities.

3.7 Allocation

Practitioners often use risk tolerance criteria for hazard scenarios or hazardous events in the belief that it is easier to calculate their risk rather than the overall risk of a facility. However, risk tolerance criteria for individual hazard scenarios or hazardous events have no meaning by themselves [24]. They must be derived by allocating or apportioning overall facility risk to the scenarios or events by dividing facility risk tolerance criteria by the estimated numbers of scenarios or events. However, the estimated numbers are uncertain unless all scenarios and events are identified, in which case, their overall risk can be aggregated to produce an estimate of facility risk that can be compared with facility risk tolerance criteria [25].

Of course, reliance solely on meeting overall facility risk tolerance criteria may result in the inequitable distribution of risk across a facility. While the overall facility risk criterion may be met, there may be processes, areas, units, process modes, people, etc. that bear the brunt of the risk resulting from the disproportionate allocation of risk across the facility. Consequently, overall risk determination should be accompanied by the allocation of the overall risk tolerance across a facility, particularly to all receptors as ultimately that is what matters.

Both individual and group risk tolerance criteria can be allocated within a facility. However, it is also possible to allocate group risk across multiple facilities as group risk is a function not only of facility hazards but also the population exposed to risks under the control of a company, i.e. it depends on the risk exposure of all receptors from all facilities. Thus, if a company has ten identical facilities at different locations, the individual risks are the same for all the facilities because they are a function only of the exposures to people from the risks at the single facility. However, the group risk for the company is 10 times the group risk for a single facility since group risk is evaluated for all individuals in the population at risk which is the same for all 10 facilities in this example. Individual risk cannot be allocated across multiple facilities.

Some companies set group risk tolerance criteria for the entire company and allocate risk across all their facilities. Ultimately, allocation of national risk tolerance criteria for an entire industry could be considered where each facility receives an appropriate allocation. This has been proposed for the nuclear industry in the UK.

3.8 Scaling

Some companies adjust, or scale, their group risk tolerance criteria to reflect the number of operations and their size [8]. Individual risk criteria usually are not scaled as justifying different individual tolerable risk levels for different operations within the same company is controversial. Scaling of group criteria addresses inconsistencies that would otherwise arise. For example, a company may operate 10 facilities and another company a single similar facility. If both companies used the same overall tolerable risk criterion, the company with 10 facilities would have to operate them all at one tenth the risk of the company with the single facility to meet the criterion. More specifically, in simple terms, if a small company operates one facility in which the individual fatality risk for onsite personnel is 1×10^{-3} per year and ten people work in the facility at any time, the group risk is 1×10^{-2} single fatalities per year and the company can expect to experience one fatality every 100 years. However, if a large company operates one hundred such facilities, that company faces a company-wide, or enterprise, group risk of one fatality per year. Similarly, it is not logical to assign the same group risk tolerance criterion to a large facility as for a small facility. Either the small facility would be allowed to operate with undue risk or more stringent requirements than necessary would be imposed on the large facility. Furthermore, if a company operated a facility and determined that its group risk was twice the tolerable value, theoretically, the company could split the facility into two and meet the facility risk tolerance criterion without actually having reduced risk at all.

The basis for scaling should be a facility metric that addresses the importance of the facility to the people it serves. However, the people it serves may not be the people it affects. Societal factors enter into the picture. However, a possible simple importance measure for a facility is the value of the products it produces as measured by their sale price. Consequently, overall company group risk tolerance criteria could allocated to individual facilities by the ratio of the value of the products from the facility to the total value of products produced by all the company's facilities.

More complex scaling metrics may be a function of additional facility metrics such as the hazardous material inventory, number of employees, size, capital investment, production volume, value of the facility to the community, and/or national strategic importance. Failure to scale group risk criteria likely will result in higher risk being tolerated for small facilities and large amounts of risk reduction being required for large facilities.

3.9 Aggregation of Risk from Different Casualty Types

The exclusive use of *fatality* risk criteria for people is not completely satisfactory. Process safety incidents can and do produce injuries as well, which are often much more numerous than fatalities for catastrophic accidents (Table 2). For the twelve accidents contained in the table, the average ratio of injuries to fatalities is 25 and the range of the ratio is about 2 to 140. If the Bhopal and Toulouse accidents are eliminated as outliers, the average ratio drops to about 8, with a much narrower range. Unfortunately, insufficient data are available to classify the injuries by severity or by receptor (worker versus member of the public). All of the incidents listed are fire and/or explosion scenarios, except Bhopal, which involved a toxic release and has the highest ratio of injuries to fatalities. Clearly, injuries can be a significant component of the harm caused by catastrophic accidents.

At some point, a certain number of injuries of a particular severity becomes equivalent to a fatality in the eyes of many people. Perhaps the ratio is 10 or more. It involves a value judgment. Given the number of injuries that can occur in catastrophic accidents, it would seem prudent to address them in risk analysis and to develop risk tolerance criteria for them.

Unfortunately, often the focus in risk analysis is exclusively on fatalities and any accompanying injuries are not addressed. For example, PHA is used to identify possible hazard scenarios for a process, and the severities of the scenario consequences are identified and categorized according to ranges and types of casualties. However, for scenarios that produce fatalities, the extent of accompanying injuries is usually not considered. If the average number of injuries that accompanies a single fatality is about 10, as seems to be the case from the record of catastrophic incidents, and 10 injuries are equated with a fatality, the actual risk is about double what is actually estimated. This may not be of undue concern recognizing other uncertainties in risk analysis. However, there may be cases where the ratio of injuries to fatalities is much higher and the risk could be increased by an order of magnitude or more. Non-fatal health effects for people can be incorporated into the risk evaluation using the concept of equivalences in which risks are combined for different casualty types using weights. For example, 10 severe injuries may be equated to one fatality. Of course, this is subjective and different people may use different equivalences. However, it does allow a more inclusive definition of risk to be employed. Risk tolerance criteria that have been developed by regulators such as the UK HSE are based on workplace fatality rates and do not include injuries.

PHA studies often do identify scenarios that result in injuries of different severities. Therefore, it is appropriate to assign risk tolerance criteria to them. Of course, a value judgment must be made in doing so. Risk tolerance criteria for injuries can be developed by using the equivalence concept and using the tolerable frequency of a fatality as a reference point. For example, if ten incapacitating injuries are equated to one fatality, and the tolerable frequency of a fatality is 1×10^{-3}, the tolerable frequency of a single incapacitating injury would be set at 1×10^{-2} (risk = frequency x severity of consequence, so equivalent tolerable frequency = $1 \times 10^{-3} / (1/10) = 1 \times 10^{-2}$. A lesser injury category could be defined, for example, as hospitalization for a certain period, and the tolerable frequency reduced by another order of magnitude to 1×10^{-1}. Although such criteria are convenient and easy to assign, the equivalences are debatable. An alternative approach is to analyze injury data for workers in industry and set risk tolerance criteria in the same way as they have been set for fatality risk. However, that has not yet been done. It is also possible to set proportionately more restrictive criteria for more serious consequence severities to account for aversion to high severity consequences.

Distinctions between the degree of damage are of lesser concern for other receptors such as equipment and property. For example, equipment damage is converted into monetary terms and aggregated for all levels of damage produced by an incident to provide an overall measure of the severity of equipment damage. Typically, separate criteria are established for different consequence severity levels for all types of receptors of concern.

3.10 Process Age

Often, older processes pose higher risks than newer processes, owing to aging factors in the former and improved technology in the latter. New facilities may be held to a higher standard than existing facilities and more stringent risk tolerance criteria used, for example, lower by an order of magnitude. Thus, companies may use more than one set of risk tolerance criteria according to the remaining life of their processes. Of course, this practice would mean that higher risk is being tolerated for older processes. Some justification is provided by the shorter remaining life of older processes than for newer processes with the bulk of their life remaining. However, if older processes are to be operating indefinitely, they must be retrofitted to meet current risk tolerance criteria.

3.11 Risk Qualities

The risks of different hazards may be the same quantitatively but they may be perceived differently by the public owing to differences in the *qualities* of risk [12]. These qualities are risk attributes that influence people's views on the tolerability of risk and include:

- Voluntary vs. involuntary nature of the risk, e.g. rock climbing versus living next to a hazardous facility.
- Number of people affected at one time, i.e. the incident size or scale.
- Fear of the risk. For example, fear of the type of casualty (e.g. toxic exposure versus fire exposure).
- Familiarity with the risk. Unfamiliar risks are less tolerable. For example, nuclear radiation versus chemical toxicity versus automobile crashes.
- Control over the risk. The more control people have over risk, the more risk they may be willing to tolerate.
- Trust in those controlling the risks, i.e. companies and regulators.
- Uncertainty in the risk. Not knowing the extent of the risk.
- Geographical concentration of the risk. Distribution of the risk across many members of the public is usually viewed as preferable to concentrating the risk with a few people.
- Perceived effects if the risk is realized, e.g. community and economic impacts.

Not enough is known currently to address all these different risk qualities in setting risk tolerance criteria. However, some of them have been addressed in a regulatory context. For example, risk tolerance criteria for facility personnel are usually set higher than for members of the public to account for the voluntary / involuntary nature of the risk [9]. Incident size has also received considerable attention in the form of F-N curves that incorporate aversion to high consequence incidents [26]. Criteria for the UK nuclear industry have been set lower than for process facilities to account for public aversion to nuclear risk [20].

Typically, risk tolerance criteria are reduced by an order of magnitude to accommodate differences in risk tolerance for different risk qualities. It is logical that there should be reductions, but the factors of ten frequently used, while seemingly sensible, are actually arbitrary. However, currently, there is no empirical basis for other reduction factors. Reduction factors of this type can be used to set lower risk tolerance criteria for certain classes of receptor. They can be addressed by aggregating risks within individual receptor classes and comparing the risks with the applicable risk tolerance criteria for the receptor class.

Ultimately, the public is the arbiter of public risk tolerance criteria so risk perceptions and the underlying values that drive them should be considered. Group risk is a subset of societal concerns but it is the only aspect which has so far been incorporated into formal decision making on the control of risk. Societal concerns include socio-economic, political, and other factors.

4. Issues in Using Risk Tolerance Criteria

4.1 Matching Calculated Risk With Risk Tolerance Criteria

The type of criteria and risk measures used must be specified and the entity to which they apply must be defined. Unfortunately, some practitioners do not address these issues and their analyses are ill-defined producing questionable and possibly erroneous results. The type and form of risk estimates must be the same as those of the risk tolerance criteria used. A common error is to use individual risk criteria but calculate societal risk measures which can lead to more stringent risk reduction measures than needed [27].

When using risk tolerance criteria from regulators or industry guidance as reference points for establishing company risk tolerance criteria, it is important to understand what type of risk criteria are being used so that corresponding risk measures can be calculated for comparison with them. For example, individual risk is calculated in the UK for a hypothetical person who is in some fixed relation to the hazard, e.g. the person most exposed to it, or a person living at some fixed point or with some assumed pattern of life; while, in the Netherlands, it is calculated for the public as location risk for a receptor that is present 24/7 out of doors with no protection [28]. In classical QRA, location risk is combined with a weighted population distribution to calculate societal risk [2]. In LOPA, the actual consequences usually are defined for specific receptors for each hazard scenario since typically the scenarios originate in a PHA so that the calculation of societal risk is performed differently than in QRA [27].

4.2 Calculation of Individual and Group Risk

Most facilities comprise multiple processes within which hazard scenarios and hazardous events can occur. For receptors located within a process, while the risk to them is usually dominated by hazardous events that originate within the process, a contribution to the risk may be made by hazardous events that originate in other processes. For example, toxic releases from one process may result in employee impacts in an adjacent process. Usually, process hazard analysis identifies impacts on receptors wherever they are located. Consequently, risk analysis for those scenarios will provide the risk contribution from the process to the overall facility risk. However, the individual risk to receptors calculated from the scenarios for the process is not a complete measure of individual risk either for the receptors within the process or those in other processes that are impacted. In order to correctly determine individual risk for receptors within the process, contributions from hazard scenarios and hazardous events in other processes must be included in the cumulative risk measures. Similarly, the correct determination of individual risk for receptors within other processes must include contributions from hazard scenarios and hazardous events in the process being studied together with those from any other processes that may contribute to the risk. Consequently, an aggregation of risk for scenarios across all processes at a facility is needed using a risk model that includes hazard scenarios and hazardous events for all the processes at the facility.

A common misconception is that group risk applies only to scenarios that involve more than one casualty. Group risk is actually calculated beginning with the consideration of single casualties. For example, the risk of a fatality of an individual person at a facility may be calculated and that is a measure of individual risk. However, the risk of a single fatality for any person at the facility may also be calculated. It is the sum of the risk of all possible single fatalities. Both individual and group risk should be allocated across facilities.

Group risk can be calculated for the public, for facility personnel, or for both groups combined. Regulatory group risk tolerance criteria usually are established only for the public. In using such criteria as reference points for setting company criteria, it is important to know what group risk is being calculated. Usually, a distinction is made in PHA between impacts on the public and facility personnel. Therefore, group risk can be calculated from the hazard scenarios for either the public or facility personnel, or both combined together. The risk measure calculated must be paired with the correct risk tolerance criterion. If PHAs do not make this distinction, group risk will be calculated for facility personnel and the public combined. This may have the effect of requiring more stringent risk reduction measures than necessary.

Some practitioners calculate societal risk but compare risk estimates with risk tolerance criteria for individuals. Obviously, this practice is incorrect and can lead to more stringent risk reduction measures than needed.

4.3 Conflicting Criteria

Companies are motivated to reduce risks for moral / ethical, legal / regulatory, financial / business, and other reasons. These motivators affect different types of receptors. Risk tolerance criteria need to be set for each type of receptor, e.g. facility personnel, members of the public, the environment, property, etc. It is possible that risk reduction requirements for different types of receptors may conflict. For example, an attempt to reduce risk to employees by moving toxic gas cylinders out of a process area to the plant fence line may increase public risk. Such issues can be addressed with a risk model that includes all relevant hazard scenarios so the risks to different types of receptors can be optimized together making tradeoffs, as appropriate. Some time ago, other authors expressed the similar view that in modeling major accidents, there is no alternative but to elaborate all the events and consequences that may result, including second order effects (i.e. consequences of consequences) [29].

4.4 Changes in Risk

Public group risk can change as a result of changes in the population exposed to risks from a facility, e.g. new housing developments, which are beyond the control of a company, as well as changes at the facility which, of course, are within the control of the company. Typically, individual risk can change only as a result of changes at a facility, unless sources of risk external to the facility change, for example, the installation of a new facility next to an existing facility located in an industrial park. When increases are anticipated in onsite and offsite populations, and the number of processes at a facility, part of the facility risk allocation should be reserved to accommodate the changes.

4.5 Transient Populations

Some people may be exposed to the hazards from a facility for a limited period of time. For example, contractors at a facility may be onsite for a short duration and members of the public may be present for a limited time when traveling through the area. While it is unlikely that any of these individuals will be exposed to risk at the maximum tolerable levels for individuals, they contribute to the onsite and offsite populations used in estimating group risk. Care must be exercised to ensure that these receptors are addressed in group risk estimates whenever they may have a significant impact on risk. The time these receptors are exposed to risk is taken into account using time-at-risk factors.

4.6 Delayed effects

In process safety accidents, usually casualties occur within a short time of the initiating event. However, there may be circumstances in which casualties are delayed. The Bhopal accident is a case in point [30]. The issue has been addressed in the context of group risk tolerance criteria [31] where it was recommended that delayed and early fatalities should be treated identically for the purposes of risk assessment.

4.7 Overlapping Hazard Zones

Process facilities may be co-located, for example, in an industrial park. In such cases, accidents within any one of the facilities may have a hazard range or zone that encompasses adjacent facilities. Hazard range and hazard zone are, respectively, the distance from, or the area surrounding, the source of a hazard that is impacted if the hazard is realized, for example, the distance at or area within which blast effects extend from the point of origin of an explosion. The hazard ranges and zones from each facility may also extend into the surrounding community and there may be multiple overlapping risk footprints. Thus, accidents in one facility may contribute to the risks within other facilities, and all co-located facilities contribute to the risk to the public.

Consequently, in the same way that individual process units at a facility may contribute to individual and group risk from the facility, neighboring facilities may contribute to the individual and group risk for facility personnel at each facility, and individual and group risk to the public may have contributions from all facilities. These contributions may have a significant impact on risk, particularly group risk to the public. Cross-company collaboration on process safety for neighboring facilities may facilitate such considerations [32].

4.8 Uncertainties in Risk Estimates

Uncertainty is the state of knowledge of a situation where the factors influencing the situation are known but their effects on the situation cannot be described precisely. Ignorance can also be a factor, but it is not the same as uncertainty. It means a lack of awareness of the factors that influence the situation. Uncertainties in estimated risk are likely most significant for high-consequence, low-frequency events, and are particularly important when risk estimates are close to risk tolerance criteria as they may be sufficient to make the risk intolerable.

Various forms of uncertainty enter into the calculation of risk [2, 33]. The models used in risk analysis are imperfect representations of the real world and result in *modeling uncertainties* in risk estimates. The data used in the calculations are often sparse or not known very well and result in *data or knowledge uncertainty*. Modeling uncertainty and knowledge uncertainty are also known by risk analysts as aleatorical and epistemological uncertainty, respectively. Comparisons of risk estimates with risk tolerance criteria must address these uncertainties to the extent possible so that decisions on risk reduction err on the side of protecting receptors. There are two approaches to addressing uncertainties:

- Perform uncertainty analysis
- Use the precautionary principle

Uncertainty analysis is used to quantify knowledge uncertainties. Modeling uncertainties are difficult to quantify. In risk analysis, point values of input data are used to calculate point values for risk. However, these values represent just one point from distributions that reflect uncertainties in the input variables [34]. The point values used by most practitioners are viewed as best estimates, i.e. likely they are viewed as the mean of the input variable distributions and they are used with the intention of calculating a best estimate of the risk. However, these best estimates of risk are not the highest values of risk that are possible and they may be optimistic depending on the form of the distributions. In uncertainty analysis, the input variable distributions are propagated through the risk analysis model to obtain a distribution for the calculated risk using techniques such as Monte Carlo analysis [35, 36]. This enables high percentiles of the distribution, such as the 90th or 95th, to be used for decision making instead of the best estimate of risk [37]. Of course, this will force higher levels of safety than would otherwise be required but that is the cost of accounting for uncertainties. An alternative approach is to employ best estimates for initial decision making but check the risk distributions to ensure the range is not so wide that it invalidates the decisions being made. Another approach that has been used is to employ more stringent risk tolerance criteria when there is thought to be greater uncertainty in risk estimates. However, there is no guarantee that the chosen criteria are sufficiently stringent to account for the uncertainties. A proposal has been made to address uncertainties within the specification of risk tolerance criteria [38].

Sensitivity analysis can be used in combination with uncertainty analysis. Sensitivity analysis is first used to identify those aspects of the risk model, and its input data, to which the calculated risk is most sensitive [36]. This information enables uncertainty analysis to focus on addressing uncertainties in those sensitive aspects of the risk model and its input data.

The precautionary principle can also be used to deal with hazards that are subject to significant uncertainties. In the context of major hazards, the precautionary principle can be expressed as, "Where there are threats of serious or irreversible harm, lack of full scientific certainty shall not be used as a reason for postponing cost-effective measures to control the risk." When uncertainties exist, assumptions are made to produce a conservative risk estimate for decision making. The greater the possible harm and the greater the degree of uncertainty, the more conservative are the assumptions made. Often, these assumptions entail using a credible worst case or a worst case possible. The assumptions can be applied to the models employed or data used. In the case of data, these are sometimes called *safety factors*.

Risk analysts may be ignorant of some matters that impact the determination of risk. For example, risk analysis depends on the identification of a complete set of hazard scenarios. However, no hazard analysis method can guarantee complete identification of hazard scenarios so risk analysts are necessarily ignorant of all scenarios that should be included in the analysis. Participation by people with a range of technical expertise and experience and the use of peer review are measures used to counteract ignorance. However, the possibility always exists that phenomena or situations unknown to anyone may arise. Aspects of the Buncefield accident illustrate this point [39].

One can speculate on the uncertainties in risk estimation. Estimates of consequence severities are probably no better than within a factor of 2, if calculated, and 5, if estimated qualitatively. Estimates of frequencies are probably no better than within a factor of 10, if calculated, and 50, if estimated qualitatively, owing to data uncertainties. Risk estimates are probably no better than within a factor of 10 owing to modeling uncertainties. Therefore, an overall uncertainty factor of at least 200 can be anticipated. The typical maximum range between de manifestus and de minimis risk tolerance values is 1,000 so these uncertainties are a major issue.

5. Conclusions

Various issues affect the development and use of risk tolerance criteria which should be approached with care. Not only do risk tolerance criteria help to determine the extent of harm that is viewed as tolerable, but also they influence the allocation of resources and the technologies used in facilities.

The focus in this paper has been on developing and using numerical risk tolerance criteria for use in decision making on risk. Such criteria are used with engineering analyses of risk. While decision making on the tolerability of risk must be informed by scientific considerations it must also address ethical, social, economic, political, technological and other concerns.

6. References

[1] Aven, T., Misconceptions of Risk, *Wiley*, 2010.

[2] Guidelines for Chemical Process Quantitative Risk Analysis, *Center for Chemical Process Safety / American Institute of Chemical Engineers*, New York, NY, 2nd Edition, 2000.

[3] Baybutt, P., Gibson, S. B., Lamb, R. S., and Leverenz, F. L., "Safety, Risk and Reliability Criteria in the United States," Annual Symposium of the Safety and Reliability Society, Safety Criteria - Requirement, Establishment and Application, Southport, England, September, 1983.

[4] ALARP Suite of Guidance, United Kingdom Health and Safety Executive, http://www.hse.gov.uk/risk/theory/alarp.htm.

[5] ANSI/ISA—84.00.01—2004 Part 1 (IEC 61511-1 Mod), Functional Safety: Safety Instrumented Systems for the Process Industry Sector — Part 1: Framework, Definitions, System, Hardware and Software Requirements.

[6] ANSI/ISA—84.00.01—2004 Part 2 (IEC 61511-2 Mod), Functional Safety: Safety Instrumented Systems for the Process Industry Sector — Part 2: Guidelines for the Application of ANSI/ISA—84.00.01—2004 Part 1 (IEC 61511-1 Mod).

[7] ANSI/ISA—84.00.01—2004 Part 3 (IEC 61511-3 Mod), Functional Safety: Safety Instrumented Systems for the Process Industry Sector — Part 3: Guidance for the Determination of the Required Safety Integrity Levels — Informative.

[8] Guidelines for Developing Quantitative Safety Risk Criteria, *Center for Chemical Process Safety / American Institute of Chemical Engineers*, New York, NY, 2009.

[9] Bottelberghs, P. H., "Risk Analysis and Safety Policy Developments in the Netherlands," *Journal of Hazardous Materials*, 71, pps. 59-84, 2000.

[10] Reducing Risks, Protecting People, United Kingdom Health and Safety Executive Books, 2001.

[11] Cox, D. C., and Baybutt, P., "Limit Lines for Risk," *Nuclear Technology*, Vol. 57, pps. 320 - 330, June 1982.

[12] Ball D. J., and Floyd, P. J., Societal Risks, United Kingdom Health and Safety Executive report, 1998.

[13] Griffiths, R. F., "Problems in the Use of Risk Criteria," in: *Dealing with Risk, Planning, Management and Acceptability of Technological Risk*, R.F. Griffiths (Editor), Wiley, New York, 1982.

[14] Carter, D. A., "The Scaled Risk Integral - A Simple Numerical Representation of Case Societal Risk for Land Use Planning in the Vicinity of Major Accident Hazards," Proceedings of the 8th International Symposium on Loss Prevention and Safety Promotion in the Process Industries, pps. 219 - 224, Vol. II, Antwerp, Elsevier, 1995.

[15] CD212, Consultative Document on Proposals for Revised Policies to Address Societal Risk Around Onshore Non-nuclear Major Hazard Installations, United Kingdom Health and Safety Executive, 2007.

[16] Societal Risk and the Consideration of Technical and Policy Issues, Status Summary of 23 Issues, UK HSE, February 2010 , http://www.hse.gov.uk/societalrisk/technical-policy-issues.pdf.

[17] HID's Approach to ALARP Decisions, SPC/Permissioning/39, Annex, http://www.hse.gov.uk/foi/internalops/hid_circs/permissioning/spc_perm_39.htm.

[18] Federal Register, Volume 57, Number 36, pages 6355 to 6417, February 24, 1992, Final Rule, Process Safety Management (PSM) of Highly Hazardous Chemicals, 29 CFR 1910.119.

[19] Baybutt P., "Using risk tolerance criteria to determine safety integrity levels for safety instrumented functions," *Journal of Loss Prevention in the Process Industries*, 25 (6), pps 1000 - 1009, 2012.

[20] The Tolerability of Risk from Nuclear Power Stations, 2nd edition, *HSE Books*, 1992.

[21] Marshall, V. C., Major Chemical Hazards, *Halsted Press*, 1987.

[22] Baybutt, P., "On Setting Group Risk Tolerance Criteria for Hazardous Facilities," to be published.

[23] The Setting of Safety Standards, *HM Treasury*, June 28, 1996.

[24] Baybutt, P., "Risk Tolerance Criteria and the IEC 61511 / ISA 84 Standard on Safety Instrumented Systems," *Process Safety Progress*, accepted for publication.

[25] Baybutt, P., "On the Allocation of Risk Tolerance Criteria," submitted for publication.

[26] Wu-Chien, J. S., and Apostolakis, G., "On Risk Aversion in Risk Acceptance Criteria," *Reliability Engineering*, pps. 45–52, Vol. 2, No. 1, 1981.

[27] Baybutt, P., "Risk Tolerance Criteria for Layers of Protection Analysis," *Process Safety Progress*, Vol. 31, No. 2, pps 118–121, June, 2012.

[28] Ale, B. J. M., "Tolerable or Acceptable: A Comparison of Risk Regulation in the United Kingdom and in the Netherlands." *Risk Analysis*, pps. 231-241, Vol. 25, No. 2, 2005.

[29] Fischhoff, B., Lichtenstein, S., Slovic, P., and Derby, S. L., Acceptable Risk, *Cambridge University Press*, 1984.

[30] Broughton, E., "The Bhopal Disaster and Its Aftermath: A Review," *Environmental Health*, 4 (1), 2005.

[31] Griffiths, R. F., "Discounting of Delayed Versus Early Mortality in Societal Risk Criteria," *J. of Loss Prevention in the Process Industries*, Vol. 7, No. 5, pps 432–438, 1994.

[32] Reniers, G., Multi-Plant Safety and Security Management in the Chemical and Process Industries, *Wiley*-VCH, 2010.

[33] Baybutt, P., "Uncertainty in Risk Analysis", Conference on Mathematics in Major Accident Risk Assessment, *University of Oxford*, July 1986.

[34] Baybutt, P., Cox, D. C., Denning, R. S., Kurth, R. E., Fraley, D. W., and Heaberlin, S. W., "The Treatment of Uncertainties in Risk for Regulatory Decision Making," International ANS/ENS Topical Meeting on Probabilistic Risk Assessment, Port Chester, New York September, 1981.

[35] Cox, D. C., and Baybutt, P., "Methods for Uncertainty Analysis: A Comparative Survey," *Risk Analysis*, Volume 1, p. 251, 1981.

[36] Baybutt, P., and Parry, G., "Uncertainty and Sensitivity Analysis," in *PRA Procedures Guide: A Guide to the Performance of Probabilistic Risk Assessments for Nuclear Power Plants* (NUREG/CR-2300), January 1983.

[37] Baybutt, P., "Procedures for the Use of Risk Analysis in Decision Making," in *Heavy Gas and Risk Assessment II: Proceedings of the Second Symposium on Heavy Gases and Risk Assessment*, Frankfurt Am Main, May 25-26, 1982 , pps 261-277, S. Hartwig, (Ed.), Springer, 1983.

[38] Henselwood, F., and Phillips, K.G., "The Development of Risk Criteria for High Severity Low Frequency Events," *Process Safety Progress*, Vol. 28, Issue 1, pps. 11–14, March 2009.

[39] Atkinson, G., Cusco, L., Grant, S., and Painter, D., "Buncefield: Vapour Generation, Dispersion and Explosion - The Investigator's Story," Symposium Proceedings Hazards

XXII - Process Safety and Environmental Protection, Liverpool, UK, Institution of Chemical Engineers, April 2011.

Table 1. Consequence Endpoints and Measures

Endpoint	Measure
Release	Quantity
Effect distance / area / zone	Distance / area impacted
Physical effects	Heat radiation level, blast overpressure, toxic exposure
Impacts	Personnel casualties, environmental damage, property damage

Table 2. Impacts of Catastrophic Accidents

Accident	Fatalities	Injuries	Ratio
Oppau, explosion, 1921	500 - 600	2,000	3 - 4
Feyzin, fire and explosion, 1966	18	81	5
Flixborough, vapor cloud explosion, 1974	28	36	1.3
Beek, explosion and fire, 1975	14	107	8
Mexico City, fire and explosions, 1984	500 - 600	5000–7000	10 - 12
Bhopal, toxic vapor cloud, 1984	4,000- 20,000	550,000	28 - 138
Norco, explosion, 1988	7	42	6
Pasadena, vapor cloud explosion, 1989	23	314	14
Sterlington, explosion, 1991	8	120	15
Toulouse, explosion, 2001	29	2,500	86
Skikda, explosion, 2004	30	70	2
Texas City, fire and explosion, 2005	15	170	11

Note: Data are from multiple sources on the internet.

Figure 1: F-N Limit Line

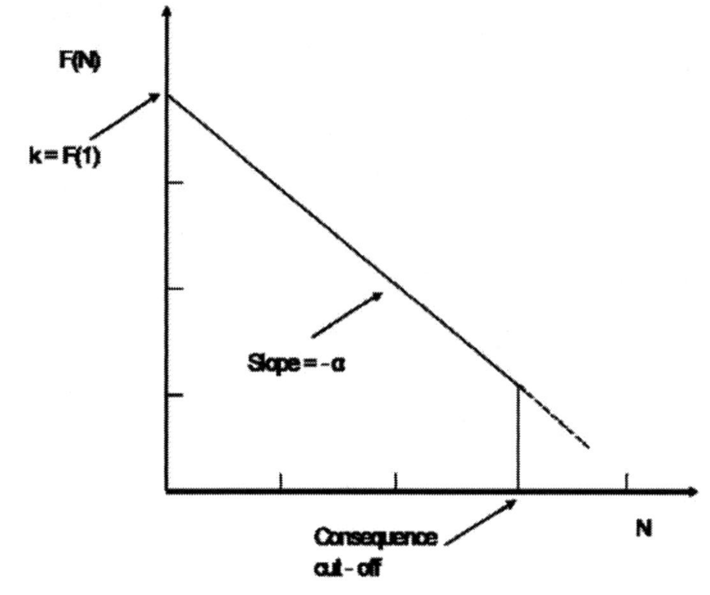

Figure 2: The ALARP Principle

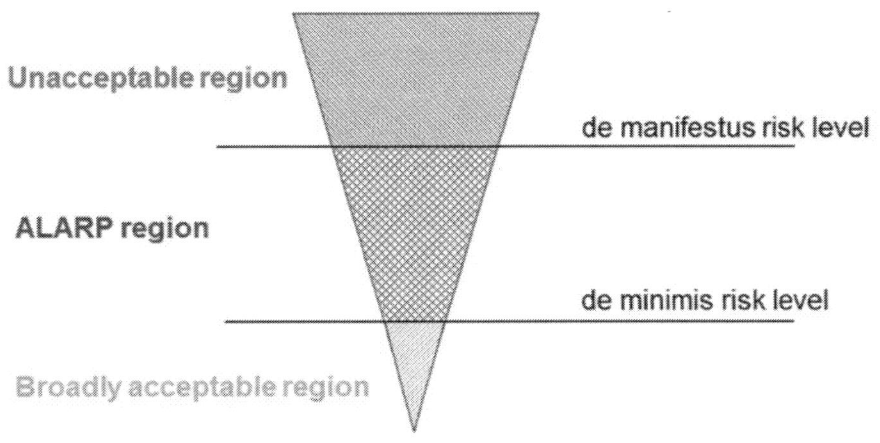

Figure 3: The ALARP Principle Used for F-N Limit Lines

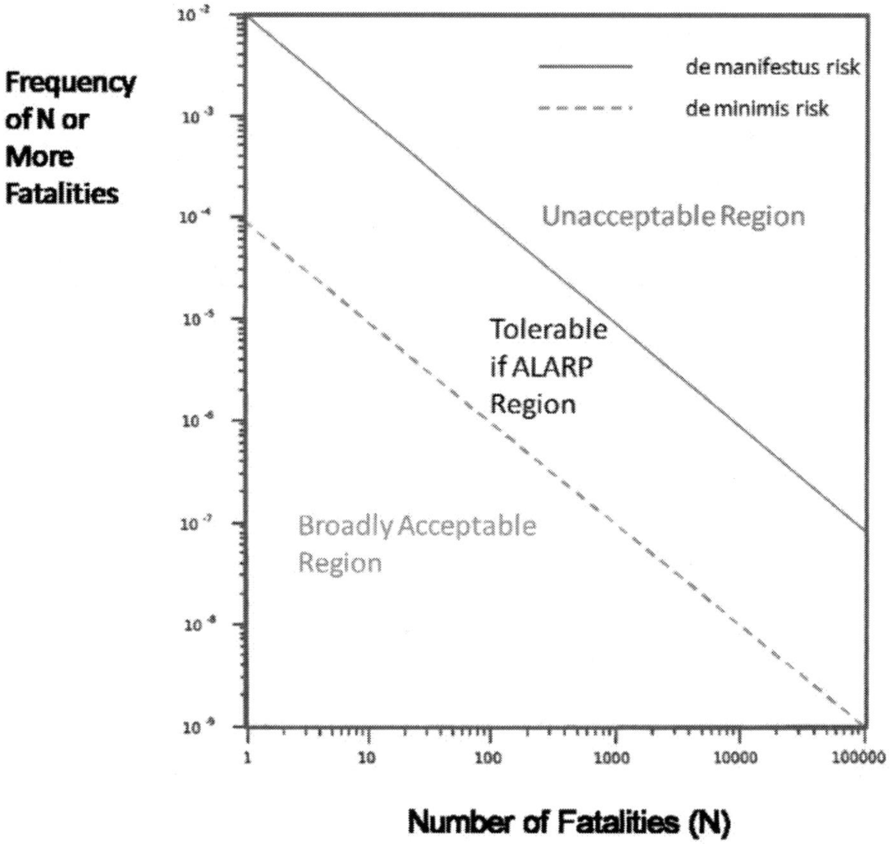

Beyond the 2 Million Safe Hours

Rao Akula,
Senior HSE Manager, WorleyParsons
125 W. Huntington Drive, Arcadia CA 91007
Rao.akula@worleyparsons.com

Prepared for Presentation at
American Institute of Chemical Engineers
2013 Spring Meeting
9th Global Congress on Process Safety
San Antonio, Texas
April 28 – May 1, 2013

UNPUBLISHED

Beyond the 2 Million Safe Hours

Rao Akula, PE (Chemical Engineer, CA), AIChE, STS
Senior HSE Manager, WorleyParsons
125 W. Huntington Drive, Arcadia CA 91007
Rao.akula@worleyparsons.com

Keywords: Arctic, WorleyParsons, Engineering/Process Safety, safety performance indicators

Abstract

In September 2012, WorleyParsons Arctic Business Unit, part of WorleyParsons Western Operations in Southern California, celebrated completion of 2 million hours without any recordable injuries/incidents for a major Alaskan Client. In the highly visible and challenging world of hydrocarbons projects in "brownfield" settings (with inherent simultaneous ongoing operations, and process safety issues), such accomplishments are definitely significant. This paper presents hindsight perspectives from a company manager in this context.

1. Introduction

WorleyParsons is a Global Engineering Services company (*www.worleyparsons.com*). Their Arctic Business Unit, part of their Western Operations in California, has been managing a Program "of the Projects, by the Projects, for the Projects", for many major oil companies in Alaska, for a long time. This paper analyzes a recent safety accomplishment for one of their clients, and presents the hindsight perspectives of the program management, over an eight year period covering a cluster of projects delivered to date.

2. The Perspectives

2.1 Be Pragmatic

As highly productive project groups pursue high quality "fit for purpose" and "cost effective" deliverables, contradictions between the passionate safety commitments by some and the casual non-compliance by others can be perplexing. Also, there can be an inherent tendency to not "over-do" anything in the name of efficiency, or if it is not a "billable" and "approved" activity (in the contractors' world). The Arctic Business Unit mentioned above has detailed safety management procedures and practices in place to ensure structure and consistency across the program. Project teams should remember that safety requirements can be an auditable function for verification of compliance with pre-defined expectations and deliverables. Also, safety culture is a value, not a deliverable –

it is OK to overdo values, and safety prevention measures are a necessary part of this value.

2.2 Delivering Safe Projects

For contracting companies serving major clients, delivering safety in projects is a condition for doing business. However, delivering safety in complex projects can often seem more of an art than a science. We have seen impressive safety records on some projects obliterated by the "weakest link" in the safety management process. No one comes to work thinking that "I am going to hurt myself today", yet safety incidents continue to occur. We should not underestimate the importance of a strong and responsive safety culture to prevent safety incidents. Common sense in application of the procedures and honest appreciation of the complexity involved is important. As individuals, we all bring in varying levels of knowledge, risk awareness, tolerance, and judgment. Therefore, open and honest communication about what we know, what we don't know, use of courtesy and consideration necessary to compensate for others' shortcomings (like in defensive driving) are very important. It is also useful to see preventive safety measures as actions addressing risks, not necessarily as someone's "short comings" (even if they appear to be). Many of us may have witnessed passionate discussions about safety, and noticed people turning "defensive" when the conversations are misunderstood. The complexities of the context and communications involving safety discussions make the safety delivery process so challenging. As a major contractor for several large Alaskan clients, WorleyParsons gets to contribute (and offer improvements) to their world of safety requirements, as well as safety culture. For many contractors, success can in fact depend on the ability to add value to their client's safety requirements as well as safety culture.

2.3 Leading vs. Lagging Indicators

The world of safety management uses two common terms for measuring its effectiveness - "leading indicators" and "lagging indicators". Leading indicators are measurements made before an "incident" occurs, lagging indicators are measurements made after the fact (past performance). The law has clear definition for "recordable injury" incidents – which we all strive to prevent. Completion of 2 million hours without a recordable injury is a lagging indicator. The recording of any potential safety concern is a leading indicator. Some other examples of leading indicators used by WorleyParsons Arctic Business Unit include meetings addressing safety such as Engineering/Process safety meetings, HSE Steering committee meetings (where we discuss status of previously recorded unsafe conditions, "near miss" reporting, suggestions for improvements, newcomer safety orientations, drills to check preparedness for emergencies, etc. These tools are useful to fuel and reinforce the project teams' passionate commitment to their HSE values. Good safety management (whether personal or in groups/projects) always includes plenty of sensitivity for leading indicators.

2.4 *Celebration of Accomplishments*

Celebration of accomplishments is an important way to reinforce and sustain them. Also, we celebrate accomplishments that are difficult to achieve. The higher the difficulty, the better we should be feeling about the accomplishments.

2.5 *What Next?*

The periods following celebrations can be the most vulnerable for safety incidents. If anyone on a project team has a question or a doubt regarding any matter that connects with their personal safety, potential safety issues that can arise during their projects' execution (even after handover to the client), no one should ever hesitate to bring them forward. "Engagement'" of the team and open, honest communication are thus very critical in safety matters. As the volume of the safety conversation increases, Project Management must weigh and balance the resource needs, pragmatically process the inputs received, and prioritize potential actions for implementation. Delivering on HSE expectations requires exceptional responsiveness to identified safety concerns. In addition, several other critical factors come to mind to ensure success. Some of these include "walking the talk" by the management, minimizing gaps between reality and expectations by striving for continuous improvement (of procedures/expectations), and sustaining creativity/innovation by maintaining a mutually respectful culture that accommodates individual styles/differences, while being consistent with application of enforcement/audit measures.

Celebrations like completion of millions of safe hours are lagging indicators in personal safety. There is so much more to manage when we think of "Engineering/Process safety" required to ensure no harm to our client's assets and the environment.

2.6 *Engineering/Process Safety*

In order to separate out "personal" safety issues (like seat belts and handrails) from the "technical" type (like complex and impersonal matters like pressure, materials, ratings etc.), we use the term "Engineering Safety" for projects while all the engineering disciplines are designing/developing the designs. I have seen "Process Safety" described in several ways; but one that would be easy to connect with "is about keeping the hydrocarbons inside the pipes, vessels, etc., where they belong". Also, a good way to describe the culture to maintain Engineering/Process Safety excellence would be to maintain a "healthy sense of vulnerability" as a team, which can make technical issues personal and keeps personal issues technical (meaning respectful/responsive). The Engineering/Process Safety controls used in the Arctic BU include several structured design/engineering tools such as hazard management plans, formal design reviews, assurance reviews, and controlled stage gate completion checks at various stages of projects' development. However, these controls would not be successful, if the "execution" aspects (like multiple contractors, suppliers, vendors, disciplines, roles/responsibilities, organizational and regional cultures, technologies, contracting strategies, etc.) were not accounted for.

Conclusion

It is a sweet feeling to be successful in safety delivery. Project teams should keep on celebrating their successes, keep their minds on leading indicators, and promote their HSE culture to the fullest extent possible.

Acknowledgements

There is an old saying that victory has 100 fathers and defeat is an orphan. The author insists that the accomplishment discussed in the paper belongs to everyone in these project teams – not just to the management. Special acknowledgements are also due to Mr. Mark Anzelon (the Program's deputy Engineering Manager), and Mr. Allen Frame (the Program's Engineering Manager), the engineering disciplines, and all the volunteers who lead the safety agenda forward at WorleyParsons.

About the Author – Rao Akula

Rao Akula is a Senior Manager for HSE and Risk Management in WorleyParsons Western Operations' Arctic Business Unit. Rao is a Registered Professional Engineer (Chemical Engineering) in California (CH 4832), with a Masters in Computer Information Systems from California State University of Los Angeles. Rao is also a full member (139736) of the American Institute of Chemical Engineers and a Safety Trained Supervisor (IEX03356). Rao joined WorleyParsons in 2007 and has worked in a wide variety of settings in the industry since completion of his Bachelor's Degree in Chemical Engineering from Andhra University, Visakhapatnam, India in 1977. Rao's past employers include Engineers India, New Delhi (1977-82), Kuwait National Petroleum Company, Mina Abdulla Refinery, Kuwait (1982-90), Brown & Root/KBR, Alhambra, California (1991-99), State of California Department of Toxic Substances Control, Glendale & Sacramento, California (2000-2006), and BP, Long Beach, California (2006-07).

The Importance of a Team-Based Approach to Overpressure Scenario Development

Neil Prophet / Principal Author
ioMosaic Corporation
93 Stiles Road | Salem, NH 03079
prophet.n.tx@iomosaic.com

Gene Gorski / Author
ioMosaic Corporation
93 Stiles Road | Salem, NH 03079
gorski.g.nh@iomosaic.com

Prepared for Presentation at
American Institute of Chemical Engineers
2013 Spring Meeting
9th Global Congress on Process Safety
San Antonio, Texas
April 28 – May 1, 2013

UNPUBLISHED

The Importance of a Team-Based Approach to Overpressure Scenario Development

Neil Prophet / Principal Author
ioMosaic Corporation
93 Stiles Road | Salem, NH 03079
prophet.n.tx@iomosaic.com

Gene Gorski / Author
ioMosaic Corporation
93 Stiles Road | Salem, NH 03079
gorski.g.nh@iomosaic.com

Keywords: Pressure Relief, Overpressure Protection, Hazard Identification

Abstract

With the increased regulatory focus on pressure relief and flare systems (PRFS) design basis, many companies are in the process of updating their existing relief systems design documentation. A very important aspect of any relief systems analysis is the correct identification of overpressure scenarios, and one way to ensure successful and accurate overpressure scenario development is to conduct this process as a team. This approach is currently in effect at many major oil and petrochemical companies including BP, Shell, Chevron Phillips, Bayer, etc., even though current relief systems, recognized and generally accepted good engineering practice (RAGAGEP), does not call for this approach.

The benefits of a team-based approach to overpressure scenario development include participation and consensus from all affected disciplines: engineering, operations, instrumentation and control, maintenance, and process safety. Unique insights can be provided through operational experience, and there are opportunities to confirm correct operating and design conditions.

In the same way that a process hazard analysis (PHA) is conducted using a team approach; so should the relief systems rationale.

This paper outlines the basic steps involved in overpressure scenario identification, the way that a team approach can be implemented, and the benefits of a team-based approach to overpressure scenario development.

1. Background

The provision of overpressure protection is a typical requirement for any ASME code-stamped equipment. When considering overpressure protection, the primary regulatory drivers in the US are:

i. OSHA Standard 1910.119 Process Safety Management of Highly Hazardous Chemicals
ii. ASME Section VIII (Pressure Vessels)
iii. ASME Section I (Boilers)
iv. ASME B31.1 and B31.3 (Piping)

When designing overpressure protection systems, the established codes and practices that are commonly considered include:

i. API Standard 520 Part I[i]
ii. API Recommended Practice 520 Part II[ii]
iii. API Standard 521[iii]
iv. API Standard 526
v. API Standard 2000
vi. NFPA-30
vii. CCPS, '*Guidelines for Pressure Relief and Effluent Handling*'[iv]

It is interesting to note that while these documents are considered RAGAGEP, they do not discuss the need for a team approach for overpressure scenario identification and development.

When conducting design or evaluation studies of existing relief systems, there are a number of commonly established steps in any project:

- Define scope, basis, and project guidelines
- Gather project and process data
- Develop overpressure scenarios for each protected system
- Determine relief requirement for each applicable scenario
- Calculate relief device capacity for each applicable scenario
- Identify any relief system deficiencies and formulate mitigation options
- Generate report

For new relief systems, the last three steps listed above are substituted with the following steps:

- Determine the minimum relief area which will satisfy all applicable scenarios
- Select an appropriately-sized relief device with proper materials of construction
- Design relief system piping system to avoid temperature concerns, downstream collection system concerns, vibration concerns, (for relief valves: excessive inlet line pressure loss or discharge line built-up backpressure), discharge consequences.
- Generate PRFS design information (Process Safety Information)

This paper focuses on the "development of overpressure scenarios for each protected system" step. A thorough hazard identification is essential for the overall study to be successful.

As Trevor Kletz once stated "Are we sure that we have identified all the major hazards and all the ways they can occur? What has not been identified can neither be assessed nor mitigated". It is therefore imperative that the overpressure scenario development stage receives the attention it merits.

2. Overpressure Scenario Identification

When considering overpressure protection, prudent evaluation of pressure-relief systems involves much more than just relief devices.

A "Relief System" should include, but is not limited to:

- All interconnected vessels, equipment, and process lines in the system
- Relief device
- Relief device inlet line
- Relief device discharge line
- Relief header
- Vent containment system (cyclone, separator, catch tank, scrubber, stack)
- Flare knockout pot
- Flare and stack
- A variety of associated indicators, sensors, and alarms used to indicate whether the process equipment is functioning properly and whether the relief systems are functional or functioning as designed
- External factors such as human error, control systems failure, fire exposure
- A consideration of the area at and around the atmospheric release point

Depending on the extent of the study scope, each of these components should be considered during the overpressure scenario development stage.

API Standard 521 provides very thorough guidance on overpressure contingency analysis, determination of relieving rates, and disposal systems. API Standard 521 lists the following upset conditions, which may result in overpressure:

- Closed outlets on vessels
- Inadvertent valve opening
- Check-valve leakage or failure
- Utility failure
- Electrical or mechanical failure
- Loss of fans
- Loss of heat
- Reflux failure
- Abnormal heat input from reboilers

- Heat exchanger tube failure
- Transient pressure surges
- Plant fires
- Process changes/chemical reactions

Using this list as a basis, it was possible to develop a matrix of overpressure scenario, cross-referenced to specific vessel types, as shown in Table 1:

Table 1. Common Contingencies Cross-Referenced to Type of Equipment

	Closed Outlets	Cooling Failure	Reflux Failure	Accumulation of Noncondensables	Mixing of Volatile and Hot Materials	Failure of Automatic Controls	Abnormal Heat or Vapor Input	Internal Explosions	Chemical Reaction	Hydraulic Thermal Expansion	Exposure to External Fire	Ruptured heat exchanger tube	Power / Utility Failure	Inadvertent Valve Opening	Transient Pressure Surges	Inbreathing/Outbreathing
Blower	☑												☑			
Boiler	☑										☑		☑			
Column - Packed	☑	☑	☑			☑	☑				☑		☑	☑		
Column - Trayed	☑	☑	☑			☑	☑				☑		☑	☑		
Compressor - Centrifugal	☑												☑			
Compressor - Positive Displacement	☑												☑			
Dryer	☑				☑	☑				☑	☑		☑	☑		
Filter	☑					☑				☑	☑		☑	☑		
Fired Heater	☑												☑			
Heat Exchanger - Aerial	☑					☑					☑		☑			
Heat Exchanger - Concentric Pipe	☑					☑				☑	☑		☑			
Heat Exchanger - Electric	☑					☑				☑	☑		☑			
Heat Exchanger - Other	☑					☑				☑	☑		☑			
Heat Exchanger - Plate and Frame	☑					☑				☑	☑		☑			
Heat Exchanger - Shell and Tube	☑					☑				☑	☑	☑	☑			
Pipe Segment	☑									☑			☑			
Pump - Centrifugal	☑												☑			
Pump - Positive Displacement	☑												☑			
Reactor	☑	☑				☑			☑		☑		☑	☑		
Storage Tank	☑					☑					☑		☑	☑		☑
Storage Vessel	☑					☑					☑		☑	☑		
Turbine	☑												☑			
Vessel	☑					☑					☑		☑	☑		

For each relevant scenario, a narrative should be developed stating whether a scenario is applicable or not, and the reasoning behind a decision. A standard overpressure scenario analysis form should be developed which addresses each of these scenarios, as well as listing the protected equipment, and overpressure protection, for each system.

It is common practice for designers of pressure-relief systems to require extensive process and equipment information. OSHA 1910.119 Section D outlines data that should be available to support a process hazard analysis study, and the requirements are similar for a relief systems design study. In order to fully conduct a relief systems design study, a full range of supporting process safety documentation should be available, such as that shown in Table 2. Additionally, piping isometrics, relief devices and vessel design parameters, and piping and instrument drawings (P&IDs) should be field verified to ensure an accurate evaluation.

Table 2. Typical Relief Systems Design Study Data Requirement

	General Data Requirements	
Process Design and Description	Piping and Instrumentation Diagrams (P&ID) Heat and Material Balances (H&MB) Process Flow Diagrams (PFD) Process Safety Flow Diagrams (PSFD) Process descriptions / operating procedures Plot plans / elevation plans	
Utility and Piping Design	Utility operating conditions (electrical, instrument air, cooling water, steam, etc.) Electrical one-line diagrams Piping designations and ratings Insulation designations and ratings	
	Data Requirements	
	Required Information	**Data Source Hierarchy**
Fluid and Mixture Properties	Thermophysical properties	1. DIPPr database using modified PR EOS 2. Company generated data 3. Estimates based on structure
	Reaction kinetic models	1. Company provided adiabatic calorimetry data 2. Open literature data 3. Externally generated adiabatic calorimetry data
Pressure Relief Devices	Manufacturer / model number Inlet / outlet / discharge area sizes Opening pressure and temperatures	Relief Device Information: 1. Maintenance records 2. Relief device specification sheets 3. Original design basis 4. P&ID 5. Valve Tag Inlet / Outlet Piping Details: 1. Existing isometric drawings 2. Field sketches
Fixed Process Equipment (General)	MAWP, MAWT, and vacuum rating Design conditions Equipment Dimensions	1. U-1A forms 2. Mechanical drawings 3. Equipment specification sheets 4. Operating Manuals 5. P&IDs 6. Nameplate
Vessels	Liquid levels	1. Operating procedures 2. P&IDs 3. Equipment design drawings

General Data Requirements		
Process Design and Description	Piping and Instrumentation Diagrams (P&ID) Heat and Material Balances (H&MB) Process Flow Diagrams (PFD) Process Safety Flow Diagrams (PSFD) Process descriptions / operating procedures Plot plans / elevation plans	
Utility and Piping Design	Utility operating conditions (electrical, instrument air, cooling water, steam, etc.) Electrical one-line diagrams Piping designations and ratings Insulation designations and ratings	
Data Requirements		
	Required Information	**Data Source Hierarchy**
		4. Level alarm set-points 5. Level-gauge tapping locations (from equipment design drawings)
	Elevation	1. P&ID 2. Equipment design drawing
	Insulation type, thickness, firegrade status	1. Maintenance records 2. Equipment design specification 3. P&IDs
Heat Exchangers	Design type Rated and normal duty Tube ID / length	1. U-1A forms 2. Heat exchanger specification sheets 3. P&ID 4. Nameplate
Heaters / Steam boilers	Tube Design Pressures Furnace design duty Boiler dimensions and design duty	1. Heater / Boiler specification sheets 2. U-1 Forms 3. P&ID 4. Nameplate
Rotating Process Equipment (General)	MAWP, MAWT Design conditions	1. Equipment specification sheets 2. P&ID 3. Equipment nameplate
Centrifugal Pumps	Pump capacity curve, rated capacity, and installed impeller size Suction Conditions	1. Performance curves 2. Pump specification sheets 3. Maintenance records (installed impeller and corresponding curve) 4. P&ID 5. Nameplate
Centrifugal Compressors	Compressor capacity curve and rated capacity Suction conditions Isentropic or polytropic efficiencies	1. Performance curves 2. Compressor specification sheet 3. Original design data 4. P&ID 5. Nameplate
Positive Displacement Pumps	Pump casing MAWP / MAWT, design conditions Rated capacity	1. Pump specification sheets 2. P&ID 3. Nameplate
Reciprocating Compressors	Compressor manufacturer/model Cylinder type (double acting, etc.), diameter Stroke length, Rod diameter, Piston displacement, Engine speed, Volumetric efficiency	1. Compressor specification sheets 2. Original design specification 3. P&ID 4. Nameplate

	General Data Requirements	
Process Design and Description	Piping and Instrumentation Diagrams (P&ID) Heat and Material Balances (H&MB) Process Flow Diagrams (PFD) Process Safety Flow Diagrams (PSFD) Process descriptions / operating procedures Plot plans / elevation plans	
Utility and Piping Design	Utility operating conditions (electrical, instrument air, cooling water, steam, etc.) Electrical one-line diagrams Piping designations and ratings Insulation designations and ratings	
	Data Requirements	
	Required Information	**Data Source Hierarchy**
Turbines	Exhaust casing MAWP / MAWT, design conditions, Steam throughput	1. Turbine specification sheets 2. P&ID 3. Nameplate
Control Valves	Sizes (inlet / outlet / port) Manufacturer and model number Fail safe position	1. Control valve data sheets 2. Vendor data 3. Nameplate

Of particular importance at the overpressure scenario development stage is the following information:

- Piping & Information Diagrams
- Process Flow Diagrams
- Heat & Material Balance
- Equipment design pressures, operating pressures, and elevations
- Relief systems set pressures and specifications

2.1 Overpressure Scenario Development Preparation

Unlike conducting a PHA, it is most efficient for an individual engineer to develop a set of narratives prior to conducting the team meeting. These draft overpressure scenario analysis forms (narratives), developed by the engineer should be considered as a 'strawman' for discussion, based on the information available to the engineer. The intent of the team-based meetings is then to ratify, or amend, these forms, to gain consensus on which overpressure scenarios are applicable, and which are not.

In addition to the draft overpressure scenario analysis forms, a visual representation of each protected system should be developed. An example is shown in Figure 1. This visual aid should show the protected equipment, equipment connectivity, pressure sources, pressure-limiting devices, process and utility lines leading into and out of the (evaluated) protected equipment. The visual aid will help to expedite the team-based overpressure scenario review meeting, which occurs upon completion of the individual draft work.

The visual aid can be created by using a portion of the PFD, one or more P&IDs, and / or a hand sketch. Color-coding can further enhance the effectiveness of these visual aids. For example, the following color scheme can be used for highlighting lines and valves:

- Red - PSV's
- Green - Feed lines to vessels or equipment
- Purple - Discharge lines
- Yellow - Control valves
- Blue - water, steam, air, or nitrogen (utility lines)

Figure 1. Visual Aid for Each Protected System

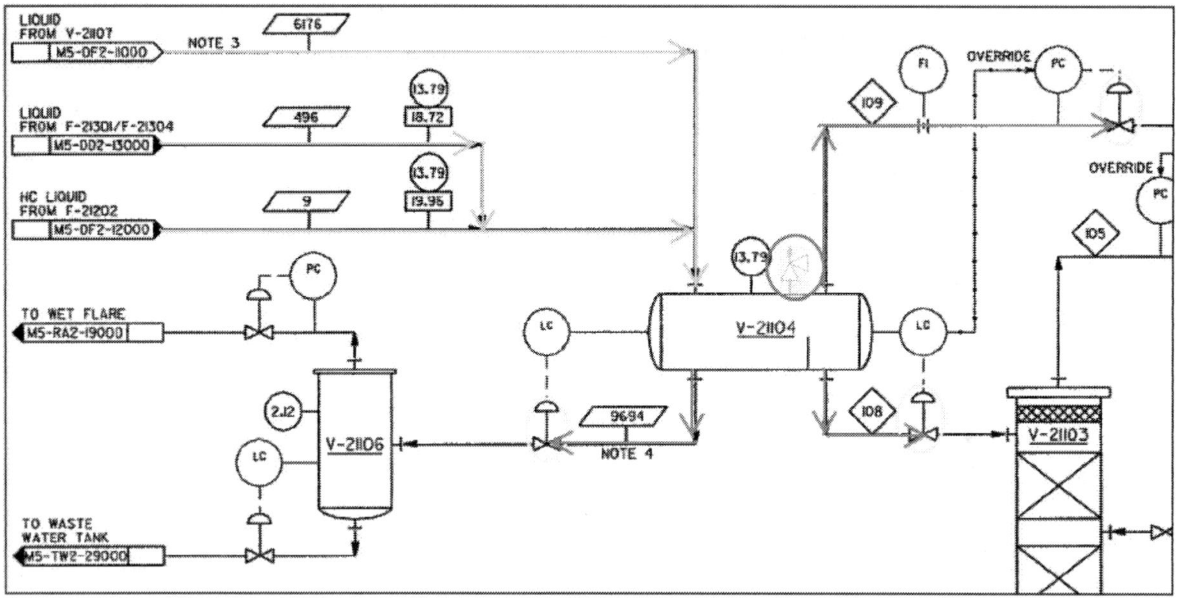

2.2 Team-Based Overpressure Scenario Review Meeting

The team-based overpressure scenario review meeting should be conducted similarly to that of a process hazard analysis. Participants should include:

- Facilitator
- Scribe
- Pressure Safety Specialist
- Process Knowledgeable Person
- Plant Operator
- Plant Process Engineer
- Project Engineer

(At a minimum, the team must include a pressure safety specialist and a process knowledgeable person to be effective.)

In addition to the main participants, other participants are likely to be required on a less-than-full-time basis, with one participant potentially providing more than one type of expertise. These "as needed participants" typically include:

- Controls and Instrumentation
- Electrical, Mechanical
- Maintenance
- Process Safety

Attendance should be taken at the start and end of each morning and afternoon session, noting full-time and part-time participants, the date and the session. The Scribe is responsible for keeping accurate attendance lists and matching the attendance lists with evaluated equipment.

Team-based overpressure scenario review meetings should assess each equipment item within the scope, and aim to cover a specific section of plant. At the start of the scenario analysis for a unit, the process knowledgeable person should briefly describe the unit's design intent and modes of operation.

Meeting progress can be expedited by sharing the draft overpressure scenario development forms (narratives) with meeting participants for their review prior to the meeting. Team members should be provided with both the relevant draft overpressure scenario development forms, and the visual aid that was assembled to help facilitate discussion.

Providing team members with a copy of the project guidelines used would also be very useful. A good project guidelines document addresses the following concepts:

- Document the relief systems evaluation and design philosophy
- Address gaps and inconsistencies in existing design methodology, guidelines or standards
- Serve as a condensed design guide/best practices document addressing all relevant codes and standards
- Identify the form, layout, and content of all project reports and deliverables

The facilitator should lead the team-based overpressure scenario review meeting, one piece of equipment (or protected system) at a time. The facilitator should use the draft overpressure scenario development forms and visual aids to lead the meeting participants through the review to determine if all credible scenarios have been identified and if the identified scenarios are correctly considered credible. A good facilitator keeps the review moving (limiting side discussions) and keeps the participants engaged. The facilitator decides when discussions should be tabled, rescheduled for a smaller group, or expanded to include one or more of the additional attendees.

As the review of each piece of equipment proceeds, the scribe should update the draft overpressure scenario development form based on team member comments and team consensus. The scribe should keep a record of all issues that require resolution, or tabled discussions that require follow-up action. Specific input from operations, or process engineering, can help resolve previously identified data discrepancies, and should be noted on the forms.

2.3 Improving Meeting Efficiency

Any meeting involving multiple participants for extend periods of time can be expensive, and difficult to organize. It is therefore in the team's interests to ensure that the team-based overpressure scenario development meeting runs as smoothly and efficiently as possible.

This can be achieved in a number of ways:

Availability of accurate process and mechanical data: it is essential that the engineer developing the draft overpressure scenario development forms has access to accurate data. If not, the process of identifying overpressure scenarios will be hindered prior to the team meeting, as well as during the team meeting. Throughout the overpressure protection study, a consistent design basis should be used.

Accurate draft overpressure scenario development forms: The progress of the team-based overpressure scenario development review meeting will be directly influenced by how little or how much revisions are required to the draft overpressure scenarios development forms. It is therefore beneficial that the engineers developing these draft forms spends as much time as is required to generate an accurate set of overpressure scenarios. Common errors include not analyzing every single control valve affecting a system, failing to trace the source of pressure far enough upstream, and not considering every equipment item in a 'protected' system.

Effective use of audio-visual equipment: As with a PHA, where the PHA worksheets can be projected using a data projector for all the participants to see, so can the draft overpressure scenario development forms. This will ensure the entire team can follow the meeting progress, and ensures agreement on conclusions reached. An adept scribe can even utilize two computers and two data projectors, where the draft overpressure scenario development form is shown on one, and the visual aid is displayed on the other.

Advanced preparation from all team members: If time permits, the draft overpressure scenario development forms should be shared with team members in advance of the team-based overpressure scenario review meeting. This will enable team members to review and develop comments and questions prior to the meeting. In addition, if time permits, it is useful to have an "example session" with all team members in advance of the team-based overpressure scenario review meeting. This will familiarize team members that have not previously participated in the review process so that they can better self-prepare for the future meeting.

Provision of P&IDs: During the entirety of the meeting, sets of P&IDs and visual aids should be available to each team member.

Grouping of similar or identical equipment: In facilities where similar or identical systems are present, and one system has already been reviewed in entirety, the team may elect to consider similar or identical systems 'by difference', thereby increasing the pace of the review meeting.

Pace of meeting: A simple system can be reviewed within a matter of minutes by an experienced team, whereas a complex system may take two or three hours. It is important that the team does not set too ambitious progress goals, so that the review meeting is not rushed. Additionally, the facilitator should recognize the need for frequent breaks in order to help the team stay focused.

2.4 Finished Product

The outcome of the team-based overpressure scenario development meetings will be a finalized set of overpressure scenario development forms (narratives). These forms will be part of the facility's relief systems design basis and, just like finalized PHA worksheets they should not be edited without consensus from the review team.

It should be realized that the finalized overpressure scenario development forms are only part of the relief systems design basis. Calculations are still required to determine required relief rates and flow capacities for each applicable overpressure scenario identified.

2.5 Conclusions

There is increased regulatory focus on pressure relief and flare systems design basis, and many companies are in the process of updating their existing relief systems design documentation.

Those companies that adopt a team-based approach to overpressure scenario development exhibit a proactive commitment to following recognized and generally accepted good engineering practices. In addition, they will have increased confidence in the quality of their relief systems design documentation, and the overpressure protection process will have increased understanding and acceptance from multiple plant disciplines.

While this step in a relief systems study is not identified as RAGAGEP, in current relief systems documentation, it is only matter of time before this process is identified as such.

3. References

[i] "Sizing, Selection, and Installation of Pressure Relief Devices in Refineries" 8th Edition, API Standard 520 Part I (2008).

[ii] "Sizing, Selection, and Installation of Pressure Relief Devices in Refineries" 5th Edition, API Recommended Practice 520 Part II (2003).

[iii] "Pressure-relieving and Depressuring Systems" 5th Edition, ANSI/API Standard 521 (ISO 23241) (2007).

[iv] "Guidelines for Pressure Relief and Effluent Handling", 1st Edition, AIChE/CCPS (1998).

Additional References not cited:

[v] "Guidelines for Hazard Evaluation Procedures", 3rd Edition, AIChE/CCPS (2008).

Process Safety as a Career Foundation

Andrew Stein
Smith & Burgess
5700 NW Central Dr. Ste. 301 Houston, TX 77092
Andrew.Stein@smithburgess.com

Jason White, P.E.
Smith & Burgess
5700 NW Central Dr. Ste. 301Houston, TX 77092
Jason.White@smithburgess.com

Prepared for Presentation at
American Institute of Chemical Engineers
2013 Spring Meeting
9th Global Congress on Process Safety
San Antonio, Texas
April 28 – May 1, 2013

UNPUBLISHED

Process Safety as a Career Foundation

Andrew Stein
Smith & Burgess
5700 NW Central Dr. Houston, TX 77092
Andrew.Stein@smithburgess.com

Jason White, P.E.
Smith & Burgess

Keywords: Process Safety, Young Professionals, Career

Abstract

The foundation of one's profession can have an exponential effect on the lifelong development of their careers. There may be no stronger launching point than the lessons learned as a young engineer specializing in process safety. A young engineer building a foundation on process safety may have the opportunity to work on and visit a magnitude of different processes and environments. One may also see similar processes operated in different manners. However, a vast range of exposure is not the only valuable material in the development of a career. In a relatively short period of time, one specializing in process safety can learn to understand how safety can sometimes become an afterthought for many process engineers. This can ultimately lead to costly or even life-threatening consequences that must be addressed reactively. With a foundation in process safety, one is of the mindset to address safety issues in a proactive manner, and thus ultimately becoming a more efficient engineer.

1. Introduction

Every young engineer leaving school to begin their career has various ideas of what they would like to accomplish. Process safety offers young engineers a tool that can form a platform for multiple career paths. In order to understand the opportunities a young engineer can create for themselves, a glance into how the variety of lessons learned in the first few years of process safety can help create opportunities and influences.

2. Process Safety as a Career Foundation

2.1 Diversified Process Experience

Process safety applies across many different fields of process engineering and gives young engineers the opportunity to diversify their process knowledge.

2.1.1 Process Safety in Refining

A young process safety engineer often has the opportunity to work in oil and gas refineries to develop the necessary safety documentation. A process safety engineer working on refinery documentation will not be limited to one unit/process, but will gain exposure to various refining units/processes.

Even brief experiences with new process units offer new learning opportunities. When evaluating a new process for upset conditions, a process safety engineer must understand how minor changes can affect the system. For example, the exact same refinery could drastically change the process to maximize its diesel production and this could have an effect on the entire analysis. It is differences like this that offer opportunities to grow as an engineer.

2.1.2 Process Safety in Petrochemicals

When a process safety engineer is working with a petrochemical facility, there is an opportunity to learn about the effects specialty chemicals can have on a process unit. Many specialty chemicals create unique hazards that must be understood to correctly understand process upsets. For example, when dealing with a component that may decompose when exposed to heat, it may be insufficient to use a standard external fire overpressure analysis. The actual required protection from overpressure may be controlled by the rate of decomposition.

Each chemical facility will offer learning experiences that are unique because when dealing with a unique process, unique considerations must be made regarding safety.

2.1.3 Safety Engineers Don't Get Caught in the Weeds

Unlike process engineers working to optimize units, engineers in process safety spend much of their time learning and understanding processes. Very rarely would a young engineer in process safety spend more than a few days on a single problem where a unit process engineer may spend multiple weeks working towards optimizing a single heat exchanger. This lack of time spent working on the tiniest of details allows for far greater exposure to the industry and provides a much steeper learning curve for engineers to develop.

2.2 Safety Applies to All Business Levels

An engineer can decide to take a technical career path, a managerial career path, or a combination of both. However, safety is a factor no matter what career path is taken.

2.2.1 Safety as a Plant Engineer

A plant unit engineer will be responsible for maintaining and enhancing their respective unit in order to continue to make products with profit in mind. While doing this, the unit engineer must know that the process is also safe. After a young engineer spends their first few years in process safety, no decision/change would be approved or made without considering process safety. This creates an engineer with an understanding of the safety impacts that can make proactive safety decisions rather reactive recovery decisions.

2.2.2 Safety as an Engineering Manager

Later into an engineer's career, management may be the chosen career path. Although it may take time and various positions, this manager will be faced with many decisions. It is a fact that many of these decisions will be critical safety decisions that could affect a capitol project, a process unit, a small company, a corporation, or an industry.

Decisions that can have that type of effect should always be made with safety in mind. An engineering manager that spent their first few years in process safety could make these decisions with confidence based on the lessons learned early in their career.

2.2.3 Safety as a Corporate Engineer

Engineers that progress into the corporate engineering ranks must also have a sound understanding of the safety impacts changes can have on an organization. Corporate safety directives and initiatives must be well thought out based on both the impacts they will have on a facility's operations and the value added (process safety can create value even if profit is not quantifiable).

2.3 Safety as an Afterthought

Many process engineers' main concerns lie with creating and maintaining more efficient processes in order to make them more profitable. This is key for any industry, as no industry would survive without innovation; however, this thought process may result in engineers having tunnel vision on that goal. This in turn can cause process safety to be an afterthought instead of something at the forefront of the decision making process. When process safety is at the forefront of any decision, it has the potential to prevent costly and life threatening incidents.

Conclusion

There are hundreds of different ways to begin a career as a chemical engineer, process safety being one of them. Beginning a career in process safety offers young engineers an opportunity to build a solid foundation in a diverse way. How an engineer evolves over their career will depend on their knowledge gained and work ethic, but lessons learned by a young engineer in process safety could greatly enhance this career.

Process Safety Management: 21 Years Plus or Minus

--What I Wish I'd Known Then and What We Can't Afford to Forget Now!

Alan C. Brackey
Valero
One Valero Way, San Antonio, TX 78249-1616
alan.brackey@valero.com

Keywords: Process Safety Management, PSM, Elements, Systems, Challenges

Abstract

This paper provides practical thoughts and insights on the essence of Process Safety Management (PSM), distilled from the author's 34 years in the industry, the past 19 of which have been spent solely in PSM practice. It focuses on several key elements, which make up the "heart and soul" of PSM and comments on the importance of and a few of the challenges presented by each of these elements. It offers some thoughts on key focus areas as PSM continues to move forward and briefly touches on a few of the frustrations that can be encountered in PSM practice. A key takeaway relates to the importance of never losing a sense of one's own vulnerability with regard to future PSM events.

Background/Introduction

For the first ten years of my career, starting as a production supervisor in 1979 and later moving into other areas of process and project engineering, I barely knew what a material safety data sheet (MSDS) was and I'm not sure the term process hazards analysis (PHA) had even been invented. However, during that time, especially as we reached the mid to late 1980's, and following the occurrence of several major incidents having significant impact on the chemical, petrochemical and refining industries, the wheels of change were already in motion. Through the early 1990's, culminating with the issuance of the OSHA Process Safety Management (PSM) standard and the EPA Risk Management Program (RMP) rule (i.e., the "PSM regulations"), our industries embarked upon a path that would forever change the way we viewed, understood, and responded to the challenge of chemical process safety, defined in its simplest terms as the notion that highly hazardous (flammable, explosive, toxic, reactive) materials must at all times remain contained in the piping, vessels, equipment, and systems designed, at least in part, for that very purpose.

Shortly after the promulgation of the OSHA PSM standard in 1992, while working at a major chemical manufacturing facility, I was afforded the opportunity to become increasingly involved in a number of significant PSM-related activities at my site as well as on a company-wide basis. During that time, I had the exceptionally good fortune to meet, interact with, and continually learn about this emerging concept of chemical process safety from some of the most capable, knowledgeable, and experienced engineers and managers in the company, many of whom were already involved in a number of industry activities that were, in retrospect, precursors to what we all now refer to as PSM.

As has been recognized with the creation of the Process Safety Management Mentoring (PSMM) program now offered as a part of the American Institute of Chemical Engineers' (AIChE) annual Global Congress on Process Safety, we in the process safety community once again find ourselves in the middle of a significant generational transition. This time, the transition involves a handoff from those of us whose careers have straddled the issuance of the OSHA and EPA PSM regulations, to those making up the next generation of process safety practitioners, many of whom have never known a time when these regulations were not a part of their everyday lives.

The intent of this paper is to briefly outline a few practical thoughts and insights on what I believe to be the essence of PSM. I'll highlight the importance of several key elements, which in my experience, have often been viewed as the "heart and soul" of PSM, and then mention a few of the challenges and opportunities presented by each of these elements. I'll also touch on a few of the frustrations that can be encountered in PSM practice. I'd like to begin by commenting on what I believe to be the single most important takeaway for readers of this paper and for process safety practitioners everywhere.

The Most Important Takeaway

Over the years, a number of significant process safety events have occurred in the chemical, petrochemical, and refining industries. The AIChE Center for Chemical Process Safety (CCPS) publication *"Incidents that Define Process Safety"*[1] is an excellent source of information regarding several of these events. The scope of this paper does not include describing or even commenting on any of these events. Nonetheless, I did want to quickly note a few other incidents which have directly impacted me and/or others with whom I've closely worked over the years:

- A small dust explosion led to an equipment fire in a fiber handling system. No one was injured and the event ended up not being much more than a minor interruption. However, later on, one of the managers was in a meeting banging his shoe on the table saying such a thing would never happen again, when the plant alarm sounded and the same thing had happened again.

- An explosion in a reactor piping system led to a major hydrocarbon release, formation of a large vapor cloud that found an ignition source, ignited, and exploded. This event resulted in multiple fatalities and several serious injuries as well as catastrophic equipment damage, and required a multi-year, nine-figure rebuild.

- A reactor explosion occurred while a unit was down for maintenance. This event also resulted in multiple fatalities and several serious injuries as well as catastrophic damage and significant offsite impact, and also required a multi-year, nine-figure rebuild.

- An explosion and fire in a compressor train during a unit startup resulted in one serious injury as well as significant operational impacts.

- A large hydrocarbon release from a failed piping system found an ignition source, resulting in a major fire, significant domino effects and several serious injuries. The entire plant was down for several months. The unit rebuild took about a year.

Perhaps most importantly - for every one of these events, as bad as they were, each one of them could have been even worse.

So for me, the most important takeaway is, quite simply, that we must never lose sight of the fact that bad things don't always happen to somebody else. Admittedly, if you weren't involved in any of these particular incidents, then as far as you're concerned, they really did happen to somebody else. But from my perspective, they didn't happen to somebody else, they happened to me (or at least to people close enough to me that I felt it could just as easily have been me)!

I recall a comment from Glenn Erwin on the U.S. Chemical Safety and Hazard Investigation Board (CSB) BP Texas City video[2], in effect that "if you think it can't happen to you, you may be setting yourself up to be next". I believe it to be absolutely critical that none of us ever lose this sense of our own vulnerability. One concern that is often expressed in this regard is that in the absence of any recent low-probability high-consequence events, the possibility of any such event occurring sometime in the future may quickly fall off everyone's radar screen.

Heart and Soul

As already mentioned, the early 1990's witnessed the promulgation of the OSHA Process Safety Management (PSM) standard, 29 CFR 1910.119, as well as the EPA Risk Management Program (RMP) rule, 40 CFR Part 68. A total of 14 "working elements" (parts) are included in the OSHA standard. The prevention program requirements of the EPA rule are very similar in nature to those contained in the OSHA standard. Outside the U.S., certain other process safety regulations (e.g., Seveso II in the European Union, COMAH in the United Kingdom) are in place. This paper will focus on key elements contained within the U.S. OSHA and EPA PSM regulations.

Without question, ***all*** the PSM elements are important and must work together to provide the multiple layers of protection that we rely upon. However, over the years, certain elements have often come to be viewed as the "heart and soul" of PSM. Not surprisingly, these same elements continue to present some of our most significant PSM challenges. In the discussion that follows, I have briefly highlighted these key elements, some of the systems required to manage them, and a few of the challenges they continue to present:

- Process Safety Information (PSI) - PSI generally comprises much of the key technical and engineering knowledge relative to process chemicals, technology, and equipment. Given that most facilities already have much of their PSI in place, one of the most resource intensive technical activities still being worked is the development (and maintenance) of relief system design and design basis information. The advent of flare header and flare recovery systems has added an even more complex dimension to the required analyses. Maintaining the accuracy of the analyses in light of process changes can present a daunting task. One thing I wish I had more fully appreciated 20 years ago is the significant effort required to properly manage pressure relief device programs across the board, from defining applicable relieving scenarios and designing the required relief systems all the way to inspecting and maintaining those systems and devices, and then effectively documenting everything along the way! Another significant PSI challenge relates to the development of an appropriate and consistent rationale for defining safe limits for each process and its associated equipment (also refer to the operating limits discussion below).

- Process Hazards Analysis (PHA) - PHAs are intended to provide a comprehensive, structured review of relevant PSI and procedures in order to postulate and recommend means for addressing significant loss of containment scenarios having the potential to impact people, the environment, or in some cases, the business. One of the greatest challenges I believe still being faced by many PHA practitioners is the need to achieve consistency of results from study to study across the facility, as well as across the organization. Over the past 10-15 years, the increased utilization of semi-quantitative risk analysis techniques (e.g., layer of protection (LOPA) and safety integrity level (SIL) analyses) and "risk matrices" has further elevated the need for consistency across the entire PHA spectrum, from scenario and consequence determination, to safeguard identification and risk assessment, to development of appropriate risk reduction or mitigation recommendations and actions.

- Operating Procedures (OP) - Effective operating procedures do not just happen - they are the result of focused efforts on the part of a cross section of highly knowledgeable employees, most importantly those personnel closest to the operating unit itself, as well as technical support staff assigned to that unit. Risk is often greatest during process startup. Clear, effective startup (and shutdown) procedures are essential to risk reduction. As with safe limits under PSI, a significant OP challenge relates to the development of an appropriate and consistent rationale for defining operating limits for each process, including the point(s) at which the process should be transitioned from normal operations to emergency operations. A detailed discussion of operating limits and safe limits is beyond the scope of this paper. One source of additional information on this topic may be found in Section V of the "*OSHA Voluntary Protection Program (VPP) Application Supplement for the Sites Subject to the Process Safety Management (PSM) Standard*"[3].

- Management of Change (MOC) and Pre-Startup Safety Review (PSSR) - Especially for mature and well developed PSM programs, the elements of MOC and PSSR are often viewed as two of the most critical PSM elements. Once the other elements of the program have been established, it falls in large measure to MOC and PSSR to ensure that the program doesn't "drift". For this reason, MOC and PSSR may touch the broadest plant populations of any of the PSM elements. Over the past several years, it has become increasingly evident that a "one size fits all" system may not provide the optimal choice for all MOC/PSSR applications. Certain variations on the standard MOC/PSSR themes (e.g., for safety system bypassing, procedural modifications, hot taps, leak repairs, etc.), so long as all technical and regulatory requirements are satisfied, may enhance the effectiveness and efficiency of these critical change management systems.

- Mechanical Integrity (MI) - MI, closely tied to the concept of mechanical reliability, is universally recognized as one of the most impactful elements of a facility's PSM program. Some might view it to operate in its own separate universe, however in reality its importance to and interrelationships with the other PSM elements are unquestioned. MI touches nearly every piece of process equipment in the facility. One key MI challenge is related to the need to define, document and implement appropriate inspection, testing and preventative maintenance (ITPM) systems for all safety critical equipment (SCE) and instrumentation (SCI). One thing I wish I had more fully appreciated 20 years ago is the significant technical complexity associated with many SCI systems and the key role they play in providing essential layers of protection for many of our units and processes. Another significant MI challenge more generally related to all SCE ITPM systems (e.g., fixed equipment, rotating equipment, pressure relief devices, etc.) is the need to define and develop a functional understanding of recognized and generally accepted good engineering practices (RAGAGEP) that may be applicable to a given situation. Finally, the utilization of effective maintenance management systems (usually computer-based and generally incorporating appropriate deficiency management capability) is absolutely essential to the success of any ITPM program.

- Management Systems/Performance Audits - Any discussion of audits should encompass more than just the regulatory required triennial PSM compliance audits. Over the past 20 years, it has been shown that strong, effective monthly, quarterly, or annual management system reviews (including periodic monitoring of selected performance metrics) will drive continuous PSM program improvement, reduce the number and nature of findings from the 3-year compliance audits and, most importantly, reduce the likelihood of significant process safety incidents. There are any number of ways to conduct these periodic management system reviews, but if your facility or organization is not already doing so, you might begin by focusing on the current status of a few of the key elements and systems noted above, then broaden, deepen and set frequencies for future reviews based on the results of your initial reviews. In addition, the CCPS publication *"Guidelines for Auditing Process Safety Management Systems"*[4] offers a detailed compilation of current guidance related to auditing virtually all system components of a comprehensive process safety management program. The CCPS publication *"Guidelines for Risk Based Process Safety"*[5] also discusses and describes several important aspects of management review and continuous improvement.

Frustrations

Without question, the practice of PSM can sometimes become quite frustrating. I believe there are probably several reasons for this, a few of which may include:

- The OSHA and EPA PSM regulations are performance-based rather than prescriptive. They state *what* needs to be done, but for the most part, leave it up to the users to decide *how* to do it. This leaves room, in many cases, for a great amount of misinterpretation and inconsistency.

- Many of us tend to believe that regulatory compliance should be viewed as a starting point, not the finish line. Unfortunately, as noted above, no two people can ever seem to completely agree on how to define every aspect of PSM compliance. Issues can arise with respect to such topics as what a PHA revalidation should "look like", how operating sites should deal with "pie in the sky" internal company policies and standards, etc.

- As with many technical endeavors, the PSM "devils are in the details", and those details touch most every aspect of the entire operation.

- The desire on the part of some to want to try to "score" everything. Scores based on well-defined objective criteria can drive positive change. On the other hand, scores that are primarily based on ill-defined subjective criteria can become counterproductive. In those cases, the focus can often shift away from true progress, results and outcomes to simply trying to figure out how to game the system to make the scores look good.

- The fact that we can really never prove a negative, in other words prove that something we did (or money we spent) actually caused something bad ***not*** to happen. It's similar to the challenge that the U.S. Department of Homeland Security faces every day. They have to be right every time. The bad guys only have to be right once.

Looking Ahead

I believe there are a few things that we need to stay focused on as we continue to move forward:

- I believe that we should continue to emphasize appropriate risk based approaches to all of our key process safety activities. A detailed discussion on the topic of risk is beyond the scope of this paper. The CCPS publication *"Guidelines for Risk Based Process Safety"*[5] (referenced above) provides a wealth of practical insights and information on this topic. The basic premise is that in the real world, there is no such thing as unlimited resources. Therefore it is incumbent upon each of us to ensure that we remain focused on prioritizing and resolving issues according to technically sound assessment models that appropriately address all aspects of risk to personnel, the community, the environment, and the business.

- I believe that we must work to develop, implement and maintain effective, functional systems that are not any more complicated than they need to be to get the job done. One consistent takeaway for me, in having spent nearly 20 years as a "boots on the ground" PSM practitioner and having participated in numerous PSM audits along the way, has been that, in most cases, simple systems work better than complicated ones. PSM is by its very nature a complicated business and the more we can do to simplify the systems that are required to manage it, the more likely they will be to produce the desired outcomes.

- I believe that we must never allow ourselves to be satisfied with anything less than the full and proper execution of these systems and activities. There are no perfect systems. There are holes (gaps) in every one, no matter how good they appear to be or how long we've been using them. You may have heard of what is often referred to as the "Swiss cheese" model. Think of each of our process safety management systems as slices of Swiss cheese overlaid on each other. No matter how good it may be, every single slice has some holes in it. The idea here is that given the number of specific systems (slices) that comprise our overall PSM program, the probability of having any one set of holes line up all the way through the stack should be significantly reduced. We must continue striving to identify and address any weaknesses in any of our systems and thereby minimize the overall likelihood of having several independent systems either fail or be failed simultaneously.

Concluding Thoughts

I believe that over the past 20+ years, the chemical, petrochemical and refining industries have made significant progress toward materially improving their process safety performance. I believe we have a far better grasp on knowing and understanding that which we may not have fully known or understood 20 years ago. But we're not there yet. And the bar will continue to be raised. From time to time over the years, it has been said that PSM should ease up, take a break, that 80% is good enough, etc., etc. Several years ago, one of my more astute colleagues wondered if the 80/20 rule really held in process safety, why not consider air to be an inert gas!

Admittedly we live and work in a world that allows for very little margin of error. It's not easy. It's often tough to prove that something we did really made a difference. Failure, of course, is often immediately obvious. We must never lose a sense of our own vulnerability. I believe that our motivation must come from the satisfaction in knowing that yes, <u>the things we do, do in fact make a difference</u>, and that difference is that lives are saved, livelihoods are preserved, and our licenses to operate are continually renewed.

References

[1] *"Incidents that Define Process Safety"* by John Atherton and Frederic Gil, a joint publication of the AIChE Center for Chemical Process Safety and John Wiley and Sons, Inc. (2008)

[2] *"Anatomy of a Disaster - Explosion at BP Texas City Refinery March 23, 2005"*, U.S. Chemical Safety and Hazard Investigation Board Safety Video (issued March 2008)

[3] *"Voluntary Protection Program (VPP) Application Supplement for the Sites Subject to the Process Safety Management (PSM) Standard*", U.S. Department of Labor, Occupational Safety and Health Administration

[4] *"Guidelines for Auditing Process Safety Management Systems"*, a joint publication of the AIChE Center for Chemical Process Safety and John Wiley and Sons, Inc. (2011)

[5] *"Guidelines for Risk Based Process Safety"*, a joint publication of the AIChE Center for Chemical Process Safety and John Wiley and Sons, Inc. (2007)

Some Lessons Learned When Communicating With the Neighbors

Richard N. Knowles, Ph.D.
The Safety Sage
Richard N. Knowles and Associates, Inc.
6083 Bahia Del Mar Circle
Unit 564
St. Petersburg, FL 33715
RNKnowles@aol.com
www.SafetyExcellenceForBusiness.com

Prepared for Presentation at
American Institute of Chemical Engineers
2013 Spring Meeting
9th Global Congress on Process Safety
San Antonio, Texas
April 28 – May 1, 2013

UNPUBLISHED

Some Lessons Learned When Communicating With the Neighbors

Richard N. Knowles, Ph.D.
The Safety Sage
Richard N. Knowles and Associates, Inc.
6083 Bahia Del Mar Circle
Unit 564
St. Petersburg, FL 33715
RNKnowles@aol.com
www.SafetyExcellenceForBusiness.com

Keywords: Outreach, Clean Air Act, Public Outreach, Worst Case Scenarios, Hazard Communications

Abstract

This paper combines 35 years of direct, plant managerial experience with processes handling highly hazardous materials and energies with a subsequent 17 years of consulting and personal reflection to understand how the leadership roles evolved from one of sleepless nights worrying about the plant into one where sleep came easier knowing that the personnel could be trusted to operate the plant safely at all times. The changes did not occur overnight. Although the change was slow, the steps were deliberate, with trust among everyone in the organization built over time. This paper shares some of the key lessons from my days as one of the plant managers working with the Kanawha Valley community sharing our core value: "ensuring the safety, health and welfare of our employees and of our surrounding communities". Once we recognized what the community wanted, in the Safety Street effort, we successfully shared our sound hazards analyses, the importance of layers of protection and risk reduction efforts with our community.

Introduction:

Process safety is a core technology for those operating a chemical plant using hazardous materials. Sloppy operations with leaks and explosions are totally unacceptable. We have no right to be careless, hurting our people, the neighbors and the environment.

Having been a plant manager in plants using highly hazardous materials, like phosgene, anhydrous ammonia, chlorine, oleum, sulfur trioxide, methyl amines, hydrogen cyanide, etc., I placed high value on process safety management, the engineers who did the analyses, the managers who had to make tough decisions, and the operating people and mechanics who followed the procedures, carried out the process inspections, analyses, maintenance and hazards recommendations. It takes a fully functioning operating team to do the whole process hazards work thoroughly and well and to ensure the safety, health and welfare of everyone inside and outside the plant.

In order for all of this to come together and to work, I had to help to create a culture of openness, honesty, respect, integrity, listening, sharing, supporting each other and doing the things we'd agreed needed to be done. Excellence in process safety management is truly a team effort, so that we can operate with the levels of safety and process integrity that are needed. It is built on a foundation of a culture of total safety excellence.

This is an effort that also reaches out into the communities surrounding our plants. Our neighbors need to have the confidence that we are operating in a way that will protect them from the potential problems we could face. We have to lower the probability and severity of an incident or release so low that it is basically never happens. This is not only and issue of our health and safety, it is also an issue of our freedom to operate and avoid more excessive regulations. These are issues of survival for our industry.

The citizens living in the communities around our plants want to talk with people in authority, learn about what is going on in the plant and establish a relationship of trust together. I spent a number of evenings in my Plant Manager assignments in Niagara Falls, NY and Belle, WV visiting the neighbors, having coffee together and talking about things that were important to us to build a sustainable working relationship with them. Many managers worry that if they talk with the neighbors that they'll tell him/her how to run the plant. I never found anyone that wanted to run the plant. What I did find was the neighbors were vitally interested in what we were doing to protect them and the environment.

Over the years I learned the three top things that get the neighbors really upset.

- Do something that hurts their children, like the methyl isocyanate release in Middleport, NY a few weeks before the disaster in Bhopal, India in 1984. I know some of those people. Both the company and the regulators treated them with indifference. It was a frightening, maddening experience. Many became instant, formidable advisories against the Chemical industry. If it had been my kids, I would have been with them.
- Do something that threatens their ability to have healthy children. This was a big factor in the Love Canal mess where someone did a sloppy study, later proven wrong, that there was chromosome damage in the adults. I saw the impact that this had on the community, and it was devastating.
- Have a release of something that puts spots on their cars and campers. This really upsets a lot of people and you have to paint lot of cars.

Fortunately I didn't do the first two, but I have had to paint a lot of cars and talk with angry neighbors. I did this a few times before it occurred to me that I was getting beaten up and I didn't cause the release. So I insisted that the supervisors who ran the process that had the release go out and talk to the neighbors. Their performance got a whole lot better very quickly.

Our neighbors don't want to run our plants. But, they are well aware that we can hurt them and want to know what we are doing to prevent incidents and releases.

It is important to remember the historical climate the Country was in during the 1970's, 1980's and 1990's. The Clean Air Act became effective December 17, 1963 and was amended in 1970, 1977, and 1990. This last part became effective 1999 and required the sharing of Worst Case Scenarios with our communities. Earth Day was April 22, 1970. The EPA was formed in December 2, 1970. The Resource Conservation and Recovery Act (RCRA) passed congress December 13, 1976. Love Canal came onto the scene in July 1978. Then we had Three Mile Island on March 28, 1979. Those of us in the industry during this period felt these huge changes and worried deeply about our freedom to operate. As a plant manager in a highly responsible, big chemical company, I felt that we needed to build bridges with our communities to survive in the long run. I received huge support from DuPont in this effort.

The Challenge:

As an example of the importance of excellent process safety management and sharing this with our neighbors, I want to relate the story of our work in the Kanawha Valley in West Virginia where, in early 1992. All the chemical plants were asked by the Emergency Response and Community Advisory Groups from Kanawha and Putnam Counties to share our worst-case scenarios. Even though this was before the Clean Air Act requirements, community groups could ask for this information under RCRA.

There were 13 chemical plants from 8 different companies in the greater Charleston, WV area. There were 300,000 people who lived in the Valley under the brow of the mountains, so any release would be held in the Valley and not dispersed quickly. All the managers were deeply concerned about the request, and were struggling with how to respond it. Several wanted to just ignore it. Others felt we should try to develop a credible response, but the pathway was not clear. None of us wanted to be shut down. We were the biggest employers in the area with high paying jobs, but we could not hide behind this as had happened in years past. One of the reporters in the Charleston Gazette, Ken Ward, was aggressively reporting on the industry and criticizing us just about everyday. The whole attitude of the community had shifted away over the years from support for the industry to one of fear and distrust.

I was the plant manager for the DuPont Plant in Belle, WV from April 1987 to February 1995. We handled 14 different chemicals that were toxic enough and in large enough quantities that a release could have a devastating impact on both the neighbors and ourselves. For example, we had a 110-feet tall, 100-feet in diameter tank with a capacity for 40,000,000 pounds of anhydrous ammonia. Everyone in the community saw this everyday. The other Kanawha Valley plants also had chemicals of concern to the neighbors. For example, Union Carbide had large quantities of methyl isocyanate stored within several hundred feet of the neighbors, and after Bhopal, everyone was on edge.

I knew that we had to find a way to respond to this request. I had learned a lot when I was the DuPont plant manager in Niagara Falls, NY. There, I had managed our way through a public relations crisis over the DuPont dump, Necco Park, in 1983 where I'd found that being open with our employees, the neighbors and the media, sharing information about the hazards and our remediation efforts and helping everyone see how important it was for all of us to be together in a safer community really helped. Managing a chemical plant with a dump problem like this within 3 miles of Love Canal was an educational experience! We did so well in healing the mistrust and hurt from Love Canal that we were given a permit to build a 660-ton/day chlorine/caustic plant in downtown Niagara Falls in 1985. I knew that being proactive, open, honest and showing the community that we cared was extremely important for our long-term future to stay business in the Kanawha Valley.

As managers we all struggled with the request and our companies also struggled with this first-of-its-kind request. What we did in the Kanawha Valley would impact the entire US chemical industry. We were all members of the Chemical Manufacturers Association (now the American Chemical Council). I was on the CAER Task group and became the communications link between the industry, the CMA and the local plants. From my previous experience as manager of the DuPont Plant in Niagara Falls, NY, I had learned how to work in the community as we managed our own dumpsite not far from Love Canal.

As the managers were struggling with this challenge, I was also actively engaged with several local environmental groups, building on my Niagara Falls experiences to build bridges with the neighbors. During one of these conversations, Mrs. Mildred Holt, one of the leaders of Citizens Concerned About Methyl Isocyanate, looked at me kindly and said, "I know you guys can kill me! I want to know what you are doing to prevent it." She opened up my eyes and gave us the key. All the managers, including me, were very aware of the large quantities of toxic materials we handled and didn't know how to talk to the community about them with scaring everyone to death. But because we were all doing a lot on process safety and building layers of protection into our systems, we could all talk about the good things we were doing in process hazards management and control; we were all doing a lot. This was one of our key lessons, and it came because of the relationship I'd developed with Mrs. Holt and the environmental community.

Building on her insights, we formed two committees. One committee to develop the technical details of our plans, and to work with EPA and industry experts to develop the worst-case scenario definitions and maps, and another committee to develop the plans for our communications processes. Both of the committees were chaired by community people because we knew that if the industry people led these, that there would be a credibility problem. Art Burk of DuPont, a highly skilled process safety expert, coordinated the technical committee's work with the industry and the CMA, while I helped lead and coordinate the communications work. Our committees were composed of industry people, neighbors, environmental activists and occasionally EPA people who were trying to figure out how to do this worst-case scenario planning and communications along with us in the Valley. Based on my Niagara Falls' experiences, I knew that we had to be as open and honest as possible and do this in a way that would keep things in perspective. All of us in the industry were doing really good process hazards work and we didn't want to sensationalize things and scare the neighbors.

Our committees met from 1992 to June of 1994 where our work culminated in Safety Street, our 2-day event to share everything with the community. There were many unknowns for us in this work.

For example the Technical Committee struggled with issues like;

- No one had done this before so how do we do it? The EPA didn't know how to do it either.
- Who should be involved in the Committee? Plant process safety experts? Local company people and/or from higher up in the companies? First responders? Regulators from WV and the EPA? Environmentalists? Where do we draw the line between who is in and who is out?
- Who should be the Chair? Paul Hill from the National Institute for Chemical Studies in Charleston, WV was chosen for this.
- What chemicals should be included? What quantity and toxicity should determine the choices?
- How big should the zones of impact be for possible deaths and then decide on even larger zones for injuries?
- How should we model the exposure levels?
- What wind speeds did we need to use to calculate the dispersion diagrams?
- For the ammonia scenario, how do we calculate the dispersion and rate since 40,000,000 pounds would take a long time to evaporate because of its high heat of vaporization? If it spilled into the Kanawha River flowing next to the tank would that make a difference?
- What would happen to the ammonia tank if an airplane from the little airport up the river ran into it? We concluded that the double-walled tank with each tank built of 1-inch thick steel with each weld fully x-rayed was so strong that a small plane would just cause a little dent.
- If there was a hydrogen explosion of one of the Horton Spheres, how large should the impact zone be modeled? When the windows blow out or the walls collapse?
- How do we co-ordinate what we were doing in the Kanawha Valley with the rest of the chemical industry. We concluded that working with the CAER (Community Awareness and Emergency Response) Task group would be best.

Some examples the Communications Committee struggled with were;

- Who should be on the committee? Plant people? Neighbors? EPA people? State regulators? Environmentalists?
- Who should be the Chair? M. Mary Frances Bleidt, a strong and dedicated community leader was asked to serve.
- How do we communicate something as complex and technical as this to the community?
- How do we keep things in perspective and not mislead the community people? We decided that each company would do their own technical analyses internally following the Committee's guidelines. We'd share the work among all the committee people except the worst-case scenario maps that we would share together at Safety Street, our name for the event.
- How do we keep the process of developing all this in perspective for the community?
- When should we plan to share all this information?

- When should we plan for the event? Where?
- Do we use a big auditorium or hall?

The questions went on and on taking us 2 years of 2-3 meetings a month to work it all out. We agreed that the scenario presentations by each plant would follow the same format in a four-page brochure. This way it would be the most easily understood as people heard about them.

- The name of the chemical
- The quantity involved.
- The toxicity of the chemical.
- The plant's 5-year release history of that chemical; those that went beyond the fence line as well as the others.
- Then our worst case scenario plume maps

As we progressed, we kept everything open for all to see except for the calculations for the worst-case scenarios themselves. These we wanted to share internally first and try to keep things in context. All during the communications process, I had 9th grade students from the local junior high school involved to see the whole process. After all, this work was to build a better future for them, and I'd found that the adults behaved better if the students were watching. There was a lot more openness and honesty among us when the students were there.

Safety Street began on June 3rd, 1994. There was no dress rehearsal. We were going to share the scenarios and hope for the best. We had an all-day meeting in the Charleston Convention Center. As I sat alone in the front row waiting for my turn to present the ammonia scenario I was really nervous. When the students who had been involved all through the communications process came and sat with me it was a huge boost. The Convention Center auditorium was filled with about 700 people. About a third came from all over the country from the industry to see what would happen. About a third were community people and the remaining third came from the plants. The morning presentation included the background presented by the local Kanawha-Putnam emergency response people, an EPA person who supported what we'd done, the hospitals who shared their emergency plans and finally two scenarios. I shared the first scenario on the DuPont ammonia tank since it was biggest plume and could injure people 30 miles away; it is a huge footprint. While the ammonia plume was the biggest, impacting the most people, I pointed out that this was a hypothetical case and the in the 75 year history of our plant we had never had a release that had impacted the neighbors. This was very important to keep things in perspective. After Rhone-Poulenc shared their methylisocyanate scenario everyone adjourned to the big Convention Center hall where each plant had their own displays set up and the plant managers and their key staff people and engineers were there to talk with people and answer questions during the afternoon. We used this open approach so that everyone who wanted to could talk directly with the plant managers and no one could dominate and take over the communications processes.

The second day, Safety Street took place in the 3-story Town Center Mall, where each plant had their displays set up in various places throughout the Mall so all the shoppers could talk with us. During the day, the plant managers probably talked with 7,000-8,000 people on a personal basis, getting our stories to them as accurately as we could. We had good news articles in the NY Times, the Charleston Gazette, on CNN and in other media about a community that had come through tough times and was on the mend. The environmental people told the community that the plants were trying to get better. The hospitals and first responders were there as well with their strong stories about their preparedness and response capabilities.

Holding the second day in the big shopping mall was the idea of Ms. Mary Frances Bleidt who led the Communications Committee. She felt that we had to be as open as possible and take our stories to where the public gathered. She called me about 3 weeks before the event and told me that the community people would not go to the Convention Center on a nice June day and that we had to move all our displays to the Mall. I had to sit down for a bit since this was really putting us out there. I knew that she was right. At first the managers hesitated, talked with their wives and agreed that we needed to go to the Mall. We were there from about 9AM to 9 PM. This was another great learning and we were given this guidance by the community.

In Safety Street, we told a community of about 300,000 people 29 ways we could kill them and trust went up. The rest of the chemical industry began to model what we'd successfully accomplished. Building on the communications work in the Kanawha Valley, the next big effort was in the Houston Ship Channel Area where a similar process worked well.

The Lessons Learned:

We learned that being open and honest with everyone in our plants and the community and co-creating the whole Safety Street process with them was the best thing we could have done. Because of Mrs. Holt's help, we learned a key lesson: we learned that the details of our process safety management efforts, especially our process hazards analyses, were of interest to the public and our engineers learned how to talk about their work in a way they could be understood. Staying involved in the process of this sharing effort for two years helped build credibility and trust among all of the participants, the larger Kanawha Valley community and the chemical industry.

On a personal level, my plant leadership role had evolved from one of sleepless nights worrying about the plant into one where sleep came easier knowing that the personnel could be trusted to operate the plant safely at all times, and we had the support of the community. We confirmed with everyone, inside and outside of the plants, how critical the work of the process hazards engineers and plant personnel really is, and that everyone has a role in making things as safe as they can be.

Process Safety Management Mentoring: Developing Leaders

Clyde W. Young
John M. Campbell & Co.
1215 Crossroads Blvd. Suite 230 Norman, OK 73072
clyde.young@jmcampbell.com

John Kanengieter
John M. Campbell & Co.
1215 Crossroads Blvd. Suite 230 Norman, OK 73072
john.kanengieter@jmcampbell.com

Prepared for Presentation at
American Institute of Chemical Engineers
2013 Spring Meeting
9th Global Congress on Process Safety
San Antonio, Texas
April 28 – May 1, 2013

UNPUBLISHED

Process Safety Management Mentoring: Developing Leaders

Clyde W. Young
John M. Campbell & Co.
1215 Crossroads Blvd. Suite 230 Norman, OK 73072
clyde.young@jmcampbell.com

John Kanengieter
John M. Campbell & Co.
1215 Crossroads Blvd. Suite 230 Norman, OK 73072
john.kanengieter@jmcampbell.com

Keywords: mentoring, leadership, management, competency, culture, commitment,

Abstract

The first pillar of Risk Based Process Safety Management is "Commitment to Process Safety." A formalized mentoring system can ensure workforce involvement, compliance with company and regulatory requirements, increase the competency of personnel and enhance the process safety culture of the entire organization. Within this element there are several essential features that lead to a more effective process safety culture.

Providing strong leadership is critical for any organization that strives to manage the risk associated with the activities associated with process safety. Leadership is a skill that is not necessarily intuitive to managers and mentors. Leadership is a skill that can be learned. We will specifically look at the roles, skills, and signature styles required to lead effective teams.

Another essential feature for an effective culture is to defer to expertise. This paper will describe how a proven model of leadership development can provide the tools needed for leaders/mentors in the oil, gas, chemical and allied industries to prepare the next generation of process safety professionals to be effective. This model has been successful in creating functional teams that work across traditional departmental lines and outside of a strict hierarchical structure.

An effective process safety culture requires leadership throughout the organization. Our industry will benefit from deferring to expertise from outside the traditional business environment.

Commitment to Process Safety

In 2012, the 8th Global Congress on Process Safety introduced a new symposium, Process Safety Management Mentoring (PSMM), which is focused on sharing career experiences to help assure continuity of the profession. Hopefully, this sharing will allow all organizations to enhance their process safety culture and effectively manage the activities in their processes. The role of Mentor involves teaching, helping, protecting, challenging, motivating, guiding, coaching, listening, and providing career guidance.

A mentor/mentee relationship should be fostered through a formalized program with specific roles, goals and responsibilities. A formalized mentoring program can ensure workforce involvement, compliance with company and regulatory requirements, increase the competency of personnel and enhance the process safety culture of the entire organization.

Over the last several years, significant resources have been devoted to examining the issue of process safety culture, and strong leadership has been cited as a key element to enhance a process safety culture. Study of major accidents within the oil, gas, chemical and allied industries have found that the safety culture of organizations is often proposed as the problem, and development of a culture of process safety as the solution. Presentations at symposia similar to PSMM point to enhancing culture and providing leadership as necessary to address breakdowns in process safety management systems.

The first pillar of the Center for Process Safety (CCPS) *Guidelines for Risk Based Process Safety* (RBPS) is "Commit to Process Safety." Supporting this pillar is the element "Process Safety Culture", which is defined as, " the combination of group values and behaviors that determine the manner in which process safety is managed." One of the four essential features of process safety culture is "strong leadership."

Leadership

What is "leadership"? It has been described as "organizing or influencing a group to achieve a common goal". This would intimate that the leader is a boss or manager, but is a manager necessarily an effective leader? There is considerable literature about leadership. This literature includes quotes about leadership, how to find "natural" leaders and how to develop leadership skills. There are workshops about leadership and even university degrees in leadership. If there are so many resources dedicated toward understanding and teaching leadership, why is leadership listed as something that needs to be enhanced in symposia, papers and reports that deal with managing process safety in high hazard activities? It may be because leadership and culture are considered human factors. When associated with process safety, they are known as factors that can lead to loss of the standards of consistently reliable human performance. These standards are relied on as part of an organization's defenses against process safety incidents.

Every person working in the oil, gas, chemical and allied industries should perform their jobs under the guidance of a process safety management system. CCPS defines a management system as a "formally established and documented set of activities designed to produce specific results in a consistent manner on a sustainable basis." Producing specific results in a consistent manner all the time requires that all personnel perform at a high level. If culture is defined simply as "the way we do things around here", this is influenced greatly by leadership. But leadership doesn't reside in the role of one person. Leadership needs to be imbedded within the organization with every person. This is a skill that can be learned by all and dependence on one individual with authority or one person who might be considered a "natural" leader can lead to failure of the system.

The National Outdoor Leadership School (NOLS) has, over the last 46 years, become the leader in wilderness education. Founded in 1965 by legendary mountaineer Paul Petzoldt, NOLS takes students of all ages on remote wilderness expeditions and teaches them technical outdoor skills, leadership, and environmental ethics. What NOLS teaches cannot be learned in a traditional classroom or on a city street. It takes practice to learn outdoor skills and time to develop leadership.

NOLS has shown that leadership is a skill that can be learned and practiced. This is done through the evolution of judgment, personal responsibility, and awareness of group needs, key leadership traits through practical experience and timely feedback. NOLS expeditions are conducted with the recognition and management of risk as critical to both the development of leadership and to the safety and health of students and staff. The curriculum and leadership model stress that successful risk management stems from good judgment based on experience, training and knowledge, much like risk management in the hydrocarbon processing industry.

NOLS has worked with NASA for many years, building teams for the International Space Station and space shuttle missions. Other organizations that have embraced the NOLS model for leadership include Google, , the U.S. Naval Academy, Interagency Wildland Firefighters, and some of the top business schools in the US.

Using this leadership model, functional teams have been developed that work across traditional departmental lines and outside of a strict hierarchical structure. This model may be foreign to high hazard industries. But if the CCPS guidelines for RBPS are to be followed, it is necessary to defer to expertise. Expertise from outside the industry may be the best way to develop leaders in a process safety culture.

This leadership model recognizes four leadership roles and seven essential leadership skills to be practiced within each role. The roles and skills are listed here and will be discussed in further detail to explore how each fits into a process safety management system and the activities associated with the PSM system.

Leadership Roles

Designated leadership: A designated leader will take responsibility for the team and determine how the team will achieve the stated goals. Working in teams has become commonplace within the industry. In most cases, a person is assigned as team leader or supervisor. In other words, somebody may have to make a decision and this role will fall to the designated leader. Within a PSM system, the elements for management of change (MOC), process hazards analysis (PHA), operating procedures, emergency response, compliance audits, operational readiness, contractor management and incident investigation have steps for authorization. It is the designated leader who will take this role.

Active followership: The active followership role supports and follows the designated leader. Active followership includes participating in group decisions and providing input and seeking clarity to ensure the team is always progressing effectively. Active followers in a PSM system may act as MOC review, PHA, incident investigation and emergency response team members. An active follower should always keep the team goals primary and provide wise counsel without being disagreeable. Every member of the team that is not the designated leader fills the active follower role. It is the responsibility of each team member.

Peer leadership: Peer leadership means that all members of the team work together and provide support for each other to meet the goals of the entire team. Each team member will see what needs to be done and do it. Peer leadership should be found in elements dealing with training, competency, stakeholder outreach and workforce involvement. Mentoring is at the heart of peer leadership. There is an old adage in our business, "I taught you everything you know, but I didn't teach you everything I know." Personnel exhibiting true peer leadership characteristics will take every opportunity to teach and mentor others for the good of the organization.

Self leadership: Self leadership is exhibited when each person takes care of themselves so that they can help take care of the team and every person shows personal initiative and character. "That is not my job," is not acceptable. Ask an employee who is responsible for their safety and most will reply that they are. Working within an effective process safety culture requires that each person should be aware of competency gaps and their contributions to the safety of the process, and show willingness to pick up the slack when necessary.

Leadership Styles

No matter which role a leader may have within the management system, each person will exhibit a specific leadership style. The one signature style is the leadership style that works best for each individual. This style is based on tapping into the leader's personality, learning preferences, passions, strengths and experiences. Within an organization, there will be several different leadership styles exhibited by team members. As team members learn the signature style of each other, communication is more effective and decisions will be more consultative. Consultative decisions generally lead to higher quality decisions and this is very important in process safety.

In general, leadership styles fall within one of four quadrants and individuals may shift within these quadrants, depending on the situation. The four quadrants are described as:

Architects and analysts are information and opinion seekers, good at analysis and process observations, prefer to make decisions based on facts, prefer as much information as possible and can come out with "out of the box" solutions that work. People with this style provide an essential function within a team when they provide process checks and alternate solutions. Taken too far, Analysts/Architects can get lost in the process, always seeking more information and decision-making can get bogged down in "analysis paralysis".

Drivers provide information and opinions, make decisions easily, take feedback well and drive for a decision. Drivers also make decisions without input from others, move too quickly and make mistakes, seem impersonal and can initiate conflict with this style. People with this style are necessary to ensure that the team will make a decision and strive for far reaching goals.

Relationship-Masters work well in teams, are excellent at building and sustaining the team; build support, consensus, and commitment and seek feedback. They display high regard for others wishes, viewpoints and actions. They may not push back and can put too much emphasis on relationships that the team falls behind. It may be necessary to have them outline expectations and ask them what they think and want when they are designated leaders.

Spontaneous motivators voice their ideas passionately and are great at motivating others. They possess a sense of mission or vision and are good at solving problems. Sometimes these motivators can be challenged by objectivity and can create high emotions within a team through challenging and confronting. Spontaneous motivators can keep a team moving forward and focused on the goals.

Teams perform many of the activities associated with process safety management. Process hazard analyses are conducted by teams with a designated leader and active followers. A team of subject matter experts should validate operating procedures. Incident investigations are conducted by teams. Emergencies are managed through an incident command team. Compliance audits, operational readiness reviews and coordination of the overall process safety management system require teams.

By understanding the leadership roles and the signature leadership styles of the team, a designated leader or supervisor should make higher quality decision and get buy in so that all personnel in the organization understand and are motivated to follow a plan. The most effective leaders apply one of these effective leadership styles based on the given set of circumstances. This is commonly known as situational leadership. The result will be fewer process safety incidents and more efficient operations.

Leadership Skills

There are seven skills that are critical to effective leadership. No matter which role a person might adopt, these skills must be exhibited at high levels. A short description of each skill is provided and an example of how they all come together for the success of a team will be explored.

Positive Group Behavior: High functioning teams will exhibit behaviors that are conducive to success of the team. Members of groups do not have to be friends, but they all certainly have to work together if the group is to succeed. Success in the realm of process safety is identification and management of hazards associated with the process.

Competence: It is not critical that all leaders have equal levels of competency. It is critical that leaders are aware of their competency and especially aware of their competency gaps. A competency gap may require that a designated leader transition to an active follower. Deference to expertise is an essential feature of managing process safety challenges.

Communication: There is considerable literature available about effective communication and how to enhance it. Effective communication is critical for making sound decisions about management of change, analyzing risk and conduct of operations. No matter which leadership role a person takes, communication among all team members reduces confusion and increases competency.

Judgment & Decision-Making: At some point, someone will have to make a decision about the correct course of action. It has been said that good judgment comes from bad decisions. Effective leaders learn from experience, not only their own but others', and use this experience to increase their judgment and decision making abilities. Analyzing the risk associated with high hazard industries is usually a judgment call and is enhanced by having accurate process safety and situational information available.

Tolerance For Adversity & Uncertainty: The major purpose for process safety management is to prevent catastrophic incidents. Adversity and uncertainty materialize when an incident occurs and effective leaders must rise to the occasion. Sometimes, processes deviate from normal and personnel must take corrective action to return to normal. It is almost impossible to predict when an emergency may evolve. It is at these times that a leader must tolerate the adversity and uncertainty that comes with these scenarios and exhibit the skills to assume the appropriate leadership role.

Self-Awareness: Effective leaders must know their limitations and abilities. To practice any leadership role requires that a person understand their competency level, refuse to take a role they are not qualified for and honestly communicate with others in the group. It is difficult for individuals to admit inability to fill a role. However, the consequences of ineffective leadership are too high and the likelihood of failure too high to not admit this when/if it occurs.

Vision & Action: The processes in this industry are constantly changing. Fluid flows, pipes erode, temperatures change, and it is important that these changes are anticipated and managed. Effective leaders are not content to assume that everything will stay the same and that the initial risk level will stay the same. Effective leaders visualize what could happen and take actions to manage the risk. Management is the process of dealing with or controlling things or people. This takes vision to see what is changing and action to control.

Developing Process Safety Leaders

An expedition to reach the summit of a mountain, like NOLS conducts, has risk associated with it. Apply the four pillars of RBPS to a mountain expedition. There needs to be commitment by expedition members to the goal of reaching the summit. The risks of the environment need to be understood and analyzed. The identified risks need to be managed with a system that provides resources and establishes procedures. Finally, the expedition members must learn from their own and others' experiences in order to succeed.

A process safety management system is a "formally established and documented set of activities designed to produce specific results in a consistent manner on a sustainable basis." Most organizations have a formalized system that describes how certain activities related to process safety are to be carried out. Very few organizations have a formalized and documented set of activities for developing leaders or even describing leadership within the organization. Most address this by assigning responsibility or describing management functions. Management is not leadership. Leadership needs to be developed in a structured manner.

Structured leadership begins with establishing roles, goals and expectations of all group members. Personnel who do not understand the goal of the organization, what role they play to meet the goal and what is expected of them will be frustrated and in many cases do what they think is best without following the structure of the management system. If these actions result in no harm or even lead to some success, then the prevailing philosophy becomes "no harm, no foul." This can easily lead to normalization of deviance and eventually can lead to a catastrophic incident.

Establishing roles, goals and expectations can come from well-developed operating procedures and should always be included in the training element. Hazard assessments, safe work practices, management of change, operational readiness and emergency management elements must consider roles, expectations and be very clear about assignment of leadership roles, especially the designated leadership role.

The workforce involvement element of RBPS should clearly establish the structure of leadership within the organization. Many times there is only a simple statement that all members of the workforce will be involved in process safety. A more rigorous, structured program for leadership development and assignment of roles, goals and expectations will enhance the organization's process safety culture.

Experience has shown that the most effective process safety leaders are those who have a thorough knowledge of the management system and the interrelationship among all the elements. Failure has come from forgetting or ignoring that a small change in a process or organization has a ripple effect throughout the management system. Use of simple checklists and other tools is helpful. However, one of the most significant things an organization can do to enhance process safety leadership is to stress that active followers play a key role in insuring that all issues are addressed in a process change and that the organization maintains a sense of vulnerability at all times.

Benefits of Structured Leadership

Mentoring has become an important part of workforce development over the last few years. As the workforce in the industry ages, less experienced personnel are taking leadership roles that have traditionally been filled by older, more experienced individuals. As these more experienced people leave, there are less of them available to act as mentors. Filling this gap will not be easy. It is time for the industry to seek expertise where available and begin developing a structured leadership program that will provide leaders for the future.

Leadership competency of all personnel in the oil, gas, chemical and related industries should be developed, with appropriate competency development activities provided. Many benefits will result from this.

With structured leadership, people understand the goals and expectations of working within a PSM system. Under stress or crisis, whether it is a process emergency or a downturn in the economics of the business, resources and efforts will not be spent on trying to figure out what to do. Structured leadership allows for quick response because the human dynamic issues have been frontloaded, thus taking these off the table as a possible stressor in the moment.

The goal of any process safety management system is to understand the hazards and risk associated with the process and reduce it to as low as reasonably practicable. Effective leaders create safe and effective teams. More effective teams make better decisions. Better decisions lead to more efficient operations. The result is a learning/generative organization that is safer, more knowledgeable, dependable, and has strong bottom-line results.

Summary

Process Safety Management Mentoring (PSMM) is dedicated to sharing career experiences to help assure continuity of the profession. This goes to the heart of competency development for the profession. Significant resources are expended to develop the technical expertise for process safety professionals.

Experience has shown that many catastrophic incidents in the industry have an origin with management system failure. Many incidents are chalked up to human factors failure. Much effort is being expended in the industry to find a way to enhance the process safety culture. More effort needs to be focused on understanding that good managers are not necessarily effective leaders and that all personnel within the organization have a leadership role that will affect the culture. Leaders need to be developed rather than expecting them to naturally appear. A clear leadership development plan that includes an understanding of leadership roles, skills, and styles will equip new mentors with fundamental knowledge that sets the stage for organizational effectiveness.

E.I. DuPont de Nemours & Co., Inc. Flammable Vapor Explosion

Mark Wingard
U.S. Chemical Safety and Hazard Investigation Board
2175 K St NW
Suite 400
Washington D.C. 20037
mark.wingard@csb.gov

Prepared for Presentation at
American Institute of Chemical Engineers
2013 Spring Meeting
9th Global Congress on Process Safety
San Antonio, Texas
April 28 – May 1, 2013

UNPUBLISHED

E.I. DuPont de Nemours & Co., Inc. Flammable Vapor Explosion

Mark Wingard
U.S. Chemical Safety and Hazard Investigation Board
2175 K St NW
Suite 400
Washington D.C. 20037
mark.wingard@csb.gov

Keywords: hot work, explosion, incident, process safety

Abstract

This case study examines a November 9, 2010, explosion at an E.I. duPont de Nemours and Co. Inc., Yerkes chemical plant in Buffalo, New York which killed a contract welder and injured a foreman who were repairing the agitator support atop an atmospheric storage tank containing flammable vinyl fluoride. The U.S. Chemical Safety Board (CSB) determined that flammable vinyl fluoride (VF) vapor from interconnected, in-service process tanks flowed undetected into the tank and ignited when the welder was repairing the agitator support assembly.

1 Incident Summary

This case study examines a November 9, 2010, explosion at an E.I. duPont de Nemours and Co. Inc., Yerkes chemical plant in Buffalo, New York when a contract welder and foreman were repairing the agitator support atop an atmospheric storage tank containing flammable vinyl fluoride. The welder died instantly from blunt force trauma, and the foreman received first-degree burns and minor injuries. The explosion blew most of the top off the tank. The top and agitator assembly hung over the side of the tank supported only by a 2-foot section of the top (cover photo). The explosion caused minor overpressure damage in the tank farm area and the adjacent production building.

The U.S. Chemical Safety Board (CSB) determined that flammable vinyl fluoride (VF) vapor from interconnected, in-service process tanks flowed undetected into the tank and ignited when the welder was repairing the agitator support assembly. In February 2010, the CSB issued a "Hot Work Safety Bulletin" that summarized 11 similar fatal incidents. Like the incidents described in the bulletin, this was another example of improperly monitored hot work activities involving flammable conditions inside a container.

2 Background

2.1 *DuPont*

E.I. duPont de Nemours and Co., Inc (DuPont) is a Fortune 100 company which was founded in Wilmington, Delaware in 1802. The American chemical company operates in about 90 countries and employs more than 60,000 people. The company offers a broad range of products for industry and consumer use, including pesticides, electronics, apparel, and biomedical supplies. The company offers services as a safety resource for other corporations to evaluate and improve workplace safety.

DuPont has owned and operated the 100 acre facility in Buffalo, New York since 1921. The facility employs over 600 workers and manufactures Tedlar®, a polymer used as a film in photovoltaic panels and Corian®, used to make countertops and other durable surface products.

2.2 Maintenance Contractor

Mollenberg-Betz, founded in 1910, operates a large steel fabrication facility in Buffalo, New York, and provides piping system design, fabrication, field construction and maintenance services. It employs more than 75 professionals and skilled construction trade staff. Mollenberg-Betz had contractors working on site at the DuPont facility repairing a steam line. DuPont used Mollenberg-Betz to repair the heavily corroded steel on the agitator support structures atop tanks 1 and 2 due to their availability on site. At the time of the incident, the Mollenberg-Betz contractors had been working at the DuPont facility for two months. The welder and foreman assigned to the work had many years of piping system and tank welding experience.

2.3 Process Description

The Tedlar process converts VF into polyvinyl fluoride (PVF).VF is generated at another DuPont facility, shipped in on tanker trucks and stored at the Buffalo facility in high pressure storage tanks. VF is pumped from the storage tanks to a reactor and reacts to form PVF slurry in water and unreacted VF. After the reactor, the PVF water slurry passes through separators to remove the VF. The VF is pumped from the separators by compressors and recycles back to the reactor. After the separators, steam is injected into the PVF slurry to raise the temperature and vaporize any VF present in the slurry. The heated mixture passes through a small slurry flash tank where the residual VF is released to the atmosphere[1].

The non-combustible PVF slurry then passes to one of three insulated slurry holding tanks (numbered 1, 2, or 3) in the tank farm adjacent to the production building (Figure 1). Under normal operating conditions, slurry is pumped to slurry tank 3. Slurry tank 3 is newer than slurry tanks 2 and 1; it had been replaced in 2009 due to corrosion[2]. Slurry tanks 2 and 1 were used as overflow storage space in the event slurry tank 3 was filled and slurry tank 1 was generally kept empty of slurry.

If the slurry flash tank level is too high, hot slurry passes through the flash tank overflow line directly into slurry tank 2. A liquid trap (seal loop) on the end of the overflow line inside slurry tank 2 was designed to prevent VF vapor and steam from passing directly from the flash tank into the slurry tank. The PVF slurry is then pumped from the slurry tanks to the cpro\ production area for further processing. PVF is used in a variety of applications as a film or a surface protector due to its resistance to weathering and flammability-lowering properties.

---

[1] The Buffalo facility normally vents approximately 5 pounds per hour of VF to the atmosphere from the flash tank..

[2] Slurry tank 1 was removed after the 2010 incident.

2.3.1 Physical, Chemical, and Biological Properties of Vinyl Fluoride

VF is a colorless, highly flammable gas at ambient conditions with an ether-like odor.[3] VF vapors are heavier than air and will accumulate in low areas. VF is classified as a reasonably anticipated human carcinogen[4]. National Institute for Occupational Safety and Health recommends an exposure limit of 1 ppm as a time weighted average and a ceiling limit of 5 ppm for workers.

---

[3] The VF lower explosive limit is 2.6 percent and the upper explosive limit is 21.7 percent. The autoignition temperature is 725° °F (385 °C).

[4] According to the National Toxicology Program from the National Institute of Environmental Health Sciences in its 2011, 12th Edition of *Report on Carcinogen*

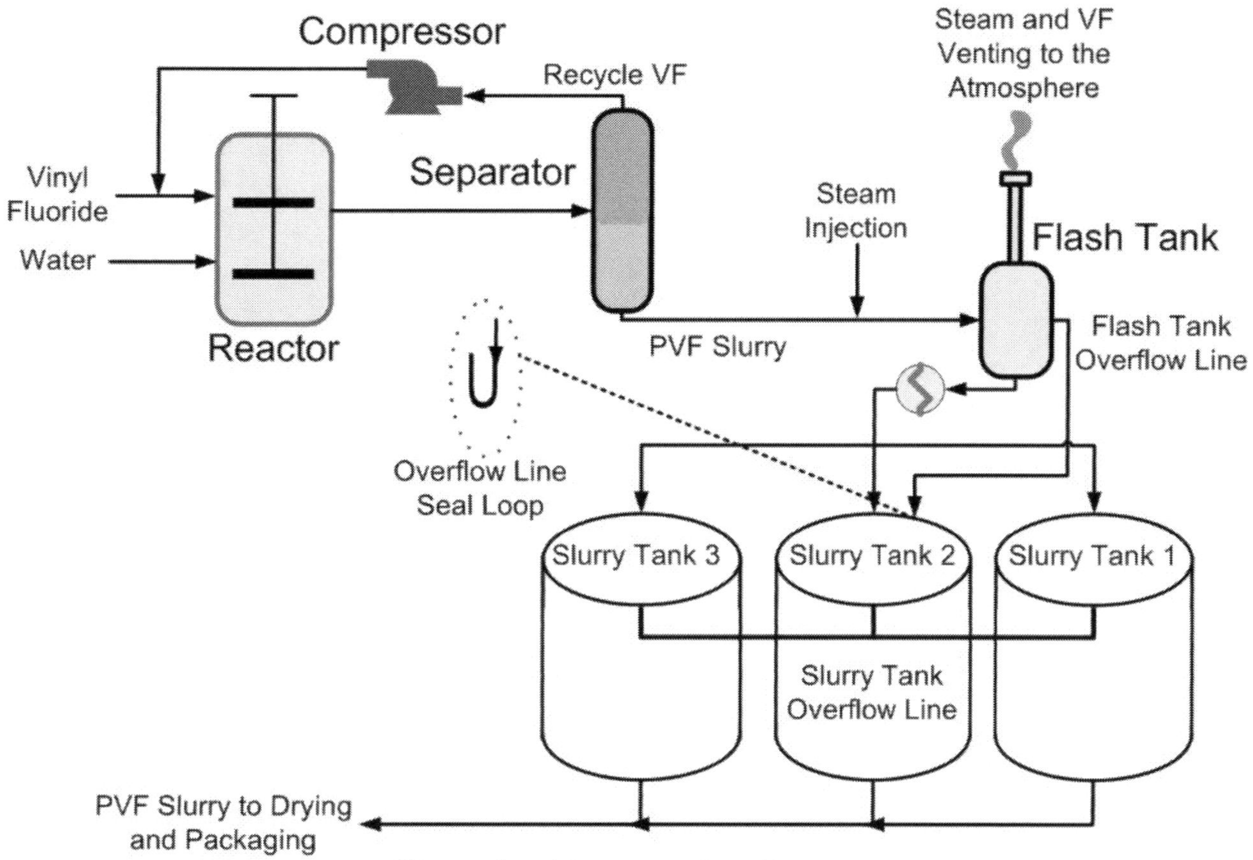

Figure 1. PVF process flow diagram

2.4 PVF Slurry Tank Farm

The flash tank and three slurry tanks are located in a tank farm adjacent to the Tedlar manufacturing building. The slurry tanks are 11 feet in diameter, 19 feet tall, and have a capacity of 10,800 gallons (Figure 2). The tank shell and top are one-quarter inch thick stainless steel. Each tank top has a hinged, unsealed steel cover on a 24-inch diameter manway and a large agitator motor drive and gearbox[5]. The agitator assembly is bolted to structural steel beams that are welded directly to the tank top.

The slurry tank feed and drain lines are equipped with isolation valves. The slurry tanks have a common overflow line attached near the top of the tanks. Blind flanges are installed on the overflow line and isolate one tank from another if one or two tanks are operating while the third is undergoing maintenance.

The 3-foot diameter flash tank vents small concentrations of flammable vapor directly into the ambient atmosphere, so the area is equipped with a flammable vapor air monitoring devices to notify personnel when explosion hazards are present. The continuous air monitors are located on the catwalk near slurry

[5] The PVF slurry tank farm process area was classified as a general purpose electrical classification area at the time of the incident at ground level, on top of the tanks, and on the tank farm elevated catwalk.

tank 3 and on the flash tank vent pipe (Figure 3). An alarm in the Tedlar control room activates if either device detects flammable vapor above the instrument set point.[6]

[6] The low alarm set point is 10 percent of the lower explosive limit (LEL) and the high alarm set point is 2 percent of the LEL.

Figure 2. PVF slurry tanks

Figure 3. Slurry flash tank (left). Agitator and open manway on tank. Second air monitor is located on catwalk near tank 3 (lower right).

3 Incident Description

3.1 Events prior to incident

The Tedlar process area conducted a planned shutdown from October 22 to November 6, 2010, ending 3 days before the fatal explosion. During this shutdown, the asbestos insulation on slurry tanks 1 and 2 was to be removed. New, non-asbestos insulation was to be installed and the slurry tanks would be externally inspected for corrosion similar to that previously found on slurry tank 3. The process was shut down on the evening of October 21 and all slurry was pumped out of slurry tanks 2 and 1.

On October 22, the slurry tanks were locked out[7] by DuPont maintenance personnel so the tanks could be cleaned and entered. On October 29, DuPont discovered damaged agitator support after the insulation was removed from the slurry tanks 1 and 2. A DuPont engineer wrote up a work order for the repair of the agitator supports and had the construction field engineer in charge of contractors begin writing a work

[7] OSHA defines lock out as "The placement of a lockout device on an energy isolating device, in accordance with an established procedure, ensuring that the energy isolating device and the equipment being controlled cannot be operated until the lockout device is removed."

scope and gather contractors and materials. DuPont hired Mollenberg-Betz to repair the agitator support based on the fact that they were already engaged in hot work on site and could be quickly made available.

The valves on the tank 1 fill and discharge lines were locked out November 1. Following the completion of DuPont site lockout and hot work safety procedures, the contractor performed the grinding and welding repairs atop tank 2 on November 1 (Figure 4), but had to delay the tank 1 work because repair materials were unavailable. DuPont personnel determined the slurry tank 1 job could be safely completed after the process restarted and scheduled the repairs for November 9, when the Mollenberg-Betz contractors would again be available.

During the tank 2 internal inspection on November 3, DuPont engineers discovered that the U-leg seal loop on the flash tank overflow line had a "fishmouth" split in the pipe (Figure 5), likely the result of the PVF water slurry freezing in the line. The engineers concluded that the slurry tank could be returned to service without repairing the split. DuPont engineers who saw the broken seal loop concluded its likely purpose was to limit steam in the flash tank from flowing into the slurry tanks. They further concluded that the operation was safe and that the broken seal loop presented no hazards. No acknowledgement was given to the fact that flammable VF gas had a pathway into the slurry tanks, an area classified as "low hazard" and not expected to contain any flammable materials. DuPont engineers scheduled a repair on this seal loop during the next planned unit outage. No management of change (Section 4.7.2) was carried out for the process to continue with this compromised seal loop.

A DuPont crew reconnected the tank 2 and 3 process piping and removed the valve locks to prepare for unit restart. Tank 1 remained out of service while awaiting the materials needed to repair the agitator support. The overflow line to all three tanks was never blinded[8], which provided a direct flow path from the vapor space in tanks 2 and 3 to the out-of-service tank 1.

[8] Blinding is the absolute closure of a pipe by fastening of a solid plate that completely isolates the interior of the pipe and is designed to withstand the maximum pressure generated within the pipe.

Figure 4. Repair weld on agitator support beam to the slurry tank 2 top plate

On November 6, DuPont restarted the Tedlar process with the valves aligned so that the PVF slurry flowed into slurry tank 3. The overflow line remained connected to all three tanks, and was not isolated or disconnected from tank 1 before the hot work was authorized. On November 8, a compressor within the Tedlar unit malfunctioned. Although unit operation without the ring compressor was listed in the Tedlar® operating procedures, and was acceptable as part of the site environmental emissions, restarting without the compressor, significantly increased the VF vapor present in the PVF slurry flowing into the slurry flash tank. The process operated in this mode until the incident.

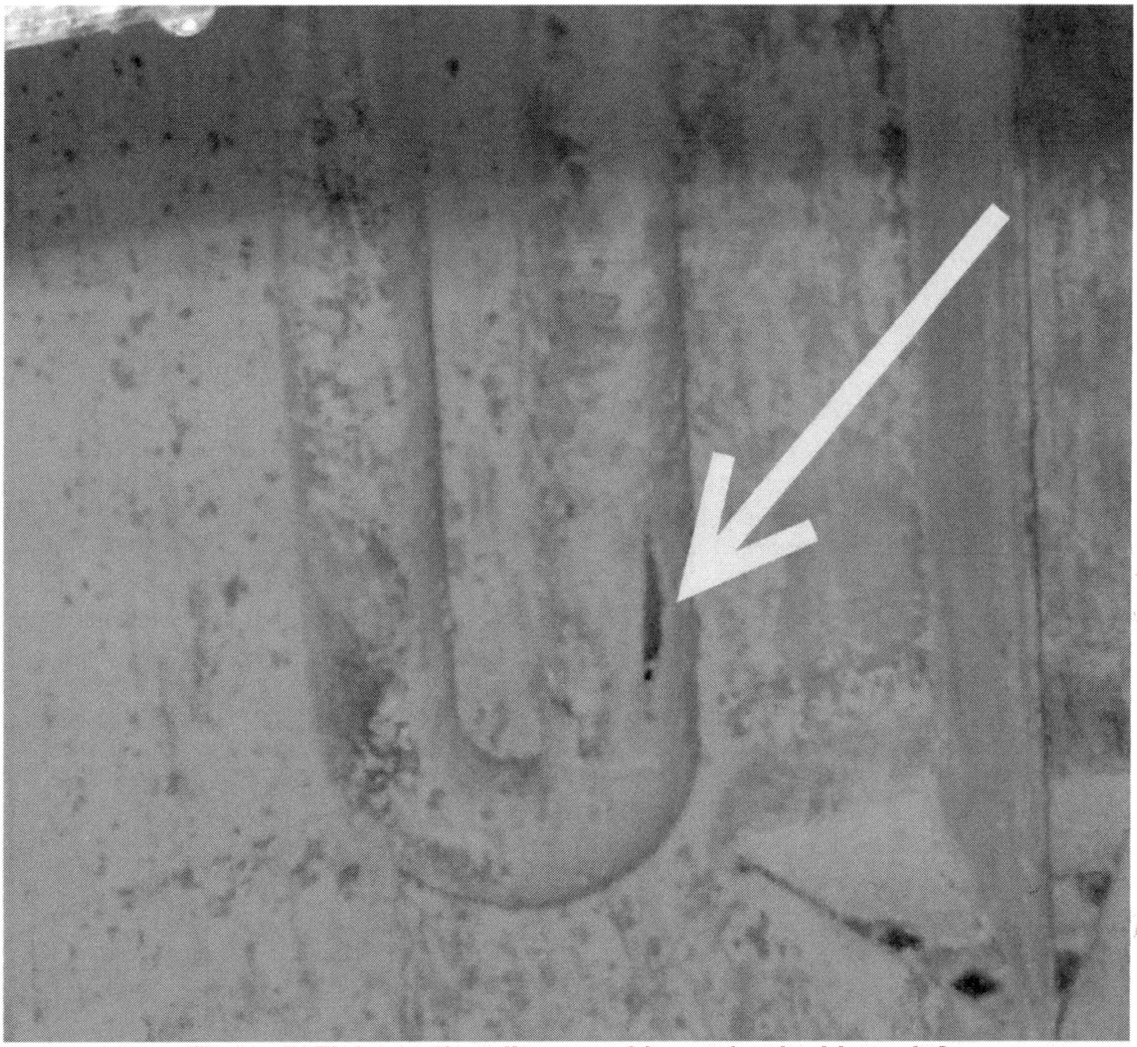

Figure 5. Fishmouth split on seal loop pipe inside tank 2

3.2 Permits in Place on Day of Incident

On November 7, 2010, a lockout card[9] completed for tank 1 work indicated that that all five valves leading to and from tank 1 and the agitator motor had been locked out. The card instructed workers to try to start the motor before beginning work, but did not reference the overflow line, which had no valves. On November 9, 2010, the contractors completed a hot work permit before starting work on slurry tank 1. The contractors did not check the lock out valves on slurry tank 1 with any DuPont employees prior to starting the hot work.

[9] A lockout card details what has been done to ensure an area has been effectively locked out for an area or a piece of equipment.

3.3 Incident Description

The morning of November 9, a DuPont lab technician tested the area around the top of the slurry tanks for flammable vapor concentration, including the region above tanks 2 and 3 and the immediate area above tank 1 where the repair work would be performed. The testing indicated there were no flammables present around the top of slurry tanks. Continuous air monitoring was also present near the top of tank 1, but the lab technician never tested the atmosphere inside the tank. The lab technician was not required to test the air inside tank 1 for flammable chemicals even though the repair work involved significant grinding and welding directly on the tank top. Nor did the DuPont construction field engineer or contractor crew ask the lab technician to test the air inside the tank.

Shortly after 9 am, the contractor crew went to the catwalk above tank 1 and started work. The contractor attached his safety harness lanyard to the agitator on tank 1, stepped out onto the tank top, and began the grinding and welding repair work. The foreman remained on the catwalk adjacent to the closed manway cover to supervise and act as fire watch. At the time of the incident the contractor was using an electric arc welder to weld metal support on to a C-channel beam on the top of slurry tank 1. The C-Channel beam was then to be welded to the agitator support.

At 11:04 am the tank 1 level transmitter in the control room recorded a sudden increase as a fire erupted inside the tank. The overpressure blew the steel manway cover off the tank, hurling it more than 100 feet. Nearly the entire 25-feet-long tank top-to-shell circumferential weld tore apart; only a segment about 24 inches long remained connected. This segment acted like a hinge as the top, agitator motor and drive assembly, and agitator shaft violently swung over the side of the tank (see cover photo). The welder died instantly in the explosion. The flash fire burned the foreman's arms and head, burst an eardrum, and scratched his eye. The fire quickly consumed the flammable vapor and self extinguished.

4 Incident Analysis

4.1 Fuel Source

The fuel source for the hot work explosion was flammable vinyl fluoride in slurry tank 1. DuPont technical engineers in process hazard analyses carried out before the incident assumed that any residual VF gas would vent safely to the atmosphere more than 10 feet above the work area through the flash tank vent. However, the defective seal loop provided a direct path for the flammable vapor and uncondensed steam to flow from the flash tank into tank 2. VF was also present in small amounts in the PVF slurry entering slurry tank 3. DuPont modeling published in July 2010 and based on data obtained from slurry samples on the Tedlar process calculated that 0.02 pounds per hour of VF would be present in the slurry going to the slurry tank. Prior to the incident, other DuPont models had stopped after the separators and made the assumption that any VF present would vent from the flash tank. DuPont did not see this as a hazard because this 0.02 pounds per hour is below the LEL for VF. DuPont did not consider that VF could accumulate in the large vapor space in the slurry tanks and reach flammable levels.

Prior to the incident DuPont calculated that the loss of the liquid ring compressor would more than double the amount of VF going to the flash tank and venting. The VF present in the flash tank had an open path to slurry tank 2 through the cracked seal loop. VF from the flash tank was the main source of flammables on the day of the incident.

The VF present in the slurry in tank 3 and from the defective seal loop in tank 2 had an unimpeded path to slurry tank 1 through the unblinded overflow line. VF, which is heavier than air accumulated to a flammable concentration in the bottom of the tank. As the steam and VF entered slurry tank 1, the steam condensed, due in part to the earlier removal of insulation from the slurry tank. The slurry tank insulation had been removed during the shutdown and not replaced and the condensing steam exacerbated the VF flow into slurry tank 1 at increased concentrations.

DuPont personnel sampled the air around the work area for flammable gas as required by the company hot work permit procedure; however, neither they nor the contractor crew recognized the possibility of VF vapor entering the tank. The sampling procedure did not require testing the atmosphere inside the tank even though the work required welding directly to the tank top. If air monitoring had been done inside slurry tank 1 prior to and during hot work, the flammable VF would have been detected and the incident would not have occurred.

4.2 Ignition Source

The ignition source came from the repair work activity of welding the agitator support on the tank top. The agitator shaft passed through an unsealed hole, approximately one-half inch, in the tank top, which provided a path for sparks to enter the tank or for flammable vapor to escape into the work area.

Welding the C-channel on the tank top using the arc welding technique increased the metal temperature above the steel melting point of 2800 °F (1538 °C). The surface of the metal would quickly get "red hot" as welding progressed, a temperature in excess of 1200 °F (650 °C), which was significantly above the VF vapor autoignition temperature of 725 °F (385 °C). The accumulated VF vapor inside the tank would ignite as soon as it contacted the extremely hot steel or if welding sparks entered the tank through the agitator shaft opening.

4.3 Hot Work Procedure

Hot work is any flame or spark producing operation including welding, grinding, and riveting. The DuPont corporate *Contractor Safety Handbook* states that a "Welding, Open Flame, and Sparking Equipment Permit" is required for flame or spark-producing activity at DuPont facilities. The two contractors who performed the hot work on slurry tank 1 completed such a hot work permit, titled the "Yerkes Work & Flame Permit," prior to starting work. The purpose of the permit was to ensure that, prior to any hot work, communication between the supervisor of the hot work and a proprietor of the area where the work was to be performed was appropriate, and to ensure the hot work was carried out safely. The permit indicated the scope of the work and listed needed safety precautions.

DuPont operating procedure "Yerkes Work & Flame Permit" provided guidance on completing the hot work permit of the same name at the Buffalo plant. The permit was required prior to the initiation of any hot work. This document was designed primarily for contractor use. As part of the hot work permitting procedure, contractors were required to complete a safety task assignment (STA) to address potential hazards in the hot work area. All employees doing the hot work were required to read this STA and sign that they had read it.

After the contractor completed the paperwork, the area proprietor[10] would sign the permit and notify the contractor of any process changes or safety information that might affect the job. This required the area proprietor to be knowledgeable about processes going on in the area. This procedure was in accordance with the guidelines set forth in the National Fire Protection standard NFPA 51B which states the individual authorizing the permit will, (Section 4.2.3)"ensure the protection of combustibles from ignition", and (Section 4.1.7) "advise all contractors about site-specific flammable materials, hazardous processes or conditions, or other potential fire hazards." The contractors had completed this same permit before doing hot work on slurry tank 2 the previous week.

In the hot work procedure, DuPont recognized that the contractor may be unfamiliar with process safety or activities in the area they would be working. DuPont determined that the construction field engineer[11] and the area proprietor would be responsible for helping the contractor understand potential hazardous conditions. On the day of the incident however, neither the construction field engineer nor the employee signing off as the area proprietor had understanding of the area, the process, or potential hazardous conditions.

4.4 DuPont Hot Work Permitting

On the morning of November 9, 2010, the contractor supervisor completed the required hot work permit for the repair work on slurry tank 1. The permit specified that the Mollenberg-Betz contractors would be burning, welding, and grinding work on the top of slurry tank 1. The requirement section of this permit asked if flammable material would be within 35 feet of the work. This section was not completed, even though the hot work was carried out within 35 feet of the slurry flash tank that was designed to vent VF to the atmosphere. The contractors were unfamiliar with the Tedlar process and the process equipment involved. The contractors did not know what the flash tank was or which chemicals were present inside it.

The safety precautions section of the permit included an item for "Lock, Tag, Try, Think" that was not checked as necessary, even though slurry tank 1 had been cleaned and locked out for the hot work. In another part of the permit the contractor supervisor indicated that lockout had been completed for this work.

The contractor supervisor also completed a safety task assignment with the construction field engineer as part of the permit. On this part of the permit the supervisor did not answer the questions "Are chemicals present? What chemicals?" and "Flammables/Combustibles present? Identify."

The "Yerkes Work & Flame Permit" requires signatures from two DuPont employees, an area proprietor and construction engineer, before contractors can begin work. DuPont's intent in structuring hot work procedures, from the corporate to the plant level, was to ensure that these individuals would understand the work and potential hazards.

In addition, the DuPont construction engineer for the slurry tank work had no working knowledge of the Tedlar process. He would not have known the effect of the loss of the compressor on increased VF in the system. On the morning of the incident he reviewed the work permit with the contractor supervisor. Since the contractor had performed hot work on slurry tank 2, the construction engineer expected the contractors

---

[10] An area proprietor is a DuPont employee who is knowledgeable about the area where any work would be done.

[11] The construction field engineer is the DuPont employee responsible for coordinating the contractor's work.

to obtain the proprietor signature from someone in the area where they were working, as required in the DuPont policies, and that the area proprietor would provide the contractors any plant-specific, process safety information necessary for the hot work.

After the construction engineer signed the permit, he obtained the proprietor signature from someone in the service department. The service department is responsible for mail, yard work, and related tasks. It was not located near nor was it related to the Tedlar process. DuPont policy states that the proprietor should be "knowledgeable" about the area they are signing off on and that the proprietor should "walk down" the area where the hot work is to be done. On the day of the incident, this was not done.

The contractors would not have known about any of the chemicals present in the Tedlar area or if they were flammable or combustible; neither did they have knowledge of the Tedlar process nor if flammable vapors might be present around the slurry tanks. Despite this they were allowed to complete the hot work permit and begin hot work without getting approval from any DuPont employee knowledgeable about the process.

Prior to this incident, the DuPont Buffalo facility allowed someone in the service department, rather than someone knowledgeable about the area, to sign off on contractor work permits. The individual in the service department who signed as area proprietor had no knowledge of the Tedlar area or any associated dangers and had been signing permits for the Mollenberg-Betz contractors for months while they did hot work in other parts of the Buffalo facility. Information gathered from CSB interviews indicated the DuPont service worker believed that he was simply releasing them to do their jobs and that the construction engineer had briefed them on the job and hazards. He was not aware that he needed to know about potential hazards or that he was not the appropriate DuPont employee to sign these permits. On the day of the incident the service department worker was still under the assumption that the Mollenberg-Betz contractors were working on the steam line and believed that was the purpose of the hot work permit.

DuPont's policies and procedures are structured to ensure that someone knowledgeable checks before hot work is performed. However, on the day of the incident, no individual with plant-specific knowledge of the Tedlar process reviewed the hot work permit. DuPont's practices had fallen short of its policies..

4.5 Lockout Procedure

DuPont's corporate *Contractor Safety Handbook* requires a lockout procedure to protect workers from injury caused by energized equipment, the opening of valves, or exposure to electrical sources. The DuPont Buffalo facility's *Yerkes Contractor Safety Handbook* states that any equipment locked for safety must be deactivated, tagged, cleared, and tried. Contractors working in a locked out area were responsible for verifying that the lockout procedure had been completed. The handbook also states that isolating piping systems in hazardous processes should include block valves or a blind flange. Piping systems were to be examined after lockout to ensure they were isolated.

DuPont Buffalo's procedure "Lockout Procedure Lock, Tag, Try, and Think" describes the lockout/tagout procedures at the Buffalo facility. This document states that "all sources of hazardous energy shall be removed or controlled prior to potential exposure to the hazards." The list of examples of hazardous energy sources does not include chemicals or fire hazards however the OSHA standard for lockout, Control of Hazardous Energy (29 CFR 1910.147) includes chemicals as energy sources for which a lockout procedure must address (Section 5.0). DuPont designated the area job representative, usually a DuPont field engineer, as responsible for securing required lockouts for contractor work. In addition, the field engineer and the contractor foreman were required to verify that the lockout procedure was done. The lockout card for slurry tank 1 did not include a requirement for blinding the overflow line; that is, tank 1 was considered completely "locked out" with the overflow open to the tank.

4.6 OSHA Citations

As a result of this incident, OSHA investigated and issued ten citations to DuPont, and eight to Mollenberg-Betz. All of the citations referenced either OSHA regulation 29 CFR 1910.147 or 29 CFR 1910.252 which regulate lockout/tagout and hot work, respectively. OSHA found several deficiencies in DuPont's lockout and hot work procedures. DuPont was fined for not including steps to isolate the slurry tank overflow line in their energy control procedure, and for not installing blinds in the overflow line.

OSHA also found that DuPont had not provided the Mollenberg-Betz contractors with specific control procedures for de-energizing slurry tank 1. OSHA fined Mollenberg-Betz because the contractors did not ensure that the lockout procedure had been completed and that the tank was not properly isolated. OSHA fined DuPont and Mollenberg-Betz for not informing their employees about the potential explosion hazards related to hot work on slurry tank 1.

4.7 Process Safety Management (PSM)

4.7.1 Process Hazard Analysis

In its process hazard analysis (PHA) for the Tedlar process, DuPont considered whether VF could ever enter the slurry tanks. In the PHA process description, DuPont stated that its separators would recover "all the VF in the slurry" and that the steam injected before the slurry flash tank "allows the remaining VF to be flashed off". DuPont told the CSB that "no tests were conducted prior to the incident to determine whether vinyl fluoride ("VF") was entering the slurry tanks".

The PHA also considered what would happen if the liquid ring compressor failed. DuPont reached the conclusion that "additional VF will leave through the flash tank vent". No consideration was given to VF making it past the flash tank to the slurry tanks.

Despite the assumption in this PHA that VF was not present in the process after the flash tank during normal operating conditions, DuPont evaluated the consequence of "more VF in slurry going to Slurry Tanks" if the steam injection failed. DuPont found that in this theoretical situation "VF left in slurry would be max 300 ppm [parts per million]" and would not pose a safety risk. In a 2009 PHA carried out on the downstream portion of the Tedlar process, DuPont considered the potential of VF in the slurry past the flash tank. DuPont determined that small amounts of VF may be present in the slurry "at a concentration below the LEL" (lower explosive limit). No recognition was given that VF below the LEL in a slurry tank with a large vapor space could accumulate until it reached explosive levels.

4.7.2 Management of Change

Management of change (MOC) is a written procedure for a company to safely regulate any changes in process chemicals, technology, equipment, and procedures. MOC requires considering impact of change on safety and health, modifications to operating procedures, and technical basis for the proposed change. MOC also requires that any employees involved in operating a process and maintenance and contract employees whose job will be affected by the change be informed of, and trained in, the change prior to start-up of the process or affected part of the process. MOC is mandated by OSHA's PSM program.

No MOC documentation or authorization was carried out for the damaged overflow line from the flash tank or for the failed liquid ring compressor. DuPont did not require a management of change if adverse safety impacts were not anticipated. DuPont engineers determined that any residual VF vapor would vent safely to the atmosphere from the flash tank above the slurry tanks. DuPont did not consider the possibility that flammable vapor could flow into, and accumulate in, the slurry tanks.

4.7.3 Mechanical Integrity

The OSHA PSM standard requires that the mechanical integrity of critical process equipment is maintained to ensure it is designed and installed correctly and operates properly. PSM mechanical integrity includes storage tanks, piping systems, and vent systems. The mechanical integrity section of PSM states:

> The employer shall correct deficiencies in equipment that are outside acceptable limits… before further use or in a safe and timely manner when necessary means are taken to assure safe operations

After DuPont personnel noticed the cracked seal loop, DuPont technical engineers decided not to repair the loop until the next turn around. If DuPont had included the slurry tanks and slurry flash tank overflow line in their PSM program they would have been required to fix the cracked seal loop before restarting the unit or ensure that the cracked seal loop posed no safety hazard.

4.7.4 DuPont PSM Classifications

The DuPont Tedlar process unit uses flammable VF in quantities exceeding 10,000 pounds; thus, the unit is covered by OSHA Process Safety Management (PSM) standard 29 CFR 1910.119. The PSM standard includes additional requirements to assess and reduce hazards associated with highly hazardous and flammable materials. DuPont included the VF storage vessels, reactors, separators, and flash tank in the PSM program but excluded the outdoor PVF slurry tank farm from the requirements of the PSM standard because DuPont concluded VF carryover into the flash tank and product storage tanks was unlikely and insignificant. DuPont further concluded that VF emissions from the flash tank would not accumulate in the tank farm area.

Prior to the incident, DuPont classified the slurry tank as a lower hazard operations (LHO) area, which is an area containing "any activity that exclusively manufactures, handles, stores, or uses any substances with low potential for death or irreversible human health effects." Post-incident, DuPont increased the slurry tank area rating to a higher hazard process (HHP), or an area containing "any activity manufacturing, handling, storing, or using hazardous substances that, when they are released, ignited, or intentionally combined, can result in death or irreversible human health effects." DuPont changed this because it modified the HHP definition to include any area that could contain more than 1 percent VF by weight or 0.65 percent by volume. DuPont did not conclude that OSHA PSM coverage extended to the slurry tanks.

The OSHA PSM standard does not apply to "Flammable liquids stored in atmospheric tanks or transferred which are kept below their normal boiling point without benefit of chilling or refrigeration." Although the slurry tanks were atmospheric vessels, they contained a flammable gas which was above its normal boiling point (-98°F, -72°C). The slurry tanks also served to agitate the slurry mixture, making them pieces of process equipment and not solely storage tanks.

In 1997, an OSHA memo about the chemical company Akzo Nobel provided guidance and serves as a letter of interpretation on how a facility should set the boundaries of PSM for a process:

> Aspects of the process which contain an HHC [Highly Hazardous Chemical] would be covered by all PSM elements, such as information, process hazard analysis and mechanical integrity. Aspects which do not contain HHC, but are interconnected or located nearby are part of the process. Such aspects may or may not be covered by the PSM standard based on whether the particular aspects could cause a HHC release or interfere with mitigating the consequences if there was [sic] a HHC release.

According to OSHA definitions, VF is an HHC. This incident clearly demonstrated that VF can enter the slurry tanks; additionally, the tanks are interconnected to the process that uses VF and are physically close to the flash tanks that remove VF. Despite DuPont's failure to correctly include the slurry tanks in the Tedlar PSM boundary limits, OSHA did not cite DuPont for failing to apply PSM to the slurry tanks.

4.8 DuPont Belle Incident

On September 20, 2011, the CSB released an investigation report[12] which detailed three incidents which occurred at a DuPont facility in Belle, West Virginia on January 22 and 23, 2010. The last of these incidents resulted in one fatality and one employee exposure to phosgene, a toxic gas.

The DuPont Belle and DuPont Buffalo incidents had a number of similarities. In the Belle report, the CSB found that DuPont practices at the Belle facility had deviated from their own policies and standards. It was determined that DuPont did not follow its own standards for the change-out of phosgene transfer hoses. This similarity was seen at the Buffalo facility where DuPont practice deviated from the hot work permitting process. The Belle report discussed DuPont Belle's PHA process and how some process hazards were not addressed. Similarly, the DuPont Buffalo PHA did not adequately consider the threat the potential threat that VF in the slurry tanks posed.

4.9 DuPont Corporate Audit

In November 2006, DuPont corporate performed a PSM audit of the DuPont Buffalo facility. DuPont awarded DuPont Buffalo a score of 99% on this audit, the highest score the auditor team had ever awarded a facility. The audit commended the Buffalo facility on their operating procedures and safe work practices and made no recommendations to improve these programs. No recommendations were made for the Buffalo facility's MOC procedure for subtle changes, stating that "all systems comply".

Another DuPont audit was carried out in October 2009. This document commended DuPont Buffalo's PHA process. It stated that the PHA's were "very well managed and executed" and are of "consistently

---

[12] Report available at www.csb.gov

high quality". These latest audits carried out by DuPont at the DuPont Buffalo facility missed many deficiencies that became apparent as a result of the November 9, 2010, incident.

5 Occupational Safety and Health Administration Standards

5.1 Lockout/Tagout

OSHA standards do not explicitly require testing the atmosphere inside a tank before performing hot work on the outside of a tank.

The OSHA Control of Hazardous Energy (lockout/tagout) standard (29 CFR 1910.147) requires positive control of hazardous energy including:

> Appropriate lockout devices or tagout devices to energy isolating devices, and to otherwise disable machines or equipment to prevent unexpected energization, start up or release of stored energy in order to prevent injury to employees.

This standard does not specifically address isolating tanks from hazards associated with flammable, combustible, or toxic material flow through pipes, but many elements of the lockout/tagout standard directly apply to assuring that tanks are properly isolated for safe confined space entry and hot work activities.

The lockout standard defines energy source as

> Any source of electrical, mechanical, hydraulic, pneumatic, chemical, thermal, or other energy.

5.2 Hot Work

The OSHA standard "Welding, Cutting and Brazing, General Requirements" (29 CFR 1910.252) includes requirements directly applicable to tank repair work. Specifically, §1910.252(a)(2)(vi)(C) prohibits cutting or welding

> in the presence of explosive atmospheres (mixtures of flammable gases, vapors, liquids, or dusts with air), or explosive atmospheres that may develop inside uncleaned or improperly prepared tanks or equipment which have previously contained such materials, or that may develop in areas with an accumulation of combustible dusts.

Furthermore, §1910.252(a)(3)(i) requires that "[a]ny pipe lines or connections to the drum or vessel…be disconnected or blanked" before welding or cutting. The overflow line, which provided the direct path for the flammable vapor to enter tank 1, was left unblinded and was not included on the lock out card for slurry tank 1.

6 Industry Standards

6.1 National Fire Protection Association

The National Fire Protection Association (NFPA) publishes consensus standards and guides about fire prevention and related subjects. Many of the standards are incorporated by reference in OSHA and industry consensus standards.

The standard "Fire Prevention During Welding, Cutting, and Other Hot Work" (NFPA 51B) provides a guidance for contractors and property managers who manage, supervise, and perform hot work. The standard is not intended to contain all necessary safety precautions and work practices involved in job-specific work, such as hot work on atmospheric storage tanks. Job-specific hot work prerequisites and practices are contained in other NFPA standards.

NFPA 51B appendices address hot work on atmospheric tanks by providing general guidance for a hot work permit program. The sample "Hot Work Permit" in Appendix A of NFPA 51B contains a checklist that includes two precautions directly applicable to the tank repair work:

- Enclosed equipment cleaned of all combustibles
- Containers purged of all flammable liquids/vapors

Appendix B summarizes more than 25 hot work fire and explosion incidents to "illustrate how such incidents occur and to emphasize the provisions of this standard." Three summarized incidents are similar to the DuPont incident in that each involved welding and other hot work on tanks and attached piping without adequate testing or other safety precautions.

Chapter 4 of the standard, "Safeguarding of Tanks and Containers for Entry, Cleaning, or Repair" (NFPA 326) contains basic precautions for performing hot work on storage tanks: "Extreme caution shall be used when work is performed on a tank or container that contains vapors related to the substances that are stored or were previously stored therein." Paragraph 4.1.6 adds, "Work on tanks or containers shall be permitted only after the characteristics…of the atmosphere within the tank or container have been determined." Isolating equipment prior to hot work is addressed in NFPA 326, paragraph 5.3.2: "The vents for the tank or container on which work is to be performed shall be isolated from the vents of other tanks or containers that might still be in service."

NFPA 326 6.1.1 states:

> To determine that an atmosphere is safe for the designated entry, cleaning, or repair work, tests for oxygen and for flammable, combustible, or other hazardous substance vapors, fumes, or dusts shall be made with an appropriate instrument as follows.
>
> (1) Before entry or re-entry
>
> (2) Before start of alterations or repairs
>
> (3) Before and during any hot work, cutting, welding, or heating operations
>
> (4) Continuously or periodically during the course of the work to be as determined by a qualified person

(5) After cleaning the interior of each tank or container to determine that the cleaning procedures have been effective

(6) After any process or activity has occurred that might change the atmosphere within the tank or container

The standard specifies that testing is needed before the start of alterations or repairs and before and during any hot work, cutting, welding, or heating operations. The OSHA hot work standard [29 CFR 1910.252] does not require air testing prior to the initiation of hot work.

6.2 American Welding Society

The American Welding Society (AWS) publishes consensus standards for welding, brazing, and thermal-cutting practices. The standard "Safety in Welding, Cutting, and Allied Processes" (ANSI Z49.1)[13] has the same purpose and scope as the NFPA 51B standard. It provides technique-specific safety criteria, detailed guidance on welder personal protective equipment such as arc-flash eye shield selection criteria, and welding fume hazard awareness. Explanatory information item E6.4, "Welding or Cutting Containers," specifically includes tanks and other large vessels in the definition of a container and warns

> All containers should be considered unsafe for welding or cutting unless they have been rendered safe…Even a water tank should be considered hazardous unless a qualified person has declared it safe to weld or cut.

"Recommended Safe Practices for the Preparation for Welding and Cutting of Containers and Piping" (AWS F4.1) discusses precautions for welding and thermal cutting on containers. The standard warns that "the container should be tested for hazardous gases, fumes, and vapors periodically to ensure that the container and work area are safe during welding." The standard is not intended for welding on "containers and confined spaces that can be entered by workers" but the DuPont tank repair work involved welding work on a tank that contained hazardous vapors; thus, the safety principles in the standard were applicable.

6.3 American Petroleum Institute

The American Petroleum Institute (API) publishes consensus standards for the petroleum industry. API Recommended Practice 2009 "Safe welding and cutting practices in refineries, gas plants, and petrochemical plants" discusses hot work in plants that deal with flammable vapors. API RP 2009 states that to effectively use a combustible gas detector, consider the characteristics of the gas (lighter or heavier than air) as well as the appropriate locations and frequency of test.

The lack of rigor and deviation from accepted practices in the permitting process and incorrect engineering assumptions resulted in VF vapor flowing into tank 1 and accumulating. When DuPont personnel discovered the damaged seal loop on the flash tank overflow line, they incorrectly concluded that it would still prevent VF vapor and steam from entering the slurry tank. In addition, they left tank 1 unblinded and still open to the overflow line when they restarted the Tedlar unit. Finally, DuPont authorized the hot work to begin without determining if the interior of tank 1 contained a hazardous atmosphere.

---

[13]ANSI Z49.1 is available as a free download at www.AWS.org (2011).

7 Other Hot Work Incidents

In February 2010, the CSB issued a hot work safety bulletin, "Seven Key Lessons to Prevent Worker Deaths During Hot Work In and Around Tanks", which discusses 11 incidents involving hot work activities where flammable vapor accumulated and ignited while repair work or process modifications were performed in or near storage tanks. Three of the incidents, which involved welding on a tank that contained flammable materials, closely parallel the incident at DuPont.

In June 2006, three contract workers were killed and one was seriously injured when a welding torch ignited flammable hydrocarbon vapors at a Partridge-Raleigh Oilfield in Raleigh, Mississippi. The contractors had been installing piping between two oil tanks. Like the DuPont incident, all of the tanks were interconnected by piping and the interconnected tanks contained flammable vapors which ignited.

In July 2008, three workers were killed and another seriously injured in an explosion at the Packaging Corporation of America fiberboard manufacturing facility in Tomahawk, Wisconsin. Workers were welding a temporary metal clamp on a damaged flange atop an 80-foot tall atmospheric storage tank. Company hot work procedures did not require sampling the atmosphere inside the tank prior to welding directly to the attached piping. The tank contained recycled water and fiber waste, which the company assumed did not pose a fire or explosion hazard. The CSB concluded that bacteria likely produced highly flammable hydrogen gas inside the tank. Sparks from the welding or the heated metal ignited the hydrogen. Two workers fell to their death and the third died after being hit by debris when the explosion blew off part of the tank top.

In February 2009, a welding contractor was killed while repairing a large crack in the bottom of a water clarifier tank at a ConAgra Foods facility in Boardman, Oregon. Water used to wash potatoes had leaked through the crack and accumulated inside the inaccessible space under the tank. Examination of a sample of the liquid indicated that bacterial decomposition of the organic matter likely produced flammable gas that accumulated inside the tank support skirt. The gas ignited when the contractor started welding the tank bottom.

This case study reiterates the key lessons of the CSB hot work bulletin to minimize fire and explosion hazards when performing hot work on atmospheric tanks, including the need to clearly understand and manage hot work activities, especially in terms of testing for a flammable atmosphere both outside and inside any tank or container where welding or grinding is done directly on the container surface.

Hot work incidents across the country continue to cause death, injury, and property damage with alarming frequency. The CSB is currently conducting a more comprehensive evaluation of hot work regulations and standards and expects to issue related safety recommendations in the near future.

8 Key Findings

1. DuPont PHAs made the incorrect assumption that VF in the Tedlar process could not reach flammable levels in the slurry tanks.

2. DuPont restarted the unit after incorrectly concluding that the defective seal loop did not increase the risk of VF vapor transfer into tank 2. Flammable VF vapor flowed directly from the flash tank into tank 2.
3. DuPont did not properly isolate and lockout tank 1 from in-service tanks 2 and 3 prior to authorizing hot work on tank 1. Consequently, flammable VF vapor passed directly from tank 2 into tank 1 through the overflow line and accumulated to a concentration above the lower explosive limit.
4. The DuPont hot work permit procedure did not require testing the atmosphere inside tank 1 for flammable vapor even though the work required welding directly to the tank top.
5. The individuals who signed off on the hot work permit were not knowledgeable in the operations and hazards of the Tedlar process.
6. The repair work created multiple ignition sources. Welding directly to the tank top increased the metal temperature inside the tank significantly above the VF vapor autoignition temperature. Welding and grinding generated hot sparks, which likely ignited the flammable vapor.
7. OSHA's hot work standard does not specifically require gas monitoring inside containers intended for hot work even though it is recommended by industry safe practice guidelines.[14]

9 Lessons Learned

1. Welding to the outside surface of a tank or container generates heat and sparks near the weld area. The extreme temperatures and sparks will likely ignite flammable materials present inside the equipment.
2. Before starting hot work activities, all process connections on tanks and similar containers should be completely isolated by closing valves, installing blanks, and disconnecting pipes to ensure that all possible and known sources of flammable materials cannot enter the container at any time.
3. The atmosphere inside any container previously containing flammables, regardless of size, should be tested before authorizing any hot work involving grinding, cutting, or welding on the outside surface and the atmosphere must be continuously monitored during the work.

10 Recommendations

[14] NFPA 326 and ANSI Z49.1

E.I. DuPont de Nemours and Co., Inc.

2011-01-I-NY R1

Develop and enforce a corporate quality control policy and procedures to require that all DuPont facilities audit their hot work permitting systems to ensure that:

- All potential explosion hazards associated with hot work activities are identified and mitigated before hot work is started
- All relevant forms required for permits are completed in accordance with corporate policies and industry standards (including NFPA 326 and NFPA 51B) before hot work is started
- Appropriate DuPont personnel officially approve hot work permits, by signature or equivalent, consistent with DuPont policies, before hot work is started

2011-01-I-NY R2 Revise corporate policies and procedures to require all process piping, vent piping, or similar connections to be positively isolated (using closed valves, blind flanges or pancake blanks) before authorizing any hot work.

2011-01-I-NY R3 Revise corporate policies and procedures to require that the atmosphere inside the container be monitored for flammable vapor prior to performing any welding, cutting, or grinding on the container surface.

2011-01-I-NY R4 Revise corporate policies and procedures to require air monitoring for flammable vapor inside the container for the duration of the hot work consistent with industry standards (NFPA 326, NFPA 51B). Create a policy for determining criteria for requiring continuous or periodical testing for the duration of hot work.

Can Fire Suppressants Promote Ignition? A Study of HFC-125 and HFC-227

Carmen H. Osorio[1], E. L. Petersen[2] and M.S. Mannan[1]

Corresponding author: carmenhelena@tamu.edu

[1]Mary Kay O'Connor Process Safety Center,
Artie McFerrin Department of Chemical Engineering,
Texas A&M University,
College Station, Texas 77843-3122, USA

[2]Department of Mechanical Engineering,
Texas A&M University,
College Station, Texas 77843-3123, USA

Prepared for Presentation at
American Institute of Chemical Engineers
2013 Spring Meeting
9th Global Congress on Process Safety
San Antonio, Texas
April 28 – May 1, 2013

UNPUBLISHED

Can Fire Suppressants Promote Ignition? A Study of HFC-125 and HFC-227

C. Osorio[1], E. L. Petersen[2] and M.S. Mannan[1]

[1]Mary Kay O'Connor Process Safety Center, Artie McFerrin Department of Chemical Engineering, Texas A&M University, College Station, Texas, USA
[2]Department of Mechanical Engineering, Texas A&M University, College Station, Texas, USA

Corresponding author: carmenhelena@tamu.edu

Abstract

C_2HF_5 (HFC-125,) and C_3HF_7 (HFC-227ea or FM-200) have gained considerable acceptance in the field of fire protection, especially after the restriction of several ozone-depleting substances under the Montreal Protocol. These two substitutes are currently included in the list of the Significant New Alternatives Policy (SNAP) Program of the U.S Environmental Protection Agency (EPA). Their desired properties such as low Ozone Depleting Potential (ODP), relatively low toxicity, electrical non-conductivity, high dispersion capabilities, and low flammability, make them two of the most-used total flooding fire suppressants at the moment. In terms of Minimum Extinguishing Concentration (MEC), C_3HF_7 and C_2HF_5, although less effective than bromine-containing compounds, are more efficient than other clean alternatives in the market. Nevertheless, the combustion properties of these agents are not well understood. In fact, previous numerical studies have shown that these compounds can act as combustion promoters under certain conditions, but there are scarce experimental data of HFCs that can be used to certify such predictions. This lack of data shows the necessity of having accurate measurements that can be used for model validation but also to better understand the role of different fire suppressants on the combustion initiation stage towards an optimal and safer application. To this end, this paper examines the effect of C_2HF_5 and C_3HF_7 on the high-temperature chemistry of methane and propane flames at different conditions. Ignition delay time measurements were obtained using a shock tube with reaction monitoring capability. Theoretical analysis was conducted using a detailed chemical kinetics mechanism that includes the hydrocarbon and fire suppressants set of reactions. Results indicate that C_2HF_5 and C_3HF_7 can actually act as ignition promoters under certain conditions. These results suggest that these fire suppressants may not be good alternatives to prevent flames, however they can still be used as fire extinguishers after the flame is well established. Finally, sensitivity analysis was carried out to identify the most significant reactions responsible for such behavior, and the results were compared with CF_3Br over a similar range of conditions.

1. Introduction

Despite that halogenated compounds have been shown to inhibit developed flames, some studies have demonstrated that they can also promote combustion at certain conditions [1-6]. Even the well-known CF_3Br (Halon 1301), as well as C_2HF_5 and C_3HF_7 (two of the most common fire suppressants alternatives), can act as both combustion inhibitors and promoters [2, 7-9]. Therefore, identification and understanding of such phenomena became extremely necessary for the safe and optimal application of such compounds. This analysis requires experimental data and validated models that help to predict the effect on different fire suppressants in a variety of systems and conditions. The chemical kinetics mechanism is also used to identify the key reactions, species and pathways, and therefore to better understand their fundamental properties.

Currently, there is an extensive research area that focuses on the HFC mechanism development; however, there are scarce experimental measurements that can be used for its validation. Considering this, the present work remedies this deficiency and focuses on providing accurate ignition delay time measurements for different fire suppressants (C_3HF_7/C_2HF_5/CF_3Br) and Fuels (CH_4/C_3H_8) systems, which can be compared directly with modeled data. The next sections describe the experimental and numerical methodology used, followed by the results, analysis and conclusions.

2. Background

To understand how fire suppressants work, it is important to have some background related to fire theory. Fire is the result of a rapid exothermic reaction between a fuel (i.e. methane) and an oxidizer (i.e. oxygen). For this process to take place, an ignition source should be present, thus the fuel can reach its ignition temperature. For the flame to be sustainable, it is necessary to have a continuous chemical chain reaction, where enough fuel and oxygen produce the required active radicals responsible for the major heat release. This combustion process can be represented through the fire tetrahedron (Fig. 1). Based on this concept, a fire can be extinguished by eliminating one of the four elements present in the process. In this work, we are focusing on understanding how the tested fire suppressants can affect the chemical chain reaction using chemical kinetics mechanisms.

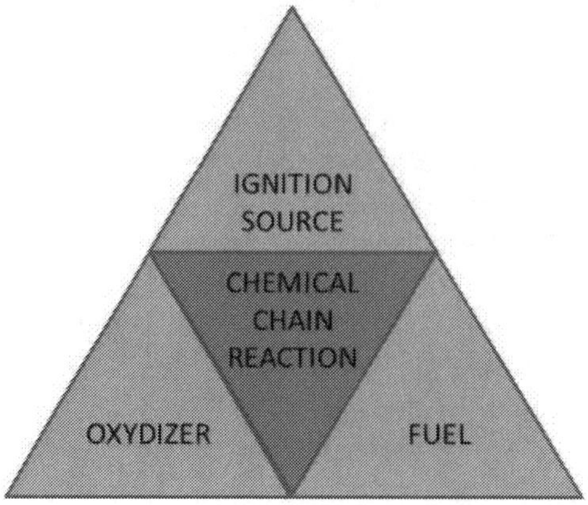

Figure 1. Fire tetrahedron for a flame.

Overall, the chemical process of combustion can be represented by individual steps called elementary reactions. To analyze the different properties of such a system, it is necessary to gather information on all of the elementary reactions that can be involved along with their kinetics parameters (A, n, Ea). This database is known as the chemical kinetic mechanism. Chemical processes are very dependent on highly reactive intermediates called free radicals such as H, OH, and HO_2, among others. In fact, elementary reactions can be categorized in four groups based on the production or consumption of the free radicals (Table 1). This classification can help to understand different phenomena. For example, generation and propagation of the flame can be attributed to branching reactions which are predominant over termination steps. It occurs when a large amount of radicals are produced due to a fast fuel decomposition.

Table 1. Classification of elementary reactions.

Initiation	One free radical is produced from stable molecules in the reactants
Propagation	One radical is consumed, and one radical-molecule is generated
Branching	Two free radicals are produced per single radical in the reactants
Termination	One radical-molecule is consumed to form stable species

3. Methodology

3.1 Modeling

Modeling is conducted using CHEMKIN 4.1 [10]. This program can solve the differential equations that result from the kinetics analysis taking into account the different kinetic parameters of all elementary reactions involved. For the ignition delay time modeling, the batch reactor routine was assumed. This model allows simulating the conditions behind the reflected shock wave in shock-tube experiments. Internal energy and volume are considered constant. Chemical kinetics mechanism consist of different sets of reactions including a recently updated H_2 and C_1-C_5 hydrocarbon mechanism [11], a widely used CF_3Br and HFC chemical kinetic mechanism, provided by Luo et al. [12-15], and a OH* chemistry mechanism [16, 17]. The final assembled chemical mechanism is validated by comparing the modeled and measured ignition delay times (τ_{ign}). To identify the most significant reactions that affect the combustion phenomena studied, sensitivity analysis is conducted using SENKIN [18].

3.2 Experimental

Ignition delay time measurements were conducted using the High Pressure Shock Tube at Texas A&M University. This apparatus has been designed to generate an incident and reflected shock wave that travels through the test mixture. This process allows the pressure and temperature to remain constant until ignition takes place. The time between the pressure rise after the reflective shock wave (time zero) and the rapid increase of OH* emission [19] is known as the ignition delay time (Figure 2). Pressure is monitored through equally spaced pressure transducers located alongside the tube and one at the endwall. Wave speed is calculated using the signal from the pressure transducers which give information of the time at which the shock wave travels from one transducer to the next. A window located at 1.6 cm from the endwall allows the light emitted after ignition to be captured. The optical signal is filtered at 310 nm (OH* wavelength) and then amplified by a photomultiplier tube detector. Pressure and emission signals are examined using GageScope software. More information about this facility is provided by Aul et al. [20].

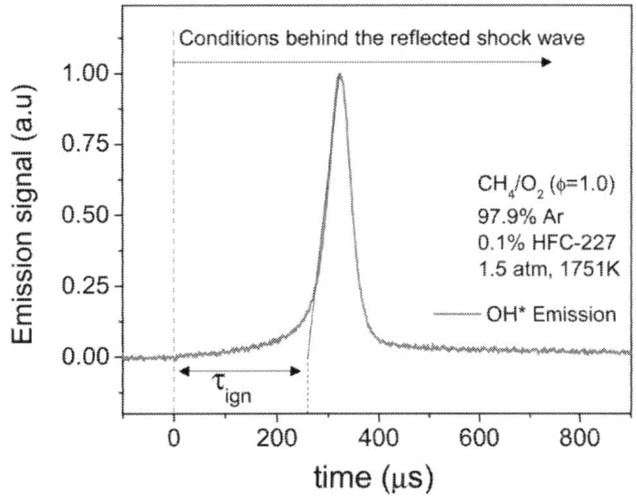

Figure 2. Definition of ignition delay time based on an OH* profile

4. Results and Discussion

4.1 Ignition delay time measurements

Different mixtures of fuel (CH_4 or C_3H_8), oxygen (O_2), and Argon (Ar) were analyzed with and without fire suppressants (i.e., CF_3Br, C_2HF_5, and C_3HF_7) at atmospheric pressure over a range of temperatures of 1350K-2200K. Figures 2 and 3 show the resulting experimental and modeled ignition delay times for the different mixtures evaluated. These results are very important because they help us to answer some general questions such as: Can a fire suppressant promote ignition? If this result is the case, at which conditions or for what fuels does it occur? Additionally, these results provide information regarding the chemical mechanism accuracy.

Experimental and modeled data (in Figure 3) show that the tested fire suppressants tend to decrease the ignition delay time in methane mixtures. Their promoting effect can be ranked as $CF_3Br < C_2HF_5 < C_3HF_7$. However, this trend is not general. Experimental data in Figure 4 show that the ignition of the propane mixtures can be slightly promoted by C_2HF_5, but retarded by the action of CF_3Br. C_3HF_7 does not show a significant effect on propane mixtures. Note that the range of practical application of the present experimental results remains strictly limited to the experimental conditions defined. This is when numerical approaches are very important since they allow predicting the combustion properties of different systems. However, such mechanism should be validated using accurate measurements, which recall one of the main objectives of this work.

Comparison between modeled and experimental data (see Figures 3 and 4) shows that the model very well predicts the effect of CF_3Br.on methane and propane; however, improvements on the HFC-chemistry are suggested for better prediction; especially for those including propane. The following section describes the results obtained from the sensitivity analysis which serves as the basis for future mechanism improvements.

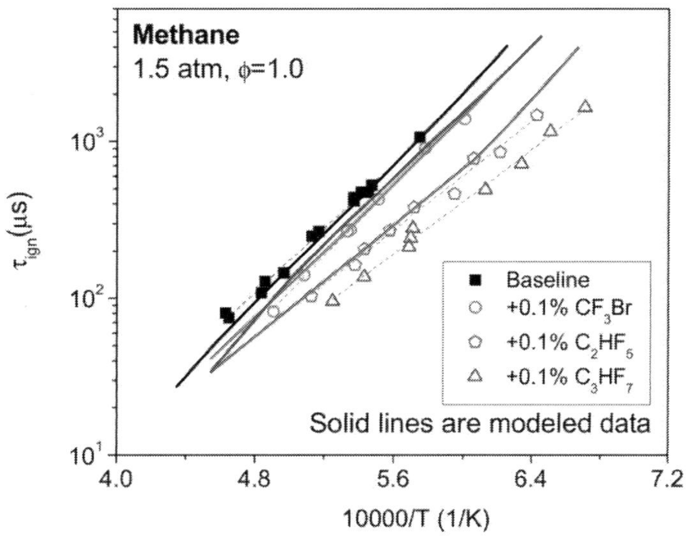

Figure 3. Experimental and modeled Ignition delay times (τ_{ign}) as a function of temperature for $CH_4/O_2/Ar$ Mixtures with and without CF_3Br, C_2HF_5, C_3HF_7.

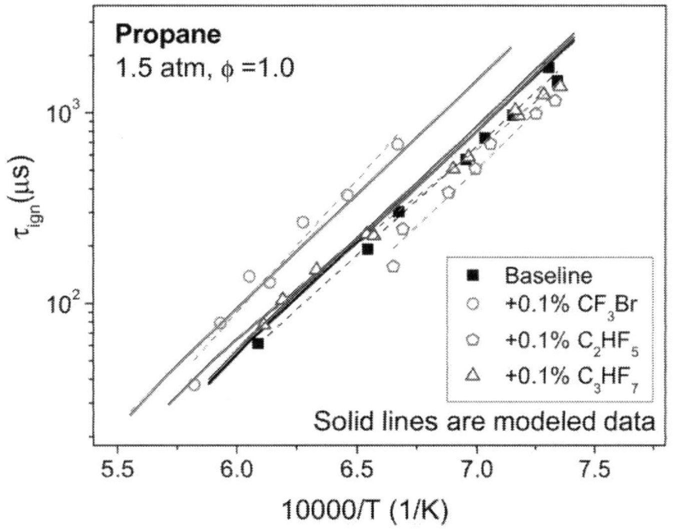

Figure 4. Experimental and modeled Ignition delay times (τ_{ign}) as a function of temperature for $C_3H_8/O_2/Ar$ Mixtures with and without CF_3Br, C_2HF_5, and C_3HF_7.

4.2 Sensitivity Analysis

Sensitivity analysis was carried out to find the key reactions responsible for the major effects of the tested agents on a specific reacting mixture. For this analysis, SENKIN [10] was used through the CHEMKIN [18] interface, which provides the reactions with the largest sensitivity coefficients. For convenience, the reaction numbering in this paper retains the same numbering used in the assembled mechanism. Tables 2 and 3 present a summary of such findings.

Results show that halogenated species can participate in promoting and inhibiting reactions that compete to give a net effect. The resulting effect depends on many factors such as the fire suppressants-fuel chemical interaction and the conditions of the system (i.e. temperature, pressure).

In terms of sensitivity coefficients, it can be seen that bromine-containing species mainly participate in strong inhibiting reactions. It can be explained due to the participation of bromine species in catalytic cycles where highly reactive radicals (i.e. H radicals) are removed from the flame reaction zone, as it was suggested by Westbrook [21]. On the other hand, fluorinated species may act as scavengers with lower radical removal efficiency. However, they can also participate in several ignition-promoting reactions. Despite all the research in this area, the flame inhibition mechanism of fluorinated species remains unclear, and further research may be conducted.

Current sensitivity analysis results can be used to improve the HFC chemistry. Particular attention is required to those reactions with significant influence, where the focus then is to find more-accurate Arrhenius kinetic parameters that lead to better predictions.

5. Summary and Conclusions

This paper examined the fundamentals of fire chemistry to understand the ignition effects of two of the most currently used HFC-clean fire suppressants (C_2HF_5 and C_3HF_7) on two common hydrocarbons. It also compared current results with the properties of the CF_3Br (Halon 1301), which is considered as the reference. This study was based on ignition delay times measurements and modeling of different, highly diluted mixtures of CH_4 and C_3H_8, with and without the fire suppressants. Our results suggest that the tested agents may not be good alternatives as ignition preventers; however, we do not deny that they can still be used as fire extinguishers. Considering the scarcity of HFC ignition delay time measurements in the literature, the current data set represents a valuable tool for model validation. Comparison between modeled and experimental data shows that the HFC-mechanism behaves well, however it can be improved. Sensitivity analysis was conducted to identify the most relevant reactions that participate at the ignition process. These results can be used as the basis for future HFC-chemical kinetics mechanism research and improvements.

Table 2. Summary of the sensitive analysis results for methane systems

Agent	Effect	Key Reactions	Net Effect
0.1% CF_3Br	Promoting	1281. $CF_3+O_2=CF_3O+O$ 57. $CH+O_2=CO+OH^*$ 1257. $CF_3+CH_4=CHF_3+CH_3$ 1737. $C_2HF+O=CFCO+H$ 1188. $CF_3Br+M=CF_3+Br+M$ 1182. $CH_3+BR=CH_2+HBR$ 1169. $CH_2O+Br=HCO+HBr$	Promotion
	Inhibiting	1166. $CH_4+Br=CH_3+HBr$ 1208. $Br+HCO=HBR+CO$ 1143. $HBr+H=Br+H_2$	
0.1 % C_2HF_5	Promoting	57. $CH+O_2=CO+OH^*$ 1375. $CHF_2\text{-}CF_3=CF_2{:}CF_2+HF$ 1453. $CF_3+CHF_2=CHF_3+CF_2$ 1737. $C_2HF+O=CFCO+H$ 1305. $CF_2+O_2=CF_2{:}O+O$ 1760. $CH_3+CF_2=CH_2{:}CF_2+H$ 1281. $CF_3+O_2=CF3O+O$ 1280. $CHF_2+O_2=>CF_2{:}O+O+H$ 1328. $CF+O_2=CF{:}O+O$ 1226. $CHF_3+M=CF_2+HF+M$ 1320. $CF_2+H=CF+HF$	Promotion
	Inhibiting	1415. $CH_3+CF_3=CH_2{:}CF_2+HF$ 1424. $CHF_2+CF_3+M=CHF_2\text{-}CF_3+M$	
0.1% C_3HF_7	Promoting	1665. $CHF+CF_2=CHF{:}CF_2$ 1764. $CHF_2+CF_2=CHF{:}CF_2+F$ 1281. $CF_3+O_2=CF_3O+O$ 57. $CH+O_2=CO+OH^*$ 1280. $CHF_2+O_2=>CF_2{:}O+O+H$ 1305. $CF_2+O_2=CF_2{:}O+O$ 1375. $CHF_2\text{-}CF_3=CF_2{:}CF_2+HF$ 1453. $CF_3+CHF_2=CHF_3+CF_2$ 1737. $C_2HF+O=CFCO+H$ 1305. $CF_2+O_2=CF_2{:}O+O$ 1760. $CH_3+CF_2=CH_2{:}CF_2+H$ 1281. $CF_3+O_2=CF_3O+O$ 1280. $CHF_2+O_2=>CF_2{:}O+O+H$ 1305. $CF_2+O_2=CF_2{:}O+O$ 1320. $CF_2+H=CF+HF$ 1453. $CF_3+CHF_2=CHF_3+CF_2$ 1257. $CF_3+CH_4=CHF_3+CH_3$ 1328. $CF+O_2=CF{:}O+O$	Promotion
	Inhibiting	176. $CH+O_2=HCO+O$ 1412. $CH_3+CHF_2=CH_2{:}CHF+HF$ 1761. $CH_2F+CF_2=CHF{:}CF_2+H$ 1776. $CH_4+F=CH_3+HF$ 1415. $CH_3+CF_3=CH_2{:}CF_2+HF$	

Table 3. Summary of the sensitive analysis results for propane systems.

Agent	Effect	Key Reactions	Net Effect
0.1% CF_3Br	Promoting	1188. $CF_3BR+M=CF_3+BR+M$	Inhibition
	Inhibiting	1146.$Br+HO_2=HBr+O_2$ 1189. $CF_3Br+H=CF_3+HBr$ 1208. $Br+HCO=HBr+CO$ 1210. $C_2H_3+Br=C_2H_3Br$ 1143. $HBr+H=Br+H_2$	
0.1 % C_2HF_5	Promoting	-	Promotion
	Inhibiting	1286. $CHF_2+OH=CHF{:}O+HF$ 1776. $CH_4+F=CH_3+HF$	
0.1% C_3HF_7	Promoting	1358. $CO+F+M=CF{:}O+M$ 1328. $CF+O_2=CF{:}O+O$ R1849. $CF_3\text{-}CHF(+M)=CHF{:}CF_2+F(+M)$	Promotion
	Inhibiting	1339. $CF+C_2H_4=>C_2H_2+CH_2F$ 1397. $CF_3\text{-}CHF+H=CH_2F+CF_3$	

References

1. Shebeko, Y.N., et al., *The influence of fluorinated hydrocarbons on the combustion of gaseous mixtures in a closed vessel.* Combust. Flame, 2000. **121**(3): p. 542-547.
2. Suzuki, A., et al., *Effect of Bromotrifluoromethane on the Ignition in Methan and Ethane-Oxygen-Argon Mixtures behind Shock Waves.* Bull. Chem. Soc. Jpn., 1991. **64**: p. 3345-3354.
3. Saso, Y., et al., *Laminar Burning Velocities of Trifluoromethane–Methane Mixtures: Experiment and Numerical Simulation.* Combust. Flame, 1998. **114**(3–4): p. 457-468.
4. Linteris, G.T. and L. Truett, *Inhibition of premixed methane-air flames by fluoromethanes.* Combust. Flame, 1996. **105**(1-2): p. 15-27.
5. Babushok, V.I., et al., *Simulation Studies of the Influence of Fluorine and Bromine Containing Fire Suppressants on Ignition Behavior*, in *Halon Options Conf. Proc.*1994: Albuquerque, NM. p. 205-216.
6. Hamins, A. and P. Borthwick, *Suppression of ignition over a heated metal surface.* Combust. Flame, 1998. **112**(1–2): p. 161-170.
7. Osorio, C.H., E.L. Petersen, and M. S. *Kinetic Modeling and Experimental Study of CF_3Br Effect on Methane and Ethane Oxidation Behind Reflected Shock Waves.* in *7th US National Technical Meeting of the Combustion Institute.* 2011. Atlanta, GA.
8. Linteris, G.T., et al., *Stirred reactor calculations to understand unwanted combustion enhancement by potential halon replacements.* Combust. Flame, 2012. **159**(3): p. 1016-1025.
9. Babushok, V.I., G.T. Linteris, and O.C. Meier, *Combustion properties of halogenated fire suppressants.* Combust. Flame, 2012. **159**(12): p. 3569-3575.
10. Kee, R.J., et al., *CHEMKIN Release 4.1.1*, 2007: San Diego, CA.
11. Donato, N.S., et al., *Ignition and Oxidation of 50/50 Butane Isomer Blends.* J. Eng. Gas Turbines Power, 2010. **132**(5): p. 051502.

12. Luo, C., *Study on the Inhibition Mechanisms of $CBrF_3$, CF_3I and C_3F_7H in Methane Fuelled Premixed Flames*, in *Department of Chemical Engineering*2010, The university of Newcastle: Australia. p. 270.
13. Luo, C., B.Z. Dlugogorski, and E. Kennedy, *Influence Of CF_3I and $CBrF_3$ on Methanol – Air and Methane – Air Premixed Flames.* Fire Technol, 2008. **44**: p. 221-237.
14. Kennedy, E.M., B.Z. Dlugogorski, and C. Luo, *Burning Velocity And Flame Structure Of Methane Flames Inhibited With Heptafluoropropane (C_3HF_7)*, in *Proceedings of the Sixth Asia-Oceania Symposium on Fire Science and Technology*2004: Daegu, South Korea.
15. Luo, C., et al., *Effect of Bromine and Iodine on Methanol and Methane Flames.* , in *Suppression, Detection and Signaling Research and Applications - a Technical Working Conference (SUPDET 2010)*2010: Orlando, Florida.
16. Healy, D., et al., *n-Butane: Ignition delay measurements at high pressure and detailed chemical kinetic simulations.* Combust. Flame, 2010. **157**(8): p. 1526-1539.
17. Healy, D., et al., *Methane/n-Butane Ignition Delay Measurements at High Pressure and Detailed Chemical Kinetic Simulations.* Energy Fuels, 2010. **24**(3): p. 1617-1627.
18. Lutz, A.E., et al., *Senkin: A Fortran Program For Predicting Homogeneous Gas Phase Chemical Kinetics With Sensitivity Analysis. SAND87-8248*, 1988: Livermore, CA.
19. Petersen, E.L., *Interpreting Endwall and Sidewall Measurements in Shock-Tube Ignition Studies.* Combust. Sci. Technol., 2009. **181**(9): p. 1123-1144.
20. Aul, C.J., DeVries, J. and Petersen, E. L. . *New Shock Tube Facility for Studies in Chemical Kinetics at Engine Conditions.* in *Eastern States Fall Technical Meeting of the Combustion Institute.* 2007. Charlottesville, VA.
21. Westbrook, C.K., *Numerical Modeling of Flame Inhibition by CF3Br.* Combust. Sci. Technol., 1983. **34**: p. 201-225.

A Focus on Fire Fundamentals Including Emergency Response Training at the National Fire Agency in Taiwan

Ronald J. Willey
Department of Chemical Engineering
Northeastern University, Boston, MA 02115
r.willey@neu.edu

Jack, Huan-Chang Hsiao, Ph.D., Director, Training Center National Fire Agency, Ministry of the Interior, Executive Yuan, ROC; No. 100, Dagong St., ZhuShan Township, Taiwan 55795, ROC, **jack@nfa.gov.tw**
Roy E. Sanders, Chemical Process Safety Consultant, Lake Charles, LA, **sanders.roy@suddenlink.net**
Arcady Kossoy, Dr. Sc. R&D Deputy Director General, Cheminform, St. Petersburg (CISP) Ltd., PO Box 25, 193232, St. Petersburg, Russia, **kossoy@cisp.spb.ru**
Chi-Min Shu, Ph. D., Center for Industry-Academia Collaboration, National Yunlin University of Science and Technology, 123 University Road, Section 3, Douliu, Yunlin 64002, Taiwan, **shucm@yuntech.edu.tw**

Prepared for Presentation at
American Institute of Chemical Engineers
2013 Spring Meeting
9th Global Congress on Process Safety
San Antonio, Texas
April 28 – May 1, 2013

A portion of this article appeared in the March 2013 issue of AIChE PSP Vol 32, No. 1 pp 2-7

A Focus on Fire Fundamentals Including Emergency Response Training at the National Fire Agency in Taiwan

Ronald J. Willey
Department of Chemical Engineering
Northeastern University, Boston, MA 02115
r.willey@neu.edu

Jack, Huan-Chang Hsiao, Ph.D., Director, Training Center National Fire Agency, Ministry of the Interior, Executive Yuan, ROC; No. 100, Dagong St., ZhuShan Township, Taiwan 55795, ROC, **jack@nfa.gov.tw**
Roy E. Sanders, Chemical Process Safety Consultant, Lake Charles, LA, **sanders.roy@suddenlink.net**
Arcady Kossoy, Dr. Sc. R&D Deputy Director General, Cheminform, St. Petersburg (CISP) Ltd., PO Box 25, 193232, St. Petersburg, Russia, **kossoy@cisp.spb.ru**
Chi-Min Shu, Ph. D., Center for Industry-Academia Collaboration, National Yunlin University of Science and Technology, 123 University Road, Section 3, Douliu, Yunlin 64002, Taiwan, **shucm@yuntech.edu.tw**

Keywords: Emergency Response Training, Fire Prevention

1.0 ABSTRACT

This article reflects on process safety fundamentals of fires and explosions for the benefit of new professionals entering the field. It provides a brief review of fire fundamentals such as the fire triangle and the respect that must be paid to the unknown ignition source. Various control methods are reviewed at the elementary level, including the reminder that the ignition source is always free. After a review of fire fundamentals, an emergency response training exercise is presented for a burning propane fire witnessed at a training facility in Taiwan. The National Fire Agency (NFA) of Taiwan has built the third largest emergency response-training center in the world. The 270 acres facility is located in central Taiwan near Taichung, Taiwan. The site has 49 separate emergency response training scenarios. Scenario pads include training for fire attack on mock-ups of chemical storage facilities, tanker trucks, power transformers, passenger airliners, and shipping vessels. Overall, this paper serves as a reminder of the safe guards used in fire safety, and the importance of training beforehand in the event of an emergency.

2.0 INTRODUCTION

This article is intended for young professionals entering the field of process safety. We will use the scenario of "fire" as our example for process safety fundamentals, and emergency response training. Let's begin with the nature of fires and some elementary considerations by any professional performing hazard analysis of process areas for fire prevention.

3.0 FUNDAMENTALS ABOUT FIRES

3.1 The Nature of a Fire and Elementary Parameters Explained

Fire in its simplest terms is an exothermic reaction. It is a reaction between an oxidizer (e.g., air, oxygen, chlorine, fluorine, or hydrogen peroxide) and a fuel (e.g., methane, hydrogen, propane, gasoline, carbon disulfide, or ether). A "third" requirement is an external source of "energy" to initiate the reaction. Typically this is a called an ignition source. The combination of these, oxidizer, fuel, and ignition source form the sides to the equilateral "fire" triangle. Eliminate any one of the three, and a fire will not initiate. An outstanding primer on fires is the textbook by Crowl and Louvar, Chapter 6 [1]. Young engineers working in the field of process safety are encouraged to purchase this reference and retain it on their desks.

Two of us (Sanders and Willey) are disciples of Trevor Kletz – a pioneer in modern process safety methods (HAZOPS for example). In honor of him, we'd like to bring forth to a younger audience a portion of his simple, yet effective, teachings in the area of fire prevention and his wise words about ignition.

The following is extracted from Trevor's "What Went Wrong"[2]

The inevitability of ignition (paraphrased for the young engineer)

> You must be familiar with the fire triangle as you begin a fire hazards analysis. Air, fuel, and a source of ignition are necessary for a fire or explosion to occur; remove one of the three, and an explosion is impossible. However, when flammable gases or vapors are handled on an industrial scale, this view, though theoretically true, is misleading. If flammable gases or vapors are mixed with air in flammable concentrations, experience shows that sources of ignition are likely to turn up as the amount of energy required to ignite a flammable mixture can be very small. As little as 0.2 mJ, or the energy produced when a one-cent coin falls 5 mm, can be enough energy to ignite a combustible mixture.

"THE IGNITION SOURCE IS ALWAYS FREE"

To further follow-up this statement, ignition sources can include the hot surfaces such as an automobile exhaust manifold located nearby the site, a hot bearing, static electricity discharge, an open flame, a discharged burning cigarette butt, lightning, or an electrical switch being turned on.

With the above "warning" written, it is prudent to include in a hazard analysis, management of change assessments, and general operation practice, the consideration of where ignition sources are present. When an external contractor arrives to a your facility, where should that pick-up truck be allowed? Are permits issued for vehicular equipment and tools capable of emitting sparks? If that contractor lights up a cutting torch, has full consideration been made to all of the consequences? Are there any potential for a fuel-air mixture? Several noteworthy fires can be traced to secondary tanks, connected but not isolated, to the tank undergoing "hot work" becoming a source of a flammable mixture. The U.S. Occupational Health and Safety Administration (OSHA) provides guidance and regulations for hot work activities that should serve as the basis for individual companies' safety programs [3].

Material safety data sheets (MSDS) or the newer label, safety data sheets (SDS), provide guidelines to help understand the nature of a flammable or combustible chemical or product. The auto ignition temperature (AIT) is a property of a material that if it is in the combustible range, will ignite upon contact with a surface at that temperature or above. The flammable limits or combustible ranges are also frequently presented as lower explosion limit (LEL) or upper explosion limit (UEL). These are the concentrations at ambient conditions in air in which the mixture is either flammable or combustible. Another set of terms is frequently used interchangeably with LEL and UEL. These are lower flammable limits (LFL) and upper flammable limits (UFL). There is a minor difference on the order of 0.1%, between the explosive and flammable limits. For example, propane has an AIT of 450°C, a LEL of 2.1% and a UEL of 9.5% [4]. Fire prevention begins by recognizing that if we have propane (fuel) plus air (oxidizer) between 2.1 to 9.5% propane, we have a potential for fire or explosion no matter what (see statement above). An additional reference source for flammability properties is a book by Yaws et al [5].

Another indicator of fire potential is the flash point [6]. This number is lowest temperature that a material can catch fire but not be self-sustaining. Generally, lower the flash point, the more flammable. It is related to vapor pressure of the material and to the method of testing (open cup or closed cup for example). Gasoline has a flash point below -43°C. Inside a closed container, at room temperature, the vapor pressure of gasoline brings it into the non-flammable range because the equilibrium mixture is well above its UFL. This is critical in maintaining safety within the world's transportation system. On the other hand, ethanol's flashpoint is 13°C and its fire point (point where a self-sustaining flame is possible) is 18°C. This presents a challenge related to storage of ethanol at ambient temperatures (20 to 30°C). The vapor-air mixture above the liquid is in the mist of the flammable limits for ethanol/air at ambient temperatures. Thus, an inert such as nitrogen must be used as a purging gas. A small flow or small positive head of nitrogen is used to keep the gas mixture within a storage tank and process vessels within the non-flammable range. This gas mixture must remain below the minimum oxygen concentration (MOC) by using nitrogen to displace the oxygen present in ambient air. Crowl and Louvar discuss methods of purging [1]. Large scale purging is one of the largest industrial applications of nitrogen either as a delivered gas/liquid or from on-site membrane separation processes.

3.2 Elementary Controls for Fire Mitigation

Fire fighting measures begin with the design of the process area. Indoor areas almost always require overhead fire sprinklers. There are two main types, wet pipe, where water is readily available at all times, and dry pipe (pressurized air is within – that trips a water valve when the pressure is released) for areas that may drop below freezing. The sprinkler head has a fusible link that typically breaks apart at 165°F (other temperature ranges are available for higher temperature areas). Water then flows out a ½" nozzle on the order of 30 gallons per minute and is dispersed by the cap of the head. Water damage is also a concern when a system goes off. Thus, access and ease of operation of control valves should also be considered in the layout of a system. Guidelines are provided by in NFPA 13 [7].

Outdoor areas, especially areas storing large amounts of flammables, may need fire monitors. These devices spray large amounts of water, several hundred gallons of minutes, into a process area. Some monitors are moveable, either manually or remotely, so that the water spray can be directed to the base of the fire. In the case of liquefied propane storage, the water spray onto the storage tank has an even more critical function of preventing one of the most severe petrochemical explosions, the boiling liquid expanded vapor explosion (BLEVE) [8].

3.3 Professional Guidelines to Consider in Fire Prevention and Design

A number of key guidelines for the control of fires exist for the young professional entering the process safety field. The Dow Fire and Explosion Index [9] provides guidance on spacing of processing units to achieve safer layouts and prevention of major catastrophes. Another Trevor Kletz axiom comes to mind. Paraphrased: "What you don't have can't leak, can't catch on fire, and can't explode" [10]. Proper hazard analysis in terms of fire and explosion prevention includes minimizing the amount of flammable and keeping inventory rotated, and away from personnel.

Fire prevention also includes reviewing materials from the National Fire Protection Association (NFPA) such as NFPA 30 [11] and NFPA 58 [12]. Is the space under assessment located inside a building, or confined space, or in an open-air situation? If inside a building or shelter, are fire sprinklers present overhead? If outside, are fire monitors required? Is there a need for flammable gas detectors? Is the primary flammable lighter than air such as methane or hydrogen? If yes, flammable gas detectors may need to be located at an elevation above ground level. If no, generally, closer to ground level is the appropriate placement as most flammable vapors are heavier than air.

Another elementary lesson in fire and explosion includes the understanding of the two major NFPA fire ratings. Unfortunately, the number system meaning is reversed and depends on which set of NFPA guidelines the user is following. The foremost is the NFPA diamond (NFPA 704) that contains four diamonds within (See Figures 1 and 2). The red diamond, upper center, contains a number from 0 to 4. Four is the most severe fire hazard. The relative ranking for flammables within the diamond is as follows [13]:

Table 1. NFPA 704 Red Diamond Flammability Ratings (adapted from [13]).

Number	Description	Example
Red - 0	Will not burn.	carbon dioxide
Red - 1	Must be heated in order to ignite. Flash point exceeds 90°C or 200°F **(NFPA 30 Class IIIB)**	mineral oil
Red - 2	Moderate heat or relatively high ambient temperature is required for ignition. Flash point between 38°C or 100°F and 93°C or 200°F **(NFPA 30 Class II and IIIA)**	diesel fuel
Red - 3	Liquids or solids that readily ignite at most ambient temperature conditions. Liquids have a flash point below 23°C (73°F) and boiling point at or above 38°C (100°F) or flash point between 23°C (73°F) and 38°C (100°F) **(NFPA 30 Class I B and I C)**	gasoline
Red - 4	Rapidly or completely vaporizes at normal temperature and pressure or readily disperses in air and readily burns. Flash point below 23°C (73°F) **(NFPA 30 Class I A)**	hydrogen, propane

The second set up NFPA rankings appear for flammable and combustible liquids and are used within NFPA 30 and NFPA 45 [14] guidelines. The boiling point is also used to segregate the hazards. These guidelines are used around the USA by fire inspectors accessing the amount of storage that flammable quantities may occupy within any enclosed room or process area. Here Class I flammable means highly flammable, and the amounts that can be stored are less than a Class II or Class III, which are considered combustibles. Table 2 presents a summary of the classifications.

Table 2. NFPA 30 Classification for Liquids [11].

Class	Description	Example
Class I A	Flash point is below 22.8°C (70°F) or ambient temperature and a boiling point below 37.8°C (100°F) **(NFPA 704 - Red 4)**	Butane
Class I B	Flash point is below 22.8°C (70°F) or ambient temperature and a boiling point above 37.8°C (100°F) **(NFPA 704 – Red 3)**	Hexane
Class I C	Flash point is between 22.8°C (70°F) to 37.8°C (100°F) and a boiling point above 100°F **(NFPA 704 – Red 3)**	Nonane
Class II	Flash point is between 37.8°C (100°F) to 60°C (140°F) **(NFPA 704 – Red 2)**	Decane
Class III A	Flash point is between 60°C (140°F) to 93.3°C (200°F) **(NFPA 704 – Red 2)**	Phenol
Class III B	Flash Point above 93.3°C (200°F) **(NFPA 704 – Red 1)**	Heat Transfer Fluid

4.0 EMERGENCY RESPONSE SHOULD A FIRE BREAK OUT

First and foremost, employees in any chemical industrial complex, must understand all of the hazardous properties of the raw materials, intermediates and finished products along with the hazards of all utilities involved in the emergency. When it comes to combustible and flammable liquids or flammable gases, they must understand the principles of flammable limits, flash points, auto-ignition temperatures and be ready to handle a potential fire. They must also understand the products of combustion of the burning materials and review for proper protection. When vinyl chloride burns the fire will emit HCL fumes. If carbon disulfide is ablaze combustion produces toxic gases and irritants including oxides of sulfur.

Emergency responders must know their personal protective equipment. NFPA 704 should be consulted for protective clothing [13]. Typically full protective clothing includes helmet, self-contained breathing equipment, bunker coast and pants, rubber boots and gloves.

Once a fire has erupted, what actions should be involved? If the fire is small enough for hand held or wheeled fire extinguishers, then what type of fire extinguisher should be used? There are three major categories, A, B, and C. “A” ratings are for trash or wood fire, “B” ratings are for extinguishing fires of combustible or flammable liquids, and “C” ratings are for electrical equipment fires [15]. Every individual within an organization should experience pulling the pin and putting out a small “B” type fire using a CO_2 extinguisher. The training should include the demarcation point as to when to “fight” or to “flee.” It is okay to flee, and this decision has saved lives!

Another key assessment in fire prevention is the location of the emergency isolation valves to stop the flow of flammables from outside sources. In a hazard analysis ask: “Should emergency isolation valves be automated? If not, can they be accessed safely in an emergency?” There are numerous examples where a fire burned uncontrollably because the proper isolation valve could not be closed.

Your assessment should include the emergency response training expected of personnel. If that response includes a fire response by personnel working in the facility, then fire training is mandatory. If an in-house fire brigade is going to be set up, then they should be given the opportunity to attend a fire training session at a professional fire training facility. Numerous professional training facilities are willing to work with private and industrial groups on fire training. Often local fire departments are willing to conduct fire response training with local industries and fire brigades for training exercises. Recent *Process Safety Progress* articles about emergency response exercises and the need for emergency response are provided in references [16-19].

4.1 Consider an Emergency Response Practice with Local Fire Agencies

Successful emergency response includes working with the local fire agencies beforehand, practicing various levels of drills. At least annually, a practice run with local fire fighters is warranted, especially in rural locations where these fire fighters may be volunteers. One of the authors (Willey) witnessed an emergency response where a million dollar a day process was brought to a complete halt because of a fume from butane lighter brought near an outside flammable detector. An operator halted plant operations who had less than one year of experience. Because of training, she hit the emergency panic button when the flammable detector alarm sounded at her control panel. Within minutes the local fire department arrived and practiced applying foam to a butane storage tank that was not burning. The exercise was then fully reviewed with plant and fire-fighting personnel on how well the group responded to this practice event. During the past 30 years, this facility has had an excellent safety record. Part of the reason can be attributed to emergency response practice with local agencies.

4.2 Fire Response Training – Practice Beforehand – An Example from Taiwan

The National Fire Agency (NFA) of Taiwan is a key part of the Ministry of the Interior and has national responsibility for fire prevention and fire fighting as well as medical response and ambulance services. It was established on March 1, 1995 to improve emergency medical services, promote professional fire prevention and fire fighting service, enhance disaster rescue efficiency, and protect civilian life & property within Taiwan. The NFA Training Center is located in central Taiwan in Zhushan, Nantou County. It occupies 109 hectares (270 acres) and is now the 3rd largest such facility in the world. The main mission of this center is to provide training to young emergency response personnel working in Taiwan who select this as their career path. A typical trainee is a person between the ages of 20 to 25 years old who has selected emergency response as a career. They attend training at the facility for a period of 6 to 18 months depending upon which authority they are responsible to. The facility provides living quarters to trainees during their training period at the facility.

Full training includes practice on a wide variety of fire scenarios reflecting real life emergencies, earthquake responses, and water rescues. The fire responses include the extinguishment of a high intensity propane fire, the entry into a completely dark building with a concurrent smoke event, the entry into a major roadway tunnel with vehicle fire occurring, the entry into an airplane that may be on fire within, the entry into a ship, the approach to a burning tank truck fire, the approach to a chemical complex under a fire condition, and the rescue of stranded persons flowing in a very powerful body of flowing water such as a strong rapids.

The Taiwanese training facility uses propane as a fuel for some of its fire evolutions. Recall the fundamentals about fires above. One must have fuel, an oxidizer, and an ignition source. As shown in Figure 3, we see a scenario area for the training used on putting out a propane fire. Propane and air can flow out through the burners without any signs of a flame. Then, an ignition switch is thrown, and very large intimidating flames evolve and become the focus of the team's attention.

In this training session, the firefighting crew of nine is manning two 2 ½ inch fire hoses in approaching the threatening inferno. Water is the medium used as the responders inch their way forward. Their tactic is to begin with a very broad spray of water (Figure 4). This water spray protects them from the heat of the flame and allows them to move closer to the fire. During this particular training session the fire actually intensifies before the firefighters approach This intensification is by design and assists on protecting the trainees for their initial fire training exercises, allowing them to learn how to approach a large propane fire appropriately. Note that there are nine crew members on these two hoses. For the novice, one is not aware of the momentous forces that exist on any curved large diameter flexible hose. At 2 ½ inches, a large hose releasing 100 gallons per minute, if not supported properly, can develop enough momentive force to break a leg of a human. It is important that emergency responders have experienced this momentum before entering a major flame up. A minimum crew is also required to handle the hose safely.

Figures 5 through 7 show the crew extinguishing the flame. As it is approached, and the flame is brought under control, the water stream narrows, and focuses towards the base of the flame. This entire exercise occurred over an absolute time period of 2 minutes. Capable supervisors, in an on-site centrally located control tower, monitor the training via closed circuit television and can instantly shut down any fire exercise if there is any emergency.

Typically, fires from pipelines or large vessels should not be extinguished unless the fuel can be shutoff and isolated. If a fire is extinguished and the fuel continues to be emitted, a cloud can form and when it reignites an explosion can occur.

4.3 Additional Fire Training Scenarios at the Taiwan Training Facility

The facility has several other fire training scenarios available. Included is a chemical area training facility. There are fires in mock up apartment building, warehouses, theaters, a commercial kitchen fire, a subway car, a large commercial airliner mock up, and much more. There are approximately 50 "simulation points" where qualified instructors can create the conditions that would warrant live firefighting. It would take a visitor a couple of days to see all responders training in each of the available mockup areas.

4.4 Additional Emergency Response Training - Swift Water Rapids Training

Taiwan, due to its geographic location is subject to many natural disasters. It experiences earthquakes, typhoons, and other natural disasters in a much higher proportion than many other locations around the world. Taiwan became interested in such a disaster-training center after a devastating earthquake in 1999.

One natural disaster is the result of its mountainous central areas and the frequency of typhoons. Swift water flows are frequent occurrences. This facility has a large flume with flow rates of 100,000 gallons per minute for the training of emergency response personnel in the rescue of swimmers and others were stranded in a rapidly flowing stream. Figure 8 shows an example of a person moving down the 100,000 gpm flume in a practice for water rescue.

The agency is looking for partnerships with other countries and international corporations. It offers 1-week training courses in swift water, tunnel fire, and aircraft fire. For those working in the process industries, there is an 8, 32, or 56 hour HAZMAT training that follows NFPA guidelines, with division into awareness, operation, and technician levels. The facility is willing to share its training facilities for the preparation of emergency response personnel associated with these groups.

5.0 CONCLUSIONS

Emergency training before an event happens makes a difference. Lack of training often contributes to the problem rather than mitigate it. Further, this training should include working with outside groups with similar interests and resources. By working together both groups learn more effective methods to mitigate emergencies. The facility in Taiwan is willing to welcoming emergency response personal from public and private affiliations worldwide. For more information please write the director, an author of this paper, Dr. Jack, Huan-Chang Hsiao, at jack@nfa.gov.tw. We also note that the United States has several excellent fire fighter training schools with large fire scenarios. Places include the Texas A&M- College Station, Texas and Louisiana State University- Baton Rouge and the reader can contact these sources as well for potential fire and emergency response training.

Figure 1. An example of the NFPA Diamond. The center red diamond is related to fire hazard of the material within. Steam is rated "0" as it is not a fire hazard.

Figure 2. A flammable solvent storage tank with a flammability rating of 3.

Figure 3. Crew approaches a propane fire.

Figure 4 Crew walks towards the flame with a cone of streaming water.

Figure 5. Roughly 20 seconds into the training session, crew now move decisively towards the fire, water stream is starting to be concentrated.

Figure 6. Fire close to extinguishment.

Figure 7. Fire extinguished, total elapsed time – 2 minutes.

Figure 8. Swift water training facility during a training session.

6.0 LITERATURE CITED

1. Crowl, D.A.; Louvar, J.F. *Chemical Process Safety Fundamentals with Applications, 3rd Ed.* Prentice Hall PTR, 2011.
2. Kletz, T. " What Went Wrong?, 5th Edition, Case Histories of Process Plant Disasters and How They Could Have Been Avoided," Butterworth- Heinemann, 2009.
3. OSHA Regulations 29 CFR 1910.252, related to hot work, welding, cutting, and brazing, Retrieved December 22, 2012, from http://www.osha.gov/pls/oshaweb/owadisp.show_document?p_id=9853&p_table=standards
4. Propane MSDS, Retrieved December 22, 2012, from http://www.airgas.com/documents/pdf/001045.pdf.
5. Yaws, Carl L.; Braker, William; *Matheson Gas Data Book* Published by McGraw-Hill Professional, 2001.
6. Flash point. Wikipedia, Retrieved December 26, 2012, from http://en.wikipedia.org/wiki/Flash_point.
7. *NFPA 13: Automatic Sprinkler Systems Handbook*, Quincy, MA, 2013.
8. Boiling liquid expanding vapor explosion, Wikipedia, Retrieved December 28, 2012, from http://en.wikipedia.org/wiki/Boiling_liquid_expanding_vapor_explosion.
9. *Dow's Fire & Explosion Index Hazard Classification Guide, Seventh Edition, 2010,* Wiley On Line Books. Retrieved December 24, 2012, from http://onlinelibrary.wiley.com/book/10.1002/9780470938195.
10. Kletz, T. "*By Accident...a Life Preventing Them in Industry*," PFV Publications, London, 2000.
11. R. P. Benedetti and P. E. May, *NFPA 30 and NFPA 30A Flammable and Combustible Liquids Code Handbook 2012 edition,* National Fire Protection Association, Quincy, MA, 2012.
12. *NFPA 58, Liquefied Petroleum Gas Code, 2011 edition,* National Fire Protection Association, Quincy, MA, 2011.
13. NFPA 704 or the Fire Diamond, Retrieved December 24, 2012 from http://chemistry.about.com/od/safetysignsandsymbols/a/Nfpa-704-Or-The-Fire-Diamond.htm.
14. NFPA 45: Standard on Fire Protection for Laboratories Using Chemicals, 2011 Edition, National Fire Protection Association, Quincy, MA, 2011.
15. What do the A B C ratings mean on Fire Extinguishers? Retrieved December 24, 2012, from http://www.brkelectronics.com/faqs/oem/what-do-the-abc-ratings-mean-on-fire-extinguishers.
16. M. Idrees and S. Aslam, "Prevent major failures and accidents with well-trained and empowered operators," Process *Safety Progress,* Volume 29, Issue 2, June 2010, Pages: 174–178, DOI: 10.1002/prs.10348.
17. R. V. Kalnins, "Emergency preparedness and response," *Plant/Operations Progress*, 6 (1987) 6–10, DOI: 10.1002/prsb.720060106.
18. J. A. Klein. "Two centuries of process safety at DuPont," *Process Saf Prog*, 28 (2009) 114–122, DOI: 10.1002/prs.10309.

19. Q. Wang, T. Ma, J. Hanson and M. Larranaga, " Application of incident command system in emergency response," *Process Saf Prog,* 31(4) (2012) 402–406, DOI: 10.1002/prs.11538.
20. Kidde Fire Trainers, retrieved December 22, 2012 www.kiddeft.com/utcfs/ws-385/Assets/Taiwan%20Site%20Sheet.pdf

Snares to LOPA Action Items

Glenn Rozmus
Smith & Burgess LLC
8700 Northwest Central Drive
Houston, TX
glenn.rozmus@smithburgess.com

Dustin J. Smith
Smith & Burgess LLC
8700 Northwest Central Drive
Houston, TX
dustin.smith@smithburgess.com

Dick A. Baum
Smith & Burgess LLC
8700 Northwest Central Drive
Houston, TX
dick.baum@smithburgess.com

Prepared for Presentation at
American Institute of Chemical Engineers
2013 Spring Meeting
9th Global Congress on Process Safety
San Antonio, Texas
April 28 – May 1, 2013

UNPUBLISHED

Snares to LOPA Action Items

Glenn Rozmus
Smith & Burgess LLC
8700 Northwest Central Drive
Houston, TX
glenn.rozmus@smithburgess.com

Dustin J. Smith
Smith & Burgess LLC

Dick A. Baum
Smith & Burgess LLC

Keywords: LOPA, RAGAGEP, IPL, PRV, SIL

Abstract

Greater numbers of action items are being generated from the LOPA process as it becomes increasingly utilized as a method for risk evaluation. The quantity and type of action items result from the combination of initiating events, conditional modifiers, and prescribed guidelines. The quality of the inputs determines whether the action items will actually provide any additional safety benefit. This paper is not a procedure for performing a LOPA analysis, but presents issues to be aware of when generating a list of initiating events, evaluation of the initiating event severities, and the influence of conditional modifiers.

1. Introduction

As a process safety consulting company, Smith & Burgess has been asked to review numerous action item lists from LOPA studies. The lists were generated based upon company standards, the guidelines of the company PSM policies, and the expertise of the facilitator. Companies are sometimes surprised at the quantity and intent of the action items. We hope to create awareness of the effects of seemingly innocuous company policies and suggest alternative strategies. Examples are given as to how the guidelines used for the LOPA study can alter the quality and quantity of action items. As a result, facility personnel will be better informed prior to and during the LOPA process and understand the effects of the inputs on the results.

2. Background

2.1 LOPA

Layer of Protection Analysis (LOPA) is a hazard analysis method used for evaluation of risk. It is more rigorous than the qualitative HAZOP method but not as rigorous as a fully quantitative risk analysis. It is normally used to evaluate select scenarios generated from a HAZOP study which require further analysis to confirm that consequences and safeguards have been properly addressed. LOPA may also be used as a basis for prioritization of action items based upon ranking of the risks.

2.1.1 Methods

There is no uniform method for performing a LOPA analysis. A well-documented procedure is presented by the Center for Chemical Process Safety (CCPS) [1]. There are a variety of software available which are available for LOPA studies. There are also various hybrid methods, created by individual facilitators or dictated by company policy. It is intended that all of the methods are substantially equivalent and achieve the same desired result.

2.1.2 Guidelines

Guidelines for performing a LOPA study are as varied as the methods used. The guidelines used for a particular study are set prior to the study either solely by the host company or from a method agreeable to both the company and the facilitator.

3. OSHA PSM Standard

3.1 Relevant Regulations

LOPA is not specifically mentioned in the OSHA PSM Standards but there are sections that refer specifically to the evaluation of hazards.

3.1.1 29 CFR 1910.119 PSM Standard

OSHA Standard 1910.119 sections (e)(1), (e)(2), and (e)(3) state that the complexity of the process affects the PHA methodology selected, lists methodologies that are appropriate to evaluate the hazards of the process being analyzed (LOPA is not mentioned), and lists the items that the process hazard analysis must address [2].

3.1.2 Enforcement Directives

Also, OSHA enforcement directive guidelines state that the process must be evaluated with an appropriate methodology according to the hazards of the system [3]. The use of LOPA analysis follows as a means to highlight high risk scenarios.

4. LOPA Action Item Snares

4.1 Company Influence

Often a company establishes the overall policies for performing a LOPA study. Since a policy is therefore in place, employees adhere to it whether it is excessively burdensome or excessively remiss. In addition, since the policy is in writing, it must be also be strictly followed since OSHA will issue citations for non-compliance to written policies.

4.1.1 Knowledge

Have you read your company PHA/LOPA policy? Even though this probably has been written by a concerted team effort by safety professionals and is well-intended, it must be followed no matter its suitability. When the LOPA analysis is scheduled it is too late to read the policy and suggest an internal review if necessary.

4.1.2 Scope

What risks are covered in your company PHA/LOPA guidelines? Are production/economic risks included? Companies who want to have economic risks identified in LOPA studies are sometimes surprised to find the scope and depth of action items generated as a result.
Are conditional modifiers included as an adjustment for the risk level? Not all companies include these for LOPA studies. Sometimes the conditional modifiers are omitted so that the risk is not minimized. However, if there is a risk of serious employee injury but there are normally no employees inside the plant site and the risk level is not modified for this situation, expect additional action items to be generated to meet the required IPLs (Independent Protection Layers).

4.1.3 No Policy

If your company has no policy concerning LOPA studies, then perhaps a verbal negotiation between the company and the facilitator will determine the guidelines for the LOPA study. This will therefore determine the quantity and quality of the LOPA action items. Without a policy there will be no consistency and without consistency the results of the study will vary.

SNARE:
A LOPA study can only be successful if a detailed and comprehensive policy is in place. The LOPA methodology is very standard but the company's determinations and agreements for the risk criteria to be used is imperative. The guidance provided from industry publications are tools for establishing the criteria but the company must provide the "hard" numbers. The quantitative portion of LOPA depends on this input. Any gray areas not addressed may lead to lengthy discussion points and result in recommendations that may not be warranted.

EXAMPLE:
Informal company guidelines were used to evaluate the severity of LOPA scenarios generated from process conditions of overpressure. This resulted in generation of a large number of action items involving a large number of additional IPLs. The first step for resolution was to recommend a RAGAGEP-based standard for evaluation of high pressure scenarios. This was found in API RP 581 [6] which lists the percentage of vessel overpressure and correlates it to the vessel and system integrity. This indicated that risks were overemphasized and that there was actually no hazard for this group of action items. The second step was to complete the action items with an alternative solution (in this case no action), based on an OSHA compliance guideline and enforcement procedure.

4.2 Facilitator Influence

Even though the facilitator may be directed by company policies and may even use specialized software, the overall quality of the LOPA study still rests with the facilitator. The ultimate goal for conducting a LOPA study is to have the team comprised of a LOPA-experienced facilitator and LOPA-trained members.

4.2.1 Scenarios

Scenarios for LOPA analyses should be the most serious consequence of all initiating events considered for individual equipment or groups of equipment. Without written guidance, the facilitator may include excessive scenarios without regard to quantity or quality of the output (action items).

SNARE:
The risk criteria as determined by the company must include the risk matrix trigger points for sending the scenario to LOPA. Different severities for safety, environment, commercial impacts or business interruption must be defined based on the company guidelines.

4.2.2 Frequency Modifiers

Has the initiating event under consideration ever happened in the history of the facility? Guidance for this frequency should be explained in the company policy. Without written guidance, the facilitator will make suggestions which may or may not be uniform from scenario to scenario and which may or may not be consistent with published guidelines for equipment failure.

SNARE:
The trap! Frequency modifiers are the qualitative portion of LOPA that must be approached with caution yet are vital to an accurate assessment of risk. Occupancy, time at risk, and ignition probability should all be factored where applicable. That's why they were developed. Caution must always been used to ensure they don't become the method to play with the numbers and reduce the risk.

4.2.3 IPLs

Independent Protection Layers (IPLs) can be interpreted with some variety. For example, the number of IPLs counted with different instrumentation interlocks may depend on whether the interlocks are on separate controllers, on separate input cards on the same controller, or on different racks on the same controller. Credit may or may not be taken for these variations.

SNARE:
A planning meeting with the LOPA facilitator to review IPL criteria has many benefits. As facilitators we have all had the training, read the books and facilitated the studies. Each company is unique and their IPL criteria will not be the same. A review between the facilitator and the company is the place for questions to be answered and resolved. Sometimes as facilitators our experience does get in the way.

4.3 Action Items Aspects

After LOPA action items have been generated, they must be addressed. With a good LOPA study this is straightforward. With a poor LOPA study there are snares that have already been tripped.

4.3.1 Quantity

A large quantity of LOPA action items may have resulted from a variety of causes. Now what? Are the action items based on analysis of totally different scenarios? Are they from the same scenarios with different contributing factors? Do the action items overlap between scenarios? Sorting through the jumble can be daunting.

SNARE:
Different initiating events can result in the same consequence, but keep in perspective that different IPL's and modifiers may exist. Each scenario has to be evaluated individually.

EXAMPLE:
A large number of LOPA action items were generated from a large number of similar scenarios as a result of company policy and facilitator guidance. Sorting through the similar scenarios and similar action items was overwhelming. The solution was to group the action items based on the root cause scenario and the specific equipment. This enabled the minimum amount of IPLs to be utilized to remedy numerous action items.

4.3.2 Quality

Will the LOPA action items improve the safety of the facility? Prior to the LOPA analysis, were applicable industry standards considered?

SNARE:
LOPA recommendations must be developed to "close the gap". Unfortunately, too often the thought is to add an IPL and move on. This may create burdensome recommendations that though they solve one problem, they create other issues. The focus should always be to prevent and detect first, working with the left side of the LOPA sheet. Eliminating or reducing the likelihood of the initiating event should always be the first objective.

EXAMPLE:
A LOPA analysis was performed on an equipment burner system without regard to RAGAGEP (Recognized and Generally Accepted Good Engineering Practices) standards. In this case this referred to industry standards for burner systems. Remedy for the numerous recommendations generated involved two steps. The first step was to compare the burner system to a NFPA standard which defines safety systems for burners [4]. After satisfying that all requirements were currently met, the second step was to complete the recommendation with an alternative solution (in this case no action). For this example, resolution of PSM action items is addressed by a PSM Compliance Guideline and Enforcement Procedure.
"OSHA considers an employer to have "resolved" the team's PHA findings and recommendations when the employer either has adopted the recommendations or has justifiably declined to do so. Where a recommendation is rejected, the employer must communicate this to the team and expeditiously resolve any subsequent recommendations of the team. An employer can justifiably decline to adopt a recommendation where the employer can document, in writing and based upon adequate evidence, that one or more of the following conditions are true:

1. The analysis upon which the recommendation is based contains material factual errors;

2. The recommendation is not necessary to protect the health and safety of the employer's own employees or the employees of contractors;
3. An alternative measure would provide a sufficient level of protection; or
4. The recommendation is infeasible."[5]

Since the burner system already met a RAGAGEP standard, the action items were not necessary to protect the health and safety of the employer's own employees or the employees of contractors and the action items could be closed.

EXAMPLE:
A company wished to resolve LOPA action items regarding installation of new pressure relief valves (PRVs). Resolution included re-evaluation of the process system to confirm or refute the hazard, PRV installation recommendations, or alternately installation of safety integrity level (SIL) instrumentation. Guidance was presented to the company for all approaches. The first guidance was to refer to ASME Section VIII [7] which states that there be no credible overpressure scenario in which the pressure exceeds 116% of the MAWP. If the process system meets this requirement then no PSVs are required.
API Standard 521 [8] states that if instrumentation is used to replace relief devices, it must be at least as reliable as a SIL 3 system. A SIL 2 system may be acceptable if some other means are used to bridge the gap to a SIL 3 system. Part of a correct instrumentation solution is a system which contains appropriate documentation.

5. Conclusion

Plan now to prepare for your next LOPA study. Refresh yourself with the OSHA PSM requirements. Read your company policies. After the LOPA study begins, the path will already be set. The action items will reflect company intentions and/or facilitator preferences. Handling the action items will then either be a meaningful effort for beneficial solutions or a less fruitful effort to justify alternatives.

6. References

[1] Layer of Protection Analysis, Center for Chemical Process Safety, New York, 2001.

[2] Occupational Safety and Health Administration, Occupational Safety and Health Standards, Subpart H, Hazardous Materials, Standard Number 1910.119, Process safety management of highly hazardous chemicals, www.osha.gov/pls/oshaweb/owadisp.show_document?p_id=9763&p_table=STANDARDS (accessed Dec. 27, 2012)

[3] OSHA Enforcement Guidelines, http://www.osha.gov/pls/oshaweb/owadisp.show_document?p_table=STANDARDS&p_id=9760 (accessed Dec. 27, 2012)

[4] NFPA 87 Recommended Practice for Fluid Heaters, 2011 Edition, NFPA, Quincy, MA.

[5] OSHA Instruction CPL 2-2.45A CH-1 September 13, 1994 Directorate of Compliance Programs, www.osha.gov/pls/oshaweb/owadisp.show_document?p_id=1558&p_table=directdire (accessed Dec. 27, 2012)

[6] Risk-Based Inspection Technology, API Recommended Practice 581, 2nd edition, September 2008 (accessed Dec. 28, 2012)

[7] 2007 ASME Boiler & Pressure Vessel Code, 2008a Addenda, VIII, Division 1, ASME, 2008 (accessed Dec. 28, 2012)

[8] Pressure-relieving and Depressuring Systems, ANSI/API Standard 521, 5th edition, January 2007, API, 2008 (accessed Dec. 28, 2012)

ADDRESSING ENABLERS IN LAYERS OF PROTECTION ANALYSIS

Dr. Paul Baybutt
Primatech Inc.
50 Northwoods Blvd., Columbus, Ohio, USA
paulb@primatech.com

Prepared for Presentation at
American Institute of Chemical Engineers
2013 Spring Meeting
9th Global Congress on Process Safety
San Antonio, Texas
April 28 – May 1, 2013

UNPUBLISHED

ADDRESSING ENABLERS IN LAYERS OF PROTECTION ANALYSIS

Dr. Paul Baybutt
Primatech Inc.
50 Northwoods Blvd., Columbus, Ohio, USA
paulb@primatech.com

Keywords: Layers of protection analysis; enabling events and conditions; time-at-risk factors; conditional modifiers; givens.

Abstract

Layers of protection analysis (LOPA) is used to evaluate the risk of individual hazard scenarios by combining initiating event frequencies with failure probabilities of protection layers. Some practitioners include events and conditions that enable the occurrence of hazard scenarios in the analysis, such as conditional modifiers, but sometimes they are excluded to ensure conservative results. However, these events and conditions, and other factors that enable scenarios, are often key parts of hazard scenarios and their exclusion from the analysis can result in overly conservative results. This paper broadens the definition of enabling events and conditions to include other factors that can have a significant impact on the risk of hazard scenarios. Such other factors include management systems to account for inadequacies in, and failure to follow, policies, procedures and work instructions; at-risk factors to account for the time period in which a process is at risk; incident outcomes to represent different possible consequences for the same initiating event; and release conditions to account for different release conditions or circumstances. Their inclusion in LOPA studies is described with examples. The determination of adjustment factors to account for their effect on scenario risk is also demonstrated.

1. Introduction

Layers of Protection Analysis (LOPA) is an analytical technique used to assess the risk of hazard scenarios for processes wherein scenario risk is defined as the couplet of frequency and severity for the scenario consequence [1, 2]. Frequency and severity can be evaluated qualitatively or quantitatively but most commonly qualitative severity and quantitative frequency estimates are used in LOPA. Scenario frequency is calculated using simplifying approximations and is determined by multiplying the frequency of the initiating event by the failure probabilities of protection layers that have met appropriate qualification criteria, including independence from other elements of the scenario [1, 2].

Hazard scenarios for processes begin with an initiating event and end with a consequence impact. The path from initiating event to worst-case consequence involves the failure of any protection layers that guard against the scenario. However, there may be other scenario elements that influence the scenario risk including enabling events and conditions, time-at-risk factors, intermediate events, incident outcomes, conditional modifiers, and given conditions. Many of these scenario elements act to reduce scenario risk because their likelihoods of occurrence are factored into the analysis through multiplication of the scenario frequency by the likelihoods which may be quite low. In some cases, scenario consequences are also impacted and risk may be increased or decreased. This article describes the role of such elements in risk analysis and describes how to address them in LOPA.

Conventionally, LOPA studies determine the risk of a hazard scenario by accounting primarily for the initiating event and protection layers. Often, only a few of the enablers and other adjustment factors described in this article are addressed. However, the events represented by these enablers and adjustment factors commonly are key parts of hazard scenarios and their exclusion from the analysis can result in overly conservative results. Enabling events and conditions are important not only because they influence the risk of hazard scenarios but also because they make scenarios possible.

LOPA was originally conceived as a simple risk analysis method that at best produces an order of magnitude estimate of scenario risk. However, it is now being used to support the determination of Safety Integrity Levels (SILs) for Safety Instrumented Functions (SIFs) to assist compliance with the IEC 61511 / ISA 84 standard on Safety Instrumented Systems (SISs) [3 - 5]. The standard requires substantial effort to support the SILs claimed for SIFs. Consequently, refinement of LOPA to support such effort is warranted. Furthermore, LOPA has evolved from its original form and current applications seek greater rigor and incorporate more detail [6].

Historically, some LOPA practitioners have not addressed enablers and adjustment factors in the belief that the uncertainties associated with them are too great and risk may be underestimated by their inclusion, and because of the effort involved. However, they are key parts of hazard scenarios and are often part of actual incidents. Their inclusion in LOPA studies arguably produces more accurate risk estimates and conservative assumptions can be made to help avoid risk underestimation. Furthermore, the effort to include them is not substantial in comparison to the overall effort required to perform LOPA.

2. Enablers

Enabling events and conditions have been defined as events and conditions that do not directly cause a scenario but are required to be present or active for the scenario to proceed [1]. A bypassed high level alarm that allows overflow of a tank is an example. This article extends that definition to include any scenario elements, in addition to the initiating event and actions of protection layers, that influence the risk of a hazard scenario, and the term *enablers* is used to encompass them all. The original definition of enabling events and conditions included modes of process operation such as startup and shutdown or the operation being in a specific phase or step [1]. These are best treated using time-at risk factors.

Conventionally, enabling events and conditions usually have been viewed as acting to reduce the frequency of a hazard scenario, or modify its consequences. They do so because their probability of occurrence is usually less than 1 and the scenario frequency is reduced accordingly to account for their role in enabling the scenario. Conservative analyses assume their probability of occurrence is 1. In practice, their probabilities of occurrence may be substantially less than 1 and they may reduce scenario risk significantly.

The broader definition of enablers includes elements that can increase the frequency, for example, lack of PM on equipment that increases its failure rate. Enablers may be categorized as:

- *Enabling events and conditions.* Must be present or active for the scenario to proceed.
- *Management systems.* Account for inadequacies in, and failure to follow, policies, procedures and work instructions.
- *At-risk factors.* Account for the time period in which a process is at risk.
- *Incident outcomes.* Used to represent different possible consequences for the same initiating event.
- *Release conditions.* Used to account for different release conditions or circumstances.
- *Conditional modifiers.* Affect the scenario consequence.
- *Given conditions (also called "givens").* Enable a scenario but are always present.

3. Enabling Events and Conditions

Enabling events and conditions do not by themselves initiate a hazard scenario but they make them possible. An enabling event may occur or an enabling condition may exist without the scenario happening every time the event occurs or the condition is present. They are sometimes called *contributing causes* or *contributing factors* and they may make possible any other element of a hazard scenario, e.g. the initiating event, a protection layer failure, or the consequence. There may be multiple enabling events or conditions for each of the other elements of a scenario and enabling events and conditions for more than one element.

Enabling events and conditions can be classified as originating with:

- Human actions (errors of omission or commission, extraneous events, and deliberate acts), for example, disabling equipment, e.g. bypassing an interlock, or overriding an inhibit condition.
- Equipment failures, for example, an alarm failure.
- External events, for example, extreme ambient conditions, e.g. low temperature; or utility failures, e.g. loss of inerting.

Enabling events usually occur prior to the initiating event, e.g. a failed or disabled alarm. They are sometimes called *latent failures*. Enabling conditions exist at the time the scenario occurs, e.g. low environmental temperature. They are sometimes called *latent conditions*.

Enabling events and conditions are addressed in LOPA by assigning a value to the probability that the event or condition will exist when the scenario occurs and multiplying the scenario frequency by it. The values are obtained by estimating the likelihood that the event or condition will have occurred or be present when the scenario occurs. Such estimates should be based on actual experience with the process as they are specific to the process and no generalized values can be provided.

Of course, in order to address enabling events and conditions they must be identified. Arguably, that should be done as part of the performance of Process Hazard Analysis (PHA) which is the most commonly used source of hazard scenarios for LOPA studies. However, enabling events and conditions are identified infrequently in PHA [7]. Consequently, LOPA teams need to question whether enabling events and conditions may be factors for the scenarios analyzed.

4. Management systems

Management system enablers are failures in the systems set up to manage safety throughout the lifecycle of a process. For example, they may be inadequate procedures, e.g. test and inspection frequencies may be set too low; no or inadequate training of personnel; inadequate skills or knowledge of personnel; failures in the execution of procedures, e.g. PM is not conducted per requirements; and mis-operation of equipment, e.g. stressing a pump by using it outside its operating limits. Fundamentally, management system enablers are failures by people. When present, they are givens for scenarios.

Management system enablers directly influence the failure rates used in LOPA studies for other scenario elements. They may increase initiating event frequencies or probabilities of failure of protection layers and they are addressed in LOPA by assigning a value by which a failure rate must be adjusted, usually in an upwards direction, so that the values are numbers greater than 1. For example, the initiating event frequency for a pump mechanical failure may be adjusted upwards to account for mis-operation of the pump outside its limits, assuming the pump failure frequency used does not already reflect the mis-operation. Similarly, the probability of failure on demand of a relief valve may be adjusted upwards to account for lack of PM on the valve, assuming the valve failure frequency used does not already reflect operation without adequate PM. The effects of operating and maintenance regimes play a similar role and they can be considered within the category of management system enablers.

The adjustment factors can be used to modify the failure data directly, or they can be identified explicitly as modification factors and folded into the overall calculation of the scenario frequency. Such adjustment factors should be based on experience with the process or the opinions of informed personnel.

In accounting for management system enablers, it is important to ensure that their effects have not already been incorporated into the frequency or probability of the event it enables. For example, if a pipe leak failure frequency includes the effects of lack of PM, or if the probability of failure of a human protection layer already addresses the skill level of the people involved, adjustment factors should not be applied as they are already incorporated into the data. However, while tabulations of human failure data may reflect various influencing factors, such as stress on people and the time available for action, usually not all pertinent factors are addressed in the tabulations. Consequently, such unaccounted factors should be considered as management system enablers if they are to be addressed in LOPA.

The same comments apply to management system enablers as for enabling events and conditions with regard to their identification.

One danger in using management system enablers is that it could institutionalize deviations from required practices by allowing some risk reduction credit to be taken even when practices are not followed exactly. Some companies may take the view that if there is a deviation from the requirements of a management system, no credit should be taken at all until the deviation is corrected. Clearly, this is a conservative position but it may have unintentional consequences. If LOPA team members are required to follow this approach, they may be unwilling to admit that management systems are not being followed, or they may claim that while there have been deviations in the past there will not be any in the future. If these claims are accepted, the LOPA results may be overly optimistic and underestimate risk. The preferred approach is to allow some justifiable credit to be taken but ensure a recommendation is made to correct the deviation promptly, or revise the requirements to comport with actual practices. Deviations from required practices that are tolerated are likely to result in more serious consequences than creating difficulties in the performance of LOPA.

5. Time-At-Risk Factors (TARF's)

Some hazard scenarios can occur only when the process is in a particular state or certain conditions exist, e.g. the process is in a particular mode, phase or step. Scenario frequencies are usually expressed on an annualized basis to match risk tolerance criteria expressed in the same form. Consequently, scenario frequencies should be adjusted for time at risk, i.e. the fraction of time the risk is present. Receptors are at risk for only this time period. If such adjustments are not made, risk may be grossly overestimated. For example, a tank that is filled 20 times per year is subject to hazard scenarios that may result in overfilling. If the cause of overfilling is failure of a BPCS level control loop with an in-service failure rate of 0.1 failures per year and each filling takes 1 hour:

> Filling failures per year = Failures per year x Fillings per year x Time spent filling
> and, Filling failures per year = 0.1 x 20 x 1/8760 = 2.3×10^{-4}

Thus, the initiating event frequency has been reduced by almost three orders of magnitude, a risk reduction that definitely should be applied. Further examples of TARFs for processes are the fraction of time:

- A piece of intermittently used equipment is in operation.
- A continuous process is in startup or some other mode of operation.
- A batch process spends in a particular step.
- A runaway reaction is possible in a batch process.
- People are at risk owing to time-of-day effects, day-of-week effects, time indoors versus outdoors

An adjustment is made to the scenario frequency by multiplying it by the fraction of time the scenario can occur. For example, for the initiating event of a pump seal leak, if the available data are for an annualized pump seal leak failure rate:

> Scenario frequency (events / year) = Pump seal leak frequency (events / year) x Hours in use / 8760 (hours in a year)

Of course, at issue is whether the data correspond to a pump that is operated intermittently and that has been annualized, or whether it corresponds to a pump that is operated continuously. Equipment in frequent use may fail at different rates than equipment used infrequently. If the data do not match the operating regime, an adjustment factor could be applied to the data, or more appropriate data sought. Note that the adjustment factor can be viewed as representing an enabler for the scenario.

Similarly, for a process that is in a particular mode of operation, e.g. startup, hazard scenarios that can occur during that mode of operation, and for which annualized frequencies have been calculated, need to be adjusted for the time at risk. For example, a process that is in startup mode once per year for 24 hours would require a time-at-risk adjustment of 1/365 for startup hazard scenarios.

Usually, TARFs are not difficult to identify, although the PHA team may not have addressed them, and they are not difficult to estimate since they depend on straightforward and readily available information. In practice, ranges of time-at-risk are usually possible and practitioners must decide if the average or end-of-the-range value should be used. Generally, the conservative choice is made.

6. Intermediate Events

Intermediate event enablers allow LOPA practitioners to account for the probabilities of different hazard scenarios that result from the same initiating event. Some events may be part of a hazard scenario but they are distinct from the initiating event, consequences and protection layers. For example, if water is introduced to a storage tank containing sulfur trioxide, a release through a relief valve may occur, but also a release owing to a corrosion failure of piping may result. These two occurrences are examples of intermediate events. Whenever there is more than one possibility for a particular intermediate event, it can be treated as an enabler representing the probability of occurrence of the specific intermediate event, provided that the events are mutually exclusive. The scenario frequency is multiplied by the probability of the intermediate event to adjust for its likelihood of occurrence. Of course, the probabilities of all such possibilities for a particular intermediate event must sum to 1. It is also entirely possible that both events may occur together, in which case an enabler would not apply.

Vessel rupture from overpressure provides another example of an intermediate event that affects the path of a hazard scenario. The probability of vessel rupture depends on various factors such as the ratio of the actual pressure reached to the design pressure for the vessel, vessel structure, vessel construction and material, vessel maintenance, and the process materials. More than one scenario is possible. In one scenario the vessel fails, presumably with serious consequences; in other scenarios the vessel does not fail but the consequences of the scenarios may still be of concern. Each of the scenarios has its own probability of occurrence and these probabilities can be used to adjust the scenario risk.

Some chemicals may pose multiple hazards, for example, hydrogen sulfide and ammonia are each both toxic and flammable. Usually, each possible hazard is addressed separately and they may be considered to be mutually exclusive in which case they can be assigned probabilities of occurrence and treated as intermediate event enablers. Mixed hazards, where one hazard is realized first followed by another, e.g. a toxic exposure followed by an explosion, usually are not modeled in order to keep matters simple. Care must be taken to avoid double counting with probability of ignition enablers.

Intermediate events are identified as part of the scenario and should have been addressed in PHA. However, it is possible the PHA addressed only one of several possibilities in which case the LOPA team should consider the additional intermediate events. Values of intermediate event probabilities may be difficult to estimate. In such cases, they may be assumed equal to 1 in order to be conservative. In some cases, consequence severities may vary with the intermediate events.

7. Incident Outcomes

Sometimes the outcomes of hazard scenarios vary. Each scenario outcome should be modeled individually and its frequency adjusted by the probability of the outcome in a similar way to that for intermediate events. Of course, the probabilities of the outcomes must sum to 1 if they are mutually exclusive. Consequence severities may vary with incident outcomes. Examples of outcomes include fire versus explosion, the type of fire, and the type of explosion.

Often, the relative probabilities of incident outcomes are not known, in which case no credit can be taken for this type of enabler. However, there may be cases in which justification can be provided for a significant difference in the probabilities of outcomes, for example, a fire that may occur with a 0.9 probability and an explosion with a 0.1 probability. In such cases, risk reduction by a factor of 10 for the explosion scenario is worth considering. Some outcomes may have a sufficiently low probability of occurrence that they can be excluded from the analysis as non-credible scenarios.

Sometimes, analysts may choose a worst-case for study. However, worst-case scenarios may vary according to the receptors at risk, for example, for employees in the immediate area of a release it may be a jet fire while for employees further away it may be a flash fire. Consequently, care must be taken when using a worst-case consequence approach.

PHA may address only one incident outcome. Other possible outcomes should be addressed by the LOPA team to ensure a full consideration of risk.

8. Release Conditions

The consequences of hazard scenarios may vary according to conditions and circumstances at the time of release. Sometimes these are referred to as *incident outcome cases*. Each such scenario may be modeled individually and its frequency adjusted by the probability of the release conditions for the scenario. Release characteristics such as hole size, location, elevation, orientation, duration, and delayed ignition may influence the scenario that occurs. Weather conditions such as wind direction, wind speed, air temperature, atmospheric stability class, and precipitation may also play an important role. To the extent that such release conditions are important, they may be accounted for in a LOPA study by adjusting the scenario risks by their probabilities of occurrence. In some cases, they may change the consequence that is possible.

Release characteristics may be difficult to determine and only a worst-case scenario may be addressed. Weather conditions are easier to address because historical data are usually available but a worst-case scenario may also be considered in lieu of analyzing variants. However, there may be certain weather conditions, such as wind direction, that have a significant impact on the scenarios possible. For example, the prevailing wind may blow towards a potential ignition source for a flammable release only 10% of the time which means that the frequency of the scenario could be reduced by a factor of 10 if credit were taken for wind direction as an enabler. It may make the difference between an explosion consequence and one in which the flammable material disperses harmlessly.

9. Conditional Modifiers

Conditional modifiers directly impact the scenario consequence. Common conditional modifiers are:

$P^{ignition}$ - Probability that a flammable / explosive material will be ignited.

$P^{present}$ - Probability that a person will be present to be exposed to a hazard. It is sometimes called the occupancy factor and represents the fraction of time personnel are exposed to the hazard.

P^{injury} - Probability that harm will occur if an individual is exposed. It is sometimes called the vulnerability.

These probabilities are used to reduce the frequency of the scenario in which harm occurs. For example, if a pump seal fire can occur during operator rounds but the operator is in the vicinity of the pump for only 30 minutes during a 12 hour shift, the frequency of the scenario in which an operator is exposed to fire should be multiplied by 0.5/12 = 0.042, assuming there are two shifts.

There can be some pitfalls with conditional modifiers [1]. During some modes of operation, such as startup, operators may always be present, i.e. $P^{present} = 1$. During the build-up to a hazardous event, more people may be present investigating the symptoms, i.e. its is likely that $P^{present} = 1$ when a release occurs. Human presence may be correlated with the cause of a hazardous event, i.e. it is possible that $P^{present} = 1$ if the person contributes to the initiating event, for example, for a release caused by the operator opening a drain valve. The initiating event, $P^{ignition}$, and $P^{present}$ may be linked. For example, the actions of the person who is present may be the ignition source, e.g. a crane operator may drop a load on the process causing a flammable release and providing an ignition source through metal-on-metal sparking (or from the crane engine). In this case, $P^{ignition} = 1$ and $P^{present} = 1$.

Various factors influence the values of conditional modifiers (Table 1). Values for various situations are available in the literature [1] but they must be used with care as they may not apply to the situation at hand. Conditional modifiers can be controlled in various ways (Table 2).

Other possible conditional modifiers include probability of sheltering, probability of escape, and probability of evacuation, although these may be incorporated into the evaluation of P^{injury}. For conservatism, many practitioners take no credit for emergency response actions unless special circumstances warrant, e.g. special provisions are made for sheltering in particularly hazardous situations and a non-zero probability of effective sheltering can be justified.

Usually, PHA teams assume the worst case for conditional modifiers for conservatism and effectively they are not identified explicitly in PHA worksheets. LOPA teams must decide if they should be addressed explicitly.

10. Given Conditions

Some apparent enablers are actually fixed aspects for a scenario. Management system enablers are an example. In contrast, other enablers may or may not be present, i.e. they are variable in nature. Given conditions are always part of the scenario and influence it. For example, a boiler house that acts as an ignition source for a release is a given condition but hot work in the process is an enabler for a fire scenario. The former is fixed and always present, unless the probability of the boiler being in a lit condition is to be addressed, while the latter is variable as hot work is not usually conducted on a continuous basis. Other examples of given conditions include the omission of safety features from the process design and locations of stationary equipment. Many given conditions do not adjust the frequency of scenarios. Rather, they make scenarios possible by their presence.

Given conditions are part of the scenario definition although they may be assumptions in PHA. LOPA team may need to clarify the assumptions.

11. Values and Use of Enabler Multipliers

As for other failure data used in LOPA, the values used to incorporate the effect of enablers on scenario risk for processes should reflect actual experience with the processes. Judgment may also be needed as often data are sparse, but the values used should be justified with process data or expert opinion.

Typically, only enablers that may impact the scenario risk by more than an order of magnitude are included in the analysis, e.g. if a disabled alarm that allows a scenario to occur is in a disabled state 10% or less of the time so that it reduces scenario likelihood by a factor of 10 or more, or if lack of PM on a vessel increases the likelihood of a corrosion failure by a factor of 10 or more. In some cases, enablers that together produce an order of magnitude risk reduction may be credited but care must be exercised as the credits taken may produce a non-conservative result owing to possible dependencies between enablers.

For enablers that represent two or more alternative scenario paths, if one path has a probability of occurrence of 0.5 or above, the enabler multiplier may be assumed to be 1 for convenience and conservatism. Generally, such multipliers are used when the effect on the scenario risk is substantial, i.e. when their probabilities are 0.1 or less.

Various enablers may combine together to reduce the risk of a hazard scenario substantially. However, multipliers for the enablers described in this article should not be used arbitrarily to meet risk tolerance criteria. The temptation to convince oneself that an extra order of magnitude risk reduction is possible by reducing the value of the multiplier for an enabler by a factor of 10, or that an additional enabler that reduces the risk to a tolerable level should be credited, must be resisted unless they can be credibly justified. All data used in LOPA must be justified and should favor conservative values.

Care must be taken not to double count enablers that have already been accounted for through scenario consequences or assumptions made in the performance of PHA or LOPA. Studies may be more susceptible to this issue when LOPA is performed by a different team or when a long time passes between the performance of PHA and LOPA. Also, enablers must not be confused with initiating events and the actions of protection layers, although they may directly impact them.

Often, PHA studies do not identify enablers, leaving the effort to LOPA teams if enablers are to be addressed. As LOPA practices evolve to address enablers, it is likely that PHA teams will begin to identify enablers. LOPA teams usually engage in more detailed discussions of hazard scenarios than PHA teams. The events that make up scenarios, including enablers, must be clearly defined and understood, and interactions between elements of scenarios must be addressed, if valid results are to be produced. The simple examples contained in this article are provided for illustrative purposes. Many actual hazard scenarios involve multiple events and event trees can be used to understand possible combinations for inclusion in LOPA studies.

12. Conclusions

Events in hazard scenarios, in addition to initiating events and the action of protection layers, can have a marked impact on their risk. Certain enabling events and conditions and some conditional modifiers are sometimes included in studies by LOPA practitioners. This article broadens the definition of enablers to include various other factors that may act and combine to make scenarios possible, including enablers relating to management systems, time at risk, intermediate events, incident outcomes, and release conditions. Factors that account for the effect of these enablers are used to multiply the scenario frequency. In some cases, the scenario consequence may also depend on the enabler. Values for multipliers must be chosen carefully and justified. In cases where data are not available, worst-case assumptions are usually employed.

13. References

[1] CCPS, 2001, "Layer of Protection Analysis," *Center for Chemical Process Safety / American Institute of Chemical Engineers*, 2001.

[2] Baybutt, P., "Analytical Methods in Process Safety Management and System Safety Engineering – Layers of Protection Analysis," in *Handbook of Loss Prevention Engineering*, J. M. Haight (ed), Wiley-VCH, 2013.

[3] ANSI/ISA—84.00.01—2004 Part 1 (IEC 61511-1 Mod), Functional Safety: Safety Instrumented Systems for the Process Industry Sector — Part 1: Framework, Definitions, System, Hardware and Software Requirements.

[4] ANSI/ISA—84.00.01—2004 Part 2 (IEC 61511-2 Mod), Functional Safety: Safety Instrumented Systems for the Process Industry Sector — Part 2: Guidelines for the Application of ANSI/ISA—84.00.01—2004 Part 1 (IEC 61511-1 Mod).

[5] ANSI/ISA—84.00.01—2004 Part 3 (IEC 61511-3 Mod), Functional Safety: Safety Instrumented Systems for the Process Industry Sector — Part 3: Guidance for the Determination of the Required Safety Integrity Levels — Informative.

[6] HSE, 2009, "Safety and Environmental Standards for Fuel Storage Sites," Process Safety Leadership Group, Final report, HSE Books, 2009.

[7] Baybutt, P., "Conducting Process Hazard Analysis to Facilitate Layers of Protection Analysis," *Process Safety Progress*, Vol. 31, Issue 3, pps 282–286, September 2012.

Table 1. Examples of Factors that Influence the Values of Conditional Modifiers

$P^{ignition}$	$P^{present}$	P^{injury}
Initiating event - if it produces or provides a source of ignition	Mode of operation	Type of event, e.g. pool fire versus flash fire
Physical properties, e.g. - Flammable and explosive limits - Physical state (gas, vapor, liquid)	Initiating event	Duration and magnitude of the exposure
Chemical properties, e.g. reactivity	Attended / unattended operation	Escape routes
Layout, e.g. proximity and location of ignition sources	Need for operator presence in an adjacent area	Ability to escape: - Detection of exposure - Time to incapacitation - Skill / knowledge - Physical ability - Availability / use of PPE
Environmental factors that impact dispersion, e.g. wind direction		

Table 2. Examples of Control Measures for Conditional Modifiers

$P^{ignition}$	$P^{present}$	P^{injury}
Hazardous area classification	Barriers	Hazardous area entry control
Ventilation	Access control	Release detection and alarms
Procedures	Exclusion areas	Escape plans
Equipment design	Procedures	Protective equipment
Release containment		Refuges
		Training

Bowtie:
Visual Aid to Reduce Spurious Trips and Nuisance Alarms

Patrick Fisher, EIT
ACM Automation Inc
700, 940 – 6th Avenue S.W.
Calgary, AB T2P 3T1
pfisher@acm.ca

Yijia Sun
ACM Automation

Fareed Ebrahim
ACM Automation
febrahim@acm.ca

Prepared for Presentation at
American Institute of Chemical Engineers
2013 Spring Meeting
9th Global Congress on Process Safety
San Antonio, Texas
April 28 – May 1, 2013

UNPUBLISHED

Bowtie: Visual Aid to Reduce Spurious Trips and Nuisance Alarms

Patrick Fisher, EIT
ACM Automation Inc
700, 940 – 6th Avenue S.W.
Calgary, AB T2P 3T1
pfisher@acm.ca

Yijia Sun
ACM Automation

Fareed Ebrahim
ACM Automation
febrahim@acm.ca

Keywords: LOPA, PHA, bowtie, alarms, SIF, interlocks

Abstract

HAZOP and Safety Integrity Level (SIL) studies encourage adding safeguards such as alarms and trip with little consideration for cost or operability of the plant once it is running. This practice results in processes that cannot be operated without bypassing a large number of the interlocks, and important alarms are lost in the hundreds or thousands of signals assaulting the operator. Many of these bypasses stay on for longer than needed, or perhaps are never turned off. This bypass condition results in a situation where, despite the best efforts of the Hazard study team, the facility is no safer than if those alarms or trips were nonexistent.

The Bowtie is a method to effectively use the SIL method while ensuring that a plant is still operable at the completion of the study. The Bowtie visualizes, on one diagram, all equipment failures which lead to similar hazardous events. Safeguards are added to reduce the overall likelihood of the hazardous events occurring. Instead of adding more alarms and trips to protect against each failure, the Bowtie allows the SIL participants to easily see which existing safeguards could be used to reduce hazardous event likelihoods. As well, the Bowtie makes it simpler for participants to visualize how increasing the reliability of one safeguard may be able to reduce the total number of trips and alarms needed for safe operation. Through visualization of the SIL method, the Bowtie emphasizes fewer, more effective safeguards to get the same SIL at a lower cost, while lowering downtime for maintenance and spurious trips.

1. Introduction

Overuse of alarms, trips and interlocks is a serious issue in facilities today. Operators are overloaded and important safety-related alarms can be easily missed amongst the influx of hundreds if not thousands of alarms during an emergency. While alarm management systems attempt to fix the issue with prioritization, it can be difficult to determine which alarms are important, and alarms which require longer-term attention may be missed. A better solution would be to reduce the amount of instrumentation connected to the alarm system in the first place. Critical alarms, trips, and interlocks are identified and added in risk studies. During a Process Hazard Analysis (PHA) or single-event layer of protection analysis (LOPA), a single cause-consequence pair is analyzed to determine the inherent risk and the approximate number of independent protection layers (IPLs) that are required to mitigate the risk. This leads to an abundance of IPLs which are only useful for a single scenario. Using a multiple-cause-multiple-consequence visualization for LOPA analysis (Bowtie), the total number of alarm-related instruments can be reduced, resulting in lower cost and reduced operator confusion.

2. Why do we need the Bowtie?

2.1 Alarm overload

The rise of computerized Distributed control systems in the past twenty years has introduced a problem that is recognizable to any person involved with plant design or plant operations: alarm overload. It is inexpensive and easy to add an alarm, so that is often the first approach taken to safeguard against a potentially hazardous event. When alarms were hard wired to a light on a literal board, the particular alarms that the operator had to see were limited by economics. With the removal of this limit, it is not surprising that there are thousands if not tens of thousands of alarms going to the control board on a daily basis. The same goes for process trips: hard wired trips were expensive, but programmed trips are inexpensive.

The assumption that there is no impact to adding more alarms, trip and interlocks is false. Alarm overload has been noted as a contributing factor to a number of process incidents, with the most notable being the BP Texas City explosion. As well, the overuse of trips and interlocks has made it difficult to run plants consistently, as a small process deviation may result in multiple cascading trips that can bring a whole facility down. Alarms and trips need to be maintained, and this can be a considerable amount of effort and money for a large facility.

The primary purpose of qualitative and semi-quantitative risk assessments (s-QRA), such as a LOPA, is to ensure that a plant can operate within the safe bounds set by the company and by society. However, there is a secondary purpose in that, if a qualitative or semi-quantitative risk assessment is done correctly, all safety-critical alarms, trip and interlocks are identified based on the scenarios chosen. In fact, if done to a sufficient level, an s-QRA will tell you the **only** alarms, trips, and interlocks that are required to operate the facility without harm to people, assets, or the environment. But how can an s-QRA be performed in a manner that ensures all risk is covered while minimizing the number of alarms, trips and interlocks required?

2. What is the Bowtie?

2.1 Bowtie as a risk visualization method

The Bowtie diagram is a visual aid to assist in identifying the need for safeguards or independent protection layers (IPLs)[1] in a risk study, and is typically presented in five sections:

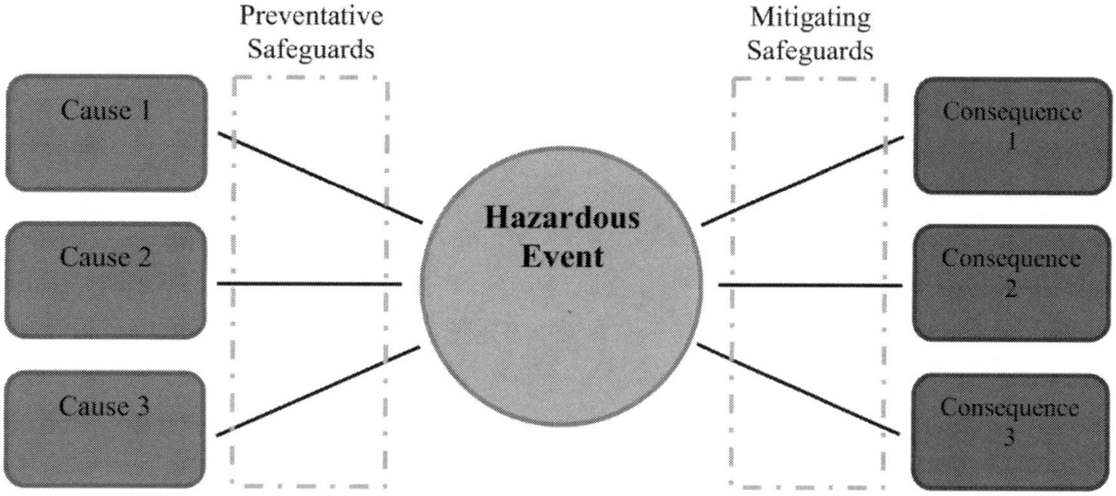

Figure 1: A typical Bowtie diagram showing three causes and three consequences

Causes are initiating events which may lead to a hazardous event if not prevented. Typical causes may range from control valve failure, to pumps running above their intended speeds. Without **Preventative Safeguards** in place, the initiating scenario may lead to a **Hazardous Event** or loss of control. Typical Preventative safeguards are alarms and trips related to level, pressure, or temperature. Typical hazardous events are loss of containment, fires, explosions or gas leaks. In the absence of mitigating safeguards, hazardous events lead to **Consequences** or risk receptors, which may include personnel injuries, financial and production impact, and impact to the environment.

Causes are represented as boxes on the left side, with lines connecting to the hazardous event at the center. The hazardous event is represented by a box in the center connected to the consequences which are represented as boxes on the right of the bowtie. Safeguards are placed as boxes or walls over the connecting line and represent a barrier for hazard progression. Safeguards which protect multiple causes or consequences are "taller" and overlap multiple connecting lines. The bowtie shape comes from the fact that, in the event of no safeguards, all causes lead to the same hazard, and the hazard may lead to any of the consequences. This multiple-cause to multiple-consequences is approximately the shape of a bow tie.
Once safeguards are added to the diagram as barriers of walls, it can become visually clear which risk are covered, which are not covered, and which may be over-covered. For example:

[1] Depending on the intent of the bowtie visualization, safeguards may not be limited to Independent Protection Layers (IPLs). For calculation of LOPA frequencies, only IPLs should be used, but for other studies all safeguards could be added. For the purposes of describing the bowtie, the more general term "Safeguard(s)" is used.

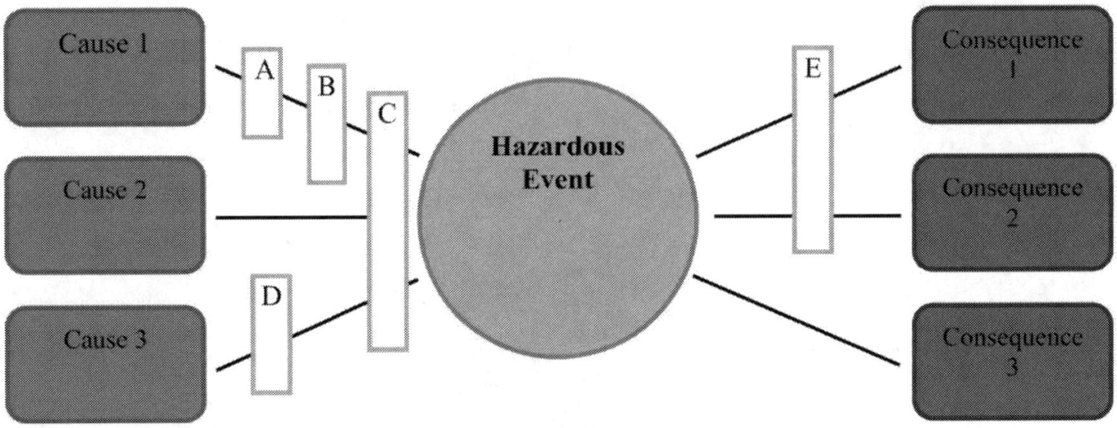

Figure 2: Standard example of a Bowtie from a LOPA

The above figure shows that the line between Cause 1 and hazardous event is packed with safeguards, while Cause 2 and Cause 3 to Hazardous event have only one and two safeguards respectively. This may or may not be ok depending on the scenario. However, we can see that between Cause 2 and Consequence 3, there is only Safeguard C. As such, if Safeguard C is bypassed or fails, there is nothing to stop Cause 2 from leading to Consequence 3. At the same time, Cause 1 is likely over-instrumented given the other causes on the diagram. Situations such as having five or more safeguards on one line while having only one on another are not unusual.

A good bowtie has only a few, large safeguards which cover all lines approximately equally. This minimizes the overall number of IPLs in a plant and spreads the safeguarding more evenly, meaning that there are fewer hidden hazards due to safeguard failure. If the safeguards are not spread evenly, the result will be many low reliability safeguards and a few very high reliability safeguards. The high reliability ones tend to get more attention and there may be too many low reliability safeguards to adequately maintain and test. Over time, risk will build up slowly without notification and hazardous events will become more likely due to complacency.

2.2 Bowtie depicting existing practices

The bowtie visualization is not intended to be an analysis tool, rather it is intended to help visualize a traditional risk assessment in such a way as to look at the bigger picture. For example, a single cause-consequence pair in a HAZOP or single-cause LOPA on a storage tank could be denoted as:

Figure 3: HAZOP or single cause-consequence LOPA Bowtie

Combining the above scenario into a multiple-cause LOPA, the bowtie would appear as:

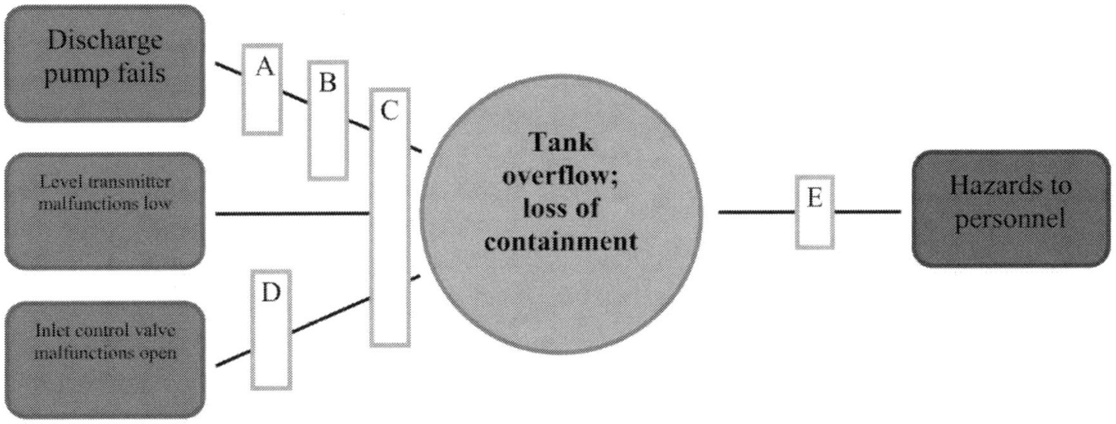

Figure 4: LOPA Bowtie

However, a full bowtie should have multiple consequences as well, which is where the bow tie shape appears:

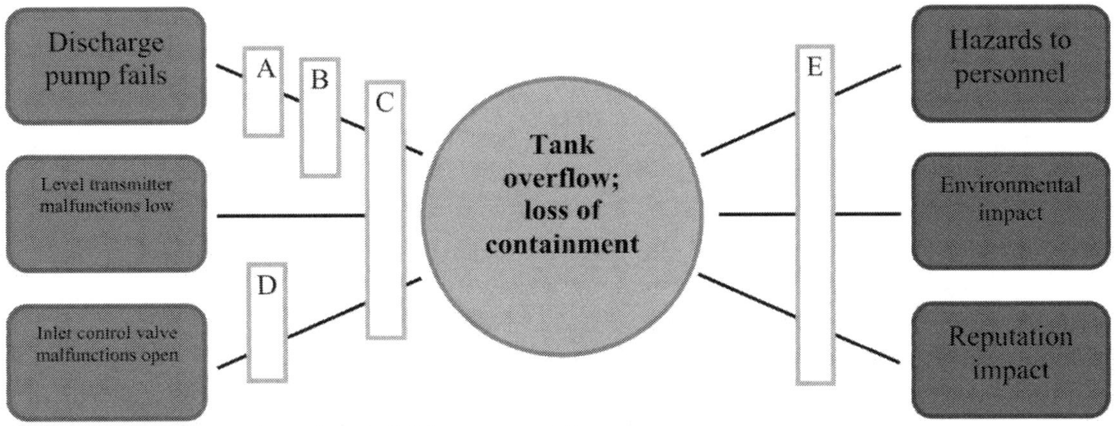

Figure 5: Big picture (multiple LOPA scenario) Bowtie

Now that the team is considering the larger picture, it can be seen that the focus thus far has been too much on specific safeguards. If the multitude of safeguards can be replaced with fewer, broader safeguards, the team could maintain the same protection while decreasing the number of alarms, trips, and interlocks the operator has to deal with. Perhaps safeguards A, B, and D could be replaced by a better safeguard that also protects against the level transmitter failure. If so, total safeguards would be reduced from five to three and the system would have better risk coverage while being simpler to operate and maintain.

2.2 Real-world analysis

An exchanger was reviewed in a HAZOP to determine the possibility of exceeding the design temperature with damage to the equipment leading to hazards to people, environment, and finances. Two causes were identified: a temperature control valve (TV) malfunction and an operator error (closing an incorrect valve). To address the TV malfunctioning closed, a low flow alarm was added and operator rounds were increased to observe heater temperature. To address operator error, an independent valve status alarm and valve tagging were used to mitigate the risk. This resulted in a small remaining risk gap which was covered by a SIL 1 loop that shuts down the unit on high-high temperature. All risks were covered through these safeguards, based on the HAZOP analysis. When plotted on a bowtie, the result was as follows:

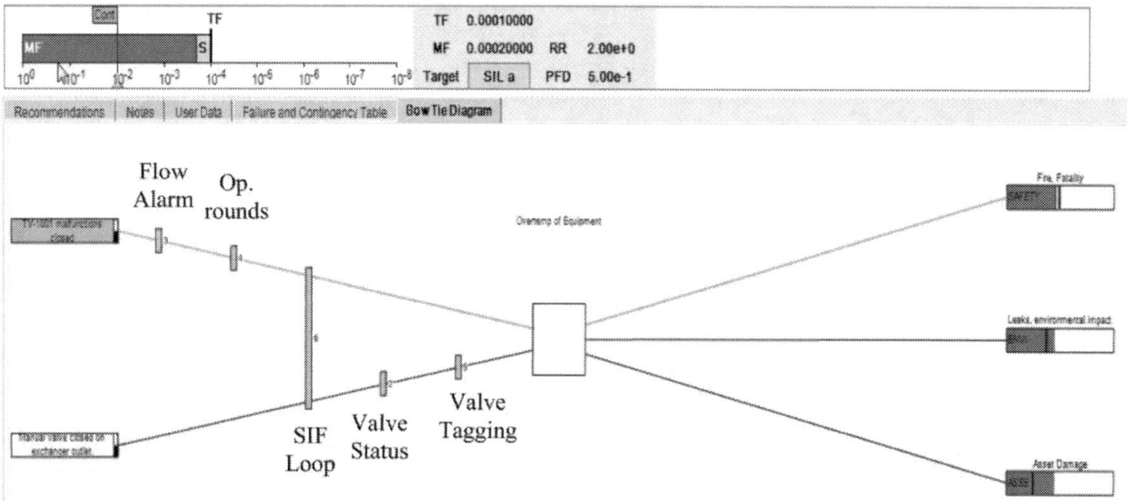

Figure 6: Bowtie resulting from a HAZOP

There were a total of five safeguards in place and the combination of the mitigated frequencies from both causes meant that there was a slight residual risk on the fatality consequence (risk shown in blue in the above diagram). Under a normal LOPA, the SIF loop reliability would be increased to a slightly more reliable SIL 1 and the gap would be covered. However, looking at the bowtie we can see that the system as a whole could be improved. Are all of those safeguards required? The team reviews the options and finds that the operator rounds and low flow alarm can actually help with both causes. These are "stretched" across both lines and the manual-valve specific safeguards are removed. The results are shown in Figure 7:

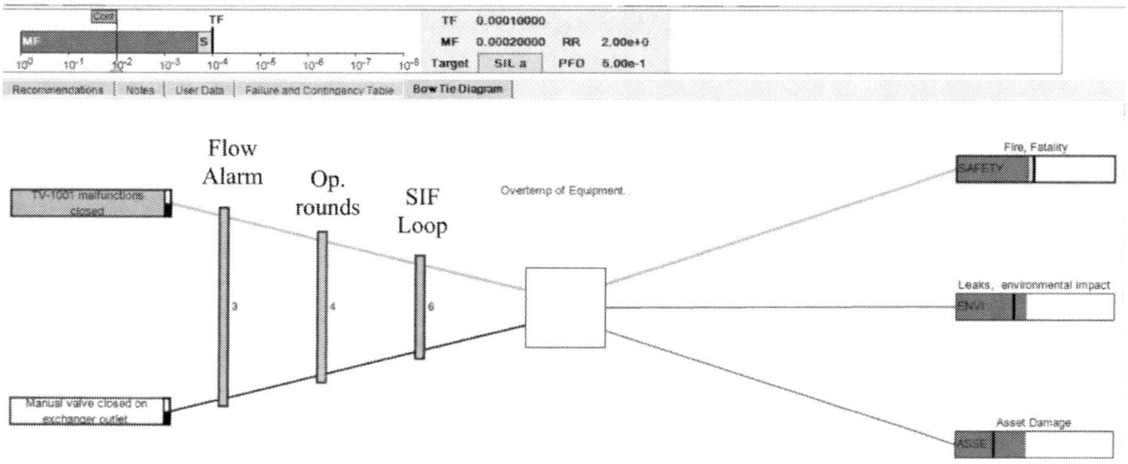

Figure 7: An unbalanced bowtie due to LOPA not looking at mitigative safeguards

The bowtie now looks "cleaner" and the risk coverage is the same. **The total number of safeguards has been reduced by 40%.** However, there are no mitigative safeguards, so if over temperature of the equipment occurs, environmental damage and hazards to personnel are almost guaranteed. To address this, a containment area with gas detection and personnel exclusion is added in the event of over temperature of equipment. Since the mitigated frequency with an additional safeguard is far lower than the tolerable frequency, the team finds that the SIF can be removed, with the resulting bowtie:

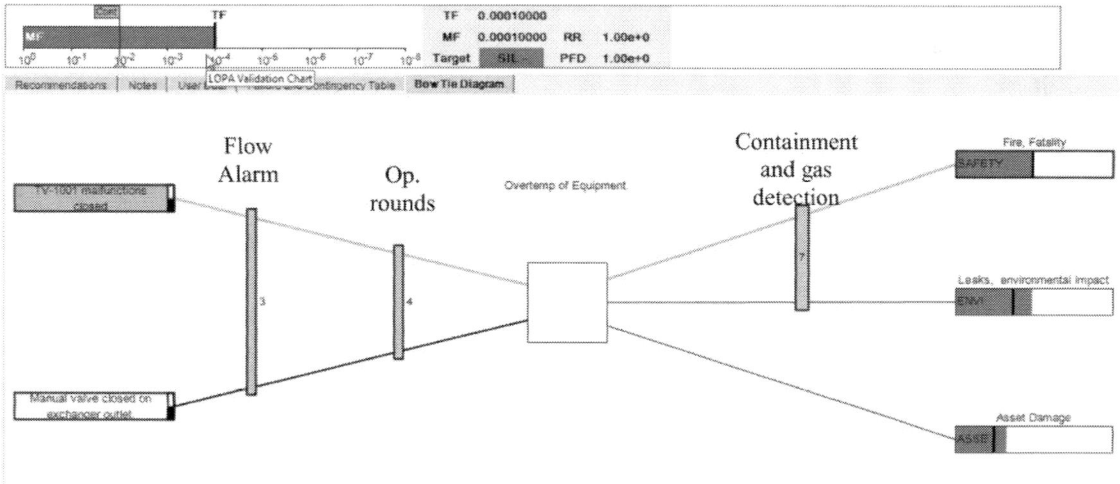

Figure 8: Completed bowtie with 40% fewer safeguards and no SIF requirement

All of the risks are now covered to an acceptable level (no colour shown on right side of bowtie), and plant now has to deal with one less safety critical procedure, one less alarm, and one less safety instrumented function. Added was a mitigation which requires minimal maintenance and will overall make the plant safer. Total lifecycle cost savings are considerable, and there are no longer any automated trips to shut down the facility.

3 Conclusion

The bowtie is an effective method for visualizing risk studies such as HAZOPs and LOPAs. HAZOPs and LOPAs do not look at the big picture, and due to the low cost of adding alarms and trips, the risk studies lead to over-instrumentation of systems. The over-instrumentation leads to systems that are difficult to operate and maintain. The bowtie takes the results of the risk assessments and plots them on an easy-to-understand diagram, complete with initiating events, safeguards, and consequences. This simple diagram can then be used to lower the total amount of safeguards required and balance the system to look at both preventative and mitigative safety systems.

The result is a system with less instrumentation, fewer safety-critical alarms, and fewer trips, all while maintaining the same level of design safety. The reduction in overall number of safeguards reduces the amount of hidden risk due to minor safeguards failing over time. The simplicity of the completed system allows operators and maintenance personnel to better understand the system and make the right decisions when hazardous events are occurring.

Development of a QRA method to calculate the risks generated by Liquefied Natural Gas (LNG) filling stations for road trucks

Albert G. Wolting
Dutch National Institute for Public Health and the Environment (RIVM)
Anthony van Leeuwenhoeklaan 9
3721 MA Bilthoven
The Netherlands
bert.wolting@rivm.nl

Luc J. Vijgen
DCMR Environmental Protection Agency
3112 NA Schiedam
The Netherlands
Luc.vijgen@dcmr.nl

Prepared for Presentation at
American Institute of Chemical Engineers
2013 Spring Meeting
9th Global Congress on Process Safety
San Antonio, Texas
April 28 – May 1, 2013

UNPUBLISHED

Development of a QRA method to calculate the risks generated by Liquefied Natural Gas (LNG) filling stations for road trucks

Albert G. Wolting
Dutch National Institute for Public Health and the Environment (RIVM)
Anthony van Leeuwenhoeklaan 9
3721 MA Bilthoven
The Netherlands
bert.wolting@rivm.nl

Luc J. Vijgen
DCMR Environmental Protection Agency
3112 NA Schiedam
The Netherlands
Luc.vijgen@dcmr.nl

Keywords: QRA, LNG filling station

Abstract

This paper describes a method to calculate risks generated by liquefied natural gas (LNG) filling stations for road trucks. An increasing number of these stations are operational in the Netherlands and they require activities such as transfer, storage, saturation and supply of LNG. Under the Dutch legislative context, facilities that handle or store large amounts of hazardous substances must carry out a quantitative risk analysis (QRA) for land-use planning purposes. The outcomes of the QRA determine the kinds of activities that are acceptable in the surroundings of the facility. The method defines which release scenarios and frequencies to use and how consequences of release scenarios must be calculated. Outcomes are presented for a generic station and some alternative designs for which calculated safety distances are in the range of 66 - 140 meter.

1 Introduction

Owners of sites where large amounts of hazardous substances can be handled or stored are required by Dutch law to provide a quantitative risk analysis (QRA) [1]. The QRA is needed for permitting and land-use planning purposes. For new hazardous activities and for renewed permit applications, a permit will only be issued if the risk that is imposed on objects around the site is acceptable. Similarly, changes in the land-use in the vicinity of the site will only be allowed if the associated risk is acceptable. Both types of decisions depend on the outcomes of the QRA.

The indicators used in assessing the acceptability of a risk are the individual risk and the societal risk. The Individual Risk (IR) is defined as the probability that during a period of one year a person will become the victim of an accident, assuming this person is in a particular location permanently and without protection. The IR is presented in the form of contours on a map with equal risks. Buildings for the general public are not acceptable at locations where the IR exceeds one in a million (10^{-6}) per year. The Societal Risk (SR) is defined as the probability per year that N or more persons outside the industry die in an accident and is calculated for various values of N. In the Netherlands, no strict acceptance criteria have been defined for SR. Instead, the competent authority must determine if the calculated level of the SR is acceptable.

The use of the QRA outcomes for public decision-making dealing with current and future land-use requires a method that is transparent, valid, robust and verifiable [2]. Only publicly accessible data maybe used since all stakeholders must be able to verify the QRA method. Likewise, the outcomes of the QRA may not depend on subjective choices of the risk analysts or on differences in software tools. Therefore, the methods to carry out a QRA are prescribed by legislation, including the prescribed use of 'the Dutch QRA handbook' [3] and the software tool SAFETI-NL® [4]. SAFETI-NL is the Dutch version of PHAST RISK® and in this version certain parameters are prescribed and are fixed and suppressed so users cannot change them.

The legislation currently applies to chemical industries, marshaling yards, pesticide storage facilities and several other hazardous activities. Liquefied natural gas (LNG) filling stations for road trucks also impose a risk on their surroundings and may have to be included in the legislation. Policymakers have yet to make the final decision.

For reasons of consistency, the QRA method for LNG filling stations follows our general method for general chemical industries [3] and our specific method for LPG filling stations [10] as much as possible. The description includes the identification of release scenarios, the associated probabilities and the calculation of the consequences of the identified scenarios. We also list a number of conditions for which the method is not directly valid. Finally, we discuss issues for further research such as the failure frequency of the hose, the probability of a Boiling Liquid Expanding Vapor Explosion (BLEVE), the failure frequency of a double-walled pressurized storage vessel and the relevance of return flow from the receiving vessel.

The project was carried out by RIVM (Dutch National Institute for Public Health and the Environment) and DCMR Environmental Protection Agency and was supervised by an advisory committee. Designers of LNG filling stations supplied technical information.

2 Activities, scenarios and frequencies

2.1 *Description of an LNG filling station*

An LNG filling station, supplying LNG to road trucks, is a relatively new development both in the Netherlands and worldwide. In fact, only a few are currently operational in the Netherlands. However, several dozen stations are being developed to create a countrywide network. No formal guidelines are available yet for the design, construction and management of Dutch LNG filling stations. However, a draft guideline is available [6]. Construction and engineering is still developing and therefore significant differences between stations pertain. An example is the transfer of LNG from the trailer truck to the storage vessel, which can be done either by using the pump on the trailer truck or by increasing the pressure in the truck vessel.[1] Another example is the LNG saturation between the storage vessel and the dispenser, which can be done either by using an in-line heater or an intermediate saturation vessel. In the case of a saturation vessel, the pressure will be 8 or 17 barg depending on the motor system of the road truck using the LNG as fuel. Due to the extensive differences between stations, a generic station has to be defined. The QRA method can be used for this generic station and some alternative designs. For atypical designs, however, the method will not be valid. Figure 1 shows a simplified drawing of an LNG filling station. For the specifics of the station, see section 2.2.1.

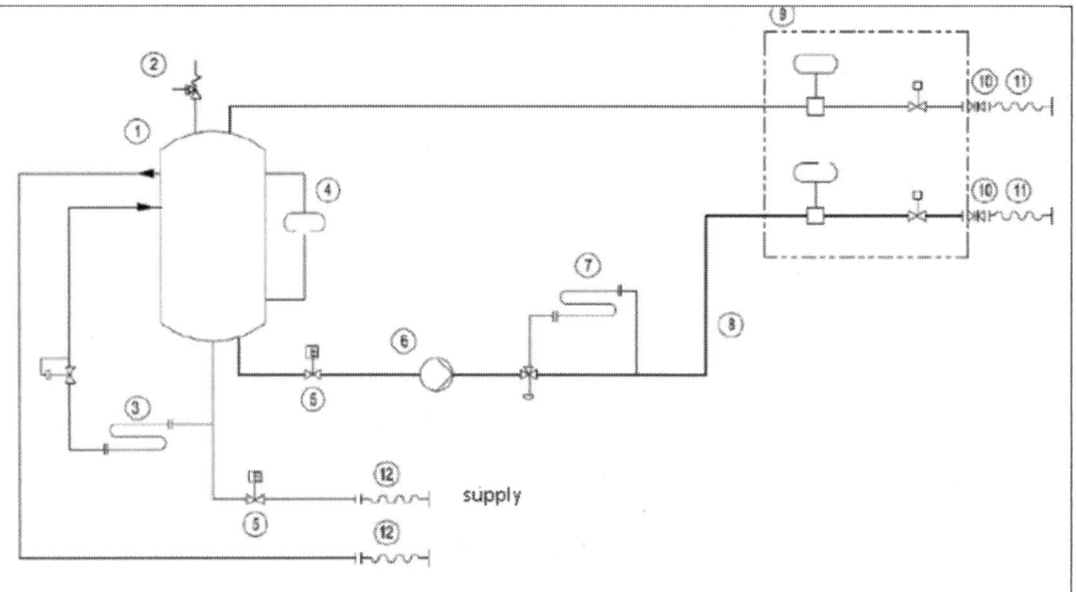

Figure 1 Simplified drawing of a generic LNG filling station.
The component descriptions are provided in Table 1.

[1] Creating a higher pressure is also known as Net Positive Suction Head (NPSH).

Table 1 Explanation of Figure 1

1	Pressurized storage vessel	7	In line heater
2	Pressure Relief Valves	8	Lines
3	Pre-saturation vaporizer	9	Dispenser
4	Level indicator	10	Break away couplings
5	Emergency shutdown	11	Dispenser hose
6	Pump	12	Supply hose 2"

In the presented simplified filling station, a trailer truck supplies LNG using a filling hose (upper hose 12). The vapor phases in the storage tank and truck vessel are connected using a vapor return system (lower hose 12). Once the storage tank is filled, the LNG is conditioned at a higher temperature and pressure using a heater (3). The filling of road trucks is done using the dispensers (9) at a pressure of 8 and 17 barg. This pressure is obtained by heating the LNG using the heater (7).

A QRA method requires the identification of the relevant activities and equipment in order to develop relevant scenarios and frequencies. **Table 2** lists the activities that can take place at an LNG station and the equipment used for these activities. It also contains examples of associated scenarios. Initially we consider most activities relevant for the QRA. Equipment containing only gaseous (and not liquefied) natural gas are not included in the QRA due to their negligible risk contribution.

Table 2 Relevance of the activities for the risk outside the LNG filling station

Activity	Associated equipment	Examples of associated scenarios
Transfer	Truck with pressure vessel	Intrinsic failure Failure due to fire Failure due to external impact
	Pump	Rupture
	Hose	Rupture
Storage	Pressure vessel	Catastrophic rupture Major continuous release (entire content in 600 s)
Saturation	Pressure vessel	Catastrophic rupture Major continuous release (entire content in 600 s)

Dispenser	Hose	Rupture
Compressed gas system	Small dimension pipes and vessels containing CNG[2]	Not relevant for off-site risk
Venting system	Vent stack	Not relevant for off-site risk

2.2 *LNG risk calculation method*

The risk calculation consists of four major parts. First loss of containment scenarios are defined for relevant activities and equipment with their probability of occurrence. In the second part, the effects of the dispersing material are calculated near the emission. The different effects that are calculated are heat load, toxic load (not for methane) and overpressure. Thirdly, the impact of these effects on humans is defined using probit relations (where a specific load results in a specific probability of death). In the last part, the probability of death for a specific effect is multiplied with the associated probability. Summation over all the different scenarios now leads to the combined risk of the site.

2.2.1 Scenario identification and frequency analysis

None of the activities listed in **Table 2** are in fact unique to an LNG filling station. For most of the activities, the release scenarios and failure frequencies are already available in the Dutch QRA handbook [3] or the method for LPG filling stations [10]. **Table 3** gives an overview of the equipment types for which we either investigated or copied existing scenarios and frequencies.

Table 3 Origin of scenario identification and frequency estimation

Equipment type	Origin of scenario identification and frequency estimation
Truck	Use of double-walled insulated vessel for which the scenario 'hot BLEVE' is discussed.
Storage	Copy of existing scenarios and frequencies.
Saturation	Copy of existing scenarios and frequencies.
Supply	Use of a composite material for the supply hose that reduces the release frequency by a factor 10.

Remark

The sudden failure of a vessel (truck vessel or storage vessel) containing a flammable liquid at a temperature well above its normal (atmospheric) boiling point resulting in a large fire

[2] CNG = compressed natural gas

ball (if ignited) is known as a BLEVE. A cold BLEVE occurs when the temperature is near the atmospheric boiling point, e.g. when the trailer truck is punctured by external impact. The term 'hot BLEVE' is used for a BLEVE at an increased temperature due to, e.g. a fire in the vicinity. The used pressure in the QRA scenario 'hot BLEVE' is related to the setting of the pressure relief valve on the truck vessel. The higher pressure influences the surface emissive power, which results in larger effect distances.

The first step in development of a QRA method is defining all the relevant QRA scenarios and modeling parameters. To this end, we compiled a spreadsheet with all scenarios and their frequencies for the generic filling station with the following specifics:

- Annual throughput of 5000 m^3
- LNG transfer from the trailer truck to the storage vessel using a liquid pump
- Total annual transfer time of 167 hours based on an off-loading flow of 500 liter/minute
- Presence of the trailer truck at the station: 1.5 times the transfer time
- Trailer truck vessel content 40 m^3 (95% filling at -126°C, results in a mass of 13640 kg), double-walled
- Storage vessel content 40 m^3 (15385 kg)[3], horizontal, double-walled
- Two saturation vessels content each 10 m^3, at 8 and 17 barg horizontal, double-walled
- One non-return valve (NRV) in the fill line between the trailer truck and storage vessel
- All pumps canned
- Dutch average weather data used

Alternative designs include e.g. a different type of LNG transfer, vertical instead of horizontal storage vessel, underground instead of above ground fill lines, in line heater instead of saturation vessels[4]. Initial risk calculations showed that for all designs only a few of the scenarios determine the risks. The IR10^{-6} distances calculated, see **Table 7**, appeared to be caused for more than 90 percent by scenarios for transfer and storage. These are the most relevant scenarios whose combined contribution to locations where the IR equals 10^{-6} per year is more than 90 percent. For brevity, we now focus on these QRA scenarios, which are listed in **Table 4**.

[3] The maximum content of the truck vessel is 95% of the net content based on the density related to the setting of the pressure relief valve (10 bara) according to ADR. This rule does not apply to the storage vessel for which PGS 33 requires adequate measures to prevent temperature increase over time.

[4] A spreadsheet (partly in Dutch) containing the in total circa 100 release scenarios and the calculation of their frequencies as well as the Safeti NL modeling parameters is available on request.

Table 4 Main QRA scenario's for a generic LNG filling station

Scenarios	Quantity (kg)	Initial failure probability (per hour)	Final probability (per year)
Transfer			
Catastrophic failure of truck due to fire in the vicinity (= hot BLEVE)[1]	13640	4.0×10^{-8}	9.5×10^{-8} ($4.0 \times 10^{-8} \times$ 167 hours x 1.5 x 0.19 x 0.05)
Catastrophic failure of truck due to external impact (= cold BLEVE)[2]	13640	9.6×10^{-10}	1.2×10^{-6} ($9.6 \times 10^{-10} \times$ 167 hours x 1.5)
Hose rupture[3] + Return flow	430 (120 x 3.4 + 5 x 4.5) 21740 (13640 + 1800 x 4.5)	4.0×10^{-7}	5.6×10^{-5} ($4.0 \times 10^{-7} \times$ 167 hours x 0.9 (operator) x 0.94 (NRV) 4.0×10^{-7} ($4.0 \times 10^{-7} \times$ 167 hours x 0.1 (failure operator) x 0.06 (failure NRV))
Storage vessel[4]			
Catastrophic rupture	15385	-	5.0×10^{-7}
Major continuous release (entire content in 600 s)	15385	-	5.0×10^{-7}

Remarks

1 The final probability of the domino scenario 'hot BLEVE' of the trailer truck includes:

- a factor 0.19 for the assessment that in 90 % of the cases of a fire in the vicinity the tank wall is cooled by the liquid, with a reduction of 0.1 due to the pressure relief valve. The probability of heating the vapor/liquid space is considered 0.1/0.9 which results in a BLEVE probability of 0.19 (= 0.1+ 0.1 x 0.9).
- a factor 0.05 for the effect of a double-walled insulated vessel. Basis for this is the analogy to an LPG trailer truck with a BLEVE coating [10]. Reasoning for this is that on one hand the double-walled vessel limits and disperses the heat transfer to the inner wall

eliminating the scenario 'hot BLEVE'. On the other hand, the temperature difference between the outer and inner wall may result in additional tension, which could facilitate tank failure. All pros and cons considered the scenario is modeled with a lower frequency, a vapor mass fraction of 0.81 (three times the flash fraction 0.27 [3]) and a failure pressure of 11.1 barg.[5]

2 The cold BLEVE of the trailer truck is modeled with a vapor mass fraction of 0.24 (three times the flash fraction 0.08) and a failure pressure of 1.36 barg (saturated liquid at -150 °C).

3 For the hose rupture (flow rate 500 l/min or 3.4 kg/s) and return flow (4.5 kg/s) the table lists for brevity, only the minimum and maximum released amounts and their associated probabilities. The method includes the four combinations of operator detection within two minutes (probability of failure 0.1) and the non-return valve (probability of failure 0.06) respectively, see section 2.2.2.

4 For the storage vessel the scenario 'hot BLEVE' is not considered relevant due to internal safety distances [6].

Values for direct ignition of natural gas vaporized from an LNG spill vary from 0.02 – 0.09 depending on the source term, see **Table 6**.

2.2.2 LNG transfer and hose rupture

During the LNG transfer there is the possibility of a hose rupture. In such a case an amount of LNG will be released from the trailer truck vessel through the pump and the hose. Additionally return flow from the storage vessel might occur. Due to its potentially large impact on the calculated risk, we investigated both issues in detail. Figure 2 shows a simplified drawing of the transfer process.

[5] The pressure relief valve (10 bara) of the truck vessel corrected with factor 1.21 [13].

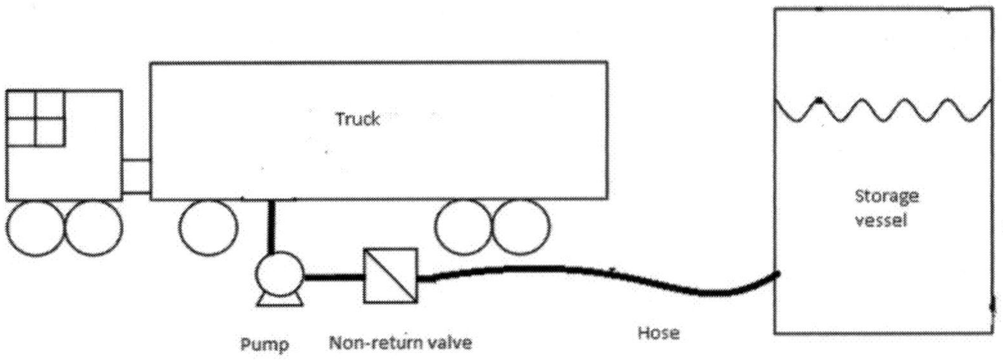

Figure 2 Simplified drawing of the process of LNG transfer from the trailer truck to the storage vessel

For LNG transfer usually increase of the pressure on the trailer truck side is required since, due to heat absorption, the pressure in the LNG storage vessel will be higher than in in the trailer truck vessel. Depending on the vapor pressure in the trailer truck, the options for transfer include:

- Transfer initiated by the liquid pump on the trailer truck (1.5 barg) to build up an overpressure. In this case, two scenarios are relevant: the failure of the pump and the rupture of the hose. A hose rupture will lead to a loss of effectiveness of the pump due to the pressure drop (and gas formation). Due to the orifices that the hose contains however the pump might to continue to run. A hose rupture should be detected by an operator within two minutes. He then trips an Emergency Shutdown Device (ESD).

- Transfer by increasing headspace pressure (maximum 7 barg) in the trailer truck vessel by vaporizing a small quantity of LNG through an onboard heater and feeding it to the gas phase above the liquid. In case of hose rupture, an automatic detection system in the filling line should indicate low pressure within five seconds. This is also known as a 'low pressure trip', which closes the ESD valves at the trailer truck and storage tank. Sufficient documentation proving the reliability that the valves close within five seconds is required.

- Transfer by combining the pump on the trailer truck (1.5 barg) and an increased pressure in the trailer truck vessel. Consequence of this may be a higher flow rate in case of failure of the pump due to the higher pressure in the vessel.

For hose rupture, the released amount depends on the source term and duration of the release. In practice it will depend on three factors: the conditions of LNG in the truck vessel (temperature and whether or not it is in equilibrium), the resistance offered by the hose and the liquid pump (if used), and the safety measures (blocking systems, shut-off valves, etc.).

Conservatively we assume for the two minutes an outflow equal to the normal pump flow. The failure of the pump is modelled as a rupture of the fill line.

The rate of the return flow and its duration depends on three factors: the tankhead (height of the liquid column in the tank) of the storage vessel, the pressure in the storage vessel (which can be higher in case headspace pressure is used) and the non-return valves installed. For our generic station the combined flow of hose rupture and return flow varies from 4.5 – 9.7 kg/s for using the pump on the truck and from 25.5 – 30.7 kg/s using headspace pressure. This approach may seem over conservative since we assumes both a fully loaded truck vessel and storage vessel. However, the net effect on the risk is limited since short-term releases, as jet fire (< 20 s) and maximum size of the flash fire, determine the IR 10^{-6}.

2.3 *Calculation of the consequences*

The second step in the development of the QRA method involves the assessment of the consequences related to the release scenarios that were defined in the previous section. When LNG accidentally is released unpressurised, near or close at ground level, an evaporating pool is formed which provides, by means of spreading and vaporization, a source of a flammable heavy gas cloud. For pressurized releases, as in this case, the liquid is expected to evaporate very quickly upon mixing with the hot air (given the very low boiling point of methane) and therefore no or very limited rainout is expected to occur. The calculation of the consequences is done with SAFETI-NL, which is a software tool from Det Norske Veritas (DNV). Because of the desire to harmonize risk analyses in the Netherlands, the use of this software tool is already prescribed by law for general chemical industries. DNV supplied the validation of the Safeti NL release models for LNG [5]. The most significant consequences of ignition of a release relate to a flash fire and BLEVE of the truck caused by a neighboring fire.

In order to be coherent with other hazardous activities, the event trees to be used for releases of flammable materials were copied from the Dutch QRA handbook (see [3], Module B, §3.4.6). For this method, we assume an outflow on concrete. **Table 5** provides an overview of the considered consequences and distinct between events for immediate ignition (e.g. fireball and jet fire) and events for delayed ignition (vapor cloud fire and explosion). Largely, the event trees of **Table 5** follow the historic development of SAFETI-NL, and some improvements are both possible and desirable. However, we do not expect that the location of the risk contours will change significantly if more realistic event trees are used.

Table 5 Consequence events

Release phase	Release type	Time of ignition	Consequence events in SAFETI-NL
Gaseous	Instantaneous	Immediate ignition	Fireball, immediate fire and immediate explosion
		Delayed ignition	Vapor cloud fire (flash fire) and vapor cloud explosion

	Continuous	Immediate ignition	Jet fire
		Delayed ignition	Vapor cloud fire (flash fire) and vapor cloud explosion

The total probability of ignition is the sum of the probability of immediate ignition and the probability of delayed ignition. The first is standardized and is shown in **Table 6**. For the probability of delayed ignition, the following rule set is used in the calculation of the IR:

1. If the area covered by the lower flammability (LFL) contour is partly outside the plant site boundary, then the cloud is assumed to ignite with maximum effect (maximum content of flammable gas). The probability for delayed ignition is 1 minus the probability for direct ignition.
2. If the area covered by the LFL is located entirely inside the plant site boundary, then the probability of delayed ignition is derived from the presence of ignition sources and their specific ignition probabilities (as a function of exposure duration).

This rule is sometimes referred to as the 'free field assumption' and is intentionally conservative. For the calculation of the SR delayed ignition depends only on the presence of ignition sources and their probabilities.

Table 6 Immediate ignition probabilities

Substance involved	**Release quantity (instantaneous release)**	**Release rate (continuous release)**	**Probability of immediate ignition**
Natural gas	0 – 1000 kg	0 – 10 kg/s	0.02
	1000 – 10,000 kg	10 – 100 kg/s	0.04
	> 10,000 kg	> 100 kg/s	0.09

3 Outcomes

3.1 *Range of safety distances*

Table 7 contains the resulting IR distance calculated for the generic filling station, see section 2.2.1, and for a number of variations.

Table 7 IR distances calculated

Situation	**IR 10^{-6} (m)**
Generic filling station	68
Annual throughput of 1000 m^3	47
Excluding catastrophic failure of truck due to	66

external impact	
Single wall truck[6]	123
Transfer by using an increased pressure (7 barg, ESD 120 seconds)	140

The lowest safety distance applies to the situation of no possibility for external impact of the truck and the highest safety distance to the situation of transfer by increasing headspace pressure.

Figure 3 shows the individual risk contours. In the Dutch context, the IR 10^{-6} /year contour is most important because various types of buildings are unacceptable or highly undesirable inside this contour (see [1] for more information).

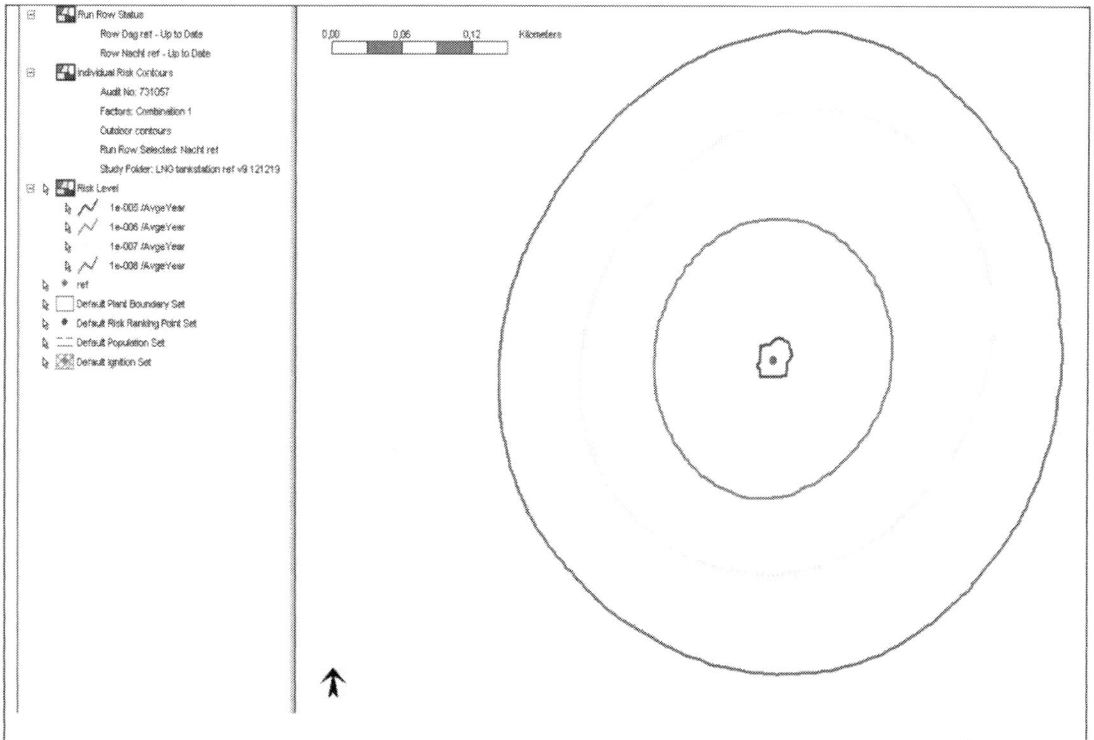

Figure 3 Individual risk outcome for the standard filling station (IR 10^{-6} in red)

[6] See section 3.4.

3.2 *Comparison to LPG filling stations*

Our method for LNG filling stations is partly based on the method for LPG filling stations [10]. That method results in a safety distance of 40 meter for an LPG filling station with an annual throughput of 1000 m^3 [12]. This distance is of the same order as the 47 meters for 1000 m^3 of LNG throughput. In the Netherland risk calculations are not performed for gasoline stations.

3.3 *Conditions for which the method is not valid*

Our QRA method is only valid within certain limits. Some cases in which a closer review of the actual situation is required are:

- The diameter of supply- or dispenser hose is larger than 2".
- The off-loading flow rate of the truck vessel is smaller than 500 l/min.
- The pressure relief valve setting of the storage vessel is larger than 10 bar.
- The storage tank is pre-saturated (no saturation vessels).

For such cases, additional research might be required.

3.4 *Issues for further research*

Based on the currently available information, the QRA method offers issues for further research. On one hand, the failure frequencies of some of the contributing major risk scenarios date back to the early seventies. Updating of those frequencies may also make it possible to reward certain safety measures taken. On the other hand, technical developments might be rewarded in the QRA. Overall, we propose as most relevant issues for further research:

- <u>Update of the failure frequency of the domino-scenario 'hot BLEVE of the trailer truck'</u>. Beside the cause of a fire in the vicinity of the truck this also involves the causes of a fire caused by an LNG leak during the LNG transfer from the truck. Currently the second cause contributes twenty times less to the risk. In contrast to Liquid Petroleum Gas (LPG), only one possible case of an LNG 'hot BLEVE' is known [14]. The available literature therefore does not exclude the possibility of a 'hot BLEVE' [9].

- <u>The frequency of the scenario 'external impact of the trailer truck' leading to a 'cold BLEVE'</u>. Update of this frequency seems also appropriate since nearly all the LNG trailer trucks have, in contrast to LPG trailer trucks, double-walled pressurized vessels. In addition, the rewarding of safety measures such as column guards and speed limits is an option.

- <u>Update of the failure frequency of the supply hose</u>. The current failure frequency (4.0 x 10^{-6} /hour) is outdated, Green and Bourne estimated it in 1972 [7], and it has several drawbacks such as a lack of differentiation between human and technical failures, and differentiation between the root causes. For improved LPG supply hoses we conservatively estimated a factor 10 reduction in frequency in 2006 [8], [10] which is currently also applied for LNG hoses. The hoses used at LNG filling stations are from a composite material that makes them more resilient than standard hoses [10].

- In case of a hose rupture research the amount of return flow from the storage vessel. Installed non-return valves or a measurement of pressure difference (Δp) which closes a valve might prevent this return flow. For both systems, sufficient documentation especially of their reliability is required.

- For a double-walled pressurized storage vessel update the frequencies and scenarios to use. The vessels on the trailer truck as well as the storage vessel are in almost all cases equipped with a double -walled. Additionally between the double walls, an insulation material, e.g. Perlite®, is used. Our Dutch QRA handbook [3] offers only a failure frequency for a single-walled pressurized storage vessel and dedicated information on the effect of a double wall on the failure frequency of a pressurized storage vessel appears to be very scarce.

4 Conclusion

Our method describes how to calculate the off-site risk resulting from activities at LNG filling stations. The specific design of the LNG station is taken into account and therefore may affect the safety distances. In generally the safety distances for LNG filling stations vary from 66 – 140 meter and are slightly larger than for LPG filling stations. This is due to the higher turnover of a LNG station that results in a higher number of risk determining LNG transfers from the trailer truck to the storage vessel.

5 References

[1] Dutch external safety degree (Besluit externe veiligheid inrichtingen), Dutch Ministry of Housing Spatial Planning and the Environment (VROM), First issue 2004, current issue 2009 (in Dutch).

[2] Framework for the adjustment of the quantitative risk assessment methodology. http://rivm.nl/Bibliotheek/Wetenschappelijk/Rapporten/2012/oktober/Protocol_aanpassing_rekenmethodieken_Externe_Veiligheid (in Dutch).

[3] Reference Manual Bevi Risk Assessments version 3.2, RIVM, 2009. http://www.rivm.nl/dsresource?objectid=rivmp:22450&type=org&disposition=inline.

[4] Det Norske Veritas (DNV),2009, the Dutch version of "*Software for the Assessment of Flammable, Explosive and Toxic Impact (SAFETI-NL)*", version 6.54.

[5] Witlox, H.W.M., Harper, M., and Pitblado, R., "Validation of Phast dispersion model as required for USA LNG Siting Applications", LNG plant safety symposium, 12th topical conference on gas utilisation AICHE Spring Meeting, April 2012.

[6] 'PGS 33 Aardgas – afleverinstallaties van vloeibaar aardgas (LNG) voor motor voertuigen': http://content.publicatiereeksgevaarlijkestoffen.nl/documents/other/PGS33_Concept_v0.1_web.pdf (in Dutch).

[7] Green and Bourne, Reliability Technology, 1972.

[8] RIVM letter 210-06, to Ministry of VROM, 2006 (in Dutch).

[9] Integ Risk Reference solution containing documents, methods and tools, for the assessment and management of emerging risks related to new and intensified technologies available for LNG regasification terminals, Stuttgart, November 01, 2010.

[10] "QRA-berekeningen-LPG-tankstations-versie-29-05-08.pdf". https://relevant.nl/download/attachments/2033212/QRA-berekening-LPG-tankstations-versie-29-05-08.pdf?version=1 (in Dutch).

[11] Assessing integrity and reliability of multicomposite LNG transfer hoses. Van Der Weijde G, Van Der Putten S Offshore Technology Conference, Proceedings 2012;3:1895-901.

[12] VROM, Gewijzigde afstanden LPG-autogastankstations (voor bestaande situaties), maart 2007, www.infomil.nl.

[13] API Recommended Practice 520, American Petroleum Institute, Washington, D.C. (Jan.2000).

[14] Planas-Cuchi, E., Gasulla, N., Ventosa, A., & Casal, J. (2004). Explosion of a road tanker containing liquefied natural gas. Journal of Loss Prevention in the Process Industries, 17, 315e321

9TH GLOBAL CONGRESS ON PROCESS SAFETY

Release Scenario Assumptions in Modeling Risk From Underground Gaseous Pipelines

Elizabeth M. Lutostansky
Air Products and Chemicals, Inc.
7201 Hamilton Blvd.
Allentown, PA 18195
lutostem@airproducts.com

Seungho Jung (jungs2@airproducts.com)
Keith Ludwig (ludwigka@airproducts.com)
Len Creitz (creitzlw@airproducts.com)
Joan Schork (shorkjm@airproducts.com)
Air Products and Chemicals, Inc

Prepared for Presentation at
American Institute of Chemical Engineers
2013 Spring Meeting
9th Global Congress on Process Safety
San Antonio, Texas
April 28 – May 1, 2013

UNPUBLISHED

Release Scenario Assumptions in Modeling Risk From Underground Gaseous Pipelines

Elizabeth M. Lutostansky
Air Products and Chemicals, Inc.
Allentown, PA
lutostem@airproducts.com

Seungho Jung, Keith Ludwig, Len Creitz and Joan Schork
Air Products and Chemicals, Inc.

Keywords: pipelines, PHAST, PHAST-Risk, SAFETI, risk, toxic, flammable, San Bruno

Abstract

Risk assessments of underground pipelines are complicated by the number of possible release conditions. Pipeline release scenarios are a function of the location of the rupture or leak, the hole size, automated shut-off valves, release direction, etc. In order to provide a realistic representation of the risk with a reasonable degree of detail, the selection of release conditions is critical. The nature of the hazard influences which assumed release conditions are appropriate.

For immediately ignited flammable releases, the peak radiation must be captured due to the time-independent nature of the radiation vulnerability model summarized in the Purple Book [1]. For time-dependent impacts, such as toxic releases, both the concentration and exposure duration are important parameters. Therefore, modeling of time-varying releases is justified.

Release direction and momentum from underground pipelines are affected by the initiating cause of the release. Larger releases and releases caused by excavation are expected to form a crater. Releases from a crater are vertical or angled above horizontal and have some degree of reduced momentum. Smaller releases, such as those initiated by corrosion, may not form a crater. Released vapor may percolate though the soil above the pipeline and be released to atmosphere with low momentum.

1. Introduction

The quantified risk assessment of underground pipelines that transport toxic and/or flammable pressurized gases is dependent on release scenarios, assumed failure frequencies, weather conditions, neighboring populations, etc. It is important to carefully define the release scenarios to be modeled. Release rates from pressurized vapor pipelines vary with time. However, risk and consequence models, such as PHAST-Risk, only model a single release rate or a series of stepped rates for dispersion or radiation calculations.

\

One key learning of our evaluation of methods for modeling flammable and toxic pipeline risk was the importance of the assumed release rate or rates. Careful consideration must be given to the flow averaging assumptions. Preferred release rate assumptions depend on depressurization time and on the nature of the hazard.

A second key learning was the importance of release momentum and direction. Releases from underground pipelines have the potential to be high momentum angled or vertical releases from a crater or low momentum releases diffusing through the soil. The release momentum has a significant impact on calculated risk.

Comparison of model results using selected assumptions with real data was possible for buried flammable pipelines. Experimental radiation data collected from a buried hydrogen pipeline was used to test assumptions for release velocity and flow rate [2]. A second comparison was done using damage evident from satellite imagery and using the numbers of injuries and fatalities caused by the 2010 natural gas pipeline release in San Bruno, California.

2. Release rate

2.1 Importance of release rate

2.1.1 Release rate characterization

The release rate from a gaseous pipeline rupture is a function of the pressure of the pipeline, the hole size and the inventory of material in the pipeline. Figure 1 shows the impact of a modest reduction in discharge area on flow rate. The dashed line shows the flow rate from a complete break of the pipeline. The solid line shows the more gradual depressurization that occurs with 25% of the original discharge area. Pinhole leaks would show even less decay in flow rate with time.

Figure 1 - Impact of Hole Size on Flow Rate

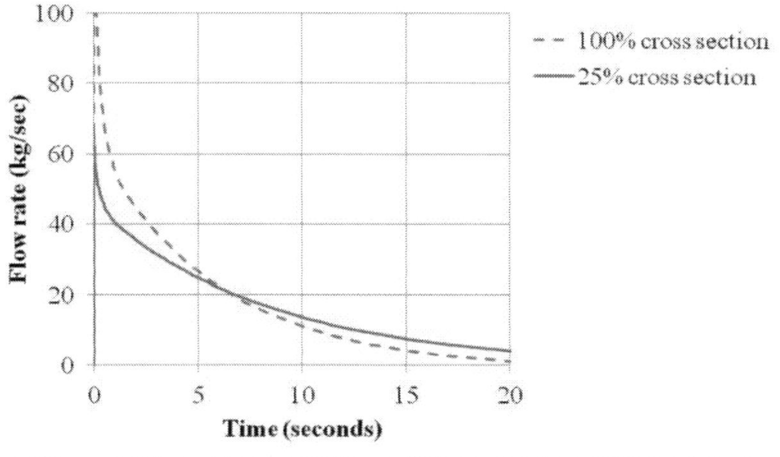

The release rate and duration may also depend on the location of the pipeline rupture relative to automated shut-off valves. Figure 2 shows the impact of release location on flow rate. A release at the center of the pipeline section (between two shutoff valves) results in the shortest depressurization time.

Figure 2 - Impact of Failure Location on Flow Rate

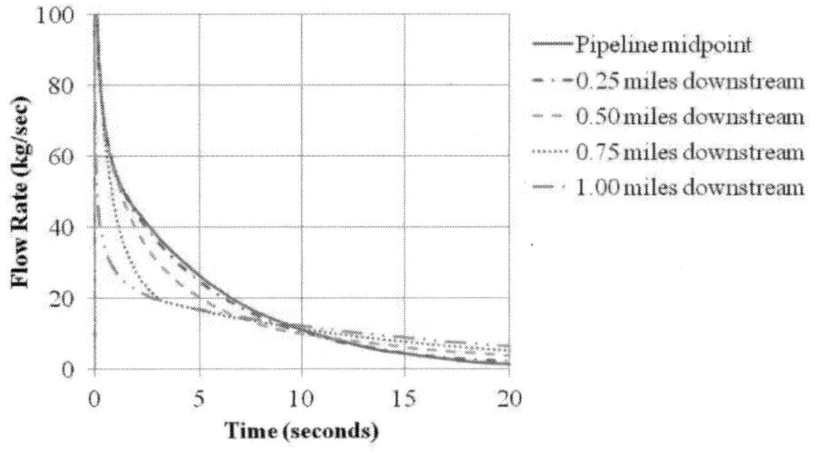

Those familiar with the long pipeline model in PHAST know that the model produces many results including voluminous time-varying reports of pressure, mass, flow rate, etc. When looking at a broad range of possible scenarios for a pipeline that extends many miles, some simplifications are required.

2.1.2 Release rate for practical implementation in quantified risk assessment

Use of the time-varying discharge rate in dispersion modeling can significantly increase computational time. Therefore, time-varying discharge rates are only used when the impact of time-varying rates is important in the assessment of the risk.

2.2 Time-independent jet fire hazard

2.2.1 Radiation vulnerability basis

The vulnerability model in the Purple Book can be used to model the radiation hazard from a jet fire [1]. The assumptions in this model include:

1. At radiation levels at or above 35 kW/m^2, vulnerability is 100%. This is based on the assumption that at this radiation intensity, clothing and buildings could ignite without an ignition source.
2. At radiation levels below 35 kW/m^2, a Probit function is used that accounts for both radiation intensity and exposure duration. A factor of 14% is applied to the dosage for outdoor exposures to account for the protection provided by clothing.

As a result of the step change in vulnerability from 100% above 35 kW/m^2 to a maximum of 14% below 35 kW/m^2, it is the high radiation zone, above 35 kW/m^2, that is typically the major contributor to the risk from radiation.

A modification factor can be applied to the 100% vulnerability above 35 kW/m^2 to account for the probability that a person may be able to escape to a safe location. This modification factor is influenced by the ability of people to exit a building and/or the ability of people to find a sheltered location. Occupants of multi-story, masonry, commercial buildings would be more likely to escape high radiant heat than occupants of wood frame structures with limited means of egress.

2.2.2 Proposed release rate basis

As a result of the assumption that the dominating radiation risk is that from time-independent exposure to radiation levels at or above 35 kW/m^2, modeling peak flow is recommended for jet fire risk. Figure 3 shows the influence of the release rate on the jet fire hazard zone, defined as the area where radiation is at or above 35 kW/m^2. Release rates 1, 5, 10, 15, and 20 seconds after the pipeline break were used to determine the jet fire hazard zone. As the flow rate decays with time, the jet fire hazard zone rapidly decreases in size. Initially, at 1 second, the hazard zone extends more than 120 m downwind. At 15 seconds, the hazard zone only extends 40 m downwind.

The PHAST default flow averaging time for the long pipeline model is to average flow over the first 20 seconds. Using this default assumption, the jet fire hazard zone extends 77 m downwind of the release point. Using the PHAST default flow rate averaging time significantly under predicts the hazard zone for a time-independent hazard.

Figure 3 – Influence of Release Rate on Jet Fire Hazard Zone

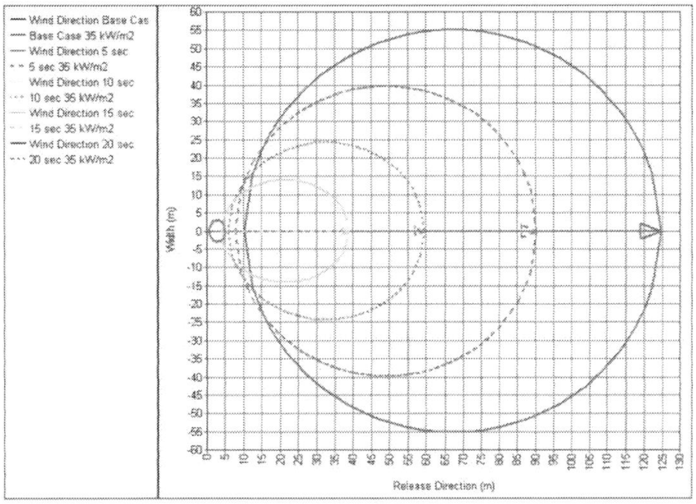

2.3 Time-dependant dispersion hazard

2.3.1 Vulnerability basis

When modeling the hazard from a dispersing toxic vapor released from a pipeline, the vulnerability is assessed using a Probit model. The Probit model relates exposure concentration and duration (i.e. the dose) to a probability of fatality. Both concentration and duration of exposure are important parameters.

When modeling the hazard of a dispersing un-ignited flammable vapor released from a pipeline, the risk is a function of the likelihood of ignition and the size and location of the flammable cloud.

2.3.2 Proposed release rate basis

Since the duration of the release is important for toxic, flash fire or explosion effects, modeling dynamic flow is recommended. However, dynamic release rate modeling may not be required in instances where the pipeline depressurization time is long enough that the flow decay does not significantly impact the calculated risk. Examples of scenarios where pipeline depressurization would be slow include pinhole leak and leaks from which the flow is less than or equal to the flow rate of material into the pipeline.

Figure 4 illustrates the impact of flow rate assumptions on the calculated hazard zone for a toxic material. The PHAST default, flow rate averaged over the first 20 seconds results in a localized hazard zone, within 20 m of the release point. The hazard zone is small due to the high averaged flow rate over the first 20 seconds. This high flow rate results in a rapid depletion of inventory. Because of the short duration of the event, the dose received downwind is not enough to result in predicted fatalities.

The duration of the release is also important to capture for indoor populations. The concentration indoors is a function of the concentration outdoors, the exposure duration and the building size and ventilation. If the duration of cloud is short, it will not last long enough for the building ventilation system to bring a hazardous amount of the material indoors.

Time-varying discharge rates give the largest hazard zones for the example toxic release. Both with 10 and 3 flow rate steps, the hazard zone extends 280 m downwind. Since there is not discernible difference in the results with 10 and 3 flow rate steps, the use of 3 steps is preferred to reduce calculation time. Averaging the flow rate over 3600 seconds results in a hazard zone extending 250 meters downwind. Since duration is an important parameter in this case, averaging over one hour may be reasonable.

Figure 4 – Influence of Flow Rate Assumptions on 1% Vulnerability for Toxic Material

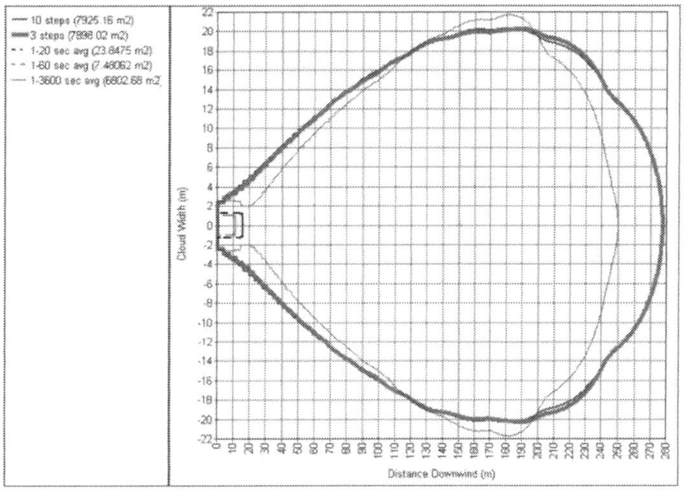

3. Release momentum and direction

3.1 Releases caused by excavation

According to the 7th report of the European Gas Pipeline Incident Data Group (EGIG), approximately 50% of pipeline incidents have been attributed to external interference [3]. These events may result in the pipe break and crater like shown in Figure 5.

Figure 5 – Example of a Pipeline Rupture Caused by External Interference

3.1.1 Release direction considerations

When an underground pipeline is contacted by excavation equipment, it is typically hit on the upper side of the pipeline. If a hole forms, it is likely to be oriented vertically or close to vertical. As shown Figure 5, the two ends of the pipeline are still oriented in a horizontal direction. High pressure vapor flowing out the two ends of the pipeline would merge together. If the two pipeline ends had been distorted and pointed in other directions, the flow out of the pipe ends could hit the crater wall or be directed out of the crater unimpinged. A horizontal release direction is not appropriate for a pipeline that is buried 3 feet or more below grade. Because of the formation of a crater, some degree of upward momentum is expected.

3.1.2 Release momentum considerations

When excavation is the initiating cause of the pipeline failure, it is expected that soil above the leak or rupture will be removed. This removal may be a result of the excavation itself or the result of the sudden large release of a high pressure vapor jet blowing out the soil above the release point.

3.1.3 Cases modeled

When modeling cases caused by excavation, the discharge is modeled in PHAST as an impinged release at an angle of 45 degrees from horizontal. Crater wall measurements from pipeline tests done show the crater wall is angled 48 to 60 degrees from horizontal [2]. Worst-case hazards are present when the plume is angled closest to grade. Therefore, a 45 degree angle was selected as the lower limit of the crater wall angle measured in the tests.

Tests done on a buried hydrogen pipeline showed that PHAST's default velocity modification factor for impinged releases, 25%, provided flame length results that were in good agreement with measured values [2]. The same degree of velocity reduction is assumed to apply to a dispersing un-ignited release.

3.2 Releases from other causes

Other causes of pipeline failures included construction defects or material failures (17%), corrosion (15%), ground movement (7%), hot-tap made by error (5%), and other or unknown causes [3].

3.2.1 Release direction considerations

Release directions for causes other than excavation are assumed to be more random in nature. In the case of a hole forming on the underside of the pipeline, one would expect a downward directed jet to push through the soil following the path of least resistance. Ultimately, the material emanates from the soil vertically.

3.2.2 Release momentum considerations

By the time that a small leak penetrates through several feet of soil, the velocity of the release is expected to be significantly less than at the release point. The Ergun equation for flow through a porous medium could be used to assess the degree of velocity reduction for uniform, well-characterized soil. Release velocities of the same magnitude at typical wind speeds ensure that the released material is modeled as a low momentum, passive release.

4. Case Study – San Bruno, California

The calculated radiation using preferred release assumptions for flammable pipelines was compared to damage following the 2010 natural gas pipeline rupture in San Bruno, California. There were 8 deaths as a result of the pipeline rupture in San Bruno. In addition, there were approximately 60 injuries and 40 homes destroyed. Figure 6 shows the damage caused in the vicinity of the pipeline rupture.

Figure 6 – Satellite Images of San Bruno Before and After Pipeline Rupture

Peak flow from the 30 inch, approximately 400 psig pipeline was modeled in PHAST. The radiation calculations were performed assuming the jet fire was angled 45 degrees from horizontal and exiting the crater with reduced momentum. The resulting 35 kW/m^2 radiation contour is shown in Figure 7 assuming the wind was blowing to the northeast.

Figure 7 – Calculated 35 kW/m^2 Radiation Contour for San Bruno Pipeline Rupture

Using the radiation vulnerability assumptions and generic day time residential population assumptions from the Purple Book [1], the calculated number of fatalities would have been much larger as shown in Table 1. However, the resulting 35 kW/m^2 contour corresponds very closely with building damage.

Table 1 – Comparison of Actual Number of Fatalities with Model Predictions

Method	Fatalities	Injuries
Actual event	8	58
Calculated based on 35 houses initially destroyed	59	NA
Calculated based on 37 houses inside 35 kW/m^2 radiation contour	62	NA
Calculated based on 35 kW/m^2 radiation contour and San Bruno population density	78	NA

Other factors influencing the consequences from this release include:

- Initial overpressure cause by pipeline rupture
- Radiation from secondary fires
- Ability of individuals to escape or find a location sheltered from the radiation
- Shielding effects of structures
- Variation in wind direction and speed

5 Conclusions

Results of risk assessment of underground pipelines are very sensitive to release assumptions. Simplifying assumptions are required to reduce computational time. These assumptions must be tailored to the nature of the pipeline hazards and pipeline parameters that influence the depressurization time of the pipeline.

The time-varying nature of potential releases is important to capture for time-dependent consequences such as building infiltration and toxic exposures. For radiation hazards, peak flow rates are critical. The default long pipeline model in PHAST uses averaged flow over the first 20 seconds of the release. Use of this default will have a large influence on the resulting assessed risk.

Causes of potential pipeline releases need to be considered as they affect release size, momentum and direction. Directing the release close to horizontal would likely be a conservative approach. However, a release from underground will have some degree of vertical momentum. The release direction and momentum have a large impact on the assessed risk. Therefore, it is important to pick a set of cases that span realistic conditions to insure that the assessed risk is representative.

6 References

[1] "Guidelines for Quantitative Risk Assessment", Publication Series on Dangerous Substances (PGS 3), CPR 18E, December 2005.

[2] Lutostansky, E., Creitz, L, Jung, S., Schork, J, Xu, Y. and Worthington, D. "Modeling of Underground Hydrogen Pipelines", 2012 CCPS conference.

[3] "7th EGIG-report, 1970-2007, Gas Pipeline Incidents", 7th Report of the European Gas Pipeline Incident Data Group, Doc. Number EGIG 08.TV-B.0502, December 2008.

Effect of Flammable Cloud Treatment on Predicted Blast Loads

J. Kelly Thomas, Ph.D.
Baker Engineering and Risk Consultants, Inc. (BakerRisk)
3330 Oakwell Court, Suite 100
San Antonio, TX 78218
KThomas@BakerRisk.com

Jihui Geng, Ph.D.
Baker Engineering and Risk Consultants, Inc. (BakerRisk)
3330 Oakwell Court, Suite 100
San Antonio, TX 78218
JGeng@BakerRisk.com

Prepared for Presentation at
American Institute of Chemical Engineers
2013 Spring Meeting
9th Global Congress on Process Safety
San Antonio, Texas
April 28 – May 1, 2013

UNPUBLISHED

Effect of Flammable Cloud Treatment on Predicted Blast Loads

J. Kelly Thomas, Ph.D.
Baker Engineering and Risk Consultants, Inc. (BakerRisk)
3330 Oakwell Court, Suite 100
San Antonio, TX 78218
KThomas@BakerRisk.com

Jihui Geng, Ph.D.
Baker Engineering and Risk Consultants, Inc. (BakerRisk)
3330 Oakwell Court, Suite 100
San Antonio, TX 78218
JGeng@BakerRisk.com

Keywords: VCE, FLACS, Blast Load, Dispersion, Equivalent Stoichiometric Cloud, Q9

Abstract

Computational fluid dynamic (CFD) analysis of flammable gas dispersion can provide a realistic evaluation of the resulting flammable gas cloud. CFD dispersion analysis can provide a more realistic assessment of the resulting gas cloud in heavily congested and/or confined environments relative to simplified dispersion analysis methods. In contrast to simplified methods, the gas cloud predicted using CFD analysis in congested/confined environments is typically complex, both in terms of shape and concentration gradients. The blast loads predicted using CFD vapor cloud explosion (VCE) analysis for the resulting flammable gas cloud are dependent upon the representation chosen for the gas cloud. Directly utilizing the predicted dispersed gas cloud, with the corresponding complexities in shape and concentration gradients, requires an extended series of analyses, since both ignition location within the cloud and the cloud location can have a strong influence on the predicted blast load. A common approach is to replace the dispersed gas cloud with a simplified representation for the purposes of performing a CFD VCE blast load assessment, such as a cubic shaped cloud with a uniform near-stoichiometric fuel concentration. This approach is very attractive since it greatly simplifies the blast load assessment. However, the relationship between the blast loads predicted using a simplified cloud representation versus those predicted for the actual dispersed cloud is not well defined.

This paper presents a comparison of the blast loads predicted using the dispersed flammable gas cloud resulting from CFD dispersion analysis versus a simplified representation of the gas cloud. The FLACS CFD code was employed for this purpose. The simplified flammable gas cloud representations used included the "equivalent stoichiometric" gas cloud parameters developed by GexCon (i.e., the Q9 approach) and utilizing the entire flammable gas volume (i.e., from the lower flammability limit up to the upper flammability limit). A congested volume typical of an on-shore process module was employed for this purpose. Blast loads were assessed at locations both within and external to the flammable gas cloud. A range of potential ignition locations were evaluated in order to define the impact of ignition location on the predicted blast load relative to that obtained with the simplified gas cloud representations. It was found that, in general, the blast loads predicted using the equivalent stoichiometric cloud are conservative relative to those predicted for the dispersed cloud. The maximum blast loads predicted using the equivalent stoichiometric clouds were much larger than those for the dispersed clouds for the cases evaluated.

1. Introduction

The "equivalent stoichiometric" (i.e., Q9) approach developed by GexCon for use with the FLACS CFD code allows a "real" flammable gas cloud (i.e., that resulting from a dispersion analysis) to be treated in a straightforward manner for the purposes of performing a FLACS VCE blast load assessment. Hansen et al. [1] discuss the development and basis of the equivalent stoichiometric cloud approach as implemented in FLACS and provide a range of comparisons with test data and alternative gas cloud representations (i.e., dispersed cloud, replacement of entire flammable volume with worst-case concentration, etc.). The basic concept of the equivalent stoichiometric cloud approach is that the non-homogenous cloud resulting from a FLACS dispersion analysis is replaced with a cloud at a uniform concentration which will give a blast that is representative of, and generally conservative with respect to, that for the dispersed cloud.

The intent of the Q9 approach is to provide a realistic worst-case blast load assessment, rather than a completely bounding result. That is, it is recognized that replacing the entire flammable gas cloud resulting from dispersion analysis with a uniform cloud of the same volume at the worst-case concentration will provide a larger blast load, and that the blast load developed in this manner will bound that for the dispersed cloud. However, since the portions of the dispersed cloud which are well away from the stoichiometric concentration (i.e., significantly lean or rich) will release less energy and will do so at a lower rate (i.e., slow flame speed), such an approach may lead to an overly conservative result.

BakerRisk utilizes the Q9 approach in most FLACS blast load analyses, although a more conservative approach (i.e., cloud volume equal to the entire flammable cloud volume with the concentration set to the worst-case) is used in some cases. The dispersed clouds resulting from dispersion analyses are not normally directly ignited to determine VCE blast loads. Hansen et al. [1] noted that this was also the normal practice for GexCon, although occasionally studies are performed where the dispersed clouds are directly ignited and an example was provided of such a study (albeit without providing actual numerical results due to confidentiality restrictions). The purpose of the evaluation described in this paper was to provide an additional comparison for conditions representative of a typical on-shore process module.

Section 2 describes the module utilized for this study, the corresponding FLACS model and dispersion cases evaluated. The results of the dispersion analyses are provided in Section 3. Section 4 discusses the model used for the blast load analyses. The results of the blast load analyses are presented in Section 5. Conclusions from this evaluation are provided in Section 6.

2. Process Module

A typical on-shore process module was employed for this evaluation. A plan (top) view of the module is given as Figure 1. The model dimensions are 36.5 m long, 17.7 m wide and 11.5 m high (volume of 7,430 m^3). The module is fully enclosed with relatively light-weight panels on all sides and the roof. A module ventilation (HVAC) system with an air flow rate of 5.4 air changes per hour (ACH) was included in the model for the dispersion analyses.

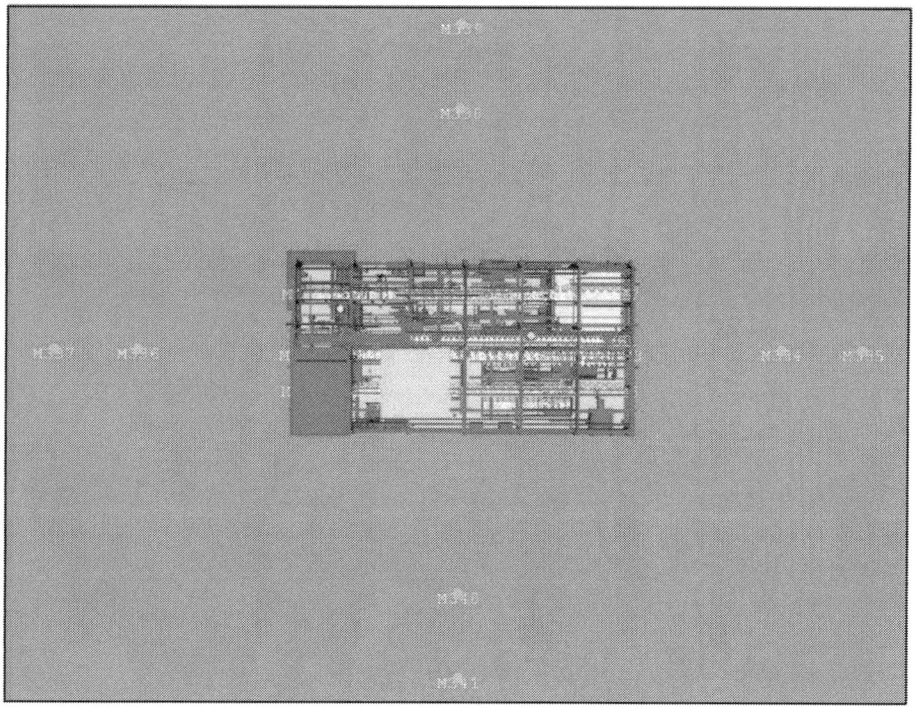

Figure 1. Module Geometry (Plan View) and Target Distribution

3. Dispersion Analysis

The background air flow induced by the ventilation system within an enclosed module strongly influences fuel dispersion and the resulting cloud size, location and concentration distribution. The ventilation system simulations indicated that the air flow patterns within the module take approximately 100 seconds to reach steady-state. The ventilation system was therefore run for 100 seconds prior to flammable gas release in each dispersion simulation. The ventilation system supply and exhaust fans were assumed to run for the entire duration of each release simulation (i.e., mitigation actions related to the ventilation systems were not considered). The averaged turbulence levels in the module from the ventilation system were used to set pre-ignition turbulence levels for the blast load analysis.

Releases of natural gas (97% methane), propane and ethylene were evaluated. Release rates in the range of 1 to 3 kg/s were found to produce the largest flammable and equivalent stoichiometric gas clouds. The results of the dispersion analyses are summarized in Table 1. As shown in this table, the flammable cloud volumes range roughly 80% to 85% of the module volume. The equivalent stoichiometric cloud volumes range from roughly 60% of the module volume (for methane and propane) up to 74% of the module volume (for ethylene); most past analyses performed by BakerRisk for enclosed process modules have shown worst-case equivalent stoichiometric cloud volumes of 50% to 60% of the enclosure volume. The larger value obtained with ethylene (74%) indicates a more uniform concentration nearer to the stoichiometric concentration. This is mainly reflective of the near neutral buoyancy of ethylene (i.e., a molecular weight of 28). This is also reflected in the ratio of the flammable cloud volume to that of the equivalent stoichiometric Q9 cloud, which ranges from 1.15 (ethylene) to 1.41 (propane).

Table 1. Dispersion Analysis Results Summary

Fuel	Volume (m^3)		Volume (%)		LFL-UFL to Q9 Volume Ratio
	LFL-UFL	Q9	LFL-UFL	Q9	
Natural Gas	6,040	4,500	82%	61%	1.34
Propane	5,860	4,160	79%	56%	1.41
Ethylene	6,300	5,480	85%	74%	1.15

Figure 2 shows, as an example, the natural gas concentration contours for a 1 kg/s release at the time when the dispersed cloud attains the largest equivalent stoichiometric cloud volume. The contour plots are sliced elevation and plan views near the module center. The contour values shown are in terms of equivalence ration, and range from the LFL to the UFL. As can be seen, due to the positive buoyancy of the natural gas, the cloud is fuel-rich near the top of the module fuel-lean near the bottom. The opposite behavior is observed for the propane due to its negative buoyancy, while a more homogeneous mixing behavior is seen with ethylene releases.

4. FLACS Model for Blast Load Analysis

The solid model used in the FLACS blast load analyses was shown in Figure 1. A total of 333 targets were evenly distributed within the module to record blast pressure. Eight additional targets were placed outside the module. The four closest external targets are located 20 m from the module corresponding module wall.

A computational mesh size of 0.65 m was used for all of the FLACS blast load analyses. This size was selected to ensure that there would be at least 10 computational cells through the cloud for the Q9 clouds. To represent pre-ignition turbulence due to the ventilation system, a length scale of 0.6 m and a relative turbulence intensity of 0.09 (characteristic velocity of 1 m/s) were employed.

The array of potential ignition source locations shown in Figure 2 and Figure 3 was evaluated in the blast load analysis. The ignition source locations evaluated are at three elevations ($z = 7$ m, 10 m and 13 m) with five ignition locations along the length of the module (5 m spacing) at each elevation. Ignition source locations off the module centerline, which are shown in these figures, were also evaluated but found to give generally lower blast loads. The same ignition locations were used for both the real (i.e., dispersed) and equivalent stoichiometric clouds.

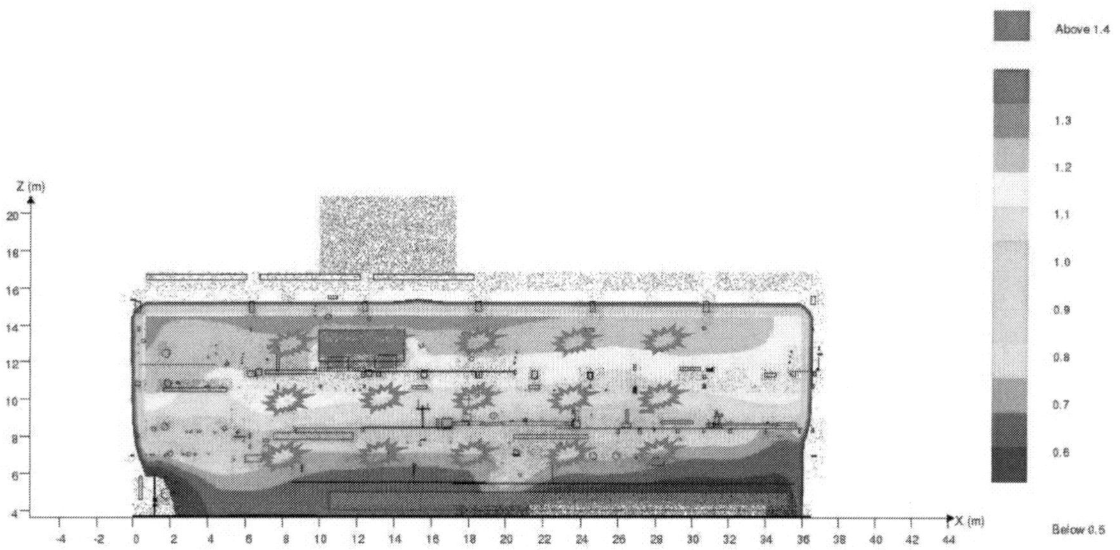

Elevation View (y= 47 m)

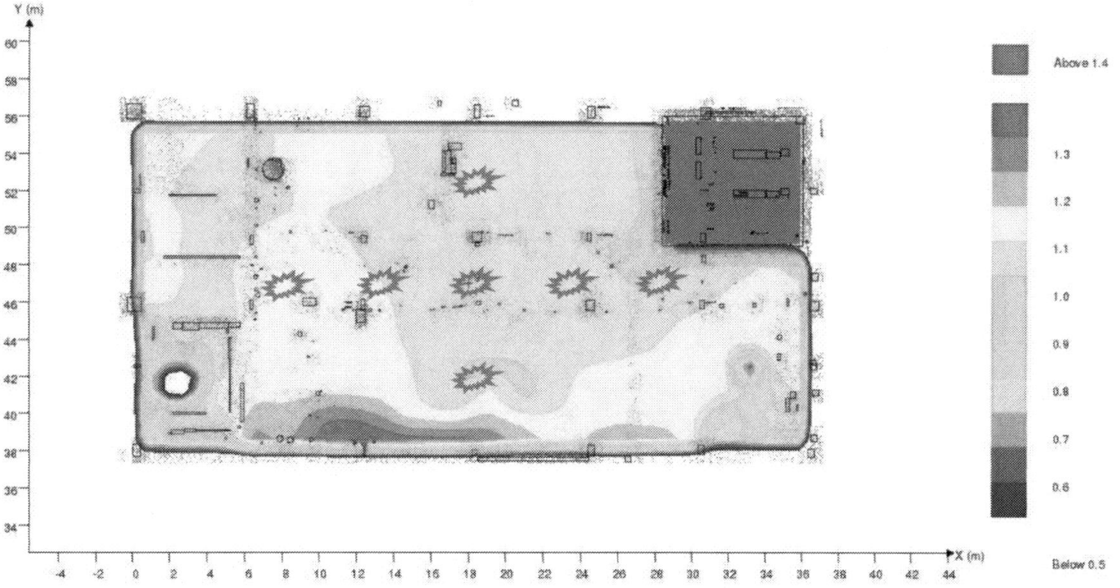

x-y Sliced Top View (z= 10 m)

Figure 2. Real (Dispersed) Flammable Gas Cloud for Natural Gas

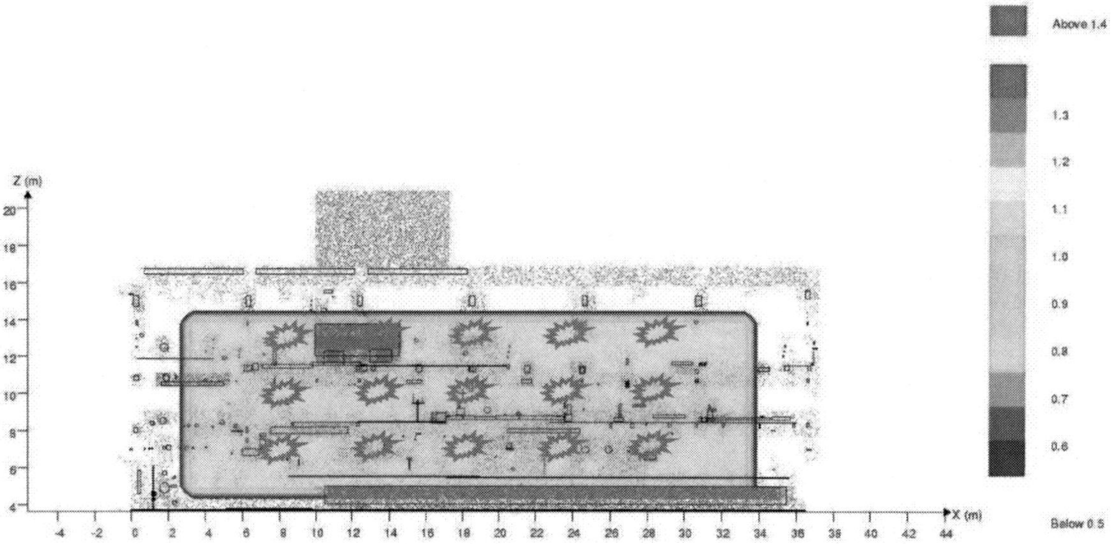

Elevation View (y= 47 m)

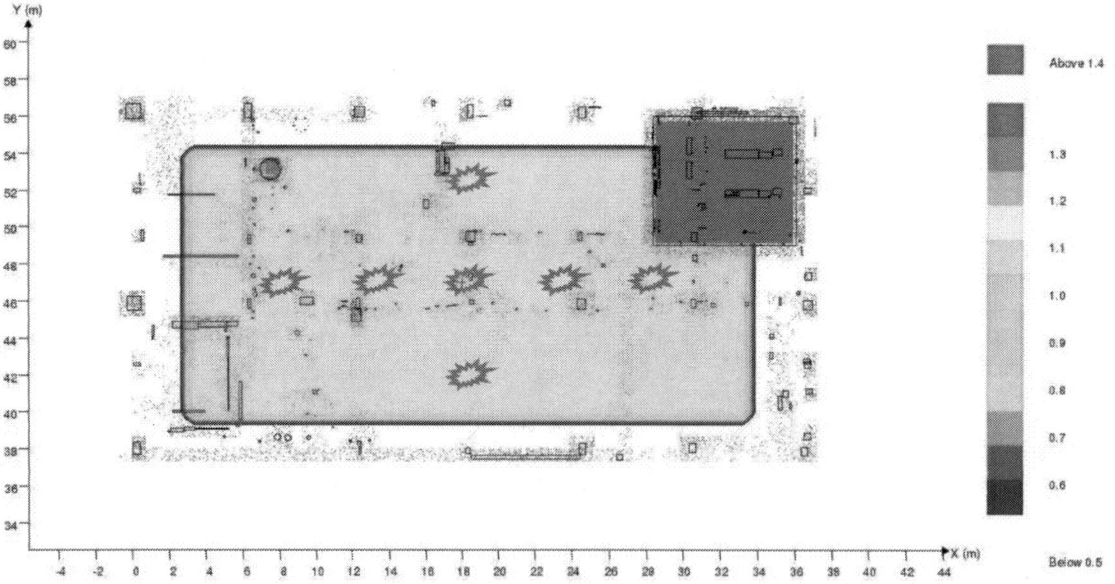

x-y Sliced Top View (z= 10 m)

Figure 3. Equivalent Stoichiometric (Q9) Cloud for Natural Gas

5. Results and Discussion

5.1 Blast Loads from Real Clouds (Natural Gas)

The average of the peak overpressures for the real (i.e., dispersed) natural gas cloud internal and external to the module for all ignition source locations considered are shown in Figure 4 and Figure 5, respectively. For the internal pressures, the values shown are average of the peak pressures recorded at all monitor locations within the module. For the external pressures, the values shown are the average of the peak pressures recorded at the 4 closest monitor locations (i.e., 20 meters outside each of the module walls). There is roughly a factor of two between the largest average internal pressure (ignition location near the center of the module at the upper elevation) and the smallest; this difference is roughly 50% for the average external overpressure.

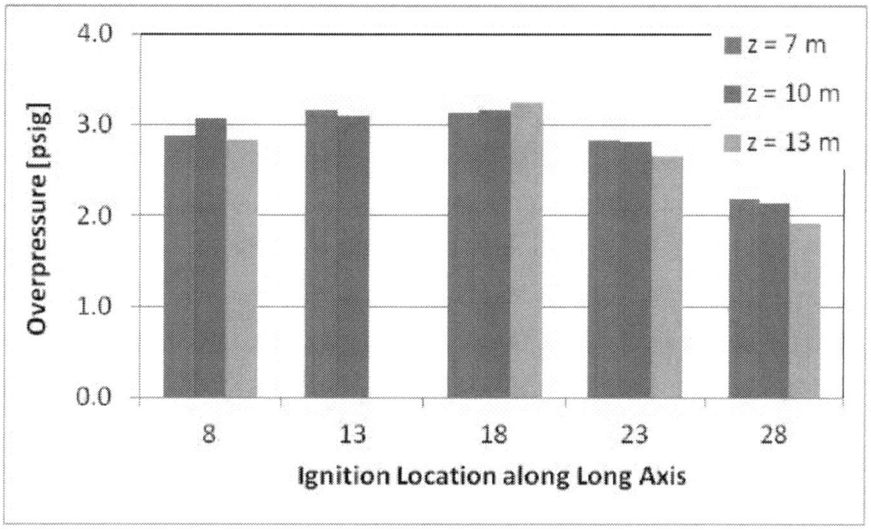

Figure 4. Impact of Ignition Location on Internal Overpressure (Real Cloud, Natural Gas)

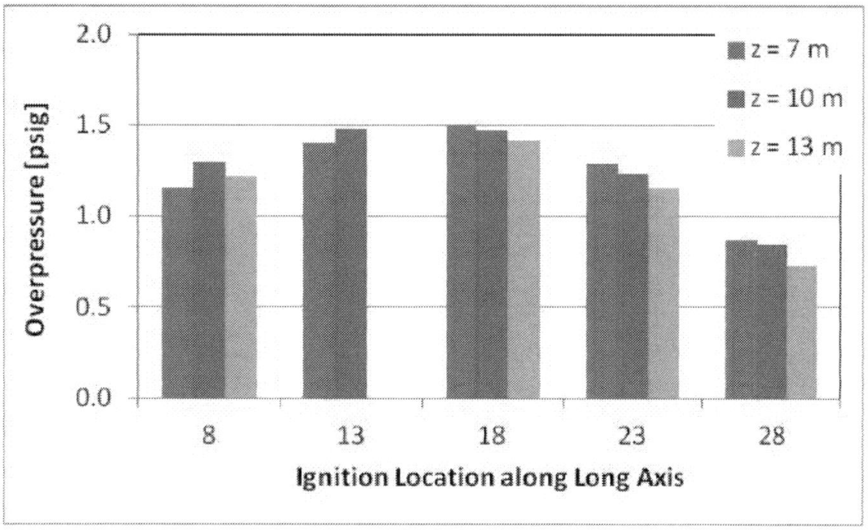

Figure 5. Impact of Ignition Location on External Overpressure (Real Cloud, Natural Gas)

5.2 Blast Load Comparison with Simplified Clouds (Natural Gas)

The maximum blast overpressures for the real (i.e., dispersed) and equivalent stoichiometric natural gas clouds internal and external to the module for the worst-case ignition elevations are shown in Figure 6 and Figure 7, respectively. The values shown are for the monitor location giving the highest peak overpressure. The maximum overpressures are much larger for the equivalent stoichiometric cloud than for the real cloud. The maximum internal overpressure with the equivalent stoichiometric cloud is more than an order of magnitude larger than that for the real cloud. However, the maximum external overpressure with the equivalent stoichiometric cloud is only 3 times larger than that for the real cloud. The maximum internal overpressure is therefore not a very good indicator of the overall blast loads, and particularly not a very good indicator of the external blast loads, as the maximum may be highly localized.

The average internal overpressure and impulse for the real cloud, the equivalent stoichiometric (i.e., Q9) cloud, and for the LFL-UFL cloud (i.e., a cloud with the same total flammable volume as that of the real cloud, but at the worst-case concentration) are shown in Figure 8 and Figure 9, respectively. The worst-case average internal overpressure for both the Q9 and LFL-UFL clouds are approximately 3 times that of the real cloud. The worst-case average internal impulse for the real cloud is slightly (less than 10%) higher than that for the simplified cloud representations, with the worst-case average internal impulse for the LFL-UFL cloud being slightly higher than that for the Q9 cloud.

The average external overpressure and impulse for the real cloud, Q9 cloud and for the LFL-UFL cloud are shown in Figure 10 and Figure 11, respectively. The worst-case average external overpressure for both the Q9 cloud is 2 times that of the real cloud, with the LFL-UFL cloud giving about 10% higher overpressures than the Q9 cloud. The worst-case average external impulse values are similar for the three cloud representations, with the worst-case LFL-UFL cloud giving slightly higher impulse values.

5.2 Blast Load Comparison with Simplified Clouds (Propane)

The average internal overpressure and impulse for the real and Q9 propane clouds are shown in Figure 12 and Figure 13, respectively. The worst-case average internal overpressure for the Q9 cloud is almost 4 times that of the real cloud. The worst-case average internal impulse for the Q9 cloud is 50% higher than that for the real cloud.

The average external overpressure and impulse for the real and Q9 propane clouds are shown in Figure 14 and Figure 15, respectively. The worst-case average external overpressure for the Q9 cloud is 3 times larger than that of the real cloud. The worst-case average external impulse for the Q9 cloud is about 20% larger than that of the real cloud.

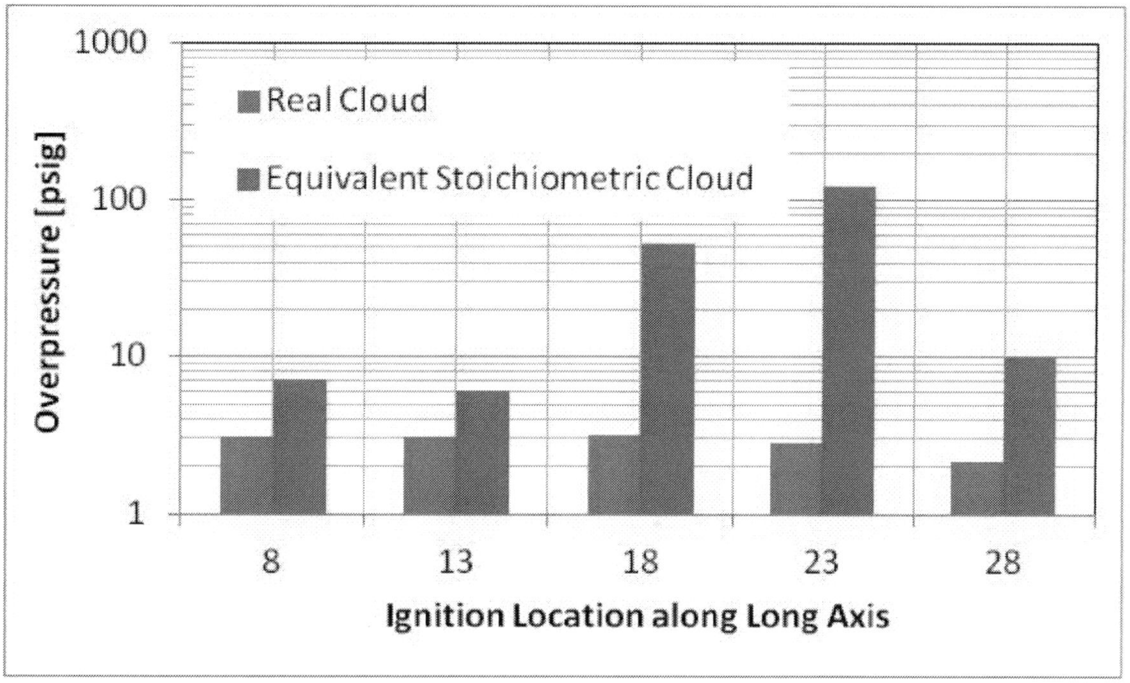

Figure 6. Internal Maximum Overpressure Comparison (Natural Gas)

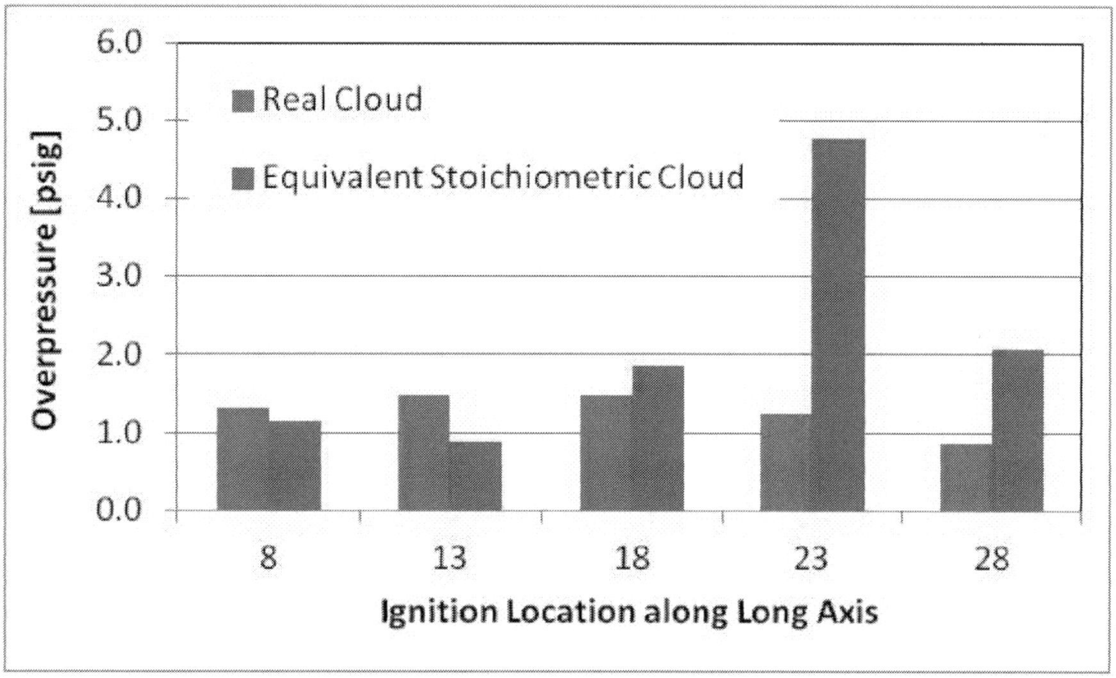

Figure 7. External Maximum Overpressure Comparison (Natural Gas)

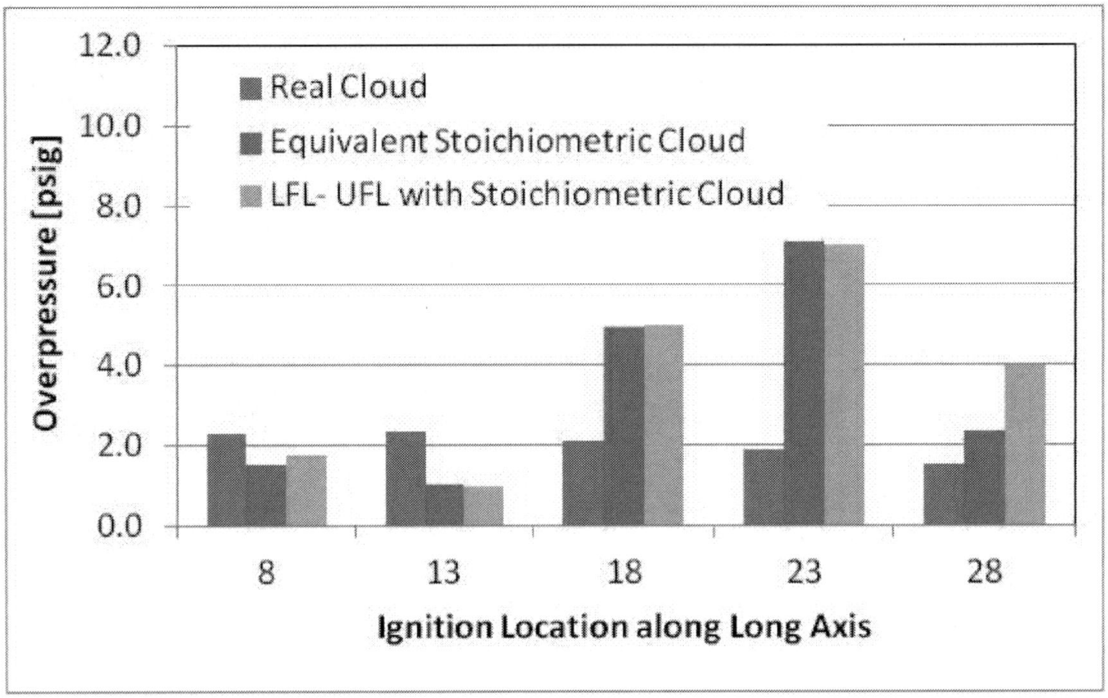

Figure 8. Internal Average Overpressure Comparison (Natural Gas)

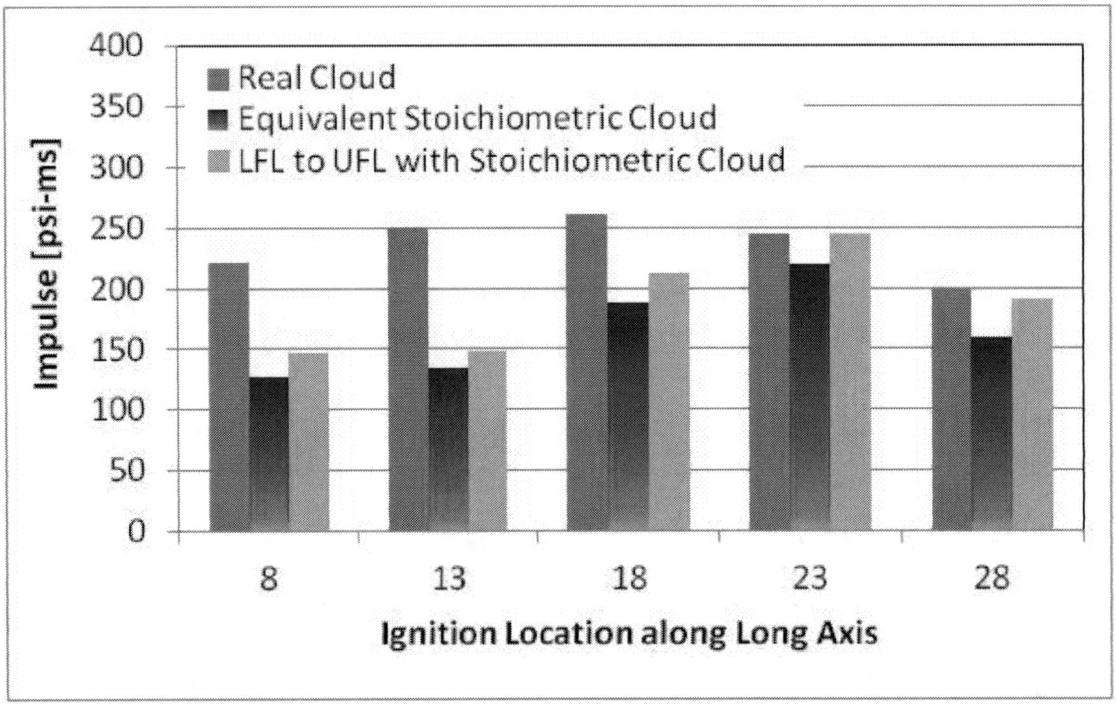

Figure 9. Internal Average Impulse Comparison (Natural Gas)

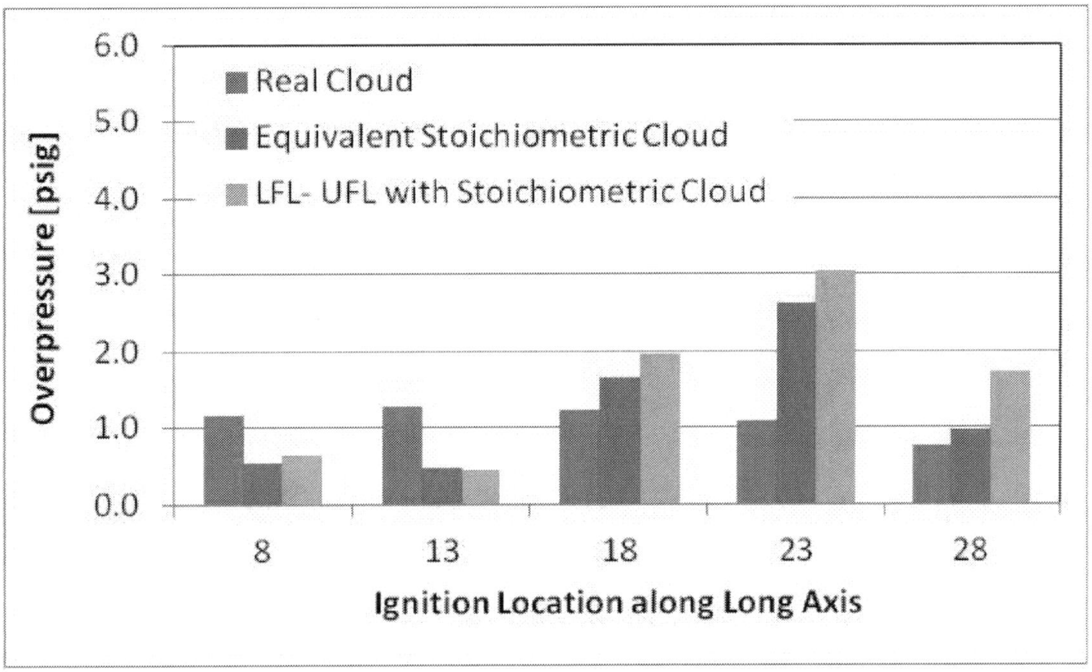

Figure 10. External Average Overpressure Comparison (Natural Gas)

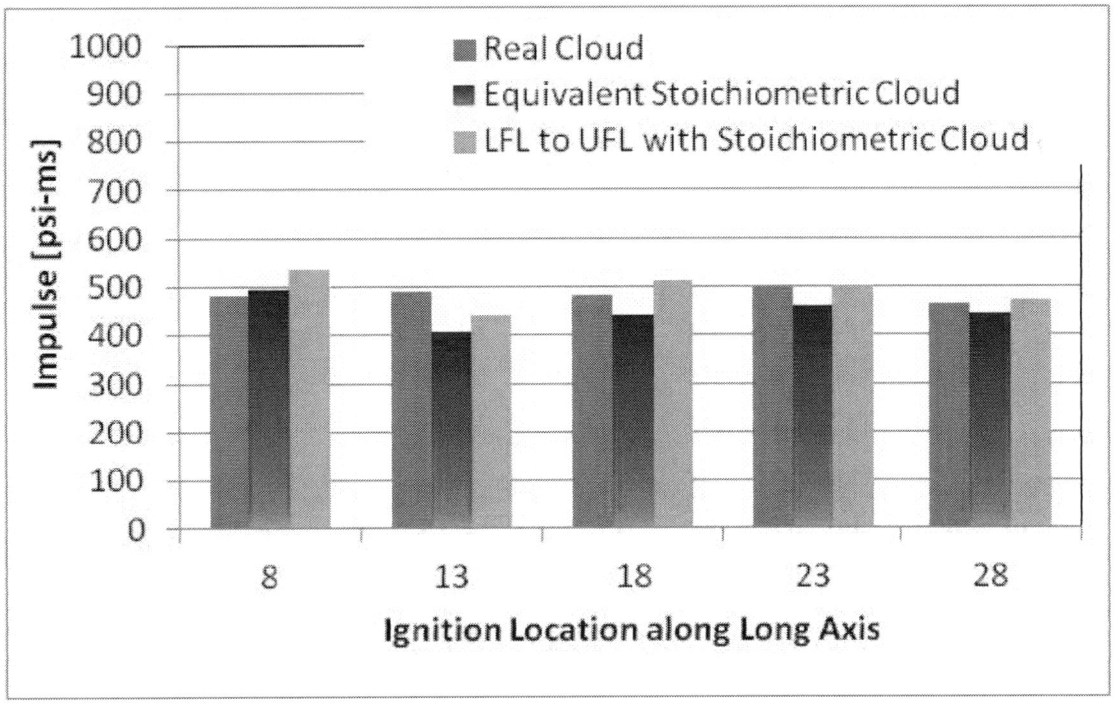

Figure 11. External Average Impulse Comparison (Natural Gas)

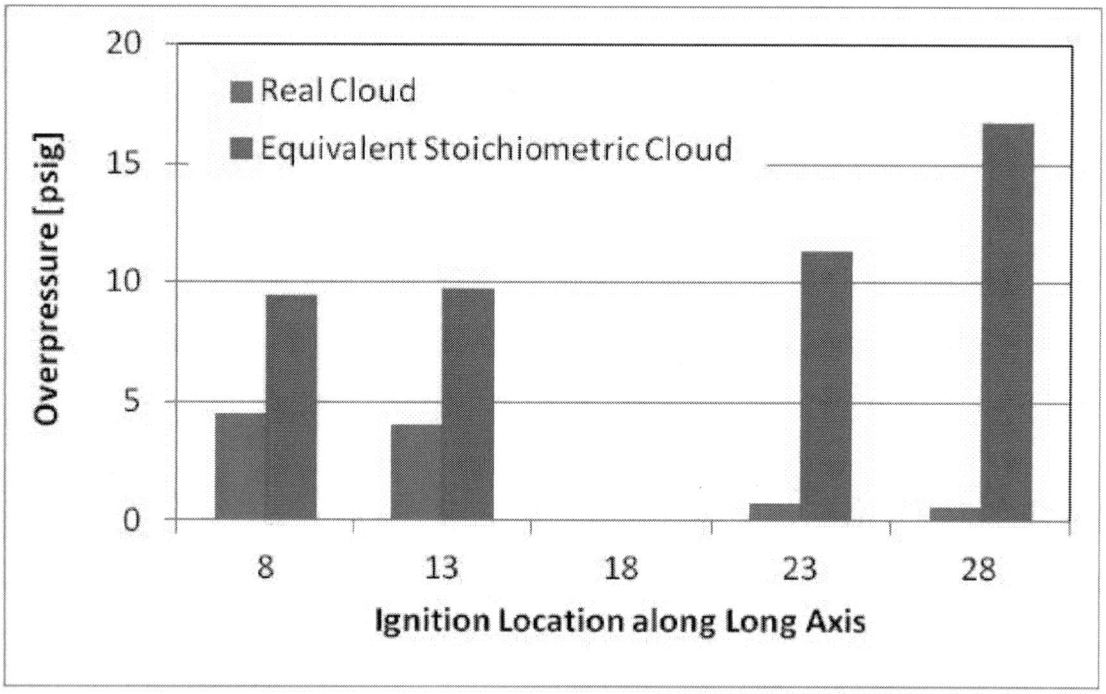

Figure 12. Internal Average Overpressure Comparison (Propane)

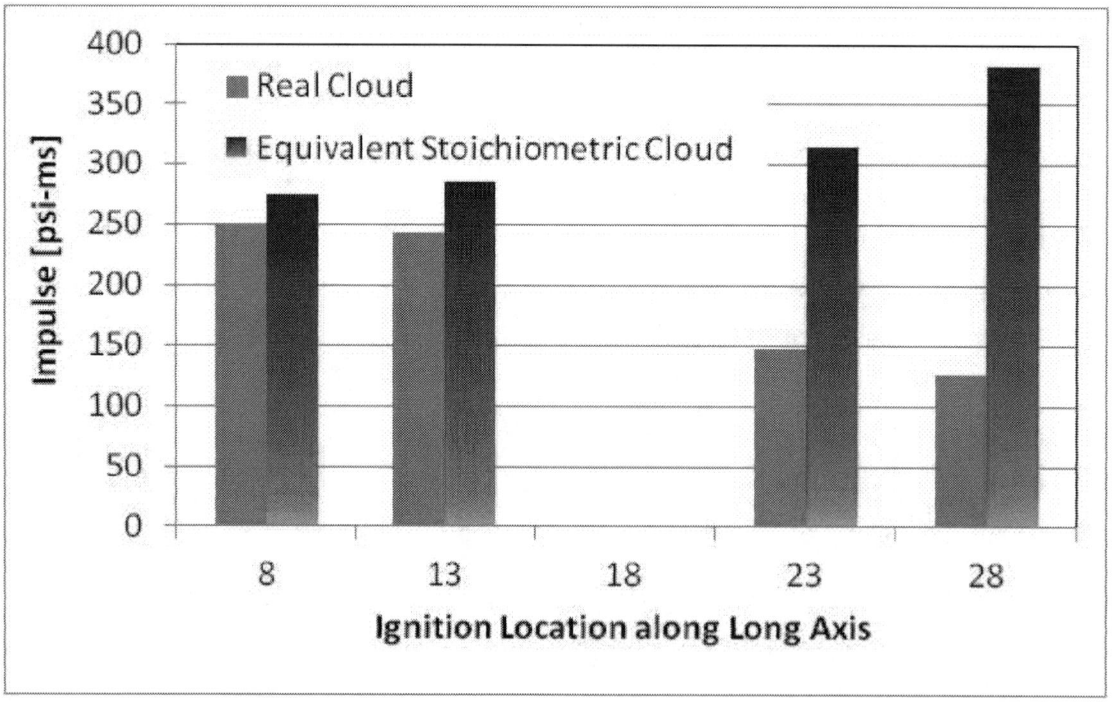

Figure 13. Internal Average Impulse Comparison (Propane)

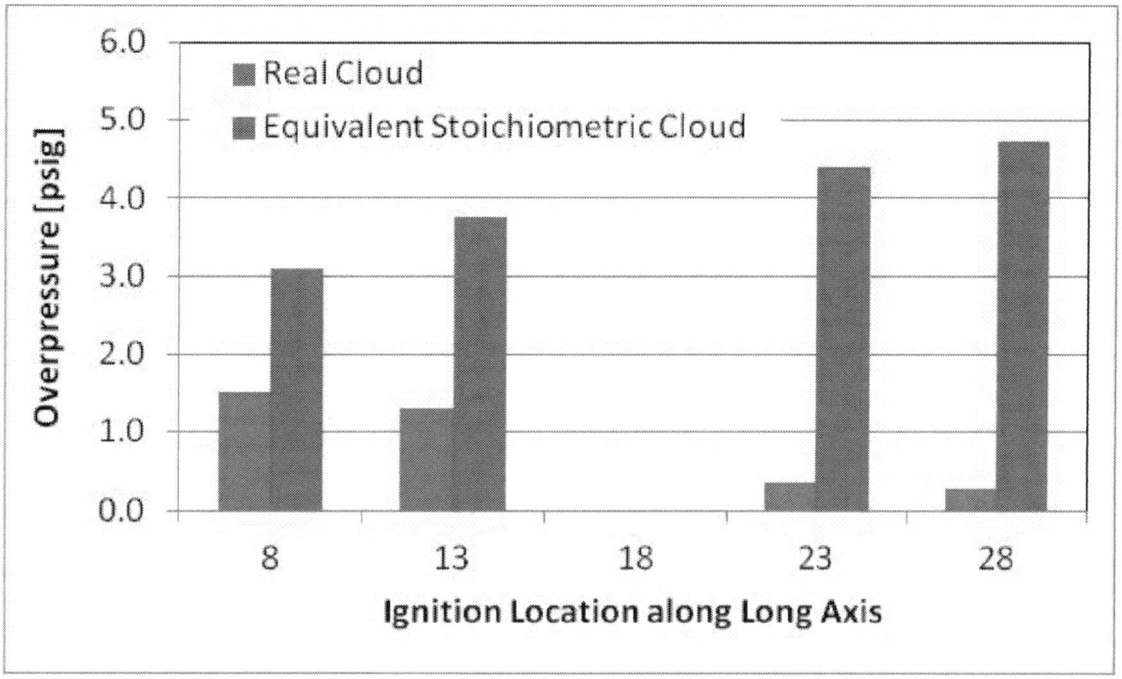

Figure 14. External Average Overpressure Comparison (Propane)

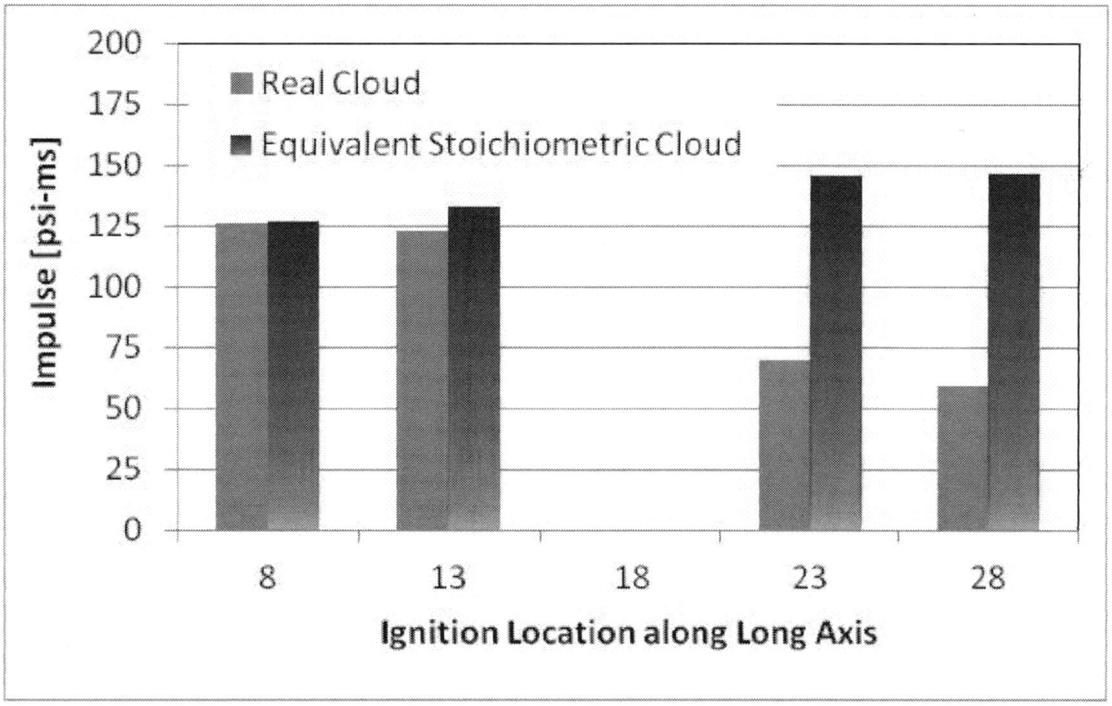

Figure 15. External Average Impulse Comparison (Propane)

5.3 Blast Load Comparison with Simplified Clouds (Ethylene)

The average internal overpressure and impulse for the real cloud, Q9 cloud and for the LFL-UFL cloud are shown in Figure 16 and Figure 17, respectively. The worst-case average internal overpressure for the Q9 cloud is 30% higher than that of the real cloud, with the LFL-UFL cloud giving about 20% higher overpressures than the Q9 cloud. The worst-case average internal impulse values are similar for the three cloud representations, with the worst-case LFL-UFL cloud giving slightly higher impulse values..

The average external overpressure and impulse for the real cloud, Q9 cloud and for the LFL-UFL cloud are shown in Figure 18 and Figure 19, respectively. The worst-case average external overpressure for the Q9 and real clouds are approximately equal, with the LFL-UFL cloud giving about 20% higher overpressures than either the Q9 or real clouds. The worst-case average external impulse values are similar for the three cloud representations, with the worst-case real cloud giving the largest maximum external impulse.

5.4 Blast Load Comparison Summary

Table 2 provides a summary of the worst-case average blast loads (over pressure and impulse) internal and external to the module for the real (i.e., dispersed) and equivalent stoichiometric (i.e.., Q9) gas clouds. As noted previously, the natural gas and propane flammable gas cloud (LFL-UFL) volumes are significantly larger that the Q9 due to significant concentration gradients (i.e., significant portions of the worst-case cloud are lean or rich), whereas this is not the case for the ethylene clouds due primarily to its near neutral-buoyancy.

Table 2. Summary of Q9 to Real Blast Load

Fuel	LFL-UFL to Q9 Volume Ratio	Ratio of Worst-Case Average Blast Load (Q9 to Real Cloud)			
		Internal to Module		External to Module	
		Overpressure	Impulse	Overpressure	Impulse
Natural Gas	1.34	3.0	0.8	2.0	0.9
Propane	1.41	3.7	1.5	3.1	1.2
Ethylene	1.15	1.3	1.0	1.0	0.9

The worst-case average internal blast overpressures for the Q9 clouds are 3 to 4 times those for the real clouds with natural gas and propane, and those external to the module for the Q9 clouds are 2 to 3 times higher than those for the real clouds. Conversely, the worst-case average internal blast overpressure for the Q9 cloud is only 30% larger than that for the real cloud with ethylene, with the worst-case average internal blast overpressures being nearly equal for the Q9 and real clouds.

The differences between the worst-case blast impulses are less dramatic than for the overpressures. The worst-case average internal impulse for propane is 50% larger with the Q9 cloud than for the real cloud. For the other fuel/location combinations, the difference between the Q9 and real cloud impulses are less than approximately 20%.

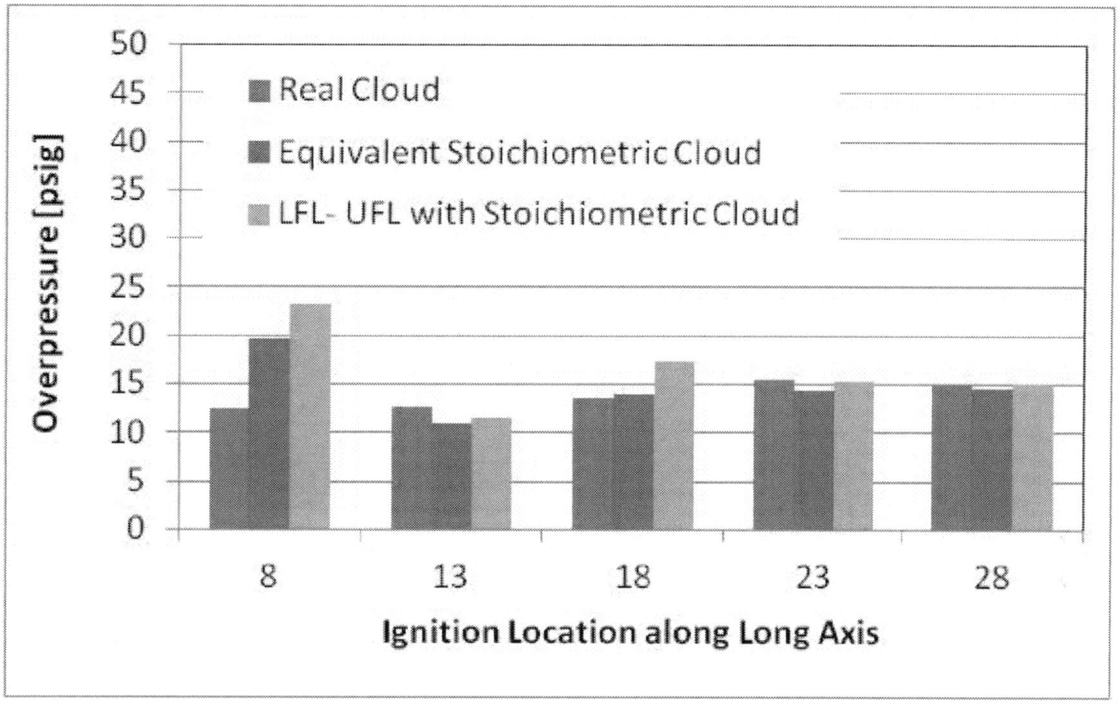

Figure 16. Internal Average Overpressure Comparison (Ethylene)

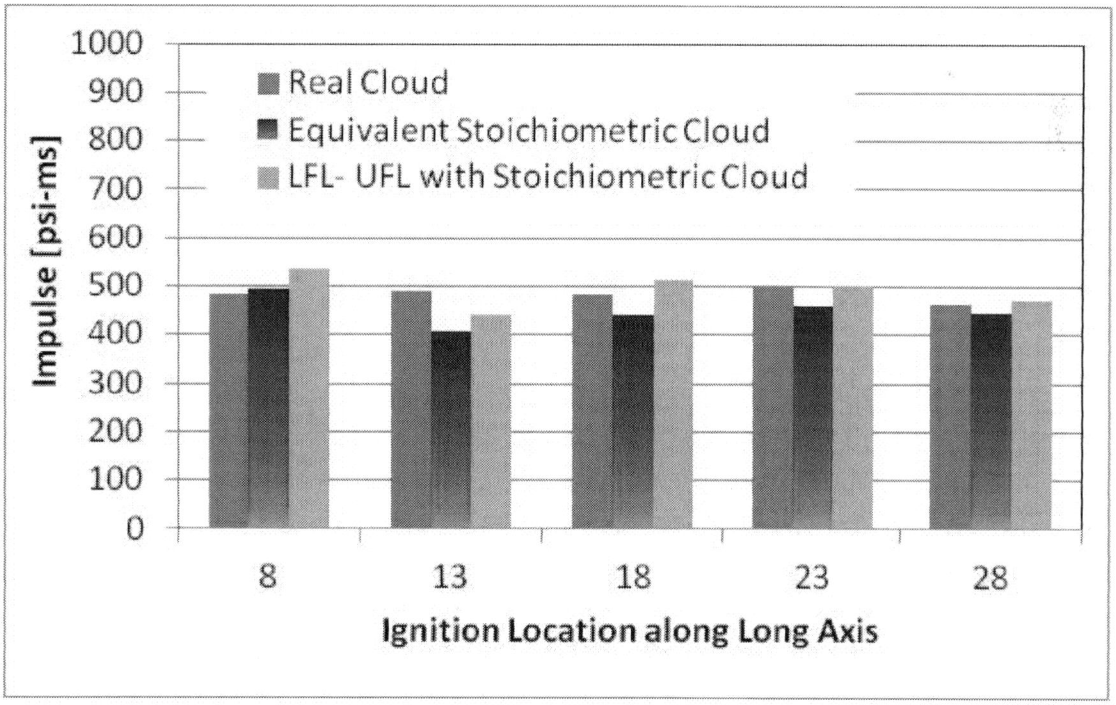

Figure 17. Internal Average Impulse Comparison (Ethylene)

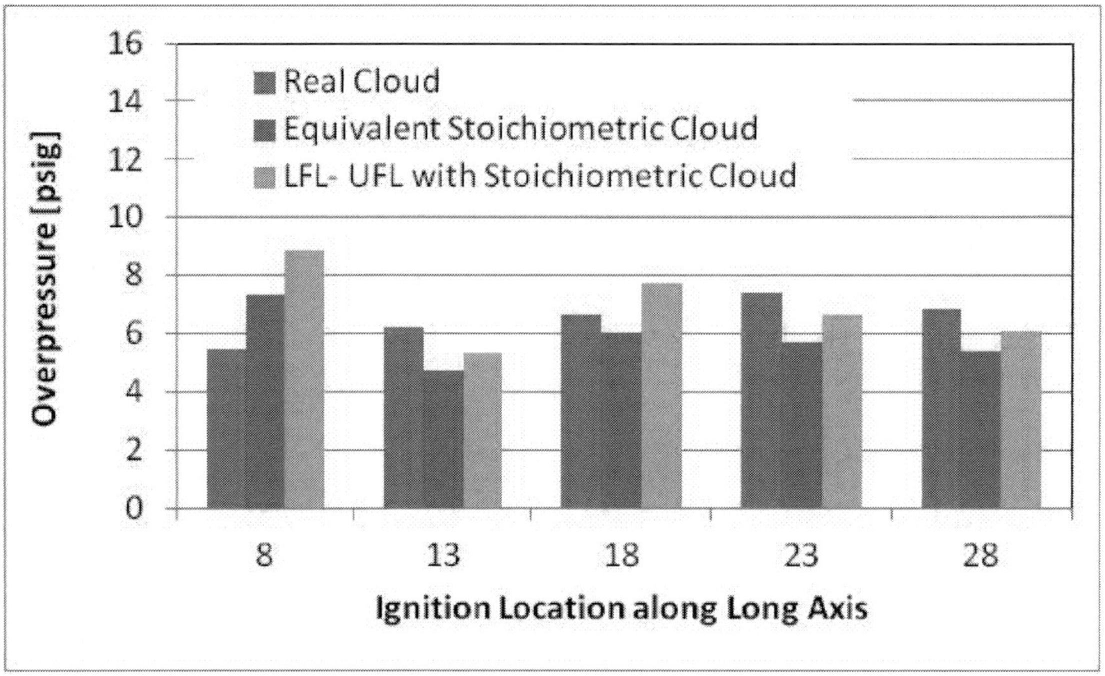

Figure 18. External Average Overpressure Comparison (Ethylene)

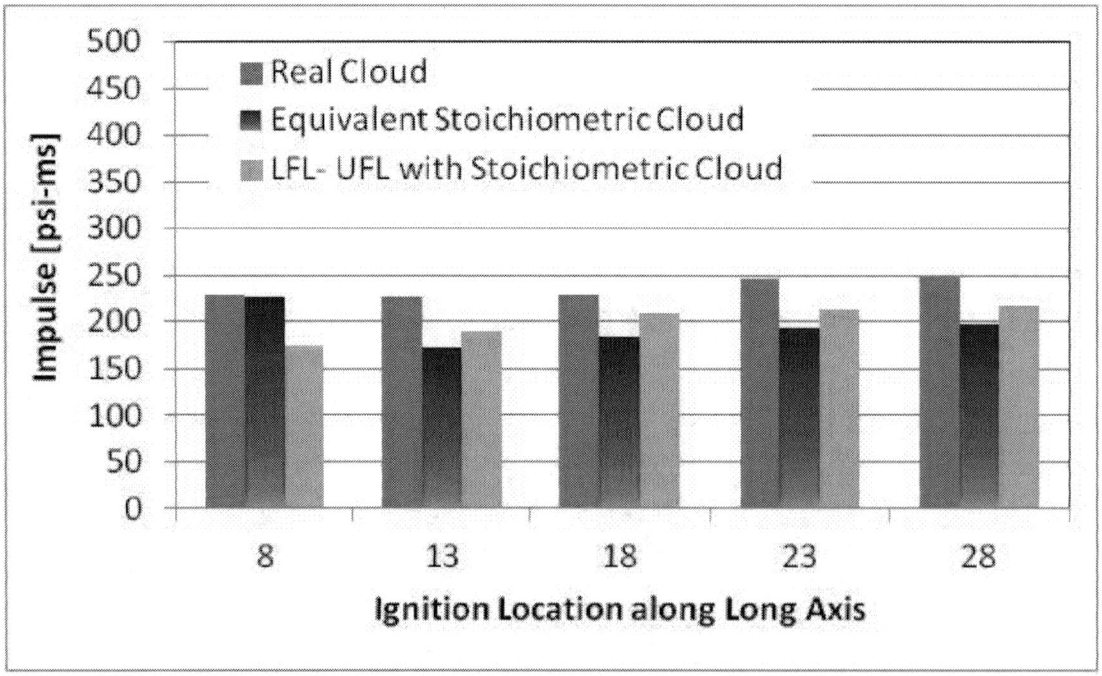

Figure 19. External Average Impulse Comparison (Ethylene)

6. Conclusion

The results of the analyses presented in this paper show that, for the conditions examined (ventilated process module, fuel release rate leading to largest flammable cloud, specified fuels, range of ignition source locations, etc.), the FLACS equivalent stoichiometric cloud (Q9) approach provides generally conservative loads with regards to those for the real (dispersed) flammable gas cloud. The maximum blast loads predicted using the equivalent stoichiometric clouds were much larger than those for the real clouds.

The natural gas and propane flammable gas cloud volumes are significantly larger that the Q9 clouds due to significant concentration gradients (i.e., significant portions of the worst-case flammable gas cloud are lean or rich). This was not the case for the ethylene clouds due primarily to its near neutral-buoyancy.

For the natural gas release, it was shown that the maximum overpressures are much larger for the Q9 cloud than for the real cloud. The maximum internal overpressure with the Q9 cloud was more than an order of magnitude larger than that for the real cloud, while the maximum external overpressure is 3 times larger. Comparing approaches based on the maximum internal overpressure may be misleading in terms of the overall blast loads as the maximum internal overpressure may be highly localized.

The worst-case average internal blast overpressures for the Q9 clouds were 3 to 4 times those for the real clouds with natural gas and propane, and those external to the module for the Q9 clouds were 2 to 3 times higher than those for the real clouds. The differences between the two approaches are much less with ethylene (i.e., due to the similarity in the Q9 and real clouds). The differences between the worst-case blast impulses are less dramatic than for the overpressures. The worst-case average internal impulse for propane is 50% larger with the Q9 cloud than for the real cloud. The difference between the Q9 and real cloud impulses both internal and external to the module for natural gas and ethylene are less than approximately 20%.

7. References

[1] Hansen, O.R., F. Gavelli, S.G. Davis and P. Middha (2011) "Equivalent Cloud Methods used for Explosion Risk and Consequence Studies," Mary Kay O'Connor Process Safety Center, 14th Annual Symposium, College Station, TX, 25-27 October 2011, pp. 78-106.

Lessons Learned from Fire in Air Supply Ducts

Al Waller
FMC Corporation
Philadelphia, PA
alvin.waller@fmc.com

Michelle Brown, CSP
FMC Corporation
Philadelphia, PA
michelle.brown@fmc.com

Prepared for Presentation at
American Institute of Chemical Engineers
2013 Spring Meeting
9th Global Congress on Process Safety
San Antonio, Texas
April 28 – May 1, 2013

UNPUBLISHED

Lessons Learned from Fire in Air Supply Ducts

Al Waller
FMC Corporation
Philadelphia, PA
alvin.waller@fmc.com

Michelle Brown, CSP
FMC Corporation
Philadelphia, PA
michelle.brown@fmc.com

Keywords: Lessons learned, fire, HEPA filter, air filter, thermal inertia

Abstract

In 2012, loss of air flow caused activation of interlocks to shut off steam and electrical heat inputs to an air heating system that was part of a drying/milling operation. Shortly thereafter, upstream filters and air ducts ignited causing more than US$500K in damage and business interruption. Investigations concluded that the interlocks functioned as designed on loss-of-airflow but were not sufficient to prevent overheating the inlet air filters (including HEPA filters) by the residual heat in the air heaters. The two main lessons learned from this incident apply to a broad range of operations in the CPI:

1) Mundane unit operations such as air filters that have combustible components must be evaluated as thoroughly as the rest of the chemical process.
2) Designers and PHA leaders must recognize the potential of thermal inertia of heaters (and coolers) to heat (or cool) a stagnant volume of fluid when flow is interrupted. Gas streams are particularly susceptible due to the low heat capacity relative to the heat capacity of a metallic heat exchanger. Interlocking the supplies (to steam and electrical heaters in this incident) is often necessary but may not be sufficient.

Other lessons "re-learned" include the importance of learning from a previous near miss, the importance of limiting escalation aspects of this fire, and the importance of preventative maintenance.

1. Introduction

A drying and milling operation experienced a fire inside its air supply ducts in 2012. Total losses exceeded US$500k, including replacement of air duct components and business interruption. An internal investigation was conducted with assistance from a consulting company by first focusing on the source of the fuel and then on the ignition mechanism.

Figure 1 below is a simplified diagram of a portion of the system; instrumentation and controls are omitted for clarity. Ambient air is pulled into the system by negative draft where it is filtered through pre-filters then HEPA filters. This filtered air is then heated via a steam heat exchanger then an electrical resistance heater to approximately 200C. This hot air then contacts damp combustible powder to convey it through a drying zone and into a cyclone. The distance between the HEPA filters and the steam heater is less than 2 meters while the distance from the electrical heater to the damp solids entry point is approximately 25 meters. Section 3.2 will more completely describe the air duct system and the process mills.

FIGURE 1 – Sketch of Air Duct for Dryer System

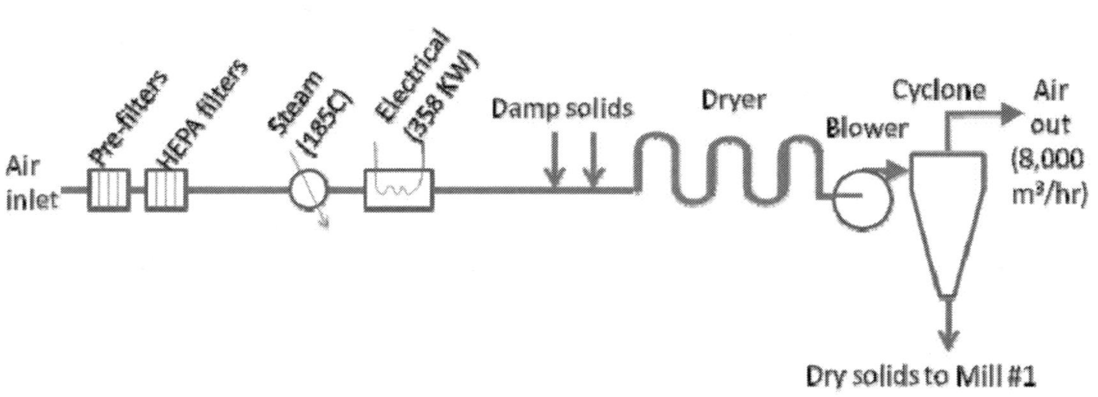

FIGURE 2 – Fire In Progress

The blower drive belts failed in September 2012 resulting in rapid loss of airflow so interlocks shut off steam and power to the heaters. Approximately 17 minutes later a portion of the air duct internally ignited.

2. "New" Lessons Learned

Forensic evidence pointed to the air filters as the source of fuel for the fire. Damage was limited near the air filters in Figure 1 plus there were no process solids or process ash upstream of the entry points for damp solids. The steam heater and electric heater were confirmed as tripped off. Calculations indicate that the residual heat (thermal inertia) was sufficient to ignite the air filters.

2.1 Filters As Fuel

The specific pre-filters and HEPA filters used at the incident site were rated for maximum continuous operation at 70C and 80C, respectively. Tables 1 and 2 list the key aspects of combustion testing of new pre-filters and HEPA filters. All components were susceptible to decomposition at 200C or less with the glass fiber and polyester component of the HEPA filter supporting a self-propagating flame in the oven test. Estimated energy release from combustion of the air filters was consistent with the damage observed from the incident.

Table 1. Stability of Air Filters

Unit	Component	Stability in 200C Oven Test	Stability in 500C Oven Test
Pre-filter	Glass fiber & polyester	Exothermic self-heating at 200C (to 600C). 47% weight loss	Exothermic self-heating (to 600C), 56% weight loss
HEPA Filter	Glass fiber & polyester	Exothermic self-heating at 160C (to 600C), self-propagating flame. 23% weight loss	Exothermic self-heating (to 600C), self-propagating flame. 39% weight loss
	Polyurethane sealant	38% weight loss	97% weight loss

Table 2. Estimated Air Filters Combustion Contributions

Unit	Component	Mass (kg)	Combustible Mass, from 500C Oven Test (kg)	Available Energy Release	
				(MJ)	(% of Total)
Pre-filter	Glass fiber & polyester	6.0	3.3	88	5%
HEPA Filter	Glass fiber & polyester	52.8	20.5	539	31%
	Polyurethane sealant	52.8	51.5	1089	63%

2.2 Residual Heat As Ignition Source

Recorded temperatures from the incident were not helpful in quantifying the residual heat remaining in the heaters after heating sources were tripped off. Recorded temperatures from a prior normal shutdown with the blower operating at 8000 cubic meters per hour were useful to calculate residual heats in the heaters as summarized in Figure 3 and Figure 4. The steam heater retains approximately 48 MJ while the electrical heater retains approximately 15 MJ.

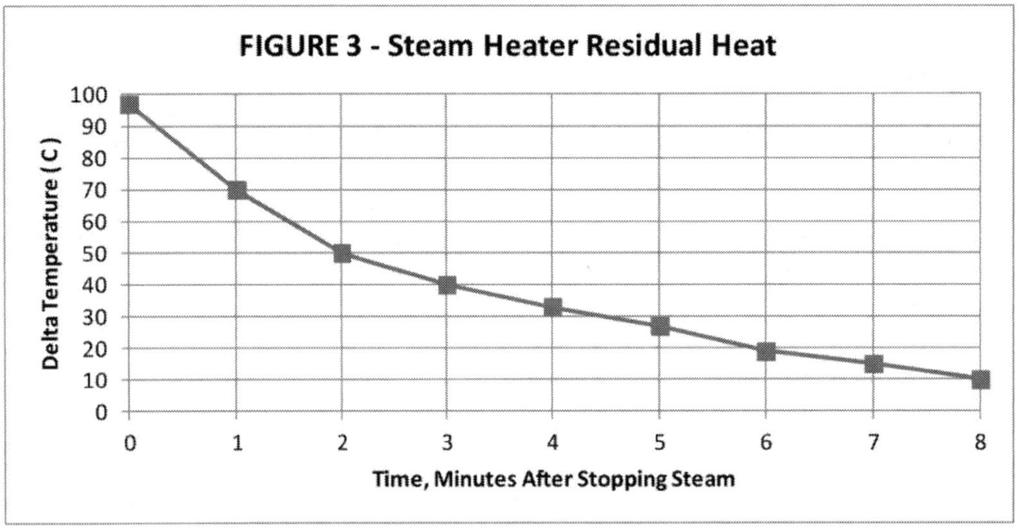

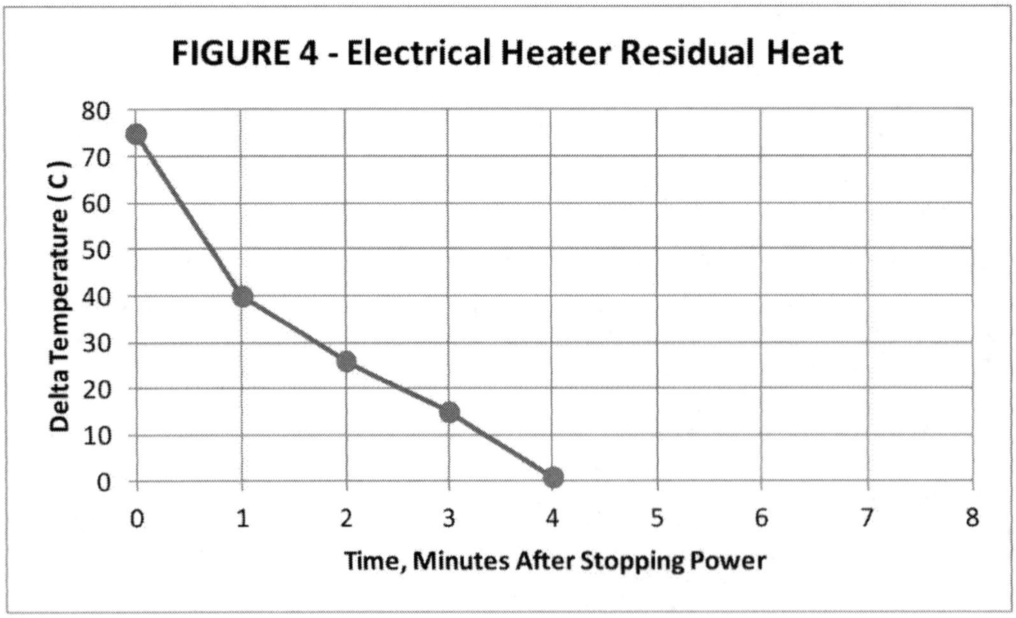

For the incident initiated by blower failure, this combined residual heat of 63 MJ can increase the temperature of stagnant air in the heater portion of the duct to well over 200C resulting in expansion driving the hot air backwards into the air filters. Alternatively, for this incident this combined residual heat could establish a convection current of dense cool air flowing forward at the bottom of the air filter duct, rising in the hot heaters zone, and hot air flowing backwards at the top of the air filter duct to the air inlet. Either incident mechanism results in excessively hot air in air filters that are rated for 70C or 80C maximum temperatures. The Standard Operating Procedures (SOP) for startup recognized the need to start the blower before heating and the SOP for normal shutdown recognized the need to keep the blower running until the air cooled down. Interlock designs recognized the necessity of tripping heating sources on loss-of-airflow. However tripping heating sources on loss-of-airflow is not sufficient to prevent the hazard of hot air reverse flow; therefore, an additional safeguard is needed. For this system, a new vertical vent duct with actuated valve was added between the heaters to open on loss-of-airflow as the primary additional safeguard to provide chimney-effect cooling of the heaters.

3. "Old" Lessons Re-Learned

In addition, lessons from this air duct fire have been refreshed multiple times and are widely known in the CPI.

3.1 Need to Investigate Near Misses and Minor Incidents

Prior to the September 2012 incident, a separate minor incident occurred during a normal shutdown. Some charring of the HEPA filters was attributed to a leaking steam heater isolation valve. The response was to repair the leaking steam valve; thus, the opportunity to investigate this precursor incident was missed.

3.2 Need to Limit Potential Escalation

The fire damage from the September 2012 incident extended beyond the Dryer air duct system (shown in Figure 1 above). The damage was much more severe to the other air filters (pre-filters and HEPA filters), other heat exchangers, and other ductwork interior surfaces as shown in Figure 5. The 3 other blowers in Figure 5 stayed on-line following the Dryer airflow failure and trip of the Dryer air heaters. Hot fumes of pyrolysis/combustion products from the Dryer air filters were pulled into the connected ductwork by these other blowers resulting in direct escalation of the fire or perhaps the Mill #2 electrical heater ignited fumes which backflashed to the air filters. While the Dryer air filters likely experienced restricted oxygen availability, the other air filters had ample oxygen to support combustion. The common air intake duct reduced the initial capital cost by a very small amount compared to separate intake ducts yet added unnecessary complications that led to major escalation. The system was rebuilt with separate air intakes.

FIGURE 5 – Sketch of Air Ducts for Complete System

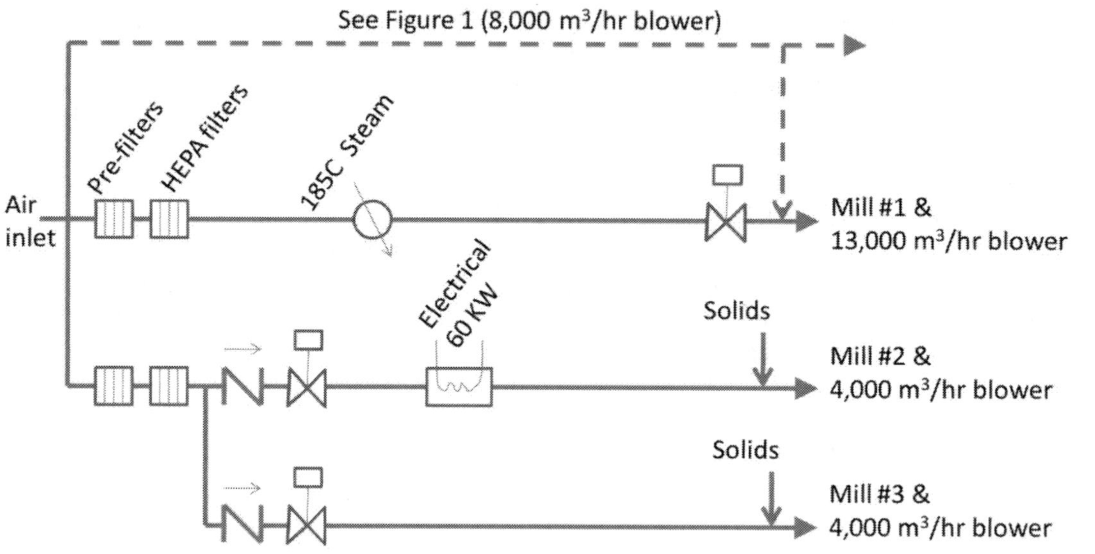

Modest external damage was also visible from smoldering combustible dust on the surfaces of some ducts due to accumulation. Housekeeping practices need to limit combustible dust accumulations below 3 mm (1/8-inch).

3.3 Need to Maintain Equipment Properly

Finally, the blowers did not have a specific installation procedure to ensure belt alignment plus they were not on a preventative maintenance program.

4. Conclusion

Those who forget the past are doomed to repeat. The main lessons learned from this incident apply to a broad range of operations in the CPI:

1) Mundane unit operations such as air filters that have combustible components must be evaluated as thoroughly as the rest of the chemical process.
2) Designers and HAZOP leaders must recognize the potential of thermal inertia of heaters (and coolers) to heat (or cool) a stagnant volume of fluid when flow is interrupted. Gas streams are particularly susceptible due to the low heat capacity relative to the heat capacity of a metallic heat exchanger. Interlocking the supplies (to steam and electrical heaters in this incident) is often necessary but may not be sufficient.
3) Near misses and minor incidents should be thoroughly investigated; an opportunity was missed at this site to prevent the major incident.

4) Keep it simple and keep areas clean. Significant escalation occurred due to a combined air duct that added unnecessary interactions between the systems. Minor escalation occurred from accumulated combustible dusts on exterior surfaces.
5) Maintain equipment properly. Rigorous installation procedures (such as belt alignment) and preventative maintenance improve both safety performance and on-stream reliability.

5. References

Additional references not cited:

[1] Kletz, Trevor. "What went Wrong? - Case Histories of Process Plant Disasters and How They Could Have Been Avoided" (5th Edition, 2009), Elsevier.

[2] DOE (U.S. Department of Energy), "Nuclear Air Cleaning Handbook" Chapter 10 – Fire Protection, DOE-HDBK-1169-2003 (2003).

Hydrogen Sulfide Exposure from Molten Sulfur – A Forgotten Hazard?

Mark J. Viz,
Sean J. Dee, Russell A. Ogle, and Brenton L. Cox
Exponent, Inc.
4580 Weaver Parkway, Warrenville, Illinois 60555
mviz@exponent.com, sdee@exponent.com, rogle@exponent.com,
and bcox@exponent.com

Prepared for Presentation at
American Institute of Chemical Engineers
2013 Spring Meeting
9th Global Congress on Process Safety
San Antonio, Texas
April 28 – May 1, 2013

UNPUBLISHED

Hydrogen Sulfide Exposure from Molten Sulfur – A Forgotten Hazard?

Mark J. Viz,
Sean J. Dee, Russell A. Ogle, and Brenton L. Cox
Exponent, Inc.
4580 Weaver Parkway, Warrenville, Illinois 60555

Keywords: molten sulfur; hydrogen sulfide; railcar; tank car; tank trailer; process safety management; PSM; hazardous materials; hazmat; PHMSA; OSHA; NA2448

Abstract

Molten sulfur is one of the most frequently transported hazardous materials (hazmat) in North America. Molten sulfur is typically transported at an elevated temperature by rail tank cars or over-the-road tank trailers as a Class 9 "Miscellaneous Hazardous Material" under the identification number NA2448. Perhaps the most prominent hazard of molten sulfur is that it is shipped at an elevated temperature, which can cause thermal burn injuries. A less obvious hazard is the potential release of hydrogen sulfide (H_2S), a highly toxic gas.

In this paper we first summarize the key features of the H_2S-molten sulfur equilibrium that are relevant to this hazard. Then we address the relative risk of H_2S exposure from an analysis of molten sulfur incident release statistics. Next, we discuss a few case studies in which vented H_2S gas from molten sulfur led to unexpected consequences for handling personnel. We close our discussion with suggestions for how to possibly prevent H_2S exposure through a combination of engineering controls, operating procedures for both loading and unloading operations, air monitoring, and personal protection equipment.

1. Introduction

Molten sulfur is one of the most widely used global feed stocks. In 2010, the global demand for sulfur reached 78.3 million tons, with elemental sulfur accounting for 66% of global production.[1] With growing demand in the Middle East and Asia, continued use of sulfuric acid, and a growing need for nickel, sulfur production is projected to grow steadily over the next decade.[2] In 2011, elemental sulfur was a $1.6 billion dollar industry in the United States, with 8.1 million tons of production.[3] Additionally, over the last thirty years there has been a significant shift in the production of molten sulfur, moving from mining of geologic sulfur deposits (Frasch process) to environmental byproduct sulfur (Claus process). This technology shift has increased the H_2S

[1] Messick, D. L., World Sulfur Outlook, The Sulfur Institute, 2010.

[2] Kitto, M., Sulfur Supply Demand Balance: The Outlook to 2015, British Sulfur Consultants, 2005.

[3] Apodaca, L. E., Sulfur, U.S. Geological Survey, Mineral Commodity Summaries, January 2012.

hazard in molten sulfur. Given the livelihood of the global sulfur market and the change in the source of the material, the need to communicate and control the hazards associated with the handling of molten sulfur is more critical than ever.

The hazard of dissolved H_2S in molten sulfur became widely recognized in 1963 with the disappearance of the molten sulfur tanker, the Marine Sulfur Queen.[4] This tanker disappeared without a distress signal off the coast of Florida with all 39 crew members. The U.S. Coast Guard searched for the vessel but was only able to recover a few life preservers and some floating debris. Reports came to light of frequent fires on the sulfur tanker and the frequent odor of rotten eggs.[5,6] Investigators from various countries performed headspace measurements in molten sulfur transportation vessels and found H_2S concentrations approaching the minimum flammable level. Recommendations were published that the dissolved H_2S content of molten sulfur should be maintained at 10 parts per million (ppm).[7]

As early as 1970, the U.S. Department of Transportation (USDOT) issued a rule under docket HM-70, that even though it was intended to address requirements for rail tank cars and tank trailers used to ship H_2S as its own commodity, clearly emphasized the hazards associated with exposure to H_2S. The comments to the federal rule state that H_2S is a "severe irritant and is highly toxic"; that exposure leads to "olfactory fatigue"; that H_2S is "heavier than air" and "tends to 'pool' rather than to dissipate into the atmosphere."[8]

Later, in the mid-1980s, the USDOT issued a Final Rule under docket HM-198 that clearly identified two significant hazards associated with the transportation of molten sulfur: the presence of unknown but potentially significant levels of H_2S and that the elevated temperature of molten sulfur could cause burns.[9] As reported in the Final Rule: "Several incidents involving the transport of molten sulfur, including three in the state of California since 1985, have emphasized the need for regulation of this material. Accidents involving spills near Barstow County and Culver City in 1986, and another in Benecia in 1985, demonstrated a lack of hazard communication. In each of these incidents, lack of swift and accurate identification of the material led to delays in handling the incidents and, in at least two instances, contributed to injuries sustained by persons on the scene."

A review of available Material Safety Data Sheets (MSDSs) for molten sulfur suggests that the range of H_2S concentrations and corresponding safe loading, unloading, and handling procedures for molten sulfur vary considerably among producers and shippers. As a consequence, the handling processes may be performed in a manner ignorant of the possible high concentrations of H_2S. According to several MSDSs, the primary hazards associated with molten sulfur are

[4] King, F.W., "Taking H_2S from liquid sulfur-why and how," Oil and Gas Journal, pp. 164-165, June 1974.

[5] Anon., "Investigations: The Queen with the Weak Back," TIME, March 8, 1963.

[6] U.S. Coast Guard, Marine Board of Investigations, "Disapearance of the SS Marine Sulphur Queen at sea on or about 4 February 1963 with the presumed loss of all persons on board," March, 1964.

[7] Lagas, J.A., "Stop emissions from liquid sulfur," Hydrocarbon Processing, pp. 85-89, October 1982.

[8] 35 FR 18919 et seq., December 12, 1970, docket no. HM-70, "Hydrogen Sulfide Gas in Cargo Tank Trucks and Tank Cars."

[9] 53 FR 17158 et seq., October 21, 1986, docket no. HM-198, "Molten Sulfur and Molten Materials Generally."

thermal burns, sulfur-fueled fires, dust explosions, exposure to H_2S, and, in the event of a sulfur fire, inhalation of sulfur dioxide.[10,11,12,13,14] The hazard of H_2S is not prominently communicated in many of these current MSDSs.

Elemental sulfur is typically transported in a molten state in railcars or tank trailers. Commonly, the H_2S hazard is reported as a potential risk, because the concentration of H_2S within the vapor space of the railcar, tank trailer or stationary storage or process vessel is specified as a range or threshold value. The actual concentration of H_2S present in the vessel headspace will likely change due to temperature variations and agitation (as occurs in railcars and tank trailers during movement), and the potential exposure will be dependent on the producer, shipper or consignee's handling procedures and best practices. The authors' own field experiences have also shown that particular to the unloading of molten sulfur railcars, unloading personnel can have a false sense of safety when opening the top fittings or manway cover to prepare a tank car for steaming. Unloading personnel occasionally believe that the solidified "crust" of molten sulfur that is usually present at the surface, especially in cold weather, will somehow prevent any significant release of H_2S. Our own investigatory experience has shown that this is not always true.

Occupational exposure to H_2S during handling can generally be categorized as an example of a low probability, high consequence risk. This is supported by a review of the Pipeline and Hazardous Materials Safety Administration (PHMSA) and Occupational Safety and Health Administration (OSHA) incident report databases, which confirms that over the last 20 years, there have been six OSHA-reported fatalities from H_2S exposure associated with molten sulfur containers and over the last 24 years, there have been no PHMSA-reported fatalities from molten sulfur releases (while the commodity is still under the jurisdiction of the federal hazmat transportation regulations).[15,16] Our collective experience based on our investigations of numerous hazmat releases is that the tendency is for successful management of low probability, high consequence hazards to lead to complacency and the occasional elimination of seemingly superfluous safeguards.

---

[10] Hovensa, Sulfur MSDS No. 1794, 2006.

[11] Tesoro, Safety Data Sheet: Sulfur, Molten, 2012.

[12] Keyera, Sulfur, Molten MSDS, 2011.

[13] Devon, Material Safety Data Sheet: Sulfur (Liquid), 2010.

[14] Marathon, Material Safety Data Sheet: Marathon Sulfur, 2010.

[15] Note that there can be overlap between the release incident reporting required by OSHA and PHMSA based on where the incident occurred and what function was being performed when the release occurred. So, for example, an accidental release of molten sulfur involving an injury or fatality that occurred while a tank car was connected to plant apparatus during unloading might be reported to OSHA and not to PHMSA even though releases during unloading can be covered by PHMSA regulations as well. Efforts to harmonize the regulations and reporting requirements promulgated by a variety of federal agencies has been the subject of federal rulemaking in dockets such as HM-223, "Applicability of the Hazardous Materials Regulations to Loading, Unloading and Storage," Docket No. PHMSA-98-4952 (HM-223), under which PHMSA issued a Final Rule on June 1, 2005.

[16] Note that at times, exposure to H_2S can result in a prolonged hospitalization that results in fatality at a date after the original paperwork related to regulatory agencies is submitted.

According to the USDOT PHMSA "Incident Reports Database," there have been 428 reported releases of molten sulfur in transportation between the years 1989 to 2012, inclusive. For all 428 reported releases, no fatalities and 19 injuries were reported. A portion of these injuries were related to exposure to fumes and odors that resembled "rotten eggs", a description commonly used for H_2S. Additionally, incident reports indicated that H_2S gas was released at various times while the commodity was in transit, being loaded or unloaded, or while being stored in transit. Although there may be some overlap between exposure incidents reported in PHMSA's Incident Reports Database and the analogous Accident Investigation Summaries from OSHA (OSHA Form 170), which are maintained in OHSA's Integrated Management Information System (IMIS), the OSHA database includes six reported fatalities from H_2S exposure associated with the handling of molten sulfur over the last twenty years.

Our review of H_2S exposure incidents is intended to act as a reminder of the importance of employing a hierarchy of safeguards when managing high consequence hazards like H_2S.

2. Technology Background

A major technological shift in the production of sulfur has resulted in an increased potential for exposure to H_2S. For most of the 20th century, the Frasch process was used to recover sulfur from underground, elemental deposits.[17] In the Frasch process, superheated water is used to melt solid deposits of sulfur which are driven to the surface with compressed air for recovery. However, with the growth of the refining industry, the supply of elemental sulfur from the desulfurizing of sour crude oil and natural gas overtook production from the Frasch process in 1970. The Claus process, which can produce elemental sulfur from feed stocks containing up to 20% H_2S, is now the preeminent production process for producing sulfur.

To convert H_2S to elemental sulfur, the Claus process uses a two-step process, first utilizing a high temperature thermal step, followed by a lower temperature catalytic step. In the thermal section, air is added to H_2S at temperatures in excess of 1800°F, to covert one-third of the H_2S to sulfur dioxide (SO_2) as shown in Equation 1.

$$H_2S + 3/2\ O_2 \rightarrow SO_2 + H_2O \qquad \text{[Eq. 1]}$$

The resulting SO_2 then reacts with the remaining H_2S in an endothermic equilibrium reaction to form elemental sulfur vapor as shown in Equation 2.

$$2H_2S + SO_2 \leftrightarrow 3/2\ S_2 + 2\ H_2O \qquad \text{[Eq. 2]}$$

To drive conversion in the endothermic equilibrium reaction shown in Equation 2, S_2 vapor is removed through condensation. As the vapor cools, the S_2 equilibrates with the S_6, and S_8 forms through multiple equilibrium reactions.

If one-third of the sulfur is consumed in Reaction 1, then the remaining two-thirds is consumed in Reaction 2 resulting in stoichiometric consumption of H_2S. However, several side reactions usually lower the conversion to SO_2 in Reaction 1, subsequently lowering the overall H_2S

[17] Greenwood, N. N. and Earnshaw, A., Chemistry of the Elements, 2nd ed., 1997.

conversion. This highlights both the importance of the air to H_2S feed ratio as a process control variable, and the need for a secondary catalytic reactor to further convert the H_2S that is not consumed in the thermal section.

The catalytic section of the Claus process is used to convert H_2S as well as side products including COS and CS_2 that are formed from side reaction in the thermal section. The catalytic section runs at significantly lower temperatures than the thermal section (inlet 400°F - 450°F, outlet 600°F - 630°F), but uses a Clause catalyst (usually aluminum (III) or titanium (IV) oxide) to covert H_2S to elemental sulfur as shown in Equations 3 and 4.[18,19]

$$2\ H_2S + SO_2 \rightarrow 3/6\ S_6 + 2\ H_2O \qquad [Eq.\ 3]$$

$$2\ H_2S + SO_2 \rightarrow 3/8\ S_8 + 2\ H_2O \qquad [Eq.\ 4]$$

The effluent of the catalytic section can be cycled through additional catalytic beds, or condensed to a final product stream at temperatures between 265°F - 350°F.

The one common feature to all of these chemical reactions is the constituent H_2S. While the Claus process can achieve 96% - 97.5% conversion with one thermal reactor, and three catalytic reactors in series, the solubility of H_2S in molten sulfur adds an additional processing constraint. The solubility of H_2S in molten sulfur actually increases with temperature up to approximately 425°F. Solubility remains approximately constant from 425°F until 630°F, where it begins to drop with further increases in temperature.[20] Thus, in the catalytic reactors and condensers, there is still an appreciable amount of H_2S dissolved in the molten sulfur. If the hot sulfur is then stored in a tank, as the tank cools, the dissolved H_2S will release into the headspace, increasing the concentration to potentially dangerous levels. If the sulfur continues to cool, then a crust will form as the surface of the product reaches temperatures below 240°F. This solid crust can serve as a physical barrier protecting personnel from potential H_2S exposure. In cases where a crust is observed, special consideration should be noted to avoid agitation that could disrupt the solid interface, resulting in a potential H_2S vapor exposure. However, it should be noted, that formation of a protective crust should not be regarded as a reliable safety indicator, since crust formation is not correlated to the concentration of H_2S in the vapor space.

The headspace concentration of H_2S that can be achieved with molten sulfur in a transportation or storage container is influenced by four primary factors: the dissolved H_2S concentration, the liquid temperature, agitation of the liquid, and the presence of any organic contaminants. To maintain an H_2S concentration below its lower flammable limit (LFL), typical specifications for the dissolved H_2S concentration range from 5 to 15 ppm. At the temperature range at which molten sulfur is typically handled, the LFL corresponds to a headspace concentration of 3.4%, or

[18] Goar, B. G. and Fenderson, S., Fundamentals of Sulfur Recovery by the Claus Process, presented at The Laurence Reid Gas Conditioning Conference, Norman, Oklahoma, March 3, 1996.

[19] Sulfur and Hydrogen Sulfide Recovery, Kirk-Othmer Encyclopedia of Chemical Technology, John Wiley & Sons.

[20] Fanelli, R., Solubility of Hydrogen Sulfide in Sulfur, Industrial and Engineering Chemistry, pp. 2031-2033, September 1949.

34,000 ppm (at room temperature the LFL for H_2S is 4.3%).[21] From the viewpoint of respiratory exposure, this is a potentially instantaneous lethal dose.

3. Hazard Communication

The MSDSs for molten sulfur identify several hazards. Common features were noted in the 19 MSDSs reviewed for this study. First, nearly all of the surveyed MSDSs noted the potential dangers associated with H_2S and listed H_2S as a material present in molten sulfur. The amount of H_2S present was usually noted as a small amount; descriptions included "trace", "< 0.1%", "< 1%", "< 500 ppm", and "< 1% (variable)". It was also common for MSDSs to include additional information about the potential hazards of H_2S exposure, the acceptable exposure limits (which even today are still a subject of debate), and to identify the need for proper ventilation during storage and handling.

In the MSDS review, there is significant confusion and room for improvement regarding the description of PPE and unloading procedures. In addition to H_2S hazards, respiratory protection also included consideration for dust hazards since molten sulfur can solidify to a yellow solid when cooled. In most MSDSs, it is noted that respiratory protection should be used, but only if the concentrations of H_2S are unknown/above regulated levels, or other engineering controls are not sufficient to control the exposure. One MSDS stated that respiratory protection was not commonly required when handling molten sulfur. While several MSDSs noted that unloading/opening of tanks should be done in well ventilated areas, it was not noted if proper ventilation is a sufficient engineering control to protect workers from potential exposure to H_2S. One MSDS included a picture of a dust mask in the recommended PPE, which is appropriate for the potential dust hazard, but not for H_2S exposure.

In addition to not specifying when a respirator should be worn, very few MSDSs had detailed recommendations for unloading procedures related to the H_2S hazard in molten sulfur. As previously discussed, most of the MSDSs recognized several key components to safe unloading: proper ventilation, respiratory protection if the H_2S concentration was unknown or above acceptable limits, and confined space warnings. Air monitoring was not commonly listed as a recommended engineering control in the MSDSs. Additionally, recommendations regarding venting of tank headspace were not usually recommended before sampling or unloading. This omission, however, could be the result of the inclusion of opening the tank car or tank trailer before steaming as part of the unloading procedure. Details such as these regarding unloading might not typically be specifically addressed in the MSDS. Other MSDSs specified venting, only if the product was excessively heated or exposed to elevated temperatures. The potential evolution of H_2S into the vapor space can be driven both by temperature increases in the presence of water or by temperature decreases which drive dissolved H_2S out of the molten sulfur. Including specific directions for when respiratory protection is required and directions for proper unloading safety controls before handling/unloading could improve the hazard communication for H_2S in molten sulfur.

[21] NFPA 655, Standard for Prevention of Sulfur Fires and Explosions, National Fire Protection Association, Quincy, MA, Section A.5.1, p. 17, 2007.

4. Review of PHMSA and OSHA Incident Data

Two meaningful sources of incident data regarding H_2S exposure from molten sulfur handling are the PHMSA Incident Reports Database and the Accident Investigation Summaries from OHSA Form 170. Although there may be some overlap in incident reporting between these two databases, in general the PHMSA database contains mandatory reports that involve the release of a regulated hazardous commodity while in transit, while being loaded or unloaded, or while in storage during transit. The OSHA database contains mandatory reports that involve incidents or accidents that have resulted in the injury or fatality of an employee. The differences between the type of incidents reported in each database and the corresponding limitations in the reports should be understood. For example, personnel may have been briefly exposed to H_2S shortly after a tank car has been opened for steaming or unloading, but if there was no release of molten sulfur, the incident would not necessarily need to be reported in the PHMSA database. In a similar manner, if a release of molten sulfur occurred after a transport vessel had been connected to plant apparatus but it did not involve the unloading equipment, this incident might only be reported in the OSHA database, and then only if a reportable injury or fatality occurred as a result. In this manner, then, the review and subsequent use of the information provided in these reports should be considered with care and properly qualified when attempting to estimate occurrence probabilities and severities.

Bulk shipments of molten sulfur in North America are typically identified by the number "NA2448." Using the PHMSA Incident Reports Database, we performed a search of reportable incidents involving NA2448 by all modes of transportation (air, water, highway and rail) over the entire range of dates for which electronic records are available. From 1989 to 2012, the database contained 428 reports of unintended releases of molten sulfur, all of which occurred either by rail or highway. In these 428 reports, no fatalities were reported and 19 injuries were reported. An interesting finding is that of the 428 reports, 43 of them were described (based on the responder's description provided in Part VI of the Form DOT F 5800.1) as being initially discovered based on the detection of a strong odor. In addition, the 43 reports that explicitly identified "odor" in the description all involved rail tank cars. Of these 43, three of them further described the odor as being that of "rotten eggs," which is the typical way that people refer to the unique smell of H_2S.

It is compelling that roughly 10% of the molten sulfur releases reported from 1989 to 2012 indicate that they were initially detected by the "odor" of sulfur, and in a few cases, the more telling odor of "rotten eggs." Many of these instances occurred when a rail tank car was technically "in transit" but was not "in movement," such as when a railcar is in a classification yard. The irony involved in the discovery of these situations is that while the inhalation hazard of the H_2S is a high consequence hazard, had it not been for personnel noticing the odor coming from these tank cars, detection of the lading loss might not have occurred.

For releases that occurred in rail transit, the most common causes were a corroded or otherwise breached rupture disk, an improperly secured manway cover (usually involving a number of loose or missing bolts) or an improperly seated gasket. These causes create concern for railcar handling personnel because unless they can detect a potential exposure by smell or see the physical indicators of lost lading (such as crusted sulfur on the outside of the tank car), they would likely approach the railcar assuming that it was completely sealed. This might lead to personnel being placed in danger if they approach an unsealed railcar that is venting high concentrations of H_2S that are not detected quickly enough for personnel to detect and identify the hazard and approach the railcar using more appropriate PPE.

The majority of releases from tank trailers involved human error on the part of the driver who failed to properly monitor the loading operation and allowed the tank trailer to be overfilled. Reports involving overfilling do not refer to sulfur or H_2S odors, likely because the loss of lading is immediately apparent by visual observation.

A review of the Accident Investigation Summaries from OSHA revealed five cases that noted an H_2S release from molten sulfur tanks during unloading.[22] A total of 16 employees were involved in the five incidents, with six total fatalities. There are several common features related to each of these incidents. First, all five incidents occurred in the fall or winter season in locations where the ambient temperatures are low. Second, none of the five incident reports makes mention of the exposed employees wearing proper respiratory protection, and two investigations noted that employees were not wearing respirators. In two of the incidents, the exposure to H_2S resulted in loss of consciousness which caused a fall for the employees who were not wearing fall protection.

The most recent OSHA report of an accident related to H_2S exposure from molten sulfur occurred in 1995. An employee was taking samples from a molten sulfur tank truck. After opening the manway cover, he reached approximately 15 inches into the tank to collect the sample. A respirator was not provided, and even though he was on a nearly nine-foot tall platform, he was not wearing fall protection. The employee was found unconscious on the gravel covered ground approximately 10 minutes after taking the sample. An autopsy listed blunt force injuries from the fall as the cause of death, but testing revealed H_2S levels up to 800 ppm using similar unloading procedures on site. OSHA maintains that H_2S was a causal factor related to the fall.

In addition to potential fatalities due to falls, H_2S can also cause poisoning or asphyxia. In 1993, three employees were unloading a molten sulfur railcar. Two employees were on a platform on top of the railcar, while the third employee was on the ground hooking up steam lines. The third employee found his two coworkers unconscious on the platform shortly after they had opened the manway cover on the railcar. The employee who opened the manway cover was found dead on arrival, the other employee on the platform was hospitalized for several days, while the third employee was examined and released.

5. Safeguards to Prevent or Control H_2S Exposures

H_2S exposure during the handling of molten sulfur can be generally categorized as a low probability, high consequence event based on an assessment of the data available from PHMSA and OSHA. However, these incidents were of high consequence due to the fatalities that have

[22] Incident reports from www.ohsa.gov. A total of 172 reports were reviewed with "hydrogen sulfide" as a key word. Five reports were related to molten sulfur tanks. Those incident reports are numbers 109097600, 109713388, 17713199, 106387822, and 1383249.

been reported. Managing this or any low probability, high consequence risk may benefit from the use of multiple safeguards for a variety of reasons.

Suggested safeguards to consider, most of which can be implemented independently or together, are:

- Closed-loop venting and treatment of H_2S vapor (e.g., adsorption systems, flares);
- Administrative controls (e.g., standard procedures and training for unloading, buddy system);
- External ventilation of unloading areas;
- Air monitoring systems with alarms;
- Respiratory protection (PPE) sufficient to the hazard level or that ensures the safety of unloading personnel when the hazard level is unknown.

In fact, some states have mandated safety procedures associated with the handling of H_2S, including specifications for personal protective equipment, venting requirements, and equipment specifications.[23] As always, the best system of safeguards will be site-specific and tailored to the capabilities of site personnel. It is also important to consider the consequences of any changes prior to their implementation, even the addition of safeguards, lest any new hazards be introduced.

6. Recommendations

The safe handling of molten sulfur requires understanding and planning regarding the potential release of H_2S. Herein, we have discussed the technical reasons for H_2S hazards in molten sulfur, reviewed accident statistics, and relevant safety documentation. The following recommendations might be considered when handling molten sulfur.

- Sulfur supply contracts, which typically include requirements for a dry weight analysis of the sulfur based on minimum or maximum amounts of ash, carbon, arsenic, selenium and possibly other elements, should also address maximum levels of H_2S. In addition, these requirements should address when and how the product sample should be taken and then analyzed, especially if it is determined that the H_2S content is strongly dependent on sample temperature and storage time.
- Documentation regarding the shipping, receipt, and unloading of molten sulfur should be maintained per regulatory and "best practices" requirements. These documents, as well as sulfur supply contracts, might include warnings regarding the potential increase in H_2S concentration during shipment.

[23] For example, California Title 8, Section 4414 and Washington Administrative Code 296-79-250.

- Development of site-specific loading and unloading procedures should be written and followed given due consideration to the assessment and management of the hazards associated with H_2S in molten sulfur. These procedures might include:
    - Forced exhaust ventilation requirements;
    - Proper temperature, humidity, and air monitoring regarding H_2S concentrations at tank inlets, outlets, and bulk concentrations within the unloading area;
    - Proper PPE for burn protection and H_2S exposure, possibly including a requirement for a SCBA respirator and an escape bottle;
    - Controlled tank headspace venting before sampling or unloading;
    - Consider fall protection requirements knowing that non-lethal exposure to H_2S can induce a loss of consciousness and potential injuries related to falls.

Finally, it is emphasized that there is typically no fixed set of safeguards that will work at every facility. A well-designed system of safeguards to protect workers from H_2S exposure should likely consider the personnel, equipment, and safety management system at that particular facility.

Calcium Stearate Deflagrations – A series of unfortunate events

Albert Ness
ABS Consulting
10301 Technology Drive
Knoxville, TN 37932
aness@absconsulting.com

Konanur Manjunath
Dow Chemical Co.
Freeport, Tx
KManjunath@Dow.com

Prepared for Presentation at
American Institute of Chemical Engineers
2013 Spring Meeting
9th Global Congress on Process Safety
San Antonio, Texas
April 28 – May 1, 2013

UNPUBLISHED

Calcium Stearate Deflagrations – A series of unfortunate events

Albert Ness
ABS Consulting
10301 Technology Drive
Knoxville, TN 37932
anesss@absconsulting.com

Konanur Manjunath
Dow Chemical Co.
Freeport, Tx
KManjunath@Dow.com

Keywords: Calcium Stearate Solids, Combustible Dust, Bulking Brush Discharge, Particle Size Distribution, Resistivity, Bonding, FIBC, Inertion

Abstract

A series of deflagrations occurred in a plant unloading Calcium Stearate from bulk bags. Each one had a different root cause and illustrates the importance of management systems for ignition control and the need to evaluate a system for many types of ignition sources. The first event was caused by static discharge from an ungrounded tote. When upgrading the system Type C bags were specified for the product and a second deflagration occurred due to static discharge from and ungrounded bag. In the third deflagrations cause was most likely a conical pile discharge. The presence of fines and coarse material plus the use of a larger cubic vessel probably enhanced the likelihood of the brush discharge.

.

1. Introduction

This paper describes three deflagrations that occurred in a bulk bag unloading system that had previously run for over six years without incident.

1.1 Product data for Calcium stearate

Table 1 lists data for the Calcium Stearate that was being used in the facility.

Table 1 Calcium Stearate Properties

Parameter	Value
D_{50}	500 microns
Minimum Ignition Energy (MIE) (< 75 μ fraction)	8 mJ
Volumetric resistivity	$14\ 10^{14}$ Ω-m

2. Unloading station description

The unloading station consisted of a bulk bag of Calcium Stearate (CaSt) hung u above a feed hopper. An Iris valve was used to control the flow rate of the CaSt into the hopper. A flexible feeding chute was attached from the hopper outlet to the lid of a 2500 L tote. Local exhaust ventilation was provided at the feed hopper and tote.

Figure 1: Unloading station for calcium stearate bulk bags

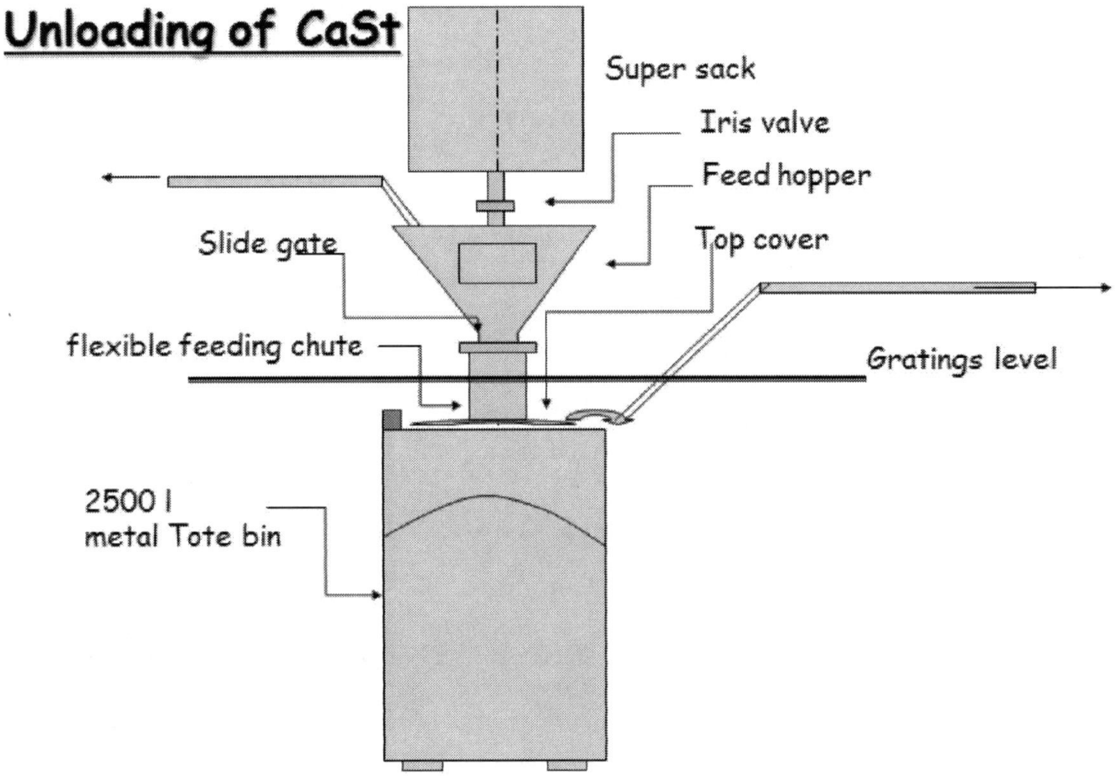

2.1 Event #1 while unloading

At the end of the loading cycle a deflagration occurred in the head space of the tote. The fire was quickly extinguished and little actual damage occurred. Investigation showed that the Feed hopper and the lid of the tote were not grounded and that the flexible chute was not static dissipative. Additionally, it was discovered that the tote vacuum system was not used. Figure 2 shows where the damage occurred.

Figure2: Unloading station at the time of event 1

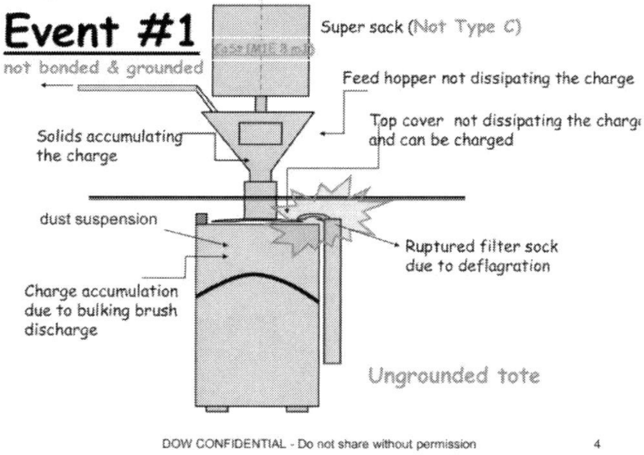

With the vacuum system inoperative, fines were able to stay suspended in the head space of the tote. A calculation of the capacitance of the lid of the lid showed that it had a capacitance of 180 pico Farads. Figure 3 shows that at that level, a static discharge energy of about 30-40 mJ for a charge of 20 EV, well above the 8 mJ MIE of the fraction under 75 microns, was possible. Therefore it was concluded that the most likely cause of the deflagration was a spark between the lid and tote ignition suspended fines in the tote.

Figure 3: Calculated electrostatic build up on process equipment

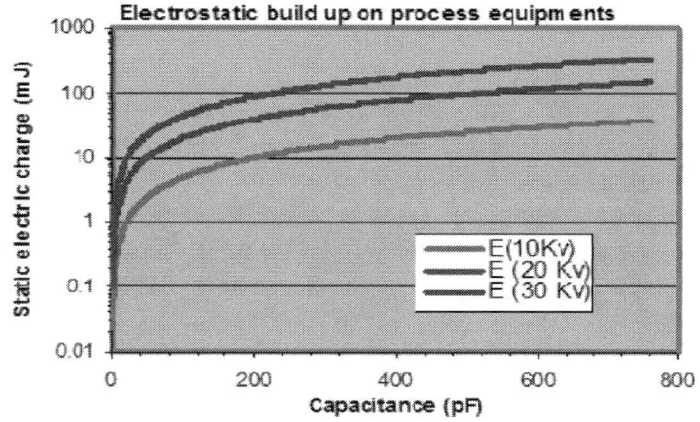

2.2 Event #2

After the first event, the metal equipment was grounded and the tote vacuum system was put back into operation. Also, it was decided to specify use of Type C bags for the bulk bags. Previously, the bulks bags were type B, that is, an all plastic bag that cannot be grounded, but has a breakdown voltage of 4kV. This allows for discharge of energy to the air. A Type C Bag has continuous conductive threads are sown into these bags, hence continuity is provided if the bag is grounded.

After starting up the improved system, a second deflagration occurred. Upon investigation, it was found that; 1) the bulk bag had not been grounded and 2) the Iris valve was not grounded due to its Teflon coating and the powder buildup on it. The failure to ground the new bulk bag is a reminder of the expression "No good deed goes unpunished". It is also a reminder of the need for a good Management of Change System. In the case of this change, the operator's training should include instruction on the hazard of NOT grounding the Type C bag.

2.3 Event #3 while unloading

Prior to restarting the unit, a ground verification system was installed to ensure the bulk bag was grounded before charging. Figure 4 shows where the damage occurred.

Figure 4: Unloading station at the time of event 3

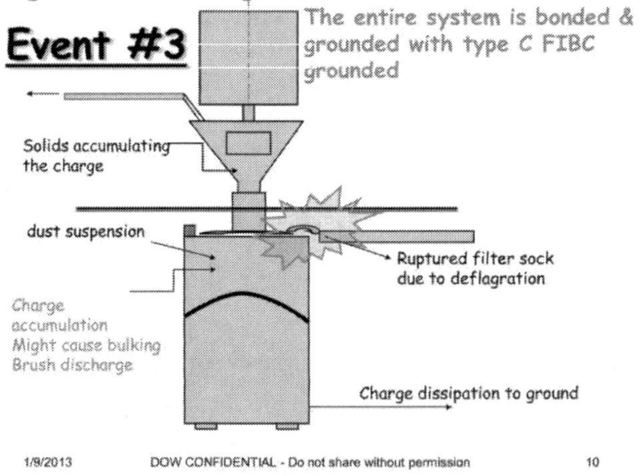

At this point, there was no obvious source of ignition. Several experts were called in and a more thorough investigation was done. Findings were:

1. The FIBC was examined for continuity of grounding. Except for the area of the bag that was damaged by the fire, the continuity was acceptable. Testing of FIBCs at the supplier found all the bags were conductive.
2. The CaSt in the FIBC from the last event was somewhat smaller than in the initial events. Table 2 shows the difference in size between the CaSt particles from the first and third events. The material from the third event was not out of spec, it was considered

acceptable material by the plant. The key point here is that normal variation in the raw material can affect the degree of hazard.

Table 2: Size Comparison

Cumulative Vol. %	Event 1	Event 3
90	800	800
50	500	425
10	200	150
% < 75 μ	2-3	~5

3. Fines collected from the top of a different tote had a D50 of approximately 10 μ and were found to have an MIE of 3 mJ.
4. The dust extraction system was again found to have been off.

2.4 Bulking or Conical Brush Discharges

Whenever solids are filled into containers, bulk bags and so on, bulking brush discharges are developed. It is the type of discharge observed on the conical pile of a heap of powder; thus it is often called conical pile discharge. Bulking brush discharges have an apparent maximum effective energy of 10-25 mJ (with respect to dust ignition), and are believed to be responsible for dust explosions in grounded silos [5, 17, 18].

Britton [17] states that bulking brush discharges cannot always be practically eliminated in equipment above about 1 meter diameter when non-conductive powders are transferred.

The following scenarios can lead to bulking brush discharges:

1. Fines in the suspension that coat the conductive wall rendering it less conductive or nonconductive. Now we have a situation of a system that is not grounded in practice. As an example the fines may be all fraction < 10 microns and the MIE for such a fraction will be much lower than the tested sample (< 74 microns). Even though this is a hypothetical situation, one needs to consider if the attrition is a factor in product processing.
2. Fines are in suspension in the free head room of the vessel and the filling continues with the mixture containing coarser particles. As the particle size increases, the equivalent energy increases. See Figure 3. This combination with fines to super fines and coarser particles is a concern as this can result in an incident.

Scenario 2 perfectly describes the situation that occurred in the third event.

2.4 Causes

The root cause of the third events is the high fines content of the CaSt in the FIBCs. The lack of dust extraction may have led to a suspension of the fines in the head space of the metal tote bin. Ignition could have been caused by either a defective Type C FIBC, or a conical brush discharge.

Testing of the existing bag was inconclusive due to the damage it sustained. Experts in static electricity and combustible solids believed that the conical pile discharge could not be ruled out.

Potential mitigation for conical brush discharges can include:

- Flow rate reduction
- Use of active neutralizers for relatively small silos.
- To minimize the probability of ignition, the first consideration
- Minimize the sub-200 mesh fines concentration to increase the MIE to greater than 20 mJ. Inertion, especially if process design or powder properties make deflagrating venting and suppression undesirable alternatives. Inerting is often considered where the powder MIE falls below the 3-10 mJ range.

In this application, inerting was chosen as the fire and explosion prevention technique

3. *Concluding remarks*

This series of events illustrates many points.

Not fully understanding the source of hazards can lead to jumping to obvious, but not comprehensive, conclusions in an investigation. It pays to consult the appropriate subject matter experts even for seemingly simple events.

No good deed goes unpunished. Changes made to improve the safety of a system can introduce new hazards. In this case the switch to Type C bags introduced a new potential ignition source. Training of operating personnel should include not only what the change is and how the procedures may change, but also on new hazards introduced.

Even on a simple system, it pays to consult with subject matter experts on potential hazards. The effect of variations of particle size on MIE and the concept of conical pile discharges, may very well not be known to process designers.

8. REFERENCES

1. Bartknecht, W., **Dust Explosions**, Berlin: Springer-Verlag, 1989.
2. Glor, M., **Electrostatic Hazards in Powder Handling**, Letchworth UK: Research Studies Press Ltd., 1988.
3. "**Recommended Practice on Static Electricity**", NFPA 77, 2003.
4. "**Recommended Practice for the Classification of Flammable Liquids, Gases, or Vapors and of Hazardous (Classified) Locations for Electrical Installations in Chemical Process Areas**" NFPA 497, 2004
5. "Recommended Practice for the Classification of Combustible Dusts and of Hazardous (Classified) Locations for Electrical Installations in Chemical Process Areas" NFPA 499, 2004
6. "Standard for the Prevention of Fire and Dust Explosions from the Manufacturing, Processing, and Handling of Combustible Particulate Solids, NFPA 654, 2006

7. Eckhoff, R.K., **Dust Explosions in the Process Industries**, London UK: Butterworth-Heinemann, 2003.
8. Britton, L.G., **Avoiding Static Ignitions Chemical Operations**, AIChE, CCPS Concept Book, 1999, Publication G-67
9. Jaeger, Norbert and Siwek, Richard; "**Prevent Explosions of Combustible Dusts**", Chemical Engineering Progress, June 1999
10. T.H. Pratt **Electrostatics ignitions of fires and explosions**. Center for Chemical Process safety. AIchE Industry technology Alliance, USA, 1997.
11. L. Britton Private communications, November , 2003
12. M. Glor **Ignition hazard due to static electricity in particulate processes**, Powder Technology 135–136 (2003) 223– 233

Effectiveness of improvised gas absorption techniques for emergency responders at releases of toxic gases

Hannes Kern
Montanuniversität Leoben
Franz-Josef Strasse 18, 8700 Leoben
hannes.kern@unileoben.ac.at

Matthias Taferner, Harald Raupenstrauch
Montanuniversität Leoben
Franz-Josef Strasse 18, 8700 Leoben
matthias.taferner@stud.unileoben.ac.at, harald.raupenstrauch@unileoben.ac.at

Prepared for Presentation at
American Institute of Chemical Engineers
2013 Spring Meeting
9th Global Congress on Process Safety
San Antonio, Texas
April 28 – May 1, 2013

UNPUBLISHED

Effectiveness of improvised gas absorption techniques for emergency responders at releases of toxic gases

Hannes Kern
Montanuniversität Leoben
Franz-Josef Strasse 18, 8700 Leoben
hannes.kern@unileoben.ac.at

Matthias Taferner, Harald Raupenstrauch
Montanuniversität Leoben
Franz-Josef Strasse 18, 8700 Leoben
matthias.taferner@stud.unileoben.ac.at, harald.raupenstrauch@unileoben.ac.at

Keywords: toxic gas release, emergency response, absorption techniques

Abstract

In September 2011 a spill of nitric acid at a galvanic company in Trieben, Austria caused the release of large amounts of nitrous fumes. An IBC tank with 1000 liter of nitric acid (53%) was flipped over by a forklift and cracked open in the top area. The content was fully released and lead to serious corrosion effects on storage racks combined with the production of critical amounts of nitrous fumes. 27 workers had to undergo medical treatment and a storage and chemical handling area of several hundred square meters was filled by nitrous fumes with concentrations of up to 100 ppm.
Emergency responders used an improvised gas absorption system to clear the area of toxic fumes. Outside the sealed area concentrations of nitrous fumes well below the AEGL-2 (4h) value for nitrous dioxide where measured. In the course of the analyses of the response the effectiveness of improvised gas absorption techniques was investigated.
For the investigations chlorine and ammonia in combination with a mobile exhauster, (10.000 m^3/h) commonly used by emergency responders in Austria were used. Different methods of gas absorption where investigated and tested upon their practical relevance.

1. Introduction

Toxic gases are commonly used in industrial and public areas, both as base products for chemical production and as technical gases for various purposes. In Europe gases like chlorine or ammonia are often used as disinfecting agents or coolant in cold storage warehouses. In February 2012 a release of several hundred kilograms of ammonia from a cold storage warehouse in Vienna (Austria) injured 62 Persons, 32 firefighters and 4 emergency medical technicians right in the busy city center of Vienna. Firefighters were able to contain the released ammonia within the warehouse and tried to absorb and disperse it with water by using water curtains and water shields. Half a year earlier, on September 5th 2011 the spillage of 1000 liter nitric acid 53% in a metallurgical production site Trieben (Austria) caused the occurrence of significant amounts of nitrous fumes in a large storage area. The nitrous fumes were produced due to the reaction of the acid with the material of storage racks and metal materials in the surrounding area. The nitrous fumes were contained in the storage area, but ventilation procedures were very critical due to the close surrounding neighborhood. At the incident in Trieben 27 workers were injured. Due to a lack of knowledge regarding improvised gas absorption techniques experiments with ammonia, chlorine and different absorption agents were carried out. The experiments were supported by the Hazmat Team of the Graz Fire Brigade.

2. Application and testing of absorption techniques

2.1 General overview on techniques for gas defense

In the field of gas defense and control different techniques are applied by emergency responders. Depending on the strategy followed by the responding crews and the situational demands there are three main categories of technical and strategic measures that can be used during a response situation dealing with toxic or hazardous gases. These methods are described in the following sections.

2.1.1 Active methods

As "active methods" all response techniques can be considered that aim to directly fight and disperse gas clouds or change their direction of spread. Most of the active response methods are based on technical devices that are able to generate large volume flows of water or/and air that are directed against the spreading gas cloud. Depending on the size of the gas release water nozzles, water monitors or different sized ventilators are used for the active response approach. Appliances like the LUF 60 (firefighting support vehicle) combine a high volume rate ventilator (90.000 m^3/h) with a circular water dispersion system and provide therefore an aerosol cloud that shows better kinetic effects than the sole air stream. For water soluble gases also absorption effects can be achieved by this method.

2.1.1 Passive methods

"Passive methods" aim to contain the released gas cloud or try to protect areas from gas by a restriction of spread. An example for such systems are, water shields normally used to protect from radiative heat during firefighting operations. Also known are water curtains mainly consisting of fire hoses with different numbers of upward directed spray nozzles. Water curtains are more flexible in application. Water shields and water curtains are available with adjustable flow rates and for different operating pressures. The effectiveness largely depends on the spilled gas type and the nozzle and flow configuration of the chosen appliance.

2.1.1 Absorption techniques

Absorption techniques for emergency responders are in general not well established. For different purposes the effects of water on water soluble gases is known and well understood. The application of knowledge from industrial processes and their applicability for emergency responders is scientifically not well reviewed. Commercial portable gas absorbers are established in some industrial emergency response crews but mostly have very low volume rates. For water soluble gases some absorption techniques concerning the use of water aerosols or spray streams are known but often improvised. For chlorine spills the effects of sodium thiosulfate were investigated and have proved to be very effective [1,2]. Other chemicals for absorption purposes regarding emergency response have not been studied very well. Effluents from absorption processes could cause hazardous situations too, but absorption in general offers the chance to reduce the three dimensional problem of a hazardous gas release to a two dimensional problem of a hazardous liquid.

2.2 Response at Trieben (Austria)

On September 5th 2011 a forklift at a metallurgical production site in Trieben (Austria) by accident dropped an IBC with 1000 liter 53% of nitric acid in the chemical storage area. The local fire brigade was called at about 11:00 am and arrived about five minutes later. The incident commander noticed during size up operations orange/brown fumes inside the cellar of the building. Most of the cellar was affected, about 200m^2 were directly contaminated with nitric acid. It is assumed that before the fire brigade was called, workers tried to dilute the nitric acid with up to 3000 liter of water, which caused severe a deterioration of the situation. About 30 minutes later the hazmat team arrived and started measurements with detector tubes. Values of around 100 ppm for nitrous fumes were measured in all areas of the cellar. The nitrous fumes occurred from the reaction of nitric acid with the material of the storage racks in the storage area. The lower parts of the storage racks were principally protected by a coating material but at some points the coating was damaged. Emergency responders over nearly 24 hours tried to contain the spill with chemical binding agents and chemical vacuum cleaners. Due to the contamination with nitrous fumes gastight suits were used which lead to huge logistical demands. The normal working time for the cleaning teams was around 45minutes to an hour, they were supplied by air hoses from a mobile compressor for respiratory air. Due to the close vicinity of the surrounding neighborhood it was not possible to vent the nitrous fumes directly in the atmosphere. A mobile exhauster was placed a spray nozzle was fixed near the exhaust opening of the ventilator. The flow rate of the water stream was around 200 l/min at 6 bar. The exhauster had an output of around 20.000 m^3/h. During the exhaust process the concentration of nitrous fumes was constantly monitored and well below 8,2 ppm (AEGL 2, 4h for NO_2) at the direct vicinity of the exhauster. The effect might have not only occurred through absorption effects but also because of a dilution effect of excess air at the input side of the exhauster.

2.3 Absorption tests with the MSA Auer COMVEX Ventilator

To investigate the potential of improvised gas absorption techniques the applicability of the MSA Auer COMVEX Ventilator was chosen as an example. The COMVEX Ventilator is used by many Austrian fire departments and offers the possibility to mount a water spray system on the ventilator that is normally used for the production of high expansion foam.

2.3.1 MSA Auer COMVEX [3]

The MSA Auer COMVEX Ventilator mainly consists of an air fan and tubing for suction and exhaust purposes. The fan has an air consumption of 10.000 m^3/h and a power of 2,6 kW. Additional to the fan, a top piece for the production of high expansion foam is available (Pic.1). The foam generating kit consists of short tube with five spraying nozzles and a mixer to be able to ad foaming agents. At an operating pressure of 4-8 bar the water consumption of the foam generating kit is around 100 l/min. Mixing rates of 1-3% are possible.

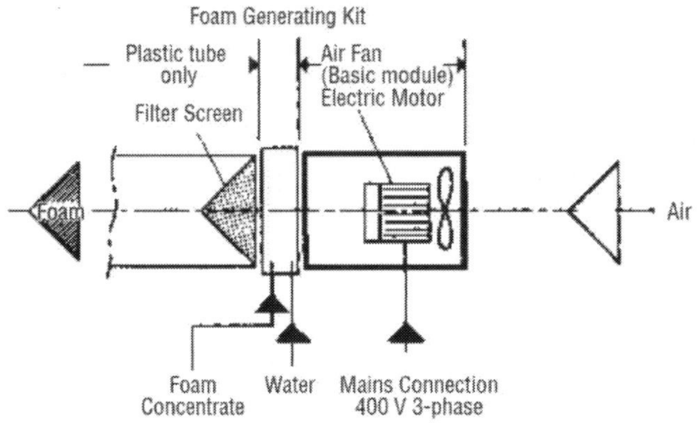

Pic. 1: Principal layout of the COMVEX Ventilator system [3]

2.3.2 Sodium thiosulfate as absorption material

Many absorption materials for technical purposes are well known. For emergency response purposes it is crucial to have absorption materials that present a low hazardous potential and have a good availability in large quantities. Logistically it is not possible to store large amounts of different absorption chemicals for specific tasks. For high risk chemicals (concerning toxicity and used quantities) it may be reasonable to foresee special measures. In case of chlorine that has a high toxic potential and is commonly used for chemical production processes and disinfection purposes, sodium thiosulfate gives good results as absorption agent in combination with water [2].

Sodium thiosulfate in an aqueous solution reduces elementary chlorine to chlorine anions. The reaction is shown in Equation 1:

$$2Na^{+} + S_2O_3^{2-} + 4Cl_2 + 13H_2O \rightarrow 2Na^{+} + 2HSO_4^{-} + 8\ H_3O^{+} + 8Cl^{-} \quad \text{[Eq. 1]}$$

One mol of sodium thiosulfate is able to reduce 4 mol of elementary chlorine and has therefore a high potential to be used as absorption agent. Sodium thiosulfate pentahydrate has a solubility of around 680 g/l at 20 °C which equals a concentration of about 2.7 mol/l.

2.3.2 The MSA Auer COMVEX as improvised absorber

The focus of the investigations was the absorption of ammonia and chlorine at concentrations around 200-300 ppm. The ventilator was equipped with a suction tube (10 m) and an equally built exhaust tube. Tests with ammonia where carried out with inlet concentrations of 250 and 290 ppm. The ammonia feed stream was provided by a bottle of pressurized ammonia, the concentration was monitored by detector tubes. At the start of the experiment a constant ammonia stream was adjusted and measured at the exhaust tube while the ventilator was operating. When constant concentration conditions were reached, the injection of water started and the gas concentration was measured again. At the first test the ammonia concentration dropped from 250 ppm to 100 ppm (33.3%). The second run lead to a concentration of 150 ppm at a starting concentration of 290 ppm (48.3%).

Pic. 2: Improvised absorption system during experiments with chlorine

Due to the good solubility of ammonia in water, the tests were only carried out with water. The second series of experiments was carried out with chlorine in the same way. The first test with chlorine was started at a concentration of 490 ppm, the injection of water caused no measurable reduction of the concentration. In the second experiment with chlorine, a 2,7 mol/l solution of sodium thiosulfate was added to the water stream by the mixer of the foam generating kit (Pic.2). A mixing ratio of 3% was chosen. The use of sodium thiosulfate resulted in a reduction of the chlorine concentration from 260 ppm to 100 ppm (61.5%).

3. Conclusion

The effectiveness of gas absorption techniques is not very well known by emergency responders in Austria. Examples from real incidents show that the availability of gas absorption techniques may improve the effectiveness of emergency response at toxic gas releases. It is clearly not possible to have ready gas absorption appliances and agents for a wide range of specific problems. Scientific research on available improvised techniques and the development of technical improvements may enhance their effectiveness. Experiments with sodium thiosulfate in cases of chlorine spills show an effectiveness of over 60% at low chlorine concentrations. Keeping in mind that the investigated ventilator/nozzle system was not designed for gas absorption in general, the results may still leave a lot of room for improvements. Future research will focus on the improvement of the effectiveness of different methods for the control of toxic gas releases.

4. References

[1] Engelhardt, F., “Chlorabsorption an fallenden Tropfen und Übertragung auf Wasservorhänge,” *Bergische Universität –Gesamthochschule Wuppertal*, Dissertation Juni 2002

[2] Bauch, M., Peine, T., “Versuche zur Erhöhung der Effizienz beim Niederschlagen von Chlor,” *Brandschutz - Deutsche Feuerwehr Zeitung*, NO. 1, 2010

[3] “MSA AUER COMVEX Portable Ventilator and High Expansion Foam Generator - Instruction Manual,” MSA AUER

9TH GLOBAL CONGRESS ON PROCESS SAFETY

Case Studies in Process Safety: Lessons Learned from Software-Related Accidents

Terry L. Hardy
Director, Safety & Risk Management
Great Circle Analytics, LLC
1238 Race Street
Denver, CO, 80206
thardy@gcirc.com

Prepared for Presentation at
American Institute of Chemical Engineers
2013 Spring Meeting
9th Global Congress on Process Safety
San Antonio, Texas
April 28 – May 1, 2013

Case Studies in Process Safety: Lessons Learned from Software-Related Accidents

Terry L. Hardy
Director, Safety & Risk Management
Great Circle Analytics, LLC
1238 Race Street
Denver, CO, 80206
thardy@gcirc.com

Keywords: Safety Analysis, Software Safety, Computing Systems, Process Safety Management

Abstract

Software and automation can play a significant role with respect to safety in the chemical process and energy production industries. Because computing systems are increasingly used to control critical functions, software may directly contribute to an accident. On the other hand, software can also be used as part of the hazard control strategy to reduce risks, and computing systems can provide valuable information to help make safety decisions. The importance of including software as part of an organization's efforts to analyze and manage hazards and risks seems clear, but for many organizations software is not effectively incorporated into process safety efforts. This paper reviews lessons learned from accidents and incidents to illustrate the potential for a software-related accident even when process safety management tools and techniques are used. This discussion is intended to provide insights to help improve process safety and software safety efforts.

1. Introduction

Process Safety Management generally refers to the application of management principles and systems to the identification, understanding, and control of process hazards to protect employees, facility assets, and the environment. Process Safety Management is focused on prevention of, preparedness for, mitigation of, response to, and restoration from catastrophic releases of chemicals or energy from a process associated with a facility. In Process Safety Management, a process is defined as any use, storage, manufacturing, handling, or the on-site movement of highly hazardous chemicals, or combination of these activities. Process Safety Management is regulated by OSHA through the Process Safety Management (PSM) Standard.

Software and computing systems are becoming essential to the operation of most industries, often through the automation of tasks, and as such can have a major impact on process safety. Software and computing systems offer tremendous advantages that encourage the use of automation. Software can allow complex operations to be performed, and can perform complicated calculations. Computing systems are reliable in the sense that there are a small number of components to perform a large number of actions. Computing systems are also flexible and can be modified to fit the application. From a safety perspective, automation can reduce the risks to operators by separating them from dangerous environments such as those in hazardous chemical processing facilities. In addition, automation can help by performing tasks that might pose a threat to humans, such as lifting large objects or operating high-speed cutting tools.

Software, computing systems, and automation have disadvantages as well. A major disadvantage to computing systems is that they often introduce increased complexity to systems, making them harder to understand, operate, maintain, and test. This complexity can also introduce new failure modes and interactions. Complex systems have a greater number of requirements, more interfaces, more subsystems, and more logical operations, increasing the potential for error. Computing systems can create a situation where no single operator can immediately foresee the consequences of a given action in the system. Computing systems can also increase coupling. Tight coupling occurs when processes are highly integrated and intrinsically time-dependent; for example, once a process has been set in motion, it must be completed within a certain period of time. Tightly coupled systems allow little time for operators to take action to prevent adverse effects. In addition, automation may require that operators think about their tasks in new ways, and may change the fundamental role of the operator from one who physically performed tasks to someone who monitors and communicates information. Finally, software will only do what a programmer has told it to do. Therefore, unlike humans, software cannot react to an abnormal situation unless specifically told to.

In spite of the fact that software is such an important part of complex systems, the analysis of hazards and risks from software has been inconsistent in the process industries. Safety analyses have historically been hardware-focused. Therefore, many analysts may not understand how to incorporate software into their system hazard analyses, and evaluators of those analyses may not understand what should be included and assessed. Organizations may be focused on compliance to regulations, which often do not address software, and as a result organizations may not properly assess or mitigate software risks. Organizations need to increase the attention given to addressing and analyzing the potential for hazards related to software and computing systems.

Software includes computer programs, procedures, scripts, rules, and associated documentation and data pertaining to the development and operation of a computer. Software can be developed by the organization implementing the system, by an outside software developer, or may be purchased as Commercial Off-The-Shelf (COTS) software. Software safety encompasses not just the software but also the computing system. A computing system includes the software and supporting hardware, sensors, effectors, humans who interact with the system, and data necessary for successful operation. Examples of computing systems include Programmable Logic Controllers (PLC) and Supervisory Control and Data Acquisition (SCADA) systems.

Through the review of hundreds of software-related accidents and incidents, and the author's personal experience in this discipline, common themes and lessons have been found.[1] This paper will use case studies to illustrate those themes and lessons learned. The discussion will center on software safety as part of a broader process safety effort and will include recommendations that can be used to improve the safety of software-driven systems.

2. Lessons Learned: Software and Computing System Related Accidents

This section discusses a number of lessons learned related to accidents and incidents where software and computing systems were contributors. These accidents are taken from detailed reports and investigative summaries from multiple industries and organizations, and are included to broadly illustrate hazards and risks in software and computing systems. Note that in discussing these accidents, this paper does not intend to oversimplify the events and conditions that led to the accidents, or place blame on any individuals or organizations. There is rarely a single identifiable cause leading to an accident. Accidents are usually the result of complex factors that include hardware, software, human interactions, environments, and procedures. The descriptions are meant to provide examples of where the analysis process failed in some way. Readers are encouraged to review the complete accident and mishap investigation reports referenced here to understand the often complex conditions and chain of events that led to each accident discussed here.

2.1 Decisions made in the acquisition and planning phases of development can profoundly affect safety

Planning typically involves trade-offs between many different facets of the program, including cost, schedule, performance, and safety. Poor planning can lead to unexpected safety consequences, and many safety decisions are actually made in the planning and acquisition phase. For example, inadequate resources may be allocated to the software safety effort. This can result in a failure to perform hazard analyses and identify safety requirements early in the program when these activities have the most impact. Therefore, safety personnel should be included in early phases of a program, especially when developing contractual requirements, and software safety considerations should be included in schedule and resource discussions.

On August 11, 1985, an Institute, West Virginia chemical production facility leaked methylene chloride and aldicarb oxime, toxic chemicals used to manufacture the pesticide Temik. Six workers were injured, and more than a hundred residents living near the facility were sent to the hospital. The toxic chemical leak occurred when a tank overheated after a steam valve failed. Due to the failure, steam leaked into the heating jacket of a storage tank containing a mixture of methylene chloride and aldicarb oxime. The methylene chloride vaporized from the heating, leaving concentrated aldicarb oxime to settle on the bottom of the tank. Operators later pumped the material from the storage tank for processing. They thought they had emptied the tank, but concentrated aldicarb oxime remained in the storage tank because a level gauge was broken and the operators performed no additional checks. The concentrated aldicarb oxime continued to be heated over several days by the steam in the heating jacket, ultimately resulting in a runaway chemical reaction days later. The high pressure from the reaction led to release of aldicarb oxime through a flare, and large amounts of the chemical were also released to the atmosphere when a burst disc failed.

The root causes of the leak were found to be the use of a tank that was not designed to hold aldicarb oxime, faulty level indicators, defective safety valves, mistaken transmission of steam to a tank, and failure of control room operators to notice critical pressure and temperature changes. In addition, a computerized monitoring system contributed to the severity of the accident. Earlier in the year, the company had installed a new air-monitoring system that could automatically identify and detect releases of hazardous gases. That computerized system could also predict whether the gases would be contained within the release area or would migrate and potentially harm the public. This information was important because it could be used to mitigate risks through evacuation or other measures if a vapor cloud appeared to be spreading. Company officials initially stated that the computer system had failed and therefore they were not informed of the lack of containment of the aldicarb oxime gas cloud. However, they later admitted that they had not purchased a version of the software programmed to detect aldicarb oxime. Had the company invested in the more expensive version of the software, they could have predicted that a cloud of aldicarb oxime would not have been contained in the production area. However, the company had only ordered the basic software model that did not contain aldicarb oxime properties.[2, 3]

2.2 Hazard causes may not include software and computing systems

Many organizations are improving their safety analyses with respect to software. However, software safety still lacks sufficient numbers of qualified practitioners, and the methods for performing comprehensive analyses are not universally known. In many cases, software causes may not be included in hazard analyses. Or if they are, the software causes may be stated in general terms in hazard analyses, such as including a generic hazard cause called "software error," instead of defining specifically what software functionality can lead to an undesirable outcome. In addition, the software hazard analyses may not pay enough attention to those cases where the software works exactly as intended, but the implemented functionality was unsafe.

On October 18, 2005, an explosion occurred at a natural gas decompression/recompression facility near Empress, Alberta. Extensive damage occurred to the facility, but no injuries were reported. The facility decompressed and conditioned high-pressure gas from a pipeline to achieve pressures and temperatures consistent with what was needed in the Natural Gas Liquids (NGL) stripping facilities. At the time of the accident the facility's "A" PLC lost communication with the NGL facility control room. Transportation Safety Board of Canada (TSB) investigators found that a processor card had failed, shutting down the PLC. When the PLC failed its outputs were de-energized. However, some of the microprocessor functions required an energized signal to initiate physical shutdown of some functions. These functions included main and auxiliary lube oil pumps, purge/damper valves, and suction and discharge valves. As a result of this condition, the pumps did not shut down and valves did not close in a safe position when the PLC failed. Operators in the NGL control room received an alarm indicating that the communications were lost. They investigated the problem and believed that the failure in communication had triggered an emergency shutdown of the facility. The operators believed that the emergency shutdown meant that all components were in a safe state. However, because the recompressor motor did not shut down, a surge event occurred in the pipe. This surge caused a pipe to break, leading to a natural gas leak. The leaking natural gas ignited, resulting in an explosion. The TSB noted that the loss of a PLC had not been identified as a safety concern in the hazard analyses prior to the accident.[4]

2.3 Risks may be underestimated or optimistically evaluated

The risk assessment process helps decision makers in their risk decisions (e.g., accept the risk without change, reduce the risk to an acceptable level through hazard elimination and mitigation, transfer the risk, or forego the activity), assists in justifying the acceptance of residual risk, and allows communication of risks throughout the organization. A risk assessment can either be qualitative or quantitative, although the emphasis in process safety is often on qualitative risk assessment. Evaluating software and computing system risk can be difficult, especially if an extensive history does not exist with the automated equipment. Risk must be carefully evaluated based on known technical factors, such as design complexity and maturity, degree of system testing, use of unproven technologies, potential for unexpected human intervention, and so on. Accidents have illustrated unfounded optimism in the use of software to assure safety.

On April 7, 2000, a pipeline failure occurred at the Chalk Point Generating Station in Maryland. The pipeline released approximately 140,400 gallons of fuel oil into surrounding wetlands, Swanson Creek, and the Patuxent River. The cost of the environmental response and cleanup exceeded $71 million. The U.S. National Transportation Safety Board (NTSB) determined that the probable cause of the accident was a fracture in a buckle in the pipe. The buckle went undiscovered because data from an in-line inspection tool were interpreted incorrectly. The NTSB stated that a contributor to the severity of the accident was inadequate operating procedures and practices for monitoring the magnitude of flow through the pipeline. When the leak occurred the 12-inch-diameter pipeline was being cleaned. An automated pipeline monitoring system existed, but this automated system could not adequately monitor pipeline conditions during this operation because the locations of the meters, pressure-sensing points, and temperature-sensing points had not been configured to be in the direct liquid flow path. Therefore, the automated system did not provide alarms or other information to operations personnel. Field personnel manually recorded tank level measurements at the upstream and downstream points, but did not use this information to evaluate whether any fuel oil had been lost. The manual monitoring procedures and lack of alarms from the automated system led to a 7-hour delay in responding to the leak. The field personnel finally realized there was a problem when the fuel pump began to cavitate. Using hand calculations the crew then realized that they had not received over 3000 barrels of fuel oil and they shut the pipeline down. Following the accident the company installed a SCADA system with software-based leak detection and radar tank gauges. In this case the operators underestimated the risks associated with a leak and overestimated their ability to detect a failure.[5]

2.4 Hazard controls may rely only on good software processes and testing

Hazard controls are devices and approaches to mitigate risk presented by a hazard by reducing either the severity of the hazard or the probability of its occurrence. Strong hazard control design follows the design order of precedence, where the first approach is to try to design out the hazard or minimize the risks through design selection. Many hazard reports will focus on the implementation of good software processes, or extensive unit testing, to prove that the design is safe. However, the software processes and testing will not help if the software design is flawed with respect to safe system operations (including hardware, software, processes, humans, and environments). In these cases the software may operate exactly as designed, but the operation may be unsafe.

On February 11, 2003, an employee of a manufacturer in Gonzales, Texas was fatally injured while performing maintenance on a reaction tank. The U.S. Mine Safety and Health Administration (MSHA) determined that the cause of the accident was a failure to close and secure a manual gate valve for a steam line and a failure to place the batch PLC in the stop mode. The company was a surface clay mill that purchased clay and blended, refined, milled and processed the material into products used in paints, inks, and grease. On the day of the accident the employee had been informed that there had been a product change in one of the batch processing systems. The employee was assigned to perform cleanup duties on a reactor tank. Two valves controlled steam entry into the tank: a manual gate valve and a butterfly valve with an automatic pneumatic actuator. The PLC controlled the functioning of the batch system based on sensors that monitored material flow. At the time of the accident the PLC was in "slurry hold" mode. In this mode the system was programmed to actuate the steam valve when the clay slurry level reached 5.5 feet. An aluminum extension ladder used by the employee caused the level sensor to falsely sense that slurry was in the reactor, which resulted in the PLC sending a command to open the steam valve. Because the manual valve had been left open, steam at 350° F then entered the tank, fatally burning the employee.[6]

2.5 Organizations may fail to consider the possibility of hazard controls not working

Hazard controls are never 100 percent reliable. Therefore, it is important for the analyst to consider what could happen if the controls fail. Consideration should be given to unexpected hardware, software, and human interactions, and a "defense in depth" approach should be used where multiple, independent methods are used to prevent a hazard from becoming an accident. Controls should be carefully evaluated to identify what could happen if they do not work as expected, especially if those controls rely heavily on software and personnel actions.

On February 18, 2009, an employee was fatally injured at a coal preparation plant in the Hunter Valley region of New South Wales, Australia. The accident occurred when 10 tons of waste rock were inadvertently released from the reject bin and fell onto the cabin of the employee's truck. At the plant, raw coal was extracted from the mine and usable coal was separated from waste rock. The waste rock was transferred approximately 2 kilometers on conveyers to the reject bin. The waste rock was then loaded from the reject bin onto trucks and hauled away. The process of loading the trucks with waste rock was controlled by a PLC system. The PLC system included truck detection sensors, traffic lights, bin capacity sensing, and remote control, hand-held transmitters used by the truck drivers. On the day of the accident the truck driver drove his truck under the reject bin delivery chute. A signal was sent from the handheld remote control to command the chute to open. The accident report stated that it was not clear whether the signal was sent inadvertently or intentionally. Opening the chute required that two of three lines of truck detection sensors be blocked in addition to a command from the remote control to assure that the truck was in the correct location. Each sensor line contained three sensors, and all three sensors had to be blocked for the entire line to be considered as blocked. At the time of the accident the truck was obscuring one line of sensors, and a second line of sensors was obscured by dirt on the lenses and therefore was not working correctly. Because two of the sensor lines were blocked and the remote control signal had been sent, the PLC automatically opened the reject bin chute door and dropped 10 tons of material on the truck cab before the driver had safely cleared the chute, resulting in the fatal injury.[7]

2.6 Testing may not provide information on subsystem and component interactions or hazard control operations

The safety verification and validation process is intended to determine that the design solution has met all the safety requirements (verification) and that the correct system is built (validation). The verification and validation process, if performed correctly, will provide evidence that risk has been reduced. Testing is an important part of the verification and validation process, and comprehensive software testing should be conducted throughout the development cycle. Testing should include not only nominal conditions based on requirements but also abnormal conditions such as improper inputs, inputs outside expected conditions, inputs exactly at and near the boundary conditions, and inputs stuck at some value. Safety-critical software must include full system integration testing of end-to-end events. That testing must include stressing of the software, and should include interactions of the software with hardware, humans, and environments. A number of accidents have occurred when no component failed in the conventional sense, but the interaction of components caused a system failure. Testing must include verifications of hazard controls, and it should assure that redundancy works when needed.

On October 24, 2002, a grinder exploded at a limestone processing plant in Foreman, Arkansas. An operator was killed when flammable waste fuel covered him and ignited. The operator had started the pump for solid waste fuel processing when the accident occurred. The MSHA stated in its report that the cause of the accident was that the safety monitoring system designed to shut off the waste fuel system pump had not been maintained so that it functioned properly. Kilns were used in processing activities at the plant, and these kilns were heated by burning coal, natural gas, and liquid waste fuel. The liquid waste fuel was delivered by truck or railcar and pumped into large storage tanks. From the storage area it was pumped through a grinder to reduce the particle size of the solids in the fuel. Two independent systems monitored and controlled the waste fuel delivery. A Foxboro Intelligent Automation Distribution Control System monitored and recorded normal operating parameters. The Foxboro also issued audible and visual alarms that were available at the plant control room. A PLC provided basic startup and shutdown of the system and responded to commands from the Foxboro. On the day of the accident the Foxboro sensed that the fuel delivery pressure was low, apparently due to blockage in the line. As designed, the Foxboro sent a command to the PLC to shut down the pumps. However, the PLC failed to respond and the pumps kept running. Three months prior to the accident this PLC had been installed; this was supposed to be a simple replacement of an older PLC of similar capability. However, the Foxboro had not been connected to the newer PLC, and the connections remained to the older non-functioning PLC. The complete system had never been tested with the new PLC. A system test had been scheduled three days prior to the accident but had been aborted when a pump failed during the test; the test had never been rescheduled. The accident report stated that the blockage may have broken free just prior to the accident. With the pumps running, the pressure elevated significantly and a "water hammer" effect caused overpressurization in the system at the grinder. The grinder was torn loose from its base, spraying fuel and pulling loose a 480-volt cable that ultimately served as an ignition source.[8]

2.7 Software change management and hazard analyses processes may not be integrated

Engineering by its very nature is an activity that requires change. Changes can come about for a variety of reasons, including the discovery of problems during development, changes in requirements, or routine upgrades of software or hardware. As the development cycle proceeds and as an organization gains more operational experience, new hazards are often uncovered and some hazards may no longer be relevant. If the hazard analysis is not updated to reflect these changes then resources may be expended on previously identified hazards that may no longer be relevant, and new hazards may not be discovered as the design matures. In addition, routine changes and modifications to software and computing systems must be analyzed for potential hazards as part of the broader hazard analysis process.

On June 10, 1999, a 16-inch-diameter pipe carrying gasoline ruptured in Bellingham, Washington. The ruptured pipeline released approximately 237,000 gallons of gasoline into a nearby creek, according to the NTSB. That gasoline then ignited, burning approximately 1½ miles along the creek. Three people died in the fire. One home and the Bellingham water treatment plant were also damaged in the accident.

The NTSB investigated the accident and determined that the probable cause of the rupture was damage to the pipe during a modification project performed in 1994. This damage weakened the pipeline, making it susceptible to rupture under increased pressure in the pipe. The NTSB also stated that inspections of the pipeline during the project were inadequate and the company did not identify and repair damage. The report noted that in-line pipeline inspection data should have prompted the company to excavate and examine that section of pipeline, but the company failed to perform such work after reported anomalies. The SCADA system computers also played a role in the accident. The SCADA system was used for operation of the pipeline, for example to open and close valves remotely as required or to operate pumps as needed. Just prior to the accident the operator was preparing to initiate delivery of gasoline to a terminal in Seattle, diverting delivery from another facility. During the process of switching delivery destinations, the pressure in the pipeline began to increase, which is a normal condition but one that required the operator to start a pump to reduce pressure. However, when the operator tried to start that pump, the SCADA system failed to execute the start command issued by the operator. The operator soon found that the SCADA system was unresponsive to any commands, something that had never happened before. The report stated that, "Had the controller been able to start the pump at Woodinville, it is probable that the pressure backup would have been alleviated and the pipeline operated routinely for the balance of the fuel delivery." Instead, the pressure in the pipe increased, and the increased pressure likely caused the damaged pipe to rupture.

The cause of computer system failure was likely a change made to the system database just prior to the accident. The NTSB accident report stated that the SCADA system administrator entered new records into the live database at the time of the accident. The system administrator however did not check the records or test the system software to see if those changes introduced any problems. The computing system problem could not be replicated after the accident and therefore the cause of the anomaly could not be definitively identified. The report stated, "The Safety Board concludes that, had the SCADA database revisions that were performed shortly before the accident been performed and thoroughly tested on an off-line system instead of the primary on-line SCADA system, errors resulting from those revisions may have been identified and repaired before they could affect the operation of the pipeline."[9]

2.8 Human-software interactions have significant safety implications that are often underestimated

Humans interact with hardware and software in a variety of ways. A number of accidents have occurred where these interactions have been a significant contributing cause. Of special concern is when a new computing system is implemented that changes what is expected of the operator in normal and emergency situations.

On August 28, 2008, an explosion occurred at a pesticide manufacturing plant in Institute, West Virginia. Two workers were killed in the explosion and eight others were injured. The U.S. Chemical Safety and Hazard Investigation Board (CSB) found that the explosion was the result of a runaway chemical reaction. The company was starting up a methomyl unit for the first time after several months of down time for maintenance to install a new computer control system and reactor vessel. Normal operations called for dissolved methomyl and waste chemicals to be fed into a preheated residue treater vessel partially filled with solvent. The treater allowed the methomyl to decompose safely, after which it was mixed with other waste chemicals and used to fuel facility boilers. On the day of the accident a methomyl-solvent mixture was added prematurely to the residue treater vessel before solvent had been added to the tank. This event occurred when operators were troubleshooting equipment problems during the startup. This mixture was supposed to be added to the vessel only after that vessel was filled with clean solvent and heated to a minimum safe temperature. An interlock system existed as part of the automatic feed control system to prevent inadvertent introduction of methomyl. This interlock was password-protected to prevent inadvertent override, but operators intentionally overrode the interlock. Bypassing the interlocks had apparently been a practice condoned by management in the past as a workaround for operational problems. Once the methomyl decomposition reaction started it could not be stopped, and the pressure rapidly rose in the vessel due to gas from the reaction, leading to the explosion. Uncrystallized methomyl existed in the tank due to equipment problems, and this material greatly increased the methomyl concentration in the residue treater, contributing to the runaway reaction.

The CSB stated that the company initiated this process startup before the company had completed critical checks, including valve lineups, equipment checkouts, and computer calibrations. The CSB said that the operating procedure for the startup was complex, but this procedure had not been reviewed or approved. In addition, the company had not performed training on the new computer system put in place as part of the maintenance. This new computer system offered significant improvements to help automatically control the operation. For example, the control system included graphical display screens that simulated the process flow. But the system was also complex and the modifications changed the way operators interacted with the system. The operators thought the screens were difficult to navigate, and responding to troubleshooting alarms was difficult. The accident report stated that, had the operators received adequate training on the new computer system, they may have been able to recognize problems in operation before the explosion.

The report faulted the company for not performing an adequate pre-startup safety review. System operators did not participate in the safety review, and review checklists showed items as completed when they were not. The CSB stated that the company had also failed to perform a thorough Process Hazard Analysis. The report stated that the Process Hazard Analysis was performed quickly because management had not allotted sufficient time for analysis. In addition, the CSB stated that the Process Hazard Analysis included invalid assumptions, and said that the team did not apply the tools properly, resulting in unmitigated accident scenarios.[10]

2.9 Support software, including models and simulations, may be as critical to safety as control software

While the focus of software safety efforts is usually on software directly controlling an application, support software may contribute to an accident or to the effectiveness of the emergency response. Examples of support software include databases used for maintenance activities, software to estimate load stability, models that provide design calculations or assurance information, and so on. A failure to analyze the hazards associated with this support software can result in unforeseen system failures.

On November 12, 2008, a 2 million gallon liquid fertilizer tank at in Chesapeake, Virginia collapsed. Two workers performing welding operations at the site were seriously injured and an adjacent neighborhood was partially flooded as a result of the accident. The CSB found that the company had not assured that welds to replace vertical joints met accepted industry standards, and the CSB faulted the company for its failure to perform inspections of the welds. The company was also faulted for not having proper procedures in place for filling the tanks following major facility modifications. In its report, the CSB also noted that the contractor hired by the company to calculate the maximum fill height had used some faulty assumptions. The maximum liquid level was supposed to be calculated in part based on the minimum measured shell thicknesses and the extent of the weld inspection (full, spot, or no radiography). The contractor used the maximum (not minimum) measured thickness, and assumed full inspection of the welds. These assumptions led to an overestimation of the allowable liquid level. The tank failed at a fill level of 26.74 feet, below the calculated maximum of 27.01 feet. The CSB also noted a number of previous overfilling accidents. The CSB found 16 other tank failures at nine facilities in other states between 1995 and 2008. These 16 failures resulted in one death, four hospitalizations, one community evacuation, and two releases to waterways. Eleven occurred due to defective welding.[11]

3. Recommendations for the Software System Safety Process

There is a great deal of information in the literature on how to improve software development processes. However, the above themes and lessons learned indicate that improving software and computing system safety is more than just providing additional training or improved software development techniques or tools. Improving software safety requires a change in the way process safety practitioners think. Software safety must be fully integrated into the process safety analysis process. And process safety practitioners must begin to ask critical questions about their software safety efforts. Some examples of these questions include the following:

- Do plans reflect how business is really done? Are plans reviewed? Do plans have unrealistic schedules or resource allocations? Are software and computing systems considering in the planning and acquisition processes? Poor or unrealistic plans may reflect an organization that does not truly place a priority on safety activities.
- Is there a convincing story that the safety analysis is complete? Is there a sufficient description of the system to support an analysis, including the software and computing system? Are computing systems treated as "black boxes"? Have software causes been thoroughly evaluated using a systematic approach, or are the software causes listed as "software error" with no further explanation? Do the hazard analyses include support software, such as design models or databases? Failure to show that the problem is being looked at systematically could indicate that there are holes in the analysis, with potentially significant problems overlooked.
- Are the hazard reports detailed enough? Are causes descriptive? Are the recommendations clear and sufficiently detailed? Does the logic make sense and is it complete? Do controls match up with the causes, showing a one-to-one or many-to-one relation? Lack of detail could be an indication of insufficient knowledge of the system, or lack of information on the system.
- Do the risk assessments have a logical basis? Have the risk assessments considered software complexity, maturity, testing, use of unproven technologies, etc.? Are assessments overly optimistic based on what is known about the computing system? A failure to provide a basis for your risk assessments may result in unrealistic assessments.
- Are the hazard controls primarily procedural rather than design changes, safety features, or devices? Is there an overreliance on alarms and signage? Do software controls only rely on good software processes? Is there an overreliance on humans and software to "save the day"? Overreliance on operational controls may indicate a weak safety design.
- Are hazard control recommendations being implemented? Can the selected hazard control strategy actually be implemented and verified? Is the control strategy so complex that it will be impossible to determine whether it will work when needed? Complex controls, overlapping control strategies, or inadequate implementation may be an indication of a weak safety design.
- Has the risk assessment truly considered the worst case? What is the basis for the likelihood levels? Is the risk analyzed only for steady-state operation, or have startups and shutdowns also been considered? Failure to provide good answers to these questions indicates a potential misunderstanding of the risk.
- Are problems found in test and design included in the hazard reports and factored into the design? Have problems and incidents been fully investigated? Does a process exist for tracking action items, including whether recommendations have been implemented? Are changes to software and computing systems properly factored into the hazard analysis? Failure to incorporate problems, changes, and corrective actions is an indication of the potential to miss serious design flaws.

These questions help to identify whether the hazard analysis process is robust. However, we must also ask questions specifically related to the use of software and computing systems in complex systems. The best questions come from real-world examples of accidents where software has been a contributor. Some examples of questions are as follows, and others can be found in the literature.[12]

- Have safety-critical software, commands, and data been identified?
- Do hazard controls for software-related causes combine good practices and specific safeguards?
- Is software and system testing adequate, and do tests include sufficient off-nominal conditions?
- Is the computing system design overly complex?
- Is the design based on unproven technologies?
- What happens if the software locks up?
- Are the sensors used for software decisions fault tolerant?
- Has software mode transition been considered?
- Has consideration been given to the order of commands and potential out of sequence inputs?
- Will the software and system start up and shut down in a known, safe state?
- Are checks performed before initiating hazardous operations?
- Will the software properly handle spurious signals and power outages?

These are by no means all the questions a decision maker should ask, and positive answers to these questions provide no assurance that an accident will be prevented. These questions should encourage critical thinking about safety and generate additional questions to provide further insight on system risk. A failure to ask these questions could mean that the potential for an accident is higher than we had assumed.

It is up to all stakeholders to look for those conditions that could lead to an accident and to recognize that the worst can happen. This means we should all express concerns about software safety management and engineering when necessary based on our knowledge, experience, and judgment, including lessons learned from accidents. We must ask questions to understand the potential for harm, to understand the steps taken to assure that the risks have been reduced, and to assure that there is proof that hazard controls are effective. We do not have to be software experts to ask questions, and in fact a lack of expertise could be an advantage when trying to understand how a system works. Most importantly, when thinking of software and computing system safety, we must think of the system and its interactions between software, hardware, humans, processes, and environments.

4. Conclusion

Process Safety Management can provide immense benefits in reducing operational risks in the chemical process and energy production industries. By proactively identifying hazards, assessing and characterizing risks, and taking actions to reduce those risks, organizations can prevent accidents and reduce the potential for death, injury, property damage, and environmental impacts. Given the importance of software and computing systems in the design and operation of many systems, software must be included as part of the broader process safety effort. However, the accidents provided in this paper, along with many others, illustrate that software is often not systematically considered as part of process safety activities. For example, software may not be identified as a hazard cause, risks may be underestimated, software-related hazard controls may be oversimplified, testing may not include sufficient safety scenarios, and system changes may not be factored into the design or hazard analysis. Process safety practitioners must take advantage of available software and computing system lessons learned, going beyond examples presented here, to improve their own safety efforts and, most importantly, to prevent accidents.

5. References

[1] Hardy, T.L., *Software and System Safety: Accidents, Incidents, and Lessons Learned,* AuthorHouse, 2012.

[2] Pool, B., "System Wrongfully Blamed in Union Carbide Leak: SAFER Fights Off Dark Cloud of Bad Publicity," *Los Angeles Times,* August 20, 1985.

[3] Schlager, N., *Breakdown: Deadly Technological Disasters,* Visible Ink Press, 1995.

[4] Transportation Safety Board of Canada, "Programmable Logic Controller Failure, Foothills Pipe Lines Ltd. Decompression/Recompression Facility, BP Canada Energy Company Empress Natural Gas Liquids Facility Near Empress, Alberta 18 October 2005," Report Number P05H0061, July 12, 2006.

[5] U.S. National Transportation Safety Board, "Rupture of Piney Point Oil Pipeline and Release of Fuel Oil Near Chalk Point, Maryland, April 7, 2000," Pipeline Accident Report NTSB/PAR-02/01, July 23, 2002.

[6] U.S. Mine Safety and Health Administration, "Report of Investigation: Fatal Other Accident (Steam Burns), February 11, 2003, Southern Clay Plants & Pits Southern Clay Prod. Inc., Gonzales, Gonzales County, Texas," Mine I.D. No. 41-00298, 2003.

[7] State of New South Wales (Australia), Department of Industry and Investment, "Fatality involving David Hurst Oldknow Ravensworth Underground Mine Coal Preparation Plant Reject bin 802 18 February 2009," May 2010.

[8] U.S. Mine Safety and Health Administration, "Report of Accident : Exploding Vessels Under Pressure Accident, October 24, 2002, Foreman Quarry and Plant, Ash Grove Cement Company, Foreman, Little River County, Arkansas," Mine I.D. No. 03-00256, 2003.

[9] U.S. National Transportation Safety Board, "Pipeline Rupture and Subsequent Fire in Bellingham, Washington, June 10, 1999," Pipeline Accident Report NTSB/PAR-02/02, October 8, 2002.

[10] U.S. Chemical Safety and Hazard Investigation Board, "Pesticide Chemical Runaway Reaction Pressure Vessel Explosion, Bayer CropScience, Institute, West Virginia, August 28, 2008," Report No. 2008-08-I-WV, January, 2011.

[11] U.S. Chemical Safety and Hazard Investigation Board, "Investigation Report: Allied Terminals, Inc. – Catastrophic Tank Collapse, Allied Terminals, Inc., Chesapeake, Virginia, November 12, 2008," Report No. 2009-03-I-VA, May 2009.

[12] Hardy, T.L., *Essential Questions in System Safety: A Guide for Safety Decision Makers,* AuthorHouse, 2011.

Technical Report on Chevron 2012 Pipe Rupture and Fire

Steve Cutchen, Lauren Grim
United States Chemical Safety and Hazard Investigation Board
1 Denver Federal Center Way
Building 20, Room A1419
Denver, CO 80225

Roger Evans, Don Holmstrom, Amanda Johnson, Dan Tillema, Kelly Wilson, Mark Wingard
United States Chemical Safety and Hazard Investigation Board

The CSB is an independent federal agency charged with investigating serious chemical incidents. The agency's board members are appointed by the President and confirmed by the Senate. CSB investigations look into all aspects of chemical incidents, including physical causes such as equipment failure as well as inadequacies in regulations, industry standards, and safety management systems. Please visit the CSB's website at www.csb.gov. For further information contact Communications Director Hillary Cohen at (202) 446-8094 or Sandy Gilmour at (202) 251-5496

Prepared for Presentation at
American Institute of Chemical Engineers
2013 Spring Meeting
9th Global Congress on Process Safety
San Antonio, Texas
April 28 – May 1, 2013

UNPUBLISHED

Technical Report on Chevron 2012 Pipe Rupture and Fire

Steve Cutchen, Lauren Grim
United States Chemical Safety and Hazard Investigation Board
1 Denver Federal Center Way
Building 20, Room A1419
Denver, CO 80225

Roger Evans, Don Holmstrom, Amanda Johnson, Dan Tillema, Kelly Wilson, Mark Wingard
United States Chemical Safety and Hazard Investigation Board

Keywords: corrosion, sulfidation corrosion

Abstract

The U.S. Chemical Safety and Hazard Investigation Board (CSB) and the California Division of Occupational Safety and Health (Cal/OSHA) released a technical evaluation report on piping samples taken from the Chevron Refinery in Richmond, California, where a hydrocarbon release and massive fire occurred on August 6, 2012. The report, prepared by Anamet, Inc., concluded the piping ruptured due to severe sulfidation corrosion, and that tested pipe samples showed a very low concentration of corrosion-inhibiting silicon.

1. Background

On August 6, 2012, a hydrocarbon release and resulting fire occurred at the Chevron U.S.A. Inc. Richmond Refinery in Richmond, California The incident occurred when a combustible hydrocarbon liquid known as "gas oil" leaked from an 8-inch pipe connected to an atmospheric crude oil distillation column in the refinery's crude unit. Workers initially noted the leak and were in the process of attempting to diagnose the source of the leak in the still-operating crude unit when the pipe ruptured catastrophically. Due to the high temperature, in excess of 600 degrees Fahrenheit, and physical properties of the material in the equipment, the gas-oil immediately formed a large hydrocarbon vapor cloud.

The CSB determined that nineteen Chevron employees were engulfed in a vapor cloud formed by the hydrocarbon release. Eighteen employees escaped before the fire started and one employee escaped without injury after the fire ensued. The fire burned for approximately 5 hours, and water was applied to the rupture site for 23 days. The incident resulted in six minor injuries. Production at the crude unit was suspended for over six months after the incident. More than 15,000 residents in the surrounding area sought treatment at area medical facilities as a result of the release and fire. Post incident on-site visual examination by Chevron, Cal/OSHA, and the CSB revealed extensive thinning of the pipe wall at the location of the rupture.

2. Sample Testing and Report

The evaluation, performed by Anamet, Inc. of Hayward, California was designed and overseen by the CSB, the California Division of Occupational Safety and Health Administration (Cal/OSHA), the United Steelworkers union, and Chevron. The evaluation was funded by the CSB. The report concludes that the 8-inch steel pipe, from a section designated as 4-sidecut which was installed in 1976, ruptured due to severe sulfidation corrosion, and that tested pipe samples showed a very low concentration of corrosion-inhibiting silicon.

The report cites wall thinning due to sulfidation corrosion as the cause of the piping failure. In crude oil distillation, the report notes, naturally occurring sulfur and sulfur compounds are available to react with steel components, particularly plain carbon steels. Corrosion rates vary according to the sulfur content of the oil being processed, temperature, and other factors, including silicon content and other materials in steel pipe.

The report concludes, "The reported service conditions of the #4 Crude Unit 4-sidecut piping, the composition of the carbon steel, presence of thick sulfide scale on the inside surfaces of the pipe, and generally uniform wall thinning indicate that sulfidation corrosion was the cause of the 8-inch 4-sidecut rupture."

The report noted that the failed pipe section was subjected to a higher corrosion rate than neighboring components due to low silicon content in the pipe. The report states that "Experience has shown that silicon in carbon steel is known to inhibit sulfidation" when the concentration is above a threshold value. Chemical analysis of the Chevron pipe showed the silicon concentration of the ruptured section to be far below this value, resulting in a significantly higher rate of sulfidation corrosion.

The report, which contains numerous photographs of the corrosion and pipe rupture, also notes the presence of an inward deformation toward the inside surface of the pipe. The CSB has determined this deformation was likely caused by a fire pike used during the plant's emergency response. A photograph depicts a tight fit between the fire pike tip and this inward deformation of the failed pipe.

3. Report Findings

The Anamet report provides the following conclusions:

1. Rupture of the 8-inch 4-sidecut resulted from wall thinning caused by sulfidation corrosion.
2. Post rupture corrosion destroyed fracture morphology that could have indicated the rupture initiation site. However, rupture likely originated in the thinnest region and initially followed a longitudinal path driven by the hoop stress. Consequently, an area of likely rupture initiation was identified.

3. The wall thickness of the ruptured section was less than all the other sections of the 8-inch 4-sidecut that were evaluated.

4. Chemical analysis showed the silicon concentration of the ruptured section was 0.01-wt%.

5. Six of twelve specimens (50%) from unique sections of the 8-inch 4-sidecut, 12-inch 4-sidecut, and 2011 12-inch samples were found by chemical analysis to have a silicon concentration of less than 0.1-wt%.

6. Deformation of a small region of the 8-inch 4-sidecut rupture edge was consistent with mechanical force applied from the outside surface of the pipe, possibly from a fire pike, such as samples E-099 and E-082-2, or other pointed object. This deformed region was located within the area of likely rupture initiation.

4. CSB Future Plans

The CSB investigation to determine the root causes of the incident is ongoing. The CSB will release a report detailing its findings and recommendations to key stakeholders later this year. The CSB, Cal/OSHA, the USW, and Chevron are cooperating under an agreement to test 4-sidecut carbon steel piping from Chevron's refinery in El Segundo, California, south of Los Angeles.

AUTHOR INDEX

AUTHOR INDEX